MECHANICAL ANALYSIS AND DESIGN

MECHANICAL ANALYSIS AND DESIGN

Arthur H. Burr

Hiram Sibley Professor of Mechanical Engineering, Emeritus
Cornell University

ELSEVIER · NEW YORK
New York · Amsterdam · Oxford

Elsevier Science Publishing Co., Inc.
52 Vanderbilt Avenue, New York, New York 10017

Sole distributors outside the USA and Canada:

Elsevier Science Publishers B.V.
P.O. Box 211, 1000 AE Amsterdam, The Netherlands

©1982 by Elsevier Science Publishing Co., Inc.
Fourth Printing, 1985.

Library of Congress Cataloging in Publication Data

Burr, Arthur Houghton, 1908–
    Mechanical analysis and design.

    Bibliography: p.
    Includes index.
    1. Machinery—Design. I. Title.
TJ230.B94   621.8'15   80-28924
ISBN 0-444-00324-X

Cover Illustration is Figure 5.30

Manufactured in the United States of America

To my father and mother

Contents

Preface

This book differs from the usual textbooks on mechanical design and offers a new approach to its teaching. The subject matter is divided according to principles of mechanics, with mechanical components serving as illustrations of the application of the principles. In several chapters one general equation is derived and then is adapted for the analyses of several kinds of components. Related equations not derived in the text are grouped in tables of reference. Analyses in greater depth and quantity follow from this principle-oriented approach.

For each component, practical considerations in its design and selection are discussed, and attention is called, when appropriate, to the modification of theory-based dimensions or chosen materials to meet manufacturing requirements or limitations. Tables of materials and standard and manufactured sizes and capacities are omitted. It is assumed that these may be found in catalogs and handbooks in the design office or classroom, as needed. Problems vary from simple computations to analyses of moderate difficulty, some of which have been a subject of recently published papers.

There is suitable and sufficient material for two semesters of university instruction, one of which may be a first, required course in mechanical design. The material is also very appropriate for a single, advanced-level course which is open to graduate students and qualified seniors, and for continuing-education courses and private reading and study.

The chapters are grouped in three parts: I—Hydrodynamic and Friction Analyses for Bearings and Torque-Carrying Components; II—Mechanical Behavior of Materials and General Design Criteria, and III—Force, Stress, and Deflection Analyses for Load-Carrying Components. The parts are outlined more thoroughly in Section 1.4. In his own teaching the author has first presented hydrodynamic and friction analyses to avoid giving the student an impression that the design course is only an extension of the course in strength-of-materials. Also, the student is quickly introduced to several interesting devices in the fields of fluid-film bearings, brakes, clutches, belt drives, fluid couplings, and torque converters. However, the three parts and, in general, their several chapters may be studied in any order. If Part II is to be studied, it, or certain sections of it, should precede Part III, since such phenomena as plastic behavior and thermal stress are introduced in Part II, then are involved in more complicated problems in Part III.

The author and several colleagues at Cornell University have used the material of this book in teaching required and elective graduate and undergraduate courses and in courses for practicing engineers. Originally intended for graduate students, the material was expanded, and the more elementary parts were used for a principle-oriented first course in mechanical design. For this first course of one semester and three-credit hours, and with topics typical of those covered when using other books in mechanical design, the following sections are indicated, with the numbers inclusive: 2.1–2.4, 2.7–2.11; 3.1, 3.2, 3.4, 3.6–3.9; 5.1–5.3, 5.5–5.7, 5.9–5.12; 6.1–6.5, 6.7–6.9; 7.1–7.7, 7.11; 8.1–8.3, 8.9, 8.10; 9.1–9.7; 10.4; 11.1, 11.5, 11.7–11.14; and 12.1, 12.3, 12.4, and 12.6–12.8. The order of presentation may be rearranged; e.g., some instructors may wish to discuss sliding-element bearings just before or after rolling-element bearings. Some sections might be omitted if they are too difficult or because of sufficient coverage in preceding courses of mechanics and materials. The foregoing listing omits some derivations but includes their applications, e.g., derivation of the Hertz equations is omitted, but their applications to cams, gear teeth, and rolling-element bearings are included, as is customary in first courses. Certainly, other topics which the instructor may consider important or interesting may be added or substituted, such as Sections 4.1–4.7 on hydrodynamic drives and Sections 7.8–7.16 on prestressing. For this first course, the instructor may wish to assign more of the numerical problems and fewer of the theory problems.

Coverage in the advanced course may depend somewhat on the interests of the students and instructor. The author's classes have included qualified undergraduates in mechanical engineering, graduate mechanical engineering students majoring in design or in thermal engineering, and graduate students in aerospace, agricultural, and nuclear engineering, and in engineering mechanics. Some of these students, of course, had not had a first course in mechanical design. Topics of common interest were readily found, and they always included topics related to energy conversion, such as pressure vessels, disks and rotors, fluid-film lubrication, thermal stresses, and the avoidance of fatigue failures. Suitable material for the advanced course, either because of its level or because it is useful material generally not given in a first course, include the following sections, with the numbers inclusive: 2.1–2.13; 3.2, 3.3, 3.5, 3.9, 3.12; 4.1–4.7; 5.4–5.8; 6.4–6.6; 7.1–7.16; 8.1–8.14; 9.7–9.11; 10.1–10.10; 11.1–11.7; 12.5, 12.9–12.11; and 13.1–13.10. The coverage may very well include a few of the sections indicated for a first course, particularly if it used a different textbook. For this advanced course the instructor may wish to assign the more theoretical problems from among those under each section. Many of the longer, more difficult problems are suitable for individual assignment and oral report, say biweekly or as term papers.

For a continuing-education course the material of the text gives excellent background reading to go with extensive lectures. Almost all of any chapter may be assigned, the more elementary parts being good for review. If practice sessions are available, some of the more difficult but practical analyses may be assigned and carried out. Topics of greatest interest in the author's courses for engineers in industry have been lubrication and some friction analyses; fatigue, stress concentration, and design of shafts; thermal properties and residual stresses; and vessels, fits, and problems of rotation. The limitation of time in these courses may prevent the study of many topics of the book, but a general description of what is available may be useful in the last lecture or two.

If the book is used for a course in advanced strength-of-materials, sufficient material and challenging problems may be found in Chapters 8 through 13.

The descriptions of the mechanisms in the book are believed adequate, but there are perhaps some for which space has not permitted sufficient explanation. The instructor may wish to discuss their operation in class. To supplement transparencies and blackboard sketches and to increase interest and understanding, the author highly recommends obtaining small-size machine components which can be handled and possibly operated. Without much cost, new and real components, or plastic models, can be obtained by writing manufacturers, at design shows, or through alumni and consulting contacts. Particularly instructive are fatigue failures and worn parts. Many can be obtained from discarded household machinery and vehicles. Faculty and students often contribute these more willingly than do the manufacturers.

There are about 75 examples which have been worked out in the text proper, and there are more than 870 problems. There are enough problems of each type for a variation in assignments over several years. The problems are placed at the end of each chapter for convenience and so that their figures may supplement those of the text. As an aid in the selection of problems, they are grouped according to the sections which they illustrate. Within each group, they generally parallel the text in that problems of theory come first, then those with numerical application, for assignment after the example in the text, which is generally numerical. However, it is sometimes preferable to assign a numerical problem before a more difficult, theory problem. About half the problems result in equations or do not have a completely numerical answer. Of the numerical examples and problems almost half are given in SI units, with the intention that they be worked out in these units. This is in keeping with the current trend for upper-level textbooks. Table 1.1 in the first chapter lists units in the SI and the British Gravitational System and relationships between them. Answers are given to about half of the numerical problems and to more than half of the others, particularly when the result is a potentially useful equation. To some long problems an answer is given to a first step only, in the expectation that an early check will prevent major errors in the final answer.

Problems may be selected for one or more objectives. There are problems which contribute to understanding the theory by requiring filling in steps in its integration, its application to a new set of boundary conditions, or its extension to the analysis of other components. There are problems in application which require the selection of appropriate equations and the determination of dimensions. Some of the problems require a physical interpretation of the results or a plotting of characteristic curves, others the determination of optimum proportions or conditions of operation. In the problems of a creative type the results may indicate the better of two arrangements, the dimensional change which will be most effective, or the feasibility of a proposal.

The book should assist engineers and engineering students in developing competence and confidence in making analyses of mechanical components and in the creative application of them to the design process. It can prepare readers to use more advanced or detailed sources of information, such as books and journals in the fields of elasticity, plasticity, lubrication, and design. References are given to many of them. Often students and engineers do not have time to study in these fields separately, and a single course and this book are suggested as a useful introduction or substitute.

The book has been developed from the author's experience in teaching undergraduate, graduate, and continuing-education courses in the Sibley School of Mechanical and Aerospace Engineering in the College of Engineering of Cornell University. This experience has been supplemented by teaching courses for shorter periods in graduate or continuing-education programs at the Pontifícia Universidade Católica de Rio de Janeiro, Brazil; the Universidad de los Andes of Bogotá, Colombia; the Indian Institute of Science of Bangalore and the NSF Summer Institute Program in India; also, the Universidad de las Américas in Cholula, Puebla, Mexico. The author acknowledges the aid and suggestions of his colleagues and students in these several places. In particular, acknowledgement is due to the staff members in the area of mechanical design at Cornell University with whom he worked closely in the courses described in the foregoing or in those closely related, namely, Professors John F. Booker, George B. DuBois, Allan I. Krauter, Richard M. Phelan, and Robert L. Wehe, and especially the late Fred W. Ocvirk. He expresses appreciation for the skilled assistance of Mrs. Gertrude Kazlauskas, who typed the material in its original form for use at Cornell University and elsewhere, and to the College of Engineering of Cornell University for some financial assistance in additional typing and copying. The author's wife, Phyllis, typed and edited much of the final manuscript. Her help, encouragement, patience, and understanding are most gratefully acknowledged.

The author will welcome comments and suggestions, as well as the indication of errors.

Tables of Reference Value

Symbols

ROMAN LETTERS

a distance from pivot or from support; moment arm; radius of hole in cylinders and plates; major semiaxis of an ellipse; linear acceleration.

A area, area of cross section; a constant; A_f, actuation factor and A_m, mechanical advantage (brakes).

b breadth, width of beam, band, or belt; base (leg) dimension of a 45° fillet weld; length of lubricant flow path; outside radius of cylinder or plate; minor semiaxis of an ellipse; semiwidth of contact area between loaded cylinders.

B constants in belt, clutch, and contact-stress theories.

c distance, distance from neutral axis to extreme fiber (beams); wall thickness of a thin bar or tube (torsion); axial velocity of wave propagation; c_d, c_r, diametral and radial clearances (bearings); c_p, specific heat.

C constant of integration, of proportionality; factors modifying endurance limit; center distance; spring index; C, basic dynamic capacity and C_s, static capacity in rolling bearings.

d diameter, the smaller of two diameters, wire diameter (springs); d_o, d_i, outside and inside diameters (hollow shafts).

D diameter, the larger of two diameters; pitch diameter (gears); mean coil diameter (springs); plate modulus $Et^3/12(1-\nu^2)$; damage (from stress fluctuation); a step ratio (impact); D_b, base circle diameter (gears).

e eccentricity; unit volume expansion; radius at an elastic-plastic boundary, a distance.

E modulus of elasticity; E_k, kinetic energy, E_p, potential energy.

f f_n, natural frequency; $f(p)$, function of some parameter p.

F force or load; F_0, F_1, loose and tight side band tensions, respectively; F_c, tension due to centrifugal force; F_n, F_a, F_s, and F_t, normal force, and

	axial, separating, and transmitted components in gears, respectively; F_1, F_2, F_3, F_4, tabulated functions of βx (elastic support).
g	acceleration of gravity, ratio of weight to mass.
G	modulus of elasticity in shear (modulus of rigidity).
h	height or elevation, head of liquid; distance, pole distance, thickness of film or band, beam depth.
hp	horsepower.
H	heat; plate deflection coefficient.
I	area moment of inertia, mass moment of inertia.
J	polar moment of inertia of area; tooth geometry factor.
k	stiffness (force per unit deflection), spring rate or spring constant, slope of force vs deflection curve; foundation modulus (force per unit length per unit deflection); radius of gyration; number of standard deviations; k_t, torsional spring constant (torque per radian of twist).
K	proportionality factor; plate stress coefficient; Wahl correction factor for spring stress; K_r, velocity factor (gears); K_t, theoretical stress concentration factor; K_f, fatigue stress concentration factor applied to endurance limit, K_f' reduced factor for limited cycles, all for normal stresses; for shear stresses add a subscript s to each.
l	length, distance between supports of a beam or shaft, length of a journal bearing, moment arm of applied force in a brake; lead of a helix; life factor with limited cycles of stress.
L	limit design factor (elastic-plastic conditions); life or number of revolutions to failure in rolling bearings; length, length of average path or periphery, length of weld; L_f, L_1, L_2, L_s, free length, initial installed length, maximum operating length, and solid length, respectively, in compression springs.
m	mass; in gears m, module and m_c, contact ratio.
m, n	lengths of beam sections.
m, n, q	radii of interior surface, fitted surface, and exterior surface, respectively, in a compound cylinder.
M	moment, bending moment; M_1, moment per unit length; M_t, torsional or twisting moment, torque; moments M_e from externally applied, M_f from friction and M_n from normal forces, respectively (brake shoes).
n	factor of safety; revolutions per minute; n', revolutions per second.
N	number of cycles or revolutions; number of teeth, of fasteners, or friction planes, of active spring coils, etc.; normal force or component; N_f, number of cycles to failure; N_L, load number in journal bearings; N_t, total number of spring coils.
p	pressure or normal force per unit area; p', unit load or force per unit of *projected* area (bearings); p_c, centrifugal force per unit area; p_f, pressure at fitted surfaces; p_i, internal pressure and p_o, outside or external pressure; p_0, pressure of fluid at source or inlet, constant pressure loading of a plate, peak pressure between contacting surfaces; p_0 and p_1, pressures where band tensions are F_0 and F_1, respectively; p, axial pitch of threads or coils; p_b, base pitch and p_c, circular pitch of gear teeth.
pwr	power.

P load or force, concentrated or total; P_1, P_2, P_s, spring forces at installed, maximum operating, and solid deflection, respectively; diametral pitch of gear or spline teeth.

q flow volume per unit time through an elemental area; notch-sensitivity factor; load or force per unit distance; q_c, centrifugal force per unit length of arc; a parameter (impact).

Q volume rate of flow; actuating force in brakes; a fictitious force for determining displacement; net fatigue strength.

r radius, fillet or notch radius; r_f, friction radius; in a curved beam r_i and r_o, radii to inner and outer surfaces, and $\bar{r}$ and r_n, radii to centroidal and neutral surfaces, respectively.

R reaction force at a support; a ratio, stress ratio (impact); R and R', respectively, minimum and maximum radii of curvature (in perpendicular planes) at a point of a surface; R_m and R_t, radii of curvature at a point on a surface, measured in the plane of the meridian and measured in a plane normal to it and to the tangential plane, respectively; $\bar{R}, R_n$, radii of spring coil measured to the centroidal axis and to the neutral axis of the wire, respectively.

s slip or fractional loss in speed; intensity of stress; stress in a wave; span, distance, or arc length; scale (graphical solutions).

S Sommerfeld number; whirl velocity (torque converter); wave stress ratio (impact); material properties in stress units: S_e, endurance limit for *infinite* life and S_f fatigue strength for *finite* life, both unprimed for mechanical parts and primed for polished specimen values; S_u, S_{ut}, S_{uc}, ultimate strength, and ultimate strength in tension and compression, respectively; S_y, yield strength; $S_{se}, S_{sf}, S_{su}, S_{sy}$, same as foregoing except strength in shear.

t time; temperature; thickness; deviation of deflection curve from a tangent elsewhere.

T period of time; temperature or temperature change; tension force per unit width in a membrane; torque; T_0, T_r, T_s, no-load, rotating member, and stationary member torques, respectively (journal bearings).

u radial displacement.

u, v, w displacements in a solid in the X, Y, Z directions, respectively; velocities in a fluid in the X, Y, Z directions, respectively.

U strain energy; U_0, strain energy per unit volume.

U, V, W velocities at the boundary of a fluid in the X, Y, Z directions, respectively; blade, fluid, and relative velocities, respectively (torque converter).

v beam deflection; velocity of a body; v, v', particle velocities in positive and negative waves, respectively; v_0, striking velocity (impact).

V volume; shear force; linear velocity, belt velocity, velocity of the pitch circle (gears), net particle velocity (impact); $\bar{V}$, velocity ratio (impact).

w plate deflection.

wk work.

W force of gravity mg or weight.

x, y, z coordinate locations, distances.

X, Y, Z coordinate directions.

$\overline{x}$	distance to a centroid of area.
$\overline{y}$	distance from a centroidal axis to a neutral axis.
Y	Lewis form factor (gear teeth).
Z	section modulus.

GREEK LETTERS

α	angle, half cone angle, apex angle; angular acceleration or deceleration; coefficient of thermal expansion; influence coefficient; a ratio, ratio of mass or moment of inertia of bar to that of attached body (impact).
β	angle; a parameter (continuous elastic support); a function of curvatures; ratio of mass or moment of inertia of bar to that of striking body (impact).
γ	angle, unit weight (weight per unit volume); shear strain or angle of distortion.
δ	interference, radial interference; deflection; δ_{st}, static deflection (from weight); δ_c, plastic deformation due to creep.
Δ	diametral interference; support misalignment; an increment.
ϵ	eccentricity ratio (journal bearings); normal strain or unit elongation; ϵ_c, creep (strain); $\dot{\epsilon}_0$, constant creep rate.
E	a ratio or group of terms.
η	efficiency; elastic constant $(1 - \nu^2)/E$; shear stress coefficient for twisted rectangular bars, correction factor for mass of elastic body (impact).
θ	angle; angle of slope in bending of beams; wrap angle of a band or belt, pressure angle in gears; angle of twist in torsion; θ_1, twist per unit of axial length.
κ, λ	functions of angular extent (shoe brakes), of curvatures (surface contact).
λ	angle, lead angle, angle-of-twist coefficient for rectangular bars; wave length; a ratio; ratio of mass or moment of inertia of striking body to that of attached body (impact).
Λ_e, Λ_p	loadings corresponding to completely elastic and to completely plastic conditions, respectively.
μ	dynamic viscosity, coefficient of friction.
ν	kinematic viscosity, Poisson's ratio.
ρ	density (mass per unit volume); radius of curvature of the neutral surface of deflected beams and plates; ratio of minimum to maximum load or stress.
$1/\rho$	curvature.
σ	normal stress; σ_x, σ_y, σ_z, stress components in X, Y, Z coordinates; σ_1, σ_2, σ_3, principal stresses in a 1, 2, 3 coordinate system; σ_r, σ_t, σ_z, radial, tangential, and axial stresses, respectively, in a cylindrical coordinate system; σ_m, σ_t, meridional and tangential stresses (shells); σ_a, σ_m, alternating and mean components of fluctuating nominal stress; σ_{st}, static stress (from weight); σ_0, initial impact stress; σ_i, maximum impact stress (energy method); standard deviation.
τ	shear stress, shear force per unit area, torsional stress; τ_{xy}, τ_{yz}, τ_{zx}, shear stresses on planes normal to directions X, Y, and Z and in the directions Y, Z, and X, respectively; τ_a, τ_m; alternating and mean components of nominal shear stress; τ_0, initial stress (impact); τ_i, τ_r, τ_t, incident,

reflected, and transmitted stresses in a torsional wave at a step, respectively.

ϕ angle; location angle; attitude angle (journal bearings); a function in torsion theory; ϕ_1, active friction angle on a drum with belt.

ψ angle, helix angle (complement of lead angle); angle between planes containing the two maximum (or two minimum) radii of curvature (surface contact); angular displacement.

ω angular velocity (radians per second); ω_0, striking velocity (impact).

Abbreviations

TECHNICAL[1]

avg	average	max	maximum
Bhn	Brinell hardness number	min	minimum, minute (time)
ccw	counterclockwise	mph	miles per hour
CI	cast iron	OD	outside diameter
const	constant	OT	oil tempered (spring wire)
cw	clockwise	psi	pounds per square inch
dia	diameter	r	radius
eff	efficiency	R_c	Rockwell hardness, C-scale
equiv	equivalent	rpm	revolutions per minute
HD	hard drawn (spring wire)	rps	revolutions per second
ID	inside diameter	SI	Système International d'Unités
kpsi	(ksi), 10^3 psi	std	standard
ln	Napierian natural logarithm	SUS	Sayboldt universal seconds
log	common logarithm (base 10)	VSQ	value spring quality (spring wire)

SOCIETIES AND ASSOCIATIONS

ABEC	Annular Bearing Engineers Committee (of AFBMA)
AFBMA	Anti-Friction Bearing Manufacturers Association
AGMA	American Gear Manufacturers Association
AISI	American Iron and Steel Institute
ANSI	American National Standards Institute

---

[1] See Section 1.5 for a discussion, also Tables 1.1 and 2.1 for abbreviations of units.

ASLE   American Society of Lubrication Engineers
ASM   American Society for Metals
ASME   American Society of Mechanical Engineers
ASTM   American Society for Testing and Materials
NASA   National Aeronautics and Space Administration (formerly NACA)
SAE   Society of Automotive Engineers
SESA   Society for Experimental Stress Analysis

MECHANICAL
ANALYSIS
AND DESIGN

1

Introduction

1.1 ANALYSIS AND MECHANICAL DESIGN: THE DESIGN PROCESS

A *machine* is a combination of mechanisms and other components which transforms, transmits, or utilizes energy, force, or motion for a useful purpose. Examples are engines, turbines, vehicles, hoists, printing presses, washing machines, and movie cameras. Many of the principles and methods of design that apply to machines also apply to manufactured articles that are not true machines, from hub caps and filing cabinets to instruments and nuclear pressure vessels. The term ''mechanical design'' is used in a broader sense than ''machine design'' to include their design. For some apparatus, the thermal and fluid aspects that determine the requirements of heat, flow path, and volume are separately considered. However, the motion and structural aspects and the provisions for retention and enclosure are considerations in mechanical design. Applications occur in the field of mechanical engineering, and in other engineering fields as well, all of which require mechanical devices, such as switches, cams, valves, vessels, and mixers.

Designing starts with a need, real or imagined. Existing apparatus may need improvements in durability, efficiency, weight, speed, or cost. New apparatus may be needed to perform a function previously done by men, such as computation, assembly, or servicing. With the objective wholly or partly defined, the next step in design is the conception of mechanisms and their arrangements that will perform the needed functions. For this, freehand sketching is of great value, not only as a record of one's thoughts and as an aid in discussion with others, but particularly for communication with one's own mind, as a stimulant for creative ideas. Also, a broad knowledge of components is desirable, because a new machine usually consists of a new arrangement or substitution of well-known types of components, perhaps with changes in size and material. Either during or following this conceptual process, one will make quick or rough calculations or analyses to determine general size and feasibility. When some idea as to the amount of space that is needed or available has been obtained, to-scale layout drawings may be started.

When the general shape and a few dimensions of the several components become apparent, analysis can begin in earnest. The analysis will have as its objective satisfactory or superior performance, plus safety and durability with minimum weight, and a competitive cost. Optimum proportions and dimensions will be sought for each critically loaded section, together with a balance between the strengths of the several components. Materials and their treatment will be chosen. These important objectives can be attained only by analysis based upon the principles of mechanics, such as those of statics for reaction forces and for the optimum utilization of friction; of dynamics for inertia, acceleration, and energy; of elasticity and strength of materials for stress and deflection; of physical behavior of materials; and of fluid mechanics for lubrication and hydrodynamic drives. The analyses may be made by the same engineer who conceived the arrangement of mechanisms, or, in a large company, they may be made by a separate analysis division or research group. As a result of the analyses, new arrangements and new dimensions may be required. Design is a reiterative and cooperative process, whether done formally or informally, and the analyst can contribute to phases other than his own.

Finally, a design based upon function and reliability will be completed, and a prototype may be built. If its tests are satisfactory, and if the device is to be produced in quantity, the initial design will undergo certain modifications that enable it to be manufactured in quantity at a lower cost. During subsequent years of manufacture and service, the design is likely to undergo changes as new ideas are conceived or as further analyses based upon tests and experience indicate alterations. Sales appeal, customer satisfaction, and manufacturing cost are all related to design, and ability in design is intimately involved in the success of an engineering venture.

1.2 ANALYSIS AND CREATIVITY: SOME RULES FOR DESIGN

In the preceding section it has been emphasized that analysis is interwoven in the design of a sound and balanced machine or device, and that it is an essential part of it. In this section it is suggested that, applied with a creative attitude, analyses can lead to important improvements and to the conception and perfection of alternate, perhaps more functional, economical, and durable products. The creative phase need not be an initial and separate one. Although he may not be responsible for the whole design, an analyst can contribute more than the numerically correct answer to a problem that he is asked to solve—more than the values of stress, dimensions, or limitations of operation. He can take the broader view that the specifications or the arrangements may be improved. Since he will become familiar with the device and its conditions of operation before or during his analysis, he is in a good position to conceive of alternatives. It is better that he suggest a change in shape that will eliminate a moment or a stress concentration than to allow construction of a mechanism with heavy sections and excessive dynamic loads. It is better that he scrap his fine analysis, rather than that he later see the mechanism scrapped.

To stimulate creative thought, the following rules are suggested for the designer and analyst.[1] The first six rules are particularly applicable for the analyst, although he may become involved with all ten rules.

1. *Apply ingenuity to utilize desired physical properties and to control undesired ones.* The performance requirements of a machine are met by utilizing laws of nature or properties of matter (e.g., flexibility, strength, gravity, inertia, buoyancy, centrifugal force, principles of the lever and inclined plane, friction, viscosity, fluid pressure, and thermal expansion), also the many electrical, optical, thermal, and chemical phenomena.[2]

---

[1] A. H. Burr, "The Principles of Machine Design," a paper presented at the Annual Meeting, ASME, New York, 1960.
[2] T. T. Woodson, *Introduction to Engineering Design*, New York: McGraw-Hill, 1966. A useful listing of "laws and effects" appears on pp. 72–82. A more extensive treatment is given in C. F. Hix and R. P. Alley, *Laws and Effects*, New York: John Wiley & Sons, Inc., 1958.

However, what may be useful in one application may be detrimental in the next. Flexibility is desired in valve springs but not in the valve camshaft; friction is desired at the clutch face but not in the clutch bearings; thermal contraction may shrink a bearing ring on a shaft, but a temperature difference between the shaft and its housing must not be allowed to cause binding at the bearings. Ingenuity in design should be applied to utilize and control the physical properties that are desired and to minimize those that are not desired.

2. *Recognize functional loads and their significance.* Functional loads are those which are inherent to the function performed by the machine or its components; e.g., centrifugal forces with cranks, gyroscopic forces with high-speed turbine disks, and fatigue loading conditions with rotating shafts. Other loading conditions that should be recognized if existent, and for which proper allowances should be made, are impact, moving loads, noncoplanar loading, and statically indeterminate loads. When several types of loading occur, their worst possible combination during a cycle of operation should be determined for its effect on stress and bearings.

3. *Anticipate unintentional loads.* Unintentional loading conditions are those not inherent to the functions of the machine. Examples are eccentric forces in a bolt due to off-center contact under the head, the rigid coupling of two shafts not in alignment, excessive vibration at operating speeds, binding due to differential temperature expansion of parts, jamming due to product failure, and general handling and shipping forces.

4. *Devise more favorable loading conditions.* A good design provides more favorable conditions for the functional and unintentional loads. Many examples may be found among machine components. A crankshaft is balanced by counterweights and by the arrangement of its several throws. An equalizer link evenly divides the actuating force applied to two band brakes. The even number of friction planes in a plate clutch avoids thrust loads external to the clutch while engaged. The helix angles of two helical gears on the same shaft may be arranged to give zero net end thrust. In planetary gear trains, two or more equally spaced planet gears give zero net radial loads on the bearings of the several shafts. On shafts, bending moments may be minimized by the location of sprockets, pulleys, and gears; unintentional bending moments may be eliminated by flexible couplings and self-aligning bearings; and variable moments of inertia may be eliminated by using three or more keyways or splines. Proper bolt tensioning will minimize fatigue conditions at the bolts. Slipping, tripping, or shearing devices are provided to prevent severe overloads. Binding at ball bearings due to assembly tolerances and axial thermal expansion of the shaft is prevented by allowing one bearing to "float" endwise. Induced drafts and fins minimize temperature problems at clutches, brakes, bearings, and worm gears.

5. *Provide for favorable stress distribution and stiffness with minimum weight.* On components subjected to fluctuating stresses, particular attention is given to a reduction in stress concentration, and to an increase of strength at fillets, threads, holes, and fits. Stress reductions are made by modifications in shape, and strengthening may be done by prestressing treatments such as surface rolling and shallow hardening. For unidirectional loading as in pressure cylinders, rotating disks, springs, and hooks, favorable prestressing may be obtained by yielding in the direction of the service load. Hollow shafts and tubing, and box, flanged, and ribbed sections give a favorable stress distribution, together with stiffness and minimum weight. Sufficient stiffness to maintain alignment and uniform pressure between contacting surfaces should be provided for crank, cam, and gear shafts, and for enclosures and frames containing bearing supports. The stiffness of shafts and other components must be suitable to avoid resonant vibrations.

6. *Use basic equations to proportion and optimize dimensions.* The fundamental equations of mechanics and the other sciences are the accepted bases for calculations.

They are sometimes rearranged in special forms to facilitate the determination or optimization of dimensions, such as the beam and surface stress equations for determining gear-tooth size. Factors may be added to a fundamental equation for conditions not analytically determinable, e.g., on thin steel tubes, an allowance for corrosion added to the thickness based on pressure. One accepts critically the special equations in catalogs and other publications unless a sound fundamental basis can be recognized. When it is necessary to apply a fundamental equation to shapes, materials, or conditions which only approximate the assumptions for its derivation, it is done in a manner which gives results "on the safe side." In situations where data are incomplete, equations of the sciences may be used as proportioning guides to extend a satisfactory design to new capacities.

7. *Choose materials for a combination of properties.* Materials should be chosen for a combination of pertinent properties, not only for strengths, hardness, and weight, but sometimes for resistance to impact, corrosion, and low or high temperatures. Cost and availability from stock are factors, as are fabrication properties such as weldability, cold formability, machinability, sensitivity to variation in heat-treating temperatures, and required finish or coating.

8. *Select carefully between stock and integral components.* A previously developed component is frequently selected by a designer and his company from the stocks of parts manufacturers, if the component meets the performance and reliability requirements and is adaptable without additional development costs to the particular machine being designed. However, its selection should be carefully made with a full knowledge of its properties, since the reputation and liability of the company suffer if there is a failure in any one of the machine's parts. In other cases the strength, reliability, and cost requirements are better met if the designer of the machine also designs the component, with the particular advantage of compactness if it is designed integral with other components, e.g., gears to be forged in clusters or integral with a shaft.

9. *Modify a functional design to fit the manufacturing process and reduce cost.* Modified shapes may be needed to produce sound castings, forgings, or weldments free of voids and cracks. Design should take advantage of the inherent characteristics of a process, e.g., a hollow box section might be specified for a weldment where a T or ribbed section would be specified for a casting. Modifications for machining may include provision for clamping points, in-line milling of bolting surfaces, and turning of concentric cylindrical surfaces without rechucking. Stock and standardized fittings, fasteners, and dimensions should be used unless otherwise justified.

10. *Provide for accurate location and noninterference of parts in assembly.* A good design provides for the correct locating of parts and for easy assembly and repair. Shoulders and pilot surfaces give accurate location without measurement during assembly. Shapes can be designed so that parts cannot be assembled backwards or in the wrong place. Interferences, as between screws in tapped holes and cored passages, and between linkages and hydraulic lines must be foreseen and prevented. Inaccurate alignment and positioning between such assemblies must be avoided, or provision must, by flexibility, be made to minimize any resulting, detrimental displacements and stresses.

1.3 PURPOSE AND SCOPE OF BOOK

One purpose of this book is to present a body of knowledge that will be useful for the analysis and design of mechanical components, particularly for performance, strength, and durability. A number of topics usually found scattered throughout several books, each of which concentrates on one discipline, have been compiled for study in this book. Although the book was written to be a textbook, the data, footnote references, and equations, some of the latter in convenient chart or tabular form, make the book useful for reference and direct application. Attention is called to Table 8.1 on cylinders, Table 9.4 on beam loadings, Table 10.3 for beams with continuous elastic support, Tables 10.4 and 10.5 on circular plates, and Tables 11.1 and 11.2 for contact stresses. Familiarity with this book should make it easier to read the single-discipline books and the papers that appear in the publications of technical societies. Hopefully, the book will interest some readers in undertaking more advanced studies in the fields of mechanics and/or mechanical design.

A second purpose of this book is to provide an opportunity for the reader to develop competence and confidence in applying available equations to the design of mechanical components and in deriving new equations as needed. For this purpose, many problems of several types are included, placed at the end of each chapter and grouped under the number of the section to which they particularly apply. The problems of application generally require computations and give results useful for design, such as dimensions. The "theory" problems may require extension of a theory for a different condition of loading and set of boundary conditions, or they may require the development of equations for a proposed device. These problems may be quite challenging but of medium difficulty, some having been the subject of papers recently published in journals. Necessarily, the statement of such a problem and the steps to be taken are outlined in more detail than they probably were to the original investigator. There are problems of a creative type, the results of which may indicate optimum proportions, the better of two arrangements, or the feasibility of a proposal. Answers are provided for selected problems, and particularly when the result is an equation of potential use. Thus study of this book and execution of a number of its problems should sharpen the reader's ability in analysis.

A third purpose of this book is to present a variety of mechanical components, including some less common but ingenious ones. They are located in sections of the book where they illustrate the applications of various principles of mechanics. Their characteristics and the practical considerations in their design are discussed. A knowledge of existing components is essential for conceptual design. A particular purpose is to show how analysis itself is part of creative engineering, as discussed in Section 1.2. Illustrations or examples of this appear among the problems as well as in the text proper, and there is some value in reading the problems without solving them. Illustrations include constant-torque and tensioning devices, self-energizing disk brakes, pivoted-pulley belt drives, and the spring-coil clutch and friction speed-governor as used in telephone dials. There are analyses for adjustable-speed fluid couplings, externally pressurized bearings, and rotating loads and sleeves on hydrodynamically lubricated journal bearings. Unique, symmetrical components created with the aid of analyses include constant-stress tanks, pressure-vessel heads, and rotating disks, also, fiber-wrapped and compound-shell cylinders. Additional illustrations of creative engineering are assemblies for beneficial prestressing, cast and welded shapes for maximum stiffness-to-weight ratios, gimbal-support rings, expansion pipe-loops, and frictionless flex-pivots, and with shafts, the minimization of bearing loads and bending moments by the arrangement and location of mounted components.

1.4 ARRANGEMENT OF TOPICS

The titles of the chapters are closely related to principles or topics of mechanics. The subtitles of the chapters indicate major subdivisions of the theory and the major components discussed. The book is principle oriented, and the general theory for a topic is developed, then applied to those components for which it is a theoretical basis. Thus based on one or a few general equations, brakes, clutches, and belts are treated together in the chapter on friction theory; pressure vessels, rotating cylinders, and interference fits are discussed in the chapter on axially symmetric loadings; and cams, gears, and rolling-element bearings are treated in the chapter on surface-contact theory. The design of shafts for strength and endurance is treated in a chapter concerned with the application of fatigue phenomena, but shaft loading and deflection, as well as the deflection and stress of many other components subject to bending, are treated in the two chapters concerned with flexure theory. By this arrangement the book differs from other textbooks on elementary and advanced mechanical design, where each component is given a separate chapter with a corresponding title.

The chapters are reasonably independent of one another, and they may be read in any order. However, their subject matter falls into three broad divisions, and they are grouped accordingly. Part I is concerned with the flow and viscosity of fluids and with dry friction and consists of Chapters 2–4 on film lubrication, friction drives, and hydrodynamic drives. Part II consists of Chapters 5–7, and general design criteria related to the engineering properties of materials are developed in these chapters. These criteria include the basic stress and strain relationships, static theories of failure, limit (elastic–plastic) analysis; fatigue, stress concentration, adaptation of the theories of failure to fluctuating stresses of various numbers of cycles and to multiple stress levels; finally, extreme-temperature properties of materials, thermal stresses, and the avoidance of harmful residual stresses and the provision for beneficial residual stresses.

In Part III which consists of Chapters 8–13, the magnitude and the distribution of stress and deflection are of prime importance. In Chapter 8, general equations are developed for axially symmetrical objects, and these equations are applied to shells, pressure vessels, disks, and rotors, including loosening of fits, elastic–plastic design, and thermal stress. Two chapters are concerned with flexure: cast, forged, and welded beam shapes to minimize weight and beam loadings in machine parts; deflection analysis by the moment area, graphical, and strain-energy methods; superposition for statically indeterminate forces and moments; continuous elastic support, curved beams and springs under bending moments, and the theory of plates, with solutions by superposition. A chapter on surface effects includes contact pressures and stresses; pitting, spalling, and wear phenomena; and durability and other considerations in the design of wheels, cams, gear teeth, and ball and roller bearings. The chapter on torsion includes axially loaded coil springs, the general equations of torsion with a few applications, and analogies for determining the stresses and twist in noncircular sections and in thin-walled, open and closed (hollow) sections. The final chapter, on impact, begins with approximate energy methods for longitudinal, torsional, and beam impact; then, it develops the more exact wave method, with application to bars, axially loaded coil springs, and shafts in torsion, including the effect of steps and finite, attached masses.

1.5 NOMENCLATURE

An extensive list of symbols follows the Table of Contents. Most of the symbols are those of "Letter Symbols for Quantities Used in the Mechanics of Solids," USAS Y10.3–1968, of the American National Standards Institute, Inc. (ANSI), formerly the American Standards Association (ASA).[3] The symbols are arranged alphabetically, first by English, then by Greek letters. Occasional departures from the Standard are made to accommodate the variety of symbols needed to cover not only solid mechanics and space, but also fluids, lubrication, and heat. Where several kinds of quantities are assigned to one symbol, they are usually from different fields and chapters, and there should be no confusion in their use.

---

[3] This particular standard is sponsored and published by the American Society of Mechanical Engineers (ASME).

Both F and P are used for concentrated force, the latter generally in the sense of "load" and normal force. Symbol P is also used in equations of the theory of elasticity, as is commonly done in standard references. Symbols, Q, R (reaction), and V (shear) are used for forces together with F and P to avoid subscripts and multiple subscripts. Symbol W is used infrequently for the force of gravity or weight. Accordingly, to avoid confusion, representation of power is made by pwr, horsepower by hp, and work by wk, in the few equations where they occur. Many equations are conveniently derived in terms of radius r, a variable, but limits of integration are usually written in terms of diameters D, dimensions that can be measured and used on specifications and drawings. Capital R is used as a dimension when two radii are needed to define a surface.

Following the table of symbols there are two listings of abbreviations used in the text. In the first listing, many of the abbreviations are from the ANSI publication Y1.1–1972, "Abbreviations for Use on Drawings and in Text." These abbreviations are written in lowercase letters, except that the first letter is capitalized when it is derived from a proper name. Plurals are not used, and periods are omitted. Following the symbols for technical terms there is a listing of abbreviations for the names of professional societies to which reference is made in the text.

1.6 UNITS AND CONVERSION

The principal physical quantities of the book are listed in Table 1.1, together with their symbols and units, and they are grouped by related quantities and units. For the units (SI) of the Système International d'Unités,[4] there are two columns, the first for the unit or its commonly used multiple, the second for the equivalent combination of base units. The latter are useful for dimensional checks of substitution in equations. Sometimes it may be more convenient when checking to substitute with prefixed units, e.g., N/mm^2 for MPa. However, answers or specifications should not be so given, since units with prefixes, except the base unit kg, are not to be used in denominators.

In the last column, equivalents to the SI unit are given in the British Gravitational System (BGS), usually with the length unit in the convenient inches of machine measurements. Currently, the units are being called U.S. Customary Units for contrast with SI units. In the row for stress, for example, we find the normal and shear stress symbols σ and τ, respectively, the SI unit MPa (megapascal), its equivalent in base units, 10^6 kg·m/s², and its equivalent in BGS units, 145.04 lb/in². Pressure p shares these same units. The abbreviation psi is used for lb/in² within the text and problem statements, as are the popular abbreviations rpm for the letter symbols r/min and rps for the letter symbols r/s. Some of the equivalents are taken from published tables of conversion factors; others are derived or calculated.

Almost half of the numerical examples and problems of the text are stated in SI units, and it is intended that they be worked in these units. Their answers when *multiplied* by the British equivalent may serve as a check on reasonableness for those whose judgement is based on experience with British units. Since chart and table data on physical properties may be available in British units only, the equivalents of the last column will also be useful for conversions to SI units. This is done by *dividing* the data in British units by the equivalent of the last column, e.g., if the yield strength is 45 500 psi, then its SI value is 45 500/145 = 314 MPa.[5]

The equivalents are given with some precision and were calculated from even more precise data. However, when they are used with engineering data of the usual impreci-

---

[4] For details see, e.g., ANSI Z210.1–1976, ASTM E380–76, ASME Guide SI–1, Special Publication 330 of the National Bureau of Standards, or SAE J916a.

[5] *Multiplying* factors for conversion from British to SI units may be found in many books, e.g., ASME Guide SI-1, loc. cit.

TABLE 1.1. SI Units of the Text and Equivalents

Quantity	Quantity Symbol[a]	Unit Symbol	Base Units (if different)	Equivalents in British Gravitational Units
Length	l, r, D	mm	10^{-3} m	0.039 37 in
				(1 in is 25.4 mm exact)
		m		3.280 84 ft
Area	A	mm^2	10^{-6} m^2	1.5500×10^{-3} in^2
		m^2		10.7639 ft^2
Volume	V	mm^3	10^{-9} m^3	61.024×10^{-6} in^3
		m^3		35.315 ft^3
		liter (liquids only)	10^{-3} m^3	0.2642 gal
Section modulus	Z	mm^3	10^{-9} m^3	61.024×10^{-6} in^3
Area moment of inertia	I	mm^4	10^{-12} m^4	2.4025×10^{-6} in^4
Time	t	s		
Velocity				
linear	v	m/s		39.370 in/s, 3.281 ft/s
angular	ω	rad/s		
Volume flow	Q	m^3/s		35.315 ft^3/s
		liter/s	10^{-3} m^3/s	15.850 gal/min
Acceleration				
linear	a	m/s^2		39.37 in/s^2, 3.281 ft/s^2
angular	α	rad/s^2		
gravity (standard)	g	9.806 65 m/s^2		386.09 in/s^2, 32.174 ft/s^2
Mass	m	kg		68.52×10^{-3} lb·s^2/ft [slug]
				5.710×10^{-3} lb·s^2/in
Density	ρ	kg/m^3		1.9403×10^{-3} (lb·s^2/ft)/ft^3
		(H$_2$O, 10^3 kg/m^3 or 1 kg/liter)		9.3573×10^{-8} (lb·s^2/in)/in^3
Mass moment of inertia	I	kg·m^2		8.851 in·lb·s^2
Force	F, P	N	kg·m/s^2	0.2248 lb
Weight[b] of 1 kg	9.806 65 N	.		2.2046 lb
Unit weight[b,c]	$\gamma, \rho g$	N/m^3	(kg·m/s^2)/m^3	3.6840×10^{-6} lb/in^3
Moment, torque	M, T	N·m	kg·m^2/s^2	8.8508 lb·in
Impulse, momentum		N·s	kg·m/s	0.2248 lb·s
Stiffness	k	N/mm	10^3 kg/s^2	5.710 lb/in
Stress and pressure	σ, τ, p	MPa(10^6 N/m^2)	10^6 kg/(m·s^2)	145.04 lb/in^2
Viscosity[d]	μ	Pa·s, (N/m^2)·s	kg/(m·s)	145.04 μreyn, [(10^{-6} lb/in^2)·s]
Energy	E_k, E_p	J, (N·m)	kg·m^2/s^2	0.737 56 ft·lb
Work	wk			$0.947\,82 \times 10^{-3}$ Btu
Power	pwr	W, J/s, N·m/s	kg·m^2/s^3	
		kW	10^3 kg·m^2/s^3	737.56 ft·lb/s, 1.341 hp
Temperature	t	°C		$(1.8\,t + 32)$°F

KEY: *base units:* m—meter, kg—kilogram, s—second, in—inch, ft—foot, lb—pound;

    derived units: J—joule, N—newton, Pa—pascal, W—watt;

    other units: °C—degree Celsius, °—degrees of plane angle; h—hour, min—minute; rad—radian;

    prefixes: μ—micro(10^{-6}), m—milli(10^{-3}), k—kilo(10^3), M—mega(10^6), G—giga(10^9)

[a] A more complete list of the symbols of the book follows the Contents.

[b] Force to restrain 1 kg against free fall at standard gravity. Weight is not recognized in the SI, but weight, unit weight, and its equivalent are useful when the product ρg occurs, and the unit weight of the material is available in lb/in^3 or lb/ft^3.

[c] To find the *density* in kg/m^3, divide unit weight in N/m^3 by 9.806 65 m/s^2, or more directly, divide lb/in^3 by 36.13×10^{-6} ($= 3.6840 \times 10^{-6} \times 9.806\,65$).

[d] See also Table 2.1.

sion, the equivalents or the calculated final result should be rounded off to a number of significant figures which is no more than in the least accurate of the numbers in the calculation. It is incorrect to imply by many digits a precision in the result that does not exist. When several zeros precede the decimal point, it is unlikely that they indicate precision. Some background knowledge is needed. The yield strength of 45 500 psi in the preceding paragraph is probably only accurate within ± 3% because of variations within the material and between lots and because of the limitations of measurement. Thus to convert 45 500 psi to MPa, we divide by 145 or 145.04, then round off the quotient of 313.70656 (given by an eight-place electronic calculator) to 314 or even to 315.[6] An exception occurs when a small difference between two much larger numbers is needed, as in some gear-tooth calculations. Subtraction should be made before any rounding off of the larger numbers.

---

[6] Guides to the use of SI units give rules and examples for rounding off converted values, e.g., the ASME Guide SI-1, loc. cit.

I

HYDRODYNAMIC AND FRICTION ANALYSES FOR BEARINGS AND TORQUE-CARRYING COMPONENTS

2

Film Lubrication
Sliding Element Bearings

2.1 CHARACTERISTICS OF FLUID LUBRICANTS

VELOCITY AND SHEARING FORCES

Most fluids tend to "wet" and adhere to solid surfaces. When fluid flows through a stationary circular tube, the velocity of the fluid at the walls is zero, and the velocity of the fluid in the center is maximum, provided it is below a certain critical value such that there is streamlined or laminar flow (Fig. 2.1(a)). Likewise, in a fluid between a stationary plate and one moving with velocity U (Fig. 2.1(b)), or between a rotating journal and a stationary bearing (Fig. 2.3), the velocity is zero at the stationary surface and equal to U at the moving surface. A plot of velocity u against distance y across the film is known as the *velocity profile*. The inverse of the slope at any point, $du/dy = \lim_{\Delta y \to 0} \Delta u/\Delta y$, is the *velocity gradient*, which has the unit of s^{-1}.

Experience shows that the larger the plate velocity U and the smaller the film thickness h (Fig. 2.1(b)), the greater is the force required to move the plate. Likewise, within the film on an element of fluid (Fig. 2.2), the shearing force would appear to be related to the velocity difference, du, of its surfaces and the distance, dy, between them. This velocity difference requires a continuous distortion or shearing of the element, and a force is required to do this. In 1668, Isaac Newton conjectured that the shearing force, F, and the shearing force per unit of area, τ_x, are directly proportional to $d\gamma/dt$, the time rate of shear distortion or angular deformation γ (Fig. 2.2). If μ is a coefficient of proportionality, then

$$\frac{F}{A} = \tau_x = \mu\frac{d\gamma}{dt} = \mu\frac{d}{dt}\frac{dx}{dy} = \mu\frac{d}{dy}\frac{dx}{dt} = \mu\frac{du}{dy} \tag{a}$$

Thus the unit shearing force is either

$$\tau_x = \mu\frac{du}{dy} \qquad \text{or} \qquad \tau_x = \mu\frac{\partial u}{\partial y} \tag{2.1}$$

The term du/dy, or its partial-derivative form $\partial u/\partial y$, is the velocity gradient. For many oils we may take μ as independent of velocity, as did Newton, and hence as constant

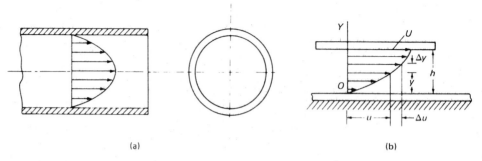

(a)   (b)

FIGURE 2.1. Fluid velocity (a) through a circular tube and (b) between plates, one stationary and one with velocity U.

in Eqs. (2.1) for any given temperature and pressure. Some oils have additives that decrease μ and, hence, the increase of shearing force F with increase of velocity. They are known as non-Newtonian oils.

Example 2.1

An *unloaded* journal of diameter d and length l rotates at speed n' centrally within its bearing, separated from it by a film of thickness $c_d/2$, shown exaggerated in Fig. 2.3, where c_d is the diametral clearance or initial difference in diameters of bearing bore and journal. The velocity gradient is constant when film thickness is uniform. Derive a formula for torque.

Solution

The area, surface velocity, and gradient are, respectively, $A = \pi dl$, $U = \pi dn'$, and $du/dy = \text{constant} = U/(c_d/2) = 2\pi dn'/c_d$. From Eq. (a)

$$F = A\tau_x = A\mu\frac{du}{dy} = (\pi dl)\mu\frac{2\pi dn'}{c_d} = \frac{2\pi^2\mu d^2 ln'}{c_d} \tag{b}$$

and torque is

$$T_0 = \frac{d}{2}F = \frac{\pi^2\mu d^3 ln'}{c_d}. \tag{2.2}$$

This equation was first presented by Petroff in 1883, and it bears his name. It will be used later in a torque ratio for *loaded* bearings.   ////

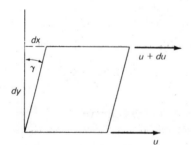

FIGURE 2.2. Shear distortion γ of an element by a difference du in velocity over a thickness dy.

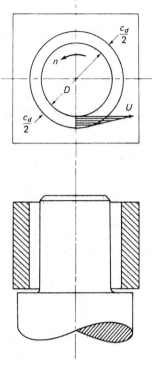

FIGURE 2.3. Fluid velocity between bearing block and concentric journal, typical of unloaded, vertical, guide bearings.

VISCOSITY UNITS

The proportionality term μ in Eqs. (2.1) is known as the *dynamic viscosity, the absolute viscosity*, or simply as the *viscosity*. Its units may be found from Eq. (2.1), solved for μ. Since the unit of du/dy is s^{-1} and $\mu = \tau_x/(du/dy)$, the unit of viscosity is a stress unit times the time unit. The units, together with those for related terms, are given for three systems in Table 2.1. The cgs system is included because many tests have been reported in it. In the earlier systems, the basic viscosity units were given a name, poise, after J. L. M. Poiseuille, and reyn, after Osborne Reynolds, a pioneer in lubrication analysis. In the SI, it is simply Pa·s. They are frequently reported in units that give more convenient numbers, the centipoise (cP), the mPa·s (equivalent of 1 cP), or the microreyn (μreyn). For equations, care must be taken to substitute values in the correct unit, usually Pa·s (the mPa·s value $\times 10^{-3}$) or reyn (the μreyn value $\times 10^{-6}$). Viscosity values at 20°C are given in Table 2.1 for several substances; however, at the temperatures of continuous operation, which are higher, the viscosities of oils are of the order of 1 to 5 μreyn.

Direct measurement of viscosities is difficult. For lubricating oils, the viscosities are commonly determined by observations made of the time required for a fixed quantity of oil to flow out of a cylindrical container through a short tube of small diameter. The most common instrument used in measuring viscosity is the Saybolt Universal Viscometer, and the time of flow is reported as *Saybolt Universal Seconds* (SUS). The SUS are commonly converted to a kinematic viscosity ν in centistokes (cSt), which is 10^{-2} cm²/s, by the equation

$$\nu = 0.22(\text{SUS}) - \frac{180}{(\text{SUS})} \tag{2.3}$$

TABLE 2.1. Viscosity Units and Equivalents

System of Units	Kinematic Viscosity ν	Density ρ	Dynamic Viscosity μ	Equivalents of μ	Typical Values of μ at 20°C (68°F)
SI	m²/s or mm²/s	kg/m³	Pa·s N·s/m² or mPa·s (10^{-3} Pa·s)	10^3 cP 145.04 μreyn cP 0.145 04 μreyn	air 0.0179 mPa·s water^a 1.0 mPa·s oils SAE 10, 110 mPa·s SAE 30, 338 mPa·s
cgs	stoke (St) cm²/s or centistoke (cSt) mm²/s	g/cm³	poise (P) dyn·s/cm² or centipoise (cP) (10^{-2} P)	 mPa·s 0.145 04 μreyn	 water^a 1.0 cP
IPS	in²/s	lb·s²/in⁴	reyn (lb/in²)·s or microreyn (μreyn) (10^{-6} reyn)	6895 Pa·s 6.895×10^6 cP 6.895 mPa·s	air 0.0026 μreyn water 0.145 μreyn oils SAE 10, 18 μreyn SAE 30, 56 μreyn

^{a} Nominal value. More exactly 1.002 mPa·s (cP).

This viscosity is kinematic because it is free of force or mass. Force for the flow comes from the decreasing head of oil being tested and, hence, is a function of oil density. The denser the oil, the greater the force and the less the flow time (SUS), hence, by Eq. (2.3) a smaller kinematic viscosity. To obtain the dynamic or absolute viscosity μ, it is necessary to multiply the kinematic viscosity by the density ρ, which in units of g/cm³ has the same numerical value as the relative density. Its values for petroleum oils are commonly listed in tables under "specific gravity," and for a temperature of 60°F (15.56°C).[1] It is thus convenient to take the listed value, correct it for the temperature at which the SUS value was obtained, and multiply it by ν from Eq. (2.3) to obtain the absolute viscosity μ in centipoises at that temperature, i.e., μ (cP) $= \rho_t$(g/cm³) $\times \nu$ (cSt), where ρ_t is the density corrected for temperature.[2] The numerical value of the result is also in the convenient mPa·s unit of the SI, and it may be converted to μreyns by multiplication with 0.145 (Table 2.1). If data are available in the basic SI units of kg/m³ for ρ and m²/s for ν, then μ comes out in the basic Pa·s of Table 2.1.[3]

The density at any temperature t in °F may be determined from the equation[4]

$$\rho_t = \rho_{60} \frac{1}{1 + \alpha(t-60)} \qquad (2.4)$$

where ρ_{60} is the density at 60°F (15.56°C) and α is the coefficient of expansion of oil (about 0.42×10^{-3} °F^{-1}). The range of relative density of oils at 60°F is 0.87 to 0.91, and if tables are not available, an average value of 0.89 may be used. At a typical operating temperature of 150°F (65.5°C); the average value of ρ_t is 0.86.

[1] D.F. Wilcock and E.R. Booser, *Bearing Design and Application*, New York: McGraw-Hill, p. 398, 1957.
[2] (g/cm³) $\times 10^{-2}$ cm²/s $= 10^{-2}$ (g·cm/s²) (s/cm²) $= 10^{-2}$ dyn·s/cm² $=$ cP.
[3] $\mu = \rho\nu =$ (kg/m³) (m²/s) $=$ kg·m/s² (s/m²) $=$ (N/m²)·s $=$ Pa·s.
[4] See Prob. 2.6 for an alternate form of the equation.

VISCOSITY AND OTHER PROPERTIES

The relationship between temperature and either Saybolt seconds or kinematic viscosity may be plotted as a straight line on charts with special scales, known as the ASTM Standard Viscosity-Temperature Charts for Liquid Petroleum Products D341.[5] For a new oil only the viscosities at two well-spaced temperatures need be determined experimentally and a straight line drawn between them. Straight lines for dynamic viscosity μ vs temperature may be plotted on the kinematic charts with minor error as shown in Fig. 2.4.

FIGURE 2.4. Dynamic or absolute viscosity vs temperature for SAE-numbered oil. (After J. Boyd and A. A. Raimondi, *J. Appl. Mech.*, Vol. 18, *Trans. ASME*, Vol. 73, p. A300, 1951.

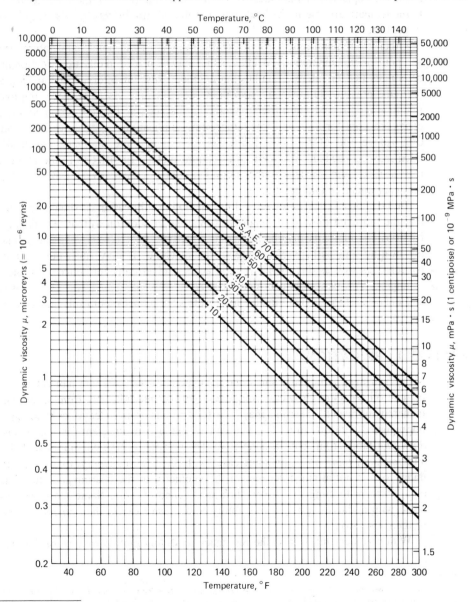

[5] Available in several sizes and ranges from the American Society for Testing and Materials, 1916 Race Street, Philadelphia, Pennsylvania. 19103.

The change of viscosity with temperature is large, and a viscosity index (VI) is used to indicate the relative magnitudes of the change for different oils.[6] A high index (around 100) indicates a smaller change than a low index (around 0).

An oil is usually selected for a viscosity that will be satisfactory at maximum operating loads. At light loads the temperatures are usually less and viscosities greater than at normal loads unless controlled by heaters before the oil is fed to the bearings. At starting, particularly of vehicles and aircraft at low atmospheric temperatures, the viscosity tends to be very high, and an oil with a low slope on the viscosity-temperature chart, i.e., one with a high VI value, should be used.

Pressure increases oil viscosity by appreciable amounts when it is above several thousand lb/in^2. This is significant in ball and roller bearings and in gears and cams, but it is usually neglected in slider bearings with the assumption that the viscosity decrease due to the temperature rise in the loaded area of the oil film offsets the viscosity increase due to pressure.

Other properties limited in oil specifications are flash-point temperatures (volatility) and pour-point temperatures. Additives that may be placed in oils are oxidation inhibitors, detergents (as for internal combustion engines), rust inhibitors, load-carrying additives (extreme-pressure lubricants), pour-point depressants, viscosity-index improvers, de-foaming agents, and emulsifiers. Synthetic oils such as diesters, polyglycols, and silicones are used for high and low temperature applications, e.g. above 250°F and below 0°F. Above 400–500°F, solid lubricants such as graphite, molybdenum disulfide, and mica may be needed. Air, inert gases, liquid metals, or silicone in the absence of air, may also be used. Water or other liquids in which a shaft operates may be used as the lubricant under certain conditions.

In the choice of the lubricant and the method of lubrication, it should be kept in mind that the lubricant may have important functions in addition to that of reducing bearing friction. The lubricant may be needed as a medium for the removal of heat generated in the bearing and perhaps elsewhere in the machine. It may flush out worn-off metal particles, as well as dirt, moisture, and acids inadvertently introduced into the machine or engine. The same lubricant may be used to lubricate gear teeth, cams, piston rings, and clutch surfaces. And, finally, lubricants such as petroleum oils and greases inhibit corrosion.

BOUNDARY LUBRICATION

Many bearings, particularly small and infrequently used ones, are lubricated with grease or with oil fed through wicks from a reservoir or intermittently from an oil can. "Permanently lubricated" bearings may consist of porous bronze sleeves or bushings, factory-impregnated with oil. In all these cases the supply is insufficient for continuous-film or *thick-film* lubrication, but oil adheres to the surfaces to give conditions called *thin-film* and *boundary lubrication*. In some cases with plentiful oil supply, thick-film lubrication under load is not attained because of the geometry of the bearing surfaces. The ability of a lubricant to cling to a given surface is called *oiliness*. Some lubricants, including those with additives, possess this ability to a greater extent than others, independently of their viscosity.[7]

With boundary lubrication the resistance to sliding motion may be of the order of 10

---

[6] D.F. Wilcock and E.R. Booser, *Bearing Design and Application*, New York: McGraw-Hill, pp. 401–403, 1957.

[7] For details on oiliness, thin-film and boundary lubrication, and dry friction see M.D. Hersey, *Theory and Research in Lubrication*, New York: John Wiley & Sons, Inc., Chapt. XIII 1966. See also, D.D. Fuller, *Theory and Practice of Lubrication for Engineers*, New York: John Wiley & Sons, Inc., Chapts. 10 and 11, 1956.

to 100 times greater than in thick-film lubrication. Calculations of torque and power loss, as with dry friction, are then based upon an empirical *coefficient of friction* μ, which is a function of the lubricant, materials, and surface finish, and sometimes of speed and pressure. Load capacity and dimensions may be based on an allowable unit load or average pressure p' corresponding to a type and speed of bearing. Frequently, capacity and dimensions are based on the assumption that wear is proportional to the rate of doing work or producing heat on a unit of projected area. Thus for a design equation, the product $p'V$ in the expression for power is set equal to a constant, where p' is the unit load or pressure based on the projected area and V is the surface velocity of the rotating member. Determined by experience, this constant may be 35 000 to 50 000 for journal bearings and 10 000 for thrust bearings, where p' is in psi and V is in ft/min.

2.2 PRESSURE–VELOCITY RELATIONSHIP: ESTABLISHMENT OF FLOW IN BEARINGS

An element of fluid within a film is, in general, subjected to forces on all sides. For clarity, in Fig. 2.5 only the X components are shown. These components are the normal forces on the $dy\,dz$ surfaces and the shear forces on the $dx\,dz$ surfaces. Inertia and gravity forces are not included because they are relatively small and negligible in most cases. There are no shear forces on the $dx\,dy$ surfaces, since adjacent elements may be considered to have the same X component motion, at least in plane and cylindrical bearings. With a summation of forces and cancellation of certain terms, there remains from the equation $\Sigma F_x = 0$,

$$\frac{\partial \tau_x}{\partial y}\, dy\, dx\, dz - \frac{\partial p}{\partial x}\, dx\, dy\, dz = 0$$

which, together with Eqs. (2.1), gives

$$\frac{\partial p}{\partial x} = \frac{\partial \tau_x}{\partial y} = \mu\, \frac{\partial^2 u}{\partial y^2} \tag{2.5a}$$

Likewise, by consideration of the Z components

$$\frac{\partial p}{\partial z} = \frac{\partial \tau_z}{\partial y} = \mu\, \frac{\partial^2 w}{\partial y^2} \tag{2.5b}$$

where w is the Z component of velocity. In the Y direction normal to one of the bearing surfaces, there is no corresponding equation since the change of pressure across a thin film, $\partial p/\partial y$, is negligible.

If the slope of the velocity profile or its inverse, the gradient $\partial u/\partial y$, is everywhere constant, $\partial^2 u/\partial y^2 = 0$, and from Eq. (2.5a), $\partial p/\partial x = 0$. Hence there can be no pressure buildup within the bearing above that at its boundaries, and no load can be supported.

FIGURE 2.5. Forces in the X direction on an element of fluid.

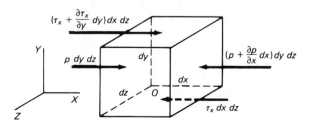

Hence we conclude that a necessary condition for pressure and support is that $\partial p/\partial x \neq$ 0 and that $\partial u/\partial y \neq$ constant, i.e., that the gradient and the slope of the velocity profile must vary across the thickness of the film.

Three methods for establishing a variable slope are in common use.

1. Fluid from a pump is piped to a space at the center of the bearing, developing pressure and forcing fluid to flow outward through the narrow space between parallel surfaces. This is called a *hydrostatic* bearing or an *externally pressurized bearing*.

2. One surface rapidly moves normal to the other, with viscous resistance to displacement of the oil. This is a *squeeze-film bearing*.

3. By positioning one surface so that it is not quite parallel to the other, then by relative sliding motion of the surfaces, lubricant is dragged into the converging space between them. This in effect is a built-in pump. It is a *wedge-film bearing*, and the type generally meant when the word *hydrodynamic bearing* is used. Positioning of the surfaces usually occurs automatically when the load is applied if the surfaces are free of certain constraints.

Under dynamic loads the action of a bearing may be a combination of the foregoing. Hence general equations will be derived and then illustrated by a study of the preceding three methods.

2.3 DERIVATION OF A GENERAL EQUATION RELATING PRESSURE TO VELOCITIES: REYNOLDS' EQUATION

Let a thin film exist between two moving bearing surfaces 1 and 2, the former flat and lying in the X-Z plane, the latter curved and inclined (Fig. 2.6). Component velocities u, v, and w exist in directions X, Y, and Z, respectively. At any instant, two points having the same x,z coordinates and separated by a distance h will have absolute velocities which give the following set of boundary conditions:

$$y = 0 \quad u = U_1 \quad v = V_1 \quad w = W_1$$
$$y = h \quad u = U_2 \quad v = V_2 \quad w = W_2$$

(2.6)

The terms $\partial p/\partial x$ and $\partial p/\partial z$ are independent of y in a thin film, and $\partial p/\partial y = 0$. Rear-

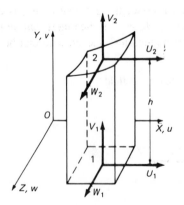

FIGURE 2.6. Component velocities at the two surfaces of a thin film.

rangement and successive integrations of Eq. (2.5a) with respect to y give

$$\frac{\partial u}{\partial y} = \frac{1}{\mu} \frac{\partial p}{\partial x} y + C_1 \qquad \text{and} \qquad u = \frac{1}{\mu} \frac{\partial p}{\partial x} \frac{y^2}{2} + C_1 y + C_2$$

and from the conditions of Eqs. (2.6),

$$C_2 = U_1 \quad \text{and} \quad C_1 = -\frac{1}{\mu} \frac{\partial p}{\partial x} \frac{h}{2} + \frac{U_2 - U_1}{h}$$

Thus

$$u = -\frac{1}{2\mu} \frac{\partial p}{\partial x} (hy - y^2) + \left[U_1 + \frac{y}{h} (U_2 - U_1) \right] \tag{2.7a}$$

Similarly,

$$w = -\frac{1}{2\mu} \frac{\partial p}{\partial z} (hy - y^2) + \left[W_1 + \frac{y}{h} (W_2 - W_1) \right] \tag{2.7b}$$

Each equation shows that a velocity profile consists of a linear portion, the second term to the right of the equals sign, and a parabolic portion which is subtracted or added depending upon the sign of the first term. For velocity u the second term is represented in Fig. 2.7 by a straight line drawn between U_1 and U_2. Since $-(hy - y^2)/2\mu$ is always negative, the sign of the first term is the opposite of the sign of $\partial p/\partial x$ or $\partial p/\partial z$, which are the slopes of the pressure vs position curves. In Fig. 2.7 note the correspondence between positive, zero, and negative slopes of the pressure curve above and the concave (subtracted), straight, and convex (added) profiles of the velocity curves below.

The flow q_x normal to and through a section of area $h\, dz$ is next found (Fig. 2.8(a)). By substitution for u from Eq. (2.7), integration, and application of limits,

$$q_x = \int_0^h u(dy\, dz) = -\frac{h^3}{12\mu} \frac{\partial p}{\partial x} dz + \frac{U_1 + U_2}{2} h\, dz \tag{2.8a}$$

FIGURE 2.7. Pressure and velocity profiles for a thin, nonuniform film with parallel surface velocities U_1 and U_2.

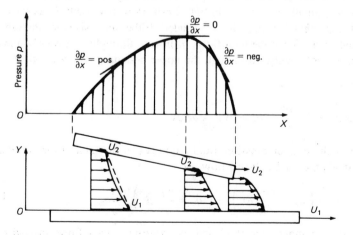

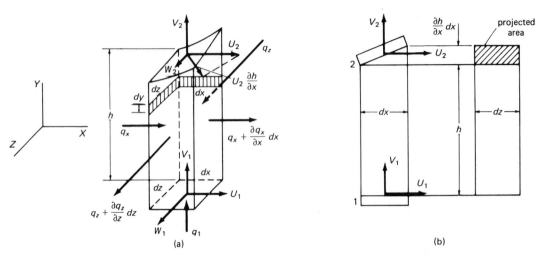

FIGURE 2.8. (a) Flows through a stationary element of space between the two moving surfaces of a thin film. (b) Flow velocities V_1 and V_2 through areas at 1 and 2, respectively, and of U_2 through the projection of the area at 2.

Similarly, through area $h\,dx$,

$$q_z = \int_0^h w(dy\,dx) = -\frac{h^3}{12\mu}\frac{\partial p}{\partial z}dx + \frac{W_1+W_2}{2}h\,dx \tag{2.8b}$$

Note that these flows are through areas of elemental width. Second integrations $\int q_x$ and $\int q_z$ must be made to obtain total flows Q_x and Q_z through a bearing slot.

Figure 2.8(a) represents an elemental geometric space within the fluid, at any instant extending between the bearing surfaces but remaining motionless; through its boundaries oil is flowing. A positive Y velocity V_1 of the lower bearing surface pushes oil *inward* through the lower boundary of the space and gives a flow q_1 in the same sense as the inward flows q_x and q_z. Surface velocities U_1 and W_1 cause no flow through the lower boundary, since the surface is flat and in the X-Z plane. Hence $q_1 = V_1\,dx\,dz$.

Because the top bearing surface is inclined, its positive velocity V_2 causes *outward* flow $V_2\,dx\,dz$. Furthermore, positive velocities U_2 and W_2 together with positive surface slopes $\partial h/\partial x$ and $\partial h/\partial z$ cause *inward* flow. In Fig. 2.8(a) there is shown a velocity component $U_2\,(\partial h/\partial x)$ normal to the top area, that may be taken as $dx\,dz$ because its inclination is very small in bearings. In Fig. 2.8(b) flow at velocity U_2 is shown through the projected area $(\partial h/\partial x)\,dx\,dz$, which is shaded. Either analysis gives the same product of velocity and area. Hence the total flows q_1 inward through the lower boundary of the geometric space and q_2 outward through the upper boundary are, respectively,

$$q_1 = V_1\,dx\,dz$$

$$q_2 = V_2\,dx\,dz - U_2\frac{\partial h}{\partial x}dx\,dz - W_2\frac{\partial h}{\partial z}dz\,dx \tag{2.9}$$

Continuity with an incompressible fluid requires that the total inward flow across the boundaries equals the total outward flow, or

$$q_x + q_z + q_1 = \left(q_x + \frac{\partial q_x}{\partial x}dx\right) + \left(q_z + \frac{\partial q_z}{\partial z}dz\right) + q_2 \tag{2.10}$$

For the case of a compressible fluid and gas bearings, mass flows instead of volume flows would be equated. Density and a relationship between it and pressure must be introduced.[8] With substitution from Eqs. (2.8) and (2.9) into Eq. (2.10), selective differentiation, and elimination of the product $dx\,dz$, there results

$$V_1 = -\frac{1}{12}\frac{\partial}{\partial x}\left(\frac{h^3}{\mu}\frac{\partial p}{\partial x}\right) + \frac{U_1+U_2}{2}\frac{\partial h}{\partial x} + \frac{h}{2}\frac{\partial}{\partial x}(U_1+U_2)$$

$$-\frac{1}{12}\frac{\partial}{\partial z}\left(\frac{h^3}{\mu}\frac{\partial p}{\partial z}\right) + \frac{W_1+W_2}{2}\frac{\partial h}{\partial z} + \frac{h}{2}\frac{\partial}{\partial z}(W_1+W_2)$$

$$+\,V_2 - U_2\frac{\partial h}{\partial x} - W_2\frac{\partial h}{\partial z}$$

With rearrangement,

$$\frac{1}{6}\left[\frac{\partial}{\partial x}\left(\frac{h^3}{\mu}\frac{\partial p}{\partial x}\right) + \frac{\partial}{\partial z}\left(\frac{h^3}{\mu}\frac{\partial p}{\partial z}\right)\right] = (U_1-U_2)\frac{\partial h}{\partial x} - 2(V_1-V_2)$$

$$+\,(W_1-W_2)\frac{\partial h}{\partial z} + h\frac{\partial}{\partial x}(U_1+U_2) + h\frac{\partial}{\partial z}(W_1+W_2) \tag{2.11}$$

The last two terms are nearly always zero since there is rarely a change in the surface velocities U and W, the "stretch-film" case. The stretch-film case can occur when there is a lubricating film separating a wire from the die through which it is being drawn. Reduction in the diameter of the wire gives an increase in its surface velocity during its passage through the die.

This basic equation or hydrodynamic lubrication was developed for a less general case in 1886 by O. Reynolds.[9] More general forms, such as Eq. (2.11), have been needed and thus have been derived by others.[10] As is common practice, we shall refer to Eq. (2.11) and its reduced forms in any coordinate system as *the Reynolds' equation*. Equation (2.11) transformed into cylindrical coordinates is

$$\frac{1}{6}\left[\frac{1}{r}\frac{\partial}{\partial r}\left(r\frac{h^3}{\mu}\frac{\partial p}{\partial r}\right) + \frac{1}{r^2}\frac{\partial}{\partial \theta}\left(\frac{h^3}{\mu}\frac{\partial p}{\partial \theta}\right)\right] = (R_1-R_2)\frac{\partial h}{\partial r} - 2(V_1-V_2)$$

$$+\,(T_1-T_2)\frac{1}{r}\frac{\partial h}{\partial \theta} + \frac{h}{r}\left\{\frac{\partial}{\partial r}\left[r(R_1+R_2)\right] + \frac{\partial}{\partial \theta}(T_1+T_2)\right\} \tag{2.12}$$

where the velocities of the two surfaces are R_1 and R_2 in the radial direction, T_1 and T_2 in the tangential direction, and V_1 and V_2 in the axial direction across the film. Again, for most bearings many of the terms may be dropped, and particularly those terms which imply a stretching of the surfaces.

2.4 HYDROSTATIC OR EXTERNALLY PRESSURIZED BEARINGS

Lubricant from a constant displacement pump is forced into a central recess as shown in Fig. 2.9(a). The lubricant then flows outward between the bearing surfaces, developing pressure and separation and returning to a sump for recirculation. The surfaces may be cylindrical or spherical, or flat with circular or rectangular boundaries. If the surfaces

---

[8] P.R. Trumpler, *Design of Film Bearings*, New York: The Macmillan Company, Sec. 34, 1966, or O. Pinkus and B. Sternlicht, *Theory of Hydrodynamic Lubrication*, New York: McGraw-Hill, 1961.

[9] O. Reynolds, "On the Theory of Lubrication and Its Applications to Mr. Beauchamp Tower's Experiments," *Phil. Trans. Roy. Soc. (London)*, Vol. 177, pp. 157-234, 1886.

[10] Derived in the form shown by J.F. Booker, "Analysis of Dynamically Loaded Journal Bearings: The Squeeze Film Considering Cavitation," Ph.D. Thesis, Cornell University, Ithaca, New York, pp. 151-160, 1961.

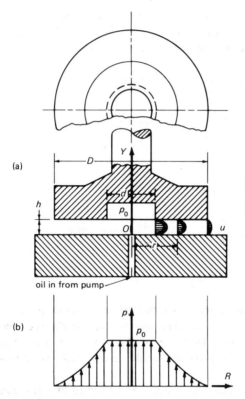

FIGURE 2.9. A circular, hydrostatic (externally pressurized) film. (a) Velocity profiles u. (b) Pressure distribution p.

are flat they are usually guided so that film thickness h is uniform, giving zero values to $\partial h/\partial x$, $\partial h/\partial z$, $\partial h/\partial r$, and $\partial h/\partial \theta$ in the Reynolds' equations. These appear in and cancel out the terms containing the surface velocities, an indication that the latter do not contribute to the development of pressure. This is expected from the discussion following Eq. (3.5). Hence with μ considered constant, Reynolds' equation, Eq. (2.11), is reduced to

$$\frac{\partial^2 p}{\partial x^2} + \frac{\partial^2 p}{\partial z^2} = 0 \tag{2.13}$$

If the pad is circular as shown in Fig. 2.9 and the flow is radial, then $\partial p/\partial \theta = 0$ from symmetry, and Eq. (2.12) is reduced to

$$\frac{\partial}{\partial r}\left(r \frac{\partial p}{\partial r} \right) = 0 \tag{2.14}$$

Equation (2.14) is readily integrated, and together with the boundary conditions of Fig. 2.9, namely $p = 0$ at $r = D/2$ and $p = p_0$ at $r = d/2$, there results,

$$p = p_0 \frac{\ln(D/2r)}{\ln(D/d)} \tag{2.15}$$

The variation of pressure over the entire circle is illustrated in Fig. 2.9(b), from which an integral expression for the total load P may be written and the following expression obtained for p_0,

$$p_0 = \frac{8P}{\pi} \frac{\ln(D/d)}{(D^2 - d^2)} \tag{2.16}$$

An equation for the radial flow velocity u_r may be obtained by putting Eqs. (2.1) and (2.5a) into cylindrical coordinates, or it may be obtained by substituting r for x and $U_1 - U_2 = 0$ in Eqs. (2.7),

$$u_r = -\frac{1}{2\mu}\frac{\partial p}{\partial r}(hy - y^2)$$

Substitution for $\partial p/\partial r$, obtained by differentiation of Eq. (2.15), gives

$$u_r = \frac{p_0(hy-y^2)}{2\mu r \ln(D/d)} = \frac{4P(hy-y^2)}{\pi\mu r(D^2-d^2)} \tag{2.17}$$

the latter term obtained by substitution for p_0 from Eq. (2.16). The total flow through a cylindrical section of total height h, radius r, and length $2\pi r$, is

$$Q = 2\pi r\int_0^h u_r \, dy = \frac{\pi p_0 h^3}{6\mu \ln(D/d)} = \frac{4Ph^3}{3\mu(D^2-d^2)} \tag{2.18}$$

This is the minimum oil delivery required of the pump for a desired film thickness h. Let V be the average velocity of flow in the line, A its cross-sectional area, and η the mechanical efficiency of the pump. Then the power requirement of the pump is

$$pwr = \frac{(p_0A)V}{\eta} = \frac{p_0(AV)}{\eta} = \frac{p_0Q}{\eta} \tag{2.19}$$

In U.S. customary units, this becomes $hp = p_0Q/6600\eta$, where p_0 is measured in psi and Q is measured in in^3/s.

If the circular pad of Fig. 2.9 is rotated with speed n' about its axis, the tangential fluid velocity w_θ may be found from Eq. (2.7b) by substituting $W_1 = 0$, $W_2 = 2\pi rn'$, and for $\partial p/\partial z$ the quantity $\partial p/\partial(r\theta) = (1/r)\,\partial p/\partial\theta$. But since $h = $ constant, $\partial p/\partial\theta = 0$. Thus

$$w_\theta = \frac{y}{h}(2\pi rn') \quad \text{and} \quad \tau = \mu\frac{\partial w_\theta}{\partial y} = \frac{2\pi\mu rn'}{h}$$

The torque required for rotation is

$$T = \int r\,dF = \int r\tau\,dA = \frac{4\pi^2\mu n'}{h}\int_{d/2}^{D/2} r^3\,dr$$

whence

$$T = \frac{\pi^2\mu n'}{16h}(D^4 - d^4) \tag{2.20}$$

If over a portion of a pad the flow path in one direction X is short compared with that in the other direction Z, as shown in Fig. 2.10, the flow velocity w and the pressure gradient $\partial p/\partial z$ will be relatively small and Eq. (2.13) may be approximated by $\partial^2 p/\partial x^2 = 0$, i.e., parallel flow is assumed for a distance b through each slot of approximate width l. Integration of the differential equation, together with the use of the limits $p = p_0$ at $x = 0$ and $p = 0$ at $x = b$ gives the pressure distribution. Integration $\int_{-l/2}^{+l/2} q_x$ from Eqs. (2.8) gives the flow Q across one area bl. The slot equations for the one area are

$$p = p_0\left(1 - \frac{x}{b}\right) \quad \text{and} \quad Q = \frac{p_0 lh^3}{12\mu b} \tag{2.21}$$

The force or torque to move a hydrostatic bearing at slow speeds is extremely small,

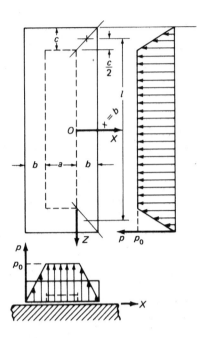

FIGURE 2.10. Pressure distribution over a pressurized, narrow, rectangular bearing pad.

less than in ball or roller bearings. Also there is no difference between starting friction and moving friction. For use in comparisons, a "coefficient of friction" is defined for rotating bearings as the tangential moving force at the mean radius of the active area divided by the normal load applied. Hydrostatic bearings are used for reciprocating platens, for rotating telescopes, for thrust bearings on shafts, and in test rigs to apply axial loads to a member which must be free of any restriction to turning. Journal bearings support a rotating shaft by a different method, but the larger bearings will have a built-in hydrostatic lift for the shaft before it is rotated, to avoid initial metal-to-metal rubbing.

To give stability to a pad, three or four recesses should be located near the edges or in the corners, as shown in Fig. 2.11. However, if one pump is freely connected to the recesses, passage of all the fluid through one recess may occur, tipping the pad and giving no flow or lift at an opposite recess. Orifices must be used in each line from the pump to restrict the flow to a value well below the displacement of the pump, or a separate pump can be used to feed each recess.

Air or inert gases are used to lift and hydrostatically or hydrodynamically support relatively light loads through flat, conical, spherical, and cylindrical surfaces. Unlike oils, air is nearly always present; it does not contaminate a product being processed by the machine; its viscosity *increases* with temperature; and its use is not limited by oxidation at elevated temperatures. Its viscosity is much lower, giving markedly less resistance to motion at very high speeds. Air and gases are compressible, but the equations derived for incompressible fluids may be used with minor error if pressure differences are of the order of 5 to 10 psi. Some applications are given in the problems.[11]

Example 2.2

Design an externally pressurized bearing for the end of a shaft to carry 1000-lb thrust at 1740 rpm, with a minimum film thickness of 0.002 in using SAE 20 oil at 140°F, pumped against a pressure of 500 psi. The overall dimensions should be kept low because of space restrictions. Assume pump mechanical efficiency is 90%.

---

[11] Probs. 2.21–2.23, 2.30, 2.74 and 2.75.

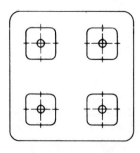

FIGURE 2.11. A pressurized, square bearing pad with corner recesses and narrow sills.

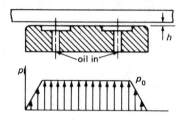

Solution

The choice of a recess diameter d (Fig. 2.9) is a compromise between pad size and pump size. Since a small outside diameter is specified, a relatively large ratio of recess to outside diameter may be tried, giving a 500-psi uniform pressure over a large interior area. Let $d/D = 0.6$. From Eq. (2.16),

$$p_0 = \frac{8P}{\pi}\frac{\ln(D/d)}{D^2-d^2}, \qquad\qquad 500 = \frac{8(1000)\ln 1.67}{\pi D^2(1-0.6^2)}$$

$$D^2 = \frac{(8000)(0.510)}{\pi(0.64)(500)} = 4.060, \quad D = 2.015 \text{ in}$$

$$d = 0.6D = 1.21 \text{ in}$$

From Fig. 2.4, the viscosity of the oil is 3.0×10^{-6} reyn and from Eq. (2.18), the pump must deliver at least

$$Q = \frac{\pi p_0 h^3}{6\mu\,\ln(D/d)} = \frac{\pi(500)(0.002)^3}{6(3.0\times10^{-6})(0.510)} = 1.37\,\frac{\text{in}^3}{\text{s}} = 0.356\,\frac{\text{gal}}{\text{min}}$$

This is a rather high rate for a small bearing. It can be decreased by using a smaller recess or lower pressure, together with a larger outside diameter, or by using a more viscous oil. Also smoother and squarer surfaces will reduce the film thickness requirement, which if halved to 0.001 in would decrease Q to one-eighth, or 0.17 in³/s.

The input power to the pump is, by Eq. (2.19)

$$pwr = \frac{p_0 Q}{6600\eta} = \frac{(500)(1.37)}{(6600)(0.90)} = 0.115 \text{ hp}$$

The rotational speed is $n' = 1740/60 = 29$ rps, and the torque to rotate the bearing is, by Eq. (2.20),

$$T = \frac{\pi^2\mu n'}{16h}(D^4-d^4) = \frac{\pi^2(3.0\times10^{-6})(29)(2.015^4-1.21^4)}{16(0.002)} = 0.384 \text{ lb}\cdot\text{in}$$

The power to rotate at the bearing is

$$pwr = \frac{Tn}{63\ 000} = \frac{(0.384)(1740)}{63\ 000} = 0.0106 \text{ hp}$$

The mean radius of the section of the pad where film shear is high is $r_f = (2.015 + 1.21)/4 = 0.806$ in. At this radius, we may imagine a tangential, concentrated friction force, $F = T/r_f = 0.384/0.806 = 0.476$ lb. The ''coefficient of friction'' is the tangential over the normal force, or $f = 0.476/1000 = 0.000\ 476$.

If lubrication were indifferently provided, with no recess, and the coefficient of friction f were 0.05 at a radius $r_f = 0.67$ in. (from Eq. 3.29, $r_f = (2/3)D$), the power requirement would be

$$pwr = \frac{Tn}{63\ 000} = \frac{frPn}{63\ 000} = \frac{(0.05)(0.67)(1000)(1740)}{63\ 000} = 0.925 \text{ hp}$$

This is eight times the power lost together at bearing and pump in the externally pressured bearing.   ////

2.5 SQUEEZE-FILM BEARINGS

With fluctuating loads, the film is constantly changed in thickness, and some lubricant is alternately pushed out and partly drawn back in at the edges of the bearing or loaded area. Together with make-up oil supplied through correctly located grooves, a parabolic velocity profile with changing slope is obtained (Fig. 2.12), and load-carrying ability is developed without the sliding motion of the film surfaces. The higher the velocity, the greater is the force developed. The squeeze effect may occur on surfaces of all shapes, including shapes that are flat and cylindrical.

For an easy example, the case of a flat circular bearing ring and shaft collar is chosen (Fig. 2.13), and the relationship between applied force, velocity of approach, film thickness, and time is determined. In Reynolds' equation, all surface velocities except V_2 will be zero, and by symmetry $\partial p/\partial \theta = 0$. With the upper surface approaching at a velocity V, $V_2 = dh/dt = -V$. Thickness h is independent of r and θ but a function of time t. Equation (2.12) becomes

$$\frac{1}{6}\left[\frac{1}{r}\frac{\partial}{\partial r}\left(r\frac{h^3}{\mu}\frac{\partial p}{\partial r}\right)\right] = \frac{h^3}{6\mu}\left[\frac{1}{r}\frac{\partial}{\partial r}\left(r\frac{\partial p}{\partial r}\right)\right] = -2V$$

Thus $(\partial/\partial r)(r\ \partial p/\partial r) = -12\mu Vr/h^3$, and by integration twice with respect to r,

$$r\frac{\partial p}{\partial r} = -\frac{6\mu V}{h^3}r^2 + C_1 \quad \text{and} \quad \frac{\partial p}{\partial r} = -\frac{6\mu V}{h^3}r + \frac{C_1}{r}$$

whence

$$p = -\frac{3\mu V}{h^3}r^2 + C_1 \ln r + C_2$$

The boundary conditions are $p = 0$ at $r = D/2$ and at $r = d/2$. Substitution, simultaneous solution for C_1 and C_2, and resubstitution of these values gives for the pressure,

$$p = \frac{3\mu V}{4h^3}\left[D^2 - 4r^2 - (D^2-d^2)\frac{\ln(D/2r)}{\ln(D/d)}\right] \tag{2.22}$$

The total force developed at a given velocity and given film thickness is found by

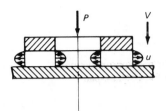

FIGURE 2.12. Velocity profiles in a squeeze film. Annular plate approaching with velocity V.

integration over the surface of the force on an elemental ring, or

$$P = \int_{d/2}^{D/2} p(2\pi r \, dr) = \frac{3\pi\mu V}{32h^3}\left[D^4 - d^4 - \frac{(D^2-d^2)^2}{\ln(D/d)} \right] \tag{2.23}$$

If the force is known as a function of time, the time for a given change in film thickness may be found from Eq. (2.23) by the substitution of $-dh/dt$ for V, the separation of variables, and integration between corresponding limits t', t'' and h', h'', thus

$$\int_{t'}^{t''} P \, dt = -\frac{3\pi\mu}{32}\left[D^4 - d^4 - \frac{(D^2-d^2)^2}{\ln(D/d)} \right]\int_{h'}^{h''} \frac{dh}{h^3}$$

If P is a constant of value W, such as obtained by a weight,

$$t'' - t' = \frac{3\pi\mu}{64W}\left[D^4 - d^4 - \frac{(D^2-d^2)^2}{\ln(D/d)} \right]\left[\left(\frac{1}{h''}\right)^2 - \left(\frac{1}{h'}\right)^2 \right] \tag{2.24}$$

The boundary condition for a solid circular plate at $r = 0$ is different, namely, $\partial p/\partial r = 0$. Use of this beginning with the equation preceding Eq. (2.22) gives

$$p = \frac{3\mu V}{4h^3}(D^2-4r^2), \qquad W = \frac{3\pi\mu V D^4}{32h^3}, \qquad \text{and}$$

$$t'' - t' = \frac{3\pi\mu D^4}{64W}\left[\left(\frac{1}{h''}\right)^2 - \left(\frac{1}{h'}\right)^2 \right] \tag{2.25}$$

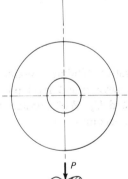

FIGURE 2.13. Squeeze-film bearing consisting of support ring and shaft collar.

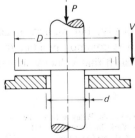

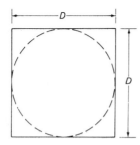

FIGURE 2.14. Circular-pad approximation with increased pressure for a square-pad squeeze-film bearing.

Flat plates of other shapes are not so readily solved. If the length is much greater than the width, it may be treated as a case of unidirectional flow, as was done for the hydrostatic bearing of Fig. 2.10. For a square plate with sides of length D (Fig. 2.14), Fuller[12] takes the average pressure to be 4/3 of that on a circular plate of diameter D, to allow for the increased length of path of the corners:

$$(p_{sq})_{avg} = \frac{4}{3} (p_{circ})_{avg} = \frac{4}{3} \left(\frac{P}{\pi D^2 / 4} \right)$$

$$= \frac{4}{3} \left(\frac{3\pi \mu V D^4}{32 h^3 (\pi D^2 / 4)} \right) = \frac{\mu V D^2}{2 h^3}$$

$$P_{sq} = (p_{sq})_{avg} D^2 = \frac{\mu V D^4}{2 h^3} \tag{2.26}$$

The action of fluctuating loads on cylindrical-bearing films is more difficult to analyze. Squeeze-film action is important in cushioning and maintaining a film in linkage bearings such as those joining connecting rods and pistons in a reciprocating engine. Here, the small oscillatory motion does not persist long enough in one direction to develop a hydrodynamic film. Theoretical and experimental studies,[13] as well as suggested design procedures,[14] are reported in the literature. See also Section 2.13 on dynamic loads.

2.6 WEDGE-FILM THRUST BEARINGS

If the surface of a pad has a certain shape or tilt, relative "sliding" motion between it and a supporting flat surface will force oil between the surfaces and develop load-carrying capacity. In Fig. 2.15 the surface velocities are U_1 and U_2. With constant viscosity μ and with $\partial h / \partial z = 0$, Reynolds' equation, Eq. (2.11), becomes

$$\frac{\partial}{\partial x} \left(h^3 \frac{\partial p}{\partial x} \right) + \frac{\partial}{\partial z} \left(h^3 \frac{\partial p}{\partial z} \right) = 6\mu \, (U_1 - U_2) \frac{\partial h}{\partial x} \tag{2.27}$$

This complete equation has not been solved analytically, but numerical analysis and digital computers may be used for particular cases. It is common practice to assume no side leakage, i.e., a bearing of infinite dimension l such that velocity w and $\partial p / \partial z$ are

[12] D. D. Fuller, *Theory and Practice of Lubrication for Engineers*, New York: John Wiley & Sons, Inc., p. 136, 1956.
[13] O. Pinkus and B. Sternlicht, *Theory of Hydrodynamic Lubrication*, New York: McGraw-Hill, 1961.
[14] R. M. Phelan, *Fundamentals of Mechanical Design*, third edition, New York: McGraw-Hill, pp. 471–478, 1970.

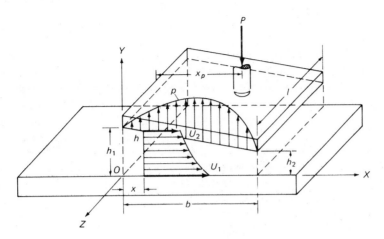

FIGURE 2.15. Pressure and velocity profiles for a hydrodynamic (wedge-film) thrust bearing with flat surfaces.

zero. Equation (2.27) is then simplified to

$$\frac{\partial}{\partial x}\left(h^3\frac{\partial p}{\partial x}\right) = 6\mu\,(U_1-U_2)\,\frac{\partial h}{\partial x} \tag{2.28}$$

Integrating once,

$$\frac{\partial p}{\partial x} = 6\mu\,(U_1-U_2)\left(\frac{1}{h^2}+\frac{C_1}{h^3}\right)$$

For the bearing of Fig. 2.15 with a film thickness at entrance of h_1 and at exit of h_2 (shown greatly exaggerated), let the inclination be $\alpha = (h_1 - h_2)/b$. Then $h = h_1 - \alpha x = h_2 + \alpha(b - x)$ and $\partial h/\partial x = -\alpha$. Hence,

$$p = 6\mu(U_1-U_2)\int\left[(h_1-\alpha x)^{-2}+C_1(h_1-\alpha x)^{-3}\right]dx + C_2$$

$$= \frac{6\mu(U_1-U_2)}{2\alpha(h_1-\alpha x)^2}\,(2h_1-2\alpha x+C_1) + C_2$$

Boundary conditions $p = 0$ at $x = 0$ and at $x = b$ are utilized to obtain

$$p = \frac{6\mu(U_1-U_2)\alpha x(b-x)}{(2h_1-\alpha b)(h_1-\alpha x)^2} = \frac{6\mu(U_1-U_2)\alpha x(b-x)}{(2h_2+\alpha b)[h_2+\alpha(b-x)]^2} \tag{2.29}$$

where the latter is in terms of the minimum film thickness h_2. The total load P is found by integration over the surface.

Machining or mounting the pads within the tolerances required of the very small angle α is difficult to attain, and thus the pads are usually pivoted. The relationship between pivot distance x_p (Fig. 2.15), and the other variables may be found by taking moments about one edge of the pad.[15] Since side leakage does occur, correction factors for the derived quantities have been determined experimentally and are available.[15]

The theory for flat pads indicates that maximum load capacity is attained by locating the pivot at $x_p = 0.578b$, but that there is no capacity if the motion is reversed. For

[15] P.R. Trumpler, *Design of Film Bearings*, New York: The Macmillan Company, pp. 57–102, 1966. See also D.F. Wilcock and E.R. Booser, *Bearing Design and Application*, New York: McGraw-Hill, Chapt. 11, 1957.

bearings with reversals, a natural location is the central one, $x_p = 0.50b$, but flat pad theory indicates zero capacity for this location. However, bearings with central pivots and supposedly flat surfaces have been operating successfully for years. A complete explanation was not available until Raimondi and Boyd[16] theoretically showed that a pad with a crown or convex surface of large radius could develop as large a load capacity with a central pivot as a pad with a theoretically flat surface pivoted at its optimal location. A crown may be unintentionally developed in three ways—by the scraping and lapping process used in an attempt to make the surface flat; by flexing of the loaded pads about their pivot; and by expansion of the surface, heated by the oil film being sheared. The crown of maximum capacity has a height of about half the minimum film thickness.[16] Since the latter is sometimes as low as 0.0003 in, a crown of half that amount may easily develop and is perhaps less easily controlled.

2.7 JOURNAL BEARINGS—ECCENTRICITY AND PRESSURES

In a bearing supplied with sufficient oil, the cylindrical journal, which is that part of a shaft inside the bearing, will rotate concentrically within the bearing under no load, but will move to an increasingly eccentric position as the load is increased, thus forming a wedge film that gives support. The eccentricity e is measured from the bearing center O_b to the shaft center O_j (Fig. 2.16). The maximum possible eccentricity equals the radial clearance c_r, or half the initial difference in diameters, c_d, and it is of the order of one-thousandth of the diameter. It will be convenient to use an *eccentricity ratio*, defined as $\epsilon = e/c_r$. Then $\epsilon = 0.0$ at no load, and ϵ has a maximum value of 1.0 if the shaft should touch the bearing under extremely large loads.

The film thickness h varies between $h_{max} = c_r (1 + \epsilon)$ and $h_{min} = c_r (1 - \epsilon)$. A sufficiently accurate expression for intermediate values is obtained from the geometry shown in Fig. 2.16 where the journal radius is r, the bearing radius is $r + c_r$, and θ is measured counterclockwise from the position of h_{max}. Distance $OO_j = OO_b + e \cos \theta$, approximately, or $h + r = (r + c_r) + e \cos \theta$, whence

$$h = c_r + e \cos \theta = c_r(1 + \epsilon \cos \theta) \tag{2.30}$$

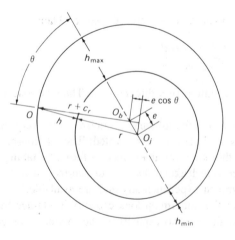

FIGURE 2.16. Relationship between film thickness and eccentricity in a hydrodynamic journal bearing. (Plotted for $\epsilon = 0.50$.)

[16] A.A. Raimondi and J. Boyd, "The Influence of Surface Profile on the Load Capacity of Thrust Bearings with Centrally Pivoted Pads," *Trans. ASME*, Vol. 77, pp. 321–330, 1955.

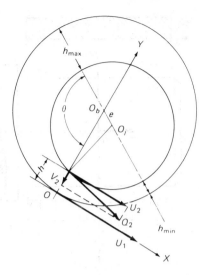

FIGURE 2.17. Surface velocity components in a journal bearing.

The rectilinear coordinate form of Reynolds' equation, Eq. (2.11), is convenient for use here. If the origin of coordinates is taken at any position O on the surface of the bearing (Fig. 2.17), the X axis is a tangent, and the Z axis is parallel to the axis of rotation. Sometimes the bearing rotates, and then its surface velocity is U_1 along the X axis. The surface of the shaft has a velocity Q_2 making with the X axis an angle whose tangent is $\partial h/\partial x$ and whose cosine is approximately 1.0. Hence components $U_2 \approx Q$ and $V_2 = U_2 (\partial h/\partial x)$. With substitution of these terms, Reynolds' equation becomes:

$$\frac{1}{6}\left[\frac{\partial}{\partial x}\left(\frac{h^3}{\mu}\frac{\partial p}{\partial x}\right) + \frac{\partial}{\partial z}\left(\frac{h^3}{\mu}\frac{\partial p}{\partial z}\right)\right]$$

$$= (U_1 - U_2)\frac{\partial h}{\partial x} + 2V_2 = (U_1 + U_2)\frac{\partial h}{\partial x} = U\frac{\partial h}{\partial x} \tag{2.31}$$

where $U = U_1 + U_2$. The same result is obtained if the origin of coordinates is taken on the journal surface with X tangent to it.

Reynolds assumed an infinite length for the bearing, making $\partial p/\partial z = 0$ and endwise flow $w = 0$. Together with μ constant, this simplifies Eq. (2.31) to

$$\frac{\partial}{\partial x}\left(h^3\frac{\partial p}{\partial x}\right) = 6\mu U\frac{\partial h}{\partial x} \tag{2.32}$$

He obtained a solution in series, which was published in 1886.[17] A. Sommerfeld in 1904 found a suitable substitution that enabled him to make an integration to obtain a solution in closed form. The result was

$$p = \frac{\mu U r}{c_r^2}\left[\frac{6\epsilon(\sin\theta)(2 + \epsilon\cos\theta)}{(2 + \epsilon^2)(1 + \epsilon\cos\theta)^2}\right] + p_s \tag{2.33}$$

where p_s is the pressure at $\theta = 0°$ for a 2π-bearing. For a π-bearing, $p_s = 0$.

This result has been widely used, together with experimentally determined end-leakage factors, to correct for finite bearing lengths. It will be referred to as the *Sommerfeld solution* or the *long-bearing solution*.

[17] O. Reynolds, "On the Theory of Lubrication and Its Applications to Mr. Beauchamp Tower's Experiments," *Phil. Trans. Roy. Soc. (London)* Vol. 177, pp. 157–234, 1886.

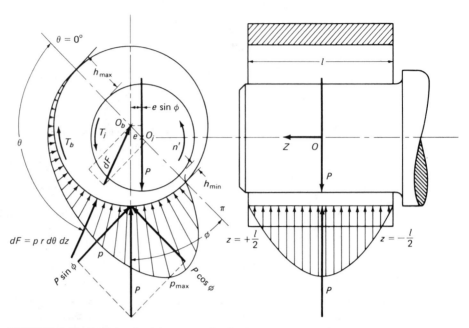

FIGURE 2.18. Radial and axial pressure distributions and resultant loads acting on the converging film of fluid in a journal bearing.

Modern bearings are generally shorter than those used many years ago. The length-to-diameter ratio is often less than 1.0. This makes the flow in the Z direction and the end leakage a much larger portion of the whole. Michell in 1929 and Cardullo in 1930 proposed that the $\partial p/\partial z$ term of Eq. (2.31) be retained and the $\partial p/\partial x$ term be dropped. Ocvirk[18] in 1952 by neglecting the parabolic, pressure-induced flow portion of the u velocity (Eq. (2.7a) and Fig. 2.7), obtained Reynolds' equation in the same form as proposed by Michell and Cardullo, but with greater justification. This form is

$$\frac{\partial}{\partial z}\left(h^3 \frac{\partial p}{\partial z}\right) = 6\mu U \frac{\partial h}{\partial x} \qquad (2.34)$$

Unlike Eq. (2.32), Eq. (2.34) is easily integrated, and it leads to the *load number,* a nondimensional group of parameters, including length, which is useful in design and in plotting experimental results. It will be used here in the remaining derivations and discussion of the principles involved. Results will be presented in chart form for the solution of design problems. It is known as the *Ocvirk solution* or *the short-bearing approximation.*

If there is no flexure or misalignment of the shaft and bearing, h and $\partial h/\partial x$ are independent of z and Eq. (2.34) may be integrated twice to give

$$p = \frac{3\mu U}{h^3} \frac{\partial h}{\partial x} z^2 + \frac{C_1}{h^3} z + C_2$$

From boundary conditions $\partial p/\partial z = 0$ at $z = 0$ and $p = 0$ at $z = \pm l/2$, Fig. 2.18,

$$p = -\frac{3\mu U}{h^3}\left(\frac{l^2}{4} - z^2\right)\frac{\partial h}{\partial x} \qquad (2.35)$$

[18] F.W. Ocvirk, "Short-Bearing Approximation for Full Journal Bearings," Technical Note 2808, Nat, Adv. Comm. for Aeronautics, Washington, D.C. 1952. See also G.B. DuBois and F.W. Ocvirk, "The Short Bearing Approximation for Plain Journal Bearings," *Trans. ASME*, Vol. 77, pp. 1173–1178, 1955.

The slope $\partial h/\partial x = \partial h/\partial(r\theta) = (1/r)\,\partial h/\partial\theta$ and from Eq. (2.30), $\partial h/\partial x = -(c_r\,\epsilon\,\sin\theta)/r$. Substitution into (2.35) gives

$$p = \frac{\mu U}{rc_r^2}\left(\frac{l^2}{4} - z^2\right)\frac{3\epsilon\,\sin\theta}{(1+\epsilon\,\cos\theta)^3} \tag{2.36}$$

This equation indicates that pressures will be distributed radially and axially somewhat as shown in Fig. 2.18, the axial distribution being parabolic. The peak pressure occurs in the central plane $z = 0$ at an angle

$$\theta_m = \cos^{-1}\frac{1-\sqrt{1+24\,\epsilon^2}}{4\,\epsilon} \tag{2.37}$$

and the value of $p_{\max}$ may be found by substituting θ_m into Eq. (2.36).

2.8 JOURNAL BEARINGS—LOAD, ATTITUDE, AND TORQUES

In Fig. 2.18 the pressures shown are those exerted by the bearing upon the *converging* oil film taken as the "free body." These pressures are normal to the film surface along the bearing, and the elemental forces $dF = pr\,d\theta\,dz$ can all be translated to the bearing center O_b and combined into a resultant force. Retranslated, the resultant P shown acting on the film must be a radial force passing through O_b. Similarly, the resultant force of the pressures exerted by the journal upon the film must pass through the journal center O_j. These two forces must be equal, and they must be in the opposite sense and parallel. In the *diverging* half of the film, beginning at the $\theta = \pi$ position, a negative (below atmospheric) pressure tends to develop, adding to the supporting force. This can never be very much, and it is thus neglected. The journal exerts a shearing torque T_j upon the entire film in the direction of journal rotation, and a stationary bearing resists with an opposite torque T_b; however, these are not equal. A summation of moments on the film, say about O_j, gives $T_j = T_b + Pe\,\sin\phi$ where ϕ, the *attitude angle*, is the smaller of the two angles between the line of force and the line of centers. If the bearing instead of the journal rotates, and the bearing rotates counterclockwise, the directions of T_b and T_j in Fig. 2.18 reverse, and $T_b = T_j + Pe\,\sin\phi$. Hence the relationship between torques may be stated more generally as

$$T_r = T_s + Pe\,\sin\phi \tag{2.38}$$

where T_r is the torque from the rotating member and T_s is the torque from the stationary member.

Load P and angle ϕ may be expressed in terms of the eccentricity ratio ϵ by taking summations along and normal to line $O_b O_j$, substituting for p from Eq. (2.36), and integrating[19] with respect to θ and z. Thus from Fig. 2.18

$$P\cos\phi = 2\int_0^{l/2}\int_0^{\pi}(pr\,d\theta\,dz)\cos\theta = \frac{\mu U l^3}{c_r^2}\frac{\epsilon^2}{(1-\epsilon^2)^2}$$

and

$$P\sin\phi = 2\int_0^{l/2}\int_0^{\pi}(pr\,d\theta\,dz)\sin\theta = \frac{\mu U l^3}{c_r^2}\frac{\pi\epsilon}{4(1-\epsilon^2)^{3/2}}$$

---

[19] F.W. Ocvirk, "Short-Bearing Approximation for Full Journal Bearings," NACA TN 2808, Nat. Adv. Comm. for Aeronautics, Washington, D.C., 1952.

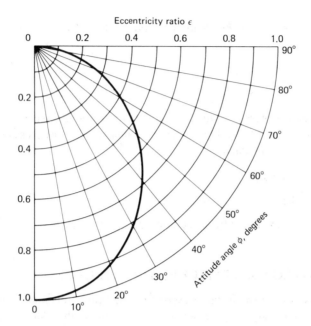

FIGURE 2.19. Path of journal center O_j with change of eccentricity. Eccentricity ratio plotted radially against angular position (attitude angle), Eq. (2.40).

whence

$$P = \sqrt{(P \cos \phi)^2 + (P \sin \phi)^2} = \frac{\mu U l^3}{c_r^2} \frac{\epsilon \left[\pi^2(1-\epsilon^2) + 16\,\epsilon^2\right]^{1/2}}{4(1-\epsilon^2)^2} \tag{2.39}$$

and

$$\phi = \tan^{-1} \frac{P \sin \phi}{P \cos \phi} = \tan^{-1} \frac{\pi \sqrt{1-\epsilon^2}}{4\,\epsilon} \tag{2.40}$$

With increasing load, ϵ will vary from 0 to 1.0, and angle ϕ will vary from 90 to 0°. Correspondingly, the position of minimum film thickness, $h_{\min}$, and the beginning of the diverging half, will lie from 90 to 0° beyond the point where the lines of force P intersect the converging film. The path of journal center O_j as load and eccentricity are increased is plotted in Fig. 2.19 for the configuration and rotation of Fig. 2.18. The fraction containing the many eccentricity terms of Eqs. (2.39) is equal to $Pc_r^2/\mu U l^3$, and although it is not obvious, the eccentricity, like the fraction, nonlinearly increases with increases in P and c_r and with decreases in μ, l, U, and rotational speed n'.

It is important to know the direction of the eccentricity so that parting lines and the holes or grooves that supply lubricant from external sources may be placed in the region of the *diverging* film or where the resistance to entrance is low. The center O_b is not always fixed; e.g. at an idler pulley, the shaft may be clamped, fixing O_j, and the pulley with bearing moves to an eccentric position. A rule for determining the configuration is to draw the fixed circle, then sketch in the circle of the movable member such that the wedge or converging film lies between the two force vectors P acting upon it. The wedge must point in the direction of the surface velocity of the rotating member. This configuration should then be checked by sketching in the vectors of force and torque in the directions in which they act on the film. If the free body satisfies Eq. (2.38), the configuration is correct.

Oil holes or axial grooves should be placed so that they feed oil into the diverging film or into the region just beyond where the pressure is low (Fig. 2.18). This should occur whether the load is low or high, hence, by Fig. 2.19, the hole should be at least in the quadrant 90–180° beyond and not infrequently, in the quadrant 135–225° beyond where load P is applied to the film. The 180° position is commonly used for the hole or groove since it is good for either direction of rotation, and it is often a top position and accessible.

The shearing force dF on an element of surface $(r\, d\theta)\, dz$, by Eq. (2.1), is $(r\, d\theta)\, dz \cdot \mu$ $(\partial u/\partial y)_{y=H}$, where either zero or h must be substituted for H. The torque is $r\, dF$. If it is assumed that the entire space between the journal and bearing is filled with the lubricant, integration must be made from zero to 2π, thus

$$T = \mu r^2 \int_{-l/2}^{+l/2} \int_0^{2\pi} \left(\frac{\partial u}{\partial y}\right)_{y=H} d\theta\, dz = \mu r^2 l \int_0^{2\pi} \left(\frac{\partial u}{\partial y}\right)_{y=H} d\theta \qquad (2.41)$$

The short-bearing approximation assumes a linear velocity profile such that $(\partial u/ \partial y)_{y=0} = (\partial u/\partial y)_{y=h} = (U_1 - U_2)/h$. Use of this approximation in Eq. (2.41) will give but one torque, contrary to the equilibrium condition of Eq. (2.38). However, the result has been found to be not too different from experimentally determined values of stationary-member torque T_s. Hence proceeding to use Eq. (2.41), with h from Eq. (2.30), integrating, substituting $c_r = c_d/2$, $r = d/2$ and $U_1 - U_2 = \pi d\, (n_2' - n_1')$, where n_2' and n_1' are the rotational velocities in rps, there results

$$T_s = \frac{\mu r^2 l (U_2 - U_1)}{c_r} \int_0^{2\pi} \frac{d\theta}{(1 + \epsilon \cos \theta)}$$

$$= \frac{\mu d^2 l (U_2 - U_1)}{c_d} \frac{\pi}{(1 - \epsilon^2)^{1/2}}$$

$$= \frac{\mu d^3 l (n_2' - n_1')}{c_d} \frac{\pi^2}{(1 - \epsilon^2)^{1/2}} \qquad (2.42)$$

Dimensionless *torque ratios* are obtained by dividing T_s or T_r by the no-load torque T_0 of Eq. (2.2), first setting $n' = n_2' - n_1'$. Thus

$$\frac{T_s}{T_0} = \frac{1}{(1 - \epsilon^2)^{1/2}} \qquad (2.43)$$

2.9 JOURNAL BEARINGS—FLOW AND TEMPERATURES

The axial flow out an end of a journal bearing, i.e., the oil "loss" at plane $z = +l/2$ or $z = -l/2$ in Fig. 2.18, may be determined by integration of Eq. (2.8b) over the pressure region of the annular exit area, substituting $r\, d\theta$ for dx. Thus since $W_1 = W_2 = 0$,

$$Q = \int_{\theta_1}^{\theta_2} q_z = -\frac{r}{12\mu} \int_{\theta_1}^{\theta_2} h^3 \frac{\partial p}{\partial z}\, d\theta \qquad (2.44)$$

To determine flow Q_H out the *two* ends of the *converging* area or hydrodynamic film, $\partial p/\partial z$ is obtained from Eq. (2.36), h from Eq. (2.30), and $Q_H = 2Q$ from Eq. (2.44). The limits of integration may be $\theta_1 = 0$ and $\theta_2 = \pi$, or the extent may be less in a partial bearing. However, Q_H is more easily found from the fluid rejected in circumfer-

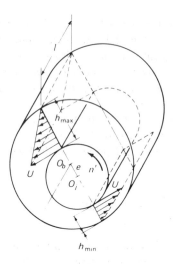

FIGURE 2.20. Velocity profiles at locations of maximum and minimum thickness, short-bearing approximation.

ential flow. With the linear velocity profiles of the short-bearing approximation (Fig. 2.20), and with Fig. 2.16 and Eq. (2.30), the flow is seen to be

$$Q_H = \frac{1}{2} U h_{\max} l - \frac{1}{2} U h_{\min} l = \frac{1}{2} U l [(c_r + e) - (c_r - e)]$$

or

$$Q_H = U l e = \frac{U \, l \, \epsilon \, c_d}{2} = \frac{\pi d (n_2' - n_1') \, l \, \epsilon \, c_d}{2} \tag{2.45}$$

where c_d is the diametral clearance. Although it is not directly evident from this simple result, the flow is an increasing function of load and a decreasing function of viscosity, indicated by the eccentricity term ϵ.

At the ends of the *diverging* space in the bearing, negative pressure may draw in some of the oil previously forced out. However, if a pump supplies oil and distribution grooves keep the space filled and under pressure, there is an outward flow. This occurs through a cylindrical slot of varying thickness, which is a function of the eccentricity. The flow is not caused by journal or bearing motion, and it is designated *film flow Q_F* by Wilcock and Booser.[20] It is readily determined if a central source of uniform pressure p_0 may be assumed, as from a pump-fed partial annular groove. Instead of starting with Eq. (2.44), an elemental flow q_z from one end may be obtained from the flat slot Eq. (2.21) by writing $r \, d\theta$ for l and $(l - a)/2$ for b, where the new l is bearing length and a is the width of the annular groove. Then $Q_F = 2\int_{\theta_1}^{\theta_2} q_z$, where θ_1 and θ_2 define the appropriate angular positions, such as π and 2π, respectively. Additional flow may occur through the short slots which close the ends of an axial groove or through a small triangular slot formed by chamfering the plane surfaces at the joint in a split bearing.[20]

Oil flow and torque are closely related to bearing and film temperature and, thereby, to oil viscosity, which in turn affects the torque. Oil temperature may be predicted by establishing a "heat balance" between heat generated and heat rejected. Heat H_g is generated by the shearing action on the oil; heat H_{oil} is carried away in oil flowing out the ends of the bearing; and by radiation and convection, heat H_b is dissipated from the

[20] D. F. Wilcock and E.R. Booser, *Bearing Design and Application*, New York: McGraw-Hill, pp. 213–218, 1957.

bearing housing and attached parts, and heat H_s from the rotating shaft. In equation form,

$$H_g = H_{oil} + H_b + H_s \tag{2.46}$$

The heat generation rate H_g is the work done by the rotating member per unit time (power loss) converted to the heat units. Thus if torque T_r is in lb·in and n' in rps, the heat generation rate is

$$H_g = \frac{(T_r/12)(2\pi n')}{778} = \frac{T_r n'}{1486} \text{ Btu/s} \tag{2.47}$$

Now T_r and T_0 [Eq. (2.2)] vary as d^3 and n' (except for a small effect on T_r through Eqs. (2.38) and (2.43) of ϵ, a function of n'). Therefore, H_g varies as d^3 and approximately as $(n')^2$. Hence large diameter and high-speed bearings generally require a large amount of cooling, which may be obtained by a liberal flow of oil through the space between the bearing and journal. Flowing out the ends of the bearing, the oil is caught and returned to a sump, where it is cooled and filtered before being returned. The equation for the heat removed by the oil per unit of time is

$$H_{oil} = c_p \gamma Q(t_{out} - t_{in}) \tag{2.48}$$

where c_p is the specific heat of the oil, about 0.48 Btu/(lb·°F) for petroleum oils, and γ is the unit weight of the oil in lb/in^3. The flow Q in Eq. (2.48) may consist of the hydrodynamic flow Q_H, Eqs. (2.45), film flow Q_F, and chamfer flow Q_c as previously discussed, or any others which may exist. The heat lost by radiation and convection may often be neglected in well-flushed bearings.

There is evidence that the outlet temperature t_{out} represents an average film temperature that may be used to determine oil viscosity for bearing calculations, at least in large bearings with oil grooves that promote mixing. The average film temperature is limited to 160° or 180°F in most industrial applications, although it may be higher in internal combustion engines. Higher temperatures occur beyond the place of minimum film thickness and maximum shear. They may be estimated by an equation given by Wilcock and Booser,[21] and they are of interest in severe applications where they may approach the softening temperature of the bearing metal.

In self-contained bearings, those lubricated internally as by drip, waste packing, oil-ring feed, or oil bath (immersion of journal), dissipation of heat occurs only by radiation and convection from bearing housing, connected members, and shaft. Studies of pillow block and pedestal-mounted bearings by Lasche, Karelitz, Fuller, and others have been directed toward experimentally obtaining overall dissipation coefficients K for still air and for moving air. These dissipation coefficients are used in an equation $H_b = KA(t_b - t_a)$, where A is some housing or bearing surface area or projected area, t_b is the temperature of its surface, and t_a is the ambient temperature.[22] Fuller presents the increment of oil-film temperature above surface temperature t_b as a function of the type of lubrication. The type giving the largest interior agitation of the oil has the smallest increment. The increment is roughly equal to $(t_b - t_a)$ with oil-ring lubrication, more than double for waste pack, and less than half for oil bath.

Radiation and convection coefficients are available for flat and cylindrical, stationary and rotating surfaces. Convection coefficients increase considerably at and above certain

[21] D. F. Wilcock and E. R. Booser, *Bearing Design and Application*, New York: McGraw-Hill, p. 222, 1957.
[22] D. D. Fuller, *Theory and Practice of Lubrication for Engineers*, New York: John Wiley & Sons, Inc., pp. 242–249 and the references cited therein, 1956.

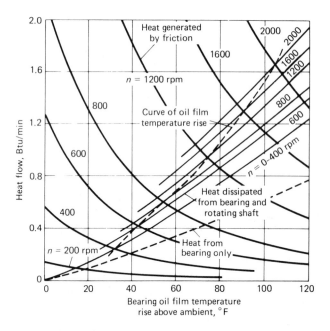

FIGURE 2.21. Heat balance diagram calculated for an unloaded bearing and a short shaft at various speeds. Bearing 1.375 in diameter by 1.50 in long and a heat-dissipating shaft length of 3.0 in. (From A. H. Burr, *Trans. ASLE*, Vol. 2, p. 239, 1960. Reprinted by permission of the American Society of Lubrication Engineers.)

critical speeds for which the surface generates its own turbulent fanning action. Burr[23] used these radiation and convection coefficients and heat transfer theory to develop equations for the temperatures at two bearings and along the shaft between them and overhung on one, which is a typical configuration. Figure 2.21 is a heat balance diagram calculated for a bearing and relatively short shaft, tested without load so that oil flow is negligible. Plotted against temperature, a heat-generated curve and a corresponding total heat-dissipated curve intersect to define the equilibrium temperature for a particular speed. Temperatures for several speeds are joined by a dashed line, which checked well with test measurements. At any temperature, the distance between a total heat-dissipated line and the dotted H_b, the bearing-only curve, represents the heat H_s dissipated by the shaft. It is seen that for the higher speeds, heat dissipation from the shaft was greater than from the bearing surface. For load tests on the same bearing, oil at 15 psi was supplied, and it was necessary to include H_{oil} in the heat balance since its contribution to heat dissipation was of the same order of magnitude as that from the housing and shaft together.

2.10 LOAD NUMBER AND DESIGN FOR BEARING SIZE

A transformation of the short-bearing load equation, Eq. (2.39), into a useful nondimensional form may be made. One substitution is a commonly used measure of the intensity of bearing loading, the *unit load p'*, which is the load P divided by the projected bearing area ld, thus

$$p' = \frac{P}{ld} \tag{2.49}$$

[23] A. H. Burr, "The Effect of Shaft Rotation on Bearing Temperatures," *Trans. Amer. Soc. Lubrication Engineers*, Vol. 2, No. 2, pp. 235–241, 1960.

where l is the bearing length, d is the nominal bearing diameter, and p' has the same units as pressure. The surface velocity sum, $U = U_1 + U_2$, is replaced by $\pi d(n_1' + n_2')$ $= \pi dn'$, where $n' = n_1' + n_2'$ is the sum of the rotational velocities. Also, c_r may be expressed in terms of the more commonly reported diametral clearance c_d, where $c_r = c_d/2$. Let the nondimensional fraction containing ϵ in Eq. (2.39) be represented by E. By transposition of Eq. (2.39), followed by substitution,

$$E = \frac{Pc_r^2}{\mu Ul^3} = \frac{(p'ld)(c_d/2)^2}{\mu(\pi dn')l^3}\frac{d}{d} = \frac{1}{4\pi}\frac{p'}{\mu n'}\left(\frac{d}{l}\right)^2\left(\frac{c_d}{d}\right)^2 = \frac{1}{4\pi}N_L$$

where N_L, called the *load number* or the *Ocvirk number*, is the product of the three nondimensional ratios. Then, from the preceding equation and the definition of E by Eq. (2.39),

$$N_L = \frac{p'}{\mu n'}\left(\frac{d}{l}\right)^2\left(\frac{c_d}{d}\right)^2 = 4\pi E = \frac{\pi\epsilon[\pi^2(1-\epsilon^2) + 16\epsilon^2]^{1/2}}{(1-\epsilon^2)^2}$$

$$= \frac{\pi^2\epsilon(1+0.621\epsilon^2)^{1/2}}{(1-\epsilon^2)^2} \tag{2.50}$$

The load number puts together the dimensions and parameters over which the engineer has a choice. Equation (2.50) indicates that all combinations of unit load p' or P/ld, viscosity μ, speed sum n', length-diameter ratio l/d, and clearance ratio c_d/d that give the same value of N_L will give the same eccentricity ratio ϵ. Eccentricity ratio is a measure of the proximity to failure of the oil film since its minimum thickness is $h_{min} = c_r(1 - \epsilon)$. Hence the load number is a valuable design aid. Also it has been very useful in tests to indicate parameters to choose so that their combinations give a good spread of points on curves of eccentricity, peak pressure, torque ratio, and oil-flow factor.

Eccentricity ratio ϵ and thickness ratio h_{min}/c_r are given as functions of load number N_L in Fig. 2.22. The curve "short-bearing approximation" is a plot of Eq. (2.50). Results of extensive experimentation[24] are plotted as the "experimental" curve. The trend of the latter supports the analysis and vice versa. However, the actual eccentricities are larger. This is as expected since the approximation made in the short-bearing theory neglects the pressure term of the velocity equation, Eq. (2.7a) and the pressure-induced flow in Eq. (2.8a). With this extra boost the same flow occurs through a narrower slot at the minimum thickness position. For design purposes, use of the experimental curve is recommended. Data for plotting the experimental curve were obtained on bearings with l/d ratios between ¼ and 1, i.e., on "short" bearings. However, subsequent tests showed that the same experimental curve may be used to determine eccentricity ratios when l/d is between 1 and 2 provided the value 1.0 is substituted in the load number equation, Eq. (2.50), for $(d/l)^2$, which is equivalent to omitting this term. This is done with less accuracy if N_L values are less than 20, but the error is "on the safe side" for design purposes.

Example 2.3

A journal bearing is to be designed to fit in a limited axial space; hence its length will be made half of its diameter. Use of a moderately high load number of 30 will save further space. A common diametral clearance-to-diameter ratio for a short bearing is

[24] G. B. DuBois, F. W. Ocvirk, and R. L. Wehe, "Experimental Investigation of Eccentricity Ratio, Friction, and Oil Flow of Long and Short Journal Bearings with Load Number Charts," NACA TN 3491, Washington, D.C., 1955. or G. B. DuBois and F. W. Ocvirk, "The Short Bearing Approximation for Plain Journal Bearings," *Trans. ASME*, Vol. 77, pp. 1173–1178, 1955.

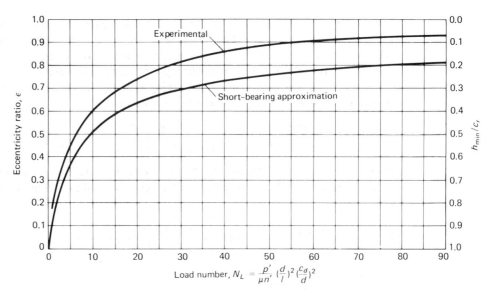

FIGURE 2.22. Eccentricity ratio and minimum film-thickness ratio vs load number for short bearings ($l/d \leq 1.0$). For design, use the experimental curve. For longer bearings use $l/d = 1.0$ when calculating the load number. (After G. B. DuBois, F. W. Ocvirk, and R. L. Wehe, "Experimental Investigation of Eccentricity Ratio, Friction, and Oil Flow of Long and Short Journal Bearings with Load-Number Charts," Tech. Note 3491, Fig. 1(c), Nat. Adv. Comm. for Aeronautics (now NASA), Washington, D.C., (1955.)

0.001. The speed is 3000 rpm and the load is 2500 lb. Oil of SAE 20 specifications, operating in the bearing at 160°F, is proposed. Determine the bearing dimensions and the minimum film thickness.

Solution

From the assumptions, $P = 2500$ lb, $l/d = 0.50$, $c_d/d = 0.001$, $N_L = 30$, $n' = 3000/60 = 50$ rps. Viscosity is $\mu = 1.97 \times 10^{-6}$ reyn, from Fig. 2.4, and $t = 160°F$.

Dimensions. From Eq. (2.50),

$$N_L = \frac{p'}{un'}\left(\frac{d}{l}\right)^2\left(\frac{c_d}{d}\right)^2, \quad 30 = \frac{p'}{(1.97\times10^{-6})(50)}\left(\frac{1}{0.50}\right)^2 (0.001)^2,$$

whence unit load is $p' = 739$ psi. From Eq. (2.49) and $l = 0.5d$,

$$ld = \frac{P}{p'}, \quad 0.5d^2 = \frac{2500}{739}, \quad d^2 = 6.77, \quad d = 2.60 \text{ in}, \quad l = 1.30 \text{ in}$$

The diametral clearance is $c_d = (c_d/d)d = (0.001)(2.60) = 0.0026$ in. Nominal "bearing size" $d \times l$ might be rounded off to $2\frac{5}{8}$ in $\times$ $1\frac{5}{16}$ in. in common fractions.

Film thickness. From the chart (Fig. 2.22), the eccentricity ratio from the experimental curve corresponding to $N_L = 30$ is $\epsilon = 0.815$. Hence

$$e = \epsilon c_r = (0.815)\left(\frac{0.0026}{2}\right) = 0.001\ 06 \text{ in}$$

and the minimum film thickness is

$$h_{min} = c_r(1-\epsilon) = c_r - e = 0.001\ 30 - 0.001\ 06 = 0.000\ 24 \text{ in} \qquad ////$$

The eccentricity must be limited, since when its value approaches 1.0, there is danger of failure. Surface asperities and foreign particles in the oil may break through the thin film, scoring metallic surfaces. Also,, because of slope deflection and misalignment of the shaft, the eccentricity will be greater at one end of the bearing.[25] With no film, metal-to-metal contact will cause high temperatures and probable seizure of the soft-bearing lining of low-melting temperature. If the bearing is a harder metal, such as bronze, both bearing and shaft may be scored. The eccentricity e is more directly related to the height of surface asperities than is the eccentricity ratio ϵ, and eccentricity e is used by some as the criterion of failure. There may be a less direct but important relationship of bearing size to asperity heights, bearing out of roundness, and the expense of replacement, thus making the ratio ϵ and its corresponding load number more conservative criteria for general use.

The choice of load number or eccentricity to use in a design must depend upon a number of things, such as accuracy and uniformity of bearing and journal, their surface finish, flexibility and alignment, and operation and maintenance practice and conditions, in other words, on experience under similar conditions. The suggestions which follow are for guidance only. G. B. DuBois has suggested that load numbers of 30, 60, and 90, corresponding to eccentricities of 0.82, 0.90, and 0.93, be considered the upper limits for "moderate," "heavy," and "severe" loading, respectively. The lowest range, up to $N_L = 30$, may be suitable for bearings in general, with the higher ranges useful for carefully manufactured and operated bearings. Thus operation of large diameter bearings at $\epsilon = 0.95$ under ideal conditions has been reported. Another guide may be based on the common practice of using clearance ratios c_d/d between 0.001 and 0.002. With a typical minimum film thickness ratio $h_{\min}/d = 0.00015$ and an average clearance ratio of 0.0015, the eccentricity is $\epsilon = e/c_r = (c_r - h_{\min})/c_r = 1 - 2\,h_{\min}/c_d = 1 - 2(0.000\,15d)/0.0015d = 0.80$, corresponding to $N_L = 28$. If one determines experimentally an average load number $(N_L)_f$ at which failures occur, then a "factor of safety" might be defined as $(N_L)_f/(N_L)_d$, where $(N_L)_d$ is the design number. A ratio of load numbers is suggested because it is related to unexpected changes or uncertain values of load, viscosity, speed, and clearance.

2.11 FURTHER DESIGN CONSIDERATIONS WITH LOAD NUMBER: ALTERNATE METHODS

Peak pressures may cause failures in the bearing metal, particularly under repeated loads such as occur in connecting-rod bearings. Fatigue cracks initiate radially and spread circumferentially and longitudinally, causing separation or "spalling." From Eqs. (2.36), (2.37), and (2.49) an equation for the ratio $p_{\max}/p'$ may be derived and written in terms of the load number. It is plotted in Fig. 2.23. Experimental points are scattered with some points directly on the analytical line and other points on both sides of the analytical line.

High temperatures, which are related to torque and oil flow, cause deterioration of oil and bearing material. They lower viscosity and increase eccentricity. Torques may be plotted against load number (Fig. 2.24) in the form of ratios to the no-load or Petroff torque T_0 of Eq. (2.2). The stationary-element torque ratio T_s/T_0, given by Eq. (2.43), is a function of ϵ and hence N_L only and plots as a single line. With the bearing as the

---

[25] G. B. DuBois, F. W. Ocvirk, and R. L. Wehe, "Properties of Misaligned Bearings," *Trans. ASME*, Vol. 79, pp. 1205–1212, 1957.

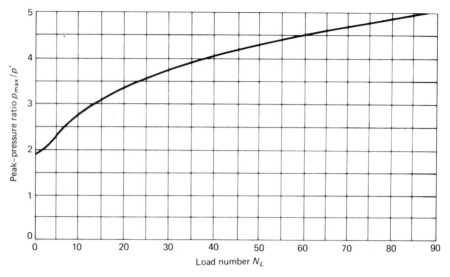

FIGURE 2.23. Peak-pressure ratio vs load number [from Eqs. (2.36), (2.37), and (2.50)].

stationary element, torques are readily measured on a test machine. Converted to the ratio T_s/T_0, experimental values are found dispersed on both sides of the analytical curve, checking it fairly well.[26] The rotating-element torque ratio T_r/T_0 must be obtained analytically by adding to T_s/T_0 the load-couple ratio $P(e \sin \phi)/T_0$, from Eq. (2.38), using Eq. (2.40) or Fig. 2.19 for attitude angle ϕ. Using Eqs. (2.39) for P it is possible to make the addition on Fig. 2.24 to give T_r/T_0 vs N_L but with l/d as an additional parameter. This is done for l/d values of ¼, ½, and 1. Torque T_r is found by the multiplication $T_0(T_r/T_0)$. It is the torque that does work and gives the heat generated, Eq. (2.47).

FIGURE 2.24. Friction-torque ratios vs load number. Torques are T_0 at no-load, T_r on rotating element, and T_s on stationary element. (After DuBois, Ocvirk, and Wehe, NACA TN 3491—reference in caption of Fig. 2.22.)

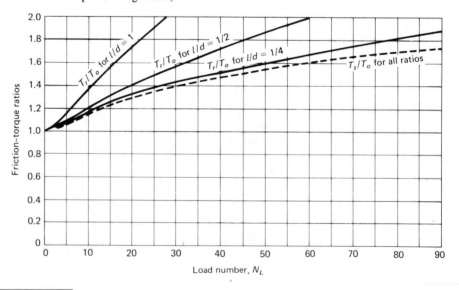

[26] G. B. DuBois and F. W. Ocvirk, "The Short Bearing Approximation for Plain Journal Bearings," *Trans. ASME*, Vol. 77, pp. 1173–1178, 1955.

Example 2.4

Continue Example 2.3 to determine peak pressure, heat and power loss, hydrodynamic oil flow, and flow to maintain a temperature of 160°F.

Solution

Peak pressure. From Fig. 2.23, $p_{max}/p' = 3.75$, whence

$$p_{max} = 3.75p' = 3.75(739) = 2770 \text{ psi}$$

Torque, heat, and power loss at bearing. The no-load (Petroff) torque is, from Eq. (2.2),

$$T_0 = \frac{\pi^2 \mu d^3 l n'}{c_d} = \frac{\pi^2 (1.97 \times 10^{-6})(2.60)^3 (1.30)(50)}{0.0026} = 8.53 \text{ lb} \cdot \text{in}$$

From Fig. 2.24, the torque ratio for the rotating member, corresponding to $l/d = 0.5$ and $N_L = 30$, is $T_r/T_0 = 1.57$. Hence

$$T_r = 1.57T_0 = 1.57(8.53) = 13.4 \text{ lb} \cdot \text{in}$$

and the loss, from Eq. (2.47), is

$$H_g = \frac{T_r n'}{1486} = \frac{(13.4)(50)}{1486} = 0.451 \text{ Btu/s } (0.638 \text{ hp})$$

Hydrodynamic oil flow. The flow, by Eqs. (2.45), is

$$Q_H = \pi d n' l \epsilon c_d / 2 = \pi (2.60)(50)(1.30)(0.815)(0.0026)/2 = 0.563 \text{ in}^3/\text{s}$$

Flow to maintain a temperature of 160°F. For SAE 20 oil, relative density $\bar{\rho}_{60} = 0.885$, specific heat capacity $c_p = 0.48$ Btu/(lb. · °F), coefficient of volume expansion $\alpha = 0.000\,42$, unit weight of water at 60°F = 0.0361 lb/in³. The relative density of the oil at 160°F is, from Eq. (2.4),

$$\bar{\rho}_{160} = \bar{\rho}_{60} \frac{1}{1 + \alpha(t-60)} = 0.885 \left[\frac{1}{1 + (0.000\,42)(160-60)} \right] = 0.85$$

and $\gamma = (0.85)(0.0361) = 0.0307$ lb/in³. We shall assume relatively little cooling from housing and shaft, a safe side assumption since it will give a higher calculated capacity requirement for the pump and allow for other uncertainties. We shall also assume that oil-cooling equipment will be available to give a controlled reentry temperature of 120°F. Then from Eqs. (2.46) and (2.48),

$$H_g = H_{oil} = c_p \gamma Q(t_{out} - t_{in})$$

By substitution and solution for Q,

$$0.451 = (0.48)(0.0307)Q(160 - 120) \quad \text{and} \quad Q = 0.765 \text{ in}^3/\text{s} \qquad ////$$

In the preceding example the flow Q required to limit the temperature was larger than the hydrodynamic flow Q_H from the converging space. Flow can be increased by an increase in the inlet pressure, by grooves or slots in the diverging region, and by an increase in clearance. The latter through a rapid increase in load number, Eq. (2.50), increases the eccentricity and decreases the minimum film thickness but at a slow rate (Fig. 2.22). Flow Q_H is increased linearly with c_d and a bit more by the increase in eccentricity [Eqs. (2.45)]. Furthermore, torque and heat generated may be decreased by the increase in clearance. An alternative to an increased flow of oil is to carry away more heat in the same flow by additional outside cooling of the oil, i.e., decreasing the inlet temperature in Eq. (2.48).

Corresponding to the Ocvirk load number N_L that follows from the short-bearing approximation, there is the Sommerfeld number S from the long-bearing approximation. From Eq. (2.33) and integrations similar to those in Section 2.8, Sommerfeld determined a load equation

$$P = \frac{\mu U l r^2}{c_r{}^2} \frac{12\pi\epsilon}{(2 + \epsilon^2)(1 - \epsilon^2)^{1/2}} \tag{2.51}$$

The integration was made from zero to 2π. Equation (2.33) indicates a negative pressure in the region from π to 2π, but no negative pressure greater than the atmospheric value can be maintained. The theory also indicates an eccentricity in a direction normal to the load direction. However, with proper correction factors the Sommerfeld solution for a "2π-bearing" has been and is extensively used. The Sommerfeld number follows from a rearrangement of Eq. (2.51), similar to the rearrangement made to obtain Eq. (2.50), thus

$$\frac{(2+\epsilon^2)(1-\epsilon^2)^{1/2}}{12\pi\epsilon} = \frac{\mu U l}{P} \left(\frac{r}{c_r} \right)^2 = \mu \frac{(\pi d n')l}{p'ld} \left(\frac{d}{c_d} \right)^2$$

whence

$$\frac{(2+\epsilon^2)(1-\epsilon^2)^{1/2}}{12\pi^2\epsilon} = \frac{\mu n'}{p'} \left(\frac{d}{c_d} \right)^2 = S \tag{2.52}$$

As might be expected for an "infinite-length" bearing, S is independent of a length ratio. When l/d in the Ocvirk load number is taken equal to 1.0, as was found necessary when $l/d > 1.0$, it is seen that $S = 1/N_L$. Charts of bearing characteristics as a function of S, as determined from the infinite-bearing equations, have been extensively used when modified by experimentally determined end-leakage factors.

In 1958, Raimondi and Boyd[27] published an extensive set of charts and tables based upon digital computer solutions of Reynolds' equation for a "finite bearing," using both terms on the left-hand side of Eq. (2.11). It is assumed that negative pressure areas give no support. Curves are plotted against Sommerfeld number S for l/d values of ¼, ½, 1, and ∞. They are given for full journal bearings and for "partial bearings" of an angular extent as low as 60°, where the closely fitting portion is reduced to minimize bearing torque, or for ease in assembly and maintenance.

2.12 BEARING CAPACITY WITH ROTATING LOADS AND SLEEVES

For journal bearings Reynolds' equation in the form of Eq. (2.31) was found to be a function of the sum of the two surface velocities of the fluid film, namely, $U_1 + U_2$, for which sum the symbol U was substituted. In the derivation of load number N_L, the substitution $U = \pi d n'$ was made, where $n' = n_1' + n_2'$, the sum of the rotational velocities of the two surfaces. Load number was found to be inversely proportional to n' and hence to $n_1' + n_2'$.

It is important to note that velocities U_1 and U_2 were taken relative to the line of action of load P, which was considered fixed in Figs. 2.17 and 2.18. If the load rotates, the same physical relations occur between the film surfaces and load provided the surface velocities are measured relative to the line of action of the load. Thus all previous

[27] A. A. Raimondi and J. Boyd, "A Solution for the Finite Journal Bearing and Its Application to Analysis and Design, Parts I, II, and III," *Trans. Am. Soc. Lubrication Engineers*, Vol. 1, No. 1, New York: Pergamon Press, pp. 159–209, 1958. Also for the 360° bearing in J. E. Shigley, *Mechanical Engineering Design*, third edition, New York: McGraw-Hill, 1977. Reprinted for the 120° bearing in *Standard Handbook of Lubrication Engineering*, J. J. O'Connor and J. Boyd, Eds., New York: McGraw-Hill, pp. 5–38 to 5–56, 1968.

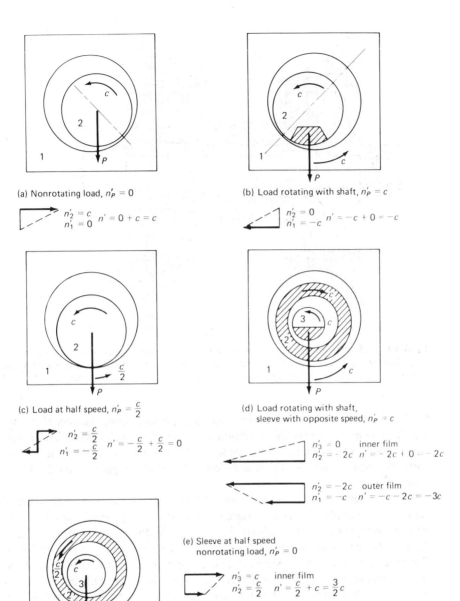

FIGURE 2.25. Effect of load and sleeve rotations on fluid film capacity. Values indicated for n' (without subscript) are to be used in the load-number calculation, Eq. (2.50) or Fig. 2.22. Load capacities (other factors remaining unchanged) are proportional to the areas of the vector diagrams.

equations apply if the rotational velocities n_1' and n_2' are measured relative to the rotational velocity n' of the load.

Several examples are given in the schematics of Fig. 2.25.[28] Member 1 is the block, and it is fixed or nonrotating and represents the support of the machine. At least one other member is driven at a rotational speed c relative to it. The load source and the two

[28] Additional examples were given by A. O. DeHart, "Which Bearing and Why," Design Engineering Conference, Philadelphia, Pennsylvania, May 25–28, 1959. Printed as ASME paper 59–MD–12 by the American Society of Mechanical Engineers, New York.

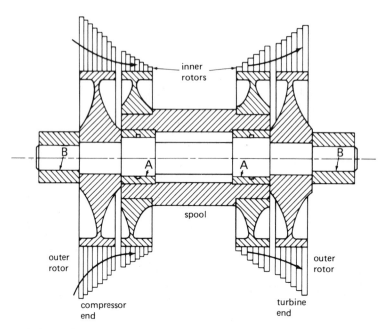

FIGURE 2.26. Jet-engine compressor driven by gas turbine. Outer (low pressure) rotors rotating with shaft at one speed, inner (high pressure) rotors on spool rotating at a higher speed, speeds controlled by gas flow. Film capacities similar to those of Fig. 2.25(e). See also Prob. 2.83.

fluid films are not necessarily at the same axial position along the central member. Thus Fig. 2.25(e) may illustrate the condition in a type of jet engine which has a central shaft with bladed rotors turning at one speed in bearings B (Fig. 2.26), and a hollow spool with additional rotors turning at a different speed and supported on the first shaft by bearings A.

Vector diagrams and the sum of the relative velocities are shown for each film. The sum $|n'|$, a multiple of c, is a measure of the load capacity, which equals c in a standard bearing arrangement (Fig. 2.25(a)). If $n' = 2c$, as for the inner film of Fig. 2.25(d), then for the same load number and eccentricity, the load capacities p' and P are doubled. The area in the diagram formed by drawing a straight line between the tips of the vectors is also a measure, since it represents the quantity of fluid flow with the Ocvirk assumption of a linear velocity profile. Thus the zero "capacity" of the bearing in Fig. 2.25(c) can be explained by the zero net area and net flow of fluid. This situation has been of some concern for crankpin bearings of four-stroke-cycle engines. The same zero net flow can occur between the bushing of an idler gear and the shaft that supports it if they turn with opposite but equal-magnitude velocities relative to a nonrotating load on the gear. The high capacity of the films in Fig. 2.25(d) is also explained by the vector diagram areas, e.g., the area for the outer film is three times that at the bearing in Fig. 2.25(a), indicating a three times greater "pumping action" and load capacity.

The analyses of Fig. 2.25 give some ideas on relative capacities that can be attained and indicate the care that must be taken in determining n' for substitution in the load number equation. However, it should be noted that the load numbers and actual film capacities are not a function of n' alone. The diameters d and lengths l of the two films may be different, giving different values to $p' = P/ld$ and to $(d/l)^2$ in the load number, but they may be adjusted to give the same load number. Also, a load rotating with the shaft (Fig. 2.25(b)) appears to give the bearing the same capacity $|-c|$ as the bearing of Fig. 2.25(a). However, unless oil can be fed through the shaft to a hole opposite the

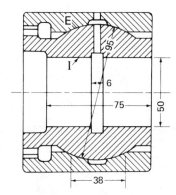

FIGURE 2.27. Self-aligning bearing with film lubrication of nonrotating spherical seats provided by a rotating load (Example 2.5).

load, it will probably be necessary to feed oil by a central annular groove in the bearing so that oil is always fed to a space at low pressure. With pressure dropping to the oil-feed value at the groove in the converging half, the bearing is essentially divided into two bearings of approximately half the l/d ratio. Since d/l is squared in the load-number equation, each half of the bearing has one-fourth the allowable unit load p', and the whole bearing has one-fourth the load capacity P of the bearing in Fig. 2.25(a).

Example 2.5[29]

In a certain shaking device, an off-center weight provides a centrifugal force of 26 000 N, rotating at 3600 rpm. This force is midway between the ends of the shaft, and it is shared equally by two bearings. Self-alignment of the bushing is provided by a spherical seat, plus loosely fitting splines to prevent rotation of the bushing about the axis of the shaft (Fig. 2.27). Oil of 10.3 mPa·s viscosity will be provided for lubrication of the interior surfaces at I and the exterior surfaces at E. The diametral clearance ratio is 0.0015 at both places, and the central annular groove at I has a width of 6 mm. Determine the load numbers and minimum film thicknesses at I and E.

Solution

Surfaces I. Relative to the load, the velocity of the bushing surface is $n_1' = -3600/60 = -60$ rps and that of the shaft is $n_2' = 0$. Hence $n' = n_1' + n_2' = -60 + 0 = -60$ rps. Each bearing, carrying $26\ 000/2 = 13\ 000$ N, is divided by the oil groove into two effective lengths of $(75 - 6)/2 = 34.5$ mm, so $l/d = 34.5/50 = 0.69$ and $P = 13\ 000/2 = 6500$ N. Unit load $p' = 6500/(34.5)(50) = 3.768$ N/mm^2 $(3.768 \times 10^6$ Pa$)$, $\mu = 10.3 \times 10^{-3}$ Pa·s.

Hence, the load number is

$$N_L = \frac{p'}{\mu n'}\left(\frac{d}{l}\right)^2\left(\frac{c_d}{d}\right)^2 = \frac{3.768 \times 10^6}{(10.3 \times 10^{-3})(60)}\left(\frac{1}{0.69}\right)^2(0.0015)^2 = 28.8$$

From Fig. 2.22, $h_{min}/c_r = 0.19$, and since $c_r = c_d/2 = d(c_d/d)/2 = 50(0.0015)/2 = 0.0375$ mm, then $h_{min} = (0.19)(0.0375) = 0.0071$ mm.

Surfaces E. Since the spherical surfaces are narrow, they will be approximated by a cylindrical bearing of average diameter 92 mm, whence $l/d = 38/92 = 0.413$. Unit load becomes $p' = 13\ 000/(38)(92) = 3.72$ N/mm^2 $(3.72 \times 10^6$ Pa$)$. Both stationary surfaces have a velocity of -60 rps relative to the rotating load, and $n' = n_1' + n_2' = -60 - 60$

---

[29] This problem is due to G. B. DuBois.

$= -120$ rps. Hence,

$$N_L = \frac{3.72 \times 10^6}{(10.3 \times 10^{-3})(120)} \left(\frac{1.}{0.413}\right)^2 (0.0015)^2 = 39.7$$

Since $c_r = (92)(0.0015)/2 = 0.069$ mm, then from Fig. 2.22,

$h_{\min} = (0.14)(0.069) = 0.0097$ mm.

The film is developed and maintained because the rotating load causes a rotating eccentricity, i.e., the center of the bushing describes a small circle of radius e about the center of the spherical cavity. The wedge shape formed by the film of oil rotates with the load, always pointing in the direction opposite to that of the motion of the load, and in effect, supporting it. Although the two surfaces of the oil film have no *absolute* tangential motion, they have a tangential motion *relative* to the load. Because of a complete film of oil, the extremely small oscillations of alignment can occur with negligible friction or binding.   ////

2.13 DYNAMIC LOADS AND SPECIAL BEARINGS

The foregoing analyses, except for the squeeze film, have been for equilibrium conditions, e.g., after a journal center has reached a stable position under a constant load. During starting after an idle period, and without a hydrostatic lift, there will be metal-to-metal rubbing until an oil film is established, after which the journal center seeks its equilibrium position. It is during start-up and shut-down, or during a failure of oil supply or other loss of film that wear or possibly seizure occurs. This, plus embeddability and conformity and the possibility for replacement are the reasons for using a soft, low-melting-temperature metal in the reparable bearing.

During running at constant speed there are several dynamic conditions that will cause the journal center O_j to orbit or whirl about bearing center O_b (Fig. 2.28). The *clearance circle* is drawn with radius c_r on O_b as a center. If a large disturbance or an unstable dynamic condition allows the eccentricity vector e to equal c_r, the film is destroyed. For analyses, force equilibrium equations for the shaft, like Eq. (2.39) and the two equations preceding it, must include not only a steady or variable external load P and the integral of film pressure but also an inertia load as a function of the mass supported and the radial acceleration $\ddot{e}$ and tangential acceleration of the journal center. There is a squeeze-film action, and the radial-flow velocity term V in Reynolds' equation is involved. The elastic stiffness and the viscous damping of the film and also the vibration characteristics of the entire bearing-shaft system may enter the analysis.

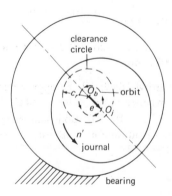

FIGURE 2.28. Orbit or whirl of journal center O_j about bearing center O_b.

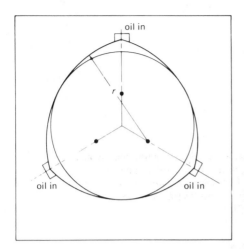

FIGURE 2.29. Three-lobe bearing to minimize whip or whirl.

Sternlicht[30] discusses several dynamical phenomena which are common in turbomachinery. *Synchronous whirl*[31] is generally caused by a slightly unbalanced rotating load, and it occurs at shaft speed, orbiting in a circle in vertical-axis bearings or in an ellipse in horizontal-axis bearings. In *half-frequency whirl* the shaft center theoretically whirls at half the speed, but actually a bit less because of side leakage. With an increase in shaft speed the eccentricity increases continuously to eventually destroy the bearing by contact. It is a fluid-dynamics phenomenon and a self-induced vibration. Furthermore, resonance of the bearing-shaft-rotor system with the half-speed frequency can join to give *resonant whip*. This occurs when the rotational speed is about twice the system's first (lowest) bending critical frequency of vibration, and the system resonates at the first critical frequency. A different *critical speed* action occurs when disturbances cause the shaft rotor to vibrate as a rigid mass on or within the film, which acts as elastic support and damper. This is of concern in gas bearings and in liquid bearings with floating sleeves, and when the clearance ratio is high and fluid viscosity low. A small unbalance can produce high amplitudes, for the resistance is low. With proper design and regard for resonant frequencies and operating speeds, the fluid film can reduce *force transmission* to the supporting structure, with dissipation of vibrational energy by viscous action of the fluid.

Journals are most vulnerable for the initiation of whirl when the external loads are light and eccentricities low. This is particularly true for vertical-axis rotors where the bearing acts principally as a guide. Several devices have been used to minimize whip. Wilcock and Booser[32] list six of these devices in order of increasing effectiveness. *Elliptical bearings*[33] are made by placing shims between two halves and machining to an enlarged, cylindrical bore, then removing the shims and placing the halves over the shaft. Axial, oil-feed grooves are cut along the two lobes. *Pressure bearings* have an annular groove in the upper half terminating in a radial-plane step or dam, which builds pressure in the diverging film. *Longitudinal-groove* bearings have four or more axial grooves feeding oil. *Three-lobe bearings* consist of three arcs of circles, with oil-feed grooves along the lobes (Fig. 2.29). *Pivoted-shoe bearings* have three or more shoes that

[30] B. Sternlicht, "Stability and Dynamics of Rotors Supported on Fluid-Film Bearings," *J. Eng. for Power, Trans. ASME,* Vol. 85, pp. 331–342, 1963. For a further mathematical treatment and charts by B. Sternlicht, see *Mechanical Design and Systems Handbook,* H. A. Rothbart, Ed, New York: McGraw-Hill, Sec. 12, pp. 73–108, 1964.

[31] The italicized names are Sternlicht's designations.

[32] D. F. Wilcock and E. R. Booser, *Bearing Design and Application,* New York: McGraw-Hill, pp. 254–280, 1957.

[33] The italicized words are the titles given by Wilcock and Booser.

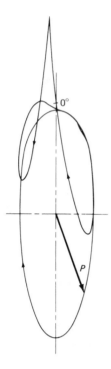

FIGURE 2.30. Cycle of loading represented by rotating vector P in a connecting-rod journal (J. F. Booker, *J. Lubrication Technology, Trans. ASME*, Vol. 93, p. 171, 1971.)

seek their own best working position. They are like those in Fig. 2.15 but have cylindrical surfaces. The pads develop large centering forces under no-load, and the interruption of continuity in the bearing surface also helps. A *nutcracker-type bearing* has one of its halves movable, and it is pushed against the shaft by a piston pressurized by oil either from an external source or by oil tapped from the loaded side of the bearing. Some of these designs are desirable for the additional cooling that they provide. Symmetrical bearings like the three-lobe and pivoted-pad ones seem more suitable for vertical shafts.

Another set of dynamical problems arises in bearings for reciprocating machines, of which the internal-combustion engine has been most studied. Its wristpin, crankpin, and main bearings are subject to complicated changes in loading during any one cycle as shown in Fig. 2.30. A combination of wedge-film and squeeze-film action ensues. It is likely that sometime during the cycle the load will rotate for an interval at and near half the rotational speed of the journal as in Fig. 2.25 (c). Then the wedge action no longer supports a film and eccentricity increases rapidly, giving a squeeze action that provides support long enough for the relative speeds of load and journal to change and restore a wedge film.

If oil-film thicknesses can be kept above and peak pressures below safe values, then provisions can be made for the other operating conditions. The mathematical difficulties of an analytical solution have required the development of several numerical methods. These have been discussed by Campbell and associates.[34] Booker[35] in a graphical method, which may also be programmed on a digital computer, determines the resultant velocity of the journal center relative to the bearing by consideration of two components, one for a stationary bearing and nonrotating journal along a squeeze path under the action of the

---

[34] J. Campbell, P. P. Love, F. A. Martin, and S. O. Rafique, "Bearings for Reciprocating Machinery: A Review of the Present State of Theoretical, Experimental, and Service Knowledge," *Proc. Inst. Mechanical Engineers*, Vol. 182, Part 3A, pp. 51–74, 1967–1968.

[35] J. F. Booker, "Dynamically Loaded Journal Bearings—Mobility Method of Solution," *J. Basic Engineering, Trans ASME*, Vol. 87, pp. 537–546, 1965.

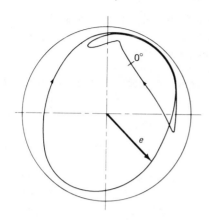

FIGURE 2.31. Journal displacement (rotating eccentricity vector e) for connecting rod of Fig. 2.30. (J. F. Booker, *J. Lubrication Technology, Trans. ASME*, Vol. 93, p. 172, 1971.)

load, and a second component to take account of the angular velocities and wedge-film action of the journal, bearing, and load. Maps of the path of the journal center can be constructed. Figure 2.31 is a displacement diagram obtained by Booker[36] for the loading of Fig. 2.30, with the short-bearing approximation as a basis. Comparisons between solutions made by several methods show excellent qualitative agreement (in orbit shape), but some differences in quantitative values. Campbell describes a *quick method* to estimate the maximum film thickness by considering only predominately squeeze intervals of the load cycle and the squeeze resistance.

Solutions can be obtained experimentally on a dynamic similarity test machine,[34] by which any cycle of loads can be applied by an electrohydraulic servoloading mechanism. DeHart and Harwick[37] report on many aspects of bearing practice in the automotive industries, such as oil holes and grooves and their location, flow rates, bearing geometry, finish, and distortion.

Gas bearings using air or more inert gases as the lubricant and film find application for lightly loaded high-speed shafts. Since torque varies as speed, power loss, Eq. (2.47), varies as the square of speed, and a fluid of greater viscosity gives prohibitive power and heating losses at speeds of 20 000 to 100 000 rpm. Gas is a compressible fluid with significant variations in viscosity and density as the temperature and pressure change throughout the film. Solutions for film thickness require a more general form of Reynolds' equation than Eq. (2.11); even so, some variations must be ignored in order to obtain a workable equation. Then it is necessary to devise numerical methods for computer solution. Trumpler[38] derives a very general form, then makes the necessary approximations. He reproduces computer-developed design charts and examples from the work of H. G. Elrod, Jr. and co-workers at the Franklin Institute.[39] Again, it is possible to plot against a "bearing number" and from the charts, to calculate eccentricity, attitude angle, and power loss.

Elastohydrodynamics is concerned with the interrelation between the hydrodynamic action of films and the elastic deformation of the supporting materials. This, for example, accounts for the ability of an oil film to support the load transmitted between convex

[36] J. F. Booker, "Dynamically Loaded Journal Bearings: Numerical Application of the Mobility Method," *J. Lubrication Technology, Trans. ASME*, Vol. 93, pp. 168–176 and errata p. 315, 1971.

[37] A. O. DeHart and D. H. Harwick, "Engine Bearing Design: 1969," *SAE Trans.*, Vol. 78, Paper 690008, pp. 69–82 in Sec. 1, 1969.

[38] P. R. Trumpler, *Design of Film Bearings*, New York: The MacMillan Company, pp. 233–252, 1966.

[39] H. G. Elrod, Jr. and S. B. Malanoski, "Theory and Design Data for Continuous-Film, Self-Acting Journal Bearings of Finite Length," Franklin Institute Interim Rep. I-A2049-17, 1962.

gear teeth. A short treatment is given by Walowit and Anno,[40] who also discuss plasticity, metal working, and foil bearings. Another grouping is furnished by *tribology,* which Moore[41] defines as "the science and practice of friction, lubrication, and wear applied to engineering surfaces in relative motion." A handbook is available[42] which presents these factors in concise and pictorial form for direct engineering application to many machine components.

PROBLEMS

Section 2.1

2.1   A weightless, flat, circular plate of outer diameter D and inner diameter d moves on a larger, flat horizontal surface, separated from it by a film of lubricant of uniform thickness h and viscosity μ. Derive expressions for (a) the force to move the plate linearly with velocity U and (b) the torque to rotate the plate about its vertical central axis with an angular velocity of n' rps.

2.2   A weightless conical pivot of base diameter D and apex angle 2α rotates in a conical seat with angular velocity of n' rps, separated from it by a uniform film of oil of thickness h and viscosity μ. Derive an equation for the torque to rotate.

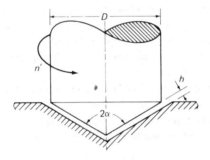

2.3   The cylindrical top guide bearing for a vertical shaft (Fig. 2.3) is 1.5 in diameter by 2 in long. The shaft rotates at 1160 rpm, diametral clearance is 0.002 in, and SAE 10 oil is used at a temperature of 140°F. Calculate the torque at the bearing.

2.4   The measurement of shaft journal diameters and bearing bore diameters by micrometers or other mechanical gages is subject to small errors such that their difference, which is the diametral clearance, may have an error that is almost as large as the clearance itself. Hence for research purposes on a given bearing, where friction torque is readily measured, the clearance may be calculated indirectly from torque at no load. Do this for a bearing where the following data are available: nominal diameter 2½ in, length 2 in, speed 3600 rpm, friction torque 6.58 lb·in, oil SAE 10 at 150°F.

2.5   Same as Prob. 2.4 except for the data: diameter 75 mm, length 75 mm, speed 890 rpm, torque 1260 N·mm, oil SAE 20 at 50°C. *Ans.* 0.118 mm.

2.6   The equation for relative density of oil is frequently given as $\bar{\rho} = \bar{\rho}_{60} - 0.000\,35\,(t - 60)$. Basically, this is the same as Eq. (2.4) but with a small approximation. Justify.

2.7   A sample of mineral oil has a relative density of 0.891 at 60°F, and the Saybolt viscometer

[40] J. A. Walowit and J. N. Anno, *Modern Developments in Lubrication Mechanics*, New York: John Wiley & Sons, Inc., 1975.
[41] D. F. Moore, *Principles and Applications of Tribology*, Oxford: Pergamon Press, 1975.
[42] M. J. Neale, Ed., *Tribology Handbook*, London: Butterworths, 1973.

readings are 195 SUS at 100°F and 38 SUS at 210°F. (a) Determine the absolute viscosities in reyn and in Pa·s at the test temperatures. Plot these points lightly on the ASTM chart of Fig. 2.4 and read the viscosity at 175°F. *Ans.* $\mu = 0.76\ \mu$reyn. (b) At this oil temperature calculate the no-load torque and horsepower for a journal 3 in long by 4 in diameter, with a diametral clearance of 0.006 in, rotating at 2000 rpm.

2.8 Same as Prob. 2.7(a) except that the oil has a relative density of 0.879 measured at 75°F, and the SUS viscometer readings are 329 s at 100°F and 54 s at 210°F. Find the viscosity at 150°F.

2.9 A marine engine develops 250 hp at 100 rpm propeller speed to push a boat at 15 mph. The efficiency of the propeller is 70%, i.e., 70% of the power is used in forward motion and 30% in churning the water. The shaft diameter is 4.5 in. The thrust bearing on the propeller shaft consists of several flat-faced collars on the shaft, bearing against stationary rings in the housing, with net-bearing surfaces 5 in inside and 8 in outside diameter. Assume that the thrust load is equally divided between them. The collars are partially immersed in oil, but the flat surfaces will not support thick-film lubrication. Hence use the design method described at the end of Section 2.1, taking velocity V at the mean diameter. Calculate the thrust force, the number of collars for the shaft, and the horsepower lost in bearing friction if the coefficient of friction is 0.05. *Ans.* $N = 2.43$, use 3.

Section 2.2

2.10 To confirm that gravity and inertia forces may generally be neglected in the equilibrium equation for Fig. 2.5, take a film of oil 1 in × 1 in × 0.001 in thick and calculate the pressures required on one edge (a) to lift it and (b) to accelerate it at 1000 ft/s². Compare these with the typical 100 to 5000 psi pressures developed in the film.

Section 2.3

2.11 (a) Derive an equation for the flow Q of a viscous fluid through a tube of bore diameter d under a constant pressure difference p_0 over a length l. Start by writing the equation of equilibrium between applied and resisting forces acting on a concentric core of fluid of any radius r. (b) Adapt the result to a vertical tube under a constant head h of oil, solving it for μ/ρ. *Ans.* $\mu/\rho = \pi ghd^4/128lQ$. What measurements and kind of instrument does this suggest for measuring viscosity?

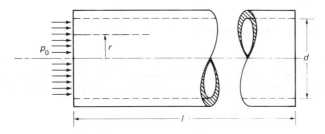

2.12 Derive in detail the equations for u and q_x, Eqs. (2.7a) and (2.8a), respectively.

2.13 Derive a special form of Reynolds' equation by assuming that the parabolic velocity term of Eq. (2.7a) and the pressure-induced flow term of eq. (2.8a), i.e., those terms containing $\partial p/\partial x$, have a negligible overall effect. This was the assumption made by F. W. Ocvirk in deriving his useful short-bearing approximation, Eq. (2.34). Also take velocities $U_1 = V_1 = W_1 = W_2 = 0$.

Section 2.4

2.14 Derive an equation in cylindrical coordinates for radial flow corresponding to Eqs. (2.5) by considering the equilibrium of forces on an elemental annulus of a cylindrical disk (instead of the rectangular parallelpiped of Fig. 2.5). From this equation derive the equation for u_r preceding Eq. (2.17).

2.15 For the circular-pad hydrostatic bearing derive the pressure equations, Eqs. (2.15) and (2.16) in detail, performing the integrations or differentiations at each step.

2.16 Same as Prob. 2.15, but for radial velocity and flow Eqs. (2.17) and (2.18), starting with Eqs. (2.7) and (2.15).

2.17 For the hydrostatic thrust bearing of Example 2.2, write an equation for the pressure variation p and integrate over the area to obtain total load P. Compare it with the given load as a check on the dimensions which were determined.

2.18 Redesign the bearing of Example 2.2 with the object of decreasing the flow requirement. Try $d/D = 0.4$ and SAE 30 oil, the maximum pressure and other conditions remaining the same. What is gained and what is lost?

2.19 A shaft carries a 10 000 lb thrust load and rotates at 600 rpm, supported by a hydrostatic thrust bearing with an outside diameter of 4 in and a recess diameter of 2 in. A pump with a constant delivery of 1.0 in³/s is available. The oil temperature will vary between 110° and 150°F depending upon outside conditions. Select a suitable SAE-numbered oil such that the film thickness will never be less than 0.0016 in. What will be the oil pressure in the recess? What are the bearing torques and the "coefficients of friction" at 150°F? At 110°F?

2.20 Same as Prob. 2.19 except the load is 21 000 N, diameters are 75 mm and 40 mm, the label on the pump reads 20 cc/s displacement, and minimum film thickness is to be 0.04 mm. *Ans.* SAE 30; 8.34 MPa, 0.112 N·m and 0.19×10^{-3} at 150°F, 0.336 N·m and 0.56 $\times 10^{-3}$ at 110°F.

2.21 Air at 68°F is used in a hydrostatic bearing to support a centrifuge weighing 10 lb and rotating at 75 000 rpm. (a) If the inlet air pressure is 10 psi, the ratio of recess to outside diameters 0.6, and the film thickness 0.001 in, what should be the dimensions and air flow in ft³/min, assuming the air to be incompressible. *Ans.* 1.43, 0.86 in, 0.136 ft³/min. (b) What is the torque for rotation and the coefficient of friction based on average ring diameter? *Ans.* 0.0013.

2.22 Same as Prob. 2.21 except inlet air pressure is 5 psi.

2.23 To provide almost frictionless motion for sliding an object in any direction, a table with several recessed, air-supported balls is proposed. One ball is shown, with dimensions in millimeters. If the load per ball is 20 N and an average film thickness of 0.015 mm is

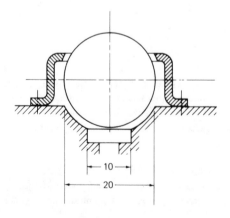

desired, what inlet air pressure and air flow are required at 20°C? Assume that the air is incompressible and that projected areas may be used in the flat-pad equations. *Ans.* 0.118 MPa, 16.8×10^{-6} m³/s.

2.24 The *stiffness* of a film-lubricated bearing is the rate of change of load capacity with film thickness. It is of interest where external loads may suddenly change. Determine its equation for a circular-pad hydrostatic bearing.

2.25 Derive equations for the axial load and the fluid flow at inlet pressure p_0 in the hydrostatic, conical bearing. *Ans.* $P = \pi p_0 (D^2 - d^2)/8 \, \ln(D/d)$, $Q = 4Ph^3(\sin \alpha)/3\mu(D^2 - d^2)$.

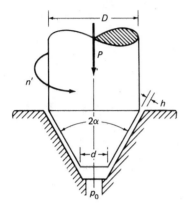

2.26 The idea is proposed that the capacity of a hydrostatic bearing might be increased by making the bearing surface slightly concave, thus "trapping" the oil. Start an investigation of this with a spherical cavity as shown. (a) Write an equation for the film thickness h, using a series expansion and retaining terms suitable for a very small concavity. (b) From Reynolds' equation obtain an equation for p in the form of an integral and constants of integration and write the boundary conditions for their evaluation.

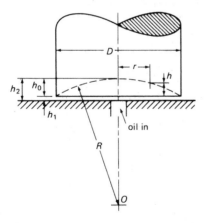

2.27 Derive a complete set of equations for the rectangular pad of Fig. 2.10, including pressure and flow [Eqs. (2.21)], supported load P, and the force to slide the pad at a velocity U. Take $c = b$.

2.28 The square pad shown here and in Fig. 2.11 is one of several which are each to support a load of 150 000 lb in a telescope. It may be considered to have a central, uniformly loaded section and four long slots or sills with a flow path of 3 in, similar to the two sills of Fig. 2.10. (a) Calculate the lift-off pressure and the central pressure p_0 to support the load hydrostatically. The contribution of the sills may be found by integration of the varying pressure over a changing width, or by averaging. (b) Calculate the flow capacity required for the pump to maintain a film thickness of 0.004 in with SAE 20 oil at 100°F. (c) If the pump is of the constant displacement type, what will be the film thickness if the temperature drops to 32°F? If this gives dimensional problems, what control do you suggest? (d) What force is required to slide the plate along each pad, located at a mean radius of 20 ft, if the rotational velocity is 1 revolution each 24 hours? The recesses are ¼ in deep. What is the value of the "coefficient of friction"? (e) What is the power requirement of the pump for each pad?

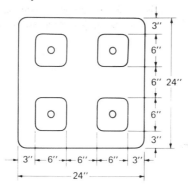

2.29   The rotating parts of a large radar antenna weigh 320 000 lb and are supported by 16 hydrostatic pads with the dimensions shown, located at an average diameter of 153 in. A 0.005 in film thickness is to be maintained at 100°F by oil with a Saybolt viscosity of 750 s. The maximum rotational speed will be 0.5 rpm. (a) With the suggestions of Prob. 2.28 as a guide, calculate the pool or recess pressure before and after the bearing ring is lifted off the pads, the pump requirement in gal/min, and the maximum "coefficient of friction." (b) Calculate the film stiffness dP/dh where $h = 0.005$ and the flow calculated in (a) remains constant. (The dimensions are those of a Haystack Hill installation, "A Hydrostatic Bearing for Haystack," described by G. R. Carroll in ASME Paper 62-WA-299, ASME, New York, 1962.) Each rectangular pad within the pool is 1 in x 3½ in.

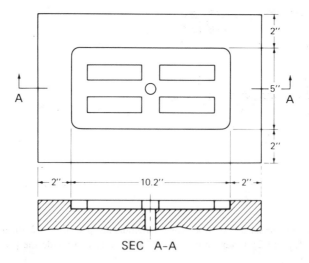

SEC A–A

2.30   Modification is planned for a heavy, 200-lb jig with a base 12 in × 10 in so that one machinist can easily and accurately position it on a drill table. Shop air at 80 psi is available, so a central 8 in × 10 in recess fed by air through several holes is proposed. How much will the jig be lifted? (The incompressible flow equations will be accurate enough for this purpose.) This much air may give trouble with the blowing of chips, etc. If the air is throttled to 5 or 10 psi, will this suffice? What friction force will hold the jig in place when the air is shut off?

2.31 Shown is a shaft journal which has been lifted hydrostatically an amount g to avoid metal-to-metal contact before rotation begins. Let c_r be the radial clearance when the journal and bearing are concentric. Journal radius r and bearing radius $r + c_r$ are drawn from journal center O_j and bearing center O_b, respectively, and $O_b O_j = e$. (a) Write an approximate equation for film thickness h. Ans. $h = c_r - e \cos \theta$. (b) Assume no axial flow and adapt Reynolds' equation, Eq. (2.11), to this case by substituting $rd\theta$ for dx. Obtain an equation for p in the form of an integral and constants of integration and write the boundary conditions for their evaluation. You will probably not find the solution for the integral in your integral tables, so stop here.[43]

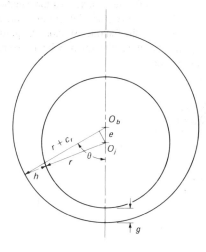

Section 2.5

2.32 Derive the pressure and load equations, the first two of Eqs. (2.25), for the squeeze-film case of a flat circular plate without a hole. Also, what is the value of the ratio p_{max}/p_{avg}?

2.33 A 4-in-diameter circular pad is supported on a platen by a 160°F film of SAE 20 oil, 0.001 in thick. It is loaded by a weight of 3800 lb. How soon does the film thickness decrease to 0.0001 in? What are the velocity of approach and the maximum oil pressure at the smaller thickness? Is pressure independent of thickness?

2.34 A 125-mm-diameter circular pad is supported on a platen by a 20°C film of SAE 40 oil, 0.025 mm thick. It is loaded by a weight of 25 000 N. In what period of time does the thickness decrease to 1/10 of its original value? 1/100? 1/1000?

2.35 A flat, narrow rectangular plate with dimensions l by $2b$ ($b \ll l$) is used as a squeeze-film bearing, with approach velocity V. From Reynolds' equation derive expressions for pressure, force, and time t to change film thickness from h' to h'' under constant load W. Verify that the rate of flow from the edges of the bearing equals the rate of fluid displacement, $2Vbl$. Ans. $p = (6\mu V/h^3)(b^2 - x^2)$, $W = 8\mu V l b^3/h^3$, $\Delta t = (4\mu l b^3/W)[(1/h'')^2 - (1/h')^2]$.

2.36 A 10 in × 2 in rectangular pad is lubricated with SAE 20 oil at 140°F. A steady load of 2000 lb is applied when the oil thickness is 0.002 in. From the answers to Prob. 2.35 determine the thickness after 1 s and 10 s and the maximum pressure.

2.37 If the pump supplying oil to the antenna bearing of Prob. 2.29 is stopped and a check valve at the bearing prevents backflow through the lines, how many seconds will it be before the

---

[43] For a solution, see D. D. Fuller *Theory and Practice of Lubrication for Engineers*, New York: John Wiley & Sons, p. 110, 1956. His book also develops the load capacity equation and charts and gives several examples and practical details of application of hydrostatic lifts (see pp. 107–125).

film thickness is reduced to 0.0005 in at 100°F, at 70°F (oil is approximately SAE 40)? This can be an order-of-magnitude quick calculation based on the sills, using the narrow rectangular pad equation of Prob. 2.35, but estimating an additional time for flow across the sills of the 0.005-in-thick-surface layer of oil within the pool, basing the estimate on relative areas.

2.38   The idea is proposed that the capacity of a circular squeeze-film bearing might be increased by making the bearing surface slightly concave, thus "trapping" the oil. Investigate this for a conical concavity, as shown. (a) From Reynolds' equation derive an equation for pressure. Check it in the limiting case of a flat surface against the equation of the text. *Ans.* $3\mu V[D^2(h_2 - 2\alpha r) - 4r^2(h_2 - \alpha D)]/4(h_2 - \alpha r)^2 h_1^2$. (b) Compare the maximum value of pressure for the conical surface with that for the flat one. Without integration of the pressures over the surface, what do you conclude about the relative load-carrying abilities for the same diameter and minimum film thickness?

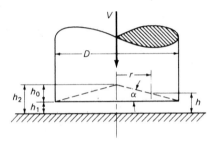

Section 2.6

2.39   Determine an equation for the load P carried by a tilted thrust pad (Fig. 2.15), when the pressure distribution is given by Eq. (2.29).
Ans. $P = 6\mu Ub^2 l[\ln(h_1/h_2) - 2(h_1 - h_2)/(h_1 + h_2)]/(h_1 - h_2)^2$.

2.40   (a) Determine the location x of the maximum pressure for the thrust pad of Fig. 2.15. Show that $x/b > 0.5$. *Ans.* $x = h_1 b/(h_1 + h_2)$. (b) Derive an equation for the maximum pressure.

2.41   (a) Set the answer to Prob. 2.39 in terms of tilt ratio $m = h_1/h_2$ and derive the equation that determines the tilt for maximum load capacity for a given minimum clearance h_2. *Ans.* $-2(\ln m)/(m - 1) + 2/(m + 1) + 1/m + 2(m - 1)/(m + 1)^2 = 0$. (b) Obtain a numerical solution for m by computer for this transcendental equation.

2.42   From Eqs. (2.1) and (2.7) obtain an integral expression for the shearing force dF when $U_1 = U$ and $U_2 = 0$. Note that unless $dp/dx = 0$ its value is different at the two surfaces $y = 0$ and $y = h$. When pressure is integrated over the surfaces, the two tangential forces are also different. Explain this for the thrust bearing of Fig. 2.15 by sketching free bodies of the pivoted pad, oil film, and horizontal plate, as necessary. On which plate is the shearing "friction" increased by a component of pressure? For calculating power loss, which surface and force should be used?

2.43   Design a thrust bearing for a ship's propeller shaft of 8½ in diameter. It will consist of pads similar to that of Fig. 2.15, arranged in a circle and with circular and radial boundaries. An inner radius of 5 in and an outer radius of 8 in seem feasible, with an arc length of 3 in at the average radius to give an almost "square" pad. The space between pads should be about one-fourth of the area of the annulus. Minimum film thickness should be about 0.0010 in, and the h_1/h_2 ratio for maximum load capacity is 2.19, obtained by pivoting at $0.58b$ downstream. The load is 25 000 lb, rpm 120, and oil temperature is maintained at

125°F. (a) By a layout or otherwise, determine the number of pads. (b) Make the customary assumption that the pads are rectangular, and by reference to Prob. 2.39 or otherwise, determine if the dimensions are feasible and what oil to use.

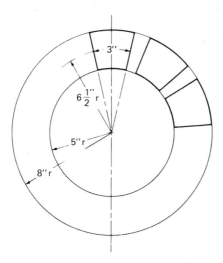

2.44 Similar to Prob. 2.43, but for a turbine with a thrust load of 20 000 N, speed 7500 rpm, oil temperature 62.5°C, shaft 37 mm diameter, pads of 44 and 84 mm diameter, and minimum film thickness of 0.015 mm. *Ans.* μ = 17.7 mPa·s, SAE 20.

2.45 Derive an equation for pressure distribution in a flat-surface tilting-pad thrust bearing like that of Fig. 2.15 but one that is relatively narrow so that the endwise flow is not negligible. Hence in Reynolds' equation drop the first $\partial p/\partial x$ term on the basis that the pressure-induced flow may be negligible, and retain the second $\partial p/\partial z$ term. Examine the result for reasonableness at various positions. For this narrow-bearing method, what must be a characteristic of the function chosen for film thickness h?

2.46 Assume the sinusoidal surface shown for a thrust pad. Write its equation and test its suitability for the narrow-bearing theory of Prob. 2.45. Derive the pressure equation and plot it for the central plane $z = 0$ when $h_1 = 2h_2$. *Ans.* $p = 3\pi\mu Un(l^2 - 4z^2)(\sin \pi x/b)/4b(m + n \cos \pi x/b)^3$, where $m = (h_1 + h_2)/2$ and $n = (h_1 - h_2)/2$.

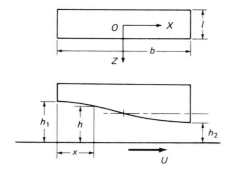

Section 2.7

2.47 In Fig. 2.17 take the origin of coordinates O to be on the *journal* surface, with the X axis tangent to it. Draw the vectors and show that Reynolds' equation reduces to Eq. (2.31).

2.48 (a) Write the reduced forms of Reynolds' equation that apply for the solution of the two cases shown. (b) By inspection of these, state for each case what combinations of values $+U$, $-U$, or 0 when given to U_1 and U_2 will result in zero load capacity, single capacity, and double capacity, corresponding to 0, $|U|$, and $|2U|$, respectively. *Note:* The right-hand side of Reynolds' equation must be a negative *number* to develop positive pressure and load capacity.

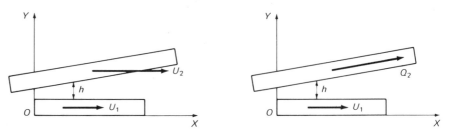

2.49 For an "infinite-length" journal bearing write the steps and the equations, including the boundary conditions, which lead to the pressure equation, Eq. (2.33), but stop short of the actual integration.

2.50 With a clockwise-rotating journal and a downward load on it when one exists, sketch with exaggerated thickness the position and shape of the oil film, the relative shapes and sizes of the velocity profiles at the $\theta = 0°$, peak pressure, and $\theta = 180°$ positions, and the distribution of pressure radially and longitudinally. Do this for the theoretical cases of a bearing with no load, an $180°$ Ocvirk "short-bearing," and a Sommerfeld $360°$ "long-bearing." *Note:* Eccentricity in the latter is always at $90°$ to the load. See the statement following Eq. (2.51).

2.51 Transform Eq. (2.36) to give the value of a pressure function, $4prc_r^2/\mu Ul^2$, for the short-bearing approximation. Show that it is nondimensional. For a bearing with length equal to diameter, show that the form of the function is the same (but with different values) as that which one obtains from the long-bearing approximation. (a) Assume $\epsilon = 0.70$ (or 0.60 or 0.80 or 0.90 if assigned) and plot the two functions from 0 to $180°$ in the central radial plane $z = 0$. (b) For the axial planes in which the peak pressure occurs (from the plot in (a) or by Eq. 2.37 for one case), plot pressure function vs. z/l from -0.5 to $+0.5$. (c) Explain why the "short-bearing" peak appears to be the higher of the two.

2.52 Derive Eq. (2.37) for the angle at which peak pressure occurs in a journal bearing. What is the value of this angle when $\epsilon = 0.70$? What is the corresponding ratio of peak pressure to unit load? Unit load p' is the load P [Eqs. (2.39)] divided by projected bearing area ld or $2lr$, i.e., $p' = P/ld$.

Section 2.8

2.53 Of interest under dynamic conditions, Section 2.13, is the stiffness or "spring constant" k of an oil film. Dynamic problems are prevalent under conditions of low load, where, depending on one's purpose, the stiffness is defined either as load P divided by the deflection of the journal center, e, or as load P divided by the component of deflection in the direction of the load. (a) For both definitions derive k as a function of eccentricity ratio ϵ. *Ans.* for the second definition, $k = (\mu Ul^3/c_r^3)[\pi^2(1 - \epsilon^2) + 16\epsilon^2]/16\epsilon(1 - \epsilon^2)^2$. (b) To study the effect of ϵ, observe that when ϵ is small (say 0.1 or less), certain terms may be neglected. Do the two stiffnesses increase, decrease, or remain nearly constant with increase of ϵ? How does the load resistance change?

2.54 Reproduce the curve of pressure vs length in Fig. 2.18, and sketch over it the approximate curve that results if the load is unchanged and an annular groove, as shown in Fig. 2.27, is cut all around the inside surface of the bearing, in order to supply more oil to the central portion. Pressure in the groove will be the same as at the oil-supply hole and much less than the peak film pressures, and, in essence, we have two bearings of half-length. Also, observe the effect of length in Eqs. (2.39). Is a groove and a greater oil supply always a good thing in comparison with a single oil hole in the diverging, low-pressure area of the film?

2.55 A clockwise-rotating shaft supported by other bearings supports in turn a bearing and rod, as shown in the no-load position. Sketch the shaft and show the position that the bearing will take when a horizontal load to the left is applied. Label the eccentricity. Show by a polar diagram how the pressure varies. Indicate a good location for an oil hole.

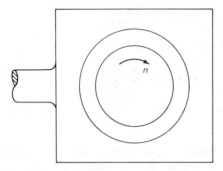

2.56 An idler pulley with a pressed-in bronze bushing is placed over the stationary shaft, and the pulley rotates counterclockwise while loaded downward. Sketch the running position of the pulley relative to the shaft, superimpose a polar diagram of pressure distribution, and on both views show a well-located oil hole and how the oil may be led to it from a pump.

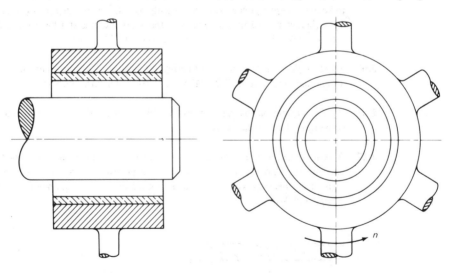

2.57 For demonstration and experiments of the effect on oil film of changes of load, speed, viscosity, and flow, there may be constructed a machine consisting of a shaft adjustable in speed and rotating in ball bearings, a constant-displacement pump with bypass valve to control oil flow into the bearing, a heater for the oil, and a bearing block with a means for loading it. (For details see J. E. K. Foreman, ''The Design and Development of a Transparent Bearing Testing Machine for the Experimental Investigation of Plain Journal Bearings,'' M.S. Thesis, Cornell University, Ithaca, N.Y., 1952.) The block is shown in its unloaded position. The block is freely movable through its clearance space, and it is made of a clear plastic so that the amber film of oil is visible. If the block is loaded, the film is seen to be broken by triangular voids or pockets of vapor and air, as shown in one view. (a) Are these in the converging or diverging part of the film? What do they tell about the direction of shaft rotation and of load on the block? Sketch a radial-plane view showing the displacement of the block relative to the shaft. (b) In additional pairs of side and radial-plane views sketch to show the effects on h_{min}, attitude angle, and void length of decreasing the flow of oil through the oil hole, of increasing the load, and of increasing the shaft speed.

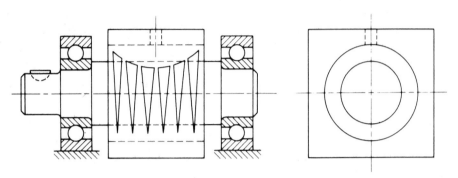

2.58 Derive the differential equations for torque T_r at the rotating shaft and T_s at the stationary bearing, basing them on the Sommerfeld (long-bearing) equation for pressure. Omit the actual integration.

Section 2.9

2.59 What is the minimum flow rate and power capacity required of a pump of 85% efficiency to keep the convergence space filled with oil with an entrance pressure of 30 psi in a bearing 10 in diameter × 10 in long and with the shaft rotating at 1200 rpm, a diametral clearance ratio of 0.0015, and an eccentricity ratio of 0.80?

2.60 Same as Prob. 2.59, but with bearing dimensions 150 mm diameter × 200 mm long and an entrance pressure of 0.20 MPa. *Ans.* 10.2 liter/min, 39.8 W.

2.61 By the integration indicated in Eq. (2.44), confirm Eqs. (2.45) for the oil rejected in the converging half of the oil film.

2.62 A bearing has a central annular groove of width t into which oil is pumped, developing a pressure p_0. Derive equations for the axial pressure distribution and for total flow out the ends of the bearing when the shaft is not loaded and the diametral clearance is c_d. *Ans.* $Q = \pi p_0 d c_d^3 / 24\mu(l - t)$.

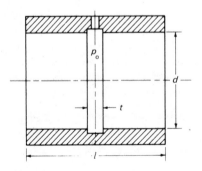

2.63 A piston has a central annular groove which feeds oil at pressure p_0 to lubricate the piston and cylinder walls. Piston velocity is W and diametral clearance is c_d. Starting with Reynolds' equation, determine equations for pressure and flow along both right and left clearance spaces. At what piston velocity is the net flow zero in one of the spaces?

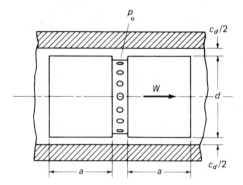

2.64 Same bearing as in Prob. 2.62, except that it is loaded and the eccentricity ratio is ϵ. (a) Derive an equation for that part of the endwise flow due to pressure p_0 in the groove. *Ans.* $Q = \pi p_0 d c_d{}^3(1 + 1.5\epsilon^2)/24\mu(l-t)$. (b) What is the largest possible ratio of flow in a heavily loaded bearing to one with no load, due to the oil-groove pressure only?

2.65 For the bearing of Prob. 2.60 what is the additional flow and pump power if an annular groove 10 mm wide is cut midlength of the bearing and the oil has a viscosity of 10 mPa·s, (a) under no-load, (b) with the eccentricity ratio of 0.80? The answers to Probs. 2.62 and 2.64 may be used.

Section 2.10

2.66 The peak pressure ratio p_{max}/p' is plotted against load number in Fig. 2.23. From equations already developed, show that the ratio is a function of N_L and ϵ only and hence of one or the other alone.

2.67 The stationary-member torque ratio, T_s/T_0, Eq. (2.43) is a function of ϵ only, so it may be plotted against N_L by Eq. (2.50). Show that the rotating-member torque ratios T_r/T_0 are functions of ϵ and also l/d and must be plotted for separate values of l/d when N_L is the abscissa, as in Fig. 2.24.

2.68 Planned clearances may be affected by manufacturing tolerances and thermal expansion or contraction relative to the steel shaft. How much does a 50% increase and a 50% decrease in clearance change the load numbers? If $N_L = 45$ what are the new and old eccentricities? Same if $N_L = 15$. An increased clearance gives a decrease in basic torque T_0, Eq. (2.2), which in turn decreases oil temperature. Hence is there a partial compensation which decreases the *change* in eccentricity? Is there a similar compensation when the clearance is *decreased*?

2.69 A shaft rotating at 900 rpm loads a bearing to 200 lb. Use SAE 10 oil at 130°F. Design the bearing for a conservative load number of 20, a length to diameter ratio of 0.75, and a diametral clearance to diameter ratio of 0.0015. Determine length, diameter, and clearance, also minimum film thickness.

2.70 The idler gear is loaded as shown and rotates at 2400 rpm. For stability with a single bronze bushing, pressed in place, a length-to-diameter ratio of 2.0 is chosen. Operating temperatures are high in the machine, estimated at 230°F for the oil at exit. Choose SAE 60 oil and a clearance ratio of 0.002, and because of limited space, a load number of 45. Determine dimensions and minimum film thickness *Ans.* $p' = 878$ psi, $1^5/_{16}$ in $\times$ $2^5/_8$ in, 0.0026 in clearance, 0.000 16 in min thickness.

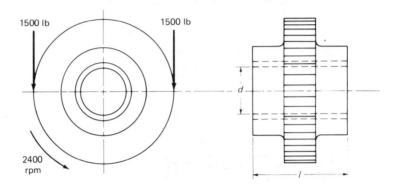

2.71 Same as Prob. 2.70 except two loads of 6000 N each, a load number of 30, and SI units throughout.

2.72 The bearing and journal with dimensions shown carry a maximum load of 4000 lb. The shaft may rotate at either 3600 rpm or 1800 rpm. The exit temperature of the oil is maintained at 150°F. Determine the oil to use if the load number is not to exceed 30 at any time.

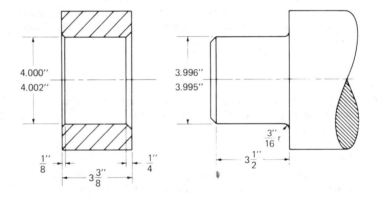

2.73 A centrifugal water pump at 1200 rpm has two journal bearings each carrying a 50-lb load. To simplify the construction and to avoid possible contamination with oil, water is being considered as the lubricant. At the highest temperature expected, 120°F, its viscosity is 0.090 μreyn. Choose a clearance ratio of 0.002 and a load number of 35. The shaft diameter is 2 in, chosen from other considerations. How long must each bearing be? What are the unit loading and minimum film thickness?

2.74 A journal bearing is to support 24 lb from a shaft rotating at 6000 rpm, using air of viscosity 0.0026 μreyn as the lubricant. Allowable unit loading is 5 psi, length-to-diameter ratio 1.2, and clearance ratio 0.0008. For small pressures the air may be treated as for incompressible fluids. Determine the dimensions, minimum film thickness, and torque required for rotation. *Ans.* 2.0 $\times$ 2.4 in; 0.000 29 in; 0.050 lb·in.

2.75 For the design of an air-lubricated journal bearing to carry a 5-lb load at 21 000 rpm, certain criteria are proposed: a minimum film thickness of 0.000 20 in to minimize surface finish expense, an average pressure (unit load) of 7 psi (range 5 to 10 psi in paragraph following Example 2.2), and a conservative load number of 10. Viscosity of air at the operating temperature is estimated to be 0.0030 μreyn. Determine the required c_d from minimum film thickness and eccentricity ratio, then determine the other dimensions. Does this result in reasonable ratios?

Section 2.11

2.76 Consider a journal bearing 2 in diameter by 1 in long with a diametral clearance of 0.004 in, with the journal rotating at 3600 rpm and the load 600 lb. The SAE 40 oil rejected out the ends of the converging film is the principal source of cooling. (a) What should be the exit temperature of the oil if the minimum film thickness is not to be less than 0.0005 in? *Ans*. 159°F. (b) What are the oil flow and the heat generated, and what should be the maximum temperature of the oil entering the bearing? Assume $\bar{\rho} = 0.86$. *Ans*. 129°F.

2.77 Example 2.4 was left with an imbalance between hydrodynamic flow Q_H and the flow Q required to keep the temperature at 160°F. (a) Try, as suggested following Example 2.4, an increase in clearance c_d. Since this may result in a greater increase in Q_H than decrease in H_g, start with a value of c_d which will increase Q_H to about $Q_H + 0.6(Q - Q_H)$. Evaluate all the other changed values and comment. (b) Determine a reduced value of t_{in} that will reduce required flow Q to the available Q_H.

2.78 It might be decided that the heat imbalance in Example 2.4 could be ignored. Assume that the bearing is built and operated with the oil and with the dimensions as designed in Examples 2.3 and 2.4. Take some higher temperature such as 170°F and with the lowered viscosity recalculate H_g and H_{oil}. Plot these together with H_g and H_{oil} for 160°F on a heat vs temperature diagram and read a predicted operating temperature at the intersection of straight lines through the plotted points.

2.79 A bearing is 2 in diameter by 1.5 in long with a diametral clearance of 0.003 in, loaded with 1800 lb by a shaft turning at 2400 rpm. SAE 30 oil enters with a temperature of 120°F. Find the effective operating temperature of the oil film, assuming that all the heat is removed in the hydrodynamic flow of oil. (For a class effort, it is suggested that each student or a pair be assigned different temperatures, such as 140°, 150°, 160°, 170°, and 180°F and that the heat-balance diagram be constructed on the blackboard.)

Section 2.12

2.80 A bearing for a shaft carrying a rotating load is provided with a "floating sleeve," as shown. (a) Assume that the sleeve rotates at half speed and calculate the relative load capacities of the two films, all other things being equal. Compare with Fig. 2.25(b) for no

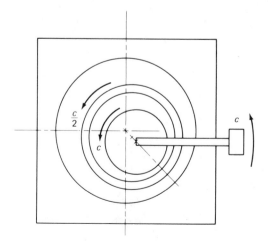

sleeve. (b) Examine the equation for load number and sketch a couple of schemes by which the capacity of one film might be changed to equal that of the other. If c_d/D, μ, and N_L are the same, what should be the D and l relationship for equal load P?

2.81 For Fig. 2.25(a) and (b) sketch free-body diagrams of the film, with force and torque vectors. Check the position of the journal center by the rule of Section 2.8 and Eq. (2.38). What special interpretation must be given to "rotating" and to "stationary" for Fig. 2.25(b)?

2.82 With the aid of sketches compare the load capacities for a stationary load and both shaft and bearing rotating with equal speeds (a) in the same direction, and (b) in opposite directions.

2.83 One type of turbine-driven compressor for a jet engine consists of outer rotors and shaft rotating in fixed bearings B at 4000 rpm (Fig. 2.26), and inner rotors or spool rotating at 6000 rpm in the same direction and supported by the shaft at bearings A. Relative motion occurs only at the surfaces indicated by arrows for bearings A and B. (a) For bearings A, the total load, which is equally divided between them, is the weight of the inner rotors totalling 4000 lb, to which is added the effect of an acceleration of $2g$ (i.e., 8000 lb), possible when pulling out of a dive. A ⅜ in wide oil groove midway of the length of each bearing A will be necessary. The film temperature of the SAE 10 oil will be held at 235°F. A suitable bearing diameter, determined from space and strength considerations, is 7.50 in, and a diametral clearance to diameter ratio of 0.002 is proposed. A minimum film thickness of 0.0010 in is desired. Determine load number and total length for each bearing. *Ans.* 42, 6⁵⁄₁₆ in. (b) Design bearings B for a diameter of 5 in. The outer rotors and shaft together weigh 4000 lb. Because tolerances can be more easily held than at A, reduce c_d/d to 0.0015 and h_{min} to 0.000 75 in. Although the bearings are in a cooler location, estimate the same film temperature since there will be less oil flow without the annular groove.

3

Friction Theory and Applications
Brakes, Clutches, and Belt Drives

3.1 INTRODUCTION

In contrast with bearings, friction is a useful and necessary physical property in brakes, clutches, belts, and traction drives. Thus in brakes and clutches, through suitable linkages and control, force and pressure are applied between surfaces to develop the friction forces needed to accelerate at desired rates, drive at constant speed, perhaps slip to prevent overloading, decelerate, stop, and hold. This is done with components of many sizes, from those in office and computing machines with torque capacities of a fraction of a lb·in and engagements occurring several times a second to those in heavy construction and mining hoists with capacities of several million lb·ft and low-operation frequency. Friction surfaces are built into many ingenious devices to synchronize the speeds of gears before sliding them into engagement, to change speed ratios and directions by locking or releasing gears in automatic transmissions, to give infinitely variable speed changing, to overrun speeds or limit torque, to limit slip between the axles driven by a differential mechanism, to multiply actuating forces as a mechanical servomechanism, and to govern speeds of rotation.

Brakes regulate speed by absorbing some or all of the kinetic energy of motion, decreasing the speed, and if desired, stopping and holding. They are used on hoists, on production machines to shorten the cycling time and for safety, and on vehicles from prams to planes. Engagement of a *friction clutch* gives a common speed to members previously rotating at different speeds. Some shock is avoided by initial slipping, and again there is an exchange and loss of energy. A *positive clutch*, one with jaws or toothlike splines, may only be engaged at small relative speeds. With a few exceptions, the word clutch is reserved for a readily engaged and disengaged component.

The word *coupling* is used for a semipermanent connection, generally through keys, set-screws, and bolts. A *rigid coupling* allows machines and shafts to be manufactured, shipped, and purchased in separate units, followed by assembly. A *flexible coupling*, in addition, allows small axial, angular, or lateral displacements, depending upon its design, thus accommodating small misalignments between the two shafts joined. There may be some torsional flexibility, either incidental or designed with consideration of vibration or shock. *Universal joints* are couplings that allow larger and varying angles between the shafts. Friction is only incidental in couplings, and they will not be discussed here. The *fluid* or *hydraulic coupling* is a principal subject of Chapter 4. In all couplings and clutches, torque is transmitted without significant change.

Continuous *belts* or rope over two or more flat, crowned, or grooved cylindrical *pulleys* or *sheaves* transmit power between shafts, usually with a change in torque and speed. A *belt drive* is, in a way, a flexible coupling between well-separated shafts, since it gives some torsional flexibility and allows a small amount of axial and angular misalignment. If the belt is readily loosened or removed, it becomes a clutch. If the belt is flat, it can limit overloads by slipping or leaving the pulleys. *Conveyer belts* support and move materials and products over distances.

Traction drives include elevator rope drives, locomotives and automotive vehicles where friction forces move the wheels along rails or road, and transmissions where disks under pressure drive by friction, replacing gears. By a shift in the axial position or shaft angle of a disk in contact with a cone or toroidal-surfaced wheel, effective diameters and hence output speed are varied in a continuous manner. Elevator and rope drives are discussed in Section 3.12. For the other traction drives a principal phenomenom and design consideration is the intensity of pressure over a very small area. Hence further mention of them is postponed to Chapter 11, on surface contacts, in particular Fig. 11.5 and Probs. 11.15 and 11.20 and its figure, plus a brief discussion of the effect of tangential friction forces at the end of Section 11.4.

The friction forces are commonly developed on inside or outside cylindrical surfaces in contact with shorter pads or shoes, or with a longer band or belt; on the flat sides of disks squeezed between pads or other disks; and on the exterior and interior surfaces of short frustrums of cones. Several of the combinations are used for brakes and clutches, and the band and cylinder pair is used in belt drives as well. Hence only a few different analyses are needed, with surface geometries influencing the equations more than the functions of the components. Also, there are common operating problems, such as pressure distribution and wear, temperature and heat dissipation, centrifugal force, different effectiveness in the two directions of rotation, and time of action. The several friction devices are effectively analyzed and studied together.

3.2 WORK, TORQUE, AND MOTION IN BRAKE AND CLUTCH SYSTEMS

Consider a machine or vehicle that has parts with masses m and weights mg $= W$, translating at time t_1 with velocities v_1 at elevations h_1, and connected parts with mass moments of inertia I rotating at angular velocities ω_1. Parts may be moving at different velocities. Thus the hoist of Fig. 3.1 lowers a mass and rope by rotation of the rope-drum shaft, motor shaft, and intermediate gear shafts. At time t_1 the total kinetic energy is $[\Sigma mv_1^2/2 + \Sigma I\omega_1^2/2]$, and the potential energy is ΣWh_1. If at time t_1 a brake is applied, then, in general, at time t_2, values will have been reduced to v_2, ω_2, and h_2, and to $[\Sigma mv_2^2/2 + \Sigma I\omega_2^2/2]$ and ΣWh_2. During the time $t_2 - t_1$ the brake has done work wk_B; rolling friction, bearing friction, and air resistance have done work wk_R; and the prime mover may have been "motored" such that work wk_M has been done to drive it. The total work equals the change in energy, or

$$wk_B + wk_R + wk_M = \Sigma m(v_1^2 - v_2^2)/2 + \Sigma I(\omega_1^2 - \omega_2^2)/2 + \Sigma W(h_1 - h_2) \qquad (3.1)$$

where the summations are made of products taken for different masses at their corresponding speeds or positions.

The work required of the brake to stop, slow, or maintain speed on an immovable surface is found by solving Eq. (3.1) for wk_B. Work wk_B indicates the mechanical energy transformed into heat at the brake or brakes. This can be used for the prediction and

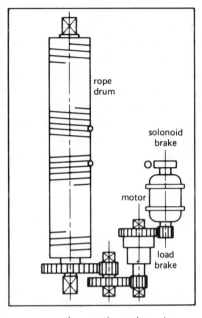

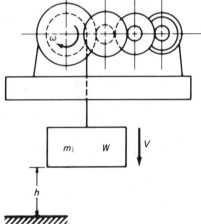

FIGURE 3.1 Hoisting arrangement on a crane trolley.

limitation of temperatures. The torque requirement of the brake may be found from wk_B. If ψ is the angular displacement of its drum,

$$wk_B = \int_{\psi_1}^{\psi_2} T \, d\psi = T(\psi_2 - \psi_1) \tag{3.2}$$

where the last term is valid only if torque may be assumed constant. If torque is constant, linear decelerations a and angular decelerations α are constant. Then, velocities v, displacements s, and times t are related by the equations of motion such that

$$v_2 = v_1 - a(t_2 - t_1) \tag{a}$$

and

$$s_2 = s_1 + v_1(t_2 - t_1) - (a/2)(t_2 - t_1)^2 = s_1 + \frac{v_1 + v_2}{2}(t_2 - t_1). \tag{b}$$

Similar equations relate angular velocities ω and displacements ψ to deceleration α and times t. These equations are applicable to braking only. Displacements are positive in the direction of motion, and decelerations a and α are positive numbers.

Example 3.1

A hoist is similar to the diagram shown in Fig. 3.1 but has only two shafts, one with a motor and brake and the other with the rope drum, with the two shafts geared together for a speed reduction ratio of 5. The "inertia" of the electric motor is given as $Wr^2 = 6.5$ lb·ft². The steel rim of the larger gear has an average diameter of 14.4 in, a width of 2 in, and an average thickness of 0.62 in. The rope drum weighs 277 lb, with a radius of gyration of 4.41 in. The grooves of the drum have 10 in pitch diameters, on which is wrapped a $^7/_{16}$-in-diameter hoisting rope with a breaking strength of 7.2 tons and a weight of 0.31 lb/ft. The maximum extended length of the rope is 200 ft. Determine brake work, torque, and rope tension when stopping within 4 ft a load of 2500 lb being lowered at 10 ft/s.

Solution

Energies will be found and used in Eq. (3.1).

Kinetic energy of translating parts

Velocity: $V_1 = 10$ ft/s, $v_2 = 0$

Weights: $W_{load} = 2500$ lb, $W_{rope} = 200(0.31) = 62$ lb

Energy: $E_k = \sum \dfrac{m(v_2^2 - v_1^2)}{2} = \dfrac{(2500 + 62)(10)^2}{2(32.2)} = 3978$ ft·lb

Kinetic energy at rotating drum shaft

Velocity: $\omega_1 = \dfrac{v_1}{d/2} = \dfrac{10}{10/(2 \times 12)} = 24$ rad/s, $\omega_2 = 0$

Weights: $W_{drum} = 277$ lb, $W_{rim} = \gamma \pi D b t = 0.28\pi(14.4)(2)(0.62) = 15.7$ lb

Inertia: $I_{drum} = mk^2 = \left(\dfrac{277}{32.2}\right)\left(\dfrac{4.41}{12}\right)^2 = 1.16$ ft·lb·s²

$I_{rim} = mk^2 = \left(\dfrac{15.7}{32.2}\right)\left(\dfrac{14.4}{2 \times 12}\right)^2 = 0.175$ ft·lb·s²

$I_{gear} + I_{shaft} = I_{rim} + 0.075(\text{estimated}) = 0.25$ ft·lb·s²

Energy: $E_k = \sum \dfrac{I(\omega_1^2 - \omega_2^2)}{2} = \dfrac{(1.16 + 0.25)(24)^2}{2} = 406$ ft·lb

Kinetic energy at rotating motor shaft

Velocity: $\omega_1 = (\text{speed ratio})(\omega_1)_{drum} = 5(24) = 120$ rad/s, $\omega_2 = 0$

Inertia: $I_{motor} = \dfrac{Wr^2}{g} = \dfrac{6.5}{32.2} = 0.202$ ft·lb·s²

Like weight W, Wr^2 is an easily conceived magnitude, here, 6.5 lb at 1 ft or 26 lb at 6 in, and it is commonly given in catalogs as a measure of the moment of inertia of a rotating part. The smaller brake drum, pinion, and shaft may add 10% more inertia, or

Inertia: $I_{add} = 0.10(I_{motor}) = 0.020$ ft·lb·s² (estimated)

Energy: $E_k = \sum \dfrac{I(\omega_1^2 - \omega_2^2)}{2} = \dfrac{(0.202 + 0.020)(120)^2}{2} = 1600$ ft·lb

Potential energy. For the 4-ft descent, neglecting change in rope length

$$E_p = \Sigma W(h_1 - h_2) = (2500 + 62)(4) = 10\ 250 \text{ ft·lb}$$

Work at brake. From Eq. (3.1), neglecting wk_R and wk_M, and substituting the preceding energy summations,

$$wk_B = 3978 + (406 + 1600) + 10\ 250 = 16\ 230 \text{ ft·lb}$$

Torque required at brake drum

Displacement: $\psi_2 - \psi_1 = (\text{speed ratio})(h_1 - h_2)/(d/2) = 5(4)/(5/12) = 48 \text{ rad}$

Torque, Eq. (3.2): $T = wk_B/(\psi_2 - \psi_1) = 16\ 230/48 = 338 \text{ lb·ft}$

Rope tension. From Eqs. (b) and (a), respectively

Stopping time: $t_2 - t_1 = 2(s_2 - s_1)/(v_1 + v_2) = 2(4)/10 = 0.8 \text{ s}$

Deceleration: $a = (v_1 - v_2)/(t_2 - t_1) = 10/0.8 = 12.5 \text{ ft/s}^2$

The tension at the top end of the rope is

$$F = W + ma = W\left(1 + \frac{a}{g}\right) = (2500 + 62)\left(1 + \frac{12.5}{32.2}\right) = 3560 \text{ lb}$$

The factor of safety is $F_s/F = 7.2(2000)/3560 = 4.05$ ////

The torque in Eq. (3.2) is the total for all brakes. In a vehicle, if identical brakes and actuating wheel cylinders are used at each wheel, the braking torque and work at each wheel is theoretically the same, except in skidding stops. To avoid individual-wheel skidding, vehicle brakes and cylinders are usually selected to give "balanced" braking at maximum deceleration. The wheel retarding force is made proportional to the normal force on the pavement. This normal force equals the weight on the wheel plus or minus a dynamic force due to deceleration of the mass of the vehicle, the center of which is some distance above the pavement. Because of variable mass distribution in both passenger cars and trucks, and because the skidding force and hence the possible maximum deceleration varies with pavement conditions, the "balance" must be a compromise. For passenger cars with the engine in front, the brakes are selected for a 60:40 or slightly smaller ratio of front-to-rear-wheel braking torque at all values of deceleration. Fortunately, the front wheel brakes are more exposed and have a higher cooling rate.

Term wk_M in Eq. (3.1) represents the work done by the machine in driving an idling engine or other energy-consuming device such as a hydrodynamic brake. In case the engine or motor drives the machine during the braking period, the equivalent energy value may be substituted for wk_M with a negative sign. For a truck, wk_R, called parasitic drag, may be estimated with the aid of factors for the calculation of power loss from various resistances.[1] If values are given in hp at different speeds, an average value, hp_{avg} may be chosen, and $wk_R = 550\ hp_{avg}\ (t_2 - t_1)$ in ft·lb when t is measured in seconds. In many machines, such as slow-speed hoists in open areas, these resistances are negligible. In any case, neglecting them will give a brake design "on the safe side."

[1] E.g., "Truck Ability Prediction Procedure-SAE J688," SAE Recommended Practice, *SAE Handbook,* Society of Automotive Engineers, New York (pp. 37.02–37.04 in the 1978 edition-Part 2; also P. G. Hykes and C. A. Herman, "Brake Drums," *SAE Trans.*, Vol. 68, pp. 336–345, 1960.

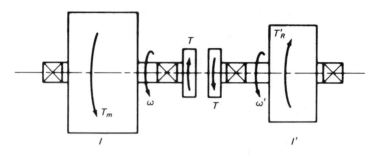

FIGURE 3.2 Shafts during clutching, with rotational velocities $\omega > \omega'$.

The work-energy relationship for clutch operations is more complicated than for most brake operations. In the clutching of two masses initially rotating freely in the same direction with different speeds, the higher-speed mass supplies energy not only for friction work at the clutch but also to increase the speed and kinetic energy of the lower-speed mass. In general, the work done by friction and other resistances equals the net change of energy *for the system*. This general relationship also holds when the braking is against a movable mass to which some kinetic energy can be transferred. When an airplane has landed and is slowing down on the deck of an aircraft carrier it is changing the speed of the carrier, however small the amount, but the brake work is affected in larger amount, depending upon the relative velocities.

A derivation will be made for the case shown in Fig. 3.2, omitting T_m and T_R' and assuming torque T at the clutch to be constant throughout the engagement period. If at initial time $t_1 = 0$, $\omega_1 > \omega_1'$, then for the shaft on the left, $-T = I(d\omega/dt)$. For the shaft on the right, $T = I'(d\omega'/dt)$. By the integration of each equation,

$$\omega = \omega_1 - \frac{T}{I}t \quad \text{and} \quad \omega' = \omega_1' + \frac{T}{I'}t \tag{c}$$

Relative motion ends at time t_2 when $\omega_2 = \omega_2'$, or when

$$\omega_1 - \frac{T}{I}t_2 = \omega_1' + \frac{T}{I'}t_2$$

whence

$$t_2 = \frac{\omega_1 - \omega_1'}{T}\frac{II'}{I + I'} \tag{3.3}$$

which is the time required for clutch engagement. By substitution of t_2 in either of Eqs. (c) the final, common velocity becomes

$$\omega_2 = \omega_2' = \frac{\omega_1 I + \omega_1' I'}{I + I'} \tag{3.4}$$

This equation can be obtained directly from the principle of conservation of momentum.

The displacements of masses I and I' after time t are, respectively,

$$\psi = \int_0^t \omega \, dt = \int_0^t \left(\omega_1 - \frac{T}{I}t\right) dt = \omega_1 t - \frac{Tt^2}{2I} \tag{d}$$

and

$$\psi' = \int_0^t \omega' \, dt = \int_0^t \left(\omega_1' + \frac{T}{I'}t\right) dt = \omega_1' t + \frac{Tt^2}{2I'} \tag{d'}$$

The work up to any time t is the torque times relative displacement or

$$wk = T(\psi - \psi') = T(\omega_1 - \omega_1')t - \frac{T^2(I + I')}{2I\,I'}\,t^2 \tag{e}$$

The total clutch work at the completion of engagement at time $t = t_2$ becomes

$$wk_c = \frac{1}{2}\frac{I\,I'}{I + I'}\,(\omega_1 - \omega_1')^2 \tag{3.5}$$

Equation (3.5) may also be obtained from the condition that work equals the net change in energy of the system. Thus $wk_c = \Delta E_k - \Delta E_k'$, where ΔE_k is the loss of kinetic energy by mass I and $\Delta E_k'$ is the gain by mass I' in arriving at the final velocity ω_2.

It is instructive to examine the special case of an initially stationary shaft, where $\omega_1' = 0$, which is brought up to speed by clutching it to a driving shaft whose speed is maintained at a constant value $\omega_2 = \omega_1$, either by power applied or by having $I \gg I'$. The velocity change in I' is ω_1, and the gain in its kinetic energy is $I'\omega_1^2/2$. An equal amount of energy is converted into work at the clutch, indicated by Eq. (3.5) for $\omega_1' = 0$ and $I \to \infty$. Thus the energy that must be supplied by the driving shaft is their sum $I'\omega_1^2$ or twice the gain by I'.

In a more general case, a motor or engine torque T_M acts upon the unprimed shaft (Fig. 3.2) and a work load or resistance torque T_R' acts upon the primed shaft.[2] This is typical for the clutches in automatic transmissions of motor vehicles. It may be shown that

$$wk_c = \frac{T(\omega_1 - \omega_1')^2}{2A} \tag{3.6a}$$

where

$$A = \frac{T - T_M}{I} + \frac{T - T_R'}{I'} \tag{3.6b}$$

Sometimes an inertia load I'' and a torque resistance T_R'' act on a shaft with speed ω'', with this shaft geared to the clutch output shaft with speed ω' (see the figure for Prob. 3.13). If the speed ratio $\omega'/\omega'' = \Omega$ and the efficiency of the gearing or other transmission is η, then equivalent inertia and torque which should be included in the I' and T_R' of Fig. 3.2 and Eq. (3.6b), illustrated in Prob 3.13, are

$$I_{equiv}' = \frac{I''}{\eta\Omega^2} \quad \text{and} \quad T_{equiv}' = \frac{T_R''}{\eta\Omega} \tag{3.7}$$

3.3 SHORT CONTACTS ON THE CYLINDRICAL SURFACES OF DRUMS

A block or short shoe can be guided to move radially against a cylindrical drum, as in Fig. 3.3(a). A normal force P develops a retarding force μP on the drum, where μ is the coefficient of friction and the required actuating force is $Q = P$. An unwanted cocking force $R = (\mu P)(a/b)$ develops at the ends of the guide, as seen from the free body of the block and rod. In Fig. 3.3(b), a lever or arm with a block at B and pivoted on a pin at

---

[2] See Prob. 3.11 and 3.15 and the sketch and reference for the latter; also Z. J. Jania, "Friction Clutch Transmissions," *Machine Design*, Vol. 30, pp. 137–143, Dec. 11, 1958. For a study of several cases involving a single mass, see L. P. Ludwig, "Engagement Characteristics of Wet-Type Clutches," *SAE Trans.*, Vol. 67, pp. 391–400, 1959.

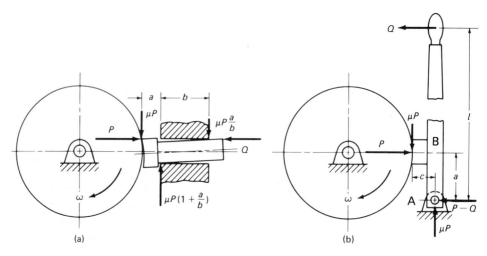

FIGURE 3.3 Development of friction force externally on a drum (a) by radial motion of the block in guide and (b) by the swinging motion of pivoted lever or arm.

A gives better guidance, together with a decreased actuating force, $Q = Pa/l - (\mu P)c/l$. Use of two or more symmetrically located arms, as shown in Fig. 3.4, provides a better balance of normal forces on the drum and its shaft and results in increased space utilization and torque capacity. In Figs. 3.3(b) and 3.4 the moment of the friction force about *A* adds to the moment of *Q* and aids in actuation. If drum rotation is reversed, the friction moment hinders actuation.

In Fig. 3.4, the central arm or flybar rotates the two curved arms that are pivoted to

FIGURE 3.4 Development of friction forces internally on a drum by pads on arms actuated by centrifugal force. Configuration similar to flybar type of return-speed governor in a telephone dial. (Sketch adapted from a telephone dial designed by Bell Laboratories and manufactured by Western Electric.)

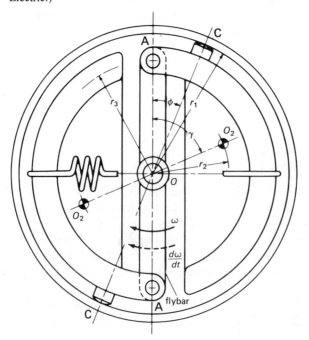

it at points A. Small studs or pads at points C are forced against the internal drum surface by lever action from the centrifugal and other inertial forces that are centered at O_2, and these forces and the friction force increase as the square of speed. With the drum fixed, the device is a speed governor, typical of those used in telephone dials, automatically clutched in for the return stroke to give uniform time intervals between electrical pulses. A more heavily constructed device, used as a *centrifugal brake*, limits speed to a safe value when a load is being lowered by a jack or winch.

With the drum free to rotate, the device becomes a *centrifugal clutch* that allows a spring or motor driving the flybar to take on the torque load gradually as speed increases, with the clutch slipping until 75 or 90% of full speed is reached. Thus a simple motor with low starting torque can be used to start loads with high initial torque. Centrifugal clutches are designed with radial-moving blocks as well as pivoted arms, as in Prob. 3.19.

Example 3.2

Fig. 3.4 represents the mechanical governor of a flybar type of telephone dial. Derive the differential equation for velocity ω in clockwise rotation. This equation will lead to a study of the uniformity of dial speed during switching operations (Probs. 3.22 through 3.25).[3]

Given are the force F_s in the retracting spring, the mass m of each arm, and the mass moment of inertia I_2 of each arm about its center of mass O_2. An enclosed, windup or power spring behind the dial drives the flybar shaft through gears to give it a higher speed for greater effectiveness (see sketch for Prob. 3.23). Assume the torque from the spring to be constant for the single revolution of the dial. It is multiplied by the gear ratio (about 1/20) and decreased by the inertial torque required to accelerate the gears and flybar. Thus the torque available to move the two arms is expressed as $T = H - K(d\omega/dt)$, where H is the spring torque times the gear ratio and K is determined from the inertia of the flybar and the gears on shafts turning at several speeds.

Solution

Figure 3.5 is a free-body diagram of one arm, using the dimensions from Fig. 3.4. Equations of equilibrium may be written either summed to zero or summed to inertial terms on the right-hand side. The former is chosen, and the reversed effective inertial-force vectors are shown on the diagram. About governor center O,

$$\Sigma M_0 = R_x r_3 - (\mu P)r_1 - \left(mr_2 \frac{d\omega}{dt} \right) r_2 - I_2 \frac{d\omega}{dt} = 0 \qquad \text{(a)}$$

Since $R_x r_3 = T/2$, where T for the two arms is defined in the problem statement, and $(mr_2^2 + I_2) = I_0$, which is the mass moment of inertia about O, Eq. (a) becomes

$$\frac{H}{2} - \mu Pr_1 - \left(I_0 + \frac{K}{2} \right) \frac{d\omega}{dt} = 0 \qquad \text{(b)}$$

About arm pivot A,

$$\Sigma M_A = Pr_3 \sin \varphi - \mu P(r_1 - r_3 \cos \varphi) + F_s r_3$$

$$- (mr_2\omega^2)r_3 \sin \gamma - \left(mr_2 \frac{d\omega}{dt} \right)(r_2 - r_3 \cos \gamma) - I_2 \frac{d\omega}{dt} = 0 \qquad \text{(c)}$$

---

[3] For a design study of a *drive-bar* type of governor and a comparison with the *fly-bar* type, including experimental data and a chatter analysis, see W. Pferd, "A Governor for Telephone Dials—Principles of Design," The Bell System Technical Journal, Vol. 33, No. 6, pp. 1267–1307 November 1954.

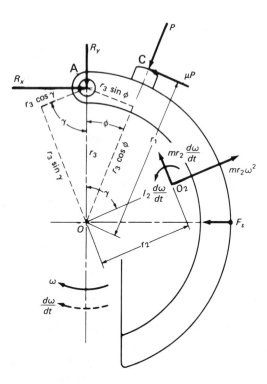

FIGURE 3.5 Free-body force diagram of right-hand arm of Fig. 3.4.

With a collection of terms and substitution of I_o, Eq. (c) becomes

$$P[r_3 \sin \varphi - \mu(r_1 - r_3 \cos \varphi)] + F_s r_3$$

$$- (mr_2 r_3 \sin \gamma)\omega^2 - (I_0 - mr_2 r_3 \cos \gamma) \frac{d\omega}{dt} = 0 \qquad (d)$$

For easier handling, let

$$s = r_3 \sin \varphi - \mu(r_1 - r_3 \cos \varphi), \; c = mr_2 r_3 \sin \gamma, \quad \text{and} \quad J = I_0 - mr_2 r_3 \cos \gamma \qquad (e)$$

Then Eq. (d) becomes

$$Ps + F_s r_3 - c\omega^2 - J \frac{d\omega}{dt} = 0 \qquad (f)$$

The unknown force P may be eliminated by multiplying Eq. (b) by s and Eq. (f) by μr_1 and adding, thus

$$\frac{sH}{2} + \mu r_1 r_3 F_s - \mu r_1 c\omega^2 - \left(sI_0 + \frac{sK}{2} + \mu r_1 J \right) \frac{d\omega}{dt} = 0 \qquad (g)$$

By rearrangement,

$$\frac{d\omega}{dt} + \frac{\mu r_1 c}{sI_0 + sK/2 + \mu r_1 J} \omega^2 = \frac{sH/2 + \mu r_1 r_3 F_s}{sI_0 + sK/2 + \mu r_1 J} \qquad (h)$$

whence

$$\frac{d\omega}{dt} + B\omega^2 = A \qquad (3.8a)$$

or

$$\frac{d\omega}{A - B\omega^2} = dt \qquad (3.8b)$$

where

$$A = \frac{sH/2 + \mu r_1 r_3 F_s}{sI_0 + sK/2 + \mu r_1 J} \quad \text{and} \quad B = \frac{\mu r_1 c}{sI_0 + sK/2 + \mu r_1 J} \qquad (3.8c)$$

The separation of variables in Eq. (3.8b) allows separate integration of the two sides. For the left-hand side, two solutions will be found in integral tables, depending on whether B is a positive or negative quantity. Both solutions should be tried and one rejected as physically unreasonable. ////

3.4 LONG SHOES ON CYLINDRICAL SURFACES

The capacity of a drum is further increased by using two long shoes, each lining subtending arcs of 90 to 150°. With longer surfaces in contact, and for the same pressure and rate of wear, the total normal and friction forces are greatly increased. However, they cannot be assumed concentrated at a central point on the drum surface, and an analysis for the force distribution must be made. Subsequently, a point will be located where they may be considered concentrated for analysis of brake effectiveness.

Pressure p in terms of angular position may be determined by assuming that the lining of a shoe is of uniform thickness and elastic, and that the drum and shoe backing are rigid. In Fig. 3.6 the lining is shown before compression (on the left) and after compression (on the right) with the distance between the dashed line and the drum surface representing the amount of compression (exaggerated in the sketch). Then a small radial deflection δ produces a pressure $p = k\delta$, where k is the stiffness constant of the lining.[4]

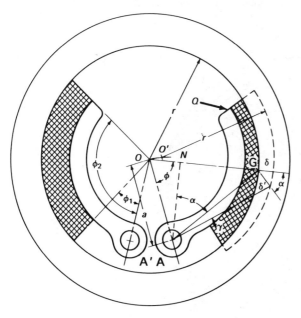

FIGURE 3.6 Long-shoe brake with lining exaggerated, that of left shoe in contact only with drum, that of right shoe compressed an amount δ at location angle ϕ.

---

[4] A different assumption which yields the same equation is that the work of friction is proportional to the volume of material removed by wear, with the lining maintaining a circular profile. In Fig. 3.6. this is the volume between the dashed line and the drum surface.

For the right-hand arm of Fig. 3.6, pivoted on the anchor pin at A, $\delta = \delta' \cos \alpha = \gamma\, AG$ $\cos \alpha$, where γ is the small angle of shoe rotation. Furthermore, $AG \cos \alpha = AN = a$ $\sin \phi$, where a is the distance AO between the anchor pin and drum centers, and ϕ is the angle between the line of centers AO and the radial line to the element. Thus $\delta = \gamma\, AN = \gamma a \sin \phi$. Stated differently, radial deflection equals the angular motion γ multiplied by the normal distance AN from the anchor pin to the radial line. Hence for an arm-mounted shoe, the pressure is

$$p = k\delta = k\gamma\, AN = k\gamma a \sin \phi \tag{3.9}$$

Pressure is maximum when $\sin \phi$ is maximum. By substitution of p_{max} and $\sin_{max} \phi$ in Eq. (3.9), $k\gamma a = p_{max}/\sin_{max} \phi$; hence Eq. (3.9) may be written as

$$p = \frac{p_{max} \sin \phi}{\sin_{max} \phi} \tag{3.10}$$

In the usual case, the extent of the shoe is such (as in Fig. 3.6) that $\phi_1 < 90° < \phi_2$, and $\sin_{max} \phi = 1.0$, whence $p = p_{max} \sin \phi$. The pressure distribution is most uniform on a given length of lining if it is placed symmetrically about the 90° line. In general, there are two shoes to a drum, operating on the same or separate anchor pins. Hence point A may or may not be vertically below the drum center O, and all angles, including the 90° angle, must be measured from the line of centers AO, whatever its orientation. Negative values of ϕ_1 and values of $\phi_2 > 180°$ are useless, and may even prevent retraction of the shoe.

The element dA of the cylindrical surface in Fig. 3.7 is $dA = br\, d\phi$, on which the normal force is $dP = pbr\, d\phi$. From the moment of the frictional force $\mu\, dP$ about the

FIGURE 3.7 Free-body force diagrams for two shoes of an internal brake. Actuating force Q, anchor-pin reactions R, and distribution and relative magnitude of pressure and friction forces.

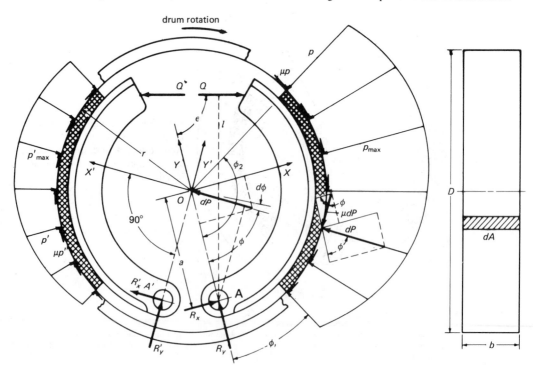

drum center O, the braking torque on the drum from one shoe is

$$T = \int_{\phi_1}^{\phi_2} r(\mu \, dP) = \mu br^2 \int_{\phi_1}^{\phi_2} p \, d\phi \tag{a}$$

By substitution from Eq. (3.10), integration, and replacement of r by $D/2$,

$$T = \mu br^2 \frac{p_{max}}{\sin_{max} \phi} \int_{\phi_1}^{\phi_2} \sin \phi \, d\phi = \frac{\mu bD^2}{4} \frac{p_{max}}{\sin_{max} \phi} (\cos \phi_1 - \cos \phi_2) \tag{3.11}$$

Solved for pressure, this equation becomes

$$\frac{p_{max}}{\sin_{max} \phi} = \frac{4T}{\mu bD^2(\cos \phi_1 - \cos \phi_2)} \tag{3.12}$$

Designate X and Y axes, with Y taken along AO. The sum P_x of the X components of all the distributed normal forces acting upon the shoe, as well as the sum P_y of the Y components, are quantities that will simplify further analysis. In Fig. 3.7 the X and Y components of dP are $\sin \phi \, dP$ and $\cos \phi \, dP$, respectively. Since $dP = pbr \, d\phi$ and with Eq. (3.10),

$$P_x = \int_{\phi_1}^{\phi_2} \sin \phi \, dP = \frac{brp_{max}}{\sin_{max} \phi} \int_{\phi_1}^{\phi_2} \sin^2 \phi \, d\phi \tag{b}$$

By integration,

$$P_x = \frac{brp_{max}}{2 \sin_{max} \phi} [(\phi_2 - \phi_1) + \sin \phi_1 \cos \phi_1 - \sin \phi_2 \cos \phi_2] \tag{c}$$

With substitution from Eq. (3.12),

$$P_x = \frac{\kappa T}{\mu D}, \tag{3.13a}$$

where

$$\kappa = \frac{(\phi_2 - \phi_1) + \sin \phi_1 \cos \phi_1 - \sin \phi_2 \cos \phi_2}{\cos \phi_1 - \cos \phi_2} \tag{3.13b}$$

and $(\phi_2 - \phi_1)$ is measured in radians. Similarly,

$$P_y = \int_{\phi_1}^{\phi_2} \cos \phi \, dP = \frac{brp_{max}}{\sin_{max} \phi} \int_{\phi_1}^{\phi_2} \sin \phi \cos \phi \, d\phi \tag{d}$$

whence

$$P_y = \frac{brp_{max}}{2 \sin_{max} \phi} (\cos^2 \phi_1 - \cos^2 \phi_2) \tag{e}$$

With substitution from Eq. (3.12),

$$P_y = \frac{\lambda T}{\mu D}, \tag{3.14a}$$

where

$$\lambda = \cos \phi_1 + \cos \phi_2 \tag{3.14b}$$

Angles ϕ_1 and ϕ_2 are measured from the line of centers AO, and this line, in general,

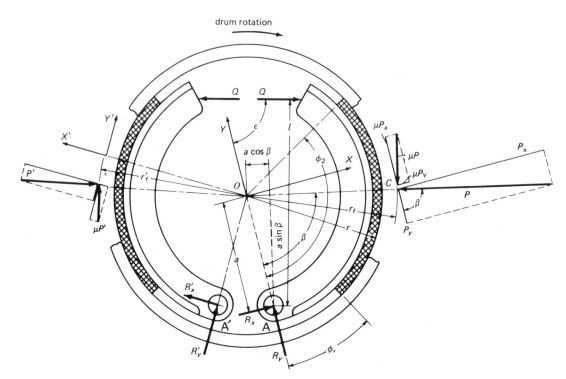

FIGURE 3.8 Resultant normal force P and friction force μP, replacing distributed forces of Fig. 3.7.

is not vertical. The angles are measured in the direction that makes $0 < \phi_1 < \phi_2 < 180°$.

Components P_x and P_y define and locate a resultant force P (Fig. 3.8) by the equations

$$P = \sqrt{P_x{}^2 + P_y{}^2} \quad \text{and} \quad \beta = \tan^{-1} P_x/P_y = \tan^{-1} \kappa/\lambda \tag{3.15}$$

That P is a normal force may be seen by considering all the elemental normal forces dP that have been translated to the drum center O. This locates their resultant P at center O, from which it may be translated outward along the radius that makes an angle β with AO, to act normal to the drum surface.[5]

Since μ has been taken as constant, the elemental friction forces $\mu \, dP$ are everywhere proportional to and perpendicular to forces dP; hence they will integrate into a resultant friction force equal to μP and perpendicular to P. It will act at a point C located at a friction radius r_f such that the product of r_f and μP equals the torque T on the drum (Fig. 3.8). Thus

$$r_f = \frac{T}{\mu P} = \frac{T}{\mu \sqrt{P_x{}^2 + P_y{}^2}} = \frac{D}{\sqrt{\kappa^2 + \lambda^2}} \tag{3.16}$$

Point C is called the "center of pressure," and the last terms of Eqs. (3.15) and (3.16) show that its location is a function of geometry only. If the elemental friction forces shown in Fig. 3.7 are translated to act at line OC of Fig. 3.8, all will lie outside the drum surface. Hence their resultant μP lies outside the drum surface. Note in Fig. 3.8

[5] A graphical method for determining and locating resultant forces is outlined by G. A. G. Fazekas, "Graphical Shoe Brake Analysis," *Trans. ASME*, Vol. 79, pp. 1322–1328, 1957. See also R. J. Auman, "Brake Design," *Machine Design*, Vol. 28, pp. 135–142, June 14, 1956.

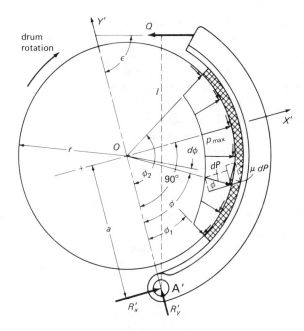

FIGURE 3.9 Forces on one shoe of an external brake

that the X component of friction force μP is normal to P_y and equal to μP_y. Likewise the Y component is normal to P_x and equal to μP_x.[6]

The distributed force system of Fig. 3.7 and the concentrated force system of Fig. 3.8 are equivalent, and the latter will be used in the following analyses. The effectiveness of a shoe is determined by the moments about its pivot A of the externally applied force Q, the normal force P, and the friction force μP. The respective moments are

$$M_e = Ql \tag{3.17a}$$

$$M_n = P(a \sin \beta) = P_x a = \frac{\kappa a}{\mu D} T \tag{3.17b}$$

$$M_f = (\mu P)(r_f - a \cos \beta) = (\mu P)\left(\frac{T}{\mu P} - a \frac{P_y}{P}\right)$$

$$= T - a(\mu P_y) = \left(1 - \frac{\lambda a}{D}\right) T \tag{3.17c}$$

The summation of moments is

$$\Sigma M_A = M_e - M_n \pm M_f = 0 \tag{3.18}$$

Torque T is always a positive quantity. The sign of the friction term M_f depends upon the direction of the drum rotation and the position of the shoe. Although the normal-force moment M_n always opposes the external, operating moment M_e, the friction moment M_f may aid or oppose M_e, hence there will be an increase or decrease in the effectiveness of the shoe. The pressure diagrams of Fig. 3.7, which are drawn to scale, indicate how much more effective one shoe may be than another to which the same actuating force Q is applied.

The analysis of an external shoe (Fig. 3.9) gives the same equations for pressure,

[6] For another treatment, see D. H. Offner, "Generalizing the Analysis of Shoe-Type Brake-Clutch Systems," *J. Engineering for Industry, Trans. ASME*, Vol. 91, pp. 694–701, 1969. This includes the case of a "floating" shoe, one with a short link between arm pivot and anchor pin, which allows some self-adjustment at installation and for wear.

torque, forces, and moments, Eqs. (3.9) through (3.18). For use of the moment equation, Eq. (3.18), the following rule may be applied for any external or internal shoe. *If to bring about an increase of pressure, the arm and shoe are rotated about the pivot in the same sense as the rotation of the drum, use the positive sign before M_f; if the rotations are opposite, use the negative sign.*[7] Thus, for the right-hand shoe of Fig. 3.8 the sign is positive and for that of Fig. 3.9, the sign is negative.[8]

The pin reactions are best determined by sketching a free-body diagram with the forces P and μP and their components, shown acting in the correct direction. Then, the sense of the reaction components R_x and R_y may be taken as positive and their values and true sense determined by a summation of forces in the X and Y directions.

Drum torque is the sum of the torques for each shoe, and there are usually two shoes. For shoes differently orientated relative to drum rotation, such as those in Fig. 3.8, there are differences in the friction forces and torques and sometimes in the externally applied moments, shoe dimensions, and lining material. The two shoes will be identified by priming the symbols of the shoe whose rotation for braking action is counter to drum rotation. Substitution into Eq. (3.18) of the values of M_e, M_n, and M_f from Eqs. (3.17) gives the following solution for torque:

$$T = \frac{Ql}{\dfrac{a}{D}\left(\dfrac{\kappa}{\mu} + \lambda\right) - 1} \quad \text{and} \quad T' = \frac{Q'l'}{\dfrac{a'}{D}\left(\dfrac{\kappa'}{\mu'} - \lambda'\right) + 1}. \tag{3.19}$$

Let the ratio of these torque be $\tau = T/T'$. Then

$$\tau = \frac{T}{T'} = \frac{Ql}{Q'l'}\left(\frac{\dfrac{a'}{D}\left(\dfrac{\kappa'}{\mu'} - \lambda'\right) + 1}{\dfrac{a}{D}\left(\dfrac{\kappa}{\mu} + \lambda\right) - 1}\right) \tag{3.20}$$

The total drum torque T_D is

$$T_D = T + T' = T\left(1 + \frac{1}{\tau}\right) = T\left(\frac{1+\tau}{\tau}\right) \tag{3.21}$$

whence

$$T = T_D\left(\frac{\tau}{1+\tau}\right) \quad \text{and} \quad T' = T_D\left(\frac{1}{1+\tau}\right) \tag{3.22}$$

The torque requirement T_D for each brake or clutch is determined from the speed change and load considerations (see Section 3.2). One procedure used to determine drum and shoe dimensions is to sketch an arrangement of the shoes and operating linkage, and then assume reasonable values for the ratio $a/D = (\frac{1}{2})(a/r)$, friction coefficients, shoe extent ϕ_1 and ϕ_2, as well as the ratio $Ql/Q'l'$ if it is not equal to 1.0. From these, values of κ, λ, τ, T, and T' may be calculated in this order. Allowable pressure p_{max} is

---

[7] Because the sense of shoe rotation is the sense of the operating moment M_e on the shoe, and the sense of drum rotation is the sense of the torque portion of the friction moment on the shoe, which torque is the T term of the expression, $M_f = T - a(\mu P_y)$, in Eq. (3.17c).

[8] Usually the numerical value of M_f is positive. For an external shoe (Fig. 3.9) if ϕ_2 is small and a/r is large, the resultant vector μP may pass between pivot point A and the drum surface. Then the numerical value of M_f becomes negative, since $r_f < a \cos \beta$ in Eq. (3.17c). The product of this negative value and the negative sign outside of M_f in Eq. (3.18) is a positive value, giving to the friction moment a positive sense, correctly expressing the aid given to the operating moment M_e.

determined from materials and service conditions. Lining width b may be expressed as a proportion of drum diameter D from space and alignment considerations and then D may be determined from Eq. (3.11) or Eq. (3.12). Dimensions a and b may now be calculated from their assumed ratios to diameter, then force components P_x and P_y, Eqs. (3.13) and (3.14), and moments M_n and M_f, Eqs. (3.17b) and (3.17c). Finally, the required operating moments M_e are determined from Eq. (3.18) and the anchor pin reactions by free bodies. If dimensions and forces are unsuitable, dimensions may be adjusted and further calculations made.

Example 3.3

Design a shoe brake for the hoist of Example 3.1, using a cast-iron drum and molded-asbestos blocks or shoes with a maximum pressure of 150 psi, based on wear considerations.

Solution

From Example 3.1 the estimated brake work is 16 230 ft·lb per stop and the torque is 338 lb·ft = 4060 lb·in = T_D. Brake drum speed is 120 rad/s or 1146 rpm. An external shoe brake of open construction is desirable for hoists, using either blocks or shoes as in Fig. 3.9, pivoted below and actuated by a linkage over the top where there is room for it. (See sketch for Probs. 3.38 and 3.39.) From Fig. 3.9, it appears that there will be adequate space if a/r is made 1.25 or $a/D = 0.625$. Take $\mu = 0.35$, $\phi_1 = 30°$, $\phi_2 = 150°$, and equal actuation moments, so that $Ql/Q'l' = 1.0$. In radians, $\phi_2 - \phi_1 = 2\pi(150-30)/360 = 2.094$. Then, from Eqs. (3.13b), (3.14b), (3.20), and (3.22),

$$\kappa = \kappa' = \frac{(2.094)+(0.500)(0.866)-(0.500)(-0.866)}{0.866 - (-0.866)} = 1.709$$

$$\kappa/\mu = 1.709/0.35 = 4.88 \quad \text{and} \quad \lambda = \lambda' = 0.866 + (-0.866) = 0$$

$$\tau = 1.0\frac{(0.625)(4.88-0) + 1}{(0.625)(4.88+0) - 1} = \frac{4.05}{2.05} = 1.976$$

$$T = 4060\left(\frac{1.976}{1 + 1.976}\right) = 2695 \text{ lb·in}$$

$$T' = 4060\left(\frac{1}{1 + 1.976}\right) = 1365 \text{ lb·in}$$

A width-to-diameter ratio $b/D = $ ¼ is satisfactory and typical. Then in Eq. (3.11), $bD^2 = D^3/4$, and for the shoe with the greater torque, with $p_{max} = 150$ psi as given,

$$T = 2695 = \frac{0.35D^3}{16}\frac{150}{1.0}(0.866 + 0.866)$$

from which $D^3 = 474$ and $D = 7.80$ in. Then shoe width is $D/4 = 1.95$, or 2 in for standardization. The drum width should overlap, so make it 2¼ in, and a rim thickness of ¼ in should suffice, since there is a central supporting disk.

Before calculating the several forces it would be wise to check the assumed drum inertia and the temperature rise. The weight of the cast-iron rim and its inertia are

$$W = \gamma(\pi D_{avg})bt = (0.26)\pi(7.55)(2.25)(0.25) = 3.47 \text{ lb}$$

$$I = mk^2 = \frac{3.47}{32.2}\left(\frac{3.775}{12}\right)^2 = 0.0107 \text{ ft·lb·s}^2$$

In Example 3.1, the I for the entire brake drum plus pinion gear and shaft was estimated to be 0.020 ft·lb·s², which now seems reasonable. For the single, maximum-stopping condition of Example 3.1, lasting only 0.8 s, we shall assume that all the energy or brake work is absorbed in the brake rim. The heat equivalent is (16 230)/778 = 20.9 Btu. The specific heat c of cast iron is 0.117 Btu/(lb·°F). Then, heat = $c \, \Delta T \, W$ or 20.9 = 0.117(ΔT)(3.47), and ΔT = 51.5°F. The friction material specified can withstand temperatures over 700°F.

A more severe heating may occur if the entire 200 ft length of rope with load is being played out while held by the brake to a constant speed of 10 ft/s, the initial velocity in Example 3.1. This would require 20 s, with work done by load and rope at a maximum rate of (2500 + 62)10 = 25 620 ft·lb/s. The heat will have time to flow to other parts of the brake and into the air, and heat transfer data and analysis are needed to predict temperatures. Instead, to limit heating and wear, a product μpV is commonly used, which if divided by 33 000 becomes hp/in². Velocity V is measured in ft/min, so here V = 10 × 60 = 600 ft/min and μpV = (0.35)(150)(600) = 31 500 ft·lb/min/in². Manufacturers' tables commonly allow only 10 000 for molded asbestos, so that for this brake and load the velocity should not be allowed to exceed 190 ft/min or 3.2 ft/s for more than a few seconds. This is a typical speed for general hoisting operations. For the continuous lowering of loads over long periods of time, as when lowering drill pipe into deep oil wells, an auxiliary speed-limiting brake, usually of the hydrodynamic type, may be used (see Chapter 4).

The forces on the higher-torque arm and their locations are found by Eqs. (3.13) through (3.18). Since λ = 0, P_y = 0, and

$$P = P_x = \frac{\kappa T}{\mu D} = \frac{(1.709)(2695)}{(0.35)(7.80)} = 1690 \text{ lb}$$

$$\mu P = (0.35)(1690) = 592 \text{ lb}$$

$$\beta = \tan^{-1} \kappa/\lambda = \tan^{-1} \infty = 90°$$

$$r_f = \frac{D}{\kappa} = \frac{7.80}{1.709} = 4.56 \text{ in}$$

This locates the point C, where P and μP act, on the X axis, a distance 4.56 − 7.80/2 = 0.66 in outside the drum surfaces. Center distance a = 0.625D = 4.875 in.

Let the effective lever length l be 1.25D = 9.75 in. Then

$$\Sigma M_A = Q(9.75) - (1690)(4.875) + (592)(4.56) = 0$$

whence

$$Q = 5540/9.75 = 568 \text{ lb}$$

This is rather high, but a smaller controlling force is readily obtained by a bell crank and other levers or by a hydraulic actuating system. The resulting longer stroke makes control easier. Calculations of the forces on the other shoe and the reactions at the pivots are left for the reader (see Prob. 3.27).   ////

3.5 SELF-ACTUATION

Where friction aids in actuation of the shoe, the shoe is said to be *self-actuating* or *self-energizing*, e.g., the right-hand shoe of Fig. 3.8 or both shoes of Fig. 3.4. If friction hinders the actuation of the shoe, we shall call that shoe *counteractuating*, e.g., the left-hand shoe of Fig. 3.8 and the shoe of Fig. 3.9. If drum rotation is reversed, the shoes exchange labels. When most of the rotation is in one direction, the self-actuating shoe will wear out sooner than the counteractuating shoe. Since both shoes will be replaced when one is worn out, the lining of the counteractuating shoe is sometimes made shorter or of different material or even arranged so that a larger external moment is applied to it. In an external shoe brake, each anchor pin can be placed so that the resultant friction force is directed through it, so that there is no friction moment. The two shoes then have equal braking effort.

At the other extreme, the pin may be placed so that the friction moment alone is sufficient to apply the brake, and the shoe is said to be *self-locking*. A shoe brought into contact with the drum immediately grabs. This is undesirable in most cases since the brake is out of the operator's control. Even a shoe designed to avoid self-locking may lock if the anchor pin is out of the adjusted position or its bearing is worn, such that toe contact only occurs. Self-locking is useful in a device known as a *backstop*, a device used to prevent reverse rotation of a shaft when such rotation would be harmful. The theoretical condition for self-locking is found by placing the external operating moment M_e equal to zero in Eq. (3.18). For a 120-deg-internal shoe symmetrical about the line $\phi = 90$ deg, and with $\mu = 0.35$, the shoe will lock when in perfect adjustment if the center distance a is approximately one-half the radius or less. With contact at the toe only, it will lock at all values of $a < 1.75r$. The latter effect is reduced, however, by the flexing of the shoe, which distributes the pressure away from the toe.

The degree of self-actuation can be measured in two ways. One measure of actuation is the actuation factor A_f, defined as the ratio of the contribution of the friction moment M_f to the entire moment to apply the shoe ($M_e + M_f$). From Eqs. (3.17) and (3.18), and for self-actuating shoes only,

$$A_f = \frac{M_f}{M_e + M_f} = \frac{M_f}{M_n} = \frac{\mu[(D/a) - \lambda]}{\kappa} \tag{3.23}$$

The actuation factor is seen to be independent of torque, and proportional to the friction coefficient and a geometry factor. Self-locking occurs when $M_e = 0$. Hence $A_f = 1.0$, and the theoretical condition to avoid self-locking is

$$\frac{a}{D} > \frac{\mu}{\kappa + \mu\lambda} \tag{3.24}$$

A second measure of actuation is the mechanical advantage A_m, sometimes called the shoe factor.[9] It is the ratio of the brake torque developed at the drum to the moment externally applied to actuate the brake,

$$A_m = \frac{T}{M_e} = \frac{T}{M_n \mp M_f} = \frac{1}{\dfrac{\kappa}{\mu} \dfrac{a}{D} \mp \left(1 - \lambda \dfrac{a}{D}\right)} \tag{3.25}$$

where the sign in the second term of the denominator is negative for self-actuating shoes and positive for counteractuating shoes. At self-locking, $A_m = \infty$. If the same moment M_e is applied to each shoe, their A_m values are added to give an A_m value for the brake.

Mechanical advantage is plotted in Fig. 3.10 for 120° linings symmetrical about the line $\phi = 90°$ and for typical center distance-to-drum-radius ratios a/r (Figs. 3.7–3.9). It is seen that the geometry of the shoe has a considerable effect on its sensitivity to changes in friction coefficient, such as the increase that occurs when the brake is wet or the decrease or fade that occurs when the brake is hot. The single internal self-actuating

---

[9] G. A. G. Fazekas, "Graphical Shoe Brake Analysis," *Trans. ASME*, Vol. 79, pp. 1322–1328, 1957.

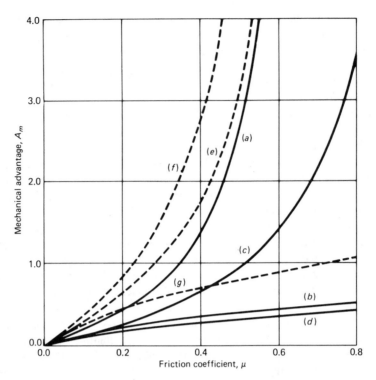

FIGURE 3.10 Mechanical advantage vs friction coefficient for 120° linings symmetrical about radial line $\phi = 90°$. All internal shoes with $a/r = 0.8$; all external shoes with $a/r = 1.2$ (Figs. 3.7 through 3.9). Internal *shoe* (a) self-actuating, (b) counteractuating. External *shoe* (c) self-actuating, (d) counteractuating. Internal-shoe *brakes* (dashed lines) (e) one shoe actuating, one counteractuating, (f) both shoes self-actuating, (g) both shoes counteractuating.

shoe (a) is more sensitive than the external shoe (c), but this sensitivity can be minimized by placing the anchor pin closer to the drum if space will allow.

Flat curves are obtained with counteractuating shoes (b) and (d). Values of A_m for two-shoe brakes are obtained by addition. They are shown for internal brakes in curves (e) through (g). All the combinations are useful, (e) for simplicity of actuation, requiring only one cam or hydraulic cylinder, (f) for minimum brake pedal effort, and (g) for maximum stability, as needed in racing cars, with the loss of mechanical advantage compensated by additional leverage in the operating mechanism or by power braking.

In the design of a self-actuating shoe, one must allow a margin of safety against self-locking. For the shoe (a) of Fig. 3.10, actuating factor $A_f = 0.51$ for the typical friction value $\mu = 0.35$. Self-locking at $A_f = 1.0$ would occur if μ were ≥ 0.68, as possible in a wet brake. An actuation factor of 0.50 is about the most for which one should design this type of shoe in a vehicle brake.[10] Where environment can be more closely controlled, higher values can be used.[11]

In a popular brake for automotive vehicles the reaction from friction on the primary shoe is the actuating force for the secondary shoe. The shoes are free to slide a small distance in the direction of the drum rotation, the secondary shoe stopping against a pin or against a block on which it can pivot. When the drum motion is reversed, the shoe

---

[10] T. P. Chase, "Passenger-Car Brake Performance," *SAE Q. Trans.*, Vol. 3, pp. 26−40, 1949.
[11] O. Von Mehren, "Internal-Shoe Clutches and Brakes," *Trans. ASME*, Vol. 69, pp. 913−924, 1947.

relationships are reversed, and the brake has the same self-actuation in both directions. When in this or other types of brakes, a shoe is not restrained to pivot about a point such as A in Figs. 3.7 and 3.8, the preceding equations do not strictly hold and other derivations are needed.

A vehicle of a certain mass traveling at a certain speed requires a definite brake torque, hence a definite friction force μP and normal force P to stop it in a given distance. The normal force is obtained by giving the necessary elastic deformations to the lining and drum. Incidental to this, there is some expansion of the oil lines and cylinders, and a small frictional resistance to the necessary motion of pedal, oil, pistons, and shoes during and after taking up about $\frac{1}{32}$ in clearance in the retracted position of the shoes. All this requires a definite amount of work to be done. In mechanical brakes, this may come partly from the energy of motion of the vehicle, which is the self-actuating effect, and the remainder from the operator's foot. The larger the self-actuation that can safely be used, the less the work required of the operator.

At the foot pedal, work may be the product of a high force and short stroke or a smaller force and a longer stroke. Both force and stroke are limited for comfortable driving, so a compromise must be made. The stroke is limited to about 5 in but some of this is needed for taking up the slack or shoe clearance, which increases between adjustments for wear. It is interesting to note that the automatic adjustment device commonly used in shoe brakes was developed during a period of increasing superhighways and a corresponding increase in speeds that demanded more work from brakes. The device eliminated periodic service shop visits for manual adjustments and decreased the allowance needed for wear in the pedal stroke, making available a longer effective stroke for the extra work of stopping from higher speeds.

3.6 DISKS AND CONES

In plate or disk clutches and brakes, e.g., Figs. 3.11 and 3.12(a), axially directed forces are applied normal to or between the flat faces of the elements to develop tangential friction forces. Usually not more than the outer 40% of radius is used, i.e., the ratio of inner to outer diameters of the friction-developing surfaces is 0.60 or greater. A ratio of 0.80 is not uncommon. Small pads of friction material instead of complete rings are also

FIGURE 3.11 Half-section view of hydraulically operated, multiple-disk clutch.

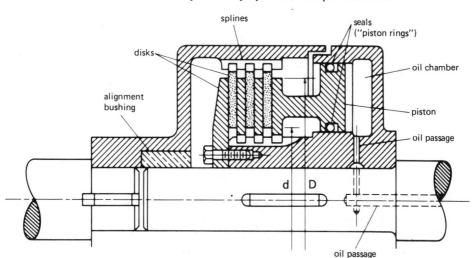

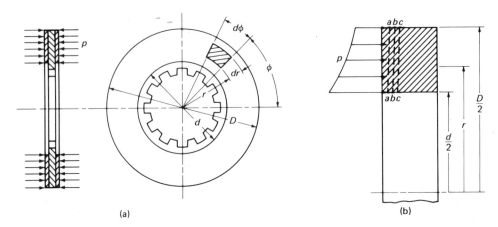

FIGURE 3.12 Single friction disk (a) initial pressure and elemental area and (b) successive profiles during wear.

used. Similar ratios are found in clutches using matching external and internal conical surfaces. It is frequently sufficient to assume that the total friction acts at the average radius of the contact area, a "friction radius," r_f. When there are N planes or places, each under the same normal force P, at which friction force μP can be developed by sliding or slipping motion (N on the driving member and N on the driven member), then the frictional torque is approximated by

$$T = N(\mu P)r_f = N\mu P \frac{D + d}{4} \tag{3.26}$$

the latter form being useful when the contact areas are bounded by rings of outer diameter D and inner diameter d. Note that $N = 6$ in the clutch of Fig. 3.11.

For a more extensive study, based on an elemental area $dA = (r\,d\phi)dr$ (Fig. 3.12(a)), the normal axial force and the torque are determined, respectively, by

$$P = \iint pr\,dr d\phi \quad \text{and} \quad T = N \iint \mu p r^2\,dr d\phi \tag{3.27}$$

For a complete (2π) ring and constant μ,

$$P = 2\pi \int_{d/2}^{D/2} pr\,dr \quad \text{and} \quad T = 2\pi N\mu \int_{d/2}^{D/2} pr^2\,dr \tag{3.28}$$

For new, accurately flat, and aligned disks the pressure will be uniform, or $p = p_0$. Integration of Eqs. (3.28), by this *uniform-pressure theory*, gives

$$P = p_0 \frac{\pi}{4} (D^2 - d^2) \tag{3.29a}$$

and

$$T = \frac{\pi N\mu p_0}{12} (D^3 - d^3) = \frac{N\mu P}{3} \frac{D^3 - d^3}{D^2 - d^2} \tag{3.29b}$$

The pressure distribution will be different after "wearing-in." If the rate of wear is assumed proportional to the work or energy conversion rate μpV, then the wear of a new clutch or brake proceeds most rapidly at the outer radius where the velocity V is greatest. Where the surface wears the most, the pressure decreases the most. The surface should wear to a shape *a-a* (Fig. 3.12(b)), such that $\mu p V$ is everywhere the same. Thereafter, wear is thought to occur uniformly at all radii, such that successive surfaces are *b-b* and *c-c*. Then $\mu p V$ remains constant and if μ may be considered constant, p is inversely proportional to V and hence to the radius r, or $p = C/r$, where C is a constant. Since $p = p_{max}$ at $r = d/2$, $C = p_{max}\, d/2$. Substitution for p into and integration of Eqs. (3.28), followed by further substitution, give by this *uniform wear theory*

$$P = \frac{\pi p_{max} d(D - d)}{2} \quad \text{and} \quad T = N\mu P \frac{D + d}{4} \tag{3.30}$$

By coincidence, this is the same result obtained in Eq. (3.26) by assuming an average friction radius. For the same torque capacity Eq. (3.30) for a "worn-in" clutch indicates a need for slightly larger areas and axial forces than does Eq. (3.29) for a new clutch. Thus it is safer to use Eq. (3.30), and together with its simplicity, it is the equation nearly always used.[12]

Example 3.4

A conveyor belt for a forge shop is to be driven through chain, gears, and a disk-type overload clutch by a 3.75 kW, 1200 rpm electric motor. The overload clutch, which is on the motor shaft, will be set for 200% of rated motor torque to protect the conveyor in case of an obstruction. Coefficients of friction are 0.40 at impending slip, 0.30 during slipping. A convenient outside diameter for the spline on the shaft will be 55 mm. Determine clutch specifications and the recovery time after the obstruction is removed. The mass of the conveyor belt and load is 22 500 kg, the ratio between motor and belt-pulley speeds is 157, the friction loss is 40%, full belt speed is 0.30 m/s, and the conveyor belt pulley diameter is 750 mm.

Solution

Velocity, $\omega = 1200(2\pi)/60 = 125.7$ rad/s,

Power, *pwr* $= 3750$ N·m/s,

Rated torque $= pwr/\omega = 3750/125.7 = 29.8$ N·m,

Slip torque $= 2 \times 29.8 = 59.6$ N·m

Choose an effective inside friction diameter of 65 mm (see proportions of disks in Fig. 3.11) and an average ratio to outside diameter of 0.70. Then $d = 65$ and $D = 93$ mm. Valuable production time would be lost if a conveyor were shut down for replacement of a clutch, so we shall use a conservative average pressure of 0.35 MPa($\approx$50 psi).

Axial force

$$P = p_{avg}(\pi/4)(D^2 - d^2) = (0.35)(\pi/4)(93^2 - 65^2) = 1216 \text{ N}$$

Numbers of friction pairs. From Eq. (3.30) of the uniform-wear theory

$$T = N\mu P(D+d)/4, \quad 59.6 = N(0.40)(1216)(0.093 + 0.065)/4$$

whence $N = 3.1$. An even number is necessary to balance out the forces internally and

---

[12] Equations (3.29) and (3.30) neglect losses in normal force, and hence in torque, caused by frictional resistance of the disks to sliding along the splines under torque. For an analysis and equations see *Mechanical Design and Systems Handbook*, H. A. Rothbart, Ed., New York: McGraw-Hill, p. 28–7, 1964, in the section on friction clutches by Z. J. Jania. See also Prob. 3.48.

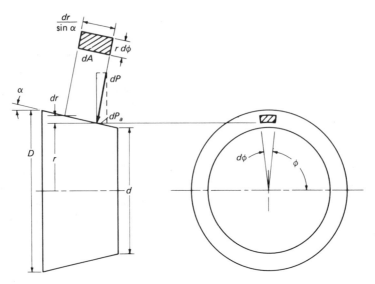

FIGURE 3.13 Elemental area of friction on a conical surface.

avoid a thrust bearing, so use two internally splined disks sandwiched between three externally splined disks, whence $N = 4$. We can now either decrease the axial force and pressure or decrease the effective outside diameter D. The former is chosen, making $p = 0.27$ MPa and $P = 942$ N.

The sliding friction is less than the static, so T becomes $(0.30/0.40)(59.6) = 44.7$ N·m. The conveyor belt pulley has $1/157$ of the motor speed, and the force at the belt is $F \approx (1 - 0.40)(44.7)(157)/(0.75/2) = 11\ 220$ N. The first factor is an estimate of the fraction of force available after friction losses, which are many due to many supporting rollers. From the relationship that impulse equals change of momentum,

$$F t = \text{mV}, \quad 11\ 220\ t = (22\ 500)(0.30), \quad t = 0.60 \text{ s}.$$

This is the time to recover full speed if the obstruction can be removed without stopping the motor. If the motor is stopped but has about 200% of torque at low speeds, desirable for conveyors, then it is an approximation to the time of recovery after stopping.   ////

On a conical surface (Fig. 3.13), $dA = (r\ d\phi)(dr/\sin\alpha)$ and $dP = p\ dA$ where α is the half-cone angle. The actuating force is the axial component dP_a of the normal force or $dP_a = dP \sin\alpha = p\ dA \sin\alpha = pr\ dr\ d\phi$. Hence

$$P_a = \int\int pr\ dr\ d\phi = 2\pi \int_{d/2}^{D/2} pr\ dr \tag{3.31a}$$

and

$$T = \int r(\mu\ dP) = \frac{2\pi}{\sin\alpha} \int_{d/2}^{D/2} \mu pr^2\ dr. \tag{3.31b}$$

With μ constant and $p = C/r$ by the uniform-wear theory, Eqs. (3.31) may be integrated to give

$$P_a = \pi C(D - d) \quad \text{and} \quad T = \frac{\mu P_a}{\sin\alpha}\frac{(D+d)}{4} \tag{3.32}$$

For the same axial force and diameters, the torque developed by a cone clutch is larger than that by a disk clutch with a single plane of slipping. The half-cone angle is commonly 12.5°, which gives a multiplying ratio of 1/sin 12.5° = 1/0.216 = 4.63. However, most "single-disk," disk clutches squeeze one disk between two plates, so N = 2. Addition of a second or third friction disk will give multiples of N = 4 or N = 6, respectively. For minimum inertia for rapid and frequent engagements, as needed in automatic machine tools, the required torque is obtained with small diameters and many thin disks, with N = 12 to 24. The driving and driven disks will often be of different metallic materials to avoid seizing. If the removal of heat by natural conduction and convection in the compact clutch is not sufficient, oil flushing and cooling is used. This reduces the friction coefficient, but the pressures and numbers of disks may be increased.

Activation of a disk or cone-type device is done in many ways. The clutch of Fig. 3.11 is closed by oil in an annular chamber, flowing from a pressure source into and through a central hole in the shaft. Note that all disks except a last one must be free to slide on their splines, internal or external, so that all surfaces carry approximately the same normal load and transmit the same torque. One or more springs may apply the actuating force in clutches and brakes that are only occasionally opened for short periods, as when shifting gears (see Prob. 3.48). Opening the disks requires a force larger than the actuating spring force, and this may be provided through levers manually, or by solenoids. A toggle mechanism may hold the clutch in engagement (see Prob. 3.46). For automobile and airplane wheels a "caliper" or yoke may span the brake disk. The caliper holds one or more hydraulic cylinders, which with friction pads put a squeeze on the disk (see Prob. 3.41). Often a major portion of the disk is exposed for cooling.

Self-actuation is not as easily attained, nor as commonly used as in shoe brakes. However, it is being advantageously used in brakes and takeoff clutches of work vehicles such as farm tractors and road-maintenance machinery. Enclosed by a wheel drum are two disks, each with friction material on their outer surfaces. Their inner facing surfaces have a series of ball ramps (see Prob. 3.49). Hydraulic cylinders force these disks into contact with surfaces on the interior of the wheel drum. Then, tangential friction forces rotate one disk relative to the other, bringing the ramp surfaces closer together, forcing a ball to roll up the ramp, further increasing the normal pressure on the friction surfaces. This self-actuation is equally effective in both directions of wheel rotation.

3.7 FLEXIBLE BANDS

A band brake or clutch has a flexible metal band or strap, usually lined with friction material, a cylindrical drum on which the band is placed, and an operating linkage to bring the band into contact with the drum (Fig. 3.15). Angles of wrap of 270 to 330° are common. A band brake is simpler and less expensive than most shoe brakes of the same capacity, but it wears more unevenly. It may require more space, but this is frequently available as at the ends of rope drums in hoists. Typical vehicle applications are parking-emergency brakes and in automatic transmissions, where they clamp certain gears in planetary trains to change the torque ratios. Band *clutches* may be used to control several of the motions in power shovels. A helical coil of spring wire, fitted over two coaxial shafts or drums, is used as the friction element of small clutches (Fig. 3.19). It is found in high-speed mechanisms because of its small size and low inertia. It can be made to grip in one direction only. Telephone dials may have a tiny coil to connect the speed governor for the return or pulsing motion and to disconnect it when the finger is winding the main spring.

The following analysis is applicable to both band and coil clutches and brakes as well as to belt and rope drives. The steps in the analysis will be similar to those previously

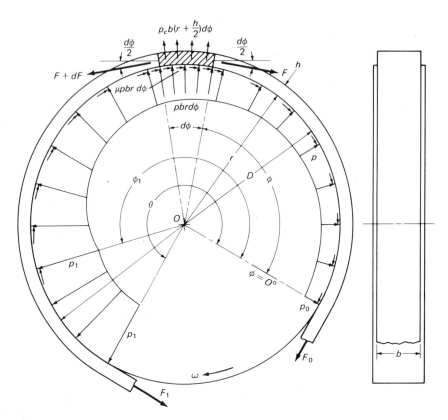

FIGURE 3.14 Free-body force diagrams of a band on a cylindrical drum surface under torque, with tight-side force F_1, loose-side force F_0, and pressure p. At top, forces on an element of band length. Pressure plotted for clockwise rotation of the drum and $\mu = 0.35$, with ϕ_1 measured from the F_0 end, as for the driving pulley of Fig. 3.21.

taken—determination of the general relationships between pressure, angular position, force, and torque from equations of equilibrium; and determination of formulas for particular types from their initial or boundary conditions.

Figure 3.14 shows a band of uniform thickness h and width b on the surface of a cylindrical drum of diameter D. Three unit forces act on the band, i.e., the drum pressure p and friction μp which vary with angular position ϕ, and a uniform centrifugal force p_c per square unit of band area that occurs when the band is rotating, as in clutches and belts.

The unit centrifugal force is given the symbol p_c because it is in the nature of a negative pressure, with pressure units. In the well-known general equation for centrifugal force, $mR\omega^2$, we substitute a mass ρh per unit of band area and an average band radius $r + h/2 = (D+h)/2$ to obtain

$$p_c = \frac{\rho h(D + h)\omega^2}{2} \tag{3.33a}$$

where all quantities are in a basic set of units. However, shaft speeds are commonly expressed in n rpm, dimensions in millimeters or inches, etc. For p_c in megapascals Eq. (3.33a) becomes[13]

$$p_c = (5.483 \times 10^{-15})\rho h(D + h)n^2 \text{ MPa} \tag{3.33b}$$

where ρ is in kg/m³, and h and D are in mm. In U.S. customary units, tables of unit weight γ in lb/in^3 are readily available,[13] corresponding to $g = 386$ in/s^2. For p_c expressed in psi, Eq. (3.33a) becomes[13]

$$p_c = (14.20 \times 10^{-6})\gamma h(D + h)n^2 \text{ psi} \tag{3.33c}$$

where h and D are measured in inches.

Within the band let the longitudinal force be F. Negligible internal moment, i.e., complete flexibility in bending, is assumed in the analysis that immediately follows.[14] For the element of band with the forces shown in Fig. 3.14, and by summation of the radial components,

$$pbr \, d\phi + p_c b(r + h/2)d\phi - F \sin(d\phi/2) - (F + dF)\sin(d\phi/2) = 0 \tag{a}$$

Now $p_c < p$, and for band clutches and belt drives, $h/2 \ll r$, and $h/2$ in the second term will be neglected. (It may be desirable to include it for some coil-spring clutches, depending upon the dimensions.) Also, in the limit, $\sin(d\phi/2) \rightarrow d\phi/2$ and $dF(d\phi/2) \rightarrow 0$. Together with the factoring out of $d\phi$, Eq. (a) becomes

$$pbr + p_c br - F = 0 \tag{b}$$

The second term is constant at constant speed. It is a force expressing the centrifugal effect on the belt tension. Let it be

$$F_c = p_c br \tag{3.34}$$

Then Eq. (b) becomes[15]

$$F = pbr + F_c \quad \text{or} \quad p = \frac{F - F_c}{br} \tag{3.35}$$

By differentiation

$$dF = br \, dp \tag{3.36}$$

By a summation of tangential components (Fig. 3.14),

$$\mu pbr \, d\phi + F \cos(d\phi/2) - (F + dF) \cos(d\phi/2) = 0 \tag{c}$$

In the limit, $\cos(d\phi/2) \rightarrow 1.0$ and

$$\mu pbr \, d\phi - dF = 0 \tag{d}$$

By substitution for dF from Eq. (3.36),

$$\mu p d\phi - dp = 0. \tag{e}$$

With a rearrangement, followed by integration within the limits indicated, we have

$$\int_{p_0}^{p} \frac{dp}{p} = \mu \int_{0}^{\phi} d\phi$$

and

$$\ln p/p_0 = \mu\phi \quad \text{or} \quad p = p_0 e^{\mu\phi} \tag{3.37}$$

The second equation gives the pressure distribution, plotted in Fig. 3.14, which increases

---

[13] For bands of spring steel, $\rho = 7830$ kg/m^3, $\gamma = 0.283$ lb/in^3; for leather belting, $\rho = 970$ kg/m^3, $\gamma = 0.035$ lb/in^3.
[14] See Sections 3.9 and 10.3 for consideration of bending moments.
[15] For rope and V-belt drives it is necessary to replace pb by q and $p_c b$ by q_c, where q and q_c are forces per unit length of rope or belt. See the discussion following Eq. (3.57).

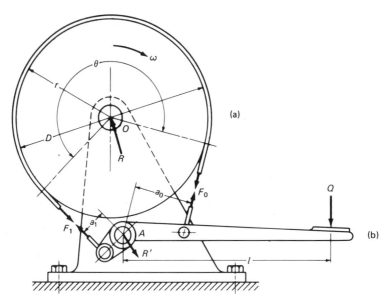

FIGURE 3.15 Band brake. (a) Free-body diagram of drum and band together. (b) Free-body diagram of operating lever.

exponentially from a minimum value p_0 at the $\phi = 0$ end to a maximum value p_1 at $\phi = \phi_1$. where ϕ_1 is the "active angle" over which slipping occurs. The maximum pressure, from Eqs. (3.37), is

$$p_{\max} = p_1 = p_0 e^{\mu\phi_1} \tag{3.38}$$

In a band clutch or brake, slipping occurs over the entire angle-of-wrap θ, and $\phi_1 = \theta$. In belt drives, however, the active angle where slipping occurs may be less than the angle-of-wrap θ. The torque developed on the drum (Fig. 3.14) is

$$T = \int_0^{\phi_1} r\,(\mu p b r\,d\phi) = \mu b r^2 \int_0^{\phi_1} p\,d\phi \tag{f}$$

similar to Eq. (a) preceding Eq. (3.11) for shoe brakes. With substitution by Eq. (3.37),

$$T = \mu p_0\, b r^2 \int_0^{\phi_1} e^{\mu\phi} d\phi = p_0 b r^2 (e^{\mu\phi_1} - 1) \tag{3.39}$$

The exponential term in the torque and pressure equations increases rapidly with angle ϕ_1. If $\mu = 0.35$, $(e^{\mu\phi_1} - 1)$ has a value of 4.2 for three-fourths of a turn of a band, and 733 for three turns with a helical coil. Hence it is possible to obtain high torques with a light pressure p_0 at one end. However, the large variation in pressure results in uneven wear.

One way of creating a pressure p_0 is to pull the band with a force F_0 at end $\phi = 0$ while its other end is held. From Eqs. (3.35) and (3.39),

$$p_0 = (F_0 - F_c)/br$$

and

$$T = (F_0 - F_c)r(e^{\mu\phi_1} - 1) \tag{3.40}$$

Another expression of torque may be obtained by taking the drum and band together as one free body, the upper one of Fig. 3.15,

$$\Sigma M_0 = 0, \quad T - F_1 r + F_0 r = 0$$

whence

$$T = (F_1 - F_0)r = (F_1 - F_0)\frac{D}{2} \tag{3.41}$$

A useful relationship between the "tight-side" force F_1 and the "loose-side" force F_0 is obtained by eliminating T from Eqs. (3.40) and (3.41), whence

$$\frac{F_1 - F_c}{F_0 - F_c} = e^{\mu\phi_1} \tag{3.42}$$

This same expression may also be obtained from Eq. (3.35) and (3.38). When there is no rotation of the band, Eq. (3.42) reduces to $F_1 = F_0 e^{\mu\phi_1}$, which is similar to the pressure relationship of Eq. (3.38).

3.8 BAND BRAKES AND CLUTCHES

The pull F_0 may originate manually or with a solenoid, hydraulic cylinder, or spring. When a band brake is manually operated, the F_0 pull is generally from a lever, as in Fig. 3.15. The F_1 end of the band may be attached to a fixed frame, to the pivot pin of the lever, or to a point out on the lever in a direction and amount to aid the operator in applying the lever, i.e., give some self-actuation, as discussed for shoe brakes in Section 3.5.

There are several requirements for locating a lever and its attachment points. In Fig. 3.15, Q is the externally applied operating force, the centrifugal effect $p_c = 0$, and the active angle ϕ_1 equals the angle of wrap θ, or $\phi_1 = \theta$. An equation of equilibrium for the lever is

$$\Sigma M_A = 0; \quad Ql + F_1 a_1 - F_0 a_0 = Ql + F_0 (a_1 e^{\mu\theta} - a_0) = 0 \tag{3.43}$$

Note that the moment $F_1 a_1$ aids the external moment Ql if $a_1 > 0$. If $a_1 = a_0/e^{\mu\theta}$, $Q = 0$, i.e., no operating force is needed once the band makes contact with the drum, and thus the band is self-locking. Because of manufacturing and environmental variations in μ, the value of a_1 should be limited to $Ca_0/e^{\mu\theta}$ where C is some fraction, perhaps 0.70 for a protected brake and less in exposed locations. Hence the first requirement is

$$0 < a_1 \leq C \frac{a_0}{e^{\mu\theta}} \tag{3.44}$$

The second requirement is that when the brake is actuated the motion given to each end of the band by the lever must have the same sense as the motion of the drum surface at the corresponding place of band tangency, except that a_1 may be zero. This locates force F_1 on the lever such that the moment $F_1 a_1$ does not oppose the operating moment Ql. Since $a_1 < a_0$ by Eq. (3.44), it also ensures that the diameter of the band is decreased to squeeze the drum.

As in shoe brakes, we may define the mechanical advantage A_m as the ratio of brake torque to external operating moment on the lever. With no band rotation and from Eqs. (3.40) and (3.43),

$$A_m = \frac{T}{Ql} = \frac{F_0 r(e^{\mu\theta} - 1)}{F_0(a_0 - a_1 e^{\mu\theta})} = \frac{D}{2a_0} \frac{e^{\mu\theta} - 1}{1 - \dfrac{a_1}{a_0} e^{\mu\theta}} \tag{3.45}$$

Equation (3.45) is plotted in Fig. 3.16 for a typical band brake.

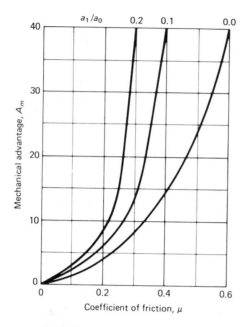

FIGURE 3.16 Mechanical advantage for a band brake with angle-of-wrap θ = 270°, a_0/D = 0.20, and three values of a_1/a_0 (see Fig. 3.15).

Example 3.5

Design the lever for a band brake with a capacity of 3600 lb·in, a drum 12 in. in diameter rotating clockwise, with the lever located beneath the drum and foot operated with a maximum force of 25 lb.

Solution

This is similar to the brake of Fig. 3.15. To use Eq. (3.44) as a guide we must know two of the three dimensions a_1, a_0, and θ. These dimensions must be determined from space considerations, so in a first trial we make a scale drawing with the pivot centered below the drum (Fig. 3.17). With a torque of 3600 lb in, the net pull is $F_1 - F_0$ = $T/(D/2)$ = 600 lb (Eq. (3.41)). If $e^{\mu\theta}$ is of the order of 4, then F_1 will be 800 lb, F_2 will be 200 lb, and the pin reaction less than 1000 lb. A ¾-in pin drawn to scale looks right, and if it and the band are 2 in wide, the bearing load is about 650 psi, reasonable for intermittent oscillation. If the pin is supported at both ends, the bending stress will be small and the shear stress in double shear, only 1100 psi.

The material around a hole is often given a diameter twice that of the hole. The lever should clear the drum enough to allow for inaccuracies and against jamming with small objects that get kicked around on floors, like machine screws and pencils. Allowing ½-in clearance, the center of the pin locates 1¼ in below the drum. The pin for attachment of the tight side of the band is conveniently placed at a radius $a_1 = 1$ in, and a tangent is drawn between the 1 in radius circle and the drum. If $e^{\mu\theta}$ is 4, attachment distance a_0 must be greater than $4a_1$ to avoid self-locking. If we draw the loose side of the band normal to the lever, we have $a_0 = 6$ in and an active angle $\theta = 225° = 3.93$ rad. Then $e^{\mu\theta} = e^{0.35(3.93)} = e^{1.37} = 3.95$ and by Eq. (3.44), choosing $C = 0.70$ as suggested,

$$a_0 = a_1 e^{\mu\theta}/C = (1)(3.95)/0.70 = 5.64 \text{ in}$$

This is close to our estimate of 6 in used to obtain θ. We are slightly "on the safe side," and rather than make adjustments now, let us use $a_0 = 6$ in to see if we obtain other reasonable dimensions. We have also satisfied the second requirement below Eq. (3.44) that all motions have the same sense, and the attachment point for F_0 moves further than

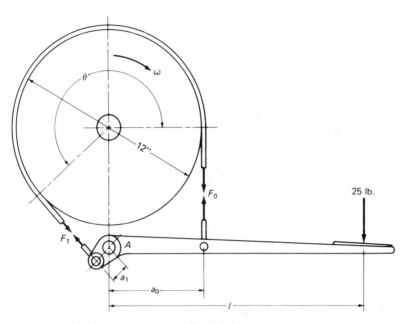

FIGURE 3.17 Layout of band brake for Example 3.5.

that for F_1, tightening the band when the foot pedal is depressed. For equilibrium,

$$\Sigma M_A = 0; \quad 25l + F_1(1) - F_0(6) = 0$$

Also, using values previously found, by Eqs. (3.41) and (3.42)

$$F_1 - F_0 = 600 \quad \text{and} \quad F_1/F_0 = 3.95$$

Solving simultaneously, we obtain

$$F_1 = 803 \text{ lb}, \quad F_0 = 203 \text{ lb}, \quad \text{and} \quad l = 16.6 \text{ in}$$

This is a reasonable length provided it locates the foot pedal in a convenient spot. The mechanical advantage is, Eq. (3.45),

$$A_m = \frac{12}{2(6)} \frac{3.95 - 1}{1 - (1/6)(3.95)} = \frac{2.95}{0.34} = 8.68$$

For reverse rotation of the drum, the same force Q of 25 lb will produce only 325 lb·in of torque. Further trials on offset locations of the pivot do not give an appreciable gain. However, the wrap angle can be increased by using a smaller a_1 (and hence a_0) and also by placing the lever and pivot higher and necessarily in front of or behind the drum, then letting pins extend under the drum for attachment to the band. An angle of 300° is possible with the bands crossing, a central slot cut in one band so that the other, narrowed down, may pass through the slot.   ////

Bands are necessarily thin for flexure and are commonly made of a steel strap faced with a woven asbestos or molded composition. This may be thin, bonded or riveted to the strap, and continuous and flexible. In large brakes of the order of 6 ft in diameter, such as used in oil well drilling operations, there may be a series of rigid blocks of a couple of inches thickness, bolted on and replaceable. They are sufficiently spaced so that the steel strap can flex between them. Special forged fittings are riveted to bands for their attachment. In automatic transmissions the bands for changing speed ranges are

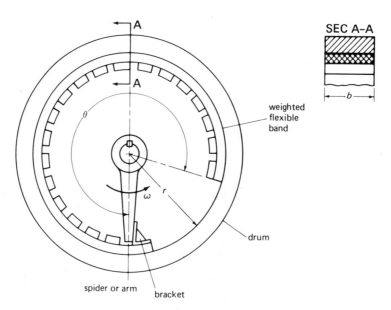

FIGURE 3.18 Centrifugal band clutch with internal weighted, flexible band. (A clutch using the same principle but with a laminated and flexible weighting member was patented by J. P. Madden of the Bethlehem Steel Corporation in 1931.)

usually of cast iron, machined and thin enough for small flexures and lubricated and cooled by the oil in the gear box.

With an internal expanding band, compressive forces are applied at the band ends to push it outward against the inside surface of a drum. Buckling should be avoided. As before, let F_1 designate the tight-side force, now at the end *toward* which the drum rotates, and F_0 designate the loose-side force, so that $F_1 > F_0$ as before. Equations (3.35) through (3.42) apply, except that the sign of the centrifugal pressure p_c should be changed wherever it appears since on a free body of the band and in the equation of equilibrium, all the forces are reversed except the centrifugal pressure. Thus Eq. (3.35) indicates that when the band rotates, the unit centrifugal force *adds* to the effect of the compressive forces along the band to increase the pressure. This can be beneficial in clutches.

Centrifugal force creates the entire loose-end pressure in a centrifugal band clutch (Fig. 3.18). At the start, an arm on the driving shaft pulls a flexible band around inside the drum. There is 100% slip until the pressure and friction between drum and band increase sufficiently to overcome starting torque, an action that can be made to occur at a favorable motor speed by determining the size of weights to be attached to the band. Slipping ceases before full running speed is reached. At the loose end $F_0 = 0$, so $p_0 = p_c$. Tension increases toward the arm end, tending to pull the band away from the drum, so pressure decreases. The torque developed is

$$T = p_c b r^2 \left(1 - \frac{1}{e^{\mu\theta}} \right) \tag{3.46}$$

3.9 THE COIL OR SPRING CLUTCH

In a spring clutch (Fig. 3.19), the contacting diameter of the coil before assembly is slightly different than that of the drums in order to give an initial interference and pressure upon assembly. If the coil is external to the drum and torque is applied to the drum in a direction such that the resulting tangential friction forces tend to wind up the coil and

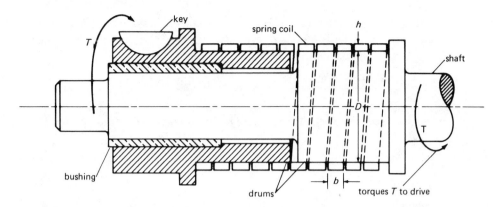

FIGURE 3.19 Elements of a coil-spring clutch.

decrease its diameter, very large forces and torques can be developed. If the torque is reversed, the friction forces tend to unwind and loosen the coil, and slipping will develop except under small torques. The drums may be attached to separate shafts, or one drum may be attached to a shaft which passes through a bushing in the other drum to support it and allow relative rotation under reversed torques, as shown in Fig. 3.19. A gear or other transmission element may be mounted on and keyed to the second drum.

An initial pressure is developed between coil and drum upon assembly, and it is uniform except near the ends of the coil. An approximate value with rectangular wire, derived in Section 10.3 as Eq. (10.27) and modified for the change due to rotation, is given by the equation

$$p_0 = \frac{8EI\Delta}{b(D + h)^4} \mp p_c \tag{3.47}$$

where E is the modulus of elasticity of the wire, I its moment of inertia of area in bending, Δ the diametral interference before assembly, D the diameter of the friction surface, h the wire height or thickness, and b the wire width (Fig. 3.19).[16] Unit centrifugal force p_c is given by Eqs. (3.33). The negative sign before p_c is used for external coils to show a loss in pressure, and the positive sign is used for internal coils to show an increase. This equation is sufficiently accurate for the usual ratios of $h/D < 0.10$. There is a corresponding longitudinal force F_0 along most of the coil, indicated by Eq. (3.35). However, there is no external moment at a free end, so the coil tends to lift from the drum near the end and contact it with a concentrated force at the very end. The end may be tapered to avoid the concentrated load and wear.

One analysis[17] assumes that the initial pressure is uniform out to the end. Then, under torque load the pressure p_0 at the end equals the initial pressure given by Eq. (3.47). Equations (3.37), (3.35) and (3.39) are used to calculate pressure, force, and torque in the direction that tightens or "winds up" the coil. A more exact analysis[18] that considers the foreshortened pressure area and a compensating concentrated force at the end gives

---

[16] Wire of approximately rectangular cross section is generally used, since it has maximum contact with the drum, hence least pressure and wear. For circular and other sections, Eq. (3.47) may be multiplied by b, and replaced by $q_0 = p_0 b = \dfrac{8EI\Delta}{(D + h)^4} \mp q_c$, where q_c is the centrifugal force per *unit length* of coil. Then q_0 is the initial contact force per unit length, which may be substituted for $p_0 b$ wherever it appears in other equations.

[17] C. F. Wiebusch, "The Spring Clutch," *Trans. ASME*, Vol. 61, pp. A-103–A-108, 1939.

[18] A. M. Wahl, "Discussion of Wiebusch's Paper," *Trans. ASME*, Vol. 62, pp. A-89–A-91, 1940.

about the same total torque. Hence the simpler equations are commonly used.[19] If N is the number of turns on one sleeve, then $e^{\mu\theta} = e^{2\pi N\mu}$, and Eq. (3.39) may be rewritten to give the torque capacity of the coil as

$$T = \frac{p_0 b D^2 (e^{2\pi N\mu} - 1)}{4} \tag{3.48}$$

Critical stresses are the tensile stress in the coil between sleeves, which from Eqs. (3.37) and (3.35) is

$$\sigma_1 = \frac{F_1}{bh} = \frac{D}{2h}(p_0 e^{2\pi N\mu} + p_c) \tag{3.49}$$

and the bending stress, derived as Eq. (10.29) in Section 10.3,

$$\sigma = \frac{Eh\Delta}{(D + h)^2} \tag{3.50}$$

induced throughout the wire at assembly. These two stresses add at the surface of the coil where it leaves the sleeve. The dynamic effects of sudden engagement add to these stresses.[20]

A reversed or negative torque, in the direction to unwind the coil, changes the sign of μ in the equations. Thus from Eqs. (3.37) and (3.8) for unwind torque T',

$$T' = \frac{p_0 b D^2}{4}(1 - e^{-2\pi N\mu}) \approx \frac{p_0 b D^2}{4} \tag{3.51}$$

when $1/e^{2\pi N\mu} << 1$. The ratio T/T', approximately $e^{2\pi N\mu}$, is a large number. Thus when p_0 is small, the spring clutch is essentially a one-way or nonreversing clutch, useful for overrunning, backstopping, and indexing with minimum backlash. With larger values of p_0 it can be used as a very stable torque-limiting clutch in the unwind direction, since T' is nearly independent of friction μ.

3.10 BELT AND ROPE DRIVES — GENERAL

A belt or rope when initially tightened around two or more drums or pulleys, then driven by the rotation of one of them, will by tension and friction transmit force and velocity to the surfaces of the other pulleys with little loss. The difference in the belt tensions on the entering and leaving sides of any pulley is the force transmitted to the pulley and is called the *net belt pull*. In the two-pulley arrangement of Fig. 3.20, the net belt pull is $F_1 - F_0$ where F_1 is called the tight-side tension and F_0 the loose-side tension. The torques developed are

$$T = (F_1 - F_0)\frac{D}{2} \quad \text{and} \quad T' = (F_1 - F_0)\frac{D'}{2} \tag{3.52}$$

in the ratio $T'/T = D'/D$.

As the force changes along the belt from F_0 to F_1 there must be a corresponding change in the strain along the belt. This requires that the belt move relative to the rigid pulley surface, a slip action commonly described as "creep." This occurs for elements (Fig. 3.14) where the change in tension dF equals the friction force $\mu p b r \, d\phi$ which is

[19] J. Kaplan and D. Marshall, "Spring Clutches," *Machine Design*, Vol. 28, pp. 107–111, April 19, 1956.
[20] J. Kaplan, "Dynamic Load Capacity of Spring Clutches," *Machine Design*, Vol. 29, pp. 105–109, April 4, 1957.

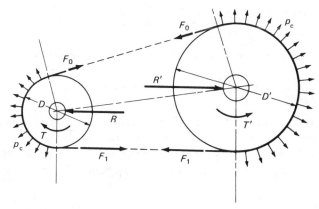

FIGURE 3.20 A two-pulley belt drive showing free-body force diagrams of each pulley together with part of the belt.

developed if slipping occurs. If the net belt pull is small, sufficient friction force to balance it is built up in a short distance, and creep occurs only along a limited portion of the pulley surface. This is the surface subtended by the active angle ϕ_1, which may be calculated from Eq. (3.42), and it is the same on both pulleys. Under idling or no-torque conditions, $F_1 = F_0$, and from Eq. (3.42), $e^{\mu\phi_1} = 1.0$, and $\phi_1 = 0°$. Experiments have shown that the active angle occurs on the leaving end of each pulley, the angles ϕ_1 and ϕ_1' in Fig. 3.21. As torque is increased the difference between F_1 and F_0 increases until $\phi_1 = \theta$, the angle of wrap on the smaller pulley, and there is creep over its entire contact surface. The corresponding torque is often taken as the capacity of the drive since a higher torque will result in a uniform relative motion superimposed upon the creep.

Creep gives a small loss in speed and power. In Fig. 3.21 the belt engages the driven pulley at E' with the surface velocity V_0 of the pulley. Creep begins at G', and the belt leaves at L' with a higher tension and in a lengthened condition, hence at increased velocity V_1. Now V_1 is the surface velocity of the driving pulley, with which the belt is in noncreeping contact from E to G. The speed loss is defined by "slip," $s = (V_1 - V_0)/V_1$ or $V_0 = (1 - s)V_1$. Since power is obtained from the product of force and velocity, the efficiency of the belt drive in percent is

FIGURE 3.21 Creep over active angles ϕ_1 and ϕ_1', with $\phi_1' = \phi_1$ and $V_1 > V_0$. Note that the active angles may be less than the wrap angles θ and θ'.

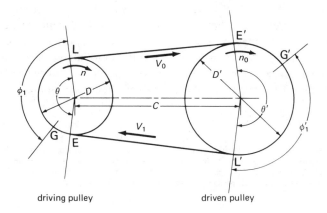

driving pulley   driven pulley

$$\eta = 100\, \frac{\text{driven power}}{\text{driving power}} = 100\, \frac{(F_1 - F_0)V_0}{(F_1 - F_0)V_1} = 100\,(1 - s) \tag{3.53}$$

In terms of the rotational velocity n of the driving pulley, that of the driven pulley is

$$n_0 = \frac{V_0}{\pi D'} = \frac{(1 - s)V_1}{\pi D'} = \frac{(1 - s)\pi Dn}{\pi D'} = (1 - s)n(D/D') \tag{3.54}$$

Creep has been found experimentally to be of the order of 1 or 2%, and overall slippage on heavily loaded drives may further increase the loss by 1 or 2%.

In calculations with belt forces the small difference $V_1 - V_0$ is neglected. The belt velocity V is taken as the surface velocity of the driving pulley. The power pwr is[21]

$$pwr = (F_1 - F_0)V \tag{3.55}$$

Equation (3.55) is solved for F_1 and taken together with Eq. (3.42) to eliminate F_0, giving

$$F_1 = \frac{pwr}{V} + F_0 = \frac{pwr}{V}\, \frac{e^{\mu\phi_1}}{e^{\mu\phi_1} - 1} + F_c \tag{3.56}$$

Inspection shows that the ratio $e^{\mu\phi_1}/(e^{\mu\phi_1} - 1)$ is smallest when ϕ_1 is largest. Hence to determine a minimum value of F_1 to transmit a given power and velocity, the active angle should equal the wrap angle θ, and Eq. (3.56) for design purposes becomes

$$(F_1)_{\min} = \frac{pwr}{V}\, \frac{e^{\mu\theta}}{e^{\mu\theta} - 1} + F_c \tag{3.57}$$

The foregoing equations beginning with Eqs. (3.33) were derived for flat bands and belts. When the belt is circular or trapezoidal in cross section (Fig. 3.22(b)), and it rides on the sides of a groove, there is a wedging action that increases the traction. The pressure and friction forces act on the sides of the belt. The products pb and $p_c b$ must be replaced by forces per unit length of belt, q and q_c, respectively. In Eq. (3.34), $F_c = q_c r$, and the normal forces on the two sides of an element of the belt are each $qr\, d\phi$. The latter have radial components $qr\, d\phi \sin \alpha/2$, where α is the angle included between tangents taken at the contact points. The first terms of the equations of equilibrium leading to Eqs. (3.35) and (3.37) in Section 3.7 are replaced by $2qr\, d\phi \sin \alpha/2 + q_c r\, d\phi$ for the radial direction, and by $2\mu qr\, d\phi$ for the tangential direction. The result is that in equations derived for flat belts, and which contain belt forces, it is only necessary to substitute for μ an *effective coefficient of friction*, $\mu' = \mu/(\sin \alpha/2)$ to make them valid for belts in grooves.

Flat leather belts are made with one, two, or three cemented layers of steer hides, called single-, double-, or triple-ply belts, respectively (Fig. 3.22(a)). For each there are two or three thicknesses, designated light, medium, and heavy. Within practical limits, any length of standard width may be specified and laced or spliced to a loop at the manufacturing plant or on the job. Flat belts of textile fabric impregnated with rubber are made continuous or endless. One or more pulleys of a drive are crowned to prevent flat belts from running off the pulleys. V-belts may have additional reinforcement from

---

[21] In SI units the unit of power is a watt, which is equivalent to a $N \cdot m/s$. Hence $(F_1 - F_0)$ must be substituted in newtons and V in m/s. In U.S. customary units (British Gravitational Units) the unit of power is the horsepower (hp), which is also used as a *symbol*. In Eq. (3.55) and others, pwr must be replaced by 550 hp if V is substituted in ft/s, or by 33 000 hp with V in ft/min. For equations in rotational speed ω or n, see the footnote to Eq. (4.6).

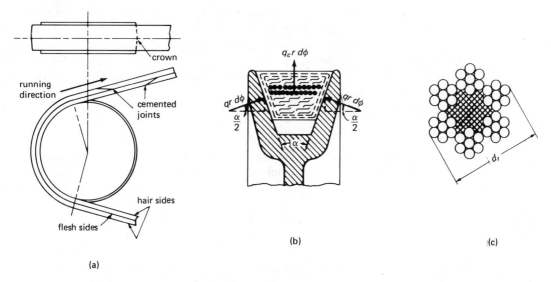

FIGURE 3.22 (a) Two-ply leather belt with beveled or scarfed cemented joints where abutting hides are joined. (b) Forces acting on V-belt. Note rubber-impregnated fabric jacket with interior tension cords above a rubber cushion. (c) Section of 6 × 7 wire rope with six strands and seven wires per strand (haulage rope).

textile or steel-wire cords (Fig. 3.22(b)). They come in a limited number of lengths and cross sections, the latter usually designated by letters A, B, C, etc. Pulleys are available with one to ten or more grooves, and the torque and power transmitted by N belts is considered to be N times that transmitted by one belt.

Wire rope is made by twisting into strands a bundle of 7 to 37 wires, usually of medium- to high-carbon steel, then twisting six or eight strands together to form the rope (Fig. 3.22(c)). Often the strands are preformed into long helices and then assembled. Rope is designated by the diameter of the circle that just envelopes the rope, by the number of strands and the number of wires in each, and by the material. A 2 in, 6 × 37 rope has more wires than a 2 in, 6 × 19 rope, but the diameter of each wire is smaller and the rope is more flexible and may be used over a sheave of smaller diameter. Under abrasive conditions the smaller wires will wear through sooner. The core or space between the strands usually contains hemp saturated with a lubricant to minimize corrosion and reduce the friction of the wires as they slide over one another, particularly during bending.

3.11 FIXED-CENTER DRIVES

In this simple arrangement, a belt is run onto or spliced over two pulleys on shafts whose centers remain at a fixed distance apart during operation of the drive. Frequently, the position of one shaft is adjustable by screws to facilitate assembly, particularly for V-section belts, and to allow convenient periodic adjustment in belt tension, since wear and stretch occur over a period of time. However, once adjusted, the center distance is fixed and the belt performs accordingly.

We shall refer to the no-load ($T = 0$) and no-velocity ($V = 0$) condition as the *standstill* or *standing* condition, with subscripts s on symbols; to the no-load but running condition as *idling,* with subscripts i; to the loaded and running condition as *driving*. The standing tension F_s is the same everywhere in the belt since there is no torque on the pulleys. With the belt idling, centrifugal forces reduce the pressure between belt and pulley without changing belt tension and length.[22] Hence $F_i = F_s$.

Under driving conditions (Fig. 3.20), approximately half of the belt will be under larger loads F_1 and F on the pulleys ($F_1 > F > F_s$) and the other half under smaller loads F_0 and F ($F_0 < F < F_s$). Let the corresponding increments of deflections be $\Delta\delta_1$ and $-\Delta\delta_0$. Suppose the stretch in placing the belt over the pulleys was $2\delta_s$. The total length is unchanged, so

$$(\delta_s + \Delta\delta_1) + (\delta_s - \Delta\delta_0) = 2\delta_s \quad \text{or} \quad \Delta\delta_1 = \Delta\delta_0 = \Delta\delta$$

Belts of different materials have different force-deflection curves, with an approximate linearity $F = k\delta$ between the minimum and maximum tensions in service. Hence

$$(F_s + \Delta F) + (F_s - \Delta F) = 2F_s$$

Now $(F_s + \Delta F) = F_1$ and $(F_s - \Delta F) = F_0$. Hence

$$F_1 + F_0 = 2F_s \tag{3.58}$$

Solutions of Eq. (3.55) for F_0, then for F_1, followed by substitutions into Eq. (3.58), give the pair of equations

$$F_1 = F_s + \frac{pwr}{2V} \quad \text{and} \quad F_0 = F_s - \frac{pwr}{2V} \tag{3.59}$$

To make F_1 minimum, F_s must be minimum, or

$$(F_1)_{\min} = (F_s)_{\min} + \frac{pwr}{2V} \tag{3.60}$$

By equating $(F_1)_{\min}$ from Eqs. (3.60) and (3.57), an expression is obtained for the minimum required tightening force,

$$(F_s)_{\min} = \frac{pwr}{2V} \frac{e^{\mu\theta} + 1}{e^{\mu\theta} - 1} + F_c \tag{3.61}$$

The tightening must actually be greater than indicated by Eq. (3.61), perhaps by a third, to allow for a period of gradual stretching of the belt before a readjustment is made. This makes the active angle $\phi_1 < \theta$. To solve for ϕ_1, one should substitute the actual value of F_s into Eqs. (3.59) and then use Eq. (3.42).

Example 3.6

A 20-hp 1750-rpm motor with a 10-in pulley drives a machine with a 30-in pulley located at a distance such that the angle-of-wrap on the smaller pulley is 153° (Fig 3.23). The belt is light, double-ply leather, 0.281 in thick by 4 in wide, and the expected coefficient of friction is 0.35. Determine: (a) the minimum required tight-side tension and (b) for a fixed-center drive, determine the required initial tensions and the resulting active angle on the smaller pulley.

Solution

(a) *Required tight-side tension.* Centrifugal "pressure," from Eq. (3.33c) and its footnote

---

[22] Actually, belt length increases by a very small amount because the reduction of pressure relieves some contraction in thickness.

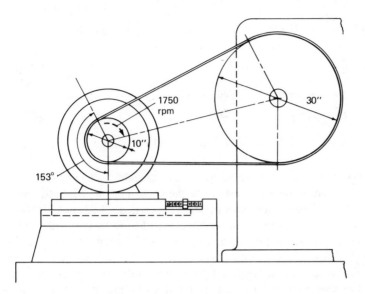

FIGURE 3.23 Electric-motor and fixed-center belt drive of Example 3.6.

$$p_c = (14.20 \times 10^{-6})\gamma h(D + h)n^2$$

$$= (14.20 \times 10^{-6})(0.035)(0.281)(10.28)(1750)^2 = 4.40 \text{ psi}$$

Belt force due to rotation

$$F_c = p_c br = (4.40)(4)(10/2) = 88.0 \text{ lb}$$

Pulley surface velocity

$$V = \pi Dn = \pi(10/12)(1750) = 4580 \text{ ft/min}$$

Net belt pull, from Eq. (3.55)

$$F_1 - F_0 = pwr/V = 33\,000 \text{ hp}/V = (33\,000)(20)/4580 = 144 \text{ lb}$$

Exponential factor, for $\theta = 153 \times \pi/180 = 2.67$ rad,

$$e^{\mu\theta} = e^{(0.35)(2.67)} = e^{0.935} = 2.55$$

Minimum tight-side tension, Eq. (3.57)

$$(F_1)_{\min} = \frac{pwr}{V} \frac{e^{\mu\theta}}{e^{\mu\theta} - 1} + F_c = (144)\frac{2.55}{2.55 - 1} + 88.0 = 325 \text{ lb}$$

(b) *Initial tensions and active angle, fixed-center drive.* Minimum standing and idling tensions, Eq. (3.61),

$$(F_s)_{\min} = (F_i)_{\min} = \frac{144}{2} \frac{2.55 + 1}{2.55 - 1} + 88 = 253 \text{ lb}$$

Initial tightening, standing, and idling tensions, made one-third larger,

$$F_t = 1.33(F_s)_{\min} = 1.33(253) = 337 \text{ lb} = F_s = F_i$$

Initial tight- and loose-side tensions, by Eqs. (3.59),

$$F_1 = 337 + \frac{144}{2} = 409 \text{ lb} \quad \text{and} \quad F_0 = 337 - \frac{144}{2} = 265 \text{ lb}$$

Initial active angle, Eq. (3.42)

$$e^{\mu\phi_1} = \frac{F_1 - F_c}{F_0 - F_c}, \quad e^{0.35\phi_1} = \frac{409 - 88}{265 - 88} = 1.81,$$

$$0.35\phi_1 = \ln 1.81 = 0.593, \quad \text{and} \quad \phi_1 = 1.69 \text{ rad} = 97.1° \qquad ////$$

Flat leather belts may be *selected* by several methods, all basically the same. One method is to use an allowable stress σ_1 of the order of $250-350$ psi, such that $F_1 = \sigma_1 bh$ in Eq. (3.57). Since F_c is also a function of bh (Eqs. (3.33) and (3.34)), Eq. (3.57) may be solved for the required cross-sectional area bh. A second method is to rate each standard thickness of belt in power capacity per inch of width at several different velocities. Several thicknesses and corresponding values of pwr/b are selected from a table. The latter are divided into the required power and multiplied by service factors to give the required belt width b. Such tables are provided by manufacturers or in handbooks.

V-belts are nearly always selected from manufacturers' catalogs with the aid of tables and charts based on the theory, with allowances for angle-of-wrap, minimum pulley diameter (bending effects), speed and centrifugal effects, fatigue life, overloads, and environmental conditions. The belts may be used singly or in multiple, side by side. Both for flat belts and V-belts, 4000 ft/min is considered an economical, good speed.

Wire rope is used for guying or staying towers, stacks, and other structures, for hauling, and for hoisting.[23] For the latter two, the rope is generally wound onto drums or guided over sheaves. In a *traction hoist*, friction of the sheave on the rope gives the driving force, and the belt equations, Eqs. (3.40)−(3.42) and Eqs. (3.35)−(3.57), may be used to determine the forces. To avoid pinching the hollow rope, the groove of the sheave is generally semicircular. The rope velocity is that of the center of this circle, the pitch circle, which coincides with the center of the rope. The effective coefficient of friction varies with material and environment, but because of the internal lubrication of rope and wires, a low value such as $\mu = 0.10$ is suggested for rope in iron or steel grooves.[24] Speeds are generally low, and the centrifugal term F_c may be neglected.

Catalogs give rope ultimate, tensile breaking strength, obtained by test. Since it is 5 to 20% less than the sum of the strengths of the wires, tested separately, the ultimate breaking stress S_u of the wires may be given as the rope strength divided by the total wire cross-sectional area. This stress, divided by a suitable factor of safety, may equal the direct stress due to maximum load and acceleration plus the bending stress. The latter is not subject to accurate calculation. A rope fails by initial breaking of its outside wires, so a better method includes the pressure based on projected contact area. If d_r is rope diameter and D is sheave pitch diameter, then from Eqs. (3.35), pressure $p = F/br = F/d_r(D/2) = 2F/d_r D$, where $d_r D$ is the *projection* of total contact area. Drucker and Tachau[25] found experimentally that p/S_u plotted against bends to failure gave typical fatigue curves. For 6×37 and 6×19 regular-lay rope the endurance limit was at $p/S_u = 0.0015$. Hence for infinite life it is logical to select wire rope by the equation

---

[23] Before individual or groups of machines were driven by their own electric motors, manila and wire ropes, sometimes a mile long, were used to deliver power to main shafts on all floors and sections of a factory or mill.

[24] The Handbook of John A. Roebling's Sons & Company gives $\mu = 0.07$ for greasy rope and 0.12 for dry rope.

[25] D. C. Drucker and H. Tachau, "A New Design Criterion for Wire Rope," *Trans. ASME*, Vol. 67, pp. A-33−A-38, 1945.

$$\frac{p}{S_u} = \frac{2F}{d_r D S_u} = \frac{0.0015}{n} \qquad (3.62)$$

where n is a factor of safety.[26] Presumably, for a traction hoist one will substitute the maximum value of F, i.e., the tight-side tension F_1.

In many cases, the actual rope selection must conform to the *minimum* standards of public building and safety codes or those of professional societies such as the ANSI and ASME.[27] In such cases, an equation such as Eq. (3.62) and the data on which it is based may be used to design for greater durability or to support recommendations for periodic inspection and replacement of the rope.

3.12 SELF-TIGHTENING DRIVES

These drives automatically maintain the necessary tensions without the allowance for stretching and without periodic retightening. Together with lower standing and idling tensions than in fixed-center drives, this gives longer belt life. Self-tightening drives require more mechanism and structure and so have higher initial cost. They are almost essential on short-center flat-belt drives to compensate for a small wrap angle and need for more frequent retightening if on fixed centers. They are seldom needed on long-center drives because of a larger wrap angle from geometry plus sag of the loose, upper side of the belt; also because the weight of the long belt helps maintain tension. In an elevator traction drive, a counterweight provides the loose-side tension on the rope.

In the *idler-pulley types,* a third pulley is forced against the loose side of the belt, either by weights and gravity or by a spring (Fig. 3.24). The extra pulley rotates freely except for the small friction of its bearings and, since it delivers no torque, it is called an idler pulley or simply *idler*. The idler is positioned so that it particularly increases the angle-of-wrap of the smaller pulley and thus the capacity of the drive. However, the reversed bending that the idler gives to the belt tends to decrease the life of the belt.

In Fig. 3.24 let W be the equivalent weight or spring force directed vertically downward through the center of the idler pulley, θ'' the angle-of-wrap on the idler pulley, and λ the angle between W and the radius at the point of belt tangency on the driving pulley side. Then the loose-side belt tension is[28]

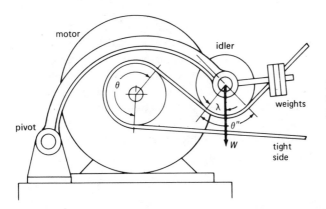

FIGURE 3.24 Loose-side belt tensioning by loaded idler pulley.

---

[26] This factor may be small, say 1.3 to 1.5. Factors in codes and catalogs are much higher, from 5 to 10, since they are applied to ultimate strength and they allow for uncalculated bending stresses.

[27] "Safety Code for Elevators, Dumbwaiters, Escalators, and Moving Walks," ANSI A17,1—1978, and yearly supplements, published by the American Society of Mechanical Engineers, New York.

[28] The derivations of Eq. (3.63) and one for $(F_0)_{min}$ are left for the reader in Prob. 3.89.

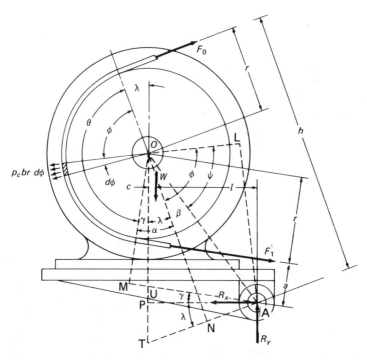

FIGURE 3.25 Belt tensioning by pivoted motor and driving pulley. The free-body diagram includes motor, platform, pulley, and a portion of the belt.

$$F_0 = \frac{W}{\sin \lambda + \sin (\theta'' - \lambda)} + F_c \tag{3.63}$$

It is seen that F_0 depends only on the force W, pulley position, and the centrifugal tension. The minimum weight or spring force W required for full utilization of the smaller angle-of-wrap on the driving or driven pulleys is determined by substituting $(F_1)_{\min}$ from Eq. (3.57) into Eq. (3.55) to obtain $(F_0)_{\min}$, then substituting this for F_0 in Eq.(3.63).[23] in Eq. (3.63).[28]

In a *pivoted-motor drive*, the motor is mounted on a pivoted platform or cradle in such a way that the moment of its weight provides the standstill tension F_s. Under running conditions the tensions become F_1 and F_0, and they are shown in Fig. 3.25, which is a free-body diagram of platform, motor, pulley, and a portion of the belt, taken together. Their combined weight W acts a distance l from the platform pivot A. The loose side of the belt is placed further from the pivot than the tight side, in order to make W more effective. Centrifugal unit forces p_c act on the belt over the arc subtended by the wrap angle θ, but are pictured only at a general location angle ϕ. Then,

$$\Sigma M_A = 0;$$

$$F_1 a + F_0 h - Wl - \int_0^\theta (p_c \, br \, d\phi) \, (AL) = 0 \tag{a}$$

The integral is evaluated as follows: setting $p_c \, br = F_c$ from Eq. (3.34) and from the triangle OLA,

$$AL = OA \sin \psi = OA \sin(\phi - \beta)$$

Then the integral becomes

$$-F_c \; OA \int_0^\theta \sin(\phi - \beta) \, d\phi$$

$$= F_c \; [OA \cos(\phi - \beta)]_0^\theta = F_c \; [OA \cos(\theta - \beta) - OA \cos(-\beta)] \qquad \text{(b)}$$

Now

$$OA \cos(\theta - \beta) = OA \cos(180 - \alpha - \beta) = - OA \cos(\alpha + \beta)$$

From triangle OMA,

$$- OA \cos(\alpha + \beta) = - OM = - (a + r)$$

From right triangle ONA,

$$- OA \cos(-\beta) = - OA \cos \beta = - ON = - (h - r)$$

Hence Eq. (b), which is the integral in Eq. (a), has the value

$$F_c \; [- (a + r) - (h - r)] = - aF_c - hF_c \qquad \text{(c)}$$

Substitution for the integral in Eq. (a) gives

$$(F_1 - F_c)a + (F_0 - F_c)h - Wl = 0 \qquad \text{(d)}$$

From Eq. (d) and Eq. (3.55),

$$F_1 = \frac{Wl}{a + h} + \frac{pwr}{V} \frac{h}{a + h} + F_c \qquad \text{(3.64a)}$$

and

$$F_0 = \frac{Wl}{a + h} - \frac{pwr}{V} \frac{a}{a + h} + F_c \qquad \text{(3.64b)}$$

At idling and standstill, respectively, F_1 and F_0 become

$$F_i = \frac{Wl}{a + h} + F_c \quad \text{and} \quad F_s = \frac{Wl}{a + h} \qquad \text{(3.65)}$$

The minimum value required of F_1 is given by Eq. (3.57). By equating this to Eq. (3.64a), a value for minimum overhang is obtained as

$$l_{\min} = \frac{(pwr/V) \, [Ea + (E - 1)(s + c\sigma)]}{W - (pwr/V)(E - 1)\sigma} \qquad \text{(3.66)}$$

where $E = e^{\mu\theta}/(e^{\mu\theta} - 1)$, $s = r + (a + r) \cos \lambda / \cos \gamma$, and $\sigma = \sin \lambda + \tan \gamma \cos \lambda$. Also,

$$h = s + (l + c)\sigma \qquad \text{(3.67)}$$

Both fixed-center and pivoted-motor drives must be designed to carry the same belt tension F_1 under maximum-load driving conditions, and this determines the belt width and thickness. This requirement is given by Eq. (3.57). The least tension $(F_s)_{\min}$ in the belt of the fixed-center drive occurs at idling and standstill and is given by Eq. (3.61). Usually this is larger to allow for stretch and this increases $(F_1)_{\min}$, according to Eq. (3.60). In the pivoted-motor drive under partial load, idling, and standstill conditions, the belt tension can be considerably lower than in a fixed-center drive. Equations (3.65) indicate this, for F_s is proportional to the overhang l. Theoretically F_s and l can be made zero by making dimension a negative with a magnitude of the order of $(\frac{1}{2})r$ to $(\frac{2}{3})$ r, determined by Eq. (3.66). This means that the tight side of the belt passes below pivot A, and its moment $(F_1 - F_c)a$ balances that of the loose side $(F_0 - F_c)h$, as in Eq. (d) preceding Eqs. (3.64). Of course, this is not practical, since there must be some initial tension at standstill to initiate gripping action and to allow for variations in the coefficient of friction. However, the motor can be cradled so that dimension a has a small negative or positive value, and the standstill tension is then small.

Example 3.7

It is proposed to arrange the belt and pulleys of Example 3.6 in a pivoted-motor drive (Fig. 3.25). The tight side of the belt will be horizontal, and the platform will be designed so that its weight and the weight of the motor, 420 lb, will act along a vertical line through the center of the motor. (a) If the tight side is 7 in above the pivot, determine the minimum overhang and compare the several tensions with those of the fixed-center drive. (b) If a standstill tension of 50 lb is satisfactory, what are the reduced values of l and a? (c) Compare the gravity moment and reaction torque in the F_1 equation, Eq. (3.64a), for cases (a) and (b).

Solution

(a) *Overhang and tensions if $a = 7$ in.* Since F_1 is horizontal, $\gamma = 0°$, $\lambda = 180° - 153° = 27°$. Since $W = 420$ lb is centered, $c = 0$. From Example 3.6, $pwr/V = 144$ lb, $F_c = 88$ lb, $e^{\mu\theta} = 2.54$, whence E $= 2.54/(2.54 - 1) = 1.65$. From below Eq. (3.66), $s = 5 + (7 + 5)(0.891)/1.000 = 15.69$ and $\sigma = 0.454$. In Eq. (3.66),

$$l_{min} = \frac{144[(1.65)(7) + (0.65)(15.69 + 0)]}{420 - (144)(0.65)(0.454)} = 8.30 \text{ in}$$

From Eq. (3.67),

$$h = 15.69 + (8.30 + 0)(0.454) = 19.46 \text{ in}$$

Tensions, Eqs. (3.64) and (3.65),

driving: $\quad F_1 = \dfrac{1}{7 + 19.46}[(420)(8.30) + (144)(19.46)] + 88 = 326 \text{ lb}$

standstill: $F_s = \dfrac{(420)(8.30)}{26.46} = 132 \text{ lb}$

idling: $\quad F_i = F_s + F_c = 132 + 88 = 220 \text{ lb}$

In comparison with the fixed-center drive, F_1 is 326 lb, the same as $(F_1)_{min}$ but less than the 410 lb after overtightening for the stretch allowance; F_s is 132 vs 338 and F_i is 220 vs 338 lb with the stretch allowance, all a considerable saving in tension.

(b) *Standstill tension F_s of 50 lb only.* W, λ, E, and σ are unchanged; s, h, l, and a have new values. First,

$$s = 5 + (a + 5)(0.891) = 9.455 + 0.891a$$

From Eq. (3.67)

$$h = 9.455 + 0.891a + 0.454l$$

From Eq. (3.66)

$$l = \frac{144\,\{1.65a + (0.65)\,[5 + (a + 5)(0.891)]\}}{420 - (144)(0.65)(0.454)} = 0.851a + 2.35$$

Also, from Eqs. (3.65) and the desired tension $F_s = 50$ lb,

$$F_s = \frac{Wl}{a + h}, \quad 50 = \frac{420l}{9.455 + 1.891a + 0.454l}$$

whence

$$l = 0.238a + 1.190$$

By equating the two expressions for l, a is found, then l and h, as follows,

$$a = -1.89 \text{ in}, \; l = 0.74 \text{ in}, \; h = 8.11 \text{ in}$$

Checks:

$$F_s = \frac{420(0.74)}{-1.89 + 8.11} = \frac{311}{6.22} = 50.0 \text{ lb (as given)}$$

$$F_1 = \frac{1}{6.22}[(420)(0.74) + (144)(8.11)] + 88 = 326 \text{ lb [same as in (a)]}$$

The standing tension F_s is 50 lb compared with 132 lb for the arrangement in (a). For a standby drive, either at standstill or idling much of the time, there may be an advantage in having the tight side of the belt below rather than above the pivot.

(c) *Gravity moment vs reaction torque.* It is of interest to compare the contributions of each to the tight-side tension given by Eqs. (3.64) in the preceding arrangements (a) and (b). In (a), $Wl/(a + h) = 132$ lb and $(pwr/V)h/(a + h) = (F_1 - F_0)h/(a + h) = 106$ lb. Although the gravity effect is not much larger, this might be called a *gravity-type drive.* In (b), $Wl/(a + h) = 50$ lb and $(F_1 - F_0)h/(a + h) = 188$ lb. Since the latter is much larger than the gravity effect, this may be called a *reaction torque drive.* ////

Two tensioning arrangements in traction rope drives are interesting. For general hauling and positioning on ships and in construction and oil-field activities, a powered, rotating pulley known as a capstan or cathead may be provided (Fig. 3.26). A man may

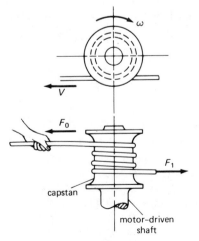

FIGURE 3.26 Loose-side tension produced manually on a capstan drive.

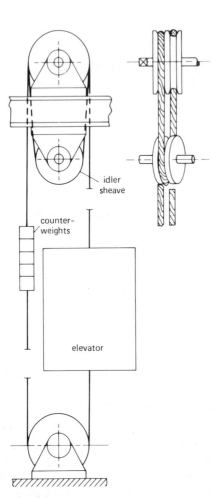

FIGURE 3.27 Tension produced by counterweights in an elevator traction drive.

quickly throw several turns of manila or light and flexible wire rope around the pulley, then by pulling on the loose end of the rope to keep it taut, he can move heavy loads with the other end. By slacking his hold to allow some slippage, he can control the speed of motion. In an elevator, a moving counterweight on one end of the rope provides loose-side tension and decreases the net drum torque (Fig. 3.27). The effective wrap angle is doubled by passing the rope from the load up and over a groove in the driving drum, then down and around a fixed-axis idler, then up and over the driving drum in a second groove and down to a counterweight. The counterweight may be proportioned so that the traction is equally effective in raising a full elevator car or in lowering an empty one.

PROBLEMS

Section 3.2

3.1   Derive Eqs. (a) and (b) for v_2 and s_2 that follow Eq. (3.2).

3.2   A hoist similar to that of Fig. 3.1 has mass moment of inertias 0.075, 0.042, 0.60, and 40 kg·m² on the motor shaft, first intermediate shaft, second intermediate shaft, and rope-drum shaft, respectively. The corresponding shaft speeds are 1750, 500, 150, and 45 rpm. The drum diameter is 600 mm, the mass on the rope is 2000 kg, and that of the rope is 75 kg. The load deceleration is to be limited to 20% of the acceleration of gravity. (a) What is the force in the rope when it is fully extended, the stopping distance, and the brake work? *Ans.* 14 290 J. (b) What torques are required of each brake if there is one at each end of the rope drum? If there is one brake only on the motor shaft? Which seems to you to be safer?

3.3 A loaded truck weighing 17 500 lb traveling at 60 mph is to be slowed to 30 mph. The average parasitic drag is 12 hp and the average motoring resistance 20 hp. The kinetic energy of the rotating parts is estimated to be 10% of that of the translating parts. The position of the load is such that more or less equal road forces occur at each braking wheel, and the coefficient of friction of tire on road is 0.80. What is the minimum time and distance for the action on a level highway and how much energy must be absorbed by the brakes? *Ans.* 1.84 s, 1.70×10^6 ft·lb.

3.4 Same as Prob. 3.3 but the action is downhill on a slope of 1:4. *Ans.* 2.62 s, 2.42×10^6 ft·lb.

3.5 The truck of Prob. 3.3 uses its brakes to maintain a constant speed of 60 mph on a long downhill grade of 1:4. At 60 mph the parasitic drag is 18 hp and the motoring resistance 25 hp. What is the rate of energy absorption at the brakes in ft·lb/s? In horsepower? If the truck has six brakes each with two shoes and a projected lining area of 52.5 in.², what is the hp/in² that is a measure of lining capacity? *Ans.* 1.01 hp/in².

3.6 (a) Derive an equation for the normal force during braking on each front and rear wheel of a four-wheel vehicle if its center of mass is a distance h above the pavement and a distance b forward from the rear wheels. The distance between front and rear wheels is l. When all wheels are skidding, what is the deceleration in terms of the coefficient of friction μ? Show that the normal forces are a function of the available friction μ. (b) If $h = 24$ in, $b = 63$ in, $l = 111$ in, and $\mu = 0.60$, what is the weight distribution ratio between front and rear wheels?

3.7 For the clutch system of Fig. 3.2 determine an equation for velocity ω_2 just after the completion of engagement by using the principle of conservation of momentum. Then derive the equation for work wk_c done at the clutch by the principle that work equals the net change of energy of the system. Omit T_M and T_R'.

3.8 By analogy to the situation for clutching, rewrite Eq. (3.5) to give the work done at the brakes of an airplane in bringing it to a stop without skidding on the deck of an aircraft carrier moving (a) in the same direction and (b) in the opposite direction. Assume a constant braking force. Make a sketch and label it with the symbols that you choose. (c) Simplify the equations by making approximations where justified.

3.9 By letting I' become infinitely large, use the clutch equations, Eqs. (3.3) through (3.5), to obtain brake equations and compare them with equations derived more directly, as a check.

3.10 Instead of "cranking" a combustion engine by an electric starting motor, it may be done by an inertia starter. This is a disk or flywheel that is slowly brought up to a required speed of rotation by a hand crank or small motor acting through gears. When up to speed, the flywheel is clutched to the stationary engine. (a) If starter and engine inertias are I and I', respectively, and the starter speed just before clutching is ω_1, write an equation for the work done at the clutch in terms of ω_2. To the extent possible, use equations already derived. (b) Compare the work done at the clutch for two designs, $I = I'$ and $I \gg I'$. On the basis of work done at the clutch, which design should be used and why?

3.11 Derive Eqs. (3.6) of the general case with torques acting on the shafts during clutching (Fig. 3.2). In the process, also show that $t_2 = (\omega_1 - \omega_1')/A$. $\omega_2 = \omega_1 - [(T - T_M)/I] \times [(\omega_1 - \omega_1')/A]$, displacement of driven mass $\psi_2' = \omega_1' t_2 + (T - T_R') t_2^2/2I'$, and relative displacement of clutch faces is $\psi_2 - \psi_2' = (\omega_1 - \omega_1')^2/2A$.

3.12 At time t_2, at the end of the clutch engagement period of Prob. 3.11, the common speed of the motor and load is below the normal speed of the motor, i.e., the speed ω_3 at which motor torque is in equilibrium with the resistance-load torque. From time t_2 until time t_3 the torque difference $(T_M - T_R')$ will act to accelerate together the two masses with combined moment of inertia $(I + I')$. Assume both torques T_M and T_R' to remain constant, and show that the angular acceleration $\alpha = (T_M - T_R')/(I + I')$, time $t_3 = t_2 + (\omega_3 - \omega_2)/\alpha$, and total displacement $\psi_3' = \psi_2' + \omega_2(t_3 - t_2) + \alpha(t_3 - t_2)^2/2$.

3.13 The ratio of input speed ω' to output speed ω'' of a speed-changing unit is $\omega'/\omega'' = \Omega$. The efficiency of the unit (power output over power input) is η. The output shaft drives a load with inertia I'' and resistance torque T_R'', sketch (a). Derive Eqs. (3.7) for an equivalent resistance torque T_{equiv}' and inertia I_{equiv}' which applied to the clutch output shaft at speed ω' will replace the driven load in a clutch-engagement analysis, sketch (b).

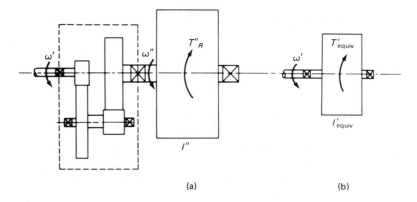

(a)   (b)

3.14 Refer to Probs. 3.10 and 3.13. Light weight is wanted in an inertia starter to make it portable (e.g., to use on small planes at airports). By what mechanism can a small inertia be made to have the equivalent effect of a larger one? Write an equation for this smaller inertia in terms of I, the equivalent, desired, driving one, carefully defining with a sketch all the symbols used.

3.15 In the development of a machine for the automatic assembly of automotive parts, the arrangement shown was proposed to drive a mechanism to intermittently transfer parts from a conveyer to the assembly device. The motor runs continuously. To effect a transfer, the speed reducer is started by engaging the clutch; when the transfer is completed, the reducer is stopped by disengaging the clutch and applying the brake. Since only one assembly machine was to be built, it was least expensive and time saving to use two conventional truck clutches, one as the clutch and one as the brake. Before the design of the assembly machine could be completed it was necessary to determine by calculations if the clutch would overheat and therefore wear out quickly; what time would be required for the transfer device to reach full speed, so that the action of other mechanisms could be synchronized; and whether or not the motor would stall, since a clutch is suddenly engaged in automatic machines.

The following properties were obtained, some from the motor and reducer manufacturers.

Motor

Nominal 1800 rpm, 20 hp.

Moment of inertia of rotor = 0.178 lb·ft·s².

Average motor torque in the expected speed range of operation = 60 lb·ft.

Speed at time of clutch engagement = 1790 rpm.

Motor must not slow down below 1440 rpm.

Speed will recover to 1750 rpm and then be constant while driving the load.

Clutch and Brake: The heavy housing is attached to the motor shaft at the clutch and to the fixed, stationary bracket at the brake. The lighter friction disks are attached to the input shaft of the speed reducer.

Housing moment of inertia $= 0.375$ lb·ft·s².

Disk moment of inertia $= 0.013$ lb·ft·s².

Assume a constant clutch torque while slipping $= 300$ lb·ft.

Speed Reducer: Double-reduction worm type.

Speed ratio, $R = 102$.

Efficiency, $\eta = 81\%$.

Equivalent moment of inertia at the input shaft $= 0.010$ lb·ft·s².

Load equivalent moment of inertia at output shaft of speed reducer $= 208$ lb·ft·s².

Frictional resistance torque at output shaft $= 2500$ lb·ft.

Couplings: Moments of inertia for the one adjacent to motor $= 0.007$ lb·ft·s²; for the one adjacent to reducer $= 0.006$ lb·ft·s².

Shaft: Neligible inertia.

Cycles of Operation: 10/min.

The following operating characteristics should be determined, not necessarily in the order listed. Refer to Probs. 3.11 through 3.13 for equations.

(a) Values of the several T's, I's, ω's, and A. *Ans.* $A = 4470$ s⁻².
(b) The minimum motor speed as a result of clutch engagement. *Ans.* 1619 rpm.
(c) Relative displacement of clutch faces.
(d) The index time lost, i.e., the time to bring the reducer input shaft from 0 to 1750 rpm.
(e) The rotation of the output shaft in degrees during this time.
(f) Heat generated at the clutch per minute.

(This problem adapted from P. West and W. D. Noon, "Determine the Operating Characteristics of a Drive and Control System for a Transfer Mechanism," *General Motors Engineering J.,* Vol. 4, No. 4, pp. 49–50, 1957; Vol. 5, No. 1, pp. 48–50, 1958; or Problem 5 in its "Reprints" booklets.)

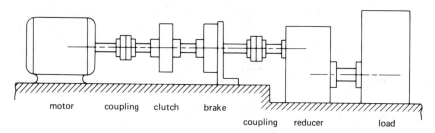

motor coupling clutch brake coupling reducer load

3.16 For the machine of Prob. 3.15, if during the braking operation with the clutch disengaged, the brake applies the same torque $T = 300$ lb·ft as did the clutch during acceleration of the load, and $T_R' = T_{\text{equiv}} = 30.3$ lb·ft as before, what is the time of stopping? *Ans.* 0.037 s. What is the rotation of the reducer's output shaft in degrees? What is the work done by the brake?

Section 3.3

3.17 A centrifugal clutch similar to that in the figure for Prob. 3.19 has eight shoes, arranged in an annular space and guided to move radially. The inner diameter of the drum is 12 in. Each shoe weighs 1.75 lb, has a projected length of 4 in and a width of 2.5 in, with its mass center at a radius of 5.25 in. The coefficient of friction is 0.35. A spring (not shown) keeps a shoe retracted until a speed of 600 rpm is attained. (a) What should be the spring force? (b) What torque and power will be developed at 1200 rpm? *Ans.* 4733 lb·in, 90.1 hp. What torque and power will be developed at 2000 rpm? Compare and comment. (c) At these speeds what are the shoe pressures based on projected area?

3.18   Same as Prob. 3.17 but the drum diameter is 250 mm, shoe width 60 mm, projected length 85 mm, radius to mass center 105 mm, and each shoe has a mass of 0.70 kg.

3.19   Determine the principal dimensions of a centrifugal clutch to be placed between a load and a 5-hp 1750-rpm induction motor, to give initial engagement when the motor speed is 800 rpm and complete engagement (no further slipping) at 1450 rpm and a torque of 225 lb·in. The maximum pressure while slipping should be low, about 15 psi, because of frequent operation. The six shoes are guided to move radially, as shown for one. They may occupy about 80% of the periphery, must lie outside a circle of 2½ in diameter, and have a width about one-fourth the drum diameter. They may be made of bronze or of cast iron ($\gamma = 0.26$ lb/in³), faced with asbestos friction material ($\mu = 0.4$). The clutch should be as compact as possible.

Suggestions: Since the dimensions of the shoes and the location of their center of mass is not known, the minimum drum inside diameter may be calculated from the maximum permissible pressure. The diameter D thus found might be increased slightly to some even or easily handled dimension, followed by calculations to see if the annular space, if filled with shoes, is sufficient for the mass needed at 1450 rpm. If so found, the shoes may be reshaped, perhaps internally, to provide a smaller mass as needed. The radius to the center of mass may be estimated. A retracting spring, such as the leaf type shown, must prevent engagement until a speed of 800 rpm is reached. The centrifugal force at 1450 rpm must equal the spring force plus the normal force needed for complete engagement. From this equality the required mass may be found. The available friction torque at rated motor speed should also be calculated and compared with the rated motor torque at 1750 rpm (which is less than motor torque at lower speeds, as in the torque curve of Fig. 4.2).

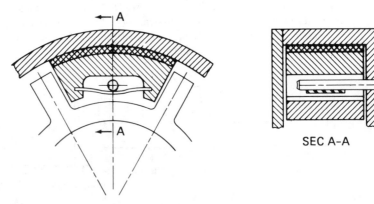

SEC A-A

3.20   As shown on the curve at (a), the power consumption of an engine fan rises at an increasing rate with speed, but the natural flow of air at high vehicle speeds provides almost all the radiator and engine cooling that is needed. Hence the centrifugal friction device shown at (b) is proposed as a mechanical means of limiting the fan speed. A drum with a 2¾-in diameter surface is keyed to the water pump shaft and riveted to a V-belt pulley which is driven from the engine. A fan plate is attached by cap screws to a part which carries the six friction blocks or shoes. This part may rotate relative to the drum, guided by the special double-row ball bearing. The shoes are held against the drum by a ribbon spring with sufficient force to prevent slipping until a speed of 2500 rpm is reached. At higher shaft speeds, the fan will rotate at 2500 rpm, since any tendency to rotate it at a higher speed will require an increased torque that is not available at the drum. This assumes that the coefficient of friction remains constant or, as is likely, decreases slightly with higher sliding velocities.

The following procedure is suggested. (a) Calculate the forces and shoe width required at 2500 rpm fan speed. The coefficient of friction is 0.40. Because of continuous slippage at higher shaft speeds, a low pressure of 10 psi during slipping seems desirable for long life. Estimate that the shoes may occupy 70% of the drum's periphery. (b) Determine centrifugal force and the required initial (installed) spring force. Take the average height of a shoe to be one-half its length. Molded-asbestos shoe material weighs 0.072 lb/in³. (c) Calculate the power loss at 4500 rpm shaft speed. Compare it with the power loss on the curve for 4500 rpm fan speed. Also calculate the μpV value, where V is the relative surface velocity in ft/min and p is pressure in psi. Compare it with the value of 10 000 often given for brakes for continuous application and poor dissipation of heat. (d) An alternative arrangement to that shown is to attach the shoe carrier to the belt pulley and/or shaft and the fan to the drum, then support and guide the drum on the shaft through a bearing. Calculate the speed at which all contact between shoes and drum will then cease. What is the advantage of this arrangement? Sketch curves of fan rpm vs shaft rpm for the two arrangements. Consider that engine speed may be high when the transmission is in the lower gears and the vehicle speed is low, and state which arrangement must be used.

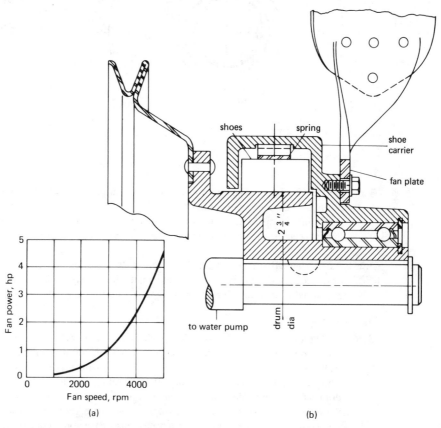

(a)

(b)

3.21 For the centrifugal governor of Example 3.2 and Figs. 3.4 and 3.5 in steady-state clockwise rotation ($d\omega/dt = 0$), sketch a free-body diagram of the left arm and derive equations for the normal force and for the angular velocity. Given the dimensions to be $r_1 = 0.46$ in, $r_2 = 0.326$ in, $r_3 = 0.36$ in, $\phi = 22.5°$, and $\gamma = 90°$. Also, $\mu = 0.15$, $F_s = 0.05$ lb, torque $T = 2 R_x r_3 = 0.0209$ lb·in, and $m = 38 \times 10^{-6}$ lb·s²/in.

3.22 Obtain the two solutions to the differential equation at the end of Example 3.2, solving for ω in terms of A, B, and t. Sketch the shapes of the two solutions and reject the one which is not physically realistic. What speed is approached as time increases? Compare the answer with the answer to Prob. 3.21 if you have solved it. *Ans.* $\omega = \sqrt{A/B} \tanh(\sqrt{AB}\, t)$.

3.23 For the dial of Example 3.2 and Fig. 3.4 with the dimensions given in Prob. 3.21, also with $I_2 = 2.48 \times 10^{-6}$ lb·in·s^2 and $T = 0.0209 - 1.31 \times 10^{-6}\, d\omega/dt$ lb·in, determine an equation for the governor speed in rad/s. If the ratio of dial speed to governor speed is 1/19.65, what is the equation for dial speed in rps. Plot dial speed against time from 0.00 to 0.30 s. What speed is approached and how uniform is it? *Ans* $n_d' = 0.727 \tanh 21.25t$ rps. (Sketch adapted from a telephone dial designed by Bell Laboratories and manufactured by Western Electric.)

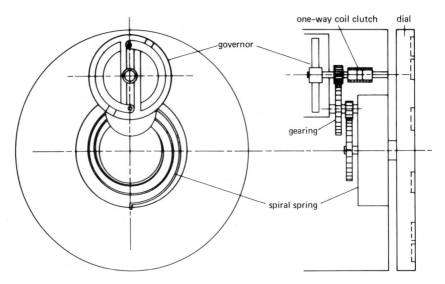

3.24 Derive equations for governor position ψ and dial position ψ_d as functions of time. Plot dial position in degrees against time from 0.00 to 0.30 s. Together with the plot in Prob. 3.23, determine the speed when the dial passes the first number, 60° from the release position. Is the speed essentially uniform? *Ans.* $\psi = (1/B) \ln \cosh \sqrt{AB}\, t$.

3.25 As an alternative to the method of Prob. 3.24, take the speed and position equations of Probs. 3.22 and 3.24, respectively, eliminate t, and solve for ω as a function of ψ, then for dial speed n' as a function of its position ψ_d. Plot for the data of the preceding problems.

Section 3.4

3.26 (a) On a drum circle of 3 in or larger diameter sketch free-body diagrams of two external shoes, pivoted below the drum and with clockwise drum rotation as in Fig. 3.9. Show equivalent concentrated forces P and μP on the arm producing the higher-torque (self-actuating) shoe, P' and $\mu P'$ on the other. Carefully locate relative to the axes and to the drum surface the points C and C' where the forces may be considered concentrated. Include the force *components* P_x, P_y, μP_x, etc. (b) For the higher-torque shoe derive an equation for the reaction force at the pivot pin in terms of the force components. (c) Same as (b), but for the lower-torque shoe.

3.27 Continuing Example 3.3, determine: (a) the forces on the arm with the lower torque and (b) the pivot reactions for both shoes. Show brake drum rotation to be counterclockwise, as seen in the end view of Fig. 3.1. Prime the symbols of the lower-torque arm. Let the angle between OA and OA' be 30°, making $\epsilon = 75°$. *Ans.* $R = 1360$, $R' = 340$ lb.

3.28 (a) From Eqs. (3.12) through (3.16) determine simplified equations for p_{max}, P, and friction radius r_f for the case of a lining symmetrical about the $\phi = 90°$ position, where the included angle is θ. Ans. $r_f = D/\kappa = r(4 \sin \theta/2)/(\theta + \sin \theta)$. (b) In a very short shoe, $T = r(\mu P)$. Plot the ratio of torques for the long shoe of (a) to that of the short shoe vs θ from 0° to 150° by 30° steps. What appears to be an advantage for long shoes in addition to lower pressure and wear? (c) Sketch a long shoe that is symmetrically hinged to its arm at the 90°, r_f location, and show the forces acting on it. What is a possible advantage for this arrangement?

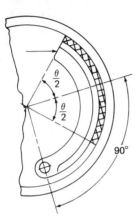

3.29 (a) Two identical shoes are used as shown in an internal brake. For $\mu = 0.30$, calculate the braking torque developed by each shoe. Note that the sense of drum rotation is counterclockwise. (b) It is proposed to make the torques developed by the two shoes equal by actuating them with a hydraulic cylinder having two diameters, as shown. Determine the necessary ratio of the two diameters and sketch the cylinder horizontally to show how it should be placed (which end for which shoe).

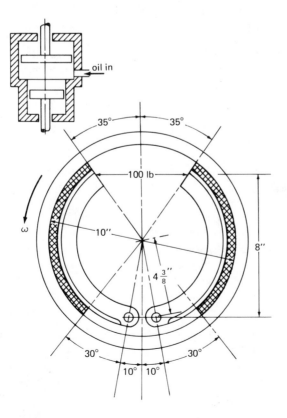

3.30 Two shoes arranged to have friction assist (self-actuation), together with separate hydraulic cylinders, have been used on automotive-vehicle front wheels, where they can be most effective during braking from forward motion, justifying the cost of the second cylinder. A view looking into a right-front brake is given with dimensions. The lining width is 50 mm, and the diameter of the hydraulic cylinders is 28 mm. The outside tire diameter is 700 mm. The coefficient of friction between the lining and drum is 0.30 and between the tire and pavement is 0.60. The mass of the car is 1400 kg, the force from which is distributed 60% on the front and 40% on the rear wheels during a stop from 100 km/h. (a) Determine forces at the wheels when skidding is about to occur, assuming that all wheels start to skid at the same time. (b) In what distance and time may the car be stopped? (c) Determine the braking torque by each front-wheel shoe. *Ans.* 432 N·m. (d) Determine maximum lining pressure, moments, actuating force, and hydraulic pressure. *Ans.* 2630 N actuating force.

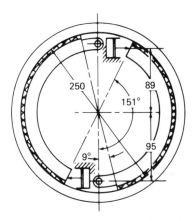

3.31 Friction assist or self-actuation for both shoes in either direction of drum rotation is obtained by allowing each shoe to "float." Each shoe butts up against one of two anchor pins, depending on the direction of drum rotation, and pivots about this pin. A lever system provides the same operating moment for either direction. (a) For the brake shown, compute the braking torque for a maximum lining pressure of 50 psi, a friction coefficient of 0.40, and a lining width of 4 in. (b) Compute the corresponding force applied by the brake cylinder against each lever.

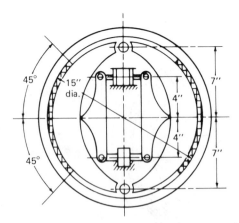

3.32 There are six two-shoe brakes on the truck of Probs. 3.3 and 3.4. Assume all brakes and shoes are equally effective. The drum diameter is 14.5 in, its width is 4 in, and the rim thickness is ⅜ in. Each shoe subtends 130° and $\mu = 0.40$. Tire radius is 19 in, and from Prob. 3.4, tangential friction force per wheel is 14 000/6 = 2330 lb. (a) Use the brake energy given as an answer to Prob. 3.4 and determine the temperature rise during the braking action if all the heat is assumed absorbed by the rim of the drum. (b) As a first approximation to the measure of the lining capacity μpV (Example 3.3), p may be taken as the normal force P divided by the projected lining area, for in many cases it will be close to the p_{max} value. Normal force P may be approximated by $T/\mu r$, as for a short shoe, where T is the torque per shoe and r is the drum radius. See Example 3.3 for units. Calculate μpV and compare its value with 100 000, which has been quoted for automobile brakes.

Section 3.5

3.33 Prove the statement following Eq. (3.25) that the mechanical advantages for the shoes subject to the same actuating moment may be added to give a mechanical advantage for the brake as a whole.

3.34 Determine the actuation factor and the mechanical advantages for the shoes of the external brake of Example 3.3, and for the brake as a whole. *Ans.* 0.735.

3.35 Determine the actuation factor and the mechanical advantage for the shoes and for the brakes of Probs. 3.29 and 3.31.

3.36 A single-shoe brake is operated through the linkage shown with a force F applied at the foot pedal. (a) Choose the direction for drum rotation to give self-actuation and draw complete free-body diagrams of all the links. Let $a = b = 5$ in, $c = e = 2\frac{1}{2}$ in, $D = 7.20$ in, $l = 15$ in, $\theta = 90°$, and $h = 5$ in. Force $F = 25$ lb and $\mu = 0.30$. (b) Evaluate the forces and determine the torque developed.

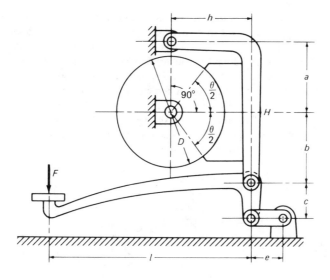

3.37 Same as Prob. 3.36 except $a = b = 125$ mm, $c = e = 50$ mm, $D = 150$ mm, $\theta = 120°$, and $h = r_f$, where the friction radius is given in the answer to Prob. 3.28. Also, the shoe is pivoted to the arm at H. Force $F = 125$ N and $\mu = 0.33$. Determine length l if a braking torque of 60 N·m is to be obtained. *Ans.* 318 mm.

3.38 An external shoe brake is actuated by a spring acting through a bell crank. The solenoid-operated release mechanism is not shown. By a yoke on one arm the two arms share the same pivot pin at A. It is proposed to use dimensions such that the braking torque is absorbed equally by the two shoes. (a) Choose the direction of drum rotation that may make this possible, sketch free-body diagrams of the bell crank and of the two arms, and from them derive an equation for the necessary ratio of arm lengths l'/l. (b) To obtain an idea of the proportions, calculate l'/l for a typical case of $D = 10$ in, $a = 6$ in, $\theta = 120°$, $b = 5\frac{1}{2}$ in, $c = 6\frac{1}{2}$ in, and $\mu = 0.35$. Obtain r_f from the answer to Prob. 3.28. Comment.

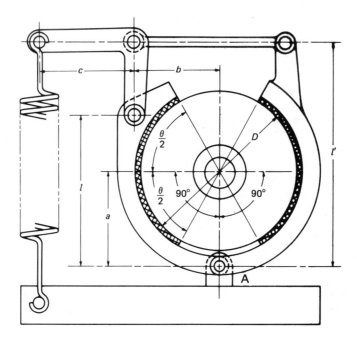

3.39 The curved arms shown with solid lines for the double-shoe brake are actuated by the same spring, initially compressed to give a force Q when the brake is closed and Q' when the brake is open. The bell crank and solenoid are used to release the brake, and they may be ignored in part (a). The drum has counterclockwise rotation. (a) For the brake closed, draw free-body diagrams for each of the loaded parts and determine an equation for the normal forces P and P' in terms of the spring force Q and dimensions a and r_f, the friction radius. (b) The brake is released by a downward pull S on the bell crank by the solenoid. Draw a free-body diagram of the right-hand lever and state why the stop is needed (and why it is adjustable). Write an equation for solenoid pull S and for the force F on the stop in terms of the spring force Q' and the dimensions.

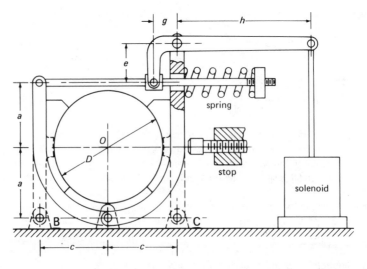

3.40 Same as Prob. 3.39 with counterclockwise drum rotation except that the shoes are "short" and the arms are straight and pivoted at B and C, as shown by the dotted lines. Let $a =$

5 in, c = 6 in, e = 2 in, g = 1 in, h = 10 in, and D = 8 in. With the brake closed, the spring force is 500 lb and μ = 0.30. With the brake open, each shoe clears the drum by $\frac{1}{32}$ in, and the spring rate is 1000 lb/in of deflection. Use numerical values on the free-body diagrams and calculate the normal forces, brake torque, solenoid pull, and force against the stop. *Ans.* Torque is 2430 lb·in, solenoid pull is 125 lb.

Section 3.6

3.41 Three circular friction pads are used on each side of the disk of a caliper-type brake, as shown. The disk is a ring with external splines so that it may slide axially within a wheel. Assume the force from each pad to be concentrated and derive equations for the braking torque and pressure in terms of the hydraulic-fluid pressure p_0 and the given dimensions.

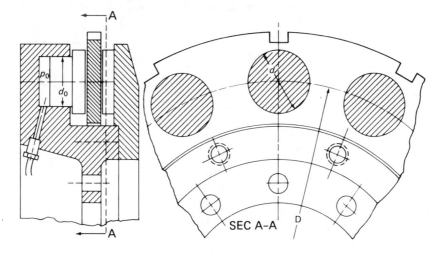

SEC A-A

3.42 Derive Eqs. (3.30) for a multidisk clutch.

3.43 Determine the torque that may be transmitted by the multidisk clutch of Fig. 3.11 after wearing-in, when the oil pressure is 75 psi and μ = 0.35. All parts are cylindrical with respect to the axis of rotation. For dimensions, scale the drawing, which is half-size. Also determine the average pressure and the maximum pressure.

3.44 In the clutch of Fig. 3.11 and Prob. 3.43, the number of disks and their composition may be changed to fit other situations in capacity, longer life, more frequent operation, oil pressure, and thinner, all-metal disks. For a torque of 1500 lb·in. transmitted with a factor of safety against slip of 2.0 after wearing-in, a low average pressure of 50 psi for frequent operation, and μ = 0.35, how many externally splined and how many internally splined disks are required? What is the maximum pressure on the disks, and what oil pressure is required? For dimensions scale the drawing, which is half-size.

3.45 Determine the numerical value of the diameter ratio $\rho = d/D$ that gives the maximum torque (after wearing-in) for a disk clutch with predetermined outside diameter D and maximum allowable pressure p_{max}. *Ans.* 0.58. Compare with ratios given in the text and suggest why these slightly higher ratios may be advantageous.

3.46 The clutch shown has a single friction plate with a splined outer surface nesting in a splined recess in the sleeve. For engagement the shifting collar is moved to the left until the short links override their central radial position. The shifting collar butts against the adjusting ring, and the links are held in the override position by a small moment of forces. The toggles thus lock the clutch in engagement, and the shifting band rides freely in the rotating collar. (a) Sketch the complete free-body diagrams of the driving plate, friction plate, toggle plate, toggle, link, and shifting collar when the clutch is engaged and the total normal force on the pressure plate is P. The toggles pivot on pins in lugs, not shown, attached to the toggle plate. (b) Unbalanced centrifugal forces must be considered in the parts of rotating devices. How does balance appear to be provided in the toggle so that it does not release itself at high speeds? (c) Why must the ring be adjustable in axial position? (Sketch courtesy of FMC Corporation, Drive Division.)

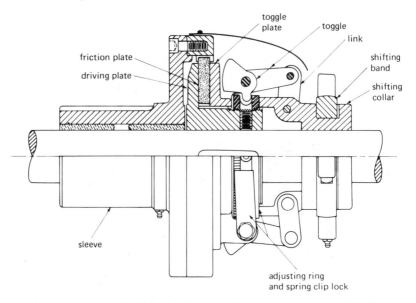

3.47 For the clutch of Prob. 3.46, let the driving and toggle plates have outside diameters of 150 mm and the friction plate an inside diameter of 100 mm, with $\mu = 0.35$. The power requirement is 10 kW at 1200 rpm. Allow a 50% margin of safety against slipping. Determine the required axial load, the average pressure, and the maximum pressure, using the uniform-wear theory. *Ans.* 2730 N, 0.35 MPa maximum.

3.48 In a multidisk clutch the frictional resistance to sliding along the splines under torque causes the effective normal force P to progressively decrease from the actuating-end pressure plate to the last pair of contacting surfaces (in the clutch of Fig. 3.11, from the piston end). This effect has been previously neglected. Start the solution of this problem by studying a so-called "single-plate" dry clutch. In the automotive type, shown here schematically, a single metal plate, faced with friction material and internally splined, as in Fig. 3.12(a), is squeezed against the flywheel by a cast-iron pressure plate, which is externally splined or keyed to

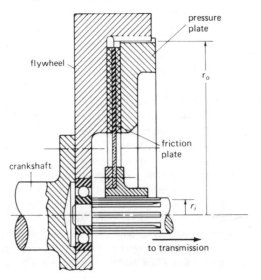

the flywheel. The friction plate slides on a shaft leading to the transmission, with splines at pitch radius r_i and friction coefficient μ_i. The pressure plate slides on splines at pitch radius r_o and coefficient μ_o. Engagement is effected by the release of springs which then act on the pressure plate with a total force P. (a) Show that the force P_1 between pressure and friction plates and the force P_2 between the plate and the flywheel are, respectively, $P_1 = P/(1 + k\mu_o/r_o)$ and $P_2 = P_1(1 - k\mu_i/r_i)/(1 + k\mu_i/r_i)$, where factor k is a constant that relates torque on a surface to the normal force on it, i.e., $T_1 = kP_1$ and $T_2 = kP_2$. (b) Write an equation in terms of P for the total torque transmitted by the clutch.

3.49 The sketch shows schematically, in three partial views, a double-disk self-actuating brake in the engaged position. It is engaged by hydraulic pressure acting on a narrow annular ring with an average radius r_a. An O-ring acts as a seal between the ring or piston and the groove or cylinder. The brake may also be engaged by tangentially placed hydraulic wheel cylinders. Frictional contact with the rotating, enclosing, wheel drum causes the two friction disks to rotate. The secondary disk strikes a stop, and the primary disk continues a short distance further, which causes rolling of the ball between each pair of ramps, with a consequent squeeze and self-induced additional normal and friction forces. For a reversal of drum or wheel motion, there is a stop such that the primary disk becomes the secondary disk. There are N balls, all at a radius r_b, and the friction lining may be considered effective at an average radius r_f. (a) Sketch a free-body diagram of a portion of the primary disk and determine equations for the normal force Q at the ramp, the normal force P at the friction surface, and the braking torque from this disk, all in terms of the applied hydraulic force H. (b) Sketch a free-body diagram of the secondary disk. Note that the radius to the stop is r_s. Determine the reaction force R at the stop and the torque developed by this disk. What is the total braking torque on the wheel drum? Check with a free-body analysis of the complete friction-disk assembly. *Ans.* $T = 2\mu H r_b r_f \tan\alpha/(r_b \tan\alpha - \mu r_f)$.

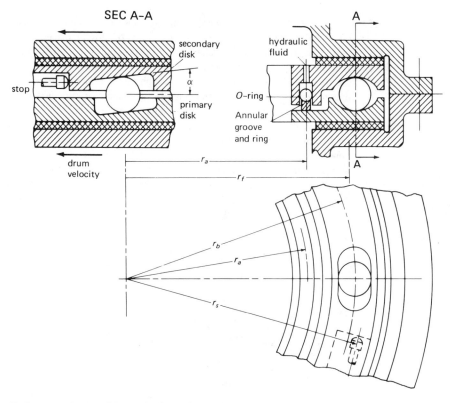

3.50 Refer to the brake of Prob. 3.49. (a) What braking torque is developed if there are no balls and ramps, but only an axial force H from the annular chamber for the hydraulic fluid? (b) Write an equation for the mechanical advantage A_m of the self-actuating brake. Reduce it to simple terms by assuming $r_f = r_b$. *Ans.* $A_m = \tan\alpha/(\tan\alpha - \mu)$. (c) If $\mu = 0.30$, at what ramp angle would the brake become self-locking? (d) If the angle is 30°, what is the value of A_m? How much change in the value of μ will cause self-locking?

3.51 A roller-ramp type of overrunning, one-directional clutch uses cylindrical rollers as wedging elements. (a) To an enlarged size sketch a free-body diagram of one of the rollers, carefully obtaining graphically the relative magnitudes and directions of the two forces and their normal and tangential components. The spring force is negligible. (b) Derive an equation for the normal component of force in terms of applied torque T, number of rollers N, and the given dimensions. *Ans.* $P_n = 2T/(ND \tan\alpha/2)$. (c) Determine the maximum value of angle α in terms of coefficient of friction μ if slipping is not to occur.

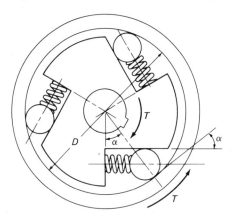

3.52 A sprag or cam type of overrunning, one-directional clutch consists of two concentric circular rings, between which there is a series of closely spaced sprags or cams. They fit loosely for one direction of relative ring rotation, allowing "free wheeling." Light springs keep the sprags in touch with the rings. A reversal of relative rotation causes a rocking of the sprags and a tightening up so that a high torque may be carried. Note that the two lines drawn from the center O of the rings to the centers of curvature O_1 and O_2 of the sprag surfaces at the contact points make an angle α, which is small. (a) Draw a free-body diagram for one of the sprags, perhaps exaggerating the angle α for clarity. Obtain graphically the relative magnitudes and directions of the two forces and their normal and tangential components. (b) Derive approximate equations for the angles between the forces and their normal components, as functions of angle α and the given dimensions. (c) Determine the maximum value of angle α in terms of the coefficient of friction μ if slipping is not to occur. *Ans.* $\alpha_{\max} \approx \mu(r_2 - r_1)/r_2$.

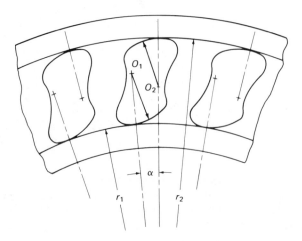

3.53 The cone clutch used in a certain application has worn out. Your company proposes to replace it with a plate or disk clutch of the same or larger torque capacity. The exact torque requirements of the application are unknown, so you measure the cone clutch and find that its friction facing has a diameter of 10 in at its large end, 9 in at its small end, with a length of 2 in measured along the conical surface. The axial spring force holding the clutch in engagement measures approximately 1000 lb. Structural and space considerations suggest the use of the same spring force and outside diameter, as well as the same friction material. Assume a reasonable inside diameter and determine the total numbers of plates you would recommend. Sketch their arrangement. What is the average pressure?

3.54 Choose a reasonable cone angle and determine the mean diameter, width, pressure, and required axial force for the lining of a cone clutch to have a torque capacity of 300 N·m when engaged infrequently at 600 rpm of the driver. The materials are molded asbestos on cast iron with friction coefficient 0.40 and an energy absorption rate limited to $\mu pV = 42$, where p is measured in MPa and V is measured in m/min. Let lining width equal 0.2 of the mean diameter for good proportion. *Ans.* 207 mm, 41.5 mm, 0.268 MPa, and 1570 N.

3.55 The friction coefficient of a clutch material is known to decrease with increase of rubbing velocity ωr within a certain range. Suppose that friction can be represented by the equation $\mu = \mu_0 - a\omega r$, where μ_0, a, and ω are constants. Consider a worn-in clutch condition where the work product μpV is constant. For a cone clutch derive the axial force and torque developed in terms of a maximum allowable pressure p_{max}. Adapt the result to a flat-plate clutch by substitution of an appropriate value for half-cone angle α.

Section 3.8

3.56 The band brake shown is operated with a foot-pedal force Q of 20 lb. If $\mu = 0.35$, what torque is developed on the drum when its sense of rotation is (a) clockwise? (b) counterclockwise? (c) What change in one dimension will make the brake self-locking? *Ans.* 2180, 233 lb·in, +0.51 in the ¾-in dimension.

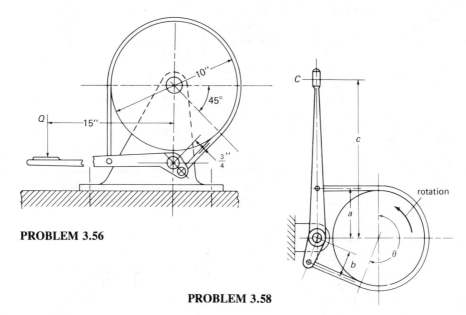

PROBLEM 3.56

PROBLEM 3.58

3.57 Same as Prob. 3.56 except that force Q is 50 N on a lever arm of 600 mm, the short arm is 25 mm, and the drum diameter is 270 mm.

3.58 What dimensional relationships must be held in order that the brake may be self-actuating, but not self-locking? What must be the direction of the force applied at C? The direction of drum rotation is indicated. Prove your conclusions.

130

3.59 A band brake must be equally effective for either direction of drum rotation, and the operating force must be as small as possible. State the criteria and sketch the arrangement of lever and band with a 270° wrap angle. For this special case derive an equation for torque in terms of symbols representing the dimensions.

3.60 With reference to Example 3.5 calculate (a) the maximum pressure on the band if it is 2 in wide, (b) the space needed below the foot pedal to allow for its motion corresponding to $1/16$-in radial clearance between band and drum, $1/32$-in compression of the lining, and $1/16$-in lining wear before adjustment, (c) the torque developed if drum rotation is reversed, and compare. *Ans.* 66.9 psi, 2.04 in, 325 lb·in.

3.61 Similar to Example 3.5 but design the brake for a capacity of 200 N·m on a drum of 150-mm diameter. The downward operating force of 75 N must be to the right of the drum, which is rotating counterclockwise. Hence try an arrangement where the ends of the band cross and the wrap angle is 300°, as suggested at the end of Example 3.5. Show a sketch of the lever and calculations.

3.62 Continue Prob. 3.61, calculating (a) the maximum pressure on the band for a width of 35 mm, (b) the motion of the end of the foot pedal when taking up a radial clearance of 1 mm between drum and band, compressing the lining 0.5 mm, and allowing for 1-mm wear of the lining, and (c) the torque developed if drum rotation is reversed, and compare.

3.63 A simple band brake may be provided on the high-speed shaft of a gear unit, one corner of which is shown here, when it is used to drive a slow, long-stroke, walking-beam, pumping unit in an oil field (Fig. 6.11). The brake holds the crank, counterweights, and walking beam when the string of pump rods is disconnected. The only place available for the brake is inside the sheave, which is driven by V-belts from an engine or motor (Fig. 6.11). The brake drum will be bolted to the side wall of the sheave. In one view with the sheave removed, there is seen above the shaft a machined boss into which a hole may be drilled to support the pivot pin of the brake arm. The major length of this arm must be offset from its pivot so that it clears the edge of the sheave. It may be locked in a holding position by a ratchet mounted on the top flat surface of the bearing housing, fastened to it by the two existing cap screws seen to the right and left of the boss.

A counterclockwise torque of 3160 lb·in must be held, with $\mu = 0.30$. Attach the F_1 end to the pivot pin to avoid sensitivity to varying environmental conditions in exposed locations. A wrap angle of 270° may be possible The operator will pull on the lever handle with a force of 30 lb. The unit is mounted on a platform 9 in high. Because of infrequent operation, a high band pressure, say 200–250 psi, may be used if necessary. The connection between band and arm should include an adjustment for initial installation and for wear. Distance from base of unit to center of sheave is 15 in.

Design the brake to the extent indicated by your instructor. A minimum requirement is a set of calculations similar to those in the preceding problems, plus a to-scale sketch on quadrille paper to check against interferences. A complete design will include the drum, band adjustment, pivot pin, arm, ratchet mechanism, and their materials.

3.64 The brake drum shown is used underneath a reel of a tape recorder to maintain a constant drag torque on the reel shaft, preventing looping of the tape in a stop from high speed. One end of the brake band is attached directly to the frame of the recorder, and the other is attached through a spring that is installed with a predetermined force. This force remains essentially constant during brake operation, since the band itself has very little stretch or movement. Alternative locations for springs are shown. (a) For each alternative A and B, write an equation for friction torque, putting it in terms of that force which remains constant (the spring end of the band). The springs chosen and the spring forces are of course *different* for A and B since the torque and the direction of reel rotation are to be *the same*. (b) Obtain the rate of change of torque with a change of the coefficient of friction. Compare the rate for the two alternatives and state which will result in the lesser change of torque from environmental changes in friction coefficient. Is the difference significant?

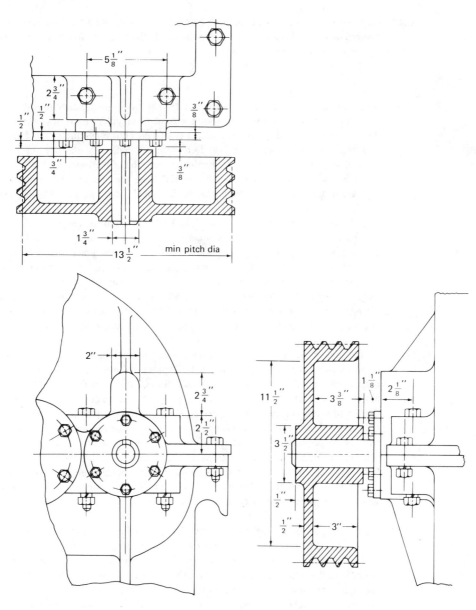

PROBLEM 3.63

PROBLEM 3.64

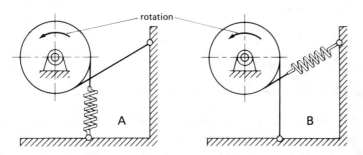

3.65 (a) Derive the torque equation, Eq. (3.46), for the centrifugal band clutch of Fig. 3.18. (b) Sketch the variation of pressure along the band. Examine the torque equation and describe the characteristics of the clutch.

Section 3.9

3.66 A coil-spring clutch, such as that of Fig. 3.19, has the following dimensions: drum diameters 0.750 in, 7 coils per drum, steel spring wire of rectangular cross section 0.033 in wide by 0.022-in-radial thickness, outside diameter of coil before assembly 0.775 in, coefficients of friction oily 0.05, dry 0.30. (a) Determine for both oily and dry surfaces the torque capacities at low speeds for both directions of rotation of the input shaft. Compare and state if suitable as an overrunning clutch. (b) Calculate the stresses. What actually limits the capacity of this clutch? To about what value should the torque rating be reduced?

3.67 A coil-spring clutch, such as that of Fig. 3.19, has drum diameters of 25 mm, each with four turns of wire 3 mm by 3 mm square, of spring steel and $\mu = 0.25$. One shaft rotates at 100 rad/s with a moment of inertia of 0.34 kg·m². The other rotates at 50 rad/s with a moment of inertia of 0.11 kg·m². Until engagement, one end of the coil is held away from its drum so there is no friction. The clutch must complete engagement in 0.25 s. (a) What torque is required of the clutch? *Ans.* 16.6 N·m. (b) With what initial pressure and interference should the coil be assembled if $E = 207$ GPa? Use the larger speed for the centrifugal effect and see footnote [13]. *Ans.* $p_0 = 66.3$ kPa. (c) What are the stresses?

3.68 At the end of Section 3.9 it is suggested that a coil clutch can be used in the unwind direction as a constant-torque slip clutch. Explore the practicality of this, relating strength, pressure, size, capacity, and friction as follows. (a) From the equations of Section 3.9 determine for rectangular wire the initial pressure as a function of bending stress. Why in this application will the tensile stress between drums be relatively small? *Ans.* $p_0 = 0.667$ $h^2\sigma/(D + h)^2$. (b) If $\sigma = 50\,000$ psi, a reasonable value for spring steels, and $h = 0.10D$, what is the pressure? What are the approximate torque capacities with drums of 1, 2, and 3 in diameters and wire widths of $0.1D$ and $0.2D$? *Ans.* 185 lb·in if $D = 3$ in and $b = 0.1D$. (c) Take $N = 6$ turns and determine the percentage change in torque with a change in μ from 0.35 to 0.10 as the surfaces become oily. (d) To minimize the variation in torque with μ, should the number of turns be large or small?

3.69 Let one end of the coil of Fig. 3.19 be enlarged in diameter and fitted *inside* a cylindrical drum. Show that this may be designed to give a slipping torque independent of the friction coefficient *and the direction of rotation*. The coil spring will be in firm contact with the drums, but there is otherwise no connection between them. (a) Neglect centrifugal effects and write equations for the torques transmitted in the two directions of rotation. Let the diameter of the external surface be D_1 and let the diameter of the internal surface be D_2. For the equation involving the internal part of the coil, the average coil diameter $(D_2 - h)$ replaces $(D_1 + h)$, that of the external part of the coil. See also Prob. 10.25. For equal torques what should be the relationship between the manufactured interferences at the two drum surfaces? (b) What is the effect of high speeds on the equality of torques? Suggest a change in the coiling and drums such that the two torques remain equal to each other regardless of speed.

Section 3.10

3.70 Derive Eqs. (3.33b) and (3.33c).

3.71 Derive Eq. (3.56) from the equations that precede it.

3.72 (a) Divide through Eq. (3.57) by belt cross-sectional area $A = bh$ to obtain stress σ_1. Solve the result for power per unit of area pwr/A. *Ans.* $pwr/A = V(\sigma_1 - \rho V^2)(e^{\mu\theta} - 1)/e^{\mu\theta}$. (b) Obtain an equation for the velocity V at which pwr/A is a maximum. Is it a linear function of stress? (c) For $\sigma_1 = 300$ psi obtain the velocity for maximum pwr/A, also for $pwr/A = 0$, and sketch the shape of the curve of pwr/A vs V.

3.73 The F_c term in the belt equations is commonly neglected for "slow-speed" drives. If $\sigma_1 = 400$ psi, in the power per unit of area, pwr/A of Prob. 3.72(a), (a) at what velocity is the error 5% if F_c is neglected? (b) what is the percentage error at 4000 ft/min? *Ans.* 2300 ft/min, 17%.

3.74 As with a belt, the performance of another flexible power-transmission component, the roller chain, is affected by centrifugal force. A few links and rollers are shown. Sketch a force diagram for a free-body consisting of one roller and two link half-lengths (a total of one pitch length p), concentrating the mass m for this length at the roller. Derive an equation for the tensile force F_c in a link due to centrifugal action in terms of chain velocity v at radius R and mass m_1 per unit length of chain.

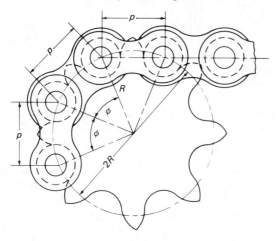

3.75 The force rating of standard roller chain has been based on a formula,

$$F = 0.273\, p^2 \left\{ 5100 - 115 V^{0.41} \left[1 + 25 \left(1 - \cos \frac{180}{N} \right) \right] \right\}$$

where p is chain pitch in inches, V is chain velocity in ft/min, and N is the number of teeth on the smaller sprocket. Now $0.273p^2$ is the projected bearing area of the pin that joins links, and 5100 is the uncorrected allowable pressure in lb/in². The term subtracted from 5100 reduces it for centrifugal and chordal effects, the latter giving a velocity variation, particularly with small numbers of teeth. (a) Derive an equation for the velocity that will give optimum horsepower rating for a given number of teeth. (b) If $N = 19$, what is the optimum velocity?

3.76 A bearing manufacturer's catalog presents a formula: total belt pull $F_1 + F_0 = 315\,000\ hp/nD$, where n is the rpm and D is the diameter of the pulley. (a) To what ratio of total belt pull to net belt pull does this formula correspond? (b) To what ratio of F_1/F_0 does it correspond and to what value of μ if the wrap angle is 150°?

Section 3.11

3.77 Starting with Eq. (3.58), derive Eqs. (3.59) and (3.61).

3.78 (a) Derive equations for the wrap angles in terms of pulley diameters and center distance C, Fig. 3.21. *Ans.* $\pi \pm 2 \sin^{-1} (D' - D)/2C$. (b) Derive equations for the length of belt, by a series expansion. *Ans.* $2C + \pi(D' + D)/2 + (D' - D)^2/4C + \ldots$.

3.79 Refer to Example 3.6 and Fig. 3.23. (a) What is the maximum stress and under what conditions does it occur? (b) Explain why the centrifugal forces affect the belt stresses even when the belt is at rest. (c) What percentage of the maximum force is the net or useful force? How does this compare with a gear drive? (d) What is the approximate force on the bearings of each pulley (neglect the cosine effect)? (e) If the modulus of elasticity E for leather is 25 000 psi, predict the slip due to the creep and the efficiency of the drive, assuming no other types of slippage or losses.

3.80 A 125 mm wide, 8 mm thick, double-ply leather belt has been selected to transmit 15 kW at a speed of 1000 m/min between two pulleys, each with a 250-mm diameter, on a fixed center distance. (a) Calculate the minimum tightening force. Assume $\mu = 0.35$. (b) Calculate the corresponding largest force and stress in the belt. (c) Calculate the total belt pull. *Ans.* 1169 N; 1.62 MPa; 2338 N.

3.81 A 75-hp motor at 870 rpm is to drive by a belt, as in Fig. 3.23, a smooth-running machine at 350 rpm in a clean plant. Use a belt velocity of about 4000 ft/min, and for good proportions take the centers of the pulleys to be a distance apart of twice the diameter of the larger pulley. Let $\mu = 0.35$ and $(\sigma_1)_{min} = 275$ psi. (a) Determine pulley diameters, rounding off to the nearest half-inch. Then, determine wrap angles from a scale drawing. *Ans.* $D = 17.50$ in, $\theta = 162.8°$. (b) Determine the required cross-sectional area for the belt, also $(F_1)_{min}$ and F_s. If tables of standard belt sizes are available, select a thickness and width. *Ans.* $bh = 4.54$ in^2, $(F_1)_{min} = 1249$, $F_s = 939$ lb, $F_c = 262$ lb.

3.82 Same as Prob. 3.81 except that the power transmitted is 10 kW, pulley speeds are 1740 and 740 rpm, belt velocity is 750 m/min, and the center distance is such that the smaller wrap angle is 157.5°. Also, reduce $(\sigma_1)_{min}$ to 1.5 MPa for longer life.

3.83 A motor pulley of diameter D_1 by a single belt drives three other pulleys of diameters D_2, D_3, and D_4, contact angles θ_2, θ_3, and θ_4, and torques T_2, T_3, and T_4, respectively. Derive an equation for the motor torque T_1 in terms of the torques of the driven pulleys. Also, write equations for the minimum wrap angle on each pulley to avoid slipping if the coefficient μ is the same on each. The wrap angles may be obtained by the proper location of idler pulleys. Neglect centrifugal forces.

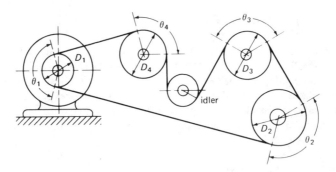

3.84 For a wedge-action belt drive, derive in detail equations for torque and the ratio of belt tensions. Start with equations of radial and tangential equilibrium, as outlined in the paragraph following Eq. (3.57).

3.85 If an existing flat-belt drive is to be replaced by a V-belt drive and the driven pulley is several times larger than the driving one, then together with the larger contact angle, it may be feasible and economical to place the V-belts directly over the existing, ungrooved, larger pulley. This is called a V-flat drive. Given a flat-pulley diameter of 35½ in. Assume the effective diameter to be at the midheight of the ½-in-high belt. Angle α, Fig. 3.22 (b), is 34°, and the speed ratio is 4:1. The center distance is made equal to the effective diameter of the larger pulley. (a) Determine the pitch diameter of the smaller, V-grooved pulley and the wrap angles (see answer to Prob. 3.78). (b) For a chosen value of F_1, the maximum belt pull, which pulley gives the smaller available net belt pull $(F_1 - F_0)$? Use $\mu = 0.15$ if needed. Which pulley, then, determines the power capacity of the drive?

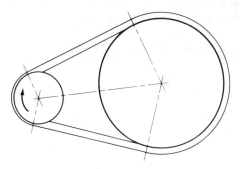

3.86 In automotive engines a single V-belt often drives the alternator and fan-water pump shafts from a sheave on the front end of the crankshaft. The wrap angles may be small. The sketch gives pitch-circle diameters and the angles between the sections of the belt. The possible simultaneous power requirements at maximum torque are 1.0 hp at the alternator and 2.0 hp at the fan and pump shaft when crankshaft speed is 800 rpm. (a) Determine the tensions in the several parts of the belt when the tightening tension is 66 lb. (b) If $\mu = 0.15$, and groove angle $\alpha = 34°$, is the wrap angle at each sheave sufficient? Neglect centrifugal effects at this low speed. The rotation is clockwise.

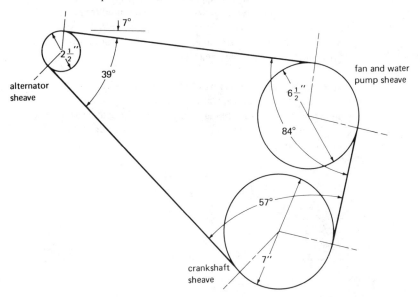

3.87 The input shaft of the gear unit of Prob. 3.63 is to be driven at about 560 rpm through parallel V-belts by an engine governed to operate at a speed of 750 rpm, for which it is rated at 20 hp. Input shaft and motor shaft centers may be placed at a minimum distance of 40 in. Shaft diameters are 1¾ and 1½ in, respectively. (a) Obtain a catalog of a V-belt manufacturer and for the approximate speed ratio, select three sheave combinations that meet the requirements for belt size, minimum sheave diameter, and center distance. (b) Using catalog-suggested overload and service factors, determine the number of belts necessary to meet the power requirements. (c) List the belts and sheaves by catalog number and from discount and net-price sheets for "original equipment manufacturers," if available, determine and list the net prices if purchase is made in quantities of 100. (d) Indicate the most economical of the three combinations.

3.88 An adjustable-speed drive consists of a wide "V-belt" and two sheaves, each adjustable in width by the same amount. The belt and sheave positions are shown at one extreme position of adjustment. (a) Copy this sketch to at least double scale, then sketch the drive in its other extreme position of adjustment. Which two cones must slide axially? Label similar dimensions. (b) Let the ratio of maximum output speed to minimum output speed be 7.0. If the input sheave is driven at a constant speed n_i of 1150 rpm, what are the maximum and the minimum speeds n_0 available at the output sheave?

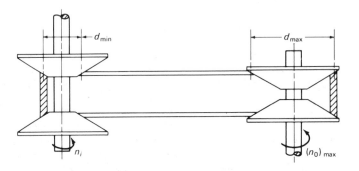

Section 3.12

3.89 (a) Derive Eq. (3.63) for the loose-side tension in a belt drive with a loaded idler pulley (Fig. 3.24). (b) From this equation, write equations for the tensions F_i when idling and F_s at standstill. (c) Derive an equation for the required (minimum) idler weight or spring force as outlined just after Eq. (3.63). *Ans.* $W_{\min} = (pwr/V) [\sin \lambda + \sin (\theta'' - \lambda)]/(e^{\mu\theta} - 1)$, where θ is the smaller wrap angle of the active pulleys. (d) Compare the tensions of (b) and (c) with those for a fixed-center drive. What are the advantages?

3.90 (a) For a better understanding of the idler-pulley drive of Prob. 3.89, sketch and label free-body diagrams of the belt at the driving pulley and of the pulley itself, each under the conditions of standstill, idling, and driving. The sample sketch shows the belt under driving conditions. (b) For a comparison with fixed-center drives, sketch and label similar diagrams for their standstill, idling, and driving conditions.

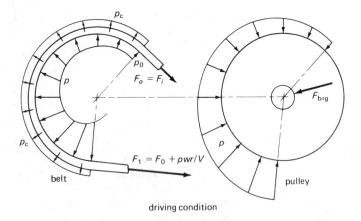

driving condition

3.91 By using a weighted idler pulley in the belt drive of Example 3.6, the angle-of-wrap on the smaller pulley is changed to 202.5° and on the larger one to 225°. The wrap angle on the idler pulley is 69° and $\lambda = 23°$ (Fig. 3.24). (a) What is the minimum effective weight required at the center of the idler pulley? (b) What are the standstill, idling, loose-side, and tight-side tensions? (c) Compare values with those of the fixed-center drive of Example 3.6 and comment.

3.92 A very thin belt of polyester film is to be used to give low turntable speeds in a disk recorder. Because a thin belt may be bent around a very small pulley without appreciable resistance, the reduction from a high-speed synchronous motor may be done with a single belt. Tension will be maintained by a spring-loaded idler pulley. (a) For clockwise rotation of the pulleys, which of the two arrangements will require the lighter spring? Why? (b) To match the capacity of a 4-W 20-rps motor, with a pulley diameter of 7.00 mm, what should be the minimum value of spring tension? Let $\mu = 0.15$ and neglect centrifugal effects. What will be the corresponding values of F_0 and F_1? *Reference:* Prob. 3.89 (c) If the allowable tensile stress is one-fifth the proportional limit of 105 MPa, and belt thickness is 0.25 mm, what should be the belt width?

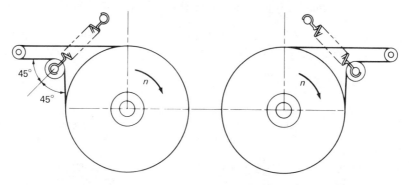

3.93 Derive Eq. (3.66) for the minimum overhang l in a pivoted-motor drive. *Hint:* After converting the equality in F_1 to the symbols following Eq. (3.66), obtain Wl (V/pwr) $= Ea + (E - 1)h$ and from Fig. 3.25 resketch and use triangles ONT, OMU, APU, and APT to obtain h in terms of s, l, c, and σ.

3.94 (a) Same as Prob. 3.90(a) except for a pivoted-motor drive. The sample sketch given with Prob. 3.90 is also valid for a pivoted-motor drive. (b) Same as Prob. 3.90(b).

3.95 Derive equations for the reaction components R_x and R_y at pivot A in Fig. 3.25. Prove that the effect of the distributed centrifugal force may be obtained by subtracting the concentrated belt tension F_c at F_0 and F_1.

3.96 For the pivoted-motor drive of Example 3.7(a) determine the bearing load at pivot A under driving conditions. *Ans.* 495 lb at 130.5°.

3.97 In the pivoted-motor drive shown, the equivalent weight acting at the motor-pulley center is 2670 N, pulley speed is 1140 rpm, dimensions are in mm, and $\mu = 0.40$. (a) Write by inspection the equation of moments about pivot A, including the centrifugal force term F_c. (b) If the maximum stress is to be 2.75 MPa, determine the required cross-sectional area bh for a leather belt. (c) Calculate the corresponding belt tensions and power transmitted.

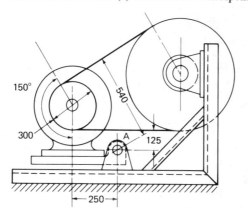

3.98 The fixed-center belt drive of Prob. 3.81 is to be converted into a pivoted-motor drive, using the same dimensions found under part (a). The motor and platform together are estimated to weigh 1250 lb, a force acting 3 in to the right of the centerline of the motor (Fig. 3.25). Force F_1 is horizontal and passes 4 in vertically above pivot A. Determine the minimum overhang l of the gravity force W, the minimum tight-side tension under driving conditions, and the standstill tension. Compare the tensions with those in the replaced drive. *Ans.* 10.43 in, 1248 lb, 451 lb, respectively.

3.99 Same as Prob. 3.98 except convert the fixed-center drive of Prob. 3.82. The weight of the motor and platform is 1225 N acting 40 mm to the right of the centerline of the motor, and force F_1 is horizontal and 60 mm vertically above pivot A.

3.100 A manila rope is wrapped three times around the capstan pulley of Fig. 3.26. The pulley has a diameter of 8 in. (a) How much force must a man exert on the rope if that part on the other side is to lift 2500 lb? Assume $\mu = 0.30$. (b) How much torque must the pulley furnish?

3.101 An elevator load of 16 000 lb is operated by the traction method. Each of six ropes passes over the upper, driving drum twice and an idler sheave once (Fig. 3.27). Coefficient μ is 0.10. (a) What is the minimum total weight required of the counterweights? (b) What is the corresponding torque to be supplied, written in terms of drum pitch diameter D? (c) As a practical matter, should there be one counterweight or six separate ones? (d) Should there be one idler drum or six separate sheaves? Why not instead wrap each rope 1½ times around the top drum?

3.102 Determine the breaking strength requirement for 6 × 37 rope to be used for the elevator of Prob. 3.101. Use the Drucker-Tachau criterion, Eq. (3.62). For this rope, tables list the wire diameter as $0.045d_r$, the metallic area of the rope as $0.40d_r{}^2$, and the average sheave diameter as $27d_r$. Assume the space is available, so use $D = 36d_r$ for rope economy, also a factor of safety of 1.5. [See footnote to Eq. (3.62).] To what ratio of ultimate load to elevator load does this correspond? *Ans.* $F_u = 59\ 300$ lb.

3.103 An elevator car weighing W with a full load of W' added is to be given an upward acceleration of $a = \gamma g$, where γ is the number of g's. The car is to have the same downward acceleration when empty. (a) For a traction drive (Fig. 3.27), what should be the weight of the counterweight so that the traction is equally effective for these extremes of car loading? Assume the mass of the rope and pulleys to be negligible. *Ans.* $\sqrt{W(W + W')}$. (b) Will there be slipping under conditions of raising an empty car, lowering a full one, or braking a full one, all at the same value of acceleration? (c) The semicircular grooves of the driving pulleys are undercut as shown to increase the effective coefficient of friction. If $\mu = 0.10$, $\theta = 135°$, $W = 4500$ lb, and $W' = 3000$ lb, what is the maximum acceleration without slipping?

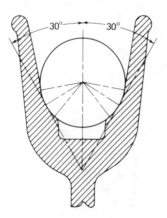

4

Hydrodynamic Drives
Fluid Couplings, Adjustable-speed Fluid Drives, and Torque Converters

4.1 INTRODUCTION

A fluid coupling or hydraulic coupling is a hydrodynamic drive that transmits power without change of torque. It is this characteristic that distinguishes it from another hydrodynamic drive, the torque converter. Magnetic couplings, electrically excited, also do not change the torque transmitted. All give smooth acceleration to the driven load and low starting torque, and overload protection without stalling to the driving motor or engine. This may change the characteristics and reduce the size required of the prime mover. By some types of couplings, output speed may be greatly decreased or the output shaft disconnected. Couplings and torque converters minimize the transmission of shock and vibration by means of the fluid or magnetic field that separates the mechanical elements. In simple couplings, the penalty for these advantages is a loss of energy, from 3.5 to 5% under full-load conditions, and much higher losses at starting and momentary overloads. There is the same loss in output speed, so the drives are not suitable when synchronization is necessary.

The torque converter and fluid coupling were patented in 1905 by Professor H. Föttinger in Germany. The converter was then developed by the Vulcan Company for the reduction of steam-turbine speeds to propeller speeds in large ships. Soon thereafter, the development of large, accurate helical gears for speed reduction made the converter unnecessary and postponed its further development. The use of Diesel engines in ships in the 1920's led to the development and installation of the simpler, fluid coupling between engines and gears to protect the latter from the inherently-high torque variations of the engines. In 1928, Harold Sinclair in England developed couplings for vehicular and industrial applications. The development and manufacture of hydrodynamic drives in the United States began in 1932 with the manufacture by the American Blower Corporation[1] of a variable-fill coupling for the control of speed in induced-draft fans. Shortly thereafter, American Blower cooperated with the Chrysler Corporation in developing a fluid coupling for automotive use.[2] Hydrodynamic drives now are best known from their

[1] Now, Industrial Products Division of American Standard, Inc.
[2] N. L. Alison, R. G. Olson, and R. M. Neldon, "Hydraulic Couplings for Internal-Combustion Engine Applications," *Trans. ASME*, Vol. 63, pp. 81–90, 1941.

use in the automatic transmissions of automobiles, and they have many industrial and marine applications as well.

4.2 COUPLING AND FLUID ACTION

The essential parts of a fluid coupling are shown in Fig. 4.1. The rotor attached to the input shaft is called the *impeller*, and the rotor attached to the output shaft is called the *turbine*.[3] Each rotor encloses a semitoroidal space, which is divided into compartments by uniformly spaced, flat, radial blades. The resulting doughnut-shaped space contains a fluid, usually a mineral oil of low viscosity. There is frequently an inner shroud or core attached to the blades to give smoother circulation of the fluid. There are bearings to pilot the turbine and impeller in relative rotation and to take axial fluid forces, and a seal to prevent leakage along the output shaft. The case or housing may have fins for air cooling, or, for heavy duty, the fluid may be continuously removed, externally cooled, and returned. One may think of the coupling as a centrifugal pump and a turbine enclosed in the same rotating case with the connecting pipe omitted.

Consider that a motor or engine applies torque to the impeller while the coupling is

FIGURE 4.1 Essential parts of a fluid coupling.

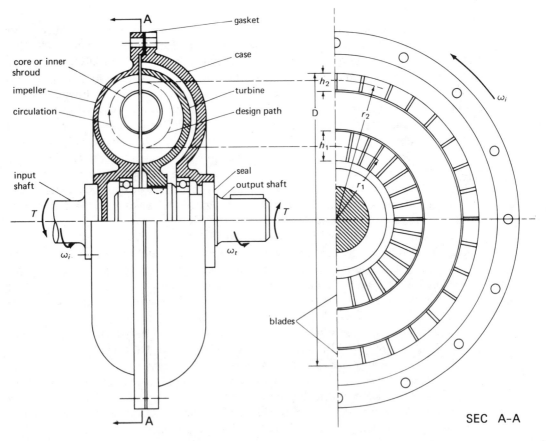

---

[3] The terminology used here is SAE Recommended Practice, "Hydrodynamic Drive Terminology," SAE J641a, SAE Handbook, Society of Automotive Engineers, Warrendale, Pennsylvania. Other commonly used names are *pump* and *primary* for the input member and *runner* and *secondary* for the output.

at rest with the turbine connected to a load. The impeller and the fluid in the pockets between its blades will rotate about the shaft axis with increasing speed and angular momentum. The connected load will prevent initial rotation of the turbine. Centrifugal forces will make the fluid flow outward in the impeller and, in the axial plane of Fig. 4.1, clockwise and across and inward through the turbine. The fluid impinges upon the stationary blades, and the resulting change in the angular momentum of the fluid creates a torque on the turbine. This torque will increase with impeller speed, until it overcomes the starting resistance of the driven machine. This may not occur until 70% of full speed has been reached. As the impeller speed further increases, the additional fluid momentum causes acceleration of the turbine until its speed is only a few percent less than the speed of the impeller. If the size of the coupling has been properly chosen to match the load, the speed difference or loss will be 3 or 4% at rated speed and load. Only when all output-end load and resistance are removed does the turbine speed come up to impeller speed. When the turbine is made to rotate faster than the impeller, as by coasting a vehicle, the direction of circulation is reversed and torque is again developed, allowing the prime mover to retard the motion.

If an overload occurs to upset the equilibrium at rated load, the turbine speed decreases, resulting in a greater "head" or difference of centrifugal pressure in the two rotors and a more rapid circulation of fluid between them. This increased flow of mass gives an increased change of momentum and more torque, and equilibrium is reestablished at a lower turbine speed. If the turbine stalls, maximum circulation and torque is developed, sometimes overcoming the resistance in a short time. If the turbine reverses, as when a load is lowered by a hoist, circulation continues in the same direction, decreasing in amount as lowering speed increases. Thus the retarding torque decreases.[4] A hydrodynamic *brake* resembles a fluid coupling, but one rotor is replaced by a set of stationary blades, so that the brake always operates under maximum-torque stall conditions. Both devices limit lowering speed, but a friction brake should be available to stop and hold the load.

4.3 SLIP, EFFICIENCY, AND TORQUE CAPACITY IN COUPLINGS

Slip is used here as the difference or the loss of speed in ratio form. Thus slip $s = (\omega_i - \omega_t)/\omega_i$, where ω_i and ω_t are impeller and turbine speeds, respectively (Fig. 4.1). Slip is often stated in percent. It follows that output or turbine speed is

$$\omega_t = (1 - s)\omega_i \tag{4.1}$$

The efficiency of the coupling is directly related to its slip. With the coupling of Fig. 4.1 taken as a free body, it is seen that the torque at the output shaft must equal the torque at the input shaft, since there is no place of attachment where a torque could be externally applied. Let the torque be T. Then, the power output and power input are proportional to $T\omega_t$ and $T\omega_i$, respectively. Their ratio is the fractional value of efficiency η, and, together with Eq. (4.1),

$$\eta = \frac{pwr_{\text{out}}}{pwr_{\text{in}}} = \frac{\omega_t}{\omega_i} = 1 - s \tag{4.2}$$

and the efficiency equals the speed ratio.

---

[4] Under certain conditions, the load might run away or the engine might stall and be driven in reverse. See W. B. Gibson, "Fluid Couplings," *Machine Design*, Vol. 32, p. 112, March 31, 1960. See also Prob. 4.8.

The torque capacity of a coupling depends upon its speed, size, fluid, and internal design. An equation for torque on a rotor will be derived from the change in angular momentum of the fluid passing through it. Consider a mass of fluid at a radius r in the coupling. It has a moment of inertia mr^2 and an angular momentum $mr^2\omega$. Torque equals the time rate of change of angular momentum or $T = d(mr^2\omega)/dt$. Now $d(mr^2\omega)/dt$ depends upon the circulation through the rotor. If there is no circulation, $d(mr^2\omega)/dt$ is zero. Let the rate for the volume of circulating flow be Q and the density be ρ. The rate of mass flow is $dm/dt = \rho Q$. Thus the expression for torque may be written.

$$T = \rho Q \Delta(r^2\omega) \tag{4.3}$$

Fluid enters and leaves the turbine at the same circulation rate Q. However, where it is about to enter the turbine, fluid has approximately the rotational-plane velocity ω_i that was imparted to it by the impeller. The mean radius[5] of effective flow, the design path of Fig. 4.1, is r_2. Fluid leaves the turbine with approximately turbine velocity ω_t, where the effective radius is r_1. Thus the net torque on the turbine, by Eq. (4.3), is

$$T = \rho Q(r_2^2\omega_i - r_1^2\omega_t) \tag{4.4}$$

For design purposes, an equation is needed that relates torque and slip to coupling speed and size. An equation is readily obtained for a series of geometrically similar couplings. Coupling size may be designated by the maximum diameter of the design path or by the maximum diameter of the flow path, D of Fig. 4.1. Then, $r_1 = K_1D$ and $r_2 = K_2D$, where K_1 and K_2 are proportionality factors. The actual magnitude of flow Q is affected by the viscosity of the fluid, the friction of the walls, and the shock and eddies as the fluid enters the space between the blades.[6] However, it is reasonable to assume that circulation Q is proportional to some tangential velocity $D\omega_i/2$ in the plane of rotation, to the slip s, and to the area (approximately $2\pi r_2 h_2$) of the passage into the turbine, which area is proportional to D^2. Then $Q = K(D\omega_i/2)sD^2 = Ks\omega_iD^3/2$, where K is a proportionality factor. This may be substituted into Eq. (4.4), together with $\omega_t = (1 - s)\omega_i$, $r_1 = K_1D$, and $r_2 = K_2D$ to give

$$T = \rho(Ks\omega_iD^3/2)[K_2^2D^2\omega_i - K_1^2D^2(1 - s)\omega_i]$$

$$= K(K_2^2 - K_1^2 + sK_1^2)\rho s(2\pi n_i/60)^2 D^5/2$$

where, finally, ω_i in rad/s, is replaced by its equivalent $2\pi n_i/60$, with n_i in rpm. Radius r_1 is usually about half of r_2, so K_1^2 is approximately $0.25K_2^2$. If use of the equation is restricted to small values of slip, say less than 0.20, then $sK_1^2 < 0.05K_2^2$, and the term $(K_2^2 - K_1^2 + sK_1^2)$ is nearly constant. With the several constants and numerals combined into a single constant C, the equation becomes

$$T = C\rho sn_i^2D^5 \tag{4.5}$$

Equation (4.5) indicates that torque is proportional to slip over a limited range, and this is supported by the curves of Fig. 4.2(a).

A numerical value for C may be calculated from a test on one or more couplings of a series. Then, Eq. (4.5) becomes useful in predicting the performance of this coupling at different speeds or in determining the dimensions of a coupling of the same proportions and fluid but for a different torque capacity. Since torque varies as the second power of input speed, power input varies as the cube. Both vary as the fifth power of dimension

[5] Frequently calculated as the root mean square of the radii to the boundaries of the passage.
[6] For a consideration of these factors, the following are suggested: R. Eksergian, "The Fluid Torque Converter and Coupling," *J. Franklin Institute*, Vol. 235, pp. 441–478, 1943; J. C. Hunsaker and B. G. Rightmire, *Engineering Applications of Fluid Mechanics*, New York: McGraw-Hill, pp. 403–413, and Bibliography p. 413, 1947.

D. Thus a large increase in torque and power is obtained with a small increase in linear dimensions.

4.4 PERFORMANCE AND APPLICATION OF FLUID COUPLINGS

Some methods of reporting coupling performance are illustrated in Figs. 4.2(a) and (b). The curve of maximum coupling torque is determined by locking the output shaft so that slip is 100%. It is also called the *drag torque* or the *stalled torque*.

The torque-speed curve of a proposed motor or engine may be plotted on a coupling chart such as Fig. 4.2(b) for a study of the operating characteristics of the combination. Here, the basis for the ordinate scale is the full-load continuous-torque rating of the electric motor, or 100%. The motor speed is 1750 rpm and the coupling slip is 3.6%. This is within the usual range of 3 to 5% slip (97 to 95% efficiency), so the coupling may properly be used with this motor. The output speed, by Eq. (4.1), is (1 − 0.036)(1750) = 1687 rpm. If the torque load drops to 50%, the motor speed is 1775 rpm, the slip is 1.75%, and the output speed is 1744 rpm. Overload speeds are similarly determined, e.g., at 200% torque, the slip is 7.7%, the motor speed is 1680 rpm, and the output speed is 1550 rpm. If the load is larger than 247%, the motor capacity, the driven machine will stall, but the motor will only drop to 930 rpm, point *S*. The coupling allows the load to stop rotating without applying the entire kinetic energy of the motor to the shafts and other parts of the transmission. If the overload continues for more than a few seconds, heating of the motor by the excessive current at this low speed will cause a thermal protective device to cut off the current.

Figure 4.2(b) is for equilibrium conditions. To understand the action at starting, let us consider the above motor and coupling to be connected to a machine which, due to friction, fluid compression, and gravity forces, requires a starting torque of 180% and a running torque of 90%, as shown in the load-torque curve of Fig. 4.2(c). In addition let us consider that the machine and its load have high inertia, and that *in comparison* the inertias of the motor and coupling rotors are negligible. The output shaft will begin to rotate or "breakaway" when the input speed is sufficient to develop a torque of 180%, namely, at 830 rpm (Fig. 4.2(b)). Up to this speed the excess of motor torque over coupling torque accelerates the motor and impeller rotors. Above this speed, the excess accelerates not only these, but also the turbine and the masses connected to the turbine. Since we are considering a case where the inertia of the motor and coupling are negligible, we may take the dotted line of Fig. 4.2 (b) to give the torque acting across the coupling upon the driven machine. From it we may read the slip at any input speed and calculate the output speed.

In Fig. 4.2(c) the scale of abscissas is laid out so that any output speed lies directly below the corresponding input speed in Fig. 4.2(b). Torque across the coupling, load torque, and slip are plotted against output speed. The difference between the two torque curves is the percentage torque that accelerates the load. The two torque curves meet at the running torque of 90%, where equilibrium is established.

Since an electric motor draws less current as the speed becomes higher, and since the acceleration to an intermediate speed is much more rapid than when the motor is directly connected to the load, there is a considerable savings in current and heating, and special starting resistances are unnecessary. A motor with a smaller starting torque may be used. Similarly, if the drive is by an internal-combustion engine, which has a low torque at idling speeds, a fluid coupling allows a considerable increase in engine speed, into its normal torque range, before the driven member starts to rotate. Conversely, low and zero output speeds at various engine torques are available without stalling the engine.

Fluid couplings are useful in decreasing the transmission of pulsations and shock, and

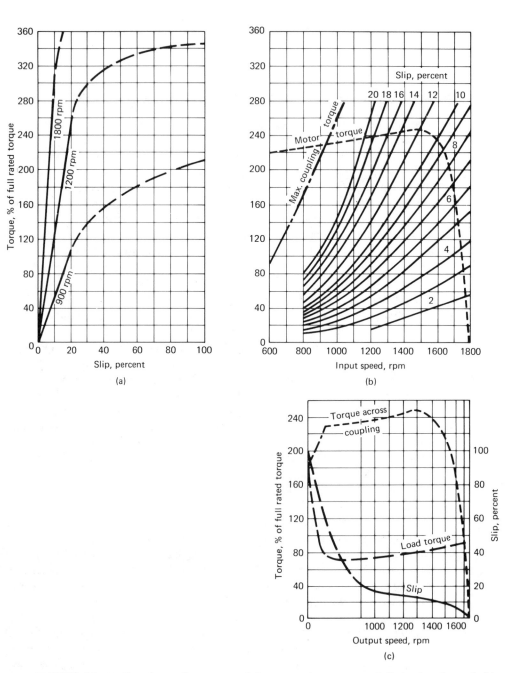

FIGURE 4.2 (a) Torque vs slip at several input speeds, constructed from the chart of (b). (b) Typical fluid coupling and induction motor performance curves. Torque vs input speed at various slips. (Chart adapted from a publication of the FMC Corporation, Drive Division, with its permission.) (c) Coupling and load torques and slip vs output speed when accelerating from zero to running speed, for motor and coupling of (b). Output speeds are plotted in line with the corresponding input speeds of (b).

in equalizing loads. One runner can oscillate about a mean position of steady rotation without greatly affecting the steady circulation of fluid within the coupling. Diesel engines give torque pulsations of large amplitudes, so the hydraulic coupling has found extensive application to Diesel engines in marine, railroad, and industrial drives, particularly, where torque pulsation would be damaging to gear teeth. Even with the coupling, however, coincidence between the prime-mover pulsation and the natural vibration frequency of the load shaft should be avoided.[7] It is difficult to start large inertia loads into motion without some shock. Hence a fluid coupling is useful in hoists, crane drives, and railway car spotters, and a lower capacity, less expensive gear reducer is permissible.[8] When two or more prime movers drive a propeller or conveyor belt, varying torque and inertia in each have less effect on the other prime movers. By draining one coupling, its engine may be disconnected for repair.

4.5 SPEED ADJUSTMENT WITH FLUID COUPLINGS

For the type of fluid coupling previously described, the output speed is determined by the torque requirements of the load and by the prime-mover capacity. However, the output speed of a fluid coupling may be controlled by adjusting the amount of fluid held in the toroidal space, thus changing the slip required to maintain a given circulation and torque. This adjustment might be made in the couplings previously described by a partial filling. In general, the adjustment is made using valves, movable scoop tubes, and pumps to rapidly remove or add fluid while the coupling continues to drive its load.[9]

Since slip is large when speed is low, the efficiency with fluid adjustment will be low. However, the *power loss* need not be high if the torque requirement of the driven member varies as some exponent of its speed, such as 2.0. This occurs for air blowers and in ship propulsion. Let the exponent on turbine speed be k, so that $T = C'\omega_t^k$, and output power is $pwr_{out} = T\omega_t = C'\omega_t^{k+1}$. The required input power is

$$pwr_{in} = T\omega_i = C'\omega_t^k\omega_i = C'(1 - s)^k\omega_i^{k+1} = C''(1 - s)^k n_i^{k+1} \qquad (4.6)$$

where substitution for ω_t is made from Eq. (4.1), and n_i is the impeller speed in rpm.[10] If torque is independent of speed, the exponent k is zero and $C' = T$. The foregoing equations give $pwr_{in} = T\omega_i$ and $pwr_{out} = T\omega_t = T(1 - s)\omega_i = (1 - s)pwr_{in}$. In general, C' may be found from the rated power and corresponding speed and slip chosen for the drive. In all cases $pwr_{out} = \eta\, pwr_{in} = (1 - s)pwr_{in}$, from Eq. (4.2), and the power loss is $pwr_{in} - pwr_{out} = s(pwr_{in})$.

Example 4.1

Two adjustable-speed coupling installations are to be compared. One is in a hoist, and the other is in a small harbor boat. For the latter, resistance to motion varies approximately as the square of velocity. For both installations, slip is 4% at the rated 100 hp and the 1000 rpm governed speed of the prime mover. If full speed is defined as the corresponding output speed, determine and compare the powers and losses at one-half and at one-tenth of full speed.

Solution

For both installations, at *rated load*, $pwr_{in} = 100$ hp, $s = 0.04$, $pwr_{out} = (1 - s)pwr_{in}$ $= (1 - 0.04)100 = 96$ hp, $n_i = 1000$ rpm, and $n_t = (1 - s)n_i = (1 - 0.04)1000 = 960$ rpm. Output half speed and one-tenth speed will occur at 480 and 96 rpm, respec-

[7] R. Eksergian, "The Fluid Torque Converter and Coupling," *J. Franklin Institute*, Vol. 235, pp. 463–471, 1943. Contains mathematics of the fluid coupling as a vibration damper, including geared systems.

[8] J. Seliber, "How Hydraulic Couplings Affect Gear and Shaft Selection," *Product Engineering*, Vol. 31, pp. 46–50, Aug. 22, 1960.

[9] N. L. Alison, R. G. Olson, and R. M. Neldon, "Hydraulic Couplings for Internal-Combustion Engine Applications," *Trans. ASME*, Vol. 63, pp. 81–90, 1941.

[10] In SI units, *pwr* is in watts, i.e., J/s or N·m/s, when the unit of ω is rad/s and the unit of T is N·m. In U.S. customary units, when T is in lb·ft, 550 *hp* should be substituted for *pwr* when ω is rad/s or 5252 *hp* for *pwr* when ω is replaced by n rpm. For equations in linear speed V see footnote to Eq. (3.55).

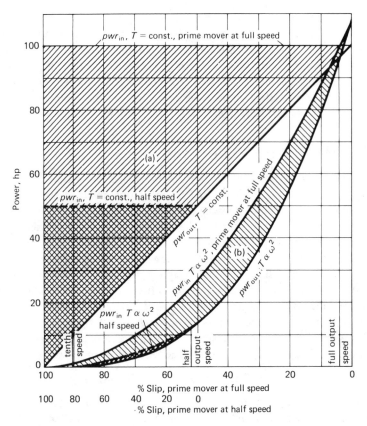

FIGURE 4.3 For adjustable-speed fluid drive of Example 4.1, power input, output, and loss (shaded areas) vs percent of full output speed for (a) constant torque load and (b) load with torque proportional to square of velocity.

tively, with corresponding slips of $(1000 - 480)/1000 = 0.52$ and $(1000 - 96)/1000 = 0.904$.

For the hoist, the load is constant during any one haul, so that torque is constant. Since prime-mover speed is held constant, pwr_{in} is a constant 100 hp (Fig. 4.3). Power for the load is $(1 - s)pwr_{in}$, and the power loss is $s(pwr_{in})$. At full, half, and one-tenth output speeds the losses are 4, 52, and 90.4 hp, respectively.

For the boat, $k = 2$, and from Eq. (4.6)[11] with rated conditions, constant $C'' = pwr_{in}/(1 - s)^k n_i^{k+1} = 100/(1 - 0.04)^2 n_i^3 = 108.5/n_i^3$. Substitution of this into Eq. (4.6) gives $pwr_{in} = 108.5(1 - s)^2$ hp, then, $pwr_{out} = 108.5(1 - s)^3$ hp. These are plotted in Fig. 4.3. The power loss is $s(pwr_{in}) = 108.5s(1 - s)^2$ hp, represented by the vertical distance between curves. At full, half, and one-tenth speeds, the losses are 4, 13, and 0.904 hp, respectively. The maximum loss is 16.1 hp, occurring at a slip of one-third or approximately two-thirds full speed. The power losses in the boat installation at reduced speeds are much smaller than those in the hoist. Of course, if the prime mover can be governed to run at some fraction, say 50%, of full speed, the power losses below half speed will be smaller in both installations. These are also shown in Fig. 4.3.   ////

The coupling allows stepless speed adjustment under load, controlled acceleration, and protection against vibration, shock, and overload. The absence of mechanical contact minimizes wear. The drive is relatively quiet, and this makes it useful for speed adjust-

---

[11] The substitutions of Footnote 10 are not needed when the torque term is not used.

ments in air-moving systems. It is bulky and comparable in diameter to the electric motor that may drive it, and it requires auxiliary cooling equipment. When torque requirements decrease rapidly with speed, as for moving air and for moving in water, as seen in Example 4.1, the power loss and corresponding cooling equipment for its dissipation may be small. They are considerably increased for so-called constant-torque drives, but under certain circumstances, the advantage may outweigh the cost and associated power loss. Examples include reciprocating pumps and conveyor drives that require speed adjustment and that are subjected to fluctuating loads and shock.

4.6 HYDRAULIC TORQUE CONVERTERS: ANALYSIS BY VELOCITY VECTORS

A torque converter (Fig. 4.4) has a ring of stationary blades, called the *reactor* or *stator*, added to a toroidal circuit like that in a fluid coupling. These blades (Fig. 4.5) are curved, and they redirect the fluid so that on entrance to the impeller it has a large component of velocity in the direction of impeller rotation. The impeller, driven by the prime mover and with blades set at an angle, adds velocity and momentum to this fluid before it leaves to enter the turbine. The turbine has bucket-shaped blades that reverse the flow of the fluid before it returns to the reactor. This reversal gives a large change of velocity and momentum to the fluid, and hence a large torque to the turbine. The output torque of the turbine equals the input torque of the impeller *plus* the reaction torque of the reactor. The need for the stationary blades and their designation is now apparent.

The action may be better followed if the reader has a clear understanding of Fig. 4.5 in which the action is shown in columns for three ratios of converter output to input speed. Each column is read from bottom to top, in the direction of the fluid circulation between the rings of blades. Rotation of the impeller and turbine rotors is to the right. Each blade is shown from entrance end to exit end, and thus, for the impeller, between radii r_1 and r_2 (Fig. 4.6). A blade profile is the intersection of the blade with the surface of a toroidal tube, generated by rotating the design path about the axis of the converter. The views in Fig. 4.5 are developed by unwrapping the intersection onto a tangential plane. If the blades were flat and radial, as in a fluid coupling, their developed profiles would be straight and perpendicular to the direction of rotation. The spaces between the rings of blades are numbered to correspond with the radii, and they are enlarged to make room for the vector diagrams. When a vector changes in value from an exit to the next entrance, it is primed at the entrance. Subscripts i, r, and t indicate impeller, reactor, and turbine, respectively.

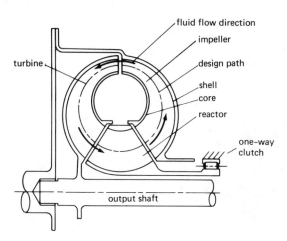

FIGURE 4.4 Two-phase single-stage torque converter. (From ''Hydrodynamic Drive Terminology,'' SAE J641a, *SAE Handbook*. Reprinted with permission. Copyright © Society of Automotive Engineers, Inc., 1974. All rights reserved.)

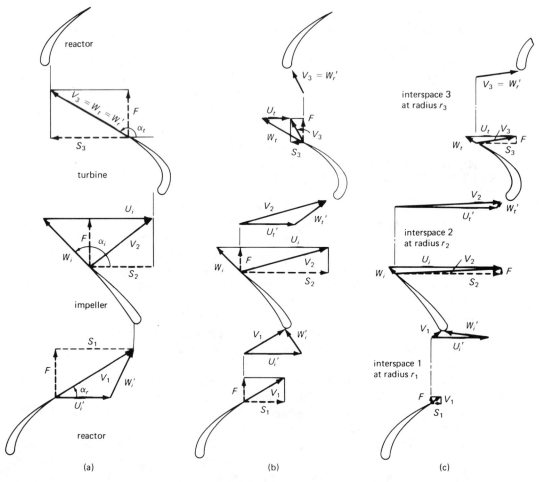

(a)　　　　　　　　(b)　　　　　　　　(c)

FIGURE 4.5 Profiles of some torque-converter blades with velocities U and fluid with velocities V, at constant impeller velocities ω_i, U_i', and U_i and different turbine velocities ω_t, U_t, and U_t', primed at the entrance end. Design-path radii r_1, r_2, and r_3, as defined in Fig. 4.6, are related by $r_3 = r_1$ and $r_2 = 2r_1$. Velocity diagrams per Example 4.2 for (a) stall or $U_t' = 0$, (b) half speed or $U_t'/U_i = 0.5$, and (c) high speed or $U_t'/U_i = 0.9$. Components of V are the constant circulation F, whirl S, and W relative to the blade. For further discussion of these SAE-recommended symbols, see Example 4.2 and the text immediately preceding it.

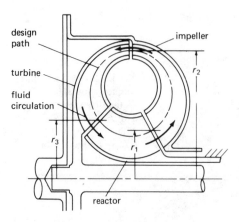

FIGURE 4.6 Design-path radii at blade entrance and exit for a single-stage torque converter.

The action is analyzed and the diagram is constructed with two sets of vectors.[12] One set of vectors consists of the absolute, *fluid velocity V*, and, shown in dotted lines, its radial-plane, tangential, *whirl component S* and axial-plane, torus-flow, *circulation component F*. Circulation component F varies with the speed ratio since it is induced by the slip, but it is essentially constant at any one ratio.[13] The other set of vectors, shown in solid lines, includes V, a *blade velocity U*, or, more exactly, the linear velocity of a point on a blade, and the *relative velocity W*, which is the velocity of the fluid relative to a blade. Thus $V = U \mp W$. Because the exit end of a blade is at a different mean radius than the entrance end, $U \neq U'$. For the impeller, $U_i = (r_2/r_1)U_i'$, and for the turbine, $U_t = (r_3/r_2)U_t'$. At an entrance, W' shows the angle at which the fluid strikes a blade.[14] The fluid is assumed to flow along a blade just before leaving it, so vector W is parallel to the blade surface at exit.

The sequence of events may now be studied, say for half speed (Fig. 4.5(b)), also with reference to Fig. 4.6. Fluid leaves the reactor with velocity V_1 and because impeller velocity U_i' at radius r_1 does not differ greatly from V_1, the fluid enters the impeller with relative velocity W_i', at only a small angle to the blade's surface. It leaves parallel to the blade at a larger radius r_2 with magnitude of W_i determined by its angle and the constant circulation component F. Because $r_2 = 2r_1$, blade velocity U_i here is twice U_i'. The resultant fluid velocity is V_2, with an increased whirl component S_2. At half speed, the turbine blade at r_2 has a velocity U_t' equal to half of U_i. However, because of the entrance angle of the blade, fluid enters smoothly with relative velocity W_t'. It leaves parallel to the blade; this together with F determines W_t. At radius $r_3 = 0.5r_2$, blade velocity is halved to U_t, which together with W_t, determines V_3 and a small and negative whirl component S_3. The change of angular momentum and torque at the turbine are proportional to $(r_2S_2 - r_3S_3)$, the terms of which add numerically. The fluid now enters with relative velocity $W_r' = V_3$ into the stationary reactor, where it is redirected toward the impeller with velocity V_1.

Example 4.2

Construct vector diagrams for the converter of Fig. 4.5. Given are the circulation vector F, the blade velocity of the impeller U_i' at the radius r_1, and $r_2 = 2r_1 = 2r_3$.

Solution

Let us first construct the diagram for the half-speed case (b), which is a general one. Since the velocity of the reactor blade is zero, V_1 is W_r. To determine V_1, one straight line is drawn from the tip of the blade as a continuation of its profile; another line is drawn upward with the length F laid off. By the completion of right triangles, V_1 and then its component S_1 are determined. Vector V_1 is redrawn above at the impeller, and U_i' is laid off from its tail in the direction of the blade rotation to establish relative velocity W_i' as their vector difference. From the exit end of the blade, vector F is drawn and a tangent to the blade profile is extended in the direction for W_i. The magnitude of W_i is found with the construction of the right triangle. Blade velocity U_i is calculated as

---

[12] The vector symbols used, F, S, U, V, and W, both primed and unprimed, correspond to the SAE Recommended Practice, "Symbols for Hydrodynamic Drives," SAE J640a, *SAE Handbook*, Society of Automotive Engineers, Warrendale, Pennsylvania.

[13] Velocity $F = Q/A$, where Q is the volume rate of circulation and A, the net entrance or exit area measured normal to F. Quantity Q is constant, and A is believed to be nearly so in most converters, obtained by decreasing the radial heights of the openings inversely as the mean radius to compensate for the increase in the circumferential width. For a fluid coupling, this may be seen in Fig. 4.1, where there is a decrease from h_1 to h_2.

[14] For a mathematical treatment, see R. Eksergian, "The Fluid Torque Converter and Coupling," *J. Franklin Institute*, Vol. 235, pp. 463−471, 1943.

$(r_2/r_1)U_i' = 2U_i'$, and it is laid off from the tip of W_i to give their resultant V_2 and its component S_2. Turbine-blade velocity U_t' is calculated as $(\omega_t/\omega_i)U_i = 0.50\ U_i$ at half speed, and it is laid off with V_2 at the blade entrance to give their difference W_t'. After one more calculation, $U_t = (r_3/r_2)U_t' = 0.50\ U_t'$, the vector construction is continued in a similar manner until completed.

The high-speed case (c) with $W_t/W_i = 0.90$ is more difficult to draw and to read because of the smaller value of F, about one-fifth by the ratio of slips. The impeller-blade velocities are the same, but $U_t' = 0.90\ U_i$. Otherwise, the construction proceeds as before. The 100% slip, stall case (a) is special only in that there are no turbine-blade components U_t' and U_t. Hence $W_t' = V_2$ and $V_3 = W_t$. The vector diagrams are combined. The magnitude of the circulation component F is arbitrarily taken as twice that at 50% slip. It may be observed that whirl components S_2 and S_3 at the turbine have a large difference at stall, and a much smaller difference at high speed, with its low torque requirement.   ////

4.7 HYDRAULIC TORQUE CONVERTERS: EQUATIONS, PERFORMANCE, AND MODIFICATIONS

Some relationships may be readily expressed. From the equation of equilibrium of torques about the axis of rotation, $T_i - T_t + T_r = 0$, and

$$T_t = T_i + T_r \tag{4.7}$$

The efficiency is output power over input power, or

$$\eta = \frac{T_t \omega_t}{T_i \omega_i} = \frac{T_t}{T_i} \cdot \frac{\omega_t}{\omega_i} \tag{4.8}$$

Efficiency is thus the product of torque ratio and speed ratio. The torque ratio may be written in terms of whirl components S, using the change of momentum equation, Eq. (4.3). If ω is the angular velocity of the fluid about the axis of rotation, then $r^2\omega = r(r\omega) = rS$. With symbols taken from Figs. 4.5 and 4.6, the net torque T_t exerted by the fluid upon the turbine is $\rho Q(r_2 S_2 - r_3 S_3)$, with due regard to the signs of S_2 and S_3. The torque required to rotate the impeller is $T_i = \rho Q(r_2 S_2 - r_1 S_1)$. The torque ratio is

$$\frac{T_t}{T_i} = \frac{\rho Q\ (r_2 S_2 - r_3 S_3)}{\rho Q\ (r_2 S_2 - r_1 S_1)} = \frac{1 - (r_3/r_2)(S_3/S_2)}{1 - (r_1/r_2)(S_1/S_2)} \tag{4.9}$$

As the speed ratio ω_t/ω_i decreases, the ratio S_1/S_2 increases (see vectors of Fig. 4.5), decreasing the denominator of Eq. (4.9). Simultaneously, the ratio S_3/S_2 decreases, then becomes negative and larger in absolute value to increase the numerator. Thus the torque ratio increases with decrease of speed ratio. Figure 4.7 shows the trends in torque ratio and efficiency. At stall, torque ratios of 2 to 3.5 are common in single-stage units like that of Figs. 4.4 and 4.5; up to 6.0 is obtainable in three-stage units (Fig. 4.9).

Also shown in Fig. 4.7 are lines for a fluid coupling. Torque ratio equals 1.0 and coupling efficiency equals the speed ratio ω_t/ω_i, by Eq. (4.2). Both torque-ratio and efficiency plot as straight lines. The efficiency of a *converter* is the product of the speed ratio and torque ratio, Eq. (4.8); hence its efficiency curve may be constructed in Fig. 4.7 by multiplication of the coupling efficiency and converter torque-ratio curves. Their largest product, the peak efficiency, usually occurs at a speed ratio between 0.6 and 0.7.

Between speed ratios of 0.8 and 0.9 the converter torque curve crosses that of the coupling, and the turbine (output) torque becomes less than the impeller (input) torque, which is not desired. Simultaneously, the converter efficiency drops below that of the coupling. These actions are confirmed by the theory. When $T_t/T_i = 1.0$, reactor torque

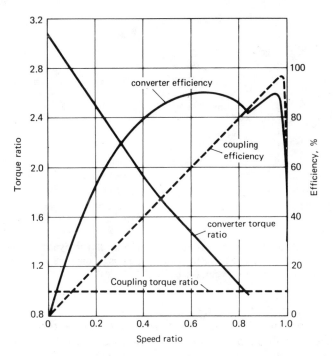

FIGURE 4.7 Torque and efficiency of a typical torque converter, similar to that of Fig. 4.4, compared with a fluid coupling. (From P. L. Fosburg, "Hydraulic Torque Converters," *Machine Design*, Vol. 28, No. 7, pp. 126–134, 1956.)

T_r is zero, by Eq. (4.7), and converter efficiency equals coupling efficiency, by Eqs. (4.8) and (4.2). When $T_t/T_i < 1.0$, Eq. (4.7) indicates that T_r is negative, i.e., that the reaction reverses direction to maintain equilibrium. This is illustrated by Fig. 4.5(c), where $U_t'/U_i = \omega_t/\omega_i = 0.90$. Relative-velocity vector W_r' shows the fluid impinging upon the *back* side of the reactor blade.

The action is improved in a converter with two mechanical phases. Above the speed ratio that corresponds to a torque ratio of 1.0, either a friction clutch automatically creates a mechanical drive or "lockup," or the reactor rotor is allowed to "freewheel" by means of an overrunning or one-way clutch between the reactor and a fixed member, as in Fig. 4.4. (See Probs. 3.51 and 3.52 for a description of the clutch.) When the torque reverses on the reactor, it rotates freely with the turbine and only slightly slower. Thus the converter becomes essentially a fluid coupling. In Fig. 4.5(c) the whirl component S_1 of the fluid delivered to the impeller is increased to something less than S_3, thus increasing T_t/T_i and η. Converter torque and efficiency curves approach those of the coupling in this phase of operation (Fig. 4.7).

The simple converter of Figs. 4.4 and 4.7 has a reasonably high and flat efficiency curve over about half its speed range. Engineers have designed many modifications to raise and widen the curve and to meet special torque requirements. Only at one speed is the entrance angle to a blade most effective. At other speeds, when the direction of the relative velocity W is quite different from the entrance angle, only half the blade may be effective in deflecting the fluid. Thus a change in the entrance angle will shift the peak of the curve to higher or lower speed ratios. Some large converters are provided with adjustable blade settings.[15] A smaller one may have two reactor rotors, differing in

---

[15] A. Lysholm, "Development of the Lysholm-Smith Torque Converter," *Trans. ASME*, Vol. 66, pp. 343–350, 1944; Vol. 68, p. 270, 1946.

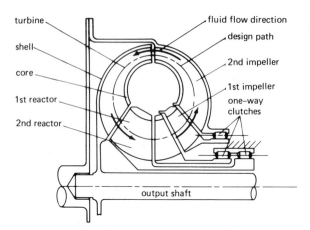

FIGURE 4.8 Four-phase, single-stage torque converter. (From ''Hydrodynamic Drive Terminology,'' SAE J641a, *SAE Handbook*. Reprinted with permission. Copyright © Society of Automotive Engineers, Inc., 1974. All rights reserved.)

blade angles, so that one freewheels at a lower speed ratio than the other. This becomes a four-phase converter with the addition of a second impeller, such that there are four functional arrangements (Fig. 4.8). With the several clutch points, a flatter efficiency curve is obtained. Further improvement is provided by two or more turbine elements or ''stages'' interspersed between impeller and reactors (Fig. 4.9).[16]

Automatic shifting of a gear transmission may allow a converter to operate in an

FIGURE 4.9 Three-member six-element single-phase torque converter with three stages. (From ''Hydrodynamic Drive Terminology,'' SAE J641a, *SAE Handbook*. Reprinted with permission. Copyright © Society of Automotive Engineers, Inc., 1974. All rights reserved.)

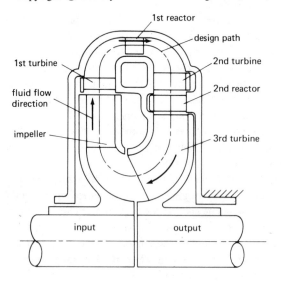

---

[16] Additional rotors may require supporting spokes that cross flow paths. The resulting drag may be minimized. See E. W. Upton, ''Determine the Angular Setting of Spokes Used in the Second Turbine of an Automatic Transmission,'' *General Motors Engineering J.* Vol. 7, No. 1, pp. 54–55 and No. 2, pp. 53–54, 1960; or Prob. 52 in its Reprints booklets.

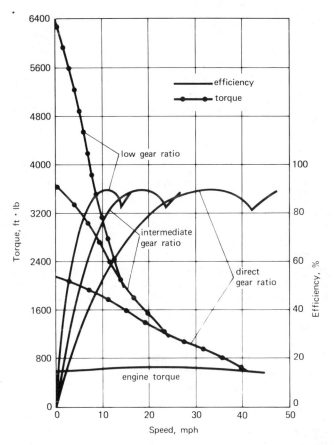

FIGURE 4.10 Torque and efficiency for a three-speed transmission in conjunction with a hydraulic torque converter. (From P. L. Fosburg, "Hydraulic Torque Converters," *Machine Design*, Vol. 28, No. 7, pp. 126–134, 1956.)

efficiency range above 90% over most of its speed range, with the gears supplying a large portion of the torque multiplication (Fig. 4.10). This is common practice in the automatic transmissions of automotive vehicles[17] and in construction machinery (e.g., bulldozers, cranes, end-loaders, and oil-field drilling rigs), where inertia loads and speeds vary over a wide range.

The ability of converters to retard the output load, as in the downhill operation of vehicles, varies and is a consideration in the choice of type for some of the above applications. The speed ratio is greater than 1.0 in the coasting vehicle. In other applications, turbine speed may be opposite in direction to impeller speed, giving a negative speed ratio. This occurs in a hoist when loads are lowering themselves. With some converter types, control can be lost as the ratio approaches −1.0. References are indicated for information on the foregoing extreme speed ratios.[18]

---

[17] A description of types and requirements may be found in J. T. Bugbee, "Talented Transmissions," *SAE Trans.*, Vol. 60, pp. 107–127, 1952 and following papers pp. 128–153, part of a SAE symposium on automatic transmissions and torque converters for vehicles.

[18] R. W. Bachman and M. W. Dundore, "Torque Converters," *Machine Design*, Vol. 32, May 12, 1960, pp. 188–198. This paper is part of a series on performance and applications by R. W. Bachman, also W. B. Gibson, H. J. Wirry, and R. C. Schneider, April 14, pp. 185–189; April 28, pp. 130–136; May 26, pp. 121–127; and June 9, pp. 178–183, all of 1960. The series specifically discusses applications to cranes, excavators, end-loaders, graders, scrapers, pusherdozers, and logging equipment.

PROBLEMS

Section 4.3

4.1   (a) At rated load, a motor running at 880 rpm delivers 2500 lb·ft of torque to a machine through a fluid coupling with a slip of 3%. What are the output speed, torque, power output, and the coupling efficiency? (b) Under an occasional overload of 4375 lb·ft, the motor speed drops to 850 rpm. What will be the approximate speed of the machine and the power output and efficiency of the coupling?

4.2   An existing fluid coupling is to be redesigned to increase its efficiency from 95 to 97% under the same conditions of operation. (a) The same geometric proportions will be used. What will be a reasonable percentage change in diameter? What are the corresponding changes in the heights of the passages and in volume? (b) If, because of some indication of turbulence, it is desired not to change heights h_1 and h_2 of the passages (Fig. 4.1) but only width and radii r_1 and r_2 in proportion, what should be a reasonable change in diameter?

4.3   A new fluid coupling is to be designed by proportioning it to an existing, successfully operating coupling. The outer passage of the existing coupling has a mean diameter of 30 in, and the coupling delivers 215 hp at 843 rpm when the input speed is 870 rpm. The new, geometrically similar coupling is to deliver 110 hp at 1500 rpm, with the same efficiency and oil. What should be its outer-passage mean diameter? What is the power required of the driving engine? *Ans.* 18.6 in, 113.5 hp.

4.4   Tests on a first unit of a series or line of geometrically similar fluid couplings show that a torque of 684 lb·in is developed at an input speed of 1150 rpm and an output speed of 1100 rpm. The oil density is 0.032 lb/in³, and the largest diameter of the flow path is 10.6 in. (a) What is the value of the constant C in Eq. (4.5) for this series of couplings with the units given? *Ans.* 1.072×10^{-3}. (b) After the tests were completed it was decided to rate all sizes of couplings for full load at a smaller slip of 3.5%. What will be the torque and the input power ratings of the first unit at 1150 rpm, at 1750 rpm? What is the output speed corresponding to 1750 rpm input? (c) For a second unit in this series of couplings, a coupling is to be designed for a rating of 100 hp at 1750 rpm and 3.5% slip. What should be its largest diameter of flow path? *Ans.* 13.05 in.

4.5   A fluid coupling with a design-path maximum diameter of 350 mm has been successfully transmitting a torque of 250 N·m at an engine speed of 1900 rpm and a slip of 3.4%. The design of a geometrically similar coupling is undertaken, to deliver 110-kW-output power to a propeller shaft turning at 450 rpm, with the same coupling efficiency and oil. (a) Estimate the required design-path maximum diameter. *Ans.* 960 mm. (b) Consider the relative sizes of the two couplings and the bases of the equation and comment on the probable accuracy of your estimate. (c) What is the numerical value of the constant C in Eq. (4.5) for the first coupling if the relative density of the fluid is 0.890 and all quantities are substituted in the basic SI units of m, kg, and s? *Ans.* 0.0397.

Section 4.4

4.6   Suppose the motor of Fig. 4.2(b) is rerated for a 20% higher full-load torque. In tabular form list the motor speed, the slip, the coupling efficiency, and the output speed of the coupling at full, half, twice, and three times the new rated torque.

4.7   (a) Suggest how the fluid-coupling principle can be used in the design of an auxiliary brake to prevent excessive speeds in a long drop (e.g., for heavy vehicles on mountain roads and lowering drill pipes into oil wells). (b) By equation and curve show how braking torque will vary with speed. (c) Suggest how to dissipate the large amount of heat that may result.

4.8   Fluid couplings are used on engine-driven hoists to pick up the load smoothly and minimize engine stalling. When a load is to be lowered, the engine speed may be throttled down to a slow speed, reducing the torque capacity of the coupling so that the load starts to fall. The operator must be careful that the engine does not stall. (a) Consider what affects the circulation of coupling fluid, and with the aid of Eqs. (4.3) or (4.4) state how torque resistance varies with the speed of descent. When is zero resistance reached? (b) What can the operator do to control the speed of descent? (c) What mechanism can the engineer include in his design to mechanically and automatically prevent reversal of the engine? Schematically sketch the components to show the location of the mechanism.

Section 4.5

4.9   Starting with Eq. (4.6), write an equation in terms of slip for power loss in an adjustable-speed coupling. For $k = 2$, obtain an equation for maximum power loss. For the coupling of the boat of Example 4.1 confirm the statement regarding maximum power loss and corresponding speed.

4.10   If the boat of Example 4.1 must frequently run at slow speeds in harbor or river operations, it would be advisable to reset the governor to hold the engine at half speed when the output speed of the coupling must be half or less. (a) Slip of this coupling at full speed and torque was 4%. If the impeller speed is halved, at what slip will the coupling produce the torque required for boat motion at half speed? (b) Obtain the output-end half speed and power requirement from Example 4.1 and derive equations for input-end power and for the power lost, both as functions of slip. Also, derive an equation for the maximum power loss and its numerical value. *Ans.* 2.01 hp. (c) Plot hp_{in} and hp_{out} on a duplicate of this chart.

4.11   (a) For the case of a torque load that is directly proportional to output speed, i.e., $T = C'\omega_t$, determine equations for input and output power, power loss, and maximum power loss. (b) For rated power of 500-kW *input*, a synchronous motor speed of 1800 rpm, and a corresponding coupling slip of 3%, determine and plot input and output power on a chart similar to that of Fig. 4.3. What is the maximum power loss? *Ans.* 129 kW.

4.12   A motor of constant speed ω_M drives an air blower through an adjustable-speed fluid coupling. Assume that, over the range of operation, air pressure and density may be considered constant and that the volume of air delivered per unit of time is proportional to the rotational velocity of the blower. Write an equation for motor power in terms of volume delivered as well as an equation for percent of maximum power in terms of percent of maximum volume. Sketch a curve of the latter equation, marking three significant points. *Ans.* $pwr/pwr_{max} = (Q/Q_{max})^2$.

4.13   A type of coupling for controlling speed and torque by variation of fluid content consists of an enclosed set of closely fitting disks or thin concentric cylinders. Torque is developed by shearing action of the fluid between the many pairs of surfaces. For example, the

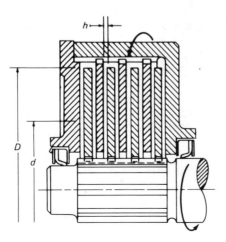

temperature of an automotive engine can control the amount of fluid in a coupling and, thereby, the fan speed (see Prob. 3.20). Assume that centrifugal action will drive the fluid outward to form a ring of inside diameter d. Also, assume that all fluid films between surfaces are of uniform thickness h, and that the viscosity is μ. Let the input speed be n_1 and the output speed n_2, both in rpm. Refer to Section 2.1 for the general torque-viscosity-shear relationships. Derive equations for slip and power loss (heat) in terms of the torque transmitted by a set of flat, circular disks of outside diameter D, as shown, with N pairs of sliding surfaces. *Ans.* $s = 960\ hT/\pi^2 n_1 N\mu(D^4 - d^4)$.

4.14 Same as Prob. 4.13 but for a device having a set of concentric, thin cylinders of length l, with a clearance space of thickness h between each pair at several diameters D_i. A ring of oil includes and extends from diameter D_M to diameter D_N. Derive equations for slip and power loss in terms of torque transmitted.

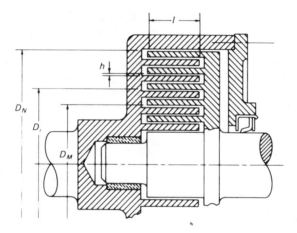

Section 4.6

4.15 For the converter of Fig. 4.5, construct the vector diagrams for quarter-speed operation, using a larger scale so that U_i' and F are laid off as 1.2 and 0.75 in, respectively (the latter intermediate between F at stall and F at half speed). The exit angles α for reactor, impeller, and turbine are 30, 135, and 150°, respectively, measured ccw from the blade-velocity vectors (Fig. 4.5(a)). *Ans.* $V_1 = 1.50$ in, $V_2 = 1.80$ in, $V_3 = 1.20$ in.

4.16 Construct a set of stall and three-quarter speed vector diagrams for a converter in which the reactor, impeller, and turbine exit angles α are 15, 120, and 135°, respectively, ccw relative to the blade-velocity vectors U. Let $r_1 = r_3 = 0.60r_2$. Take vector length U_i' to be 30 mm and length F to be 20 mm at stall, 10 mm at three-quarter speed. *Ans.* At stall, $S_3/S_2 = -0.51$, $S_1/S_2 = 1.94$; at three-quarter speed, $S_3/S_2 = 0.28$, $S_1/S_2 = 0.85$.

4.17 For a regular, nonadjustable fluid coupling (Fig. 4.1) in operation with a slip of 33%, construct a set of vector diagrams similar to those of Fig. 4.5, using the same notation. The blades are flat and perpendicular to their direction of rotation. Assume $\vec{W_i} = \vec{W_t} = \vec{F}$. Let $r_1 = 0.5r_2$ and $U_i' = 1.2$ in as in Prob. 4.15, but with the circulation vector of that problem reduced in proportion to the slips, namely, $F = 0.33$ in. (*Suggestion:* First construct V_1 or V_2 from exit conditions.) *Ans.* $V_1 = 0.85$, $W_i' = 0.55$ in.

Section 4.7

4.18 An automobile engine provides a torque of 150 lb·ft and a speed of 3000 rpm to a torque converter that is operating with a torque ratio of 3 and a speed ratio of 0.25. (a) What are the torque on the reactor and the efficiency of the converter? (b) What is the minimum amount of converter oil in gal/min that must be pumped through a cooler if the maximum temperature of the oil is to be held to 250°F? The surface area with the cooling water is sufficient to bring the oil temperature down within 10°F of a maximum water temperature of 200°F. The specific heat of oil is 0.52 Btu/(lb·°F) and a gallon of oil weighs about 7 lb.

4.19 A machine requires a torque of 595 lb·ft at a speed of 670 rpm when driven by an engine governed to run at 1735 rpm. A torque converter is selected to operate at an efficiency of 77%. What will be the power required of the engine, the heat of the converter to be dissipated, the torque and speed ratios, and the torque on the reactor? *Ans.* Torque ratio 2.00, speed ratio 0.386.

4.20 Same as Prob. 4.19, except in SI units, with a driven torque and speed of 475 N·m and 1200 rpm, respectively, engine speed of 1760 rpm, and converter efficiency of 80%.

4.21 It is seen from Fig. 4.7 that the converter torque vs speed curve may approximate a straight line over a limited range. (a) Write an equation for torque ratio τ in terms of speed ratio σ and two known points (σ_1, τ_1) and a lower one (σ_2, τ_2). Substitute for τ in the efficiency equation, and determine equations for the location and magnitude of maximum efficiency. *Ans.* $\eta = b^2/4a$ at $\sigma = b/2a$, where $a = (\tau_1 - \tau_2)/(\sigma_2 - \sigma_1)$ and $b = a\sigma_2 + \tau_2$. (b) Check the foregoing against Fig. 4.7 by taking the known points at $\sigma = 0.4$ and $\sigma = 0.8$ and calculating the maximum efficiency.

4.22 Scale the needed vectors of Fig. 4.5 and calculate, for this converter, the torque ratios and the efficiencies at stall, 0.5, and 0.9 speed ratios. Assume that the torque-ratio line is linear between the 0.5 and 0.9 points and calculate the magnitude and location of the maximum efficiency, using the answer to Prob. 4.21(a). *Ans.* At half speed, 1.44 and 72%; $\eta_{\max} =$ 81.4%.

4.23 Show the following with the aid of Eq. (4.9). (a) If $T_t/T_i = 1.0$, the relationship between whirl components is such that the torque on the reactor must be zero. (b) With freewheeling of the reactor and a larger S_1, the efficiency is increased.

4.24 If the intermediate gears of the transmission of Fig. 4.10 were eliminated, what would be the converter efficiency and torque ratio at 20 mph? At 15 mph? Extrapolate as necessary. Above what speed does the reactor appear to freewheel? Explain.

4.25 Products such as paper and wire emerge from processing machines at a uniform velocity V, as at (a). For storage and shipment, the product is usually wound onto spools or reels, of which the wound diameter sometimes increases 100 or 200% from start to finish. Various means are used to control reel speed and winding tension. One method combines a fluid coupling (not adjustable) with a single-planetary gear train. Make a study of this combination. The following steps are suggested. (a) State the change that must occur in the rotational speed n_r of the reel as the wound diameter d_w increases. If the winding tension remains reasonably uniform or tends to increase, what change occurs in the torque as the wound diameter increases? (b) In a fluid coupling, what change occurs in the torque as the output speed decreases? Is the coupling possibly useful for a reel drive? Is the practical range of speed change or slip in the coupling sufficient for the reel? If not, what must the planetary gearing do to this speed change? (c) For the planetary drive shown, rotating counterclockwise at (b), derive an equation relating the speeds n_1 of the ring gear, n_2 of the planet carrier, and n_3 of the sun gear. *Ans.* $n_3 = n_2 - (n_1 - n_2)(d_1/d_3)$. (d) Show that the objective of the planetary gearing under (b) is accomplished by connecting the ring gear to the coupling housing or impeller, the planet carrier to the turbine, and the sun gear to an output shaft that drives the reel. If they are combined into a single, rotating unit and the input speed to the unit is $n_{\rm in}$, the output speed is $n_{\rm out}$, and the coupling slip is s, show that the speed ratio of the unit is $n_{\rm out}/n_{\rm in} = 1 - s\,[1 + (d_1/d_3)]$. By how much is the slip multiplied?

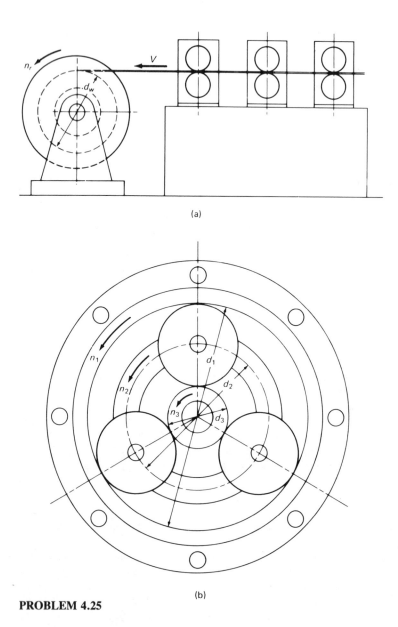

(a)

(b)

PROBLEM 4.25

4.26 (a) Determine the relationship between torque T_2 on the turbine or carrier and the output (sun gear) torque T_{out} for the unit of Prob. 4.25 (d). (*Suggestion:* Sketch a free-body force diagram for the sun gear, showing transmitted (tangential) forces F_3 at each of three teeth; then sketch, in order, diagrams for a planet pinion, the ring gear, and the carrier, labeling them in terms of F_3.) *Ans.* $T_2 = T_{out} [1 + (d_1/d_3)]$. (b) As a check, sketch a free-body diagram with the several torques acting on the housing of the unit. Show that the equation of equilibrium, together with the equations of (a), give $T_{in} = T_{out}$, necessary, of course, for equilibrium of the unit as a whole. (c) From Eq. (4.5), which holds over a limited range of slip s, we may write $T_t = T_2 = cs$, where c is a constant of proportionality for a given coupling at a given, constant impeller speed. From the equation of (a) obtain T_{out} as a function of s. *Ans.* $T_{out} = csd_3/(d_1 + d_3)$.

4.27 Refer to Probs. 4.25 and 4.26 and their answers. Usually an external speed reduction will be needed, by belt or otherwise, between motor and unit and/or between unit and reel. Let

$R = n_{out}/n_r$, where n_{out} is the output speed of the unit and n_r is the speed of the reel; also, let the wound diameter on the reel be d_w. (a) For a constant winding velocity $V = \pi d_w n_r$, coupling slip s is determined by V and R and the wound diameter d_w. Derive an equation for s. *Ans.* $s = [d_3/(d_1 + d_3)] [1 - RV/\pi d_w n_{in}]$. (b) Determine the tension P in the product being wound on the reel, first as a function of slip, then by elimination of s, as a function of one variable d_w. *Ans.* $P = (2Rc/d_w)[d_3/(d_1 + d_3)]^2 [1 - RV/\pi d_w n_{in}]$. (c) Derive an equation for the maximum (or minimum) winding tension and the diameter at which it occurs. *Ans.* $P_{max} = (\pi c n_{in}/2V)[d_3/(d_1 + d_3)]^2$ at $d_w = 2RV/\pi n_{in}$. (d) Determine the efficiency of the unit. *Ans.* $\eta = n_{out}/n_{in}$.

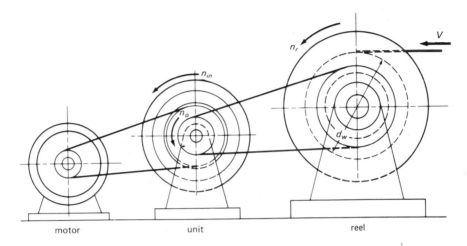

motor   unit   reel

4.28 Newspaper stock is to be started on a tube 10 in. in diameter and wound to a diameter of 30 in. A tensile force of 20 lb at the maximum diameter is desired to produce a firm roll. The paper velocity is 10 ft/s. Motor and input speed is 1750 rpm. Coupling slip should be limited to 20% for linearity. Refer to the equations obtained in the preceding three problems and determine: (a) the rpm of the reel at the maximum and minimum wound diameters; (b) the ratios R and d_1/d_3 to give 3% coupling slip at the minimum wound diameter and the same tension of 20 lb at both maximum and minimum diameters. What is then the slip at the maximum diameter? *Ans.* $R = 5.73$, $d_1/d_3 = 7.31$; (c) the coupling constant c and the maximum and minimum values of winding force and corresponding diameters? *Ans.* 26.75 lb at 15 in, 20 lb at 10 in and 30 in; (d) torques, efficiency, power required of motor, and power loss, all at the maximum wound diameter? Why is the power loss greater than that in a fluid coupling with the same slip but no attached gears?

4.29 Design a winding system for an application where initial wound diameter is 200 mm, the final diameter 360 mm, and the product speed 5 m/s. The pull at the two diameters is to be 150 N and the slip 10% at the lower diameter (a slip to minimize ratio d_1/d_3). The sun gear shaft will be directly connected to the reel ($R = 1.0$), and the necessary input speed to the coupling must be determined and provided by V-belts between motor and coupling. Determine the characteristics of the system as outlined in (a)−(d) under Prob. 4.28.

II

MECHANICAL BEHAVIOR OF MATERIALS AND GENERAL DESIGN CRITERIA

5

Stress, Strain, Strength, and Theories of Failure

General Elastic Relationships, Graphical Representation, Plasticity and Limit Design, Fatigue Strength and Stress Concentration

5.1 INTRODUCTION

One purpose of a theory of failure is to make it possible to design for a complex situation from experience with simpler situations. Structures such as pressure vessels and shafts that may be subjected to stresses of different types, from several directions and with varying magnitude, often must be designed from data obtained for the material in simpler and easily made failure tests such as the tensile test and the rotating-beam fatigue test.

This chapter is concerned with basic theory and phenomena. The general relationships of stress and strain are reviewed, including the principal stresses and graphic representation. Equations for some commonly used failure theories are derived and compared. Based on yield strength, they are the maximum shear-stress theory and the Mises criterion or maximum energy-of-distortion theory. Based on ultimate strength, they are the normal stress and the Mohr theories. Limit analysis or design based on elastic–plastic phenomena is briefly treated. Fatigue strength and theoretical stress–concentration factors are presented, together with the modifications needed for the usual conditions of manufacture and use. Suggestions are made for design changes to decrease stress concentration, including the use of a flow analogy for visualizing the change.

Chapter 6 applies the theories and phenomena of Chapter 5 to obtain equations for the design and safety of components under loads that give fatigue conditions of simple or compound fluctuating stresses. Application is made in particular to the design of shafts. The design for finite life and for several levels of stress as well as the concept of fatigue damage are also considered.

The theories based upon yield strength are said to apply to ductile materials while those theories based upon ultimate strength are said to apply to brittle materials. The distinction between ductile and brittle is sometimes taken arbitrarily at 5% total elongation in the tensile test. With the exception of some steels for tools, almost all steels, including those in a suitable hardened and strengthened condition, will have 10% or more elongation and should be considered ductile. Most copper and wrought aluminum alloys are ductile. Ordinary gray cast irons and some aluminum casting alloys are typical of brittle materials in machines and structures. In any case, for applications where a permanent set will interfere with proper operation, the use of an equation based upon yield strength is suggested.

5.2 STRAIN, STRESS, AND STRENGTH

Stress is designated by σ if it is a normal stress and by τ if it is a shear stress. Normal strain or unit elongation is designated by ϵ and shearing strain or angle of distortion is designated by γ. If an elastic condition exists or may be approximated, stresses and strains are related by certain constants of proportionality: the modulus of elasticity E, the shear modulus of elasticity G, and Poisson's ratio ν. In an orthogonal coordinate system XYZ (Fig. 5.1),

$$\epsilon_x = \frac{1}{E}[\sigma_x - \nu(\sigma_y + \sigma_z)]$$

$$\epsilon_y = \frac{1}{E}[\sigma_y - \nu(\sigma_x + \sigma_z)] \tag{5.1}$$

$$\epsilon_z = \frac{1}{E}[\sigma_z - \nu(\sigma_x + \sigma_y)]$$

and

$$\gamma_{xy} = \frac{1}{G}\tau_{xy}, \quad \gamma_{yz} = \frac{1}{G}\tau_{yz}, \quad \gamma_{zx} = \frac{1}{G}\tau_{zx} \tag{5.2}$$

In Fig. 5.1 the first subscript to a shear stress τ indicates the direction of the normal to the plane on which it acts, and the second subscript indicates the direction of the stress on this plane. Since $\tau_{yx} = |\tau_{xy}|$, etc., the three equations of Eq. (5.2) are sufficient. Moduli G and E are related by

$$G = \frac{E}{2(1 + \nu)} \tag{5.3}$$

With $E = 30 \times 10^6$ psi, and $\nu = 0.30$, as commonly used for steels, G is calculated to be 11.5×10^6 psi, or in SI units, $E = 207\,000$ MPa and $G = 80\,000$ MPa.

The unit change in volume of an element with sides of initial lengths dx, dy, and dz is called the volume expansion e, and it is seen to be the sum of the volume changes in each direction (Fig. 5.2(a)), divided by the initial volume, i.e.,

$$e = \frac{(\epsilon_x\,dx)dy\,dz + (\epsilon_y\,dy)dz\,dx + (\epsilon_z\,dz)dx\,dy}{dx\,dy\,dz} = \epsilon_x + \epsilon_y + \epsilon_z \tag{5.4}$$

Substitution of Eqs. (5.1) into Eq. (5.4), gives the following linear relationship between

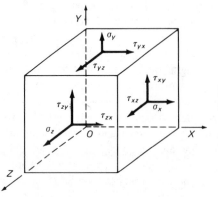

FIGURE 5.1 Normal and shear stresses at a point.

volume expansion and the sum of any orthogonal set of normal stresses,

$$e = \frac{1 - 2\nu}{E}(\sigma_x + \sigma_y + \sigma_z) \tag{5.5}$$

Strains are also related to displacements. Displacement of a particle in a body is its change of position relative to a fixed point on the body. This displacement can be expressed by components u, v, and w parallel to the X, Y, and Z axes, respectively. It often will be useful to first obtain expressions for displacements and then determine the strains and stresses, since the following relations hold:[1]

$$\epsilon_x = \frac{\partial u}{\partial x}, \quad \epsilon_y = \frac{\partial v}{\partial y}, \quad \epsilon_z = \frac{\partial w}{\partial z} \tag{5.6}$$

and

$$\gamma_{xy} = \frac{\partial u}{\partial y} + \frac{\partial v}{\partial x}, \quad \gamma_{yz} = \frac{\partial v}{\partial z} + \frac{\partial w}{\partial y}, \quad \gamma_{zx} = \frac{\partial w}{\partial x} + \frac{\partial u}{\partial z} \tag{5.7}$$

Displacement, strain, and stress relationships in polar coordinates are written and used in Chapter 8.

The work done on a body elastically is stored in it as potential energy of deformation or *strain energy*. The deflections $\epsilon_x dx$, $\epsilon_y dy$, and $\epsilon_z dz$ of Fig. 5.2(a) are the result of normal forces $\sigma_x dy\, dz$, $\sigma_y dx\, dz$, and $\sigma_z dx\, dy$, respectively, applied to the three faces of the element. The action at the positive x face is plotted in Fig. 5.2(b). The work done is the average force times its deflection, or the area of the triangle, $(\frac{1}{2})(\sigma_x dy\, dz)(\epsilon_x dx)$. By a shear force couple such as $(\tau_{xy} dy\, dz)(dx)$ in (Fig. 5.3(a)), the work done is its average value times the angular deflection $\gamma_{xy}/2$ that it causes (Fig. 5.3(a) and (b)). By the two couples on the element, and since $\tau_{yx} = |\tau_{xy}|$, the total shear strain energy is

$$(1/2)(\tau_{xy} dy\, dz)(dx)(\gamma_{xy}/2) + (1/2)(\tau_{yx} dx\, dz)(dy)(\gamma_{xy}/2) = (1/2)\tau_{xy}\gamma_{xy} dx\, dy\, dz$$

In general, the strain energy for the element $dx\, dy\, dz$ is the sum of three normal-strain

FIGURE 5.2 (a) Normal forces and deflections of an element, together with volume changes. (b) Strain energy (shaded area) with deflection $\epsilon_x dx$.

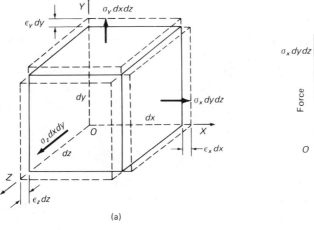
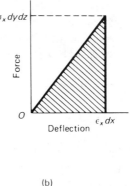

(a)   (b)

---

[1] For a derivation, see Fig. 12.15 and the text preceding Eqs. (12.27).

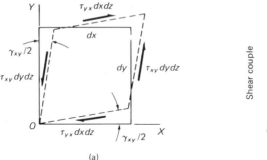
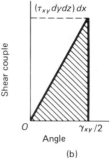

FIGURE 5.3 Shear forces and distortion of an element. (a) Angular changes. (b) Strain energy (shaded area) with angular distortion $\gamma_{xy}/2$ due to a shear-force couple.

terms and three shear-strain terms, all containing the product $dx\,dy\,dz$. The strain energy per unit volume is then the strain energy of the element divided by its volume $dx\,dy\,dz$, or

$$U_0 = \frac{1}{2}(\sigma_x\epsilon_x + \sigma_y\epsilon_y + \sigma_z\epsilon_z + \tau_{xy}\gamma_{xy} + \tau_{yz}\gamma_{yz} + \tau_{zx}\gamma_{zx}) \tag{5.8}$$

This may be expressed in terms of stress components only or in terms of strain components only by substitution from Eqs. (5.1) and (5.2). In terms of stress components, Eq. (5.8) becomes

$$U_0 = \frac{1}{2E}(\sigma_x{}^2 + \sigma_y{}^2 + \sigma_z{}^2) - \frac{\nu}{E}(\sigma_x\sigma_y + \sigma_y\sigma_z + \sigma_z\sigma_x)$$

$$+ \frac{1}{2G}(\tau_{xy}{}^2 + \tau_{yz}{}^2 + \tau_{zx}{}^2) \tag{5.9}$$

The total strain energy in an elastic body is

$$U = \iiint_V U_0\,dx\,dy\,dz \tag{5.10}$$

Example 5.1

The solid, steel, piston rod of a high-pressure hydraulic actuator has a diameter of 25 mm and a length of 375 mm, of which 295 mm is inside the cylinder at the end of the inward stroke. The inside diameter of the cylinder is 65 mm and the pressure is 70 MPa. Because of close tolerances the elastic change in the total length of the rod is needed. The changes in diameter, volume, and energy for the part of the rod inside the actuator are also needed.

Solution

The cross-sectional area for the cylinder is $(\pi/4)(65)^2 = 3318$ mm² and for the rod the cross-sectional area is $(\pi/4)(25)^2 = 491$ mm². The net piston area A on which the pressure acts is $3318 - 491 = 2827$ mm², whence the axial pull $F = pA = (70\ \text{N/mm}^2)$ $\cdot(2827\ \text{mm}^2) = 197\ 900$ N. The stresses in the rod are

$$\sigma_x = \frac{F}{A_{\text{rod}}} = \frac{197\ 900}{491} = 403\ \text{MPa} \quad \text{and} \quad \sigma_y = \sigma_z = -p = -70\ \text{MPa}$$

Strains are determined by Eqs. (5.1) and the constants following Eq. (5.3). The axial strain in that part of the rod inside the cylinder is

$$\epsilon_{x1} = \frac{1}{E}[\sigma_x - \nu(\sigma_y + \sigma_z)] = \frac{1}{207\,000}[403 - 0.30(-70-70)]$$

$$= 2.150 \times 10^{-3}$$

and the axial strain for that part of the rod outside the cylinder, with $\sigma_y = \sigma_z = 0$, is

$$\epsilon_{x2} = \frac{1}{E}[\sigma_x] = \frac{403}{207\,000} = 1.947 \times 10^{-3}$$

The total elongation is

$$u = l_1\epsilon_{x1} + l_2\epsilon_{x2}$$

$$= (295)(2.150 \times 10^{-3}) + (375 - 295)(1.947 \times 10^{-3}) = 0.79 \text{ mm}$$

Inside the actuator the radial strain in the rod is

$$\epsilon_y = \epsilon_z = \frac{1}{E}[\sigma_z - \nu(\sigma_x + \sigma_y)]$$

$$= \frac{1}{207\,000}[-70 - 0.30(403 - 70)] = -0.821 \times 10^{-3}$$

The decrease in diameter is

$$v = w = d\epsilon_z = 25(-0.821 \times 10^{-3}) = -0.021 \text{ mm}$$

The unit volume expansion is

$$e = \epsilon_x + \epsilon_y + \epsilon_z = (2.150 - 0.821 - 0.821)10^{-3} = 0.508 \times 10^{-3}$$

The initial volume of the interior portion of the rod was

$$V = (\pi/4)(25)^2(295) = 144\,800 \text{ mm}^3 \ (0.1448 \times 10^{-3} \text{ m}^3)$$

Since there is uniform straining over the 295 mm length, the total volume increase is

$$\triangle V = eV = (0.508 \times 10^{-3})(144\,800) = 73.6 \text{ mm}^3$$

Unit volume expansion may also be found from Eq. (5.5), thus

$$e = \frac{1 - 2\nu}{E}(\sigma_x + \sigma_y + \sigma_z) = \frac{1 - 2(0.30)}{207\,000}(403 - 70 - 70) = 0.508 \times 10^{-3}$$

From Eq. (5.9), the unit energy is

$$U_0 = \frac{1}{2E}(\sigma_x{}^2 + \sigma_y{}^2 + \sigma_z{}^2) - \frac{\nu}{E}(\sigma_x\sigma_y + \sigma_y\sigma_z + \sigma_z\sigma_x) + 0$$

$$= \frac{403^2 + (-70)^2 + (-70)^2}{2(207\,000)}$$

$$- \frac{0.30[(403)(-70) + (-70)(-70) + (-70)(403)]}{207\,000}$$

$$= 0.491(\text{N/mm}^2)(\text{mm/mm}) = 0.491 \text{ N} \cdot \text{mm/mm}^3 = 491 \times 10^3 \text{ N} \cdot \text{m/m}^3$$

$$= 491 \times 10^3 \text{ J/m}^3$$

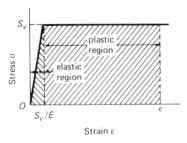

FIGURE 5.4 Idealized stress-strain diagram.

The total energy is

$$U = U_0 V = (491 \times 10^3 \text{ J/m}^3)(0.1448 \times 10^{-3}\text{m}^3) = 71.1 \text{ J} \qquad \qquad ////$$

The word *strength* will refer to the properties of a particular material, will be given the symbol S, and will have the units of lb/in² (psi) or, in SI units, N/mm², which is a MPa. Stress σ has the same units but refers to the condition at a point in a body at a particular time caused by a particular loading. If a stress exceeds a certain strength, this may constitute failure according to some theories of failure. The ratio of this strength to a lower stress is then the *factor of safety n,* and the difference between the strength and stress is the *margin of safety.* Sometimes it is necessary to use n as the ratio of failure *load* to applied load.

Strengths that will be used frequently are the *yield strength* S_y and the *ultimate strength* S_u, both obtained by the tensile test; the *endurance limit* S_e' for *infinite* life, obtained with a standard *polished specimen* in a completely reversed fatigue test, usually the "rotating-beam" test; the lower value, *endurance limit* S_e for the same material in a mechanical part as affected by its environment, surface finish, and size; and the corresponding *fatigue strengths* S_f' and S_f for any *finite* life. When the strengths are different in tension and compression, subscripts will be added, e.g., S_{ut} and S_{uc}. When the strengths are for the material in shear, a subscript s precedes, e.g., S_{sy} and S_{sf}. When there is no pronounced yield point, as with many heat-treated steels, the strength corresponding to a permanent strain of 0.2% is commonly taken as the yield strength.

An idealized stress-strain diagram (Fig. 5.4), shows the strain increasing without further increase of stress once the yield strength is reached. The inclined straight line with slope E defines the elastic region, and the horizontal straight line defines the plastic region. These straight line assumptions simplify the mathematics of elasticity and plasticity. The energy per unit volume that can be stored elastically is the area of the triangle, $S_y^2/2E$, and the additional energy absorbed plastically with a total strain ϵ is the area of the rectangle, $S_y(\epsilon - S_y/E)$.

5.3 COMPONENT STRESSES AND PRINCIPAL STRESSES

The configuration and loading of a stressed body may be such that it is convenient to determine values of normal and shear stresses in an *XYZ* coordinate system, so oriented at each point as to simplify the calculations. However, the directions chosen will not, in general, define the planes on which the maximum normal stresses or maximum shear stresses act. At each point, the particular coordinate system that defines the maximum normal stress is said to have the *principal axes* or directions 1, 2, and 3 (Fig. 5.5). The normal stresses in these directions are the *principal stresses* σ_1, σ_2, and σ_3, respectively. One of these stresses will be the maximum normal stress and another will be the minimum normal stress. Stresses σ_x, σ_y, and σ_z will henceforth be referred to as *component stresses.*

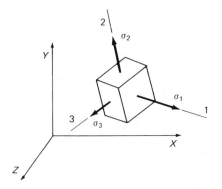

FIGURE 5.5 General orientation of principal axes.

In the general triaxial case, if the values of the six stresses τ_{xy}, τ_{yz}, τ_{zx}, σ_x, σ_y, and σ_z are known, the principal stresses may be found as the three roots of σ in the cubic equation,[2]

$$\sigma^3 - (\sigma_x + \sigma_y + \sigma_z)\sigma^2 + (\sigma_x\sigma_y + \sigma_y\sigma_z + \sigma_z\sigma_x - \tau_{xy}^2 - \tau_{yz}^2 - \tau_{zx}^2)\sigma$$

$$- (\sigma_x\sigma_y\sigma_z + 2\tau_{xy}\tau_{yz}\tau_{zx} - \sigma_x\tau_{yz}^2 - \sigma_y\tau_{zx}^2 - \sigma_z\tau_{xy}^2) = 0 \qquad (5.11)$$

The roots are σ_1, σ_2, and σ_3, representing stresses in directions 1, 2, and 3, respectively, acting normal to planes 2–3, 3–1, and 1–2, respectively. An equation for the shear stress on any plane making an angle with a principal plane shows that the shear stress is zero when the angle is zero, i.e., there is no shear on these principal planes (cf. Fig. 5.5 with Fig. 5.1).

On planes through each of the principal axes 1, 2, and 3, bisecting the angles between the other two principal axes (i.e., at 45° with each), the shear stresses have the following values, respectively,

$$\tau = \frac{1}{2}|\sigma_2 - \sigma_3|, \quad \tau = \frac{1}{2}|\sigma_3 - \sigma_1|, \quad \tau = \frac{1}{2}|\sigma_1 - \sigma_2| \qquad (5.12)$$

To distinguish them from the shear stresses on other planes, they will be called the *major* shear stresses. Absolute-value signs are used to indicate that the sign of the shear stress has no significance here. That equation of Eqs. (5.12) with the maximum numerical difference in principal stresses gives the maximum value of shear stress at the point in question.

In the so-called biaxial case, it is known that no stresses act on two of the faces in the coordinate system chosen. If these are the faces normal to Z, then $\sigma_z = \tau_{zx} = \tau_{zy} = 0$, and Eq. (5.11) may be solved to give the principal stresses

$$\sigma_1 = \frac{\sigma_x + \sigma_y}{2} + \sqrt{\left(\frac{\sigma_x - \sigma_y}{2}\right)^2 + \tau_{xy}^2} \qquad (5.13a)$$

$$\sigma_2 = \frac{\sigma_x + \sigma_y}{2} - \sqrt{\left(\frac{\sigma_x - \sigma_y}{2}\right)^2 + \tau_{xy}^2} \qquad (5.13b)$$

$$\sigma_3 = 0 \qquad (5.13c)$$

---

[2] S. Timoshenko and J. N. Goodier, *Theory of Elasticity*, 3rd ed., New York: McGraw-Hill, pp. 219–226, 1970. A. J. Durelli, E. A. Phillips, and C. H. Tsao, *Introduction to the Theoretical and Experimental Analysis of Stress and Strain*, New York: McGraw-Hill, pp. 17–21, 1958.

Equations (5.12) then become

$$\tau = \frac{1}{2}|\sigma_2|, \quad \tau = \frac{1}{2}|\sigma_1|,$$

$$\tau = \frac{1}{2}|\sigma_1 - \sigma_2| = \sqrt{\left(\frac{\sigma_x - \sigma_y}{2}\right)^2 + \tau_{xy}^2}$$

$$(5.14)$$

It is important to note that if σ_1 and σ_2 have the same sign, i.e., if they are both tensile or both compressive stresses, then one of the first two shear stresses is the maximum shear stress.[3] When there are only two principal stresses, there will be three major shear stresses, unless $\sigma_x = \sigma_y$ and $\tau_{xy} = 0$.

If the X, Y, Z axes are so chosen that shear stresses τ_{xy}, τ_{yz}, and τ_{zx} are zero, i.e., if the planes normal to the X, Y, and Z directions are free of shear, then these are the principal planes, by definition. Substitution of zero for the shear stresses in Eqs. (5.13) gives $\sigma_1 = \sigma_x$ and $\sigma_2 = \sigma_y$; and substitution of zero for the shear stresses in Eq. (5.11) gives an equation with roots $\sigma_1 = \sigma_x$, $\sigma_2 = \sigma_y$, and $\sigma_3 = \sigma_z$, as expected. When a body is shaped and loaded in such a way that some symmetry is apparent, it is usually possible to recognize planes on which no shear can exist, and thus to select the initial coordinate axes as the principal axes. This is possible, for example, at certain points in pressure vessels and in rotating uniform disks.[4]

5.4 GRAPHICAL REPRESENTATION

The relationships of Section 5.3 may be represented by three circles on a chart of τ vs σ by a method suggested by Otto Mohr. Each circle is the locus of points (σ, τ) representing the normal and shear stresses on any plane perpendicular to a plane defined by a pair of principal stresses. The angle between the radii drawn to any two points on the circle is twice the angle between the planes in the stressed body subject to the stresses represented by the points. Thus in Fig. 5.6, the angle between radii CM and CX is 2ϕ, where ϕ is the angle in the stressed body between the planes acted upon by stresses represented by points M and X.[5]

Circles are readily constructed for the biaxial case on a τ vs σ chart with origin O. If the stresses acting on one set of coordinate planes XYZ are known, then those stresses acting on planes normal to directions X and Y, namely, (σ_x, τ_{xy}) and (σ_y, τ_{yx}), the latter the same as $(\sigma_y, -\tau_{xy})$, are plotted at points X and Y, respectively (Fig. 5.6). A straight line drawn through X and Y intersects the σ axis at C, establishing the center for the circle, and the circle is drawn through X and Y. Since the angle between radii CY and CX is $2\phi = 180°$, the angle between the stressed planes is $\phi = 90°$, which, of course, is necessary for points X and Y to represent the stresses on coordinate planes.

From the construction of the circle, it is seen that the length OC equals $(\sigma_x + \sigma_y)/2$ and that radius CX, which is the hypotenuse of a right triangle, equals $\sqrt{[(\sigma_x - \sigma_y)/2]^2 + \tau_{xy}^2}$. The length OC is the first fraction term of Eqs. (5.13a) and (5.13b) and the radius is their radical term. Thus the principal stresses σ_1 and σ_2, the sum and the difference of the two terms, are represented by the points A and B, re-

---

[3] In a first course in strength-of-materials, Eqs. (5.13) and (5.14) are usually derived by sketching a right triangle with an included angle ϕ, the two legs of which are acted upon by σ_x, τ_{xy} and σ_y, τ_{xy}, respectively, and the hypotenuse is acted upon by σ_n and τ_n, respectively. Equations for the "combined stresses" σ_n and τ_n are obtained from the conditions for force equilibrium, then maximized to obtain the two principal stresses σ_1 and σ_2 and a single so-called "maximum" shear stress τ_{max}. With such a derivation, it is easy to forget that the two other shear stresses of Eq. (5.14) should be considered in order to obtain the largest value of the three.

[4] See Section 8.6.

[5] Proof of this may be found in textbooks for a first course in strength of materials.

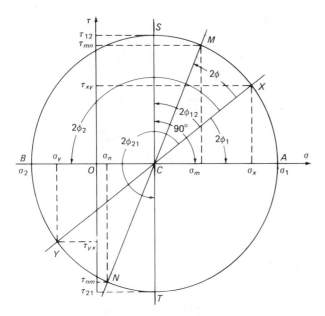

FIGURE 5.6. Construction and reading of a Mohr's circle.

spectively, the intersections of the circle with the σ axis. It is seen that these points represent the maximum and minimum values of the normal stress and zero values of shear stress, the characteristics of the stresses on the principal planes. The angles made by radii CA and CB with the radius CX are designated $2\phi_1$ and $2\phi_2$, respectively, since half their values are the angles ϕ_1 and ϕ_2 between the normals to the principal planes and the X axis in the stressed body.

Radii drawn from C parallel to the τ axis intersect the circle at points S and T, and it is seen that they represent the maximum and minimum values of shear stress, τ_{12} and τ_{21}, respectively, and that $\tau_{21} = -\tau_{12}$. Their values are confirmed by the last of Eqs. (5.14), which is the radius of the circle. In the stressed body the normal to the plane of maximum shear stress is at an angle ϕ_{12} with σ_x and at 45° with principal stress σ_1, one-half the angles $2\phi_{12}$ and 90°, respectively, on the chart. The planes of maximum shear also carry a normal stress, the abscissa of points S and T, of value $OC = (\sigma_x + \sigma_y)/2$. Only in the special case of $OC = 0$, i.e., when $\sigma_2 = -\sigma_1$, are these planes free of normal stress. This case is known as *pure shear*, and then, by Eqs. (5.12), $\tau_{max} = (\frac{1}{2})[\sigma_1 - (-\sigma_1)] = \sigma_1$.

A rapid method that avoids the double angle 2ϕ is to establish a pole P on the circle at its intersection with a vertical line through X (Fig. 5.7). To obtain the proper orientation for the planes on which the stresses act, it is necessary to use the sign convention shown on the element adjacent to P, positive for tensile stresses and for shear stresses that form a *clockwise* couple.[6] Then line PX represents the reference plane, normal to σ_x, and line PY at 90° represents the coordinate plane normal to σ_y. On any plane in the stressed body making an angle ϕ with the reference plane, the stresses are given by the circle at its intersection M with a line or ray drawn through P at an angle ϕ with PX. Proof of the construction is left for the reader.[7] For the plane perpendicular to this one, the stresses

---

[6] If shear stresses are taken positive for a *counterclockwise* couple, the point X must be taken as the pole, and that point on the circle vertically across the circle from a ray's intersection point represents the stresses, e.g., see E. P. Popov, *Introduction to Mechanics of Solids*, Englewood Cliffs, New Jersey: Prentice-Hall, Inc., pp. 297–302, 1968.

[7] See Probs. 5.19, 5.20, and 5.21.

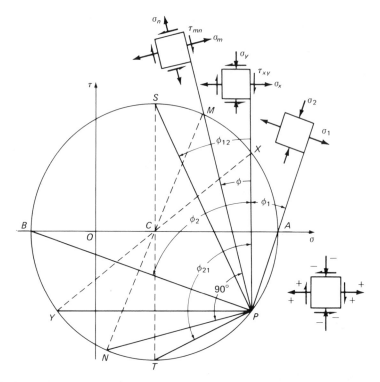

FIGURE 5.7. Alternate method of reading the Mohr's circle of Fig. 5.6.

are given at the intersection N of the circle with a ray PN drawn 90° to PM, or, more accurately, by drawing a diameter from M through the circle center C. Of particular interest are the pair PA and PB that establish the angles ϕ_1 and ϕ_2 made with the reference plane by the planes of the principal stresses, and the pair PS and PT for the angles ϕ_{12} and ϕ_{21} made by the planes of maximum shear stress.

The stresses read from the chart may be conveniently pictured with correct orientation on elements placed on extensions of the rays, as shown along PM, PX, and PA. The stresses that act on the sides parallel to a ray are read at its intersection point, say M, and those on the sides perpendicular to the ray are read at the intersection point of the perpendicular ray, in this case at N.

The drawing of circles for the biaxial case is not finished because the third normal stress σ_3, Eq. (5.13c), although zero, must be represented. The circle of Figs. 5.6 and 5.7 is reproduced in Fig. 5.8, with its center designated C_{12} to indicate that the circle represents stresses in the plane of the principal stresses σ_1 and σ_2. Added to it are the two circles defined on the σ axis by the principal stresses σ_1 and σ_3 and by σ_2 and σ_3 ($\sigma_3 = 0$). Inspection shows that these two circles have the same maximum τ values as given by the first two equations of Eqs. (5.14). Figure 5.9 is a biaxial case where σ_1 and σ_2 have the same sign. Note that the largest shear stress, the stress for use in design, is not determined by σ_1 and σ_2 but rather by σ_1 and σ_3, giving $\tau_{\max} = \sigma_1/2$.

A general case of triaxial stress is shown in Fig. 5.10. The two intersection points of the circle of center C_{12} with a diameter drawn at an angle $2\phi_1$ determine the normal and shear stresses on two planes in the body passing through axis 3 (Fig. 5.5), and making angles ϕ_1 and $\phi_1 + 90°$ with axis 1. Similarly, the intersection points on the circle of center C_{13} represent the stresses on two planes through axis 2 and making angles θ_1 and $\theta_1 + 90°$ with axis 1. The intersection points on the circle of center C_{23} represent the

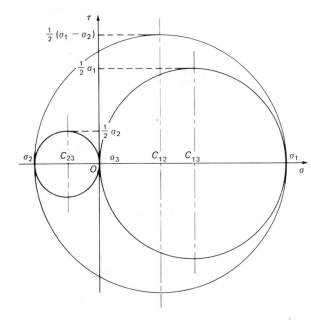

FIGURE 5.8 A complete diagram of circles for the biaxial case of Fig. 5.6. Note $\sigma_3 = 0$, and σ_1 and σ_2 have opposite signs.

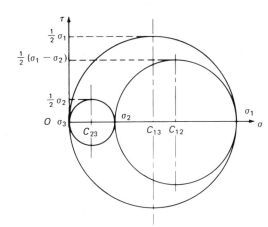

FIGURE 5.9 Mohr's circles for the biaxial case $\sigma_3 = 0$ and σ_1 and σ_2 with the same sign.

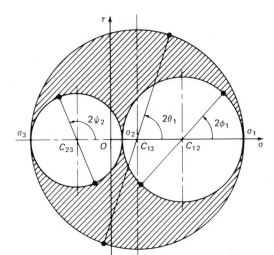

FIGURE 5.10 A general case of triaxial stress.

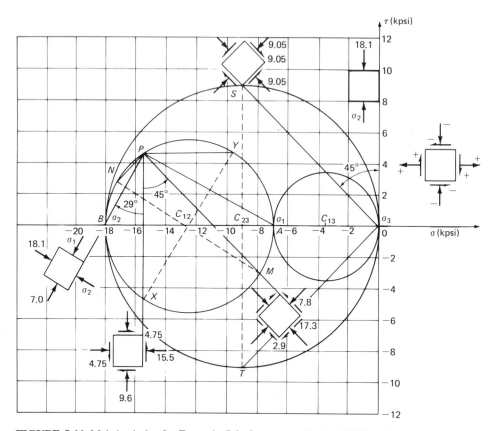

FIGURE 5.11 Mohr's circles for Example 5.2. Stresses are in kpsi (1000 psi).

stresses on two planes through axis 1 and making angles ψ_2 and $\psi_2 + 90°$ with axis 2. On planes that are inclined to all three axes the points that represent the stresses may be determined, and they lie within or on the borders of the shaded area.[8,9]

Example 5.2

At a point in a body the stresses have been calculated from external forces to be $\sigma_x = -15\,500$ psi, $\sigma_y = -9600$ psi, and $\tau_{xy} = -4750$ psi. Determine the magnitude and planes of the principal stresses and the maximum shear stresses. Also determine the stresses on a plane at 45° counterclockwise from the plane normal to X.

Solution

The given stresses are plotted to a scale of one division for 2000 psi (Fig. 5.11). The points X (−15 500, −4750) and Y (−9600, +4750) are first plotted and the center C_{12} is fixed at the intersection of diameter XY and the σ axis. The circle is drawn to determine the principal stresses, designated arbitrarily as σ_1 and σ_2. Since the third principal stress σ_3 is zero, the other two circles are drawn through zero and σ_1 and σ_2, with centers C_{13}

---

[8] O. Hoffman and G. Sachs, *Introduction to the Theory of Plasticity for Engineers*, New York: McGraw-Hill, pp. 14–15, 1953; A. Nadai, *Theory of Flow and Fracture of Solids*, Vol. 1, 2nd ed., New York: McGraw-Hill, pp. 96–99, 1950; F. B. Seely and J. O. Smith, *Advanced Mechanics of Materials*, 2nd ed., New York: John Wiley & Sons, Inc., pp. 60–64, 1952; A. P. Boresi, O. M. Sidebottom, F. B. Seely, and J. O. Smith, *Advanced Mechanics of Materials*, 3rd ed., New York: John Wiley & Sons, Inc., pp. 28–31, 1978.

[9] Mohr's circles may be used to determine strains, moments of inertia, and natural frequencies of vibration. For a general review of Mohr's circles and a bibliography see F. Y. Chen, "Mohr's Circle and Its Applications to Engineering Design," Paper 76-DET-99, ASME, September 1976.

and C_{23} on the σ axis. Pole P is located at the intersection of the first circle and the vertical line through point X. The given stresses are marked on an element along the extension of PX to show their orientation relative to other stresses to follow. The vectors indicate the type of stress, and negative signs are omitted. From P, rays are drawn through A and B to give the orientation of the principal planes, and an element with the stresses at A and B is sketched on an extension of PB. A ray is drawn through P at 45° counterclockwise from PX to give the intersection M, and 90° clockwise from PM to give the intersection N. Their values are read and line PM is chosen for the element on which the stresses are shown.

The values of the maximum shear stresses and the accompanying normal stresses are read from the points marked S and T on the largest circle. The planes on which they act are known, of course, to be at 45° with the principal planes normal to axis 2 or axis 3. However, this angle may be found by construction, taking the pole for the largest circle to be at a point representing a principal stress, in this case σ_3 or origin O, then drawing a vertical reference line through it. A ray is drawn through S, and the angle is measured between it and the reference line. Elements with stresses are shown on both OS and the reference line for comparison. It should be emphasized that these two elements lie in the plane normal to direction 1, whereas the elements previously constructed lie in a plane normal to direction 3, which is also direction Z in this problem.   ////

5.5 MAXIMUM SHEAR-STRESS THEORY OF FAILURE

This theory assumes that failure will occur by yielding when the maximum shear stress from a combination of principal stresses equals or exceeds the value obtained for the shear stress at yielding in the tensile test. In the latter, a uniaxial case, the load divided by the cross-sectional area is the principal stress σ_1, which equals S_y at yielding, while the other principal stresses are $\sigma_2 = \sigma_3 = 0$. Therefore, by Eqs. (5.12), $\tau_{max} = (\frac{1}{2})\sigma_1$ $= (\frac{1}{2})S_y$. Failure should occur by shear on planes bisecting the angles between the axis of the tensile specimen and a plane normal to it. Evidence of this is given by the shape of the break in a cylindrical tensile specimen of ductile metal. The surface of the fracture has an outer conical ring at about 45° to the axis. Also, Lueders' lines or traces of planes of slip occur at 45° in some uniaxial loadings.

The biaxial case must be studied in several parts. If $\sigma_3 = 0$ and σ_1 and σ_2 are both positive with $\sigma_1 > \sigma_2$, then $\tau_{max} = (\frac{1}{2})(\sigma_1 - \sigma_3) = (\frac{1}{2})\sigma_1$, and failure occurs when $(\frac{1}{2})\sigma_1 = (\frac{1}{2})S_y$, i.e., when $\sigma_1 = S_y$. Likewise, when $\sigma_2 > \sigma_1$, failure occurs when $\sigma_2 = S_y$. These two boundary values may be plotted as horizontal and vertical lines in the first quadrant of a σ_1 vs σ_2 diagram (Fig. 5.12). By the same reasoning, if the yield

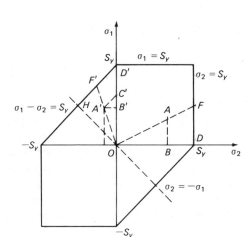

FIGURE 5.12 Maximum shear-stress theory of failure. Boundary of nonfailure stress combinations σ_1 and σ_2 for the biaxial case ($\sigma_3 = 0$).

strength under compressive stress is $-S_y$, and σ_1 and σ_2 are both negative, two similar boundary lines may be drawn in the third quadrant.

If σ_1 is positive and σ_2 is negative, then the maximum shear stress is $\tau_{max} = (\frac{1}{2})(\sigma_1 - \sigma_2)$ which is equated to the shear stress at failure in the tensile test $(\frac{1}{2})S_y$, giving $\sigma_1 - \sigma_2 = S_y$. The line representing this equation is shown in the second quadrant of Fig. 5.12. Note that the sum of the lengths of any pair of dotted lines equals S_y, i.e., $\sigma_1 + |\sigma_2| = S_y$. A similar 45° line in the fourth quadrant completes the diagram. Summarized, the equations for the four quadrants are

$$\text{I} \quad \sigma_1 = S_y, \qquad \sigma_2 = S_y, \qquad \text{II} \quad \sigma_1 - \sigma_2 = S_y$$
$$\text{III} \quad \sigma_1 = -S_y, \quad \sigma_2 = -S_y, \quad \text{IV} \quad \sigma_2 - \sigma_1 = S_y \tag{5.15}$$

Any combination of stresses within the boundary should not cause failure. If the combination of σ_1 and σ_2 occurs at A in the first quadrant, the loads or stresses need to be multiplied in the ratio OF/OA to give their values to cause failure. Thus the factor of safety is $n = OF/OA = OD/OB = S_y/\sigma_2$. In the second quadrant, $n = OF'/OA' = OD'/OC' = OD'/(OB' + B'C') = S_y/(\sigma_1 - \sigma_2)$. This is the same as dividing the shear stress at yielding in the tensile test, $S_y/2$, by the maximum shear stress $(\sigma_1 - \sigma_2)/2$. It is seen that if S_y/n is substituted for S_y in each of Eqs. (5.15), formulas for n in each quadrant are obtained. In quadrants I and III the smaller n is *the* factor of safety.

In the case of pure shear, $\sigma_2 = -\sigma_1$, and all combinations fall on the diagonal line at 45° through the origin of Fig. 5.12. The limiting shear stress occurs at point H and is $\tau = (\sigma_1 - \sigma_2)/2 = \sigma_1 = S_y/2$. Thus the maximum shear-stress theory of failure indicates that the shear strength at yielding, S_{sy}, is

$$S_{sy} = S_y/2 = 0.500 S_y \tag{5.16}$$

It may appear that a maximum normal-stress theory of failure is frequently applied to ductile materials by setting $\sigma_{max} = S_y/n$, the design stress. However, this is done when the parts are uniaxially stressed, as are rods and columns, and beams in which the transverse shear stresses are negligible. There is but one principal stress, and the maximum shear stress is $\tau_{max} = \sigma_{max}/2$. If this shear stress is equated to the shear yield strength divided by n, then $\sigma_{max}/2 = S_y/2n$ and $\sigma_{max} = S_y/n$, the same result. Thus the first procedure is equivalent to applying the maximum shear-stress theory of failure, and it simplifies the calculations a bit.

The general case of triaxial stress is not so readily diagramed, but the maximum shear stress is determined from Eqs. (5.12) or from the outer circle of Mohr's diagram (Fig. 5.10), and equated to $0.500 S_y$ to determine the limiting principal stresses.

5.6 MISES CRITERION OR MAXIMUM ENERGY OF DISTORTION THEORY OF FAILURE

It has been observed that a solid under hydrostatic, external pressure can withstand very large stresses. Under such uniform loading there is a change of volume or uniform shrinkage but no distortion or change of shape. Thus it appears that large amounts of energy can be stored by only a change of volume. When there is also energy of distortion or shear to be stored, as in the tensile test, the stresses that may be imposed are limited. In most cases some of the energy of strain is due to volume change and some is due to distortion. With the foregoing as a basis, it is proposed that failure by yielding under a combination of stresses occurs when the energy of distortion equals or exceeds the energy of distortion in the tensile test when the yield strength is reached. Test results for ductile metals more closely fit this theory than the maximum shear-stress theory. The theory is now in common use for the design of critical parts and parts where weight must be conserved.

Equation (5.9) gives the total strain energy per unit of volume. The unit distortion energy is obtained by subtracting from the total that part of it due only to volume change. Let each principal stress consist of two parts, a stress σ_v that gives only volume change, and the remainder σ', thus

$$\sigma_1 = \sigma_v + \sigma_1'$$
$$\sigma_2 = \sigma_v + \sigma_2' \tag{a}$$
$$\sigma_3 = \sigma_v + \sigma_3'$$

To cause no distortion, stresses σ_v must be the same on all faces, as in hydrostatic loading. To cause no volume change, stresses σ' must sum to zero, by Eq. (5.5). By the addition of Eqs. (a), $\sigma_1 + \sigma_2 + \sigma_3 = 3\sigma_v + (\sigma_1' + \sigma_2' + \sigma_3') = 3\,\sigma_v$, whence

$$\sigma_v = \frac{\sigma_1 + \sigma_2 + \sigma_3}{3} \tag{b}$$

By analogy to Eqs. (5.1), the corresponding strain is

$$\epsilon_v = \frac{1}{E}\left[\sigma_v - \nu(\sigma_v + \sigma_v)\right] = \frac{1 - 2\nu}{E}\,\sigma_v \tag{c}$$

The energy for volume change only results from straining an amount ϵ_v under stress σ_v in each of the three principal directions, or the energy per unit volume (Fig. 5.2 and Eq. (c)) is

$$U_v = 3\left(\frac{\sigma_v \epsilon_v}{2}\right) = \frac{3(1 - 2\nu)}{2E}\,\sigma_v{}^2 \tag{d}$$

Substitution from Eq. (b) gives the unit volume-change energy U_v in terms of the three principal stresses,

$$U_v = \frac{1 - 2\nu}{6E}(\sigma_1 + \sigma_2 + \sigma_3)^2 \tag{5.17}$$

Equation (5.9) for total energy may be rewritten in terms of the principal stresses by recalling that when the planes defined by the principal axes 1, 2, and 3 coincide with those defined by the component axes X, Y, and Z, shear stresses do not exist on the sides of the element. Thus

$$U_0 = \frac{1}{2E}(\sigma_1{}^2 + \sigma_2{}^2 + \sigma_3{}^2) - \frac{\nu}{E}(\sigma_1\sigma_2 + \sigma_2\sigma_3 + \sigma_3\sigma_1) \tag{5.18}$$

Expansion of Eq. (5.17), its substraction from Eq. (5.18), and rearrangement of terms gives the unit distortion energy U_d in the form

$$U_d = U_0 - U_v = \frac{1 + \nu}{6E}\left[(\sigma_1 - \sigma_2)^2 + (\sigma_2 - \sigma_3)^2 + (\sigma_3 - \sigma_1)^2\right] \tag{5.19}$$

In the tensile test $\sigma_2 = \sigma_3 = 0$ and at yielding $\sigma_1 = S_y$. Substitution of these values into Eq. (5.19) gives the tensile-test distortion energy at yielding,

$$(U_d)_y = \frac{1 + \nu}{6E}\left[S_y{}^2 + S_y{}^2\right] = \frac{1 + \nu}{6E}\left[2S_y{}^2\right] \tag{5.20}$$

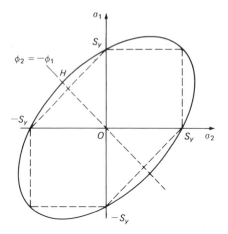

FIGURE 5.13 Mises criterion or maximum energy of distortion theory of failure. Elliptical boundary of nonfailure stress combinations σ_1 and σ_2 shown superimposed upon the boundary for the maximum shear-stress theory (dotted lines).

By the energy of distortion theory, yielding failure in the general case occurs when its distortion energy equals that at yielding in the tensile test. Hence by equating Eq. (5.19) to Eq. (5.20) a failure equation is obtained, namely,

$$(\sigma_1 - \sigma_2)^2 + (\sigma_2 - \sigma_3)^2 + (\sigma_3 - \sigma_1)^2 = 2S_y^2 \qquad (5.21)$$

Divided through by 4, one side of the equation is seen to consist of the sum of the squares of the three major shear stresses of Eqs. (5.12),

$$\left(\frac{\sigma_1 - \sigma_2}{2}\right)^2 + \left(\frac{\sigma_2 - \sigma_3}{2}\right)^2 + \left(\frac{\sigma_3 - \sigma_1}{2}\right)^2 = \frac{S_y^2}{2} \qquad (5.22)$$

The criterion for failure thus appears to be a function of shear stresses only, as might be expected from a theory based on distortion.

When the sum of the squares of the major shear stresses is less than $S_y^2/2$, there is a factor of safety n. By substitution of $(S_y/n)^2$ for S_y^2 in Eq. (5.21)

$$n = \frac{S_y}{\sqrt{[(\sigma_1 - \sigma_2)^2 + (\sigma_2 - \sigma_3)^2 + (\sigma_3 - \sigma_1)^2]/2}} = \frac{S_y}{s} \qquad (5.23)$$

where the denominator s may be considered an equivalent stress or *intensity of stress*, s.

For a biaxial stress condition, with $\sigma_3 = 0$, Eqs. (5.21) and (5.23) simplify to,

$$\sigma_1^2 - \sigma_1\sigma_2 + \sigma_2^2 = S_y^2 \qquad (5.24)$$

and

$$n = \frac{S_y}{\sqrt{\sigma_1^2 - \sigma_1\sigma_2 + \sigma_2^2}} \qquad (5.25)$$

where the intensity of stress is now $s = \sqrt{\sigma_1^2 - \sigma_1\sigma_2 + \sigma_2^2}$. Equation (5.24) is the equation of an ellipse on a σ_1 vs σ_2 chart (Fig. 5.13), having in common six points of the maximum shear hexagon shown in dotted lines, but enclosing a larger area. Thus, in general, the distortion energy theory is the less conservative one, and its use in design will give slightly smaller dimensions for most stress combinations. It has an advantage in being represented by one continuous equation, unlike the maximum shear theory where the discontinuities seem physically unrealistic.

In the case of pure shear, $\sigma_2 = -\sigma_1$, and substitution into Eq. (5.24) gives $\sigma_1^2 + \sigma_1^2 + \sigma_1^2 = 3\sigma_1^2 = S_y^2$, whence $\sigma_1 = S_y/\sqrt{3}$. The limiting shear stress, point H of Fig.

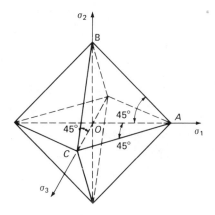

FIGURE 5.14 Octahedral shear planes leading to the Mises criterion of failure.

5.13, is $\tau = (\sigma_1 - \sigma_2)/2 = \sigma_1 = S_y/\sqrt{3}$. Thus the Mises criterion indicates that the shear strength at yielding, S_{sy}, is

$$S_{sy} = S_y/\sqrt{3} = 0.577 S_y \tag{5.26}$$

This indicated strength compares with $0.500 S_y$ by the maximum shear-stress theory, Eq. (5.16). Tests in shear show $0.577 S_y$ to be the more realistic value for yielding in shear.

The foregoing derivation based on distortion energy was first presented by H. Hencky in 1924. However, R. von Mises in 1912 had proposed Eq. (5.21) as a means of giving a continuous function to replace the several equations of the maximum shear-stress theory. The equation can also be derived from other physical principles. A. Eichinger in 1926 and A. Nadai in 1937 obtained this equation by considering the shear stress on a plane of an octahedral element within the body (Fig. 5.14), as the limiting condition for yielding failure. The traces of the plane make 45° angles with each of the principal planes.[10] G. Sachs in 1928 and H. L. Cox and D. G. Sopwith in 1937 showed the critical shear stress to be $0.577 S_y$ by considering the statistical behavior of a randomly oriented aggregate of crystals. J. Marin obtained Eq. (5.21) from Eq. (5.11) by considering that the quantities within the first two pairs of parentheses in the cubic equation must be invariant, since the values of the principal stresses must be independent of the directions X, Y, and Z.

Derivations of these and other theories, discussion, and references to original works may be found elsewhere.[11] Because of the several bases for Eq. (5.21), many engineers believe that the theory of failure that the equation represents should not be called the maximum distortion-energy theory. In the past it has sometimes been called the Mises–Hencky theory. The brief name, Mises criterion, after the man who first proposed the equation, has been increasingly used, and it will be used in this book.

Example 5.3

A thin steel spherical tank filled with a liquid of unit weight γ is supported on a ring (Fig. 5.15). The tangential stress σ_t and the meridional stress σ_m, shown on an element of the sphere, at 150° from the vertical axis, have been calculated by the equations of Prob. 8.30 to be $3.56\,\gamma R^2/t$ and $-1.70\,\gamma R^2/t$, respectively, where R is the radius of the

---

[10] See Probs. 5.38 and 5.39.

[11] J. Marin, *Mechanical Behavior of Engineering Materials*, Englewood Cliffs, New Jersey: Prentice-Hall, Inc., pp. 112–130, 1962. R. E. Peterson, *Stress Concentration Factors*, New York: John Wiley & Sons, Inc., pp. 7–9, 1974. A. Nadai, *Theory of Flow and Fracture of Solids*, Vol. 1, 2nd ed. New York: McGraw-Hill, pp. 96–99, 1950.

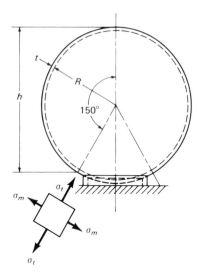

FIGURE 5.15 Spherical tank of Examples 5.3 and 5.4.

sphere, t is its wall thickness, and $t << R$. Determine formulas for the factor of safety by the theories of failure and compare.

Solution

Because of the symmetry of tank shape, loading, and support, the given normal stresses must be the same on all elements around a circle at any one depth h. Thus there cannot by any shear between these elements, and, without torque about the vertical axis, no shear on their top and bottom faces. This identifies the normal stresses as principal stresses, say σ_1 and σ_2, in the tangential and meridional directions, respectively. The third principle stress, if any, must have a radial direction, perpendicular to the plane of σ_1 and σ_2. On the inside surface the radial stress equals the pressure of the liquid γh and on the outside surface the radial stress is zero. A comparison of γh with the given stresses shows that it will be negligible when $t << R$. Hence we take $\sigma_3 = 0$. The material is ductile.

By the maximum shear theory, Eqs. (5.15),

$$n = \frac{S_y}{\sigma_1 - \sigma_2} = \frac{S_y}{(3.56 + 1.70)\gamma R^2/t} = 0.190 \frac{S_y t}{\gamma R^2}$$

By the Mises criterion, Eq. (5.25),

$$n = \frac{S_y}{\sqrt{\sigma_1^2 - \sigma_1 \sigma_2 + \sigma_2^2}} = \frac{S_y}{\sqrt{3.56^2 - (3.56)(-1.70) + (-1.70)^2}\, \gamma R^2/t} = 0.215 \frac{S_y t}{\gamma R^2}$$

By indicating a larger factor of safety, the Mises criterion is the less conservative, as is expected from Fig. 5.13. If only the maximum normal stress were considered, the factor would be $n = S_y/\sigma_1 = 0.281 S_y t/\gamma R^2$, a high and misleading factor.   ////

5.7 NORMAL-STRESS FAILURE THEORIES: THE MOHR THEORY FOR BRITTLE MATERIALS

The theories are principally applied to brittle materials, using ultimate strengths as bases. By one normal-stress theory of failure, the principal stresses, if there are more than one, are considered separately, and, of these, only the tensile stress and the compressive stress of the largest magnitudes are considered. Principal stresses of smaller magnitude are

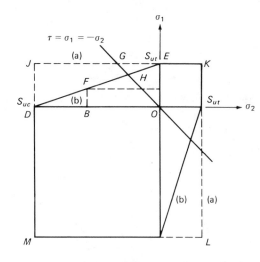

FIGURE 5.16 Theories of failure for brittle materials in the biaxial case ($\sigma_3 = 0$). (a) Normal-stress criterion. (b) Mohr theory.

assumed to have no effect on failure. Then, the equations for the envelope of safe combinations of stresses are

$$(\sigma_1, \sigma_2, \sigma_3)_{\text{tensile}} = S_{ut}, \quad (\sigma_1, \sigma_2, \sigma_3)_{\text{compressive}} = S_{uc} \tag{5.27}$$

Considered biaxially, the envelope forms the square *JKLM* shown in Fig. 5.16. It shows the condition that is most common for brittle materials, that with $|S_{uc}| > S_{ut}$, and the theory is the *normal-stress criterion*. A special case, when $|S_{uc}| = S_{ut}$, is known as the *maximum normal-stress theory*.

The dotted-line envelope in the second and fourth quadrants seems unrealistic, since in the case of pure shear, on the line $\tau = \sigma_1 = -\sigma_2$ at point *G*, the ultimate shear strength for fracture is $S_{su} = S_{ut}$, a value larger than that found in torsion tests. Otto Mohr proposed a theory (now called the *Mohr theory*) that gives smaller values, with an envelope represented by the solid lines of Fig. 5.16. The equation of the straight line in the second quadrant, considering that the numerical value of σ_2 is negative, is found by proportional triangles, $BF/BD = OE/OD$ or $\sigma_1/(|S_{uc}| + \sigma_2) = S_{ut}/|S_{uc}|$, whence

$$\frac{\sigma_1}{S_{ut}} - \frac{\sigma_2}{|S_{uc}|} = 1 \tag{5.28}$$

From the substitution of S_{ut}/n for S_{ut} and of S_{uc}/n for S_{uc} in Eq. (5.28), the factor of safety may be written[12]

$$n = \frac{S_{ut}}{\sigma_1 - \sigma_2 |S_{ut}/S_{uc}|} \tag{5.29}$$

The intersection point *H* of Mohr's boundary line and the line of pure shear represents the ultimate shear strength S_{su}. At point *H*, $\sigma_1 = S_{su}$ and $\sigma_2 = -\sigma_1 = -S_{su}$. Substitution for σ_1 and σ_2 in Eq. (5.28) gives

$$S_{su} = \frac{S_{ut}}{1 + |S_{ut}/S_{uc}|} \tag{5.30}$$

If $|S_{uc}| = S_{ut}$, $S_{su} = 0.5S_{ut}$. If $|S_{uc}| = 4S_{ut}$, typical of ordinary gray cast iron, $S_{su} = 0.80S_{ut}$. These values are more realistic than those given by the normal-stress criterion and Eqs. (5.27).

---

[12] See Section 6.2 and Eqs. (6.6)−(6.9) for the effect of stress raisers.

Example 5.4

The same information applies here as was given in Example 5.3 except that the material is gray cast iron with $|S_{uc}| = 4S_{ut}$.

Solution

This case falls in the second quadrant of Fig. 5.16 for brittle materials. By the normal-stress criterion,

$$n = \frac{S_{ut}}{\sigma_1} = \frac{S_{ut}}{3.56\gamma R^2/t} = 0.281\frac{S_{ut}t}{\gamma R^2}$$

which is smaller than $n = S_{uc}/\sigma_2$ and therefore is the factor of safety. By the Mohr theory, Eq. (5.29),

$$n = \frac{S_{ut}}{\sigma_1 - \sigma_2|S_{ut}/S_{uc}|} = \frac{S_{ut}}{[3.56 - (-1.70)(1/4)]\gamma R^2/t} = 0.251\frac{S_{ut}\,t}{\gamma R^2}$$

The Mohr theory indicates a lower factor of safety and is, therefore, more conservative when $|S_{uc}| > S_{ut}$.   ////

5.8 ELASTIC–PLASTIC CONDITIONS: LIMIT DESIGN

After the yield strength S_y is reached in a tensile-test specimen or in any uniformly stressed bar of ductile material, no increases or only small increases in load are required to overcome strain-hardening effects and to continue the elongation in relatively large amounts. The material is considered to be in a plastic state. The yielding is idealized for purposes of analysis by assuming a constant stress S_y in the plastic region of the stress-strain diagram (Fig. 5.4).

In contrast, consider ductile parts in which stress is nonuniformly distributed elastically, such as in beams and thick-walled pressure vessels. As the loading is increased, the stress to cause yielding is first reached at only the surface and location most highly stressed. With further loading, the strain continuously increases here without further increase in stress, while at adjacent places the strain continuously increases, and the stress progressively reaches and remains at the yield value S_y. The plastic flow is said to be contained as long as additional loading is required to continue straining.

A *limit* to loading is reached when continuous yielding occurs without further increase in loading, leading to ultimate collapse. Use of this limit as a basis for determining the size or safety of parts is called *limit design*. The measure of this is the *limit-design factor* L, which is the ratio of the loading for a completely plastic condition to the loading for a completely elastic condition with $\sigma_{max} = S_y$. The word "loading" has been used in a general way, for it may consist of a moment in a beam, pressure in a vessel, or rotational velocity in a disk. Expressed by a general equation, then, the limit-design factor is

$$L = \frac{\Lambda_p}{\Lambda_e} \tag{5.31}$$

where Λ_p is the loading corresponding to the completely plastic condition and Λ_e is the loading corresponding to a completely elastic condition. Values between 1.1 and 2.0 are common. A reduced, more conservative factor is based on limiting the plastic condition to some fraction of section thickness. A corresponding, elastic–plastic loading Λ_{ep} should then replace Λ_p in Eq. (5.31).

The determination of factors L is illustrated herewith by general equations for a beam

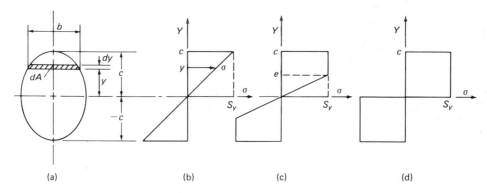

FIGURE 5.17 A symmetrical section under a bending moment. (a) Elemental stressed area. (b) Elastic stress distribution at initial yield. (c) Partially plastic distribution. (d) Completely plastic distribution.

under bending-moment loading, and in Example 5.5, for a cylindrical bar of circular section under bending-moment and under torsional-moment loadings. In Section 8.12 factors are determined for thin rotating disks of uniform thickness and for thick-wall vessels under internal pressure.

In a beam with a symmetrical section (Fig. 5.17 (a)), the internal moment that resists bending is the integral of the product of an elemental force $\sigma\, dA$ by its distance y from the neutral axis, integrated over the area of the section, i.e., $M = \int_A (\sigma\, dA)y$. In the completely elastic condition of Fig. 5.17(b), $\sigma_{\max} = S_y$ and $\sigma = (y/c)S_y$. Hence for a symmetrical section,

$$M_e = \int_A (\sigma\, dA)y = \frac{2S_y}{c} \int_0^c y^2\, dA = \frac{S_y I}{c} \qquad (5.32)$$

where $I = 2\int_0^c y^2\, dA$, the familiar moment of inertia of area. If this is unknown, $b\, dy$ may be substituted for dA and the integration carried out. Figure 5.17(c) shows the stress distribution for a partially elastic, partially plastic condition, with the division at $y = e$. For it the moment M_{ep} may be obtained by separate integrations over each part. Figure 5.17(d) shows the completely plastic condition, the stress is $\sigma = S_y$, and with this and $b\, dy$ substituted for dA in the general moment equation,

$$M_p = \int_A (\sigma\, dA)y = 2S_y \int_0^c by\, dy \qquad (5.33)$$

In beams of symmetrical cross section, limit-design factors are a function only of the shape of the section.

Example 5.5

Determine limit-design factors for a bar of circular cross section: (a) under bending moment and (b) under torsional moment.

Solution

(a) For a circular area of radius R, $I = \pi R^4/4$ and $c = R$, so by Eq. (5.32), $M_e = \pi S_y R^3/4$. From Fig. 5.18(a), width $b = 2\sqrt{R^2 - y^2}$, and by Eq. (5.33),

$$M_p = 4S_y \int_0^R (R^2 - y^2)^{1/2} y\, dy = -\frac{4}{3} S_y \left[(R^2 - y^2)^{3/2} \right]_0^R = \frac{4S_y R^3}{3}$$

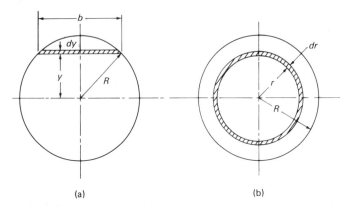

FIGURE 5.18 Circular section of Example 5.5. (a) Elemental stressed area under bending moment. (b) Elemental stressed area under torsional moment.

The limit-design factor under bending moment, by Eq. (5.31), is

$$L_b = \frac{M_p}{M_e} = \frac{4S_y R^3/3}{\pi S_y R^3/4} = \frac{16}{3\pi} = 1.70$$

(b) The torsional moment under elastic conditions is $M_t = \tau J/c$, where the polar moment of inertia J is $\pi R^4/2$. With maximum shear stress τ equal to the yield strength in shear S_{sy}, the elastic moment is $(M_t)_e = \pi S_{sy} R^3/2$. A general expression for the internal moment, with $dA = 2\pi r\, dr$ from Fig. 5.18(b), is

$$M_t = \int_0^R (\tau\, dA)r = 2\pi \int_0^R \tau r^2\, dr \tag{5.34}$$

For a completely plastic condition, τ is constant and equal to S_{sy}. Hence from Eq. (5.34),

$$(M_t)_p = 2\pi S_{sy} \int_0^R r^2\, dr = \frac{2\pi S_{sy} R^3}{3}$$

The limit-design factor under torsional moment, by Eq. (5.31), is

$$L_t = \frac{(M_t)_p}{(M_t)_e} = \frac{2\pi S_{sy} R^3/3}{\pi S_{sy} R^3/2} = \frac{4}{3} = 1.33 \qquad\qquad ////$$

For beams of rectangular cross section the limit-design factor L_b in bending is 1.50 for full yielding and 1.22 for partial yielding to one-quarter depth ($e/c = 0.75$, Fig. 5.17(c)).

Along a beam, full yielding develops at the plane of maximum bending moment, with partial yielding occurring in adjacent planes (Fig. 5.19(a)). A plastic *hinge* is said to have formed, and this simply supported beam collapses, bending or rotating at this hinge against a constant resistance moment M_p, without further increase in the applied load F_p. In continuous and other statically indeterminate beams and frames, a hinge first forms in the plane of maximum bending moment, but collapse does not occur until a second or third hinge has formed. In Fig. 5.19(b), a beam and its moment diagram are shown for the limiting elastic condition where $\sigma_{max} = S_y$, the loading is $q = q_e$, and the maximum moment just equals the maximum completely elastic moment M_e. As q is increased, a plastic region develops at the supports, which are the planes of maximum moment, until hinges are formed. The beam does not collapse because it is supported by shear forces at its ends. It has become a statically determinate beam, the end moments

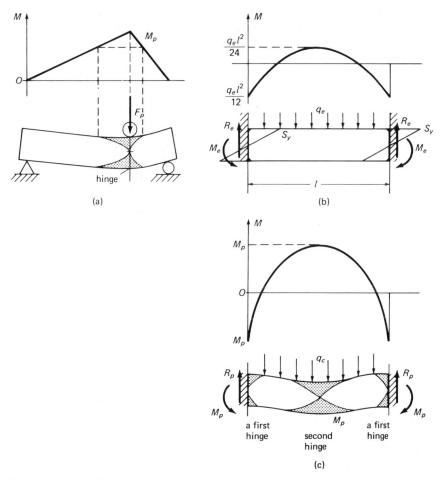

FIGURE 5.19 Collapse in beams. Shaded areas locate the plastic hinges. (a) Simply supported beam with one hinge at collapse. (b) Beam, built-in at supports, at maximum uniform load q_e for completely elastic stresses. (c) Same beam at collapse load q_c, showing three plastic hinges.

having the known, constant value M_p. Due to symmetry, a peak moment exists at mid-length, and as q is further increased, a plastic region develops there (and may have already started), until a hinge with moment of value M_p is formed and collapse occurs (Fig. 5.19(c)). Note the changes in the moment diagram.

The loading $q = q_c$ at collapse (Fig. 5.19(c)) is found from the equation of equilibrium of the segment of the beam to the right or left of the plane of the middle hinge. The ratio q_c/q_e is a measure of the overload beyond the completely elastic condition that may occur before failure by collapse. The load capacity and the ratio benefit not only from the redistribution of stresses across the sections, as measured by the limit-design factor L_b, but also from the favorable redistribution of moments along the beam, by which the peaks of the moment diagram are equalized. Since most structural beams are of ductile steel and are statically loaded, limit analysis and limit design are particularly applicable, and they have been extensively studied.[13]

With limit design, it is logical to substitute LS_y for S_y in equations for factor of safety n. Thus if only a bending stress is present, $n = L_b S_y/\sigma$. However, for multiple-stress

---

[13] P. G. Hodge, *Plastic Analysis of Structures*, New York: McGraw-Hill, 1959.

conditions, it may be necessary to *divide* each stress by the corresponding factor. Thus with reference to Example 5.5, if a bending stress $\sigma_x = \sigma$ and a shear stress $\tau_{xy} = \tau$ exist simultaneously at a point, the factor of safety obtained from Eqs. (5.14), based on the shear theory of failure is

$$n = \frac{S_{sy}}{\tau_{max}} = \frac{S_y/2}{\sqrt{(\sigma/2L_b)^2 + (\tau/L_t)^2}} = \frac{S_y}{\sqrt{(\sigma/L_b)^2 + 4(\tau/L_t)^2}} \tag{5.35}$$

The factors of safety used in limit design may be sufficiently large so that all the stresses remain elastic under normal operation. Unless an unexpected overload occurs, there may be no plastic flow and permanent set. Then, the limit-design factor essentially gives a reduction in what may be an excessively high factor of safety. There is recognition with limit design that under static loading there exists against complete failure a margin of safety above the loading that initiates yielding. The margin of safety is reduced if the material's stress-strain curve beyond the elastic range departs very much from the horizontal line of the idealized diagrams. Then, a reduced design factor may be applicable.[14] Also, the product LS_y should be less than the tensile strength. Factors are not applicable to hardened, higher-carbon steels. Nor is limit design suitable for closely fitting parts or for shafts where deflections must be limited and a permanent set is not allowable. Finally, limit design is not applicable when the stresses are alternating, a condition for fatigue failure. This includes most shafts.

5.9 FATIGUE STRENGTH

A typical rotating machine component, such as a shaft or gear, is subjected to varying or fluctuating stresses, a condition that may lead to fatigue failure. The fatigue properties of any one material are usually determined from laboratory specimens with highly polished surfaces and a diameter of 0.30 in. The specimen is loaded by a bending moment that is constant in amount and direction, and the specimen is rotated at high speed, thus inducing stresses that completely reverse every revolution from tension to compression and return.[15] After a period of time, if the stress is sufficiently high, a crack or cracks will start at the surface and, over a second period of time, progress inwardly, perhaps halfway or further through the section before a sudden complete fracture occurs through the remainder of the section. The entire fracture is "brittle," with no evidence of yielding.

The stresses corresponding to a low number of cycles to failure, such as a thousand, may not be much lower than the ultimate tensile strength. With tests initiated at somewhat lower stresses, the life is increased until with some materials such as steels, there is found a stress below which life is infinite or unlimited (Fig. 5.20). The stress corresponding to failure at some finite or limited life is the *fatigue strength* S_f', and that value which just allows infinite life is the *endurance limit* or *endurance strength* S_e'. Plots of fatigue strength vs number of cycles to failure are known as *S–N* curves. Plotting on a log-log chart allows a straight line to approximate the experimental results, and it emphasizes the "knee" at the endurance limit. For steels, the knee falls at about 10^6 cycles.

The fatigue test is a long and expensive one and the complete *S–N* curve, and even the endurance limit, may not be included with the other physical properties that are reported for the many available steel compositions and treatments. Fortunately, for most

---

[14] J. A. Van Den Broek, *Theory of Limit Design*, New York: John Wiley & Sons, Inc., 1948.

[15] Some axial-load fatigue tests are made under a fluctuating constant amplitude of strain, e.g., for bolts above the yield strain in low-cycle fatigue, where strain seems to be the most significant variable.

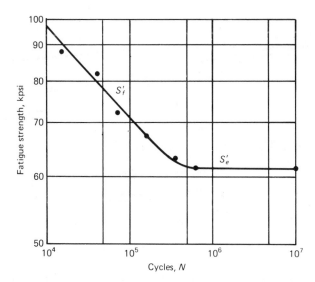

FIGURE 5.20 Typical chart of fatigue strength S_f' vs cycles of life N for a quenched and tempered alloy steel.

wrought steels the endurance limit determined by rotating-beam tests is between 0.45 and 0.60 of the tensile strength S_u, as indicated by Fig. 5.21, and $0.50S_u$ is commonly used for design purposes. However, above $S_u = 160\,000$ psi the value $0.50S_u$ should be used only if the microstructure is a tempered martensite or other stress-relieved micro-structure. Above $S_u = 200\,000$ psi, S_e' should be taken equal to 100 000 psi, for this is its maximum value in most steels. In summary, when it is necessary to make a conservative estimate of the mean endurance limit in bending for *wrought* steel, use

$$S_e' = 0.50S_u \qquad \text{when} \quad S_u \leqslant 200\,000 \text{ psi}$$
$$S_e' = 100\,000 \text{ psi} \quad \text{when} \quad S_u \geqslant 200\,000 \text{ psi}$$

(5.36)

with the foregoing limitation on microstructure when $S_u > 160\,000$ psi.

FIGURE 5.21 Fatigue strength vs life, unnotched polished specimens, wrought steel, reversed bending loads, plotted in ratio form, S_f'/S_u. (From C. Lipson and R. C. Juvinall, *Handbook of Stress and Strength,* New York: The MacMillan Company, 1963. Used with permission of The MacMillan Publishing Co., Inc., New York.)

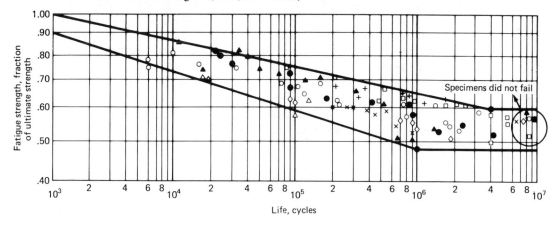

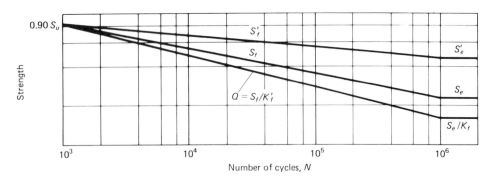

FIGURE 5.22 A method of estimating fatigue strengths S_f' (Section 5.9) and S_f (Section 5.10), and net fatigue strength $Q = S_f/K_f'$ (Section 5.11 and Example 6.1 (c)).

If only the hardness of a steel is available, then it may be converted by tables to an equivalent Brinell hardness number (Bhn), if necessary, and the tensile strength S_u estimated as 500 × Bhn. A conservative estimate for cast iron and cast steel is $S_e' \approx 0.40S_u$. Most nonferrous metals do not have an endurance limit. For aluminum alloys the fatigue strengths have been estimated to be between $0.30S_u$ and $0.40S_u$ at a commonly reported 5×10^8 cycles. The soft, wrought aluminums have the higher value. The hard, heat-treated aluminums seem to peak at a fatigue strength of about 20 000 psi.

To obtain a basis for estimating fatigue *strengths*, S_f', of steels, Lipson and Juvinall[16] took published test results for several different wrought steels, and plotted them in the form of ratios S_f'/S_u vs cycles of life N, both on a log scale (Fig. 5.21). It is seen that a reasonably safe estimate for wrought steels in bending may be a straight line drawn on a log-log chart between the points $(10^3, 0.90S_u)$ and $(10^6, S_e')$, as drawn in Fig. 5.22, where, if necessary, S_e' is estimated as $0.50S_u$. Fatigue strengths obtained in tests with reversed *axial* loading are lower than those obtained by bending. For an estimate, the S_f line may be drawn between the points $(10^3, 0.75S_u)$ and $(10^6, 0.45S_u)$.

Shear fatigue strengths obtained by alternating shear or torsion of steel specimens are seldom reported. However, studies have been made of the relationship between the torsion fatigue endurance limit S_{se}' and the bending fatigue or endurance limit S_e', as shown in Fig. 5.23. The equation of the dotted line is $S_{se}' = 0.500S_e'$, corresponding to the maximum shear theory of failure for ductile materials and Eq. (5.16). In this equation, the shear-stress and normal-stress endurance limits are substituted for the shear-stress and tensile-test yield strengths, respectively. Similarly, and corresponding to Eq. (5.26), the equation $S_{se}' = 0.577S_e'$ is written and the solid line is drawn on Fig. 5.23 to correspond to the Mises criterion of failure. The latter appears to be a good average for the experimental points, whereas the former is a safe lower boundary. Both are used to estimate the shear or torsional endurance limits of steels.

Unless interrupted by irregularities, the fatigue crack under bending is normal to the tensile stresses, i.e., perpendicular to the axis of a shaft. In torsional-shear fatigue failures the crack is again normal to the principal tensile stresses, which are at 45° angles with the shear stresses (pure shear), hence the crack is at a 45° angle to the axis of the shaft or spring wire.

---

[16] C. Lipson and R. C. Juvinall, *Handbook of Stress and Strength*, New York: The MacMillan Company, 1963.

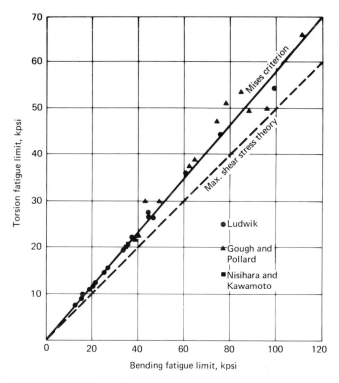

FIGURE 5.23 Comparison of torsion and bending fatigue limits for ductile materials. (Reprinted from R. E. Peterson, *Stress Concentration Factors,* New York: John Wiley & Sons, Inc., 1974, with permission of the publisher.)

5.10 MODIFICATION OF FATIGUE STRENGTH FOR DESIGN

Because of differences in surface, size, and operating environment between engineering components and highly polished, small-diameter laboratory specimens, fatigue strengths are generally lower in the engineering components. Although full-size fatigue tests with the surface and environment conditions of usage would be best, the engineer, at least in initial design, must usually estimate the combined effects. He may also wish to allow for scatter or reliability of the results reported for the laboratory specimens. One way of estimating a modified *endurance limit* S_e from the specimen value $S_e{'}$ is by the equation[17]

$$S_e = C_s C_z C_t C_r C_m S_e{'} \tag{5.37}$$

where C_s, C_z, C_t, and C_r are factors for surface, size, temperature, and reliability, respectively, and C_m is for miscellaneous effects such as plating, case hardening, internal defects, directional strength, and residual stresses. Residual stresses, which can be beneficial or harmful, are discussed in detail in Sections 7.8 through 7.16. Factors C_s, C_z, and C_r are given and discussed further in this section. Temperature effects will vary with the material, and in most cases, values of $S_e{'}$ should be obtained by tests at the particular temperature rather than by a C_t factor.

---

[17] The form of this equation is thought to be due to J. Marin, "Design for Fatigue Loading—Part 3." *Machine Design,* Vol. 29, No. 4, p. 127, February 21, 1957.

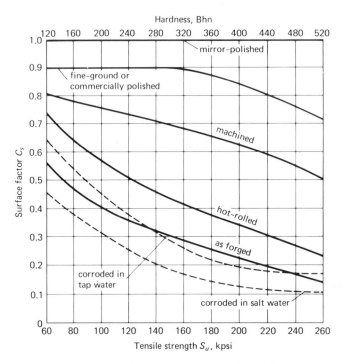

FIGURE 5.24 Surface factor C_s for modification of the endurance strength of steels according to surface finish. (From R. C. Juvinall, *Engineering Considerations of Stress, Strain, and Strength*, New York: McGraw-Hill, 1967. Used with permission of McGraw-Hill Book Company.)

An equation similar to Eq. (5.37) is not written for fatigue strengths S_f, which occur at a lower number of cycles. There are no appreciable surface and size effects under static bending loads and very little between 10^2 and 10^3 cycles; also, there is little scatter effect in the results of static tests. Thus C_s, C_z, and C_r are nearly 1.0 in value. The several effects increase progressively in some manner with an increase in the number of cycles to failure. The combined effect has been estimated for steels by a straight line on a log S–log N chart, drawn between the points $(10^3, 0.90S_u)$ and $(10^6, S_e)$ (Fig. 5.22). The line gives an estimate of S_f at any number of cycles N between 10^3 and 10^6.

Surface effect. Fatigue strength is sensitive to the condition of the surface because the maximum stresses occur here in bending and torsion and because surface irregularities, such as tool marks and scratches perpendicular to the stress direction, are stress raisers that initiate fatigue cracks. A crack once started is likely to propagate because the root of a crack is, itself, a stress raiser. Figure 5.24 gives surface factors C_s for steel parts as functions of ultimate tensile strength. The top line, for mirror-polished surfaces, corresponds to laboratory specimens and $C_s = 1.0$. The factors decrease with increasing roughness, in the order of ground, machined, hot-rolled, and hot-forged surfaces.[18] Corroded surfaces, where there is chemical action as well as roughening, have the lowest factors. Presumably, the curves are not for corrosion-resistant steels.

The surface factors also decrease with an increase of tensile strength, in some cases canceling any gain that might be expected in endurance strength. In consideration of the extra cost of obtaining higher strength by the use of alloy steels and heat treatments,

---

[18] If the strength is reported for tests on specimens with a finish other than polished, then $C_s = 1.0$ for the reported finish, and C_s must be adjusted for other finishes.

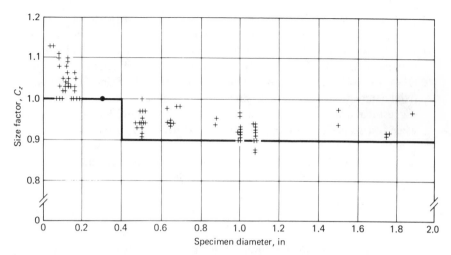

FIGURE 5.25 Size factor C_z, bending and torsional loads. Data are for steels, $S_y = 50$ to 165 kpsi; test points compiled from data reported by many investigators. (From R. C. Juvinall, *Engineering Considerations of Stress, Strain, and Strength*, New York: McGraw-Hill, 1967. Used with permission of McGraw-Hill Book Company.)

there may be good economy as well as safety in using a superior finish such as commercial polishing, fine grinding, and superfinishing, particularly along highly stressed surfaces and at stress raisers. Thus it is disturbing to see on a shaft a poorly machined fillet adjacent to a bearing seat that has been carefully ground for reasons of accuracy. Hot worked parts should have irregularities such as flash and mill scale removed. In radial aircraft engines, the forged connecting rods may be polished all over. Electroplated hard coatings may not be suitable for highly stressed steel parts subject to fatigue and corrosion unless low plating current densities and shot peening are used to partially overcome the tensile residual stresses in the coating.[19]

Size effect. The decrease in endurance limit with increasing size may be due to the probability that a larger piece is more likely to have a weaker grain or metallurgical defect at which a fatigue crack will start. In bending and torsion this decrease with increasing size may also be due to a smaller stress gradient. This means that high stresses act over a greater depth, with the greater probability of finding weak spots as well as hastening crack penetration. Because of many variables, numerical values for size effect are difficult to determine. Juvinall[20] compiled test data which compared size factors C_z with those at test specimen diameters of 0.30 in (Fig. 5.25). It would appear wise to use $C_z = 1.0$ for diameters equal to or less than 0.30 in and $C_z = 0.90$ to 0.85 for diameters between 0.40 and 2.0 in when the parts are loaded in bending or torsion. In axial loading (with no stress gradient) there is no apparent size effect up to 2 in and $C_z = 1.0$. Under *all* types of loading in the 4- to 12-in-diameter range, the factors were 0.75 or less, leading Juvinall to suggest that 0.6 to 0.75 be assumed for large parts.

Reliability factor. There is considerable scatter in the results of fatigue tests with apparently identical specimens, far more than in static tensile tests. For statistical analysis, the distribution of fatigue strengths has been approximated by the Weibull, log-

[19] For a discussion of the effect of platings, see R. C. Juvinall, *Engineering Considerations of Stress, Strain, and Strength,* New York: McGraw-Hill, pp. 339–341, 1967.

[20] R. C. Juvinall, *Engineering Considerations of Stress, Strain, and Strength,* New York: McGraw-Hill, pp. 231–233, 1967.

normal, and normal distributions,[21] the first one being the most representative and the last one, the most convenient. Stulen, Cummings, and Schulte[22] analyzed long-life fatigue strengths of several hot-worked heat-treated alloys, four of them nonferrous and the remainder AISI 4340 and 4350 steels, heat treated to give fatigue strengths from 71 000 to 97 000 psi. The standard deviations were between 5 and 8%: They state that for a fixed number of cycles, fatigue strength values have a reasonably normal distribution and suggest that a standard deviation of 8% of the long-life fatigue strengths may be assumed when test values are not available for a specific alloy, provided the material is of good quality.

A normal-distribution curve[23] gives the number of standard deviations k to be subtracted from the mean strength to obtain a desired survival rate or reliability. The reliability factor for use in Eq. (5.37) is

$$C_r = S_e/S_e' = (S_e' - k\sigma)/S_e' = 1 - (\sigma/S_e')k \qquad (5.38)$$

where σ is the standard deviation. With the suggested deviation of 8% of long-life fatigue strength, $\sigma = 0.08S_e'$, and Eq. (5.38) reduces to $C_r = 1 - 0.08k$. Corresponding to reliabilities or probabilities of survival of 90, 95, 99, 99.9, and 99.99%, the values of k are 1.3, 1.6, 2.3, 3.1, and 3.7, respectively, and Eq. (5.38) gives C_r values of 0.90, 0.87, 0.82, 0.75, and 0.70, respectively. Stulen et al. state that survival rates determined by the foregoing method can only be considered as a guide, since the actual law of distribution has not been accurately determined.

Other approaches to the use of reliability in design are being studied and published, but more data are needed. Also, the use of reliability factors depends upon the quantity of the tests and how results are reported, e.g., as minimum values or as mean values.

Example 5.6

A manufacturer stocks AISI 8740, hot-rolled steel bars in various diameters. Their use is proposed for a shaft of about 2 in diameter that is to be normalized. In a steel company's chart the tensile test properties in this size and treatment are given as 132 000 psi tensile strength, 87 500 psi yield strength, and 16.7% elongation. A hardness of 262 Bhn allows it to be finish machined after heat treatment. Estimate the endurance limit with a reliability of 99%.

Solution

$S_u = 132\ 000$ psi (910 MPa). By Eqs. (5.36),

$S_e' = 0.50S_u = (0.50)(132\ 000) = 66\ 000$ psi (455 MPa)

From Fig. 5.24, the surface factor is $C_s = 0.71$. From Fig. 5.25, a size factor of $C_z = 0.85$ should be sufficiently low. The reliability factor corresponding to 99% is $C_r = 0.82$. For normal temperatures in operation, $C_t = 1.0$. Hence from Eq. (5.37) the

---

[21] For an initial reading about each and their application to fatigue tests, the following are suggested: C. S. Yen, "Fatigue Statistical Analysis," in *Metal Fatigue—Theory and Design*, A. F. Madayag, Ed., New York: John Wiley & Sons, Inc., 1969. R. C. Juvinall, *Engineering Considerations of Stress, Strain, and Strength*, New York: McGraw-Hill, pp. 351–369, 1967. J. Y. Mann, *Fatigue of Materials*, Melbourne, Australia: Melbourne University Press, 1967.

[22] F. B. Stulen, H. N. Cummings, and W. C. Schulte, "Preventing Fatigue Failures—Part 5," *Machine Design*, Vol. 33, No. 13, pp. 159–165, June 22, 1961.

[23] R. C. Juvinall, *Engineering Considerations of Stress, Strain, and Strength*, New York: McGraw-Hill, Fig. 17.4, 1967.

endurance limit in service is estimated to be

$$S_e = C_s C_z C_t C_r C_m S_e{}'$$

$$= (0.71)(0.85)(1.0)(0.82)(1.0)(66\ 000) = 32\ 700\ \text{psi}\ (225\ \text{MPa})\qquad ////$$

Factor of safety. It should be noted that C_r as determined by Eq. (5.38) accounts only for material variables and not for variables in processing and assembly, nor for the uncertainty of loads in service, the simplifying assumptions in the analysis, and different environments and operating skills. These vary with specific manufacturing organizations, products, and service. For example, high reliability would be expected in the design of aircraft, for which material and processing are carefully controlled and checked, where service stresses are carefully determined, and where critical components are examined in periodic overhauls for the initiation of fatigue cracks. The principal mechanical components for which reliability data are readily available for design purposes are ball and roller bearings. Data could be determined and made available for other high-production components, both for the individual parts and the assembly. However, for components in general, including those designed for fatigue conditions, a factor of safety or a margin of safety appears to be necessary to allow for variables other than those of the material. Also, it is likely that many design engineers will choose to include in this factor of safety a consideration for the scatter in the material properties, rather than use a reliability factor C_r.

References. A number of books have been written on the subject of fatigue, and the reader is referred to them for further information.[24] Also, the science of fracture mechanics is being applied in studies of fatigue, such as the relative resistance of materials to crack propagation and to the rate of propagation.[25]

5.11 STRESS CONCENTRATION

Stresses at and near holes, abrupt changes of section, and other irregularities are higher than those at adjacent, uniform sections. The maximum stress that would exist at the location of the irregularity if the irregularity did not exist is designated the *nominal stress*. The ratio of the maximum of these localized stresses to the nominal stress is the *stress concentration factor*. The factors are determined mathematically, from photoelastic models, and by direct strain-gage measurements. These factors are considered to be theoretical ones and are given the symbol K_t. Values are shown in Figs. 5.26 through 5.28 for fillets of radius r that join a part of diameter d to one of larger diameter D at a step or shoulder in a bar or shaft. Nominal stress is based on the smaller diameter, d.

---

[24] A. F. Madayag, *Metal Fatigue: Theory and Design*, New York: John Wiley & Sons, Inc., 1969; C. C. Osgood, *Fatigue Design*, New York: John Wiley & Sons, Inc., 1970; H. J. Grover, *Fatigue of Aircraft Structures*, NAVAIR 01-1A-13, U.S. Government Printing Office, Washington, D.C., 1966; G. Sines and J. L. Waisman, Eds., *Metal Fatigue*, New York: McGraw-Hill, 1959; S. S. Manson, *Thermal Stress and Low Cycle Fatigue*, New York: McGraw-Hill, 1966; A. E. Carden, A. J. McEvily, and C. H. Wells, Eds., *Fatigue at Elevated Temperatures*, ASTM Special Tech. Publ. 520, American Society for Testing and Materials, Philadelphia, Pennsylvania, 1973; S. S. Manson, Ed., *Metal Fatigue Damage—Mechanism, Detection, Avoidance, and Repair*, ASTM Special Tech. Publ. 495, American Society for Testing and Materials, Philadelphia, Pennsylvania, 1971; H. E. Boyer, Ed., "Failure Analyses and Prevention," *Metals Handbook*, 8th ed., Vol. 10, American Society for Metals, Metals Park, Ohio, 1975.

[25] C. C. Osgood, "Applying Fracture Mechanics to Design," *Machine Design*, Vol. 43, pp. 102–106, August 19, 1971; Proceedings of the National Symposia on Fracture Mechanics, ASTM Special Tech. Publ., American Society for Testing and Materials, Philadelphia, Pennsylvania, e.g., No. 601, *Cracks and Fracture*, 1976 and No. 631, *Flaw Growth and Fracture*, 1977; F. Erdogan, Ed., *Mechanics of Fracture*, AMD-Vol. 19, ASME, New York, 1976; T. W. Crooker, "Fracture Mechanics—Fatigue Design," *Mechanical Engineering*, Vol. 99, pp. 40–45, June 1977.

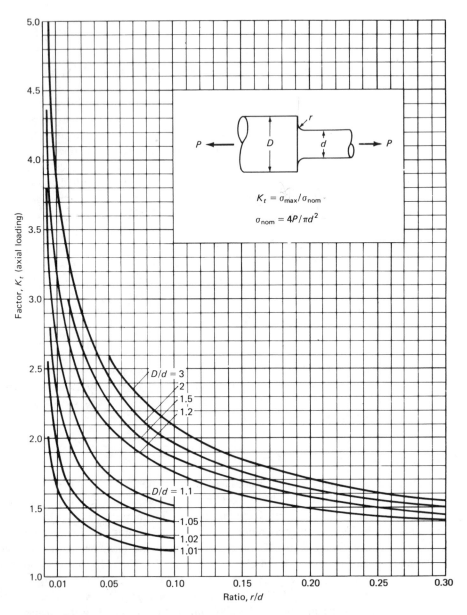

FIGURE 5.26 Theoretical stress concentration factor K_t for AXIAL loading of a stepped round tension bar with a shoulder fillet. K_t values are approximate. (Reprinted from R. E. Peterson, *Stress Concentration Factors,* New York: John Wiley & Sons, Inc., 1974, with permission of the publisher.)

Stress concentration K_t decreases with increase of r/d and decrease of D/d in each figure—for axial force P in Fig. 5.26, bending moment M in Fig. 5.27, and torsional moment T in Fig. 5.28.

Note carefully that these figures are for cylindrical members, and those for flat bars of the same proportions must be obtained elsewhere.[26] When the properties such as area and section modulus of the smallest section at the irregularity can be readily calculated, e.g., for the circular section at the bottom of a groove, the concentration factor is usually based upon it. For sections with holes, keyways, etc., it must be carefully observed whether the full or net section is the basis. Charts are available for notches, grooves, shoulder fillets, and holes in many two- and three-dimensional situations, and are also

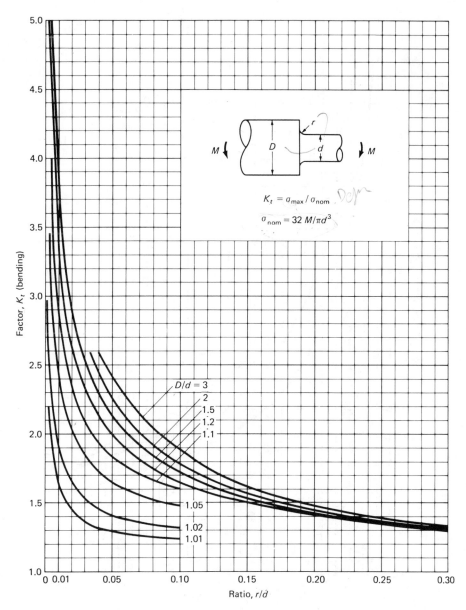

FIGURE 5.27 Theoretical stress concentration factor K_t for BENDING of a stepped round bar with a shoulder fillet, based on photoelastic tests of Leven and Hartman, Wilson and White. (Reprinted from R. E. Peterson, *Stress Concentration Factors,* New York: John Wiley & Sons, Inc., 1974, with permission of the publisher. See Source Fig. 78a for small values of r/d.)

available for machine elements such as key slots. gear teeth, interference fits, bolts and nuts, blade fastenings, crankshafts, and pressure-vessel heads.[26]

Stress concentration is a very significant factor in failure by fatigue. However, its effect on the nominal stress is not as large as indicated by the theoretical factors, particularly at very small radii. This is supported by the observation that scratches of extremely small root radii do not reduce the strength to the extent indicated by the theoretical factor.

---

[26] Probably the most complete set of charts for this and many other configurations is in the book by R. E. Peterson, *Stress Concentration Factors,* New York: John Wiley & Sons, Inc., 1974.

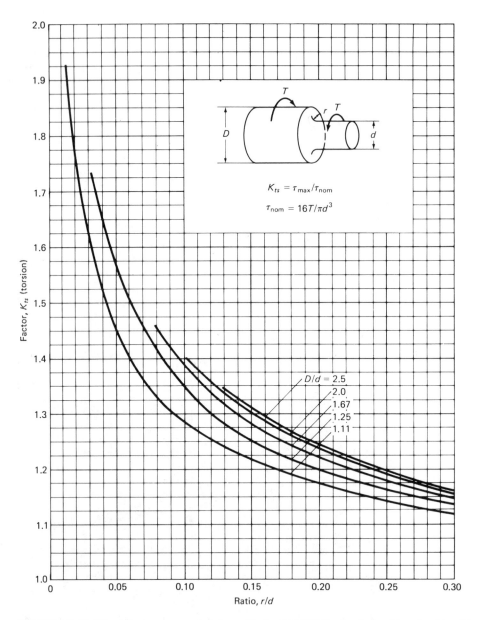

FIGURE 5.28 Theoretical concentration factor K_{ts} for TORSION of a shaft with a shoulder fillet, data from Matthews and Hooke. (Reprinted from R. E. Peterson, *Stress Concentration Factors*, New York: John Wiley & Sons, Inc., 1974, with permission of the publisher.)

A fatigue test on a grooved or filleted specimen generally gives an endurance strength which, when divided into that of a uniform specimen of the same material and minimum diameter, results in a ratio called the *fatigue* stress-concentration factor or the *fatigue notch factor* K_f, which is less than K_t. Values of K_f may be estimated from K_t by the equation

$$K_f = 1 + q(K_t - 1) \tag{5.39}$$

where q is the *fatigue notch sensitivity*, equal to $(K_f - 1)/(K_t - 1)$. Average values of q may be taken from Fig. 5.29. Its values range from 1.0 for full sensitivity at large radii and $K_f = K_t$, to $q = 0$ or no notch effect and $K_f = 1.0$ as the radius approaches

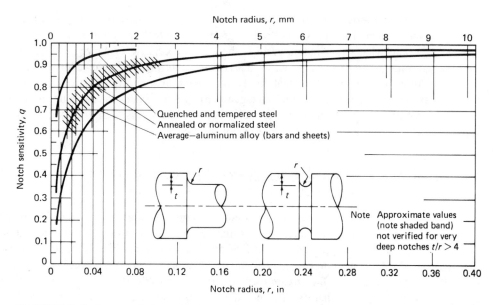

FIGURE 5.29 Average fatigue notch sensitivity q as a function of fillet or groove radius r. (Adapted from R. E. Peterson, *Stress Concentration Factors*, New York: John Wiley & Sons, Inc., 1974, with permission of the publisher.)

zero. Actually, data on very small radii are incomplete and the lines of Fig. 5.29 are not extended to zero. The higher the strength of a steel, the greater its notch sensitivity at all radii.

The K_f values obtained from Eq. (5.39) and the K_t charts are for use with the endurance limit. The notch effect in ductile materials decreases with a decrease in the number of cycles to failure until there is none in the static case. Some have assumed its effect to be negligible below 10^3 cycles. Then, one method of estimating the effect between 10^3 and 10^6 cycles has been to draw a straight line between points (10^3, $0.90S_u$) and (10^6, S_e/K_f) on a log S–log N chart (Fig. 5.22). The line gives a *net* fatigue strength, $Q = S_f/K_f'$, where K_f' is the reduced stress-concentration factor at cycles N and fatigue strength S_f (Fig. 5.22). Although it is a convenient method, it may give results on the unsafe side at the low-cycle end. Not only does $0.90S_u$ often exceed the yield strength, but the notch effect may only disappear below 10^2 cycles in bending and torsion and below a few cycles in axial loading. To account for the former, the line may be drawn through point (10^2, S_y) as in Fig. 6.14 of Section 6.10. Other methods have been presented. For example, Heywood[27] reported notch sensitivities at various cycles and tensile strengths for several materials. For normalized AISI 4130 steel Peterson[28] presented continuous curves of strength under axial loading from 1 to 10^8 cycles at different K_f values. With calculations based on Heywood's data, Juvinall[29] found good agreement with Peterson's curves between 10^3 and 10^5 cycles for specimens of 120 000 psi tensile strength under axial reversed loading. Low-cycle and finite-life fatigue is an active subject of study, and recent publications on current research should be consulted for curves to use in design.

[27] R. B. Heywood, *Designing Against Fatigue*, London: Chapman & Hall, Ltd., 1962.

[28] R. E. Peterson, "Fatigue of Metals in Engineering and Design," (Edgar Marburg Lecture), ASTM, Philadelphia, Pennsylvania, 1962, Fig. 29. An Appendix gives a method for developing similar charts for other axially-loaded, notched materials.

[29] R. C. Juvinall, *Engineering Considerations of Stress, Strain, and Strength*, New York: McGraw-Hill, pp. 260–266, 1967.

On the effect of coincident stress raisers there is very little data and agreement. For the *superposition* of a small-radius groove upon the maximum stress-intensity region of a somewhat larger-radius groove, there is photoelastic evidence that the stress-concentration factor is the product of the two separate factors.[30] However, in other cases the flow analogy may show that one kind of notch may relieve the stress from another kind, with much depending on relative geometries. For *adjacent* stress concentrations, such as a fillet next to a press fit, the factor is most likely less than a product, and, with a bit of spacing, perhaps only the larger of the two.

5.12 DESIGN TO REDUCE STRESS CONCENTRATION: THE FLOW ANALOGY

It is good design to reduce stress concentration as much as possible. There is an easily visualized qualitative analysis that may be used in the absence of theoretical and experimental data. To picture the flow of an ideal incompressible fluid, streamlines are used to divide a duct into channels of equal volumes of flow. If the duct is uniform, velocities are uniform, and the streamlines are equally spaced, as in Fig. 5.30(a). Where a duct section changes sharply, the velocities of flow increase near the corner and the channels of equal flow must narrow, the streamlines crowding together as in Fig. 5.30(b) and Fig. 5.30(c). In a stressed member of the same cross section, the increase in the stresses is analogous to the increase in velocities or inversely to the change in space between streamlines. Thus in Fig. 5.30(c), we may visualize the change from the nominal bending stress pattern at section 1 to one of higher, concentrated stress at the filleted section 2, then back to a linearly distributed stress at section 3, lower in value than at section 1.

It is obvious that the larger fillet radius in Fig. 5.30(c) decreases the crowding of the lines that took place in Fig. 5.30(b). But, also, the smaller the enlargement between ducts, the less the crowding and stress, as in Fig. 5.30(d), where $D' < D$. A groove behind the shoulder decreases the rate of change and the stress, as shown in Fig. 5.30(e). If the step is only needed for locating an assembled part, the narrow shoulder in Fig. 5.30(f) will greatly reduce the spread of the streamlines and the stress.[31] If the increased diameter is only needed to carry a larger bending moment or for the mounting of a component of larger bore, such as the sheave of Fig. 6.2, the stress concentration can be made very small by a large fillet radius of constant magnitude. If axial space is restricted, a spiral may be used or a profile, as in Fig. 5.30(g), with a radius that changes from a large value at the tangent point to a smaller one at the larger diameter. An increased fillet radius adjacent to a rolling-element bearing in a region of high nominal stress may be obtained from a spacing ring as in Fig. 5.30(h), or the shafts of Probs. 6.21 and 6.40, also by the undercut shoulder of Fig. 5.30(i). The foregoing comparisons made for the bending of shafts apply as well to axial and torsional loadings and to flat members.

The flow analogy will show that stress concentration is decreased at the end of a shaft keyway by cutting it with a rotary milling cutter rather than by an end mill; at the end of screw threads by cutting a semicircular, annular groove slightly deeper than the roots of the threads, as in Fig. 6.10, or by cutting the threads on a surface of larger diameter; at a radial hole by rounding and cold-working its mouth, or by milling or impressing transverse shallow notches on each side of the hole; in a welded butt joint by grinding flush the bead or "reinforcement"; and at lap joints by "streamlining" the fillet welds.

[30] F. W. Paul, Jr. and T. R. Faucett, "The Superposition of Stress-Concentration Factors," *J. Eng. Ind.*, *Trans. ASME*, Vol. 84, pp. 129–134, 1962.

[31] See R. E. Peterson, *Stress Concentration Factors*, New York: John Wiley & Sons, Inc., 1974. pp. 99–101 for charts giving K_t values for several narrow shoulder proportions.

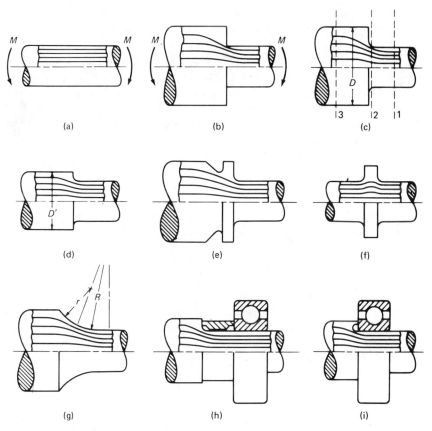

FIGURE 5.30 Flow analogy applied to minimize stress concentration at steps on shafts and other components (Section 5.12).

The flow analogy shows that on large areas of plate or sheet, openings with square corners should be avoided, with circular or oval openings used instead, as at the windows in aircraft fusilages; also, that an enlarged hole should be provided at the end of a narrow slot cut in a link or crankarm to give clamping flexibility.

A shaft is particularly vulnerable to fatigue failure at the ends of the hub of a component that has been firmly attached by pressing or shrinking. There is a combination of stress concentration where the shaft is squeezed inwardly at each end of the hub (see Fig. 8.18(a)) and a fretting corrosion with formation of red iron oxide Fe_2O_3, accompanied by roughening. The fretting is a minute rubbing against the hub by the surface fibers of the shaft as they overcome friction and elongate and compress under an alternating bending moment. Because of many variables, the strength-reduction factors generally lie in a broad range between 2 and 4, with the higher values applying particularly to larger diameter shafts and to higher strength steels. In fact, a net endurance limit over 20 000 psi is quite uncommon. However, considerable improvement is made, up to 100%, by inducing residual stresses by cold-rolling or flame-hardening the seat on the shaft (Sections 7.12 and 7.15), by cutting annular grooves on the sides of the hub to relieve the high pressure there on the shaft, and, if feasible, by increasing the diameter of the shaft seat. For details one may refer to books referenced at the end of Section 5.10.[32]

---

[32] Also, O. J. Horger in *Metals Engineering-Design*, 2nd ed., New York: McGraw-Hill, pp. 350–354, 1965.

Stress concentration and fatigue failures can be caused by faulty or uncontrolled processing and maintenance operations. These include identification marks stamped in places of high stress, clamping marks from vice jaws, removal of carbon deposits by coarse emery cloth, cracks from the heat of too-rapid grinding, incompletely blended (nontangential) fillets, die scratches, feather edges, and surface decarburization not removed after heat treatment. Internal stress raisers are nonmetallic inclusions in forgings, blow holes and voids in castings, and porosity in welds. Residual stresses from cooling, mechanical work, and assembly may be harmful or beneficial (Sections 7.8 through 7.16) and should be controlled.

PROBLEMS

Section 5.2

5.1   Derive Eqs. (5.5) and (5.9) from the equations that precede each.

5.2   Solve Eqs. (5.1) for σ_x and write the equations for σ_y and σ_z by similarity. *Ans.*

$$\sigma_x = [E/(1 + \nu)(1 - 2\nu)][(1 - \nu)\epsilon_x + \nu(\epsilon_y + \epsilon_z)]$$

5.3   Derive Eq. (5.3) for the relationship of moduli G and E. *Suggestion*: Use the equilibrium of forces on the enlarged triangular element of base unity at (b), cut out of the rectangular plate at (a), then relate strains and geometry of the distorted triangular element. The plate is stressed as shown, with $\sigma_y = -\sigma_x$, to give a condition of *pure shear*, i.e., no normal stress on the inclined plane.

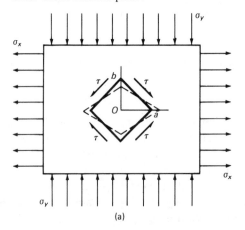

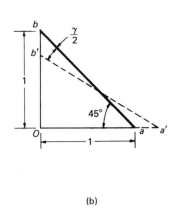

(a)   (b)

5.4   Transform Eq. (5.8) into strain components only, in a form similar to that of Eq. (5.9), which is in stress components only. The answers of Prob. 5.2 may be used. *Ans.*

$$U_0 = [E/2(1 + \nu)(1 - 2\nu)][(1 - \nu)(\epsilon_x^2 + \epsilon_y^2 + \epsilon_z^2)$$
$$+ 2\nu(\epsilon_x\epsilon_y + \epsilon_y\epsilon_z + \epsilon_z\epsilon_x)] + (G/2)(\gamma_{xy}^2 + \gamma_{yz}^2 + \gamma_{zx}^2)$$

5.5   Show that the answer to Prob. 5.4 may be further transformed into

$$U_0 = \lambda e^2/2 + G(\epsilon_x^2 + \epsilon_y^2 + \epsilon_z^2) + (G/2)(\gamma_{xy}^2 + \gamma_{yz}^2 + \gamma_{zx}^2)$$

where $\lambda = E\nu/(1 + \nu)(1 - 2\nu)$ and G and e are given by Eqs. (5.3) and (5.4). This is a form presented in books on elasticity.

5.6   In Eqs. (5.6) and (5.7), six strains are given as functions of only three displacements. Thus the strains cannot be independent. In the theory of elasticity, a relationship between strains which assures that the displacements are single valued is known as an *equation of compatibility*. In some problems only the three strains ϵ_x, ϵ_y, and γ_{xy} need be considered. Show that the equation of compatibility is then

$$\partial^2 \epsilon_x / \partial y^2 + \partial^2 \epsilon_y / \partial x^2 = \partial^2 \gamma_{xy} / \partial x \partial y$$

5.7 Metals under plastic stress as well as rubber behave like incompressible fluids, without appreciable change of volume and with the same stress in all directions. On this basis determine the value of Poisson's ratio for the plastic state. *Ans.* 0.50.

5.8 Normally, a bolt is expected to fail by a tensile fracture through the root area at the threads, since the nut height is sufficient to prevent failure by shear through the threads. However, under static overloading, large bolts with shallow threads may fail by bending and stripping of the threads. Assume that an 8% plastic elongation can occur without necking where the threads enter the nut, and that Poisson's ratio is 0.50 in the plastic range. Explain the failure with the aid of a simple calculation, using a bolt with an outside diameter of 3.000 in, 8 threads/in, and a root diameter of 2.846 in, engaging a nut with an inside thread diameter of 2.870 in.

5.9 The steel cylinder shown in cross section is held together by eight ¾-in steel bolts. A tightening force of 12 000 lb per bolt is desired. (a) One way to obtain this is to pretighten by wrench to compress the local high spots, then loosen and retighten by hand "finger tight," then turn the nut by wrench through a predetermined angle. Assume rigid flanges and that deflection occurs only in bolts and cylinder wall. If there are 16 threads per inch, through what angle should the nut be turned? *Ans.* 57.5°. (b) Other methods include the placing of strain gages on the exposed portion of the bolt, measuring the change of overall bolt length by micrometer, and possibly the diameter change. Make simple calculations to show the readings corresponding to 12 000 lb. Neglect the partial elongation of the bolt within the nut. Comment on the practicality and accuracy of these methods relative to the method under (a).

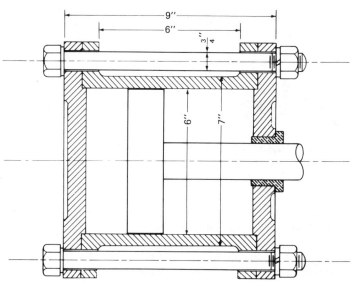

5.10 A torque-carrying member such as a gear is usually fitted on a shaft by pressing it on, called a *press fit*, which by the compression in the axial direction tends to expand the diameter of the shaft ahead of the motion. Or the hub of the member is expanded by heating, placed in position, and allowed to cool, called a *shrink fit*. Consider the possibility of a "stretch fit." Assume it is mechanically possible to put the member over an adjacent but smaller diameter of the shaft, then pull the shaft from its ends, slide the member into place, and release the pull. The interference (difference in diameter of shaft and hole) is 0.0005d for a standard, medium force fit and 0.0010d or more for a shrink fit, obtained by a selective assembly of parts with tolerances in their diameters. State if these fits are feasible for the stretch fitting of shafts where the smallest diameter at the end of the shaft is equal to or greater than 0.80 of the fitted diameter, and the strengths are: (a) S_y = 60 000 psi, typical of heat-treated, medium carbon-steel shafts, and (b) S_y = 100 000 psi, typical of alloy-steel shafts.

5.11 Several disks on a 30-mm-diameter length of the shaft of a small steam turbine have been damaged and need to be repaired and replaced as soon as possible. This can be done locally if the disks can be removed. A long press is not desirable. The shaft has threaded ends, and a nearby engineering college has a tensile testing machine with a dial graduated to 30 000 kg force. Make calculations to show if it will be feasible to try to remove the disks by shaft stretching, perhaps with the aid of a little hammering and local heating with a gas flame. See Prob. 5.10 for some data. *Ans*. It is feasible if alloy steel.

5.12 In a manufacturing operation slugs or cylinders of 25 mm length are cut from cold-drawn aluminum-alloy bars approximately 40 mm in diameter. They are to be sized to a diameter of 40.50 mm by expanding them one at a time in a heavy steel die cavity. This is done by applying a high axial pressure with a closely fitting steel punch, as sketched at (a). The modulus of elasticity of aluminum is 71 GPa. Allow for spring-back and determine the required inside diameter of the die if the material: (a) has a yield strength of 250 MPa and its stress-strain diagram can be approximated by Fig. 5.4; (b) a yield strength of 250 MPa and a straight-line approximation beyond the yield strength, as sketched at (b) with a slope such that σ/ϵ = 6000 MPa. *Ans*. 40.688 mm.

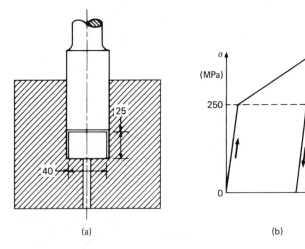

(a)   (b)

5.13 In the drilling of oil wells, the steel drill pipe or tubing must be partially supported by the swivel hook of the drawworks to reduce the force on the drill bit and minimize buckling of the pipe under its own large weight. Let F be the lifting force at the top of the well, P the force at the drill bit, w the weight per unit length of pipe, L its length, and A its cross-sectional area. (a) Derive equations for force F and deflection u as a function of a desired drill force P and of distance x from the bottom of the hole. *Ans*. $u = x(wx - 2P)/2AE$. (b) Determine the deflections at the top if $F = 0$, if $P = F$, and if $P = 0$. On sketches show the relative deflections of the pipe. (c) For a hole 10 000 ft deep, with the pipe weighing 25.25 lb/ft with couplings, and an outside diameter of 5.563 in and an inside diameter of 4.733 in, calculate the forces and deflections for the three conditions in (b). How much must the pipe be raised at its top in going from the first condition to the last condition? (d) If the desired drill-bit force is 90 000 lb, what is the supporting force F, the maximum stress in the pipe, and the distance to lift from the position of no support?

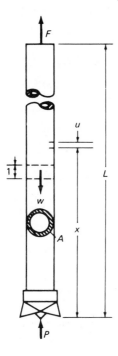

5.14 Derive an equation for the elongation of a conical rod of length L, tapering from a diameter D to a diameter d, and loaded by an axial force P at its ends. *Suggestion*: Take the origin of coordinates at the vertex of the extended cone. *Ans.* $u = 4PL/\pi DdE$.

Section 5.3

5.15 Derive Eqs. (5.13) and (5.14) for the biaxial case from the general Eqs. (5.11) and (5.12).

5.16 Derive Eqs. (5.13) and (5.14) from the equilibrium of an element, as outlined in the footnote following Eq. (5.14).

5.17 Show that $(\sigma - \sigma_x)(\sigma - \sigma_y)(\sigma - \sigma_z) = 0$ is equivalent to Eq. (5.11) when the three shear stresses are zero. What special case is this?

5.18 On the inner surface of a 3-in outside diameter, 2-in inside diameter cylinder with a high internal pressure of 5000 psi, the stresses are 13 000, −5000, and 4000 psi in the tangential, radial, and axial directions, respectively. The cylinder is also twisted about its axis, giving a tangential plane shear stress of 7211 psi in the tangential and axial directions. Determine the principal stresses and the major shear stresses from Eqs. (5.11) and (5.12).

Section 5.4

5.19 Demonstrate the validity of the pole method for Mohr's circle by proving that angle ϕ of Fig. 5.7 equals one-half angle 2ϕ of Fig. 5.6, i.e., that $\angle XPM = (\frac{1}{2})\angle XCM$ in Fig. 5.7. *Suggestion*: Construct triangles PCM and PCX within the circle.

5.20 For the pole method of Mohr's circle prove that the angle between the lines representing a plane of principal stress and a plane of maximum shear stress is 45°, e.g., $\angle APS$ of Fig. 5.7.

5.21 Prove that, if the reference plane desired for a Mohr's circle is a principal plane, and if pole P is placed at the point on the circle representing the principal stress acting on that plane, then a ray drawn from P to any point M on the circle makes the same angle with a vertical reference line through P as the plane on which the stresses act makes with the principal plane (e.g., as done at O in Fig. 5.11).

5.22 Given the stresses shown, construct the Mohr's circles and determine graphically the principal stresses and the maximum of the shear stresses with accompanying normal stresses, showing them and the given stresses on correctly oriented elements. What angle do the planes on which the largest principal stresses act make with the plane normal to the X axis? *Ans.* 13 300 psi at 75.5° counterclockwise, −7300 psi, 0; $\tau_{max} = \pm 10\ 300$ with $\sigma = 3000$ psi.

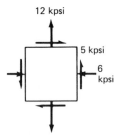

5.23 Same information as Prob. 5.22, but for the stresses shown below.

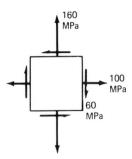

5.24 Same information as Prob. 5.22, but for the stresses shown in the figure. *Ans.* 102 MPa at 36° clockwise, 38 MPa, 0; $\tau_{max} = \pm 51$ MPa with $\sigma = 51$ MPa.

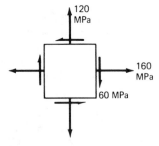

5.25 A cylindrical steel bar carries bending and torsional moments such that a tensile stress of 40 000 psi, together with shear stresses of 20 000 psi in a direction to give a counterclockwise moment, act on a surface element on its two sides normal to the axis of the bar. Sketch the element and construct the Mohr's circles. Determine graphically the principal stresses, the maximum shear stresses and accompanying normal stresses, and the stress on a plane

30° clockwise from a radial plane seen from above. What angle do the planes on which the largest magnitude principal stresses act make with a radial plane?

5.26 Same information as Prob. 5.25, but on the sides normal to the axis a compressive stress of 80 MPa and shear stresses of 110 MPa giving a clockwise moment. *Ans.* Principal stress, -157 MPa at 35° counterclockwise, 77 MPa, 0; on plane of maximum shear stress, $\tau = \pm 117$ MPa with $\sigma = -40$ MPa; on plane at 30° clockwise, $\sigma = 34$ MPa, $\tau = 90$ MPa.

5.27 Sketch freehand the three Mohr's circles for the cases shown. What are the τ_{max} values?

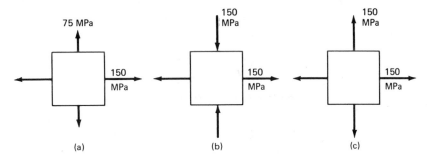

(a)   (b)   (c)

Sections 5.5 and 5.6

5.28 Construct the envelope for the maximum shear-stress theory of failure when $S_y = 40\ 000$ psi. Determine by measurements the factors of safety when: (a) $\sigma_1 = -18\ 000$ psi, $\sigma_2 = 12\ 000$ psi; (b) $\sigma_2 = \sigma_1 = 25\ 000$ psi; (c) pure shear with $\tau = 14\ 000$ psi.

5.29 Derive Eq. (5.21) step-by-step starting with Eq. (5.17).

5.30 From the Mises criterion in terms of principal stresses, Eq. (5.24), obtain an equivalent equation in terms of stress *components* σ_x, σ_y, and τ_{xy}. *Ans.*

$$\sigma_x^2 - \sigma_x \sigma_y + \sigma_y^2 + 3\tau_{xy}^2 = S_y^2$$

5.31 For the common case of a single normal and a single shear stress, as in a shaft under bending and torsion, to what does the equation of Prob. 5.30 reduce? Derive a similar equation based on the maximum shear-stress theory of failure, and compare.

5.32 (a) The energy failure theory that was first proposed was based on *total energy* and the total energy at yielding in the tensile test. Derive the relationships for this theory and an equation for the factor of safety. (b) Reduce the equations of part (a) to the biaxial case and to the case of a single normal and single shear stress. *Ans.* $\sigma_x^2 + 2(1 + \nu)\tau_{xy}^2 = S_y^2$. Compare and state if this theory is more or less conservative than the Mises criterion. What ratio S_{sy}/S_y does this give?

5.33 A maximum principal *strain* theory has been used in the past, particularly for the design of thick cylinders and guns, because of the agreement with some test results. The theory assumes that failure will occur when the maximum principal strain exceeds the maximum strain at yielding in the tensile test, namely $\epsilon_{yp} = S_y/E$. (a) Derive the equations of failure (or of the envelope) for the biaxial case when $\sigma_1 > \sigma_2$, when $\sigma_2 > \sigma_1$. (b) Carefully sketch a boundary diagram or envelope on a σ_1 vs σ_2 chart. Use $\nu = 0.30$, and derive any additional equations as needed. On the same axes sketch the relative positions of the maximum shear theory and of the Mises criterion envelopes and comment on the relative safety of the three theories. (c) For the general case of three principal stresses derive the equation for the case $\sigma_1 > \sigma_2 > \sigma_3$. Write the other two boundary equations by analogy. What is the equation for the factor of safety when $\sigma_1 > \sigma_2$ and $\sigma_1 > \sigma_3$? *Ans.* $n = S_y/(\sigma_1 - \nu\sigma_2 - \nu\sigma_3)$.

5.34   Calculate the factors of safety for a high-pressure cylinder where the stresses are 13 000 psi tangentially, -5000 psi radially, and 4000 psi axially on the inner surface and $S_y = 40\,000$ psi and $\nu = 0.30$. Compare for the maximum shear-stress, Mises criterion, and maximum strain theories (Prob. 5.33). *Ans.* 2.22, 2.56, 3.00.

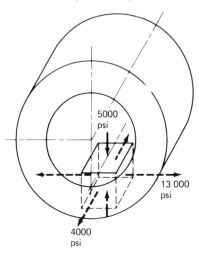

5.35   In a rotating disk of uniform thickness, at the radius of maximum radial stress, the radial stress has been calculated to be 87.6 MPa and the tangential stress has been calculated to be 154.5 MPa. The yield strength of the steel is 345 MPa. Determine the factors of safety indicated by the maximum shear-stress theory, the Mises criterion, and the maximum strain theory. (Use the answer of Prob. 5.33.)

5.36   In a thin cylinder of diameter D, thickness t, and under internal pressure p, the stresses are $pD/2t$ tangentially, $pD/4t$ axially, and negligible radially. What are the equations for the factors of safety by the maximum shear-stress theory and by the Mises criterion? *Ans.* $2.00tS_y/pD$, $2.31tS_y/pD$.

5.37   For the steel bar of Prob. 5.26 use the given answers to determine the factors of safety by the maximum shear-stress theory and by the Mises criterion if $S_y = 380$ MPa.

5.38   Determine the normal stress σ and the shear stress τ on face ABC of the octahedron of Fig. 5.14 due to a single principal stress σ_1. *Suggestion:* Let the lengths OA, OB, and OC be unity. Sketch the body $OABC$ with forces on sides OBC and ABC. Determine the stresses from the equilibrium of the force components in normal direction OG and in tangential direction GA. Note that the three forces will lie in the plane of triangle OGA and that the shear force will have the direction GA. *Ans.* $\sigma = (\frac{1}{3})\sigma_1$, $\tau = (\sqrt{2}/3)\sigma_1$.

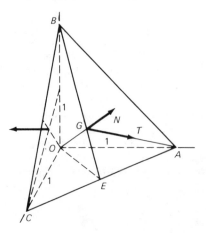

5.39   (a) Continue Prob. 5.38 to determine equations for the normal and shear stress on face ABC due to three principal stresses σ_1, σ_2, and σ_3. By analogy use the answers given to Prob. 5.38 to write the contribution of each principal stress and determine the results. *Ans.*

$$\sigma = (1/3)(\sigma_1 + \sigma_2 + \sigma_3),$$

$$\tau = (1/3)\sqrt{(\sigma_1 - \sigma_2)^2 + (\sigma_2 - \sigma_3)^2 + (\sigma_3 - \sigma_1)^2}$$

(b) Assuming that failure of ductile materials is by yielding in shear on an octahedral plane, write an equation for failure based on S_y of the tensile test. Compare it with Eq. (5.21).

Section 5.7

5.40   The cast-iron hubs of pulleys and other attached members are pressed over steel shafts to give an interference fit. For a hub with an outside diameter twice the shaft diameter and an interference of 0.001 inch per inch of shaft diameter, the maximum stresses are 10 800 psi tangentially and −6480 psi radially. The tensile and compressive ultimate strengths for cast iron ASTM 30 are 31 000 and 109 000 psi, respectively. Determine the factors of safety by the two methods for brittle materials. *Ans.* 2.87 and 2.45.

5.41   Same requirement as Prob. 5.40, but for an ANSI FN2 fit (Prob. 8.76). The stresses are 9450 psi and −4910 psi, and the material is cast iron ASTM 20 with tensile and compressive ultimate strengths of 22 000 and 83 000 psi, respectively.

Section 5.8

5.42   For a beam of elliptical cross section for which $c^2 z^2 + a^2 y^2 = a^2 c^2$, bent about its Z-axis, determine: (a) the bending moment that will just cause plastic yielding over the entire section and (b) the limit-design factor.

5.43   Same requirement as Prob. 5.42, but for a beam of square cross section with one diagonal of length h in the plane of bending. *Ans.* (a) $S_y h^3/12$. (b) 2.0.

5.44   Same requirement as Prob. 5.42, but for a circular tube of outside diameter D and inside diameter d. Express in ratio form and determine the limit-design factor in bending L_b for $d/D = 0.75$. *Ans.* 1.44.

5.45   Same tube as Prob. 5.44 but determine the limit-design factor in torsion L_t and its value for $d/D = 0.75$.

5.46   For a beam of rectangular section, with height h and width b, determine the plastic moment and the limit-design factor when plastic yielding occurs inwardly to the location $y = e$ (Fig. 5.17(c)). What are the values of the factor for yielding to one-quarter, one-half, and all the way to the center? *Ans.* $L_b = (\frac{1}{2})[3 - (e/c)^2]$, where $c = h/2$.

5.47   Same requirement as Prob. 5.46, but for a solid circular section in bending. *Ans.* 1.26, 1.49, 1.70.

5.48   Another criterion for design is a designated limit for the strain ϵ_p that occurs in the plastic region. Locate $y = e$ by a sketch of ϵ vs y and use the answer in Prob. 5.46 to show that $L_b = (\frac{1}{2})\{3 - [S_y/(S_y + E\epsilon_p)]^2\}$. If the material is steel, $S_y = 40\,000$ psi, and ϵ_p is limited to 0.001, what is the value of L_b? What is the maximum possible value for L_b?

5.49   For the beam of Fig. 5.19(b) and (c): (a) derive an equation for the loading q_c that will cause collapse and (b) derive the ratio q_c/q_e in terms of L_b (which depends upon the beam cross section and equals M_p/M_e).

5.50 (a) For the beam shown derive an equation for the load F_c that will cause collapse. *Ans.* $M_p(2l - a)/ab$. (b) If F_e is the maximum load under which the stresses are completely elastic, and $b = 2a$, for which $(M_e)_{max} = 3F_e l/16$, derive an equation for the ratio F_c/F_e in terms of L_b, the limit-design factor for the cross section, (M_p/M_e). *Ans.* $1.40L_b$.

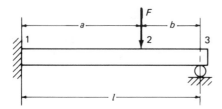

Sections 5.9 and 5.10

5.51 (a) How many hours of stress reversals are experienced by a steel shaft directly connected to a 1750 rpm motor before it reaches the usual number of cycles where endurance limit begins? *Ans.* 9.5 hours. (b) By a shaft connected through a V-belt with a reduction ratio of 3.5? (c) How many days for the latter if it drives a one-cylinder refrigerator compressor intermittently for a total of 2 hours a day? (d) How many years before the aluminum housing of the compressor is load cycled to the commonly reported cycles in the fatigue test for aluminum? *Ans.* 22.8 years.

5.52 (a) Construct and clearly label on a log S–log N chart a curve of strength vs cycles for a normalized steel with a Brinell hardness number of 275 and a highly polished surface of small diameter. Make estimates as necessary and indicate the scales of the axes. (b) The steel will be manufactured to a diameter of 1.625–1.627 in with a machined finish, and a desired reliability of 99% in the endurance strength. Determine the fatigue strength at 5×10^4 cycles.

5.53 Same instructions as Prob. 5.52, but Bhn = 450 by a quench treatment, and $d = 20$ mm, the surface is ground, and the fatigue strength at 10^5 cycles is desired. Consideration of reliability will be included in the factor of safety. *Ans.* $S_e = 550$ MPa, $S_f = 740$ MPa or $S_e = 490$ MPa, $S_f = 690$ MPa, if S_e' is limited to 690 MPa by Eq. (5.36).

5.54 The standard deviation in fatigue tests on a certain material is found to be 6%. Determine the reliability factor C_r for a 99.9% probability of survival and compare it with that given for the steels.

5.55 An examination of a fatigue fracture in a shaft may show the area of the final, sudden break to be either close to the edge, as at (a), or centered, as at (b). The surface appears to be uniform around the periphery, but it is known that material is not homogeneous and the strength of the fibers may vary as in the diagram at (c). Which type of fracture indicates the need for the greater reduction in applied load? Explain.

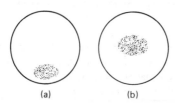

(a)   (b)

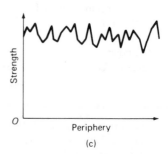

(c)

Section 5.11

5.56 Continue Prob. 5.52 but at a full fillet with a ⅛-in radius joining the given diameter to one of 1.875 in. Determine the design stress-concentration factors to use with the endurance limit for bending and torsional stress calculations. Estimate the net fatigue strength at 5×10^4 cycles in bending.

5.57 Continue Prob. 5.53 at a fillet of 0.80 mm, joining shaft sections of 20 and 26 mm diameters under torsional loadings. Estimate $S_{su} = 0.80\, S_u$ and $S_{se} = 0.577\, S_e$ (Section 5.9). Determine the net torsional fatigue strength at 10^5 cycles. *Ans.* 350 MPa or 320 MPa with S_e' limited.

5.58 A 1-in steel shaft is case hardened to a depth of 1/16 in. The tensile strength of the core is 115 000 psi, and the hardness of the case is 640 Bhn. At a fillet the stress-concentration factor is 3.2, with the concentration of stress confined to the case. At an adjacent section there is no stress concentration. Graphically or otherwise show whether fatigue failure will initiate in the case or core at these two sections. *Suggestion*: Plot the endurance strengths to scale along a shaft section, drawn double size for clarity. Superimpose lines that show the distribution of bending stress at failure for each section.

5.59 Write an equation for estimating fatigue strength S_f', the top line of Fig. 5.22. Modify it to obtain an equation for Q, the bottom line of Fig. 5.22.

5.60 An alternate and probably more accurate method for determining net fatigue strength is to use Heywood's experimental results[33] for K_f' at 10^3 cycles; e.g., for steels he found that the ratios $(K_f' - 1)/(K_f - 1)$ were 0.24, 0.36, 0.46, and 0.55 at tensile strengths of 750, 1000, 1250, and 1500 MPa, respectively. For a shaft loaded axially, with $r = 1$ mm, $d = 20$ mm, $D = 24$ mm, $S_u = 900$ MPa, quenched, tempered, and ground, and with the reliability included in the factor of safety, (a) determine the value of K_f' at 10^3 cycles. *Ans.* 1.33. (b) Recalling that the estimate for S_f' is $0.75 S_u$ at 10^3 cycles and $0.45 S_u$ at 10^6 cycles for axial loading (Section 5.9), construct the S_f/K_f' line on a log S–log N chart. What is its value at 2×10^4 cycles? *Ans.* 310 MPa.

5.61 Stress σ_1 in axial planes induces a stress σ_2 in the radial plane at a groove or fillet. (a) Show that σ_2 is required for equilibrium by sketching stresses or components on the two elements in the end view. (b) Stress $\sigma_2 = c\sigma_1$, where c is a fraction that depends on the shape of the groove. Fraction c is usually less than 0.25. One of the theories of failure predicts that yielding will begin when $\sigma_1 = S_y/m$, where m is a function of c alone. Try out the theories and derive m, then calculate the range in values of m. What experimental evidence has been given in Section 5.9 that supports the extension of this theory of failure, from yield to fatigue failure?[34] *Ans.* $\sqrt{1 - c + c^2}$.

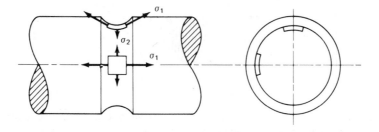

---

[33] R. B. Heywood, *Designing Against Fatigue*, London: Chapman & Hall, Ltd., 1962.
[34] In R. E. Peterson, *Stress Concentration Design Factors*, New York: John Wiley & Sons, Inc., 1953, charts are presented of mK_t, a combined factor to be used instead of K_t. However, most designers have continued to use the slightly higher and safer values of K_t alone.

Section 5.12

5.62 By using lines of flow compare the fatigue strength of a shaft of diameter D having a semicircular groove of radius r and root diameter d with that of a stepped shaft of the same two diameters D and d joined by a fillet of radius r. If stress-concentration charts of grooves are available, confirm your conclusion.

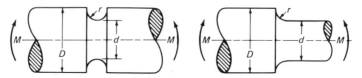

5.63 Reproduce the three shaft designs, sketch flow lines, and deduce the probable order of fatigue strength at the fillet on the smallest part of the shaft. Its radius and the adjacent shoulder diameter are the same in each design. Assume the bending moment is constant along the portion of the shaft shown.

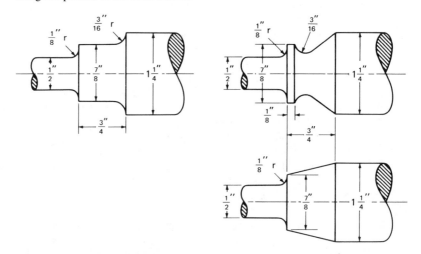

5.64 In a standard bolt, fatigue failures occur just under the head, at the shank end of the threads, and, most frequently, in line with the inner face of the nut. Sketch several special designs that will give increased strength against fatigue at the first two places. Prove your suggestions by the fluid analogy.

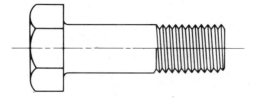

5.65 Nuts of special shape such as (1) or (2), shown in section, are used to increase the strength of bolts against fatigue failure. A standard nut (3) is relatively stiff compared with the bolt, and since stress is only developed by strain, most of the bolt force is transferred by the first few threads, creating a concentration of stress. (a) Carefully copy the three designs and sketch flow lines of stress, continuous from bolt through threads and nut to its support. These lines should show the transfer of force and concentrations of stress in bolt and nut. Which design do you like and why? (b) Suggest a change in material and a change in threads that might be made to the standard nut without changing its external shape.

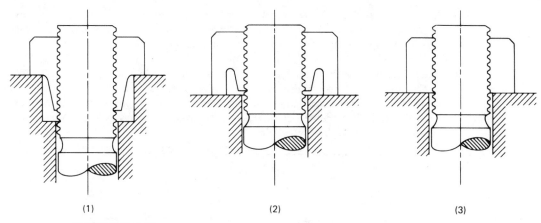

(1)   (2)   (3)

PROBLEM 5.65

5.66 Two rods may be joined and their tension adjusted by threaded rod ends and a thin socket or turnbuckle. Compare the fatigue strength at the threads with that with the standard nut (3) of the preceding problem, using flow lines to show force transfer and stress concentrations.

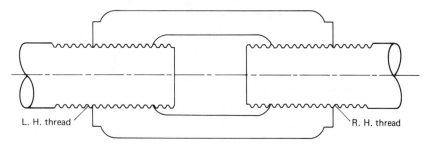

L. H. thread   R. H. thread

5.67 Put into wartime service, a large number of army tanks of new design had many early fatigue failures of pins which held the track rollers. Investigation found unblended fillets, acid pickling for cleaning, followed by brass plating, and ordinary heavy press fits. Explain the failures and suggest improvements.

5.68 Sketch an improved design for each tension member (a) aircraft rod end, by using a one-piece forging; (b) thin link, without changing its external shape; (c) bar, without changing the hole or its orientation.

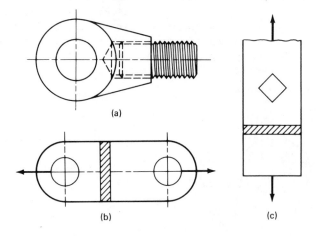

(a)

(b)   (c)

5.69 It is an anomaly that a removal of material may increase strength. Explain and sketch five illustrations, all of which may be obtained from the text, or elsewhere. Add a sketch showing how to stop a crack that has already progressed inward from the edge of a plate.

5.70 A lap joint is made by overlapping two flat bars and forming two transverse, 45° fillet welds, as shown. The dotted lines show the original surfaces and the curved lines show the depth of fusion. Reproduce and sketch flow lines to show the transfer of stress from one bar to the other. Where is the highest stress concentration? How may the stress concentration be decreased by reshaping? by grinding?

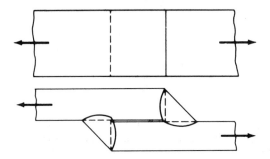

5.71 The transfer of force between a narrow and a heavy bar with two longitudinal welds, as shown, is analogous to the transfer between a bolt and a nut (Prob. 5.65). How can the parts be preshaped to better distribute the transfer and minimize the stress concentration?

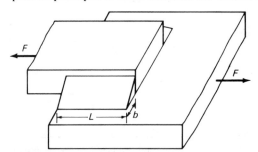

5.72 In rough and cold seas some of the first all-welded freighters developed long and sometimes fatal cracks through the plates on the deck and sides. This was contrary to experience with ships where the plates were riveted together. Why was it more likely that a crack should start and why should it progress further than in riveted plates?

5.73 A long, mill roll of 30 in diameter failed after four months of service. Examination showed that cracks had started and progressed at the relatively small, ¼ in fillet between the roll and the 10-in-diameter bearing-pin extension. The replacement cost and time was too great for a redesign and replacement. Sketch with flow lines your recommendations for machining to reshape roll and prepare pin for welding, welding itself, and processing after welding, so that the result will be much stronger than the original.

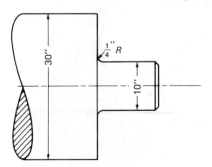

6

Design for Strength and Endurance
Types of Stress Variation, Design Equations for
Fluctuating Simple and Combined Stresses, Shaft
Design, Limited Cycles, and Multiple Stress Levels

6.1 INTRODUCTION

This chapter may be considered a continuation of the preceding one, since the theories
and relationships developed there are the bases for the design equations here. However,
this chapter may be read separately by those already familiar with theories of failure and
the basic phenomena of fatigue and stress concentration. Equations are developed for
dimensions and safety of real components subject to alternating or fluctuating forces and
moments, with corresponding simple stresses or combined normal and shear stresses.
The equations are applied to the design of shafts. Attention is called to other analyses
in the book that apply to shaft design, as well as to further practical considerations. The
final sections of the chapter are concerned with design for limited life and for operation
at two or more levels of fluctuating stress, where operation at each level contributes to
a fatigue damage.

6.2 TYPES OF STRESS VARIATION:
DESIGN FOR STEADY AND FOR ALTERNATING STRESSES IN
BRITTLE AND IN DUCTILE MATERIALS

Different patterns of stress variation with time are identified in Fig. 6.1. Figure 6.1(a)
is *steady stress*, and Figs. 6.1(b) and (c) are *alternating stresses*, which alternate between
equal tensile and compressive values with a consequent *complete reversal* of stress.
Figure 6.1(b) is *sinusoidal*, a variation commonly experienced by the fibers of a rotating
shaft when the load is steady and unchanging in direction. Figure 6.1(d) is *pulsating
stress* or *repeated stress*, varying between zero and a maximum value with each application of load, as on the teeth of gears. Figures 6.1(e) and (f) are *fluctuating stresses* or
general cases of varying stress. Figure 6.1(e) is a typical loading for an engine valve
spring, which is preloaded at installation, then further compressed when the valve is
opened. The stress variation of Fig. 6.1(f) might appear on a crankshaft.

If cyclic, Figs. 6.1(d), (e), and (f) may each be considered to result from the superposition of an alternating stress upon a steady stress. For all types, a steady stress or
mean component σ_m and an alternating stress or component σ_a may be defined, as shown

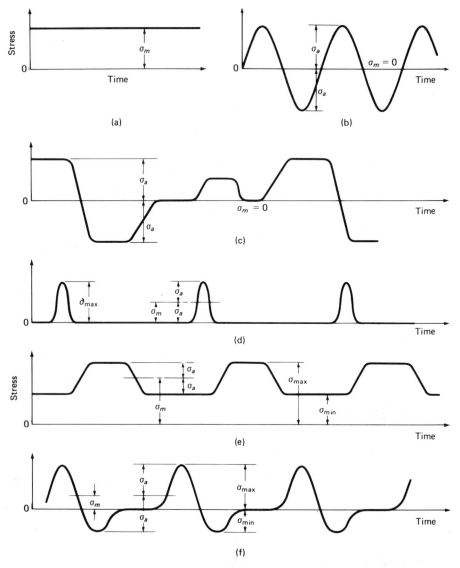

FIGURE 6.1 Types of stress variation. (a) Steady. (b) Alternating (sinusoidal). (c) Alternating (general). (d) Pulsating. (e) Fluctuating (without reversal). (f) Fluctuating (general).

on the sketches and by the equations,

$$\sigma_m = \frac{\sigma_{\max} + \sigma_{\min}}{2} \quad \text{and} \quad \sigma_a = \frac{\sigma_{\max} - \sigma_{\min}}{2} \tag{6.1}$$

Note that these stress components are independent of the shape of the stress-time curve. Also, $\sigma_{\max}$, $\sigma_{\min}$, σ_m, and σ_a are nominal stresses, as defined at the beginning of Section 5.11. Provision for the effect of stress concentration is made separately.

If the stress has a *steady* value σ_m and the material is *ductile*, with a stress-strain curve not rising much above the idealized one of Fig. 5.4, then any high localized stress such as indicated by the product $K_t \sigma_m$ can only slightly exceed S_y, and stress concentration is not considered significant in causing failure. The localized yielding that does occur does not cause significant overall deformation. Hence it is common practice to ignore the stress concentration. For a single *steady* stress $\sigma_m = \sigma_{max}$, the factor of safety in *ductile* material is

$$n = \frac{S_y}{\sigma_m} \tag{6.2}$$

If there is more than one applied stress, then the principal stresses σ_{1m}, σ_{2m}, and σ_{3m} may be determined by the methods of Sections 5.3 or 5.4 and used in a theory of failure for ductile materials as given in Sections 5.5, 5.6, or 5.8.

If the stress is *alternating*, as in Figs. 6.1(b) and (c), and the material is *ductile*, then fatigue failure is expected to occur if the product of the *fatigue* stress-concentration factor K_f and the alternating stress $\sigma_a = \sigma_{max}$ exceeds the fatigue strength. Crack initiation and propagation occur without any apparent yielding, and the final fracture looks "brittle." Hence the product $K_f\sigma_a$ must be used in design. If there is a single *alternating* normal stress σ_a, then the factor of safety for *infinite* life and *ductile* material is

$$n = \frac{S_e}{K_f \, \sigma_a} \tag{6.3}$$

where S_e is the modified endurance limit, such as obtained by Eq. (5.37). For *finite* life, where the fatigue stress-concentration factor has a reduced value, say K_f', it and S_f may be substituted for K_f and S_e, respectively, if known. If estimated by the method of Section 5.11, then the ratio S_f/K_f' is substituted, such that $n = (S_f/K_f')/\sigma_a$.

If there are two or more applied stresses, all peaking simultaneously at a place of stress concentration, then, logically, the principal stresses σ_{1a}, σ_{2a}, and σ_{3a} may be determined by Sections 5.3 or 5.4, then multiplied by stress-concentration factors, and the resulting products $K_{f1}\sigma_{1a}$, $K_{f2}\sigma_{2a}$, and $K_{f3}\sigma_{3a}$ substituted for σ_1, σ_2, and σ_3, respectively, in an equation or envelope for a theory of failure for ductile materials as given in Sections 5.5 or 5.6. Yield strength S_y is replaced by fatigue strengths S_f or S_e. There is a difficulty in that the factors K_1, K_2, and K_3 are not likely to be available for the particular directions of the principal stresses. Hence some common cases of multiple applied stresses are further treated in Sections 6.4 and 6.5.

Similarly, for single shear or torsional stresses in *ductile* materials, the equations for factor of safety are, with a *steady* stress τ_m,

$$n = \frac{S_{sy}}{\tau_m} \tag{6.4}$$

and with an *alternating* stress τ_a and shear endurance limit S_{se},

$$n = \frac{S_{se}}{K_{fs} \, \tau_a} \tag{6.5}$$

The values of S_{sy} may be estimated from S_y by Eqs. (5.16) or (5.26), based on the maximum shear stress or the Mises theory of failure, respectively. The same ratios are used to estimate S_{se} from S_e, thus $S_{se} = 0.500S_e$ or $S_{se} = 0.577S_e$. The latter is recommended in consideration of the fatigue test results of Fig. 5.23. Figure 5.28 gives values of the theoretical shear stress-concentration factor K_{ts} for a shaft fillet in torsion, and the modified fatigue factor K_{fs} may be estimated by using q in Eq. (5.39) and replacing K_t by K_{ts}.

If the material is *brittle*, a stress raiser increases the likelihood of failure under either steady or alternating stresses, and it is customary to apply a stress-concentration factor to both. Because of limited knowledge of brittle material behavior and notch sensitivity q, the full, theoretical factor K_t is recommended for general use. Thus for single normal stresses in *brittle* materials, the equations for factor of safety are, with a *steady* stress σ_m and with the ultimate tensile stress S_{ut} as the basis of failure,

$$n = \frac{S_{ut}}{K_t \, \sigma_m} \tag{6.6}$$

and with an *alternating* stress σ_a and endurance limit S_e,

$$n = \frac{S_e}{K_t \, \sigma_a} \tag{6.7}$$

If the basis for failure in shear, S_{su}, is obtained from the conservative Mohr theory (Eq. (5.30) and Fig. 5.16), then for single shear stresses in *brittle* materials, the equations are, with a *steady* stress τ_m,

$$n = \frac{S_{su}}{K_{ts} \, \tau_m} = \frac{S_{ut}}{K_{ts} \, \tau_m \left[1 + (S_{ut}/S_{uc}) \right]} \tag{6.8}$$

and with an *alternating* stress τ_a and endurance limit S_{se},

$$n = \frac{S_{se}}{K_{ts} \, \tau_a} = \frac{S_e}{K_{ts} \, \tau_a \left[1 + (S_{ut}/S_{uc}) \right]} \tag{6.9}$$

It is known that the strength of gray cast irons is not much reduced by notches, probably because of the inherent microdiscontinuities already formed at the graphite flakes. Hence use of the *theoretical* stress-concentration factor appears unnecessary for cast irons. There have been some carefully executed designs in large castings that were made possible only by reduced factors. However, there is insufficient information for estimating actual static and fatigue factors from the theoretical value K_t, and the use of K_t for cast irons is commonly recommended for ordinary design. Peterson[1] states that the use of the full factor may be partly justified as compensating, in a way, for the poor shock resistance of brittle metals. It is not possible to design in a more rational way for the shock and mishandling in transportation and installation.

Example 6.1

The rope over the sheave of the hoist shown in Fig. 6.2 carries a load of 14 kN. The material, finish, and reliability of Example 5.6 will be used for the shaft, the dimensions of which are shown in millimeters. Fillet radii are $R = 10$ mm and $r = 3$ mm. Determine the factor of safety for the following three designs: (a) The shaft is clamped at the supports 1 and 3 to prevent its rotation, and the sheave has a bronze bushing and rotates on the shaft at 2, retained by collars not shown. (b) The sheave is pressed on the shaft at 2, and sheave and shaft rotate together with bushings at 1 and 3, not shown. Service is more or less continuous. (c) Same as (b) but use is occasional, with an estimated life requirement of 50 000 cycles.

Solution

(a) This is a case of static loading of the shaft under steady loads, constant in direction. The central downward force from the two parts of the rope is $2 \times 14 = 28$ kN. It is on the safe side to assume that the 14-kN reactions are centered along 1 and 3. There are three possibly critical sections, where the moment is largest for each diameter, namely, at the steps and at midlength. We are on the safe side to take the 28-kN force concentrated

---

[1] R. E. Peterson, *Stress Concentration Factors*, New York: John Wiley & Sons, Inc., p. 13, 1974.

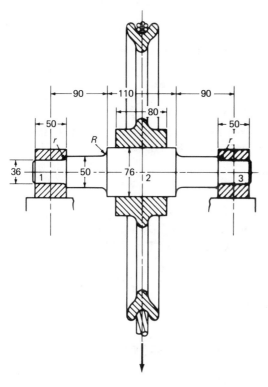

FIGURE 6.2 Shaft with idler rope sheave of Example 6.1. The dimensions are in millimeters.

at midlength, rather than distributed along the 80-mm length, in case the bushing wears to a bell-mouth shape from slightly out-of-plane rope pulls.

The moments are approximately 350, 1260, and 2030 N·m ($\times$ 10^3 N·mm) on diameters of 36, 50, and 76 mm, respectively, with corresponding section moduli ($I/c = \pi d^3/32$) of 4580, 12 270, and 41 420 mm^3. The stresses $\sigma = M/(I/c)$, are, respectively,

$$\sigma_m = \frac{(350 \times 10^3)}{4580} = 76.4 \text{ MPa}$$

$$\sigma_m = \frac{(1260 \times 10^3)}{12\ 270} = 102.7 \text{ MPa}$$

and

$$\sigma_m = \frac{(2030 \times 10^3)}{43\ 110} = 47.1 \text{ MPa}$$

The largest significant stress, 102.7 MPa, occurs in the 50-mm portion. The yield strength is 87 500 psi or 603.3 MPa, from the statement of Example 5.6. Then, the factor of safety for the shaft, Eq. (6.2), is

$$n = \frac{S_y}{\sigma_m} = \frac{603.3}{102.7} = 5.87$$

This is unnecessarily large, and for this nonrotating shaft, if the same dimensions are needed to maintain stiffness and fits, a plain-carbon, medium-carbon content, normalized steel should be considered.

(b) This is a case of alternating stresses, a complete reversal of stress with each rotation of the shaft, and the stress concentration is significant. Endurance limit is the basis for the factor of safety. Figures 5.27 and 5.29 and Eq. (5.39) will be used.

At the 3-mm fillet:

$$\frac{r}{d} = \frac{3}{36} = 0.083, \quad \frac{D}{d} = \frac{50}{36} = 1.39, \quad K_t = 1.80$$

Corresponding to $r = 3$ mm and normalized steel $q = 0.93$,

$$K_f = 1 + q(K_t - 1) = 1 + 0.93(1.80 - 1) = 1.74$$

The modified endurance limit from Example 5.6 is $S_e = 225$ MPa. The steady stresses of part (a) are now alternating stresses σ_a. By Eq. (6.3),

$$n = \frac{S_e}{K_f \, \sigma_a} = \frac{225}{(1.74)(76.4)} = 1.69$$

The factor of safety is probably higher than this and the moment and stress somewhat lower at this location because the shaft deflection tends to shift the center of bearing pressure inward, as indicated in Section 9.4 and Fig. 9.11.

At the 10-mm fillet:

$$\frac{r}{d} = \frac{10}{50} = 0.20, \quad \frac{D}{d} = \frac{76}{50} = 1.52, \quad K_t = 1.44$$

$$q = 0.97 \quad \text{and} \quad K_f = 1 + 0.97(1.44 - 1) = 1.43$$

$$n = \frac{S_e}{K_f \, \sigma_a} = \frac{225}{(1.43)(102.7)} = 1.53$$

Along the midsection, the highest stress concentration caused by the press fit occurs at the ends of the hub, where $M = (14 \text{ kN})(105 \text{ mm}) = 1470$ N·m. If we guess $K_f = 3.0$ for this intermediate-size shaft (Section 5.11), then the net endurance limit is $S_e/K_f = 225/3 = 75$ MPa (10 875 psi $< 20\,000$ psi), and the factor of safety is

$$n = \frac{S_e}{K_f \, \sigma_a} = \frac{225}{(3)(47.1)} = 1.59$$

Any uncertainty about the strength at the fit can be removed by cold-rolling or flame-hardening the 76-mm diameter. *The* factor of safety for the shaft is 1.53, as calculated at the 10-mm fillet.

(c) In part (b) the weakest section was found at the 10-mm fillet, where $K_f = 1.43$ and $\sigma_a = 102.7$ MPa. From Example 5.6, $S_u = 910$ MPa, $S_e' = 455$ MPa, and $S_e = 225$ MPa. Part (c) is a case of finite life, and the graphical method of Section 5.11 and Fig. 5.22 will be used to estimate the combined fatigue strength and reduced notch effect, i.e., the net fatigue strength $Q = S_f/K_f'$. Value $0.9S_u = (0.9)(910) = 819$ MPa, at $N = 10^3$ cycles. Value $S_e/K_f = 225/1.43 = 157$ MPa, at $N = 10^6$ cycles. The two points are plotted on closely ruled log-log paper,[2] the line Q is drawn between them, and the value at 5×10^4 cycles is read as 320 MPa. The term $Q = S_f/K_f'$ replaces S_e/K_f in Eq. (6.3), thus

$$n = \frac{Q}{\sigma_a} = \frac{320}{102.7} = 3.12 \qquad\qquad ////$$

---

[2] Figure 5.22 was constructed from the data of this problem. Paper with 1 × 3 cycles was sufficient, but usually 2 × 3 cycle paper is required.

6.3 DESIGN FOR SIMPLE FLUCTUATING STRESSES

We now consider the more general case of an alternating stress σ_a superimposed upon a mean stress σ_m, the fluctuating stresses of Figs. 6.1(d), (e), and (f). On a chart of σ_a vs σ_m, most experimental points lie above the line $MB'U$ of Fig. 6.3(a), which is drawn from the left horizontally through the laboratory specimen fatigue strength S_f' or S_e' at B' on the σ_a axis, then downward on the right to the ultimate tensile strength S_u at U on the σ_m axis. Although there has often been considerable scatter, tests on steel and aluminum specimens carefully mounted to ensure uniform axial loading indicated a fairly close linear relationship,[3] which is represented in Fig. 6.3(a) by an average line $LB'T$.

FIGURE 6.3 Limiting combinations of alternating and mean components of stress. (a) Alternating stress σ_a or $K\sigma_a$ vs mean component σ_m. (b) Upper and lower limits of stress and range of stress vs mean component σ_m.

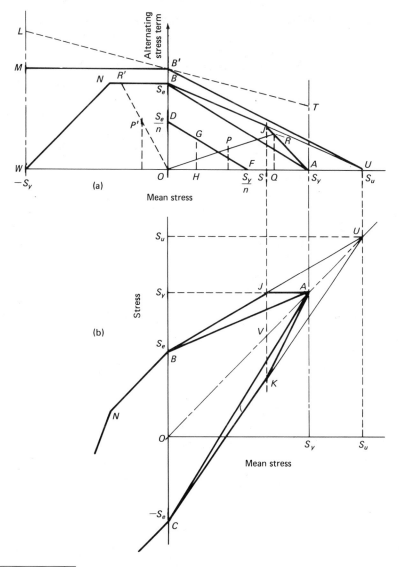

[3] G. Sines and J. L. Waisman, *Metal Fatigue*, New York: McGraw-Hill, pp. 155–158, 1959, reporting on tests by N. M. Newmark, R. J. Mosberg et al.; T. Nishihara, K. Kojima, and T. Sakurai; and M. Ros.

Point T, where $\sigma_m = +S_y$, has a σ_a value of about $0.63S_f{}'$. There is a definite reduction in alternating-stress capacity with increasing tensile mean stress, and, although not as generally accepted, there is a strong indication of an increase of capacity with increase of compressive mean stress. This is consistent with the observation that a fatigue failure requires that a crack open and spread, an action aided by tension and suppressed by compression; also, that compressive residual stresses are beneficial, tensile residual stresses harmful in fatigue (Section 7.9).

Charts for design should be based on the modified fatigue strengths S_f or endurance limits S_e (Section 5.10). Thus in Fig. 6.3(a), the line BU, known as the modified Goodman line, replaces line $B'U$, and a fatigue failure is considered a possibility for any alternating- and mean-stress combinations which plot above it. Since yielding may also constitute failure, particularly in machine parts, C. R. Soderberg proposed as a boundary the line BA, drawn to S_y, and Karpov, Lipson, Noll, and others proposed the line BJA, where the 45°-line JA limits the stress to $OQ + QR = OQ + QA = S_y$. For the same limitation against compressive yielding, line WN may be added.

The same failure lines are replotted in Fig. 6.3(b) to show the mean stress, OAU, the upper and lower limits of stress $NBJA$ and CKA, respectively, and the range of stress between them, all as functions of the mean stress. Points A, B, J, N, and U represent the same quantities as in Fig. 6.3(a), and additional points are added, C at $-S_e$ and V and K below J such that $VK = -VJ$.

The line $WNBJA$ in Fig. 6.3(a) is a reasonable boundary, and some graphical solutions are readily made. Consistent with Eqs. (6.2) and (6.3), it has been found satisfactory with ductile materials to multiply only the alternating component by the stress-concentration factor. Thus a point $(\sigma_m, K_f\sigma_a)$ is plotted, say at P, and the factor of safety equals OR/OP. If, instead, some dimensions are unknown, the stress components may be found in terms of the symbol of one dimension, say d, and in the ratio $K_f\sigma_a/\sigma_m$, d cancels out and the ratio is the slope of the line OR. The line is laid off on the diagram to locate R, and $OP = OR/n$. The dimension may be calculated from either the abscissa σ_m or the ordinate $K_f\sigma_a$ at point P.

It is often convenient to have the relationships for positive mean stresses in equation form, particularly for use with combined bending and torsional stresses (Section 6.4). This is done most readily with the continuous and conservative Soderberg line BA. With a factor of safety n, a parallel design line DF may be drawn between S_e/n on the σ_a axis and S_y/n on the σ_m axis (Fig. 6.3(a)), giving the Soderberg design triangle DOF. A point G has the abscissa $OH = \sigma_m$ and the ordinate $HG = K_f\sigma_a$. From the proportional triangles GHF and DOF,

$$\frac{HF}{OF} = \frac{HG}{OD} \quad \text{or} \quad \frac{(S_y/n) - \sigma_m}{S_y/n} = \frac{K_f\sigma_a}{S_e/n}$$

whence

$$n_{\text{ductile}} = \frac{S_y}{\sigma_m + \dfrac{S_y}{S_e}K_f\sigma_a} \tag{6.10}$$

The denominator is the distance OF, and the numerator is OA.

In brittle materials there is no appreciable yielding, and the failure line BA of Fig. 6.3 is replaced by BU. The theoretical stress concentration factor K_t is used and applied to both components, as in Eqs. (6.6) and (6.7). We may write by analogy to Eq. (6.10),

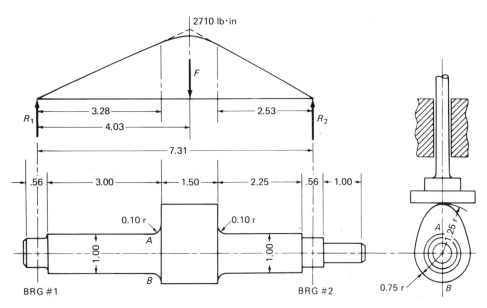

FIGURE 6.4 Camshaft of Example 6.2. The dimensions are in inches.

$$n_{\text{brittle}} = \frac{S_u}{K_t \left(\sigma_m + \dfrac{S_u}{S_e} \sigma_a \right)} \tag{6.11}$$

Many tests in fluctuating torsion on a large number of *ductile* metals show that the shear endurance limit is not reduced by a mean torsional stress provided the *maximum* shear stress is less than about 0.80 of the shear yield strength.[4] This suggests that Eq. (6.5) may be used within limits for the factor of safety. In practice, however, a triangle and equation similar to those proposed by Soderberg may be used for shear stresses, e.g., the chart for helical springs (Fig. 12.11), and the combined stress situations of Section 6.4.

Example 6.2

The camshaft (Fig. 6.4) of an intermittent-motion mechanism operates against a follower load of 1500 lb in its maximum lift position. However, a stop (not shown) limits the return motion of the follower, such that it rides free of the cam during more than half a shaft revolution. Consider the torque to be negligible. The material is AISI 8620, carburized on the cam surface, reheated and oil quenched and tempered, leaving the section at the fillet with an ultimate strength of 122 000 psi and a yield strength of 78 000 psi, and the cam surface with a hardness of R_c 62. The fillet and adjacent surfaces are fine ground. A material reliability of 99.9% is desired. Dimensions are in inches. Determine the factor of safety for the camshaft for unlimited life.

Solution

Force is exerted on the cam during less than half of each shaft revolution. Thus the shaft is given a pulsating force (Fig. 6.1(d)), with a maximum load of 1500 lb and a

---

[4] G. Sines and J. L. Waisman, *Metal Fatigue*, New York: McGraw-Hill, p. 156, 1959 (see footnote 3).

minimum of zero. Dimensions for calculation, forces, and moment diagram are shown above the camshaft. For equilibrium,

$$R_1 = \frac{(7.31 - 4.03)}{7.31}(1500) = 673 \text{ lb}$$

and

$$R_2 = \frac{4.03}{7.31}(1500) = 827 \text{ lb}$$

The moment diagram is constructed from a "peak" moment of $673 \times 4.03 = 2710$ lb·in. It is seen that the moment at the left side of the cam is larger than at the right, and it equals

$$M = 673 \times 3.28 = 2208 \text{ lb·in}$$

For a 1.00-in diameter, $I/c = \pi d^3/32 = 0.0982 \text{ in}^3$

$$\sigma_{max} = \frac{M}{I/c} = \frac{2208}{0.0982} = 22\,480 \text{ psi} \quad \text{and} \quad \sigma_{min} = 0$$

$$\sigma_m = \frac{\sigma_{max} + \sigma_{min}}{2} = \frac{22\,480 + 0}{2} = 11\,240 \text{ psi}$$

$$\sigma_a = \frac{\sigma_{max} - \sigma_{min}}{2} = \frac{22\,480 - 0}{2} = 11\,240 \text{ psi}$$

Corresponding to a ground finish, 1.00-in diameter, and 99.9% material reliability, the factors from Figs. 5.24 and 5.25 and following Eq. (5.38) are, respectively, $C_s = 0.90$, $C_z = 0.90$, and $C_r = 0.75$. From Eqs. (5.36) and (5.37), estimated values of endurance limits are

$$S_e{}' = 0.50 S_u = 0.50(122\,000) = 61\,000 \text{ psi}$$

$$S_e = C_s C_z C_t C_r C_m S_e{}' = (0.90)(0.90)(1.00)(0.75)(1.0)(61\,000) = 37\,000 \text{ psi}$$

The step in the shaft at the cam is not symmetrical. It is reasonable to assume that the stress at fiber A is principally influenced by the 1.25-in-cam radius, equivalent to a diameter $D = 2.50$ in, and the fiber at B, by the 0.75-in-cam radius, equivalent to $D = 1.50$ in. Accordingly, with $r = 0.10$ in, $r/d = 0.10/1.00 = 0.10$; at A, $D/d = 2.50/1.00 = 2.50$ and $K_t = 1.84$ from Fig. 5.27; and at B, $D/d = 1.50/1.00 = 1.50$ and $K_t = 1.72$. From Fig. 5.29. the notch-sensitivity curve for quenched and tempered steel ends at a radius of 0.08 in, indicating that available data do not justify extension of the curve and that we must assume full notch sensitivity, i.e., $q = 1.0$. Use of Eq. (5.39) is unnecessary, since the substitution of $q = 1.0$ gives $K_f = K_t$. Hence $(K_f)_A = 1.84$ and $(K_f)_B = 1.72$.

Since the stress concentration factor at A is only 7% larger than at B, we should expect fatigue failure to begin at B because the stress pulses are tensile at B and compressive at A. At B, the factor of safety is, by Eq. (6.10),

$$n_B = \frac{S_y}{\sigma_m + \dfrac{S_y}{S_e} K_f \sigma_a} = \frac{78\,000}{11\,240 + \dfrac{78}{37}(1.72)(11\,240)}$$

$$= \frac{78\,000}{11\,240 + 40\,760} = \frac{78\,000}{52\,000} = 1.50$$

The stresses at A are always compressive or zero. Factor n_A may be obtained graphically by plotting point $(-\sigma_m, K_f\sigma_a)$ at P' on a chart like the one given in Fig. 6.3(a) and extending the line OP' to intersect the failure envelope. A quickly made, approximately to scale sketch shows that the intersection R' is with the horizontal line NB. Hence the factor of safety is

$$n_A = \frac{S_e}{K_f\,\sigma_a} = \frac{37\ 000}{(1.84)(11\ 240)} = 1.79$$

If we assume that the moment at the fillet adjacent to each ball bearing is too small to make it a critical location, then *the* factor of safety for the camshaft is 1.50. If the load is well controlled and there is no impact, this factor seems quite sufficient. However, deflection should be checked, particularly since lift motion is involved. How can one make a quick, approximate calculation of the deflection of this stepped beam?   ////

6.4 FLUCTUATING NORMAL AND SHEAR STRESSES: MAXIMUM SHEAR AND NORMAL-STRESS THEORIES OF FAILURE

Soderberg[5] extended the straight-line method for fluctuating uniaxial stresses (Fig. 6.3) to the case of a fluctuating normal stress and a fluctuating shear stress (Fig. 6.5(a)). This combination may occur in shafts and crankshafts from an axial force and bending and torsional moments. Normal stress $\sigma_m \pm K_f\sigma_a$ and shear stresses $\tau_m \pm K_{fs}\tau_a$ result. The worst case occurs when σ_a and τ_a are in phase or when the frequency of one is an integral multiple of the other. Forces are shown on an elemental prism of unit thickness (Fig. 6.5(b)). A summation of forces tangent to the diagonal determines the combined shear stress τ_c, as follows:

$$-\tau_c\,dc + (\sigma_m + K_f\sigma_a)\,dy\cos\phi + (\tau_m + K_{fs}\tau_a)\,dx\cos\phi$$
$$- (\tau_m + K_{fs}\tau_a)\,dy\sin\phi = 0 \tag{a}$$

FIGURE 6.5 (a) Fluctuating normal and shear stresses on an element of a shaft. (b) Resulting forces on an elemental prism of unit thickness.

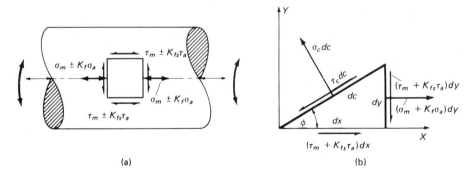

(a)   (b)

[5] C. R. Soderberg, "Factor of Safety and Working Stress," *Trans. ASME*, Vol. 52, Part 1, paper APM 52−2, 1930. The particular treatment in the following is adapted from that by R. E. Peterson, *Stress Concentration Design Factors*, New York: John Wiley & Sons, Inc., pp. 148−150, 1953.

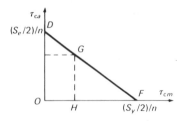

FIGURE 6.6 Soderberg design line for shear stresses, based on maximum shear-stress theory of failure.

$$\tau_c = (\sigma_m + K_f\sigma_a)\frac{dy}{dc}\cos\phi + (\tau_m + K_{fs}\tau_a)\frac{dx}{dc}\cos\phi - (\tau_m + K_{fs}\tau_a)\frac{dy}{dc}\sin\phi$$

$$= (\sigma_m + K_f\sigma_a)\sin\phi\,\cos\phi + (\tau_m + K_{fs}\tau_a)\cos^2\phi - (\tau_m + K_{fs}\tau_a)\sin^2\phi$$

$$= \frac{\sigma_m + K_f\sigma_a}{2}\sin 2\phi + (\tau_m + K_{fs}\tau_a)\cos 2\phi \tag{b}$$

The combined shear stress is separated into mean and alternating components, thus

$$\tau_c = \tau_{cm} + \tau_{ca}$$

$$= \left(\frac{\sigma_m}{2}\sin 2\phi + \tau_m\cos 2\phi\right) + \left(\frac{K_f\sigma_a}{2}\sin 2\phi + K_{fs}\tau_a\cos 2\phi\right) \tag{c}$$

Analogous to line DF of Fig. 6.3(a), a suitable design line on a τ_{ca} vs τ_{cm} chart for the maximum shear-stress theory of failure is drawn between end points $S_{se} = (S_e/2)/n$ and $S_{sy} = (S_y/2)/n$ (Fig. 6.6). This uses the value of S_{sy} of Eq. (5.16) and the first value of S_{se} following Eq. (6.5). Note that, unlike point G of Fig. 6.3(a), the ordinate here does not contain a stress-concentration factor since the necessary two factors are included in the τ_{ca} term of Eq. (c). From proportional triangles GHF and DOF (Fig. 6.6),

$$\frac{HF}{OF} = \frac{HG}{OD} \quad\text{or}\quad \frac{(S_y/2)/n - \tau_{cm}}{(S_y/2)/n} = \frac{\tau_{ca}}{(S_e/2)/n} \tag{d}$$

whence

$$\frac{1}{n} = \frac{\tau_{cm}}{S_y/2} + \frac{\tau_{ca}}{S_e/2} \tag{6.12}$$

By substitution for τ_{cm} and τ_{ca} from Eq. (c),

$$\frac{1}{n} = \frac{(\sigma_m/2)\sin 2\phi + \tau_m\cos 2\phi}{S_y/2} + \frac{(K_f\sigma_a/2)\sin 2\phi + K_{fs}\tau_a\cos 2\phi}{S_e/2}$$

$$= \left(\frac{\sigma_m}{S_y} + \frac{K_f\sigma_a}{S_e}\right)\sin 2\phi + 2\left(\frac{\tau_m}{S_y} + \frac{K_{fs}\tau_a}{S_e}\right)\cos 2\phi$$

whence

$$\frac{1}{n} = A\sin 2\phi + 2B\cos 2\phi \tag{e}$$

where

$$A = \frac{\sigma_m}{S_y} + \frac{K_f\sigma_a}{S_e} \quad\text{and}\quad B = \frac{\tau_m}{S_y} + \frac{K_{fs}\tau_a}{S_e}$$

The factor of safety n for the stressed point is its minimum value as the τ_c plane is

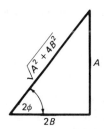

FIGURE 6.7 A relationship leading to Eq. (6.13).

given various orientations ϕ. The minimum value for n corresponds to the maximum value for $1/n$. By differentiation of $1/n$ in Eq. (e) and equating the result to zero, the corresponding value of ϕ is obtained, thus

$$\frac{d}{d\phi}\left(\frac{1}{n}\right) = 2A \cos 2\phi - 4B \sin 2\phi = 0$$

whence

$$\frac{\sin 2\phi}{\cos 2\phi} = \tan 2\phi = \frac{A}{2B}$$

This relationship is shown on the triangle of Fig. 6.7, and it is seen that

$$\sin 2\phi = \frac{A}{\sqrt{A^2 + 4B^2}}, \quad \cos 2\phi = \frac{2B}{\sqrt{A^2 + 4B^2}}$$

and, by substitution into Eq. (e), that

$$\frac{1}{n} = \frac{A^2 + 4B^2}{\sqrt{A^2 + 4B^2}} = \sqrt{A^2 + 4B^2} \tag{f}$$

With resubstitution of the values of A and B,

$$\frac{1}{n} = \sqrt{\left(\frac{\sigma_m}{S_y} + \frac{K_f\sigma_a}{S_e}\right)^2 + 4\left(\frac{\tau_m}{S_y} + \frac{K_{fs}\tau_a}{S_e}\right)^2} \tag{6.13}$$

By inversion and the factoring of S_y,

$$n = \frac{S_y}{\sqrt{\left(\sigma_m + \frac{S_y}{S_e}K_f\sigma_a\right)^2 + 4\left(\tau_m + \frac{S_y}{S_e}K_{fs}\tau_a\right)^2}} \tag{6.14}$$

If in Eqs. (5.13) and (5.14), obtained in a previous treatment of biaxial stresses, we set $\sigma_y = 0$, $\sigma_x = \sigma$, and $\tau_{xy} = \tau$, then the maximum shear stress becomes $\tau_{\max} = \sqrt{(\sigma/2)^2 + \tau^2}$, and the factor of safety is

$$n = \frac{S_y/2}{\tau_{\max}} = \frac{S_y/2}{\sqrt{(\sigma/2)^2 + \tau^2}} = \frac{S_y}{\sqrt{\sigma^2 + 4\tau^2}} \tag{6.15}$$

Equations (6.15) and (6.14) have the same form, and we may think of $\sigma = \sigma_m + (S_y/S_e)K_f\sigma_a$ and $\tau = \tau_m + (S_y/S_e)K_{fs}\tau_a$ as equivalent static stresses, each consisting of a steady component and an alternating component, the latter weighted for the effect of fatigue and stress concentration. Equation (6.14) reduces to Eq. (6.10) for a normal stress only, and to the several single-component equations of Section 6.2.

Since the torque is fairly steady on many rotating shafts which have complete reversal of bending stresses, a useful form of Eq. (6.14) is obtained by setting $\sigma_m = \tau_a = 0$; thus

$$n = \frac{S_y}{\sqrt{\left(\dfrac{S_y}{S_e} K_f \sigma_a\right)^2 + 4\tau_m{}^2}} \tag{6.16}$$

Limited experimental data[6] indicate that a steady torsional stress does not decrease the bending-stress fatigue strength, with or without notches. Hence the data do not support this equation, except to indicate that it is *on the safe side*, and therefore useful in design until a better theory or method is found.

For brittle materials somewhat similar equations are obtained by the application of stresses $K_t(\sigma_m + \sigma_a)$ and $K_{ts}(\tau_m + \tau_a)$. Forces are summed *normal* to the diagonal of Fig. 6.5(b), and the resulting expression for σ_c is separated into mean and alternating components, σ_{cm} and σ_{ca}. The design line will extend from S_e/n to S_u/n, and its equation, corresponding to Eq. (6.12), is

$$\frac{1}{n} = \frac{\sigma_{cm}}{S_u} + \frac{\sigma_{ca}}{S_e} \tag{6.17}$$

By substitution and procedure as before, the result, which should be compared with Eqs. (6.14) and (5.13), is

$$n = \frac{S_u}{\dfrac{1}{2}K_t\left(\sigma_m + \dfrac{S_u}{S_e}\sigma_a\right) + \dfrac{1}{2}\sqrt{K_t{}^2\left(\sigma_m + \dfrac{S_u}{S_e}\sigma_a\right)^2 + 4K_{ts}{}^2\left(\tau_m + \dfrac{S_u}{S_e}\tau_a\right)^2}} \tag{6.18}$$

6.5 FLUCTUATING NORMAL AND SHEAR STRESSES: MISES AND RELATED THEORIES OF FAILURE

R. E. Peterson[7] modifies Eq. (6.14) by changing the coefficient of the shear-stress term from 4 to 3, such that the Mises criterion is satisfied in each of the end- or single-component conditions, thus

$$n = \frac{S_y}{\sqrt{\left(\sigma_m + \dfrac{S_y}{S_e} K_f \sigma_a\right)^2 + 3\left(\tau_m + \dfrac{S_y}{S_e} K_{fs}\tau_a\right)^2}} \tag{6.19}$$

If the normal-stress components do not exist, and $\tau_a = 0$, Eq. (6.19) reduces to $n = (S_y/\sqrt{3})/\tau_m$, corresponding to Eqs. (5.26) and (6.4). If $\tau_m = 0$, Eq. (6.19) reduces to $n = (S_e/\sqrt{3})/K_{fs}\tau_a$, corresponding to Eq. (6.5) and the discussion following it, with $S_{se} = (1/\sqrt{3})S_e = 0.577S_e$. For all other combinations that include shear stresses, the denominator is smaller and the factor of safety is slightly larger than indicated by Eq. (6.14). Thus the equation is less conservative. In terms of equivalent static stresses the equation is

$$n = \frac{S_y}{\sqrt{\sigma^2 + 3\tau^2}} \tag{6.20}$$

which may be compared with Eq. (6.15). For the common, shaft case of $\sigma_m = \tau_a = 0$,

[6] G. Sines and J. L. Waisman, *Metal Fatigue*, New York: McGraw-Hill, pp. 157–158, 1959; M. G. S.-R. Aiyer, "Fatigue of Metals under Combined Stresses," Ph.D. Thesis, Cornell University, Ithaca, New York, 1969.

Eq. (6.19) reduces to

$$n = \frac{S_y}{\sqrt{\left(\frac{S_y}{S_e} K_f \sigma_a\right)^2 + 3\tau_m^2}}$$

(6.21)

In his presentation of Eq. (6.19), Peterson[7] separates the mean-normal component σ_m into an axial stress and a bending stress and divides the latter by a limit-design factor, L_b (Section 5.8). He also divides the mean-shear component τ_m by a limit factor L_s. This form gives an even higher value to n.

For the general biaxial case of fluctuating stress, the principal stresses are $\sigma_1 = \sigma_{1m} \pm \sigma_{1a}$ and $\sigma_2 = \sigma_{2m} \pm \sigma_{2a}$. The denominator of Eq. (5.25) is considered an equivalent stress, *intensity of stress*, or von Mises stress

$$s = \sqrt{\sigma_1^2 - \sigma_1\sigma_2 + \sigma_2^2}$$

(6.22)

Soderberg[8] defined *mean* and *alternating* intensities of stress for the biaxial case as, respectively,

$$s_m = \sqrt{\sigma_{1m}^2 - \sigma_{1m}\sigma_{2m} + \sigma_{2m}^2}$$

(6.23a)

and

$$s_a = \sqrt{(K_{f1}\sigma_{1a})^2 - (K_{f1}\sigma_{1a})(K_{f2}\sigma_{2a}) + (K_{f2}\sigma_{2a})^2}$$

(6.23b)

These stress intensities plot as a point (s_m, s_a) on a chart of σ_a vs σ_m, such as Fig. 6.3(a). The factor of safety may be found graphically as described in Section 6.3, using either the envelope *WNBJA* or the line *BA* as the limit beyond which failure will occur. Instead of the latter, the equation of the Soderberg design line (Eq. 6.10) may be used, namely,

$$n = \frac{S_y}{s_m + \frac{S_y}{S_e} s_a}$$

(6.24)

Note that K_f is omitted from Eq. (6.24) so that different factors K_{f1} and K_{f2} may be used in Eq. (6.23b) for the effect of the stress raiser in planes 1–3 and 2–3, if available.

Equations (6.22), (6.23), and (6.24) may be adapted to the case of a torsional stress $\tau = \tau_m \pm \tau_a$ and a single normal stress $\sigma = \sigma_m \pm \sigma_a$, typical of shafts. The resulting equation is

$$n = \frac{S_y}{\sqrt{\sigma_m^2 + 3\tau_m^2} + \frac{S_y}{S_e}\sqrt{(K_f\sigma_a)^2 + 3(K_{fs}\tau_a)^2}}$$

(6.25)

Although it contains the same terms as Eqs. (6.14) and (6.19), the form is different, and a comparison is difficult. For the typical shaft case where $\sigma_m = \tau_a = 0$, Eq. (6.25) reduces to

$$n = \frac{S_y}{\frac{S_y}{S_e} K_f\sigma_a + \sqrt{3}\tau_m}$$

(6.26)

[7] R. E. Peterson, *Stress Concentration Factors*, New York: John Wiley & Sons, Inc., p. 18, 1974.
[8] C. R. Soderberg, *Handbook of Experimental Stress Analysis*, M. Hetenyi, (Ed.), New York: John Wiley & Sons, Inc., pp. 454–455, 1950.

It is readily proven that Eq. (6.26) gives smaller values of n than does Eq. (6.21) except when either σ_a or τ_m is zero. Hence Eq. (6.26) is more conservative. Also, except at small ratios of the first term of the denominator to the second, Eq. (6.26) gives smaller values and is more conservative than Eq. (6.16), the equation based on the maximum shear-stress theory of failure. This is contrary to expectations from the Mises theory for the static case, illustrated in Fig. 5.13 by the Mises ellipse surrounding the shear-theory hexagon, and in Fig. 5.23 by the higher, Mises line on the torsion fatigue limit vs bending fatigue-limit chart. Also, by giving more emphasis to the static component, it is contrary to experimental evidence such as the line $B'T$ of Fig. 6.3(a) and the horizontal or nearly horizontal lines through B', such as those of Sines and Kececioglu, discussed in the next section. Hence the equations for shaft design developed in Section 6.7 are based on the less conservative sets of equations, Eqs. (6.14) through (6.16) and Eqs. (6.19) through (6.21) in preference to Eqs. (6.25) and (6.26).

6.6 FLUCTUATING NORMAL AND SHEAR STRESSES: SOME ADDITIONAL THEORY AND EXPERIMENTATION

Sines[9] observed that a change of fatigue strength with changing static stress σ_m occurred only in tests where a static normal stress acted on the planes of maximum alternation of shear stress. A net tensile normal stress decreased and a net compressive stress increased the allowable σ_a, as along line $LB'T$ of Fig. 6.3(a). Sines proposed that the permissible alternation of the octahedral shear stress τ_a be a linear function of the sum σ_m of the orthogonal or principal normal static stresses (which sum is three times the normal stress on the orthogonal plane). An equation for the octahedral shear stress is given in the answer to Prob. 5.39. The alternating components of σ_1, σ_2, and σ_3 are substituted in it. The mean components are summed to give σ_m. The relationship is then $\tau_a = A - \alpha\sigma_m$, or

$$\tau_a = \frac{1}{3}\sqrt{(\sigma_{1a} - \sigma_{2a})^2 + (\sigma_{2a} - \sigma_{3a})^2 + (\sigma_{3a} - \sigma_{1a})^2}$$

$$= A - \alpha(\sigma_{1m} + \sigma_{2m} + \sigma_{3m}) \tag{6.27}$$

The constants A and α may be determined from the coordinates of two well-separated points on an experimental line, such as $LB'T$ of Fig. 6.3(a).

For the biaxial case of Eq. (6.27), with $\sigma_{3a} = \sigma_{3m} = 0$, one alternating component σ_{1a} plots like a Mises ellipse against the other alternating component σ_{2a}, for any one constant, steady-component sum σ_m. Sines found that the very few experimental data available plotted well along the ellipses. For the common shaft situation of alternating bending stress σ_{xa} and steady torsional stress τ_{xym}, the planes of maximum alternation of shear stresses τ_a are at 45° with the axis of the shaft, and the normal (principal) stresses due to the steady torque sum to zero. The principal stresses are $\sigma_{1m} = \tau_{xym}$, $\sigma_{2m} = -\tau_{xym}$, and $\sigma_{3m} = 0$ by Eqs. (5.13), whence $\sigma_m = \sigma_{1m} + \sigma_{2m} + \sigma_{3m} = 0$. Also, $\sigma_{xa} = \sigma_a$ and $\sigma_{ya} = \sigma_{za} = \tau_{xya} = 0$, whence by Eqs. (5.13), $\sigma_{1a} = \sigma_a$ and $\sigma_{2a} = \sigma_{3a} = 0$. Hence in Eq. (6.27), the sum τ_a is $(\frac{1}{3})\sqrt{\sigma_a{}^2 + \sigma_a{}^2}$ or $(\sqrt{2}/3)\sigma_a$, and the equation becomes $(\sqrt{2}/3)\sigma_a = A$. The value of A is determined from the known point, $\sigma_a = S_e$ at $\sigma_m = 0$. Substitution gives $A = (\sqrt{2}/3)S_e$, whence Eq. (6.27) is reduced for

[9] George Sines in G. Sines and J. L. Waisman, *Metal Fatigue*, New York: McGraw-Hill, pp. 158–167, 1959; also G. Sines, "Failure of Materials under Combined Repeated Stresses with Superimposed Static Stresses," NACA Tech. Note 3495, November 1955; and G. Sines, *Elasticity and Strength*, Boston, Massachusetts: Allyn and Bacon, Inc., pp. 72–78, 1969.

this special but common case to

$$\sigma_a = S_e \tag{6.28}$$

This indicates that the allowable alternating bending stress is independent of the torsion applied, i.e., line $LB'T$ is horizontal, which is consistent with the observation made following Eq. (6.16).

Attention is called by reference to studies by Kececioglu, Chester, and Dodge[10] and by Mitchell and Vaughan.[11] Kececioglu et al. are concerned with tests on grooved specimens under different alternating-bending and steady-torque stresses. These are converted to von Mises stresses $s_a = \sigma_{xa}$ and $s_m = \sqrt{3}\tau_{xym}$ and plotted on a σ_a vs σ_m chart. The line for which an equation is formulated lies somewhat higher than the modified Goodman and is nearly horizontal through S_e and halfway to S_u to the right, thus again indicating the very small dependence of the allowable alternating stress on the torsional stress in that region. The work of Mitchell and Vaughan is concerned with the analysis of any type of failure theory and design lines on a σ_a vs σ_m chart, with the type of overloading taken into account. The modified Goodman chart with Mises stresses and the Soderberg equations for alternating bending and steady torque are derived as special cases.

For a shaft with stress concentration, it will be seen that unless the shaft is relatively short so that the alternating-bending moment is considerably less than the steady torque, the shear-stress term is often negligible relative to the normal-stress term. Thus the coefficient of the shear-stress term, or its neglect, may be of little practical consideration.

6.7 SHAFT DESIGN

The foregoing equations are most often applied to the design of shafts, axles, spindles, and pins. Sketches of complete shafts, some in assembly, may be found in Figs. 6.2, 6.4, 6.8, and 6.10 and Probs. 6.16, 6.21, 6.34, 6.36, 6.37, 6.40, and 6.41 of this chapter, as well as in Figs. 4.1, 9.1, 9.10–9.12, 9.16, 9.17, 9.29, and Prob. 9.29. Initially, the shaft may be drawn as part of an assembly of components, which will partly determine the shaft dimensions, particularly the length and the axial spacing of steps. Then forces and moments are calculated. Some diameters may be determined by the bores of standard or stock components such as gears, sprockets, pulleys, and bearings, and the economy of minimizing dimensional variations between assemblies of different capacity. Diameters may be chosen to give an apparently sufficient stiffness to the shaft, according to experience with similar assemblies. Material may be chosen from stock or warehouse availability, its treatment and finish according to the shop's equipment and skills. Thus with tentative dimensions, moments, and strength known, factors of safety may be calculated at possibly critical sections by the foregoing equations, then a decision may be made on where changes are needed.

However, the engineer may be free to make the necessary assumptions and calculate diameters directly. This may be particularly so in preliminary design, on overhung or extended lengths of shafts, on shafts carrying rotors, and where components such as gears and flanges are formed integral with the shaft. It is convenient to have the preceding equations solved for diameter d in terms of estimated mean and alternating components of bending and torsional moments, $M_m \pm M_a$ and $T_m \pm T_a$, respectively. Since, in

[10] D. Kececioglu, L. B. Chester, and T. M. Dodge, "Combined Bending-Torsion Fatigue Reliability of AISI 4340 Steel Shafting with $K_t = 2.34$," *J. Eng. Ind., Trans. ASME*, Vol. 97, pp. 748–761, 1975.

[11] L. D. Mitchell and D. T. Vaughan, "A General Method for the Fatigue-Resistant Design of Mechanical Components—Part I Graphical, Part II Analytical," *J. Eng. Ind., Trans. ASME*, Vol. 97, pp. 965–975, 1975.

general, $\sigma = Mc/I = 32M/\pi d^3$ and $\tau = Tc/J = 16T/\pi d^3$, we may substitute $32M_a/\pi d^3$ for σ_a, $16T_m/\pi d^3$ for τ_m, etc. Equation (6.14), based on the maximum shear-stress theory of failure, becomes

$$d^3 = \frac{32n}{\pi S_y} \sqrt{\left(M_m + \frac{S_y}{S_e} K_f M_a\right)^2 + \left(T_m + \frac{S_y}{S_e} K_{fs} T_a\right)^2} \tag{6.29}$$

Peterson's modification, Eq. (6.19), becomes

$$d^3 = \frac{32n}{\pi S_y} \sqrt{\left(M_m + \frac{S_y}{S_e} K_f M_a\right)^2 + \frac{3}{4}\left(T_m + \frac{S_y}{S_e} K_{fs} T_a\right)^2} \tag{6.30}$$

In these equations for d^3, the values of S_y, S_y/S_e, K_f, and K_{fs} depend somewhat on d itself. However, they may be estimated if a likely range for the diameter may be anticipated. Then, corresponding to the steel and treatment chosen, S_y and S_u may be read from mass-effect tables, and $S_e{}'$ estimated by Eq. (5.36). Size factor C_z, reliability factor C_r if desired, and surface factor C_s may be chosen, from which fatigue-endurance limit S_e is estimated by Eq. (5.37). Unless the fillet radius may be anticipated, one may take general values for notch-sensitivity factor q, about 0.85 for annealed and normalized steels and 0.95 for quenched and tempered steels (Fig. 5.29). Then, if there is freedom for the choice of the ratios r/d and D/d, K_t and K_{ts} may be read from Figs. 5.27 and 5.28, respectively, and K_f and K_{fs} calculated by Eq. (5.39). With the desired factor of safety n, Eq. (6.29) or Eq. (6.30) may now be solved for a first approximation to the diameter d. This is followed by a recalculation with more accurate values of strength factors and stress-concentration factors.

If the diameters d, D, and fillet radius r are first determined from other considerations, one may wish to use Eqs. (6.29) or Eq. (6.30) to obtain the strength required, and hence the material and its treatment. This is done by interchanging the positions of d^3 and S_y outside the radical. Before solving for S_y, one must estimate the ratio S_y/S_e, the value of which depends to some extent on the strength desired and the method of estimating S_e. As a guide the following values are offered, calculated from charts for AISI 1030, 1040, 4140, 4340, and 8740 steels in diameters ½, 1, and 2 in. Endurance limit S_e was obtained by Eq. (5.37) with no reliability factor ($C_r = 1.0$), and with size factor $C_z = 0.90$, suitable for this range of diameters. To obtain surface factor C_s, the steels when quenched and tempered were considered finished fine-ground, and the steels when normalized, finished by machining. For the hardened steels tempered to 1000°F and to 1200°F, the mean S_y/S_e value is 1.80 for the water-quenched carbon steels and 2.25 for the oil-quenched alloy steels. For the normalized steels, the mean value of S_y/S_e is 1.85 for the carbon steels and 2.10 for the alloy steels. Thus some anticipation of strength and treatment of the material is helpful in using the equations to solve for S_y. From the solution, the material and its treatment may be tentatively selected.

Adjacent to a rolling-element bearing, the fillet radius and the step in diameter depend on the bore and other dimensions of the bearing. If the shaft diameter is to be determined and followed by selection of the bearing, the stress-concentration factors must be estimated for a first calculation. Although there is considerable variation, a K_f value of 2.20 is close to the average for quenched and tempered steels and 2.07 for annealed and normalized steels with bearings of the 02, 03, and 04 series (Section 11.11) and in bore diameters from 15 mm (0.60 in) to 100 mm (3.94 in). The corresponding K_{fs} values may be 1.65 and 1.58, respectively. After a first calculation for d, a bearing, together with shaft shoulder diameter D and fillet radius r, may be tentatively selected from a bearing catalog. Then K_t and q values may be read, K_f calculated, and Eq. (6.29) or Eq. (6.30) used for a second calculation of d, or, alternatively, for a calculation of factor of safety.

A quick, preliminary calculation may be made to give the order of magnitude of the diameter in the common case of steady torque and alternating-bending moment. Here M_m = T_a = 0, and with $T_m \approx M_a$ and with typical K_f and S_y/S_e values of 2.0, the first term in Eq. (6.29) or Eq. (6.30) is 16 or more times the second. Thus with the second term neglected, and with typical values chosen for S_y, S_y/S_e, and K_f, the equation may be solved to give the order of magnitude of d. Tentative values of strength and other factors may then be chosen.

6.8 EXAMPLES OF SHAFT DESIGN

Example 6.3

General proportions, forces, and moment diagrams are shown in Fig. 6.8 for a shaft with a spur gear mounted and centered at location 1, a V-belt sheave at location 3, and rolling bearings at locations 0 and 2. The material proposed is AISI 4140, which is readily available in many bar sizes, and it will be oil quenched and tempered as necessary. The factor of safety is to be at least 2.0, which although high, includes consideration of reliability and the need for shaft stiffness at the gear. The torque is essentially steady. The bearing seats are fine ground, the adjacent fillets commerically polished, and the remaining surfaces machined. Determine the approximate diameters required for strength at the critical sections.

Solution

There appear to be four critical sections. Two are at the gear where the moment is highest and where there are stress concentrations, at the end of the keyway and at the fillet of the step. The others are at the two sides of bearing #2, where the moment is still high, the cross section is smaller, and the fillet radii are limited.

We should temper after quenching, preferably to 1200°F for greater quenching-strain relief which is important for shafts that should remain straight after machining. From steel company tables that show the effect of mass,[12] we see that the yield strength of AISI 4140 averages about 100 000 psi after tempering at 1200°F. For the quick calculation previously described and with reference to the foregoing data, we select typical values of 2.25 for S_y/S_e and for K_f, 2.0 at the gear and 2.20 at the bearing. The S_y/S_f

FIGURE 6.8 Preliminary sketch for shaft of Example 6.3, with force and moment diagrams.

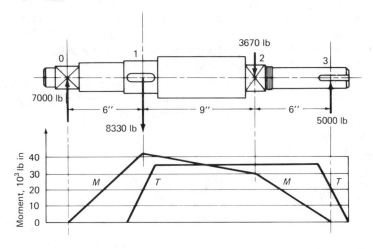

---

[12] E.g., *Modern Steels and Their Properties*, Bethlehem Steel Corporation, Bethlehem, Pennsylvania.

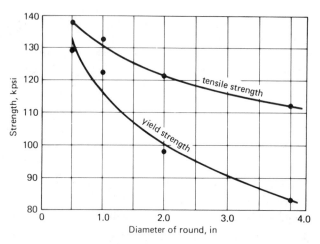

FIGURE 6.9 Strengths vs diameter of rounds, AISI 4140 steel oil quenched from 1550°F, tempered at 1200°F. (Data plotted from table in *Modern Steels and Their Properties,* Bethlehem Steel Corporation, Bethlehem, Pennsylvania, p. 148 in the edition of 1978. Data courtesy of Bethlehem Steel Corporation.)

value is based on a fine-ground finish, and since the shaft is machined at the gear, we might estimate a higher value there. However, we shall use the lower value to demonstrate that even an inaccurate estimate leads quickly to a final dimension. The fillet size is adjustable at the gear so that a round number of 2.0 is chosen for K_f. The bending moments, read from the diagram, are about 41 000 and 31 000 lb·in at the fillets. The quick calculation indicates diameters of 3.35 in at the gear and 3.15 in at the bearing. These diameters seem reasonable for the length, so we undertake further calculations based on these diameters. We plot yield and tensile strengths vs diameters in Fig. 6.9 to give easy interpolation of values.

At the gear. For $d = 3.35$ in, $S_y = 88\,000$ psi and $S_u = 115\,000$ psi. Corresponding to the latter, for a machined surface $C_s = 0.73$, and from statements in Section 5.10, the size factor C_z is about 0.80. Then by Eqs. (5.36) and (5.37), $S_e = C_s C_z (0.50 S_u) = (0.73)(0.80)(0.5 \times 115\,000) = 33\,600$ psi, whence $S_y/S_e = 88/33.6 = 2.62$.

At the shoulder a D/d ratio of 1.33 should give sufficient support for the gear, and there is room for a fair-size fillet, say $r = 0.08d$, or about ¼ in. The ring of contact between gear and shoulder will have a width of $(½)[1.33D - d - 2(0.08d)] = 0.085d$, or about 0.28 in, which is certainly sufficient. From Figs. 5.27 and 5.29, $K_f = K_t = 1.81$ at the fillet. For the end-milled keyway, we find in a standard reference[13] a K_t value of 2.14 at the end of a keyseat with a bottom radius that is (1/48) times the shaft diameter, an average for standard keyseats. Thus the radius here is about ¹⁄₁₆ in, for which $q = 0.97$ and $K_f = 1 + q(K_t - 1) = 2.11$. This is higher than the 1.81 at the fillet, and we will use it in our calculations. Hence from Eq. (6.30),

$$d^3 = \frac{32n}{\pi S_y} \sqrt{\left(\frac{S_y}{S_e} K_f M_a\right)^2 + \frac{3}{4} T_m^2}$$

$$= \frac{32(2.0)}{\pi(88\,000)} \sqrt{[(2.62)(2.11)(41\,000)]^2 + (0.75)(35\,000)^2}$$

$$= 52.94, \quad d = 3.75 \text{ in}$$

[13] R. E. Peterson, *Stress Concentration Factors*, New York: John Wiley & Sons, Inc., pp. 245–248 and p. 266, 1974.

This value is larger than the 3.35-in basis for the several factors in the calculation, and S_y in particular is lower. A calculation with $d = 3\frac{3}{4}$ in gives $n = 1.96$. Hence the diameter will be increased to $3\frac{13}{16}$ (3.8125). Then $S_y = 85\,000$ psi and $S_u = 113\,000$ psi, whence $S_e = 33\,200$ psi and $S_y/S_e = 2.56$. Keyway fillet radius $= (\frac{1}{48})(3.81) = 0.079$, $q = 0.98$, and $K_f = 2.12$. Equation (6.30) solved for the factor of safety gives

$$n = \frac{\pi d^3}{32} \frac{S_y}{\sqrt{[(S_y/S_e)K_f M_a]^2 + (\frac{3}{4})T_m^2}}$$

$$= \frac{\pi(3.812)^3}{32} \frac{85\,000}{\sqrt{[(2.56)(2.12)(41\,000)]^2 + (0.75)(35\,000)^2}} = 2.06$$

At the bearing. The radius and diameters at the fillet are related to the bearing size which is as yet unchosen. Hence for K_f the value of 2.20 previously suggested at bearings for quenched and tempered steel will be used in the first approximation. For the diameter of 3.15 in indicated by the quick calculation, $S_y = 89\,000$ psi, $S_u = 115\,000$ psi, and $C_s = 0.90$ (polished fillet). Then $S_e = C_s C_z S_e' = (0.90)(0.80)(0.5 \times 115\,000) = 41\,400$ psi and $S_y/S_e = 89\,000/41\,400 = 2.15$. Moment $M_a \approx 31\,000$ psi, and

$$d^3 = \frac{32(2.0)}{\pi(89\,000)} \sqrt{[(2.15)(2.20)(31\,000)]^2 + 0.75(35\,000)^2}$$

$$= 34.27, \quad d = 3.25 \text{ in}$$

The next larger, standard bearing bore is 85 mm (3.3465 in). There are several types of bearings, e.g., ball, roller, single-row, double-row, etc., each in several different rolling-element sizes. Therefore, a bearing suitable for the load, speed, and life expectancy can probably be found. Assume that this bearing is a medium-series, single-row ball bearing, #317. The catalog specifies a maximum fillet of 2.5 mm (0.098 in), a minimum shoulder diameter of 98 mm, and an inner ring diameter of 114 mm. We choose an average shoulder of $4\frac{1}{8}$ in (104.8 mm). Then $r/d = 0.098/3.3465 = 0.029$, $D/d = 4.125/3.3465 = 1.23$, $K_t = 2.38$, $q = 0.98$, and $K_f = 2.35$. The quenching rate and strengths at the fillet may correspond to some intermediate diameter, say $3\frac{3}{4}$ in. Then, $S_y = 85\,000$, $S_u = 113\,000$, and $S_e = (0.90)(0.80)(0.5 \times 113\,000) = 40\,700$ psi, whence $S_y/S_e = 2.09$. The width of the bearing is 41 mm (1.614 in), and the bending moment, by proportional distances is 31 075 lb·in. The factor of safety is

$$n = \frac{\pi(3.3465)^3}{32} \frac{85\,000}{\sqrt{[(2.09)(2.35)(31\,075)]^2 + 0.75(35\,000)^2}} = 2.01$$

Final dimensions are shown in Fig. 6.10. From a 5-in diameter the shaft is tapered to $4\frac{1}{8}$ in. An alternative would be a second shoulder, far enough left of the bearing so that the jaws of a bearing puller may be inserted to remove the bearing. To the right of the bearing seat there is a polished, shallow groove, a runout space for the tool that cuts the threads for the bearing-retainer nut. The groove is 3.22 inches in diameter, with a fillet radius of $\frac{1}{16}$ in. Then $r/d = 0.020$, $D/d = 1.04$, $K_t = 1.96$, $q = 0.96$, and $K_f = 1.92$. The strengths are $S_y = 88\,000$ and $S_u = 115\,000$ psi, and $C_3 = 0.73$ (machined), whence $S_e = 33\,600$ psi and $S_y/S_e = 2.61$. From the moment diagram, $M_a = 26\,000$ psi. The factor of safety is

$$n = \frac{\pi(3.22)^3}{32} \frac{88\,000}{\sqrt{[(2.61)(1.92)(26\,000)]^2 + (0.75)(35\,000)^2}} = 2.16$$

The factor of safety for the shaft as calculated is the smallest one, or 2.01.

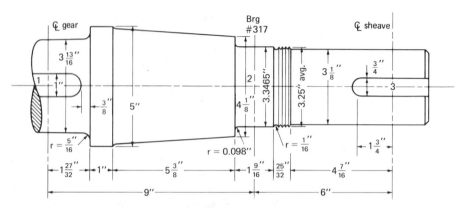

FIGURE 6.10 Partial view of completed shaft of Example 6.3.

Other considerations. It has been assumed that the press fits are light and that the reduction of strength that they cause is not any greater than the reduction of strength caused by the stress concentrations that have been considered. It has also been assumed that the effects of the press fits are not superimposed. The screw threads were ignored because they are very shallow and are cut into a raised surface.

Angular deflection (twist) of this same shaft is the subject of Prob. 12.9. The same shaft configuration, with the bearing seat slightly lengthened to accommodate a spacing ring, is used in Section 9.9 and Fig. 9.23 to illustrate a graphical method of obtaining deflections of a stepped shaft. Although no numerical values are obtained, it is seen from Figs. 9.23(d) and (e) that the slope of the shaft is small at the gear, thus providing for uniform pressure along pairs of meshing teeth. Slope and deflection are large at the pulley, but the relatively flexible V-belts can accommodate to this. Vibration may be another consideration but it is beyond the scope of this book.   ////

Example 6.4

The output shaft of a gear reducer rated at 6000 N·m torque is driven by a herringbone-type gear at its midlength which then extends axially in both directions through plain bearings (Fig. 6.11(b)). Beyond one bearing, a crank is mounted and connected by a rod to an overhead lever or "walking beam," the other end of which drives the pump rods for oil-field recovery of underground oil. For economy, the shaft is to be turned from 125-mm hot-rolled bar stock to a uniform diameter of 110 mm, except at the gear where the diameter will be 112 mm. The bending moment is maximum at midlength of the crankside bearing, and the values of it and the corresponding torque for the calculation of stresses in two fibers are listed at several crank positions in Table 6.1. As with all crankshafts the cycling is difficult to define, but the values in the table are believed to represent the extreme or most critical conditions. Fiber *A* lies 90° ahead of the crank and fiber *B* is in line with the crank. The sign of a moment is taken relative to the fiber, not

TABLE 6.1. Extreme Moments for Calculation of Stresses at *A* and *B* in Several Crank Positions (For Example 6.4 and Fig. 6.11, with Units in N·m)

Crank Position (o'clock)	Fiber A		Fiber B	
	M	*T*	*M*	*T*
12	0	0	+ 5400	0
3	+ 5050	+ 6000	0	+ 6000
6	0	0	− 4700	0
9	+ 5050	+ 6000	0	+ 6000

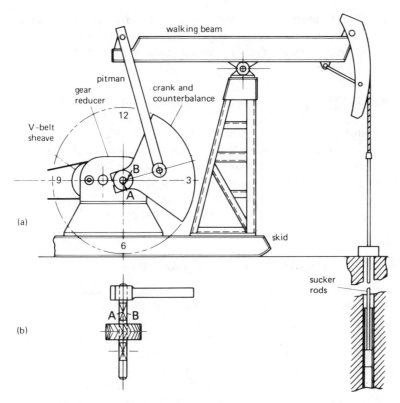

FIGURE 6.11 (a) Oil-field pumping unit with crank at 2:30 o'clock position. (b) Bottom view of output shaft of gear reducer. Example 6.4 and Table 6.1.

space, and a plus sign on a bending moment always gives a tensile stress and a negative sign always gives a compressive stress. The effective value of the bending moment is zero when the fiber lies in the neutral plane of bending. The two absolute values of the moment on fiber B differ because of inertia effects of the counterbalancing, which aids lifting the oil.

Determine the required strength and materials based on moment conditions at the bearings, a reliability of 99% on material, and a factor of safety of 2.5 to allow for shock and starting overloads and environment.

Solution

Fiber A. The bending and torque moments are in-phase and pulsating. Components are $M_m = 2525$, $M_a = 2525$, $T_m = 3000$, and $T_a = 3000$ N·m. There is no stress concentration at the bearing. For economy we anticipate normalizing only to remove mill and machining strains, and the use of a plain carbon steel, which will have the same stiffness as an alloy steel. For this size (approximately 4⅜ in) we estimate $S_e = C_s C_z C_r S_e$ $= (0.78)(0.75)(0.82)(0.5 \times 80\ 000) = 19\ 200$ psi and $S_y/S_e = 48\ 000/19\ 200 = 2.50$. From a rearranged Eq. (6.30), and since N·m/mm³ = 1000 N/mm²,

$$S_y = (1000)\frac{32n}{\pi d^3} \sqrt{\left(M_m + \frac{S_y}{S_e}K_f M_a\right)^2 + \frac{3}{4}\left(T_m + \frac{S_y}{S_e}K_{fs}T_a\right)^2}$$

$$= (1000)\frac{32(2.5)}{\pi(110)^3} \sqrt{[2525 + (2.5)(1.0)(2525)]^2 + (¾)[3000 + (2.5)(1.0)(3000)]^2}$$

$$= 243 \text{ MPa } (35\ 200 \text{ psi})$$

Fiber B. The torque is 90° out-of-phase with the moment, the latter being almost completely reversed. It appears that the pulsating torque will not contribute to any failure, and we shall make calculations for the condition $T_m = T_a = 0$, with $M_m = 350$ and $M_a = 5050$ N·m. Then,

$$S_y = (1000) \frac{32(2.5)}{\pi (110)^3} [350 + (2.5)(1.0)(5050)] = 248 \text{ MPa (36 000 psi)}$$

This is slightly larger than for fiber *A*. Ordinary 0.20% carbon steel could be used, but 1040 steel with AISI quality specifications is superior and readily available for shafting. Also, it can be flame hardened at the gear. For this size when normalized, $S_y \approx 48\,000$ psi (331 MPa) and $S_u \approx 80\,000$ psi (550 MPa). Because of this jump in strengths, there is no need for second, more accurate calculations. The choice of steel is tentative pending investigation of conditions at the gear.   ////

6.9 FURTHER CONSIDERATIONS IN SHAFT DESIGN

Attention is called to other considerations in the design of shafts, many of which may be found in the chapters appropriate to the theory or concept on which they are based. Thus Chapters 2 and 11 are concerned with the characteristics and capacities of plain journal bearings and rolling-element bearings, respectively. The choice that is made of the type and size of bearing affects the shaft, and vice versa. Chapter 12 develops methods for determining torsional stresses in noncircular sections and Chapter 13 develops methods for determining stresses from torsional impact. The development of compressive residual stresses at the surface of shafts by cold-rolling, shot-peening, and shallow-hardening, is discussed in Sections 7.12, 7.13, and 7.15. Stresses at press and shrink fits are discussed in Section 8.10. In connection with processing and assembly problems, attention is called again to Section 5.12 on the avoidance and minimization of stress concentration.

Force and deflection aspects of beam and shaft flexure are the subjects of Chapter 9. Hollow shafts give better stress distribution and weight saving. These hollow shafts and their moments of inertia are treated along with composite shafts and slotted shafts in Section 9.3 and in Probs. 9.5 through 9.8 and Probs. 9.12 through 9.15. Gyroscopic forces on shafts is the topic of Prob. 9.18. Determinate, coplanar forces and moments on shafts are treated in Section 9.4, noncoplanar forces and moments are treated in Section 9.6, and statically indeterminate forces and moments are treated in Section 9.11. Deflections of a stepped shaft are determined by the moment-area method in Section 9.8 and by graphical methods in Section 9.9. The superposition of simple, tabulated, deflection equations for more complicated load and support situations is applied in Section 9.10, particularly to overhung shafts. In Section 9.11, a statically indeterminate, stepped, three-bearing shaft with a misaligned bearing is analyzed for bearing forces and shaft deflections by both the superposition and the graphical methods.

Shaft forces are needed for bearing selection and for moment diagrams. Deflection and slope, and their reciprocal, stiffness, must be determined and controlled for various reasons. Deflections are related to the natural frequencies of shaft vibration, which should not be too close to disturbance frequencies of rotational and functional forces. These natural frequencies may often be determined by the substitution of deflections, particularly those at lumped masses, into equations of vibration theory. Elastic dimensional changes with speed and load are restricted in order to prevent misfunctioning of attached components such as linkages, cams, and roller chain and the development of dynamic forces. They are also restricted to prevent nonuniform spacing and interference with closely fitting enclosures such as stator blading in turbines and magnetic poles in motors, also to ensure uniformity in thickness and roundness of the products of machine tools and rolling mills, whether metallic or nonmetallic. At the teeth of mating gears, shaft deflection may cause chatter, and slope of one of the shafts will change the uniformity of pressure along the teeth, causing nonuniform wear and possible end-breakage of the teeth. The slope is usually largest at the journal ends of a shaft, tending to wear plain bearings to a bell-mouth shape.

Elastic matching is related to slope deflection and uniformity of pressure. It is the locating of components, and hence forces, such that the slope at one component, like a gear, matches that of a mating one. A typical example is the overhanging of the sleeve-bearing supports on the inside walls of a housing so that the forces from the shaft deflect each support into a slope that has the same direction as the slope of the shaft at its journal (see Probs. 9.16, 9.63, and 9.65). A similar improvement in design is *force matching* of components, such as gears, to minimize the net forces at bearings and elsewhere, particularly axial thrusts. This is discussed in Section 1.2 and illustrated by Probs. 9.17, 9.29 through 9.31, 9.84, and 9.85.

There are other considerations in shaft design, details of which are outside the scope of this book. These other considerations include dimensioning and tolerancing and the control of processing treatments, straightness, and finish. Avoidance of binding by thermal expansion, at assembly, and by coupling between shafts is important. Provision, at least on relatively long shafts, is made to take axial forces (thrust) from either direction at one bearing only to avoid binding forces between bearings. Separate seating surfaces are provided, at least with small differences in diameter, for each mounted component such as gears and seals, so that the axial assembly of one does not score the seat for the other and the pressing distance is not longer than necessary. Force-carrying members, such as gears, must have a light press or shrink fit on the shaft to prevent eccentricity, wobble, and fretting. Components are most accurately located in first assembly, and in reassembly after maintenance, by shoulders or by snap-rings in grooves cut on the shaft in the lathe. An exception occurs when mounting on a shaft taper, where no shoulder must interfere with the axial tightening motion. Positive locking in place is provided by nuts over threads on the shaft, possibly with sleeves as spacers, or by end plates fastened by two or more cap screws. Positive angular locking is provided by keyways and splines. Superior strength is obtained by the upset forging of gear blanks on the shaft or of flanges for mounting gears by bolts, also a flange at one end for coupling to another shaft. These avoid shaft weakness at interference fits and usually allow the use of large-radius fillets.

6.10 DESIGN FOR A LIMITED NUMBER OF CYCLES

Some machines or some of the components or mechanisms of machines operate only intermittently, or their function is one of intended short life, or the stresses are normally steady but occasionally changed or subjected to overloads or superimposed vibrations. If the expected number of cycles is well under that required to establish an endurance limit for the material, then it is economical to design for fewer cycles, basing the design on fatigue strength at limited life. Limited life is also the basis for design when there is successive cycling at several levels of stress as explained in Section 6.11.

One will deduce from Sections 5.10 and 5.11 that, for lack of sufficient test data, there is no general agreement on how to relate the effect of the surface, size, and stress concentration of actual components to the laboratory-specimen fatigue strength and theoretical values. Several methods were discussed, one by Fig. 5.22, to obtain a net fatigue-strength curve. The design method presented here is a semigraphical one, and it may be

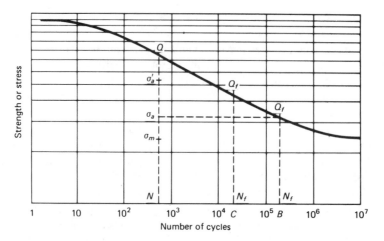

FIGURE 6.12 General curve of net fatigue strength Q, with construction for obtaining number of cycles at failure N_f, at B for complete reversal and at C when a steady stress is superimposed.

used with any line analytically or experimentally determined for net fatigue strength, whether plotted as log S vs log N or as S vs log N.

Presumably, the line for net fatigue strength Q, such as that in Fig. 6.12, is for alternating or completely reversed stresses. If the design situation is also one of complete reversal ($\sigma_m = 0$), then the alternating stress $\sigma_a = \sigma_{max}$ may be plotted on the chart at the required number of cycles N. Projection upward gives the strength Q or failure stress at N cycles, and the factor of safety against overloading is $n = Q/\sigma_a$. Also of interest is the number of cycles that would cause fatigue failure if operation were continued at the same stress beyond the design life N. This is obtained by projecting horizontally from σ_a to intersect the strength line at Q_f, below which at B is read N_f, the number of cycles to failure. With the later, one may define a life factor $l = N_f/N$, which is a measure of safety against overcycling. Because of the logarithmic scale, l will have a larger value than is apparent from the chart.

In general, the stress is fluctuating, and there is a mean component σ_m superimposed upon the alternating component σ_a. Factors n and l are not quite as simply determined. Some equivalent alternating stress must replace σ_a on the chart. An equation for it can usually be developed from a well-defined relationship between σ_a and σ_m for fatigue failure, such as the Soderberg line or the modified Goodman line, adapted for net fatigue strength Q. In Eq. (6.10), representing the Soderberg design line DF of Fig. 6.3(a), S_e/K_f must be replaced by the quotient $S_f/K_f' = Q$. With this change and a rearrangement of terms, the factor of safety may be written

$$n = \frac{S_y}{\sigma_m + \dfrac{S_y}{Q}\sigma_a} = \frac{Q}{\dfrac{Q}{S_y}\sigma_m + \sigma_a} = \frac{Q}{\sigma_a'} \tag{6.31}$$

where the denominator is an *equivalent alternating stress*

$$\sigma_a' = \frac{Q}{S_y}\sigma_m + \sigma_a \tag{6.32}$$

It consists of a reduced value of σ_m added to σ_a, the sum being presumed equivalent in its effect to a condition of complete reversal. It corresponds to the equivalent static or

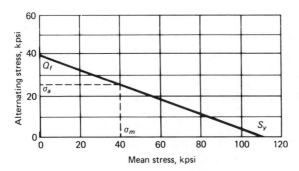

FIGURE 6.13 Graphical method for obtaining the strength at failure Q_f.

mean stress defined immediately following Eq. (6.15). It is plotted on the chart of Fig. 6.12 along the ordinate at N cycles. Then, according to Eq. (6.31), the factor of safety is the strength Q on the ordinate divided by this equivalent stress $\sigma_a{}'$.

As a part endures a larger number of cycles than N, its fatigue strength is reduced along the Q line to a value Q_f at failure. With a reduction in the value of Q, the equivalent alternating stress is reduced according to Eq. (6.32). Failure occurs when the factor of safety becomes 1.0. Substitution of 1.0 for n and Q_f for Q in Eq. (6.31) gives

$$Q_f = \frac{\sigma_a}{1 - \dfrac{\sigma_m}{S_y}} \qquad (6.33)$$

for the strength at failure, based on the Soderberg-line relationship. This failure strength Q_f may be obtained graphically by plotting the points (σ_m, σ_a) and $(S_y, 0)$ on a σ_a vs σ_m chart (Fig. 6.13), and then drawing a straight line through the points to intersect the σ_a axis at a point that has the value of Q_f. Below Q_f at C on the S–N chart (Fig. 6.12), one reads the cycles to failure N_f. The life factor is

$$l = \frac{N_f}{N} \qquad (6.34)$$

Example 6.5

A spindle less than 0.40 inch in diameter, ground at a fillet where $K_f = 2.0$, has been designed for a life of 20 000 cycles under bending stresses fluctuating between 15 000 and 65 000 psi. Minimum-strength values are $S_u = 140\ 000$, $S_y = 110\ 000$, and $S_e{}' = 70\ 000$ psi. Determine the life and safety factors based on a conservative estimate of finite life.

Solution

We choose a strength curve suggested in Section 5.11 to avoid yield in bending, i.e., a straight line through the points $(10^2, S_y)$ and $(10^6, S_e/K_f)$, and plotted on an S–log N chart (Fig. 6.14).[14] We also choose the Soderberg-line relationships Eqs. (6.31) and (6.33). From endurance-strength modification factors $C_s = 0.90$ and $C_z = 1.00$ and by Eq. (5.37), $S_e = (0.90)(1.00)(70\ 000) = 63\ 000$ psi. Then $S_e/K_f = 63\ 000/2.0 = 31\ 500$ psi. Together with $S_y = 110\ 000$ psi, the strength line is drawn on Fig. 6.14 and the strength $Q = 65\ 000$ psi read at $N = 2 \times 10^4$ cycles. From the given $\sigma_{\max}$ and $\sigma_{\min}$,

---

[14] R. E. Peterson, *Stress Concentration Design Factors*, New York: John Wiley & Sons, Inc., p. 18, 1953. This particular relationship was included by Peterson in a completely analytical solution for n and log N. See also footnote 19.

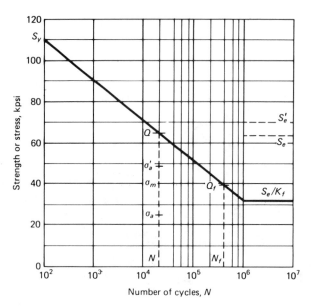

FIGURE 6.14 Stresses and cycles for Example 6.5. Net fatigue-strength line for bending assumed through $(10^2, S_y)$ and $(10^6, S_e/K_f)$ on an $S - \log N$ chart.

we obtain $\sigma_m = 40\ 000$ psi and $\sigma_a = 25\ 000$ psi. By Eq. (6.32)

$$\sigma_a' = \frac{65}{110}\ (40\ 000) + 25\ 000 = 48\ 600\ \text{psi}$$

This equivalent alternating stress is plotted along the N ordinate of Fig. 6.14, and it is seen that the factor of safety, also expressed by Eq. (6.31), is

$$n = \frac{Q}{\sigma_a'} = \frac{65\ 000}{48\ 600} = 1.34$$

By Eq. (6.33), or graphically in Fig. 6.13, the strength at failure is

$$Q_f = \frac{25\ 000}{1 - \dfrac{40}{110}} = 39\ 300\ \text{psi}$$

The corresponding cycles to failure (Fig. 6.14) is $N_f = 4 \times 10^5$, and the life factor is

$$l = \frac{N_f}{N} = \frac{4 \times 10^5}{2 \times 10^4} = 20 \hspace{4cm} ////$$

6.11 DESIGN FOR SEVERAL LEVELS OF STRESS: CUMULATIVE DAMAGE

Many components are undoubtedly designed for infinite life at maximum load when, in service, the level of the load cycles may frequently change between idling and several load levels to occasional overloads. If the loading is completely random within bounds, the rating of a component may be based on fatigue tests of the component under random-programmed loads. For situations where the loading can be predicted to occur for different numbers of cycles at each of several stress levels, as in Fig. 6.15, several methods for strength prediction have been proposed, but test results may differ widely. Nevertheless,

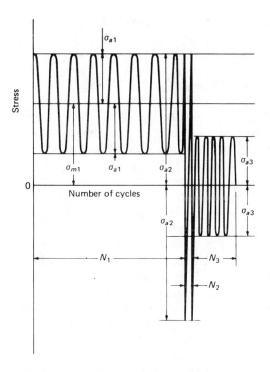

FIGURE 6.15 Designations for several levels of cyclic stress. Proportions are the same as those given in Example 6.6.

the need is great for some rational approach for use in design. A linear summation of equivalent cycle ratios, known as Miner's equation, is generally used. To apply it and to include levels of fluctuating as well as alternating stresses, the semigraphical method for limited cycles (Section 6.10) is used in the following.

It was postulated that if a stress fluctuates at a sufficiently high level between constant values $\sigma_m \pm \sigma_a$, then fracture will occur by fatigue in N_f cycles. Damage is total. If separate levels of stress fluctuation, $\sigma_{m1} \pm \sigma_{a1}$, $\sigma_{m2} \pm \sigma_{a2}$, . . . , $\sigma_{mi} \pm \sigma_{ai}$, occur for N_1, N_2, . . . , N_i cycles, respectively, as in Fig. 6.15, then each level supposedly contributes some damage. Palmgren, Langer, and Miner separately proposed that the total damage equals the sum of the separate damages D_1, D_2, . . . , D_i, where each is proportional to its ratio of cycles, N/N_f. In Example 6.5, the number of cycles to failure was found to be $N_f = 4 \times 10^5$. Thus operation for the design number of cycles, $N = 2 \times 10^4$, gives a damage proportional to $N/N_f = (2 \times 10^4)/(4 \times 10^5) = 1/20$. If operation is continued for a second and equal interval, the damage will be proportional to 2/20, and if for 20 intervals, to 20/20 = 1.0. A number of cycles N_f has been reached, there is fatigue failure, and the damage is total. If the measure of total damage is 1.0, each ratio N_1/N_{f1}, N_2/N_{f2}, etc., may be considered a fractional damage D_1, D_2, etc. Another implication is that if a certain fraction of life N_1/N_{f1} is consumed at stress interval 1, there remains a fraction of life $N_2/N_{f2} = 1 - N_1/N_{f1}$ usable at any other interval 2. Thus Miner's equation[15] for damage D_f at failure is

$$D_f = D_1 + D_2 + \cdots + D_i = \frac{N_1}{N_{f1}} + \frac{N_2}{N_{f2}} + \cdots + \frac{N_i}{N_{fi}} = 1.0 \tag{6.35}$$

The equation does not indicate that there may be a difference in damage from the order

---

[15] A. F. Madayag (Ed.), *Metal Fatigue: Theory and Design,* New York: John Wiley & Sons, Inc., 1969; for a detailed discussion of Miner's theory and others see Chapter 6 therein by H. L. Leve; see also, R. C. Juvinall, *Engineering Considerations of Stress, Strain, and Strength,* New York: McGraw-Hill, pp. 218–224, 1967, and M. A. Miner in G. Sines and J. L. Waisman, (Eds.), *Metal Fatigue,* New York: McGraw-Hill, pp. 278–289, 1959.

in which the stresses are applied or divided among themselves. It does not indicate that there may be damage to the endurance limit by a preceding, higher cyclic stress.

Miner[16] first reported tests on 22 specimens, and D_f varied from 0.61 to 1.49, with an average close to 1.0. Later, Miner and others reported even wider ranges. Applying alternating stresses on AISI 4340 steel heat treated to a hardness of R_c 28–32, with an endurance limit of about 72 000 psi, Marco and Starkey,[17] using from 2 to 20 steps between 80 000 and 120 000 psi, found that D_f averaged 1.13 for a stepped, ascending order, known as "coaxing," and averaged 0.71 for a descending order; using 2 to 10 steps from 25 000 to 50 000 psi on an aluminum alloy 76S-T61, D_f averaged 1.48 ascending and 0.73 descending. Marco and Starkey concluded that the relationship between damage and N/N_f was exponential rather than linear.

Manson et al.[18] presented a simple graphical method on an S–log N chart that recognizes the order of stress application, a lesser damage at low numbers of cycles, and a reduction of the endurance limit by preceding higher cyclic stresses, lowering the right-hand end of the fatigue-strength line. They noted that the order of stressing was most pronounced in a simple two-stress case, where the test specimen is run to failure upon application of the second, lower stress. They further state that for the two-stress cases in which the conditions of stressing are repeated many times, calculations have shown diminishing differences in the fatigue life predicted by their method and by Miner's method, approaching a magnitude that falls within the normal scatter of fatigue-test data. Most engineers use Miner's equation with $D_f = 1.0$, believing it to be reasonably conservative if the loading is random, as it probably is in most machines and vehicles.

When the sum of the damage or cycle ratios, $D = \Sigma(N/N_f)$, is less than 1.0, Miner's equation indicates that a factor of safety exists. Determination of its value requires that a strength at failure be divided by a *single* stress. Peterson suggested the following method for obtaining equivalent cycles, leading to equivalent stresses and strengths.[19] Essentially, the method increases the number of cycles at each stress level by weighted values of the numbers of cycles at all the other stress levels. Thus the equivalent number of cycles at stress level 1 is

$$N_1' = N_1 + N_2(N_{f1}/N_{f2}) + N_3(N_{f1}/N_{f3}) + \cdots + N_i(N_{f1}/N_{fi}) \tag{a}$$

For two levels of stress only, Fig. 6.16 shows the augmentation of N_1 by $N_2(N_{f1}/N_{f2})$ and that of N_2 by $N_1(N_{f2}/N_{f1})$. Let the total damage be

$$D = D_1 + D_2 + \cdots + D_i = \Sigma N/N_f \tag{6.36}$$

Then, from Eq. (a), a more convenient form of N_1' for calculation is

$$N_1' = N_1 + N_{f1}\left[\frac{N_2}{N_{f2}} + \frac{N_3}{N_{f3}} + \cdots + \frac{N_i}{N_{fi}} \right] = N_{f1}\sum\frac{N}{N_f} = N_{f1}D \tag{b}$$

With the other equivalent numbers of cycles written by analogy,

$$N_1' = N_{f1}D, \quad N_2' = N_{f2}D, \ldots, N_i' = N_{fi}D \tag{6.37}$$

[16] M. A. Miner, "Cumulative Damage in Fatigue," *Trans. ASME*, Vol. 67, pp. A-159–A-164, 1945.

[17] S. M. Marco and W. L. Starkey, "A Concept of Fatigue Damage," *Trans. ASME*, Vol. 76, pp. 627–632, 1954.

[18] S. S. Manson, A. J. Nachtigall, C. R. Ensign, and J. C. Freche, "Further Investigation of a Relation for Cumulative Fatigue Damage in Bending," *J. Eng. Ind. Trans. ASME*, Vol. 87, pp. 25–35, 1965.

[19] R. E. Peterson, *Stress Concentration Design Factors*, New York: John Wiley & Sons, Inc., pp. 17–20, 1953. It should be noted that in the book's second edition. *Stress Concentration Factors*, New York: John Wiley & Sons, Inc., 1974, Peterson omits the method and equations, including those for a limited number of cycles. He states that the latter were based on test data on polished specimens of 0.2 to 0.3 in diameter, and if the member being designed is not too far from this range, the equations may be useful as a rough guide; but that otherwise the equations are questionable since the number of cycles for a crack to propagate to rupture depends on the size of the member.

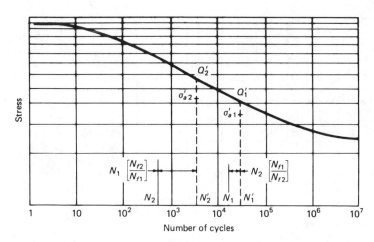

FIGURE 6.16 Equivalent cycles N', stresses σ_a', and strengths Q' for two levels of cyclic stress.

Intersections of the N_1', N_2', ..., N_i' ordinates with the strength curve (Fig. 6.16) determine the equivalent strengths Q_1', Q_2', ..., Q_i'. With these strengths, the equivalent alternating stresses σ_{a1}', σ_{a2}', ..., σ_{ai}' can be calculated by Eq. (6.32). Each equivalent stress is plotted in Fig. 6.16 below its corresponding equivalent strength, and their ratios Q_1'/σ_{a1}', Q_2'/σ_{a2}', ..., Q_i'/σ_{ai}' are calculated. Presumably, the smallest of these is *the* factor of safety, or

$$n = \left(\frac{Q'}{\sigma_a'}\right)_{min} \tag{6.38}$$

It is the smallest stress multiple, applied equally to all the levels of cyclic stress, which will cause a fatigue failure at the end of the specified number of cycles, $N = N_1 + N_2 + \ldots + N_i$. Life factor l is the reciprocal of the sum D of the damages, or,

$$l = \frac{1}{D} = \frac{1}{\Sigma(N/N_f)} \tag{6.39}$$

For a single stress level, this reduces to Eq. (6.34) of Section 6.10.

Example 6.6
To the spindle with the stress cycle of Example 6.5 and strength curve of Fig. 6.14, there are added predicted alternating stresses of $\pm$ 65 000 psi, for a total of 1000 cycles and of $\pm$ 25 000 psi for a total of 5000 cycles (Fig. 6.15). Determine the damage and the life and safety factors.

Solution
Let the cycling of Example 6.5 be designated level 1. Then N_1 = 20 000 cycles, σ_{m1} = 40 000 psi, and σ_{a1} = 25 000 psi. These two stresses were used in Eq. (6.33) or in Fig. 6.13 to determine for this single level the strength Q_{f1} at failure and thence, from the strength curve, the corresponding number of cycles to failure, N_{f1} = 4 $\times$ 10^5 cycles. The strength curve and failure condition of Fig. 6.14 are replotted on Fig. 6.17. The damage caused is $D_1 = N_1/N_{f1} = (2 \times 10^4)/(4 \times 10^5) = 0.05$.

Since there is complete reversal at stress levels 2 and 3, their failure cycles may be found by simple projections (Section 6.10 and Fig. 6.12). Thus in Fig. 6.17, from σ_{a2} = 65 000 psi, projection is made across to Q_{f2} and downward to $N_{f2} = 2 \times 10^4$ cycles. With the predicted N_2 = 1000 cycles, $D_2 = N_2/N_{f2} = 1000/(2 \times 10^4) = 0.05$. By

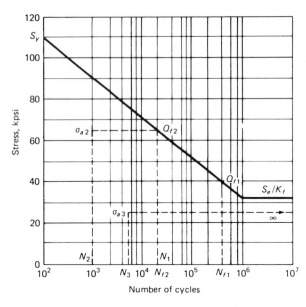

FIGURE 6.17 Cycles to failure N_f for each of the three stress levels of Example 6.6. (Line of strength and Q_{f1}, N_{f1} from Fig. 6.14.)

coincidence, level 2 with the same maximum stress of level 1 contributes the same damage as does level 1, but by complete reversal with only $\frac{1}{20}$ of the number of cycles. From stress level 3, with $N_3 = 5000$ cycles, no damage is indicated, because $\sigma_{a3} = 25\,000$ psi is below S_e/K_f in value, and no intersection occurs by horizontal projection from σ_{a3}. Thus $D_3 = N_3/N_{f3} = 5000/\infty = 0$.

The total damage is $D = D_1 + D_2 + D_3 = 0.05 + 0.05 + 0.00 = 0.10$. The life factor, Eq. (6.39), is $l = 1/D = 1/0.10 = 10$, an indication that cycling in the same proportions may be repeated for a total of ten times before failure.

The equivalent cycles needed to determine the factor of safety are, Eq. (6.37),

$$N_1' = N_{f1}D = (4 \times 10^5)(0.10) = 40\,000 \text{ cycles}$$

$$N_2' = N_{f2}D = (2 \times 10^4)(0.10) = 2000 \text{ cycles.}$$

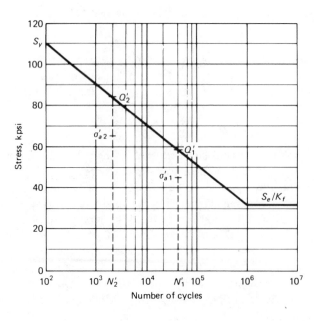

FIGURE 6.18 Equivalent cycles N', stresses σ_a', and strengths Q' for Example 6.6.

Coordinates at N_1' and N_2' on Fig. 6.18 give strengths $Q_1' = 59\,000$ psi and $Q_2' = 84\,000$ psi, respectively. By Eq. (6.32), the equivalent alternating stresses are

$$\sigma_{a1}' = \frac{59\,000}{110\,000}(40\,000) + 25\,000 = 46\,500 \text{ psi}$$

$$\sigma_{a2}' = \sigma_{a2} = 65\,000 \text{ psi}$$

These are plotted along the respective ordinates to show the ratios

$$n_1' = \frac{Q_1'}{\sigma_{a1}'} = \frac{59\,000}{46\,500} = 1.27 \quad \text{and} \quad n_2' = \frac{Q_2'}{\sigma_{a2}'} = \frac{84\,000}{65\,000} = 1.29$$

By Eq. (6.38), the smaller of these is the factor of safety, $n = 1.27$. The addition of the 1000 cycles of reversing stress to the 20 000 cycles of fluctuating stress of Example 6.5 has changed σ_a' from 48 600 psi to an equivalent 46 500 psi, N from 20 000 cycles to an equivalent 40 000 cycles, n from 1.34 to 1.27, and l from 20 to 10, all of which seem reasonable.   ////

PROBLEMS

Section 6.2

6.1   Examine the designated components of the following figures and prepare a table naming the figure number, type of loading (axial, bending, or torsional), type of cycling (alternating, pulsating, etc.), and illustrate with a sketch; also name the places of stress concentration, if any: Fig. 3.19—clutch shaft, Fig. 8.15—rotating disk, Fig. 9.6(a)—hook, Fig. 9.8(c)—leaf spring, Fig. 9.12—spindle, Fig. 9.16—crankshaft, and Fig. 9.17—shaft with large gear.

6.2   (a) In the solution to Example 6.1(a) it was suggested that a material of lower strength could be used. For a factor of safety of 2.0 and without a change of dimensions, what yield strength is required? Will the suggested material do? Will a material of even lower carbon content do? (b) What change in the 50-mm diameter is required to bring the factor of safety at the 10-mm fillet to 2.0, using the same material. *Ans.* 55 mm.

6.3   A shaft is similar to that of Example 6.1, except that the three diameters are 2, 3, and 4 in, the widths at locations 1 and 3 are 2.5 in, and for the hub at location 2, 4.5 in. Also, the dimensions 90, 110, and 90 are replaced by 4, 5.5, and 4 in, respectively. Radii are $r = \frac{1}{4}$ in and $R = \frac{3}{8}$ in. For a rope force of 4000 lb and a factor of safety of 1.6: (a) determine the yield strength required if the shaft is clamped at supports 1 and 3; (b) determine the endurance limit required if the sheave is a tight fit at location 2 and the shaft rotates in bearings at locations 1 and 3.

6.4   The reciprocating rod of a double-acting pump, operating in fresh water, receives reversing loads of 8000 lb. It has a uniform 2-in diameter except near its ends, where the rod is reduced to 1½ in, with a machined fillet of ⅛-in radius. The material is AISI 1040, normalized, with $S_u = 84\,250$ and $S_y = 53\,000$ psi. A reliability of 99.9% is wanted. What is the factor of safety? *Ans.* 1.48.

Section 6.3

6.5   Derive equations for the factor of safety n based on design lines parallel to the modified Goodman line BJ and the yield limitation line JA (Fig. 6.3 (a)). Also derive an equation for σ_m at the transition point. *Ans.* $(S_u/n)[(S_y/S_e) - 1]/[(S_u/S_e) - 1]$. Will the difference between it and the Soderberg method be larger for normalized or for tempered steels?

6.6 Immediately following Eq. (6.11) it was indicated that for torsion, $K\tau_a$ vs τ_m was a horizontal line. If the limitation for this is that the maximum stress not exceed S_{sy}/n, write the two equations for factor of safety and the equation for τ_m at the transition point.

6.7 The nominal stress varies sinusoidally between 120 and 180 MPa, the stress concentration factor $K_f = 1.65$, $S_u = 525$ MPa, $S_y = 350$ MPa, and $S_e = 225$ MPa. (a) Construct a Soderberg diagram and determine the factor of safety graphically. *Ans.* 1.54. (b) Same as (a) but use a modified Goodman diagram, with a limiting line against yielding.

6.8 Same requirements as Prob. 6.7(a) and (b), except the variation is between $-10\ 000$ and $+35\ 000$ psi, $K_f = 1.40$, $S_u = 130\ 000$, $S_y = 95\ 000$, and $S_e = 55\ 000$ psi.

6.9 A camshaft like that of Fig. 6.4 has a span between bearings of 250 mm, with a diameter of 30 mm. The cam is centrally located with a width of 40 mm, a height of 60 mm, and fillets of 3-mm radius. Its lift is 15 mm against a force of 8000 N. The material has $S_u = 1000$ MPa and $S_y = 800$ MPa, with a fillet surface machined before heat treatment. A material reliability of 99% is wanted. Determine the factor of safety.

6.10 In the train of gears shown, the load transmitted gives a nominal stress of 20 000 psi at the root of each tooth. The material is steel, normalized, the yield strength 80 000 psi, the corrected endurance limit, $S_e = 54\ 000$ psi, and the fatigue stress-concentration factor at the root, 1.60. (a) Sketch nominal stress vs time for a tooth of gear No. 1 and of gear No. 2. (b) Determine the factor of safety for each tooth. *Ans.* 2.37, 1.69.

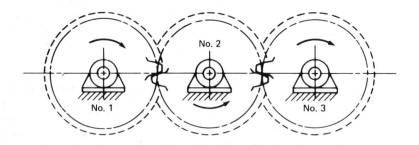

6.11 Pulsating stresses (0 to σ_{max}) occur in gear teeth, shafts with a single cam, single-acting pump rods, etc. Let this be Case 1 and compare it with alternating (completely reversed) stresses, Case 2, by the Soderberg line as follows. (a) For equal factors of safety, derive an equation for the ratio $(\sigma_1)_{max}/(\sigma_2)_{max}$. *Ans.* $2K_f(S_y/S_e)/[1 + K_f(S_y/S_e)]$. (b) An oil-quenched alloy-steel shaft with ball bearings will be subjected to pulsating loads, such as those given the shaft with a single cam of Example 6.2. How much larger may the load be than one applied during complete rotation of the same shaft? Use $K_f = 2.20$ and $S_y/S_e = 2.25$, the average values for quenched and tempered alloy steels of Section 6.7. (c) Figure 11.21 gives allowable maximum nominal stresses for gear teeth as normally loaded in one direction. Use of 0.7 of the chart values is suggested for idler gears, where the stress is alternating. What ratio is given by the answer to (a), using the data from Prob. 6.10.

6.12 Bolts are commonly pretightened. If a soft gasket is used, a portion of any variable service load is added and removed in cycles (Section 7.11). If the bolt load thus varies between 15 000 and 25 000 lb, the root diameter of the nominal $1\frac{1}{4}$ in threads is 1.075 in, the stress-concentration factor is $K_f = 3.0$, $S_e = 36\ 000$ psi, and $S_y = 63\ 000$ psi, determine the factor of safety by the Soderberg equation or determine it graphically.

6.13 Figure 12.11 shows the type of strength diagram usually plotted for helical springs, which have load and stress cycles as shown in Fig. 6.1(e). The relationship is linear and can be considered a special case of Fig. 6.3. Each maximum-stress line, extended, intersects the minimum-stress line at the ultimate shear-stress value S_{su}. For the 3.18-mm-diameter line of Fig. 12.11(a), replot the relationships on a τ_a vs τ_m diagram by a line corresponding to BU of Fig. 6.3(a), and state the numerical values of points B and U.

6.14 It is useful to have the maximum shear stress of the spring chart of Fig. 12.11(a) in terms of the ratio $\rho = P_{min}/P_{max} = \tau_{min}/\tau_{max}$. (a) Derive an equation of the 0.250-in line. *Ans.* $\tau_{max} = 43\,000/(1 - 0.653\rho)$. (b) If the spring force varies rapidly between 40 and 96 lb, what is the maximum useful stress τ_{max}?

6.15 Refer to Fig. 9.12. The actual load depends upon road conditions and the driver, but assume that the bending moment at the fillet adjacent to the left bearing varies between 4000 and 8000 in·lb for at least 10^6 cycles of rough driving. Assume a fillet radius to diameter ratio $r/d = 0.033$, and a shoulder diameter such that $D/d = 1.35$. Final values will depend on the bearing size (Section 6.7). Material is oil quenched and tempered alloy steel to give $S_u = 138\,000$ psi and $S_y = 123\,000$, ground at the fillet. Make an initial determination of the spindle diameter at the fillet for a fatigue-test reliability of 99.9% and a factor of safety of 1.6. *Ans.* 1.37 in.

6.16 A steel shaft carries two equal cams, spaced as shown. The lobe of one cam is 180° from the lobe of the other. Each cam operates a valve against a maximum force of 120 lb. Torque is negligible. (a) Sketch moment diagrams for the 0°, 90°, and 180° positions of the camshaft. Determine for the most critical fillet the mean and alternating components of bending moment relative to a fiber A, which is on the top of the shaft in the 0° position. *Ans.* $M_m = 52.5$ lb·in, $M_a = 120$ lb·in. (b) What yield strength is required of a quenched and tempered alloy steel if the factor of safety is to be 2.0 and the ratio S_y/S_e is estimated to be 2.20 (Section 6.7). Allowance for material reliability is included in the factor of safety.

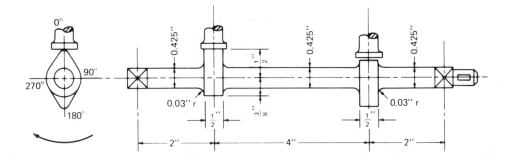

6.17 The six-lobe camshaft of Fig. 9.1 has a distance between bearing centers of 266 mm, divided into seven approximately equal spaces by the center lines of the six cams. The maximum load on a cam is 7300 N, and, with only a minor error in stress, the load may be assumed to act on one cam at a time. Note on the end view the relative angular positions of each cam. The timing is 1-5-3-6-2-4. The longest dimension of each cam is 47.2 mm, the width 16 mm, and the lift 9.2 mm. The shaft diameter is 32.5 mm, based on deflection considerations (Prob. 9.41), and the fillets have a radius of 2.0 mm. (a) Place the load at the worst positions, sketch bending-moment diagrams, and determine the maximum and minimum nominal stresses at that fillet which is maximum-stressed for fatigue failure. Neglect the torque moment. (b) Determine the factor of safety if the material is an alloy steel, carburizing grade, with $S_y = 850$ MPa and $S_e = 380$ MPa at the uncarburized fillet. *Ans.* 1.66 at fillet in maximum tensile stress, 1.51 at fillet in maximum compression.

Section 6.4

6.18 In an ASME Handbook, *Metals Engineering—Design,* in the section on Theories of Failure by Joseph Marin, there appeared the following equations for defining failure (i.e., $n = 1.0$) in terms of certain component stresses:

$$\sigma_x' = (1 - p)\sigma_x'' + S_e$$

and

$$\sqrt{[(1 - p)(\sigma_x'' - \sigma_y'') - (\sigma_x' - \sigma_y')]^2 + 4[(1 - p)\tau_{xy}'' - \tau_{xy}']^2} = \pm S_e.$$

Here σ_x', σ_y', and τ_{xy}' are maximum values and σ_x'', σ_y'', and τ_{xy}'' are mean values. Symbol p is the ratio S_e/S_y. Show that for $n = 1$, and the special case $\sigma_y' = \sigma_y'' = 0$, these are Soderberg's equations without stress-concentration factors. These, presumably, may be included in the calculations of σ_a and τ_a.

6.19 Derive Eqs. (6.17) and (6.18).

6.20 The yoke for a Hooke's coupling is forged integral with its shaft and quenched and tempered. The torque on the output shaft shown is $T_{in} (1 - \sin^2\alpha \sin^2\theta)/\cos \alpha$, and there is an induced bending moment, approximately $T_{in} \tan \alpha \sin \theta$, where T_{in} is the input shaft torque, α is the angle of misalignment of the shafts, and θ is the angle of rotation. The yoke joins the shaft with a 3-mm fillet between circular sections of 24- and 36-mm diameter. Input torque T_{in} is 400 N·m, $S_u = 1200$ MPa, $S_y = 1000$ MPa, and $S_e = 450$ MPa. Determine the factor of safety by Soderberg's equation, Eq. (6.14), if $\alpha = 25°$. *Ans.* 1.60.

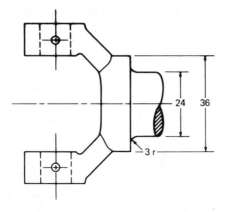

6.21 The sketch shows the proposed design of a shaft to carry two roller-chain sprockets, with the chains pulling in opposite directions. The moment diagrams have been calculated and are plotted below the sketch. Torque is 63 000 lb·in. The material is AISI 4340 oil quenched and tempered at 1000°F. From mass-effect tables, the strengths in 2- and 4-in rounds are, respectively, $S_u = 170\,000$, $S_y = 159\,500$ psi, and $S_u = 164\,750$, $S_y = 145\,250$ psi. (Data from *Modern Steels and Their Properties,* Bethlehem Steel Corp. edition of 1978.) Material reliability will be considered part of the factor of safety. Fillets are commercially polished, and the cylindrical surfaces at the fits have been cold-rolled, so that stress concentration from the fit is negligible. (a) Calculate bearing loads and check the moment diagrams. (b) Determine the factor of safety of that part of the shaft to the right of plane *A-A* according to the maximum shear-stress theory of failure of Soderberg. (For twist, see Prob. 12.10.)

6.22 Same information as Prob. 6.21 except determine the factor of safety for that part of the shaft to the left of plane *A-A*.

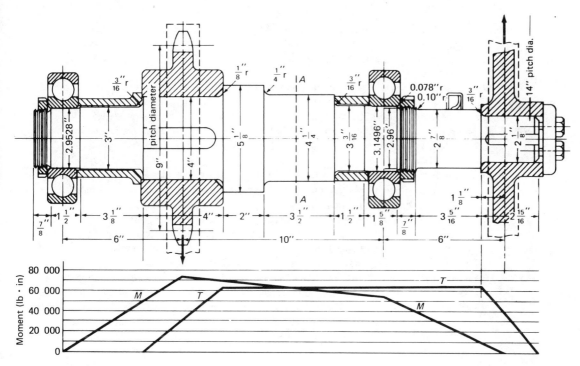

PROBLEM 6.21

Section 6.5

6.23 (a) Using Eqs. (5.13), transform Eq. (6.22) into an equation of component stresses σ_x, σ_y, and τ_{xy}. With this form, write equations for the separate stress intensities s_m and s_a. (b) From (a), derive the special, Mises-based equation for shafts, Eq. (6.25). (c) Some fatigue tests with alternating bending and steady torsion have been based on a parameter which is the stress ratio s_a/s_m. Determine the ratio in terms of σ_a and τ_m. *Ans.* $\sigma_a/\sqrt{3}\tau_m$.

6.24 A variable torsional shear stress $\tau = \tau_m \pm \tau_a$ is applied to a shaft which has planes of stress concentration. There are *no* bending stresses. Determine the principal stresses and use them in the Mises-based equation, Eq. (6.22), to determine a form for the separate stress intensities. Write an equation for the factor of safety. *Ans.* $n = 0.577S_y/[\tau_m + (S_y/S_e)K_{fs}\tau_a]$. Compare it with the equations of Soderberg and Peterson, reduced for the same case.

6.25 A torsion-bar spring is subject to torques varying between 9500 and 15 500 lb·in. The bar is attached by splines cut into enlarged ends which are blended by spiral fillets into the main bar section (Fig. 12.2). Thus the stress concentration in shear may be considered no more than 1.15. Yield strength is 190 000 psi, and the mimimum endurance limit is 85 000 psi, determined by bending-fatigue tests on similar bars, as manufactured. For a factor of safety of 1.5 what diameter is required according to the Mises-based equation of Prob. 6.24? According to the Soderberg equation? Which is more conservative?

6.26 Prove that Eq. (6.26) gives smaller values of n than does Eq. (6.21), except equal values when either σ_a or τ_m is zero. Which is the more conservative equation?

6.27 Prove that Eq. (6.26) gives smaller values than Eq. (6.16) except at small ratios of the alternating term to the steady term of the denominator. At what ratio σ_a/τ_m does the change occur? *Ans.* $\sigma_a/\tau_m = (S_e/S_y)/(2\sqrt{3}K_f)$.

6.28 For a rotating thin disk of uniform thickness, a maximum radial stress of 15 500 psi has been calculated by Table 8.1. At this same radius the tangential stress is 27 300 psi and the axial stress is zero. Both stresses are tensile. The material strengths are $S_u = 95\,000$ psi,

$S_y = 60\ 000$ psi, and $S_e = 40\ 000$ psi. The rotational speed of the disk is continuously changed from zero to a maximum and back to zero. Determine the factor of safety by the Mises criterion as adapted by Soderberg.

6.29 In a rotating thin disk the maximum radial stress is 200 MPa, the corresponding tangential stress is 352 MPa, and the axial stress is zero. The disk is used as a flywheel; its speed varies from full to 0.707 full speed, and its stresses vary as the square of speed. Material properties are $S_u = 1040$ MPa, $S_y = 700$ MPa, and $S_e = 420$ MPa. Determine the factor of safety by the Mises criterion as adapted by Soderberg. *Ans.* 1.96.

6.30 Crankshafts are generally designed for bearing capacity, cylinder spacing, and stiffness, and with reference to past design, then checked for stress. A portion of a crankshaft for a single-cylinder overhung-crank engine is shown. In this size, for a normalizing treatment, $S_u = 95\ 000$, $S_y = 60\ 000$ psi, and the fully corrected endurance limit, S_e, is 33 000 psi. Determine the factor of safety by using Soderberg's equations and investigating the following critical cases, the nominal stresses for which have been determined from an analysis of gas and inertia forces. (a) At section *a-a*, the fiber in the top dead-center position has a bending stress of 8500-psi tension, which changes to 1900-psi compression in the bottom dead-center position. (b) At section *b-b*, the fiber at *m* changes from 8000-psi tension to 1800-psi compression. (c) The maximum torque occurs at an angle of 38° beyond top dead center. The torsional stress is then 3800 psi, changing later to 900 psi in the opposite sense. At the same time, the bending stress of the fiber at *n* is 4600-psi tension, changing to 1250-psi compression when the torsional stress is minimum. Also, calculate case (c) by the equations of Peterson's modification and the Mises criterion, and compare the three results. *Ans.* 2.80, 3.04, and 3.02 in part (c).

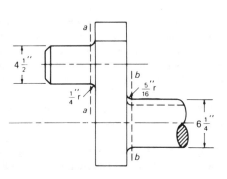
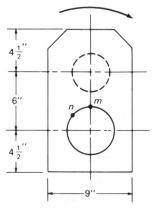

Section 6.6

6.31 Investigate the effect of a static, tensile, axial stress when superimposed on an alternating axial stress, as follows. (a) Sketch the planes of maximum alteration of shear stress, showing the shear and normal components of stress. According to Sines, what trend in the σ_a vs σ_m line should be expected? (b) Determine an equation for this line based upon Eq. (6.27) and fatigue-strength test points $\sigma_{xa} = S_e$ at $\sigma_{xm} = 0$ and $\sigma_{xa} = S_p$ at $\sigma_{xm} = \sigma_{xa}$ (pulsating). *Ans.* $\sigma_a = S_e - [(S_e/S_p) - 1]\ \sigma_m$. (c) In axial fatigue tests of a material, the endurance limit was reported to be 80 000 psi and the alternating component for endurance in pulsation was 63 000 psi. What is the equation relating σ_a to σ_m? Will this equation hold for negative values of σ_m?

Sections 6.7 and 6.8

6.32 From Peterson's equation, Eq. (6.19), derive an equation for the outside diameter of a hollow shaft in terms of the ratio of inside to outside diameters, $C = d_i/d$, and the mean and alternating components of bending and torsional moments. Compare it with Eq. (6.30).

6.33 Rewrite Eq. (6.26) of the Mises criterion to base it on the modified Goodman line. From it derive an equation for the outside diameter of a hollow shaft subject to an alternating bending moment M_a and a steady torque T_m. Let $C = d_i/d$, where d_i is the inside diameter. *Ans.*[20] $d^3 = [32n/\pi(1 - C^4)][K_f(M_a/S_e) + 0.866(T_m/S_u)]$.

6.34 The general proportions of a shaft for an idler sprocket with a constant overhung load are shown with dimensions in millimeters. The material is an oil-quenched alloy steel AISI 1340 normalized to give $S_u = 828$ MPa and $S_y = 517$ psi in the expected diameters of 75 to 100 mm. All surfaces are machined, and a reliability of 99.9% is wanted on the endurance limit. The fillet radius at *a-a* is restricted by the ball bearing, but the fillet radius at *b-b* can be a more generous, full fillet, with perhaps $r/d = 0.075$. Sketch the bending-moment diagram and estimate the minimum diameters required for the right-hand ball-bearing seat and for the shaft to the right of it, if the factor of safety is 1.60.

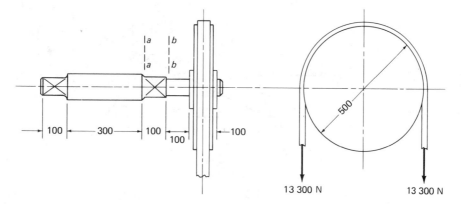

6.35 For economy it is proposed to eliminate the spacing rings for the shaft of Prob. 6.21 and, instead, locate the bearings by shoulders on the shaft. The maximum radius that the right-hand bearing will allow at the shoulder is 0.079 in (2 mm), and an average D/d value is 1.25. The stress concentration is increased, and a larger shaft diameter is required. The next standard, bearing-seat diameter is 3.3465 in (85 mm). Will this be sufficient for a factor of safety of 1.60 at the fillet immediately to the left of the right-hand bearing?

6.36 A rotating shaft carries a rotor with a mass of 6800 kg, this mass being uniformly distributed over a 500-mm length of shaft. A steady torque of 22 600 N·m is developed by the rotor and transmitted through a flexible coupling. The ratios r/d and D/d are estimated to be 0.10 and 1.50, respectively. Surface finish is "machined." Material is normalized steel with $S_u = 585$ MPa and $S_y = 345$ MPa. Sketch the shape of the moment diagram and determine d, D, and r for an endurance-limit reliability of 99.99% and a factor of safety of 1.50, using Peterson's modification of the diameter equation, Eq. (6.30).

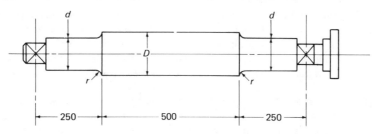

---

[20] Without the C, in the form derived by L. D. Mitchell and D. T. Vaughan, "A General Method for the Fatigue-Resistant Design of Mechanical Components—Part II Analytical," *J. Eng. Ind., Trans. ASME,* Vol. 97, pp. 965–975, 1975. See Eq. (40).

6.37 Plots of steady torque and alternating bending moment are shown below the shaft. At fillets with radii r_1 and r_2, shoulders are provided with diameters 25% larger than the adjacent diameters. A fillet ratio no larger than $r/d = 0.05$ may be used to avoid interference. The steel proposed is AISI 1040 normalized, with $S_u = 580$ MPa and $S_y = 365$ MPa. Finish is ground. The high factor of safety of 2.5 includes an allowance for reliability of endurance-limit data. Determine the minimum diameters d_1 and d_2, using Peterson's modification of the diameter equation, Eq. (6.30).

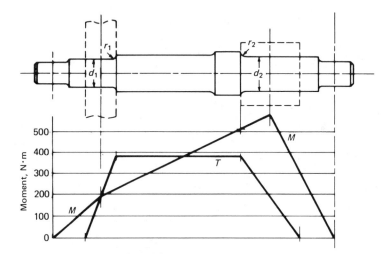

6.38 In Example 9.4 and Fig. 9.16 maximum-moment diagrams are obtained for the flywheel shaft of a punch press. One critical section for fatigue failure is at the fillet adjacent to the pinion. At this fillet the torque varies from 16 500 to -1500 lb·in, the bending moment on a worst fiber, from 20 500 to $-20 500$ lb·in, not every working stroke but over a long period because of a "hunting" gear ratio. The pitch diameter of the pinion is 6 in, and a generous fillet can be used if it is below the teeth. The heat treatment is chosen for an oil-quenched alloy steel to give a Brinell hardness less than 300, so that the shaft may be finish machined after heat treatment. The hardness is 250 Bhn, $S_u = 125 000$ psi, and $S_y = 90 000$ psi in this size. Consideration of data reliability is included in the high factor of safety chosen, 2.5. Determine the minimum-shaft diameter adjacent to the pinion based on Peterson's modification for diameter.

6.39 Maximum torque and the combined bending moment at a critical section on a vehicle rear axle, such as that of Prob. 9.26, vary between 9000 lb·in and zero, and 7700 to -7700 lb·in, respectively, relative to any fiber. There is a shoulder, and the fillet diameter is limited by the adjacent ball bearing. The material is an alloy steel, quenched and tempered to give $S_u = 180 000$ and $S_y = 160 000$ psi. The surface is commercially polished, the size of similar axles is of the order of 1½ inch, and 99.9% material reliability is wanted. The bearing seat is cold-rolled to prevent failure by fretting. Determine a diameter for the axle for a factor of safety of 1.6, using the diameter equation with Peterson's modification.

6.40 In Example 9.5 and Fig. 9.17 moment diagrams are obtained for the output shaft (the one with the larger gear) of a gear-reduction unit. The shaft has been redesigned as shown here to accommodate proper size bearings, and to minimize stress concentration and facilitate assembly. Note the large fillet at one end of the gear and the small diameter changes elsewhere, obtained by using a spacing ring. All surfaces are machined. With the axis of the shaft as the abscissa, plot on quadrille paper the combined moment diagram of Fig. 9.17(g) and the constant torque of 4060 lb·in from Example 9.5. For a factor of safety 2.0, determine the yield strength required. Can a carbon steel such as AISI 1040 normalized be used? (See Example 6.3 for K_t at a keyseat.)

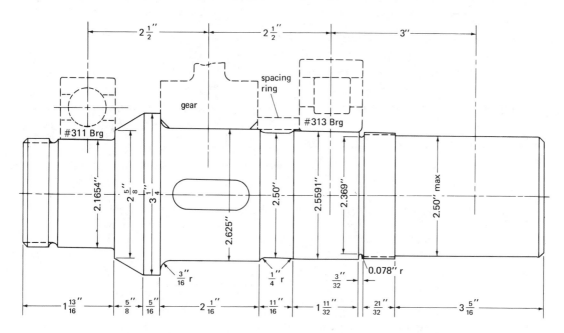

6.41 For a small, single-acting positive-displacement pump, working under a constant head, the resistance to motion of the piston will be a constant 6000 N during the working stroke and zero during the return stroke. Because of a large connecting-rod to crank-arm ratio, approximately the same forces may be assumed to act on the crankpin of the crankshaft. The crankshaft is shown at the beginning of a working stroke when the piston is above. Torque is received from the left through a flexible coupling, which prevents the transmission of external bending moments. It is likely that an oil quenched and tempered alloy steel will be required, with the fillet ground. Modify the estimated S_y/S_e factor to allow for a 99% reliability in the endurance limit. The variation in the bending and torsional stresses during rotation should be investigated for at least two points or "fibers" at the critical plane in order to determine the worst combination of stresses. A table such as the one given in Example 6.4 is suggested. Determine the minimum required value of S_y for a factor of safety of 1.6 at plane *a-a*. Dimensions are in millimeters. *Ans.* 794 MPa.

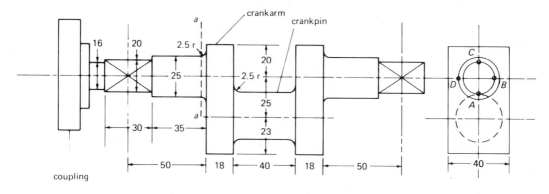

6.42 With reference to Prob. 6.41 and with the answer given there, determine the diameter required for the crankpin, which is the 40-mm length of shaft. The fillet is limited to 2.5 mm.

Section 6.10

6.43 Obtain an equation for Q from the straight line of Fig. 6.14, and revise Eq. (6.31) to give an equation for factor of safety that avoids the graphical step. *Ans.*

$$n = \cfrac{1}{\cfrac{\sigma_m}{S_y} + \cfrac{\sigma_a}{\cfrac{3}{2} S_y - \left(S_y - \cfrac{S_e}{K_f}\right) \cfrac{\log N}{4} - \cfrac{S_e}{2K_f}}}$$

6.44 On a σ_a vs σ_m chart such as that given in Fig. 6.3, an ellipse through the endurance limit and the ultimate strength may well fit the experimental points (Section 6.6). For the design line, an ellipse would be drawn through points $(0, S_e/n)$ and $(S_u/n, 0)$. Derive the equation for this line, solving it for n. Replace S_e/K_f by $S_f/K_f' = Q$ in the result, and determine an equation for an equivalent alternating stress σ_a', similar to Eq. (6.32). *Ans.* $\sqrt{(Q/S_u)^2 \sigma_m^2 + \sigma_a^2}$.

6.45 Let the modified-Goodman line BU of Fig. 6.3(a) replace the Soderberg line as a basis for Eqs. (6.31) through (6.34). Rewrite the equations, and together with a sketch justify a statement as to whether the use of the Goodman line is more conservative or less conservative than the use of the Soderberg line.

6.46 Plot net fatigue strength Q as a straight line between $(10^2, S_y)$ and $(10^6, S_e/K_f)$ on an S vs log N chart (semilogarithmic), where $S_y = 135\,000$ psi and $S_e/K_f = 21\,400$ psi. Determine the safety and life factors for a rotating shaft designed for a finite life of 10^5 cycles of stress alternating between $+30\,000$ and $-30\,000$ psi. *Ans. 1.65, 4.9.*

6.47 A missile part is to be designed for a ten-minute life pulsating in bending between zero and 450 MPa at the rate of 30 cycles per second. The stress-concentration factor is $K_f = 1.45$, and $S_y = 720$ MPa and $S_e = 380$ MPa. Plot net fatigue strength for axial loading as a straight line between $(10^2, S_y)$ and $(10^6, S_e/K_f)$ on an S vs log N chart, and determine the safety and life factors.

6.48 A thin ring cycles repeatedly between zero and a maximum speed of rotation. The only stress is a "hoop stress," equivalent to an axial stress in a straight bar, and it has a maximum value of 47 000 psi. The stress-concentration factor K_f at each of several radial holes is 2.73. Material properties are $S_y = 120\,000$ psi and $S_e = 58\,400$ psi. Construct a net fatigue-strength curve on an S vs log N chart, through the points $(1, S_y)$ and $(10^6, S_e/K)$. Determine the safety and life factors for a rated life of 10 000 cycles. *Ans. 1.59, 32.*

6.49 A shaft is to be designed for a finite life of 200 000 revolutions with a completely reversed bending moment of 1130 N·m. Material and finish have been chosen, with strengths $S_u = 800$ MPa, $S_y = 585$ MPa, and an estimated $S_e = 310$ MPa. The fatigue stress-concentration factor is estimated to be 1.75. Plot net fatigue strength Q as a straight line between $(10^3, 0.90S_u)$ and $(10^6, S_e/K_f)$ on a log S vs log N chart and determine the diameter required for a factor of safety of 1.5. *Ans. 41.3 mm.*

6.50 (a) Adapt Eqs. (6.31) and (6.32) for the common, shaft case of $\sigma_m = \tau_a = 0$, with reference to the Mises forms, i.e., Eqs. (6.24) and (6.26). *Ans.* $\sigma_a' = \sqrt{3}(Q/S_y)\tau_m + \sigma_a$. (b) A steady torque of 1580 N·m is superimposed upon the alternating bending stress in the shaft of Prob. 6.49. Determine the diameter required.

Section 6.11

6.51 In the operation of the shaft of Prob. 6.46, 10^4 cycles of 50 000 psi alternating stress will be intermixed with the 10^5 cycles of 30 000 psi alternating stress. Determine the life and safety factors for the combination.

6.52 Suppose that because of a faulty governor the ring of Prob. 6.48 is subject to 10% over-speeding for a total of 500 randomly spaced cycles, in addition to the rated 10 000 cycles. Stress is proportional to the square of the speed. Determine the safety and life factors for the combination. *Ans.* 1.42, 27.

6.53 A more careful analysis of the camshaft of Example 6.2 shows that it may be subjected to occasional overloads, giving maximum stresses of 44 000 psi for 10^4 cycles and 36 000 psi for 10^5 cycles. Plot a line through $(10^2, S_y)$ and $(10^6, S_e/K_f)$ on an S vs log N chart and determine the life and safety factors.

6.54 It is estimated that a gear tooth in the transmission of a vehicle, operating under pulsating forces (zero to maximum), will undergo maximum root stresses of 90 000 psi for 2×10^4 cycles, 80 000 for 10^5 cycles, and 60 000 for 5×10^5 cycles, all intermixed together with many cycles below 40 000 psi. The tooth and fillet surfaces are carburized after machining, giving strengths $S_y = 180\,000$ psi and $S_e = 65\,000$ psi. The stress-concentration factor is $K_f = 1.50$. Plot a line through $(10^2, S_y)$ and $(10^6, S_e/K_f)$ on an S vs log N chart (semilog) and estimate life and safety factors. *Ans.* 1.24.

6.55 Pulsating stresses in a gear tooth, as in Prob. 6.54, except 540 MPa for 5000 cycles, 460 MPa for 40 000 cycles, and 400 MPa for 200 000 cycles, with the remaining cycles at or below 300 MPa. Material strengths are $S_y = 830$ MPa and $S_e = 450$ MPa. Stress concentration $K_f = 1.50$. Estimate life and safety factors.

6.56 Data are available from NACA (NASA) reports on the frequency with which air gusts of different mean velocities and hence severities are scattered among one million gusts during flight. The number of gusts per mile for a particular plane depends on its mean wing chord. For a particular wing the loads calculated for each velocity were used together with an experimentally determined S-N curve to determine the number of cycles to failure N_f of a certain riveted lap joint between aluminum sheets (Table 6.2).[21] Only the frequency or number of gusts at each load in a total of a million gusts and the corresponding N_f are listed. The maximum loads varied from 24.6% of ultimate for the conditions at the first frequency to 86.2% of ultimate at the conditions of the last frequency. The number of gusts in the table total 494 560, and the remaining 505 440 gusts in each million gave negligible damage. The average number of gusts per mile was estimated to be 4.36 based on a mean wing chord of 11 ft.[21] Determine the fraction of life consumed by a million gusts according to Eq. (6.36), then the total life in miles of flight. *Ans.* 2.47×10^6 miles.

TABLE 6.2. Data for Prob. 6.56

Frequency	435 000	54 000	5000	500	50
Cycles N_f	2.0×10^7	1.0×10^6	4.0×10^5	1.5×10^5	6.0×10^4
Frequency	5	1	0.3	0.1	0.03
Cycles N_f	3.0×10^4	1.5×10^4	6.0×10^3	3.0×10^3	1.4×10^3

Extracted from Table 4 of L. R. Jackson and H. J. Grover, "The Application of Data on Strength under Repeated Stresses to the Design of Aircraft," NACA W-91. Originally issued October 1945 as ARR5H27 (National Advisory Committee for Aeronautics, Washington).

---

[21] For further details and an example see M. A. Miner in G. Sines and J. L. Waisman (Eds.), *Metal Fatigue*, New York: McGraw-Hill, pp. 286–287, 1959.

7

Thermal Properties and Stresses. Residual Stresses

Creep Rupture, Creep and Stress Relaxation, Stresses from Thermal Expansion, Harmful and Beneficial Residual Stresses from Assembly, Yielding, and Transformation, Bolt Tightening

7.1 INTRODUCTION

The topics of this chapter have some common characteristics that make them suitable to group together for study and application in design. If environmental and service-imposed temperatures are depressed or elevated from normal temperatures, impact and *short-time strengths* are altered, often unfavorably. Loading at elevated temperatures for a longer period, from several hours to several thousand hours, may result in *rupture,* or in *creep* and *stress relaxation,* depending upon the nature and the magnitude of loading. If at these elevated temperatures, or at normal temperatures, there are temperature differences throughout a body, it may be necessary to consider the *thermal stresses* which are induced by the restriction to expansion which the colder parts impose upon the warmer. These stresses exist independently of those stresses caused by external loading, and, when those stresses caused by external loading are superimposed upon them, may increase or decrease the maximum or critical stress.

The name *residual stress* is given to those stresses that are induced during manufacturing and assembling operations, either incidentally or by design. They may be induced by clamping together several parts during assembly; by the yielding of portions within one part during the mechanical operations of rolling, bending, cutting, or proof-loading; and by local growth, temperature differences, and yielding during casting, welding, grinding, and heat treatment. They are relieved or relaxed by creep when sufficiently heated. Like thermal stresses, residual stresses are independent of the stresses from external loading in service. When an externally induced stress and a residual stress are superimposed, the latter may have either a harmful or beneficial effect on the total stress and the resistance to fatigue failure. The discussion of residual stress will principally concern itself with the methods and the design for prestressing with beneficial effects.

In this chapter, thermal stresses are discussed only for elementary shapes. For residual stresses, the treatment is qualitative, which, for design purposes, is frequently the most valuable approach. After the derivation of additional equations in the following chapters, thermal and residual stresses are determined analytically for cylinders, tubes, and plates.

7.2 THE EFFECT OF TEMPERATURE ON SHORT-TIME MECHANICAL PROPERTIES

Short-time properties include those that can be measured at the speed of the usual tensile-machine test and by the very rapid impact test. Figure 7.1 gives some general trends that are characteristic of steels. The slopes and transition points will vary with the composition and treatment.

Yield strengths, proportional limits, and moduli of elasticity drop steadily as the temperature increases above room temperature with the proportional limits dropping more rapidly than the yield strengths. On the other hand, the ultimate tensile strength around 500°F may be higher than at room temperature. This is particularly true of plain carbon steels in the "brittle-temper" range of temperatures. With higher temperatures, the tensile strength drops rapidly. Corresponding to this initial rise in strength, there is an initial drop in the reduction of area and elongation, followed by a rapid rise at higher temperatures.

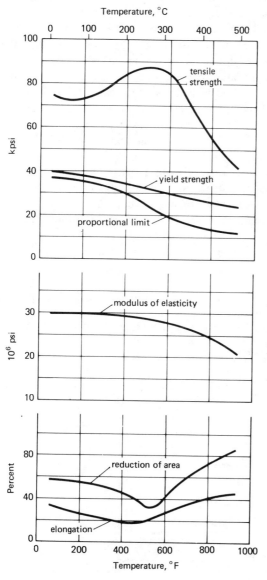

FIGURE 7.1. Tensile-test properties of a normalized 0.37% C steel at elevated temperatures. (T. O. Lynch, N. L. Mochel, and P. G. McVetty, "The Tensile Properties of Metals at High Temperatures," *Proc. ASTM,* Vol. 25, Part 2, p. 14, 1925, with permission of the ASTM, publishers, and as adapted by R. C. Juvinall, *Stress, Strain, and Strength* in Fig. 20.6, p. 436. Copyright © 1967 by McGraw-Hill, New York, and used with permission of the McGraw-Hill Book Company.)

The tensile test for temperatures below normal may be continuations of those above normal for a few or for many degrees, depending upon the materials. In a typical low-carbon steel, the tensile strength and particularly the yield strength rise with falling temperature. The ductility as measured by the reduction of area will drop slowly at first. Then at some temperature known as the energy-transition temperature, there will be a sharp drop in the reduction of area and the tensile strength. As temperatures are lowered further, the tensile strength curve may meet, then rise with the still-rising yield strength curve, and the reduction of area drops gradually to nearly zero, both indicating extreme brittleness.

Transition occurs at higher temperatures when the strain rate is increased and when there are triaxial stress components. Both are supplied by a notch in a tensile specimen, but the strain rate is further increased by impact. Thus a convenient test for the comparison of materials is the notched-bar impact test of the Charpy or Izod type. In Fig. 7.2 the fracture energy is plotted against the temperatures for steels of different carbon contents. The temperature of the energy transition for a very low-carbon steel is not only low but the transition is steep. Higher carbon steels begin to lose fracture energy at temperatures several hundred degrees above zero, but the transition is more gradual.

In general, it has been found that increases in carbon and in other interstitial impurities that prevent slip, such as phosphorous, silicon, chromium, vanadium, and molybdenum, raise energy transition temperatures whereas manganese and titanium lower it. Strain aging, which precipitates "keys" against slip, and radiation damage raise the transition temperature. Nickel, which is in solid-state solution with iron, lowers the transition temperature, and low-nickel steels are particularly useful for extremely cold applications. The high-alloy stainless steels, AISI types 302–304 (approximately 18% Cr, 8% Ni), do not have an energy transition.

In design, material should be selected with a transition temperature or range well below the operating temperature so that unavoidable dynamic overloads will not find low-energy resistance to crack propagation and fracture. Attention to this is particularly important when quench-strengthened, plain and alloy, medium-carbon-content steels are used. Sharp notches and other sources of stress concentration should be avoided. In some devices it may be possible to reduce the rate of strain application by cushioning or spring

FIGURE 7.2. Effect of carbon on the fracture-energy transition range of plain-carbon steels, measured in a notched-bar Charpy impact test. (From J. A. Rineholt and W. J. Harris, Jr., "Effect of Alloying Elements on Notch Toughness of Pearlitic Steels," *Trans. ASM*, Vol. 43, p. 1197, 1951.)

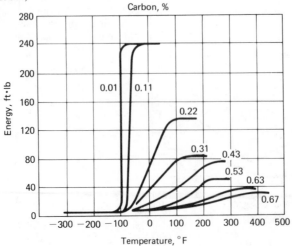

suspension. Low-carbon and unquenched steels, although of lower yield strength, may be preferred. Small grain size is desirable since it minimizes the size of microcracks that may be formed by plastic deformation. It may be necessary to use low-manganese and low-nickel alloy steels, or the more expensive stainless steel or other face-centered-cubic metals. Further details, design recommendations, and reports on the continuing studies in this area and on fracture mechanics in general may be found in many references.[1]

7.3 LONGER-TIME PROPERTIES: CREEP AND CREEP RUPTURE

When steels are held for periods measured in hours and days at elevated temperatures but at axial stresses below their short-time yield strengths at those temperatures, there is a continuing elongation which may end in a fracture called *stress rupture* or *creep rupture*. At even lower stresses and temperatures steels may elongate at a steady and much slower rate for an indefinite number of years, and the phenomenon is called *creep*. If elongation is restricted, then the initial stress will slowly decrease and this is called *stress relaxation*. Materials differ greatly in their ability to resist these effects of temperature. Whereas some alloys have been developed for use at 2000°F, lead and plastics under stress creep at room temperatures.

Creep is measured in units of strain. Each test is made at constant stress and temperature, and its progress is plotted as creep vs time, as shown in Fig. 7.3. The curves have two or three stages. In a short first or transient region, creep proceeds at a rapid but decreasing rate. The rate is constant in the second or steady-state region. One explanation is that the strain-hardening rate is balanced by a rate of recovery, due to the diffusion that occurs readily at high temperatures. At low stresses this creep continues indefinitely at a constant rate. At higher stresses, there is a third stage in which the creep rate increases rapidly and ends in fracture or rupture. The third stage may be caused by an increase in true stress because of the loss in area due to thinning and to voids formed by

FIGURE 7.3. The progress of creep and rupture in one material at one temperature but under different test stresses. The crosses mark rupture points.

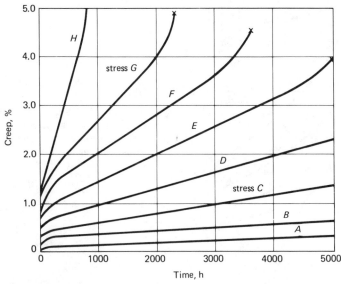

[1] F. A. McClintock and A. S. Argon (Eds.), *Mechanical Behavior of Materials*, Reading, Massachusetts: Addison-Wesley Publishing Co., Inc., pp. 557−564, 1966; A. S. Tetelman and A. J. McEvily, jr., *Fracture of Structural Materials*, New York: John Wiley & Sons, Inc., 1967. For references on fracture mechanics in general, see the footnotes at the end of Section 5.10.

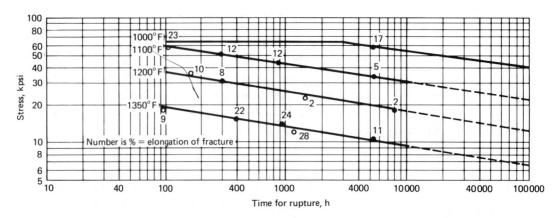

FIGURE 7.4. Applied stress vs time for rupture at four different temperatures, for a stainless-steel Type 316 + Cb. (From ''Resumé of High Temperature Investigations Conducted During 1963–65,'' The Timken Roller Bearing Company, Steel and Tube Division. Data reproduced by permission of The Timken Company.)

localized internal fracture. The elongation in a 2-in length at rupture may be in the range of 5 to 85% with steels. Ingenious testing machines and equipment have been developed for the required, very accurate measurements of creep.[2]

The rupture points can usually be represented by straight lines on a log-stress vs log-time chart (Fig. 7.4). Figure 7.4 is a chart for a steel with 0.053% C, 16.7% Cr, 14.1% Ni, 2.25% Mo, and 0.77% Cb, one of the stronger stainless steels suitable for boiler superheater tubes in the 1100°−1350°F range. The room temperature strength of this steel is 82 000 psi. Tests were run up to 8000 hours (about one year), and results were extrapolated to 100 000 hours (about 11.5 years). Stress values read from Fig. 7.4 are

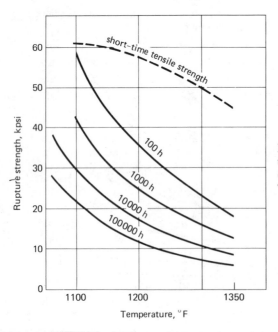

FIGURE 7.5. Rupture strengths vs temperature for the steel of Fig. 7.4. (Data reproduced by permission of The Timken Company.)

---

[2] G. V. Smith, *Properties of Metals at Elevated Temperatures*, New York: McGraw-Hill, 1950. Tests and equipment are described, together with a very complete treatment of the metallurgical aspects of creep in steels.

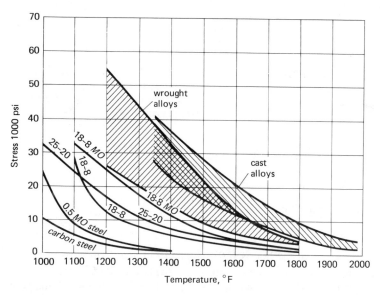

FIGURE 7.6. The 1000-hour rupture strengths of alloys containing substantial amounts of Co, W, Mo, or Cb (shaded areas) compared with austenitic stainless steels, a low-molybdenum steel, and a carbon steel. (From G. V. Smith, *Properties of Metals at Elevated Temperatures*. Copyright © 1950 by McGraw-Hill, New York. The figure appears on p. 385 of the Appendix, written by R. F. Miller. Used with permission of the McGraw-Hill Book Company.)

plotted in Fig. 7.5 as rupture strength vs temperature for the four commonly quoted lifes of 100, 1000, 10 000, and 100 000 hours. Plotted for comparison are the short-time tensile strengths. The long-time values are well under these, and from 1100° to 1350°F each of the hour curves drops about two-thirds in value.

Higher temperatures require alloys with a substantially higher total of metals such as cobalt, tungsten, molybdenum and columbium, and these alloys are sometimes called the superalloys. Their rupture strengths for 1000 hours of life fall in the shaded areas of Fig. 7.6, where they are compared with a carbon steel, a low-molybdenum steel, and with some austenitic stainless steels, the latter indicated by two numbers that are their nominal percentages of chromium and nickel, respectively. Strengths are shown for the wrought superalloys to 1800°F, and for the cast superalloys to 2000°F, the temperatures of operation of some gas-turbine blades.

7.4 APPLICATION OF CREEP DATA FOR LONG-LIFE DESIGN

Parts of vessels and machines may need to operate at elevated temperatures for ten- or twenty-year periods. Rupture can be avoided, but creep and stress relaxation cannot. The extrapolated 100 000-hour creep-rupture strength is a guide. For design, a lower stress is chosen to limit the creep or loss of initial stress to values that will avoid interference or leakage due to dimensional changes and ensure that the creep rate is well below that which would lead to rupture. Although a few laboratory tests have been carried out for ten years or more, the great majority of tests are ended within three months. Thus the problem is to *extrapolate* in time the creep data that have been obtained for a particular material in a particular treated condition, and also to *interpolate* them for stresses other than those of the tests. The following is one solution to the problem.

If ϵ_c is the creep strain plotted against time t, as for the typical curve of Fig. 7.7, then the creep rate is $\dot{\epsilon}_c = d\epsilon_c/dt$, and this has a constant value $\dot{\epsilon}_0$ along the straight portion of the curve. If this straight line is extended to intersect the strain axis at ϵ_a, then the

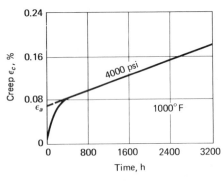

FIGURE 7.7. Creep ϵ_c vs time under a stress of 4000 psi at 1000°F. Creep rate of 0.035% per 1000 hours is the same as for the material of Fig. 7.8 at 4000 psi and 1000°F.

value of ϵ_c at any time t is

$$\epsilon_c = \epsilon_a + \dot{\epsilon}_0 t \tag{7.1}$$

In Fig. 7.7 the slope of the line is 0.035% per 1000 hours, and this is the creep rate. If this rate and those rates from curves for other stresses at 1000°F are plotted on a log $\dot{\epsilon}_0$ vs log σ chart (Fig. 7.8), a straight line may usually be drawn through the points. Similar lines are constructed from data obtained at other temperatures to complete the chart. The equation of any one line is

$$\log \dot{\epsilon}_0 = \log \dot{\epsilon}_{01} + n(\log \sigma - \log \sigma_1) \tag{a}$$

where n is the slope of the line and $(\sigma_1, \dot{\epsilon}_{01})$ is any convenient point on the line. Thus

$$\dot{\epsilon}_0 = \dot{\epsilon}_{01}\left(\frac{\sigma}{\sigma_1}\right)^n = \frac{\dot{\epsilon}_{01}}{\sigma_1^{\,n}}\,\sigma^n = B\sigma^n \tag{7.2}$$

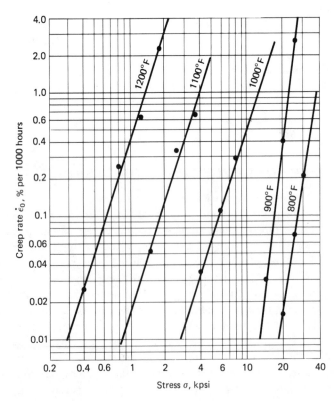

FIGURE 7.8. Log-creep rate vs log-stress chart for a "killed" carbon steel. (From data courtesy of The Timken Company.)

where n and $B = \dot{\epsilon}_{01}/\sigma_1^n$ are constants determined from the chart or from data given for any two, well-spaced points.

It is common practice to plot and speak of creep in terms of percent and creep rate in terms of percent per 1000 hours. In Eq. (7.2), and in those equations that follow, it is recommended that the rate of actual strain be used for $\dot{\epsilon}_0$ and $\dot{\epsilon}_{01}$. This is obtained by multiplying the value in percent per 1000 hours by 10^{-5}. Since strain is dimensionless, the unit of strain rate becomes h^{-1}. The unit of the constant B is $h^{-1}(lb/in^2)^{-n}$ when stress is in lb/in^2 and $h^{-1}(MPa)^{-n}$ when stress is in MPa.

Formulas other than Eq. (7.2) have been proposed, but this one is the simplest to use.[3] If one chooses a creep rate for a particular situation, and if one has available the values of n and B for the chosen material and expected temperature, then Eqs. (7.1) and (7.2) can be used to determine a design stress σ. Alternatively, if creep charts like Fig. 7.3 are available, one may make trial solutions using Eq. (7.1) and several of the curves, then interpolate for the design stress to give the chosen value of total creep ϵ_c.

The allowable stresses of the ASME Boiler and Unfired Pressure Vessel Codes have been based upon a maximum creep of 0.01% in 1000 hours, which would be 1 inch per 100 inches of pipe length in 11.4 years. This corresponds to a creep rate of 0.1×10^{-6} per hour if the small ϵ_a value in Eq. (7.1) is neglected. A Code specification for seamless tubing of 3% Cr, 1% Mo steel with a minimum tensile strength of 60 000 psi once listed design stresses of 15 000 psi at $-20°$ to 650°F, 7000 psi at 1000°F, and 1500 psi at 1200°F, and these values are typical of many other low-alloy steels.

Between materials, a comparison of the stresses to produce a given creep rate will usually place the materials in the same order and relative position as does the rupture-strength chart of Fig. 7.6. If the creep rate is 1×10^{-6} per hour, which is ten times the rate of the ASME Code, the creep-stress values will be roughly one-half the corresponding rupture strengths. Finnie and Heller list and comment on many creep and creep-rupture data.[4]

Example 7.1

From the material and test results of Fig. 7.9(a) determine the limiting stress for a part to be subjected to 950°F continuously and at constant load if the creep is limited to 1% in 100 000 hours (a) by interpolation of the curves of Fig. 7.9(a) and (b) by the construction of a chart and an equation for creep-strain rate as a function of stress.

Solution

(a) Creep rate $\dot{\epsilon}_0$, or the slope of a curve, should be obtained from well-spaced points, say $\dot{\epsilon}_0 = (\epsilon_{1400} - \epsilon_{400})/1000 \ h^{-1}$. The intercept ϵ_a with the creep axis is measured, and substituted together with $\dot{\epsilon}_0$ into Eq. (7.1) to obtain for ϵ_c, the percentage creep ($0.049 + 0.008 \times 10^{-3}t$), ($0.065 + 0.018 \times 10^{-3}t$), and ($0.084 + 0.040 \times 10^{-3}t$) for stresses of 3000, 4000, and 5000 psi, respectively. Substitution of 100 000 hours for t gives 0.85, 1.87, and 4.08%, respectively. From a plot of percent creep vs stress for these points (Fig. 7.9(b)), we read 3200 psi as the limiting stress for a creep of no more than 1% in 100 000 hours.

(b) For an equation, actual strain is preferred to percentage creep. Hence from Fig. 7.9(a), strain rates $\dot{\epsilon}_0$ of 0.008×10^{-5}, 0.018×10^{-5}, and 0.040×10^{-5} per hour are plotted against their stresses on a log-log chart (Fig. 7.9(c)). The slope of this line is n

---

[3] I. Finnie and W. R. Heller, *Creep of Engineering Materials*, New York: McGraw-Hill, p. 116, 1959. J. Marin, *Mechanical Behavior of Engineering Materials*, Englewood Cliffs, New Jersey: Prentice-Hall, Inc., pp. 338–352, 1962. R. C. Juvinall, *Stress, Strain, and Strength*, New York: McGraw-Hill, pp. 401–428, 1967.

[4] I. Finnie and W. R. Heller, *Creep of Engineering Materials*, New York: McGraw-Hill, pp. 231–254, 1959.

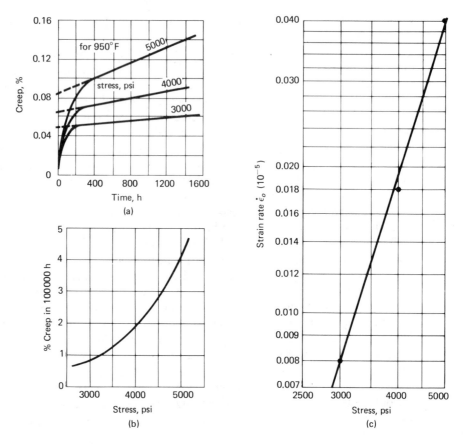

FIGURE 7.9. Example 7.1 (a) Idealized chart of creep at several stresses for a material at 950°F. (b) Interpolation for stress to give 1% creep at 100 000 hours. (c) Log strain rate vs log stress to obtain constants n and B of Eq. (7.2).

$= \Delta \log \dot{\epsilon}_0 / \Delta \log \sigma = 3.1$. For determining B, any easily read point will do. At $\dot{\epsilon}_{01} = 0.007 \times 10^{-5}$, $\sigma_1 = 2900$ psi, whence $B = \dot{\epsilon}_{01}/\sigma_1^n = 0.007 \times 10^{-5}/2900^{3.1} = 1.29 \times 10^{-18}$ h^{-1}(psi)$^{-n}$. Equation (7.2) becomes

$$\dot{\epsilon}_0 = B\sigma^n = 1.29 \times 10^{-18} \sigma^{3.1} \qquad (b)$$

which may be rewritten

$$\sigma = (7.73 \times 10^{17} \dot{\epsilon}_0)^{0.323} \qquad (c)$$

The initial effect ϵ_a must be estimated, say an average value of 0.0006, but its effect is small for a large time. From Eq. (7.1), $\dot{\epsilon}_0 = (\epsilon_c - \epsilon_a)/t$. For a creep strain $\epsilon_c = 0.01$ in a time $t = 100\,000$ h, $\dot{\epsilon}_0 = (0.01 - 0.0006)/100\,000 = 0.0094 \times 10^{-5}$ h^{-1}. From Eq. (c),

$$\sigma = [(7.73 \times 10^{17})(0.0094 \times 10^{-5})]^{0.323} = 3220 \text{ psi}. \qquad ////$$

It is not always necessary to plot strain rates against stress to obtain the creep constants n and B, as was done in Example 7.1. Frequently, only two pairs of values are available, such as the percent creep in 1000 hours for two different applied stresses. Then from Eqs. (a) and (7.2), if $(\sigma_2, \dot{\epsilon}_{02})$ and $(\sigma_1, \dot{\epsilon}_{01})$ are the two pairs

$$n = \frac{\log \dot{\epsilon}_{02} - \log \dot{\epsilon}_{01}}{\log \sigma_2 - \log \sigma_1} \qquad (d)$$

and

$$B = \frac{\dot{\epsilon}_{01}}{\sigma_1{}^n} = \frac{\dot{\epsilon}_{02}}{\sigma_2{}^n} \tag{e}$$

See Example 7.2 for an application of this.

7.5 STRESS RELAXATION

In parts which are given an initial fixed displacement, a slow decrease of stress rather than a change in length occurs at creep temperatures and times. An important example of this is tightened bolts, such as those used to connect pipe flanges and nozzles of vessels in high-temperature service. Loss of tension may require periodic retightening.

Suppose that a nut is screwed on a bolt (Fig. 7.10(a)), until the length between the nut face and the head face is less by an amount δ_i than the distance between the two surfaces which it is to connect, all parts being at the same elevated temperature. Suppose that the bolt is now stretched elastically, slipped into the U-slot, and released (Fig. 7.10(b)). If the flange and head thus joined are much stiffer than the bolt, it may be assumed that the bolt starts its service with a deflection δ_i and stress σ_i. At any later time, let the stress be σ, where $\sigma < \sigma_i$. If the bolt is removed and released, it will contract elastically an amount δ, which is less than δ_i (Fig. 7.10(c)). The difference is the plastic deformation due to creep, $\delta_c = l\epsilon_c$, where l is the bolt length and ϵ_c is the creep strain. Hence

$$\delta_c = \delta_i - \delta \tag{a}$$

or

$$l\epsilon_c = l\frac{\sigma_i}{E} - l\frac{\sigma}{E}$$

whence

$$\epsilon_c = \frac{\sigma_i - \sigma}{E} \tag{7.3}$$

FIGURE 7.10. Illustrating stress relaxation in the bolt of a tightened joint. (a) Nut positioned on bolt. (b) Initially assembled without turning nut. (c) After several years, bolt removed without turning nut, showing permanent increase δ_c in bolt length.

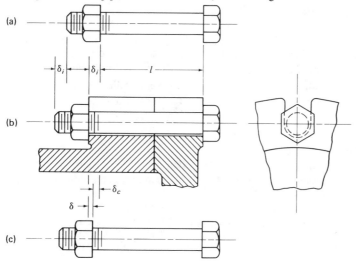

Creep ϵ_c and stress σ are functions of time, and σ_i is not. Hence the creep rate obtained by differentiation of Eq. (7.3) with time t, is

$$\dot{\epsilon}_c = -\frac{1}{E}\frac{d\sigma}{dt} \tag{b}$$

If ϵ_a of the initial creep phase is neglected, the value of the constant creep rate $\dot{\epsilon}_0$ given by Eq. (7.2) may be substituted for $\dot{\epsilon}_c$. Thus

$$B\sigma^n = -\frac{1}{E}\frac{d\sigma}{dt} \tag{7.4}$$

Rearranged and integrated,

$$dt = -\frac{1}{BE}(\sigma^{-n}d\sigma)$$

and

$$t = -\frac{\sigma^{-n+1}}{(-n+1)BE} + C = \frac{1}{(n-1)BE\sigma^{n-1}} + C \tag{c}$$

The constant of integration C is obtained from the initial condition $\sigma = \sigma_i$ when $t = 0$. The final equation becomes

$$t = \frac{1}{BE(n-1)}\left[\frac{1}{\sigma^{n-1}} - \frac{1}{\sigma_i^{n-1}}\right] \tag{7.5}$$

Solved for the stress to which a bolt must be tightened to retain a stress σ after a time t, Eq. (7.5) is

$$\sigma_i = \frac{\sigma}{[1 - BE(n-1)\,t\sigma^{n-1}]^{1/(n-1)}} \tag{7.6}$$

Because ϵ_a in Eq. (7.1) was neglected, Eq. (7.6) will indicate a larger stress than actually exists at a given time, particularly for times soon after the bolts have been put in service. For a solution to be possible the denominator must be positive. Thus material, time, and stress must be so chosen that $BE(n-1)t\sigma^{n-1} < 1$, preferably less than 0.8, to prevent excessive loss of tightness. A third useful form of the equation gives the initial stress in terms of the ratio of minimum stress to initial stress, $R = \sigma/\sigma_i$ namely,

$$\sigma_i = \frac{1}{R}\left[\frac{1 - R^{n-1}}{BE(n-1)t}\right]^{1/(n-1)} \tag{7.7}$$

Equations have been developed for the effects of creep in beams, pipe-expansion loops, columns, pressure cylinders, rotating disks, and other multiaxial stress situations. These equations may be found elsewhere.[5] An excellent discussion of the choice of design stresses and of specific applications of metals and nonmetals in various fields is available.[6]

Example 7.2

Stainless steel bolts of 12% Cr (AISI 410), each 3.5 in long, are proposed for joining flanges where the temperature will be 900°F. The available table gives only two creep-

[5] I. Finnie and W. R. Heller, *Creep of Engineering Materials*, New York: McGraw-Hill, pp. 114–202, 1959; J. Marin, *Mechanical Behavior of Engineering Materials*, Englewood Cliffs, New Jersey: Prentice-Hall, pp. 357–413, 1962.

[6] I. Finnie and W. R. Heller, *Creep of Engineering Materials*, New York: McGraw-Hill, pp. 255–327, 1959.

rate values, 1% in 100 000 hours with a stress of 13 600 psi and 1% in 10 000 hours with a stress of 24 000 psi. The modulus of elasticity is 30×10^6 at room temperature and 25×10^6 at 900°F. The plant schedule calls for bolt retightening every two years. Determine reasonable allowable stresses and the corresponding deflection at tightening.

Solution

Time $t = 24 \times 365 \times 2 = 1.752 \times 10^4$ h. The strain rates are $\dot{\epsilon}_{01} = 0.1 \times 10^{-6}$ h^{-1} with $\sigma_1 = 1.36 \times 10^4$ psi and $\dot{\epsilon}_{02} = 1.0 \times 10^{-6}$ h^{-1} with $\sigma_2 = 2.40 \times 10^4$ psi. From Eqs. (d) and (e) of Section 7.4 the creep constants are

$$n = \frac{\log 1.0 - \log 0.1}{\log 2.40 - \log 1.36} = \frac{\log 10}{\log (2.40/1.36)} = \frac{1.0}{0.247} = 4.05$$

$$B = \frac{\dot{\epsilon}_{01}}{\sigma_1{}^n} = \frac{0.1 \times 10^{-6}}{(1.36 \times 10^4)^{4.05}} = 1.82 \times 10^{-24}$$

As suggested after Eq. (7.6), to avoid unreasonable loss in tightness we shall try $BE(n - 1)t\sigma^{n-1} = 0.70$, whence the stress after two years (1.752×10^4 h) is

$$\sigma = \left[\frac{0.70}{BE(n - 1)t} \right]^{1/(n-1)}$$

$$= \left[\frac{0.70}{(1.82 \times 10^{-24})(25 \times 10^6)(3.05)(1.752 \times 10^4)} \right]^{0.328}$$

$$= 5740 \text{ psi}$$

The initial tightening stress at the operating temperature is, by Eq. (7.6),

$$\sigma_i = \frac{5740}{[1 - (1.82 \times 10^{-24})(25 \times 10^6)(3.05)(1.752 \times 10^4)(5740)^{3.05}]^{0.328}}$$

$$= 8600 \text{ psi}$$

This gives a ratio σ/σ_i of 0.667 or two-thirds. If a higher ratio is desired, then the operating stress must be reduced by using a bolt of larger diameter.

If tightening is done while hot the strain is $\epsilon = \sigma/E = 8600/(25 \times 10^6) = 0.000\ 344$ and the elongation is $\delta = l\epsilon = (3.5)(0.000\ 344) = 0.0012$ in. The coefficient of expansion of this ferritic stainless steel does not differ much from the carbon-steel flanges that it joins. Hence the same elongation is required if tightening cold, i.e., when the parts are at room temperature. However, with a larger value for E, the stress will be larger, namely, $\sigma = \epsilon E = (0.000\ 344)(30 \times 10^6) = 10\ 300$ psi. It has been assumed in the calculations that the deflection, stress, and creep of the flanges are negligible because of the large area around the bolt holes. ////

7.6 ELEMENTARY THERMAL STRESSES

Restriction to normal thermal expansion may come from external or internal sources. The rod of Fig. 7.11(a), which just fits the space between massive supports at one temperature, would, if free, expand an amount $\delta = l\alpha T$ (Fig. 7.11(b)), where l is the rod length, α the coefficient of thermal expansion, and T the temperature increase. With longitudinal expansion prevented (Fig. 7.11(c)), the stress developed is that which elastically compresses the rod the amounts $\delta = l\epsilon$, where ϵ is the elastic strain. The net

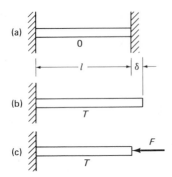

FIGURE 7.11. Illustrating thermal stress. (a) Bar assembled with zero fit at room temperature. (b) Free bar expands an amount δ with rise T in temperature. (c) Bar forced to original length to fit between walls.

displacement u is zero, or

$$u = l\alpha T + l\epsilon = l\alpha T + l\frac{\sigma}{E} = 0$$

whence

$$\sigma = -\alpha ET \tag{7.8}$$

Similarly, for a plate prevented from expansion in its plane (Fig. 7.12), the net displacements u and v in the x and y directions are zero. This, plus substitution from the elastic strain equations, Eqs. (5.1), gives

$$u = l_x \alpha T + l_x \epsilon_x = l_x \alpha T + (l_x/E)(\sigma_x - \nu\sigma_y) = 0$$

and

$$v = l_y \alpha T + l_y \epsilon_y = l_y \alpha T + (l_y/E)(\sigma_y - \nu\sigma_x) = 0$$

from which are obtained two equations

$$\sigma_x - \nu\sigma_y = -\alpha ET$$

and

$$\sigma_y - \nu\sigma_x = -\alpha ET$$

From simultaneous solution of these two equations

$$\sigma_x = \sigma_y = -\frac{\alpha ET}{1 - \nu} \tag{7.9}$$

With this two-way restraint, the stresses in the plate are larger than in uniaxial strain, Eq. (7.8), by the ratio $1/(1 - \nu)$, or if $\nu = 0.30$, by the ratio 1.43.

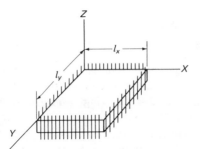

FIGURE 7.12. Dimensions of plate restricted on all four sides against thermal expansion.

Similarly, for a block that is restrained from expansion on all sides, and with the use of all three Eqs. (5.1), it may be shown that

$$\sigma_x = \sigma_y = \sigma_z = -\frac{\alpha ET}{1 - 2\nu}.$$ (7.10)

If $\nu = 0.30$, the ratio $1/(1 - 2\nu) = 2.50$.

From the preceding cases of external restraint, stresses for some cases of internal restraint may be deduced. If a part of or the entire surface of a body is quickly heated or cooled as in spot welding or in liquid quenching for heat treatment, expansion parallel to the surface is restrained by the mass beneath until the mass beneath has changed temperature. This restraint is similar to that for the plate so that the upper bound of stress is given by Eq. (7.9). It is actually less because the temperature cannot change abruptly between a surface layer and the mass beneath it. If a thin web and a heavy rim of a rotor are each uniformly heated in operation to different temperatures, or if a thin, solid web of a gear is much more rapidly cooled in a quenching operation than is the rim, displacement of the platelike web is restricted, and the upper-bound stress is given again by Eq. (7.9).

If one side of a free bar is heated to a temperature T, the bar will take a circular shape, and there will be no stresses. However, thermal stresses will develop if bowing is prevented by moments applied to the ends. If the bar is free to expand axially, the displacement of its central plane is $l(\alpha T/2)$. Each surface differs from the central plane by a temperature $T/2$. Hence from Eq. (7.8), maximum thermal stresses of value $\alpha ET/2$ may be expected, compressive on the hotter side, and tensile on the cooler side.

Similarly, one may reason that for a thin circular plate heated on one side and built in at its ends, the maximum stress will have the value $\alpha ET/2(1 - \nu)$, half that in Eq. (7.9). In a thin, circular cylinder, such as a liner used in a large diesel engine (Fig. 7.13), inherent symmetry prevents bending except near the ends, and since the effective curvature of the cylinder is small, the preceding plate formula may be applied. The accuracy of this deduction for a thin wall is checked in Section 8.14, where thermal stresses for cylinders of any thickness are derived, also in Example 10.11 of Section 10.8 for a clamped plate heated on one side. For numerous other shapes and devices there are many references available.[7]

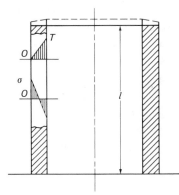

FIGURE 7.13. Thermal expansion and stress distribution in a thin cylinder heated to give an internal surface temperature T degrees higher than the external surface temperature.

---

[7] S. S. Manson, *Thermal Stresses and Low-Cycle Fatigue,* New York: McGraw-Hill, 1966.
B. E. Gatewood, *Thermal Stresses,* New York: McGraw-Hill, 1957.
B. A. Boley and J. H. Weiner, *Theory of Thermal Stresses,* New York: John Wiley & Sons, Inc., 1960.

7.7 THERMAL FATIGUE AND SHOCK: STRESS MINIMIZATION

Fatigue failures at elevated temperatures may be caused by mechanical cycling or by temperature cycling. Under the former, typical $S–N$ curves are common, with the fatigue (endurance) strength $40-60\%$ of the short-time tensile strength at the same temperature. Surprisingly, as the temperature and time period increase, the fatigue (endurance) strength increases relative to the static creep-rupture strength, and may be double it at higher temperatures.[8]

The term *thermal fatigue* usually refers to fatigue failures brought about by a cycling of temperatures which induces thermal stresses because of external or internal restraints and nonuniform heating. In one case, if the thermal stresses exceed the yield strength in a region of highest temperatures, plastic flow occurs.[9] When the temperature is reduced to an original value, the stress is reversed in direction and may cause reversed plastic flow. This may occur with each heating cycle, and the material may fail after a few hundred cycles.[10] Creep, which is a plastic yielding, may be involved in stress cycling over longer periods of time when temperatures are less frequently changed.

Thermal shock occurs when there is a single, very rapid temperature change. A surface may be rapidly cooled by quenching, and the tensile stress induced may cause cracking in a brittle material such as gray cast iron or tool steel. A surface may be quickly heated locally, as by a torch, which causes compressive stresses within the heated area. In ceramics and other brittle materials with low-heat conductivity, the temperature gradient is high where a spot is suddenly heated. Compressive stresses may approach those indicated by Eq. (7.9) and cause spalling or flaking and crumbling of the surface.[11]

Thermal stresses and failures may be minimized by proper design. Uniform sections with a minimum of geometric discontinuities, avoidance of small holes and notches, generous fillets, and good surface finish are often of more importance in minimizing thermal-stress fatigue than mechanical fatigue. Avoidance of constraints to differential thermal expansion, or provision for flexibility will minimize thermal stress. Analogous to the well-known ''floating''-bridge construction, with one end on rollers, are turbine-nozzle blades sliding within but guided by the outer shroud ring. When a component must be fixed at both ends, like heat-exchange tubes, flexibility may be provided, as by preforming the tubes in a sine wave along all or a portion of their lengths. Large structures containing small, hotter parts are best avoided or, if necessary, segmented to minimize constraints. Control of temperature during slow starting and stopping, perhaps with auxiliary heating or cooling, may avoid temperature surges and large gradients. Expansion compatibility, using metals of lower thermal expansion at points of highest temperature, properly placed insulation, and favorable residual stresses are other possibilities.[12]

7.8 DETRIMENTAL RESIDUAL STRESSES

A residual stress is one that exists without external loading or internal temperature differences on a structure or machine. It is usually a result of manufacturing or assembling operations. Sometimes it is called initial stress, and the operation, prestressing. When the structure or machine is put into service, the service loads superimpose stresses. If the

[8] B. E. Gatewood, *Thermal Stresses,* New York: McGraw-Hill, p. 137, 1957.

[9] See Section 7.14.

[10] B. E. Gatewood, *Thermal Stresses,* New York: McGraw-Hill, p. 138, 1957; also A. S. Tetelman and A. J. McEvily, Jr., *Fracture of Structural Materials,* New York: John Wiley & Sons, Inc., pp. 309–394, 1967.

[11] For details, see e.g., S. S. Manson, *Thermal Stress and Low-Cycle Fatigue,* New York: McGraw-Hill, pp. 275–312, 1966.

[12] S. S. Manson, *Thermal Stress and Low-Cycle Fatigue,* New York: McGraw-Hill, chapt. 9, pp. 365–395, 1966.

(a)

(b)

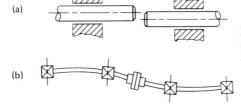

FIGURE 7.14. Offset or lateral misalignment of shafts at (a) gives residual bending stresses when assembled with a rigid coupling at (b).

residual stresses add to the service-load stresses, they are detrimental; if they subtract from the service-load stresses they are beneficial.

Only a few examples of detrimental residual stresses will be given here. One, in the *assembly* of machinery, occurs when two shafts are not in line or are a few thousandths of an inch out of parallel, and they are forced into connection by rigid couplings (Fig. 7.14). The resulting stresses in the shafts become reversing stresses when the shafts are rotated. The correction, when perfect alignment cannot be economically attained, as is frequently the case, is to use flexible couplings of a type necessary for the degree of misalignment.

The preceding case occurs with elastic stresses only, and the residual stresses are maintained by bearing constraints. In applications where *mechanical work* causes plastic yielding, stresses remain when the constraints are removed. For example, the forging of shafts and crankshafts and the cooling after forging may induce residual stresses, the equilibrium of which is changed in machining, causing some warping of the shafts. It is then common practice to straighten the shafts in a press before the final machining operation. Straightening requires a bending moment large enough to cause permanent set or yielding. This applied moment, if the material were entirely elastic, might result in stress distribution *A-A* (Fig. 7.15). Since the surface fibers are strained past their yield strength S_y, the actual stress distribution is given by the lines *B-B* if the stress-strain diagram is an idealized one (Fig. 5.4). This is similar to Fig. 5.17(c) in Section 5.8 on plastic–elastic conditions. The moment about the neutral axis of the stresses represented by the lines *B-B* must equal that represented by the line *A-A*, which equals the applied moment. The bending stress is removed elastically as the straightening moment is taken off, so that the residual stress is distributed as in lines *C-C*, these lines being obtained by "subtracting" line *A-A* from lines *B-B*. The clockwise and counterclockwise internal moments of stresses represented by the lines *C-C* must, of course, be equal for static

FIGURE 7.15. During straightening of a shaft, calculated elastic-stress distribution *A-A* and actual elastic-plastic distribution *B-B*. After straightening, residual stresses *C-C*.

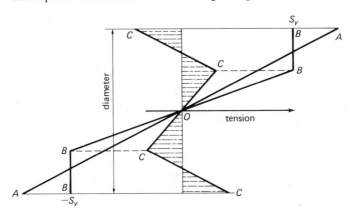

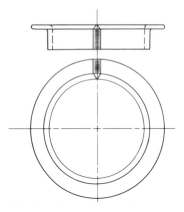

FIGURE 7.16. Flanged cylinder subject to residual stresses at weld after weld cools.

equilibrium. In rotation under steady load, alternating stresses will be superimposed, which, if sufficiently high, will start a crack in the residual tension side. Fatigue tests on automobile rear axles and on aircraft engine crankshafts which were straightened in production gave endurance limits 20 to 30% lower than nonstraightened axles and shafts.[13] Warping may be prevented by annealing before machining or after rough machining. In the case of automotive axles, the effect of the unfavorable residual stresses was overcome by shot peening after straightening.

Detrimental residual stresses commonly result from differential *heating* or *cooling*. A weld is a common example. The weld metal and the areas immediately adjacent are, after solidification or bending, at a much higher temperature than the main body of metal. The natural contraction of the metal along the length of the weld is partially prevented by the large adjacent body of cold metal as shown in Fig. 7.16. Hence residual tensile stresses are set up along the weld. Furthermore, if the weld is V-shaped, the top of the Vee will tend to contract more than the bottom. If the adjacent metal is free to move, then there will be warping with some relief of the stresses but if the adjacent metal is constrained, as in the flanged cylinder shown in Fig. 7.16, residual tensile stress will also occur in a direction across the weld.

In general, local or shallow heating which would expand the region or surface, if it were free, a distance well beyond that which the adjacent larger volume will allow causes yielding and upsetting of the heated material. This readily occurs because of the reduced yield strength at elevated temperatures. The same cooler volume prevents the upset, heated region from fully contracting during its cooling, and tensile stresses result. Thus flame-cut surfaces are in tension and weak, and grinding cracks occur because of high-tensile stresses. A general rule is that the "last to cool is in tension," although there is an exception if certain transformations of microstructure occur (Section 7.15). Methods for minimizing or reversing these stresses include annealing for stress relief and hammer or shot peening of the weakened surface (Section 7.13). Annealing requires heating mild steel to 1100°–1200°F, some alloy steels to 1600°F, then holding an hour per inch of thickness, followed by slow cooling. Some preheating of the parts to be joined may minimize the tensile stresses in welds. Electrodeposited nickel and chromium plating is in a state of tensile residual stress, and the endurance limit may be reduced by 50%. Low-current densities during disposition and shot peening before plating will minimize the reduction.

---

[13] M. Hetenyi (Ed.), *Handbook of Experimental Stress Analysis,* New York: John Wiley & Sons, Inc., pp. 523 and 554, 1950.

7.9 BENEFICIAL RESIDUAL STRESSES: PRESTRESSING

Favorable residual or initial stresses have been obtained for many years by assembling or overstressing, and more recently much attention has been directed toward cold working of surfaces and heat quenching. These developments have been accelerated by techniques in the measurement[14] of residual stresses, particularly by nondestructive X-rays. Evidence from many sources indicates that fatigue cracks start on surfaces in fields of tensile stress, not compressive stress. Hence prestressing is directed toward obtaining compressive stresses at those surfaces, including regions of stress concentration, which will be critically stressed in tension, or alternately in tension and compression, when put into service. Prestressing is widely used for improving fatigue life, particularly of those parts subjected to bending and torsion.

Applications of beneficial prestressing are discussed in what follows under the groupings of prestressing at assembly, bolt tightening, mechanically induced yielding, peening, thermally induced yielding, and transformation of internal structure. Then they are summarized in Section 7.16 according to their applicability. Throughout the sections on residual stresses the following notation is consistently used for the stress lines in the sketches, as in Figs. 7.15, 7.22, and 7.23. Lines A-A represent stresses that would occur during the manufacturing process if the material remained elastic; lines B-B represent the actual stresses with yield strength S_y in the plastic region; lines C-C represent the residual stresses after the manufacturing forces are removed, obtained by *subtracting* lines A-A from lines B-B; lines D-D represent the stresses calculated from the load imposed in service; and lines E-E represent the net stresses in service, obtained by *adding* lines D-D to lines C-C.

7.10 PRESTRESSING AT ASSEMBLY

Several examples of prestressing at assembly will be given. Flanged cylinders, such as those used on radial aircraft engines, have variable gas forces acting on the head, which subject both bolts and cylinder walls to variable tensile stresses (Fig. 7.17(b)). The initiation of cracks at the flange fillet due to stress concentration is effectively prevented by slightly tapering a matching face, as in Fig. 7.17(a), so that when the bolts are tightened to draw the faces together, the fillet is squeezed, inducing compressive stresses. Similarly, if the top leaf of an assembly of leaf springs is forged with a smaller curvature than the others (Fig. 7.18), it will be given built-in bending stresses, compressive on top and tensile below, when the U-bolts draw the leaves together. These stresses are favorable, decreasing the bending stresses imposed in service under the weight of the car.

In a single *simple* circular cylinder with high internal pressure, the distribution of tangential stress is illustrated by curve D-D of Fig. 7.19. The maximum stress is at the inner surface. The curve may be calculated by the equations of Table 8.1 and Sections 8.8 and 8.9. If the cylinder is *built up* by shrinking one cylinder over a second, then initial compressive stresses are produced in the inner cylinder and tensile stresses are produced in the outer cylinder, as shown by lines C-C. The net stress in service is represented by lines E-E, obtained by adding lines C-C to lines D-D. Stress distributions and optimum dimensions for two-layered compound cylinders are treated in Section 8.11.

---

[14] Many measurements and techniques are reported in *Proceedings of the Society for Experimental Stress Analysis*. One of the early summaries was made by O. J. Horger, *Handbook of Experimental Stress Analysis*, New York: John Wiley & Sons, Inc., pp. 459–578, 1950. For equations, see J. H. Faupel, *Engineering Design*, New York: John Wiley & Sons, Inc., pp. 659–667, 1964.

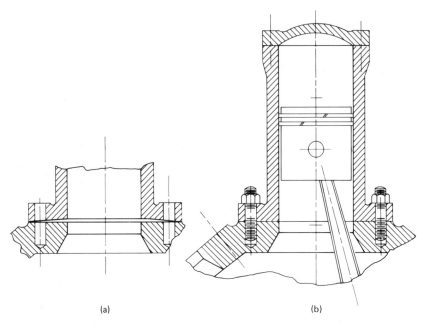

FIGURE 7.17. (a) A flange with face slightly conical or tapered (exaggerated in sketch) in a direction such that compressive-residual stresses are induced at the fillet when the flange is drawn flat by the bolts. (b) Application to a cylinder of a radial engine.

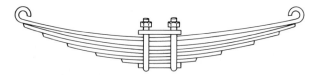

FIGURE 7.18. Top leaf of spring when clamped against others is given bending stresses which are oppositely distributed to those during service.

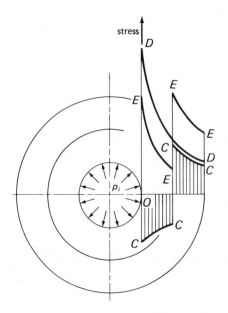

FIGURE 7.19. Compound cylinder assembled by heating and shrinking an outer cylinder onto an inner one, leaving tangential residual stresses *C-C*. In service, stresses *D-D* from internal pressure superimpose to give net tangential stresses *E-E* (plotted for radii in the ratio 3:2:1 and shrink pressure of $0.141 \times$ internal pressure).

Additional layers are used for a more even distribution of the stresses. Vessels are also made by wrapping thin plate under tension to put the inner layers in compression.

7.11 BOLT TIGHTENING

Bolt tightening is a prestressing at assembly but the stresses are tensile. Its objective, in addition to making a firm or leak proof connection, is to minimize that portion of service load that is added to the tightening load, particularly when the service load is fluctuating. The prestressing is particularly effective if the fasteners are much more flexible than the parts joined.

Figure 7.20(a) is a schematic of a head, gasket, and cylinder before assembly. Di-

FIGURE 7.20. Bolted connections. (a) Nut threaded on bolt with spring and washer before assembly and gasket set between the head and cylinder. (b) Parts assembled showing total fastener deflection δ_f and that of parts joined, δ_j. (c) Forces vs deflection, plotted in line with sketch (b), showing initial or tightening force F_i, and service-imposed force F. (d) Equilibrium of forces on head. (e) Variation of force in fastener, F_f, and at joint, F_j, as service load pulsates in value between zero and F.

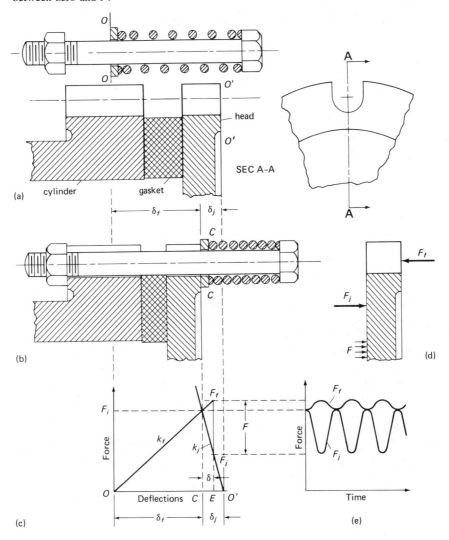

mensional differences are exaggerated for visibility. As is occasionally done to make the fastener more flexible, the bolt is shown surrounded by a coil spring. Assembly is made by applying equal and opposite forces on the nut and washer, thus stretching the bolt and compressing the spring, increasing this force until it is possible to slip the bolt into the U-slots. With assembly (Fig. 7.20(b)), the gasket is compressed, together with a smaller compression of the head and cylinder around the slots. The total deflection δ_f of the fastener is measured between the washer's initial position O-O and its final position C-C. With the faces at the nut assumed fixed, the total deflection δ_j of the parts joined is measured between the initial position O'-O' and final position C-C of the right-hand face of the head. The forces on the fastener and the joint are plotted in Fig. 7.20(c) immediately below and in line with Fig. 7.20(b). At position C the lines of force cross, and the fastener force F_f and the joint force F_j are in equilibrium. They are designated F_i, the initial or tightening force, which corresponds to the deflections δ_f and δ_j.

The stiffness constant of the fastener, which is the slope of its force-deflection line in Fig. 7.20(c), is designated k_f, and the stiffness constant of the parts joined is designated k_j. Then $F_i = k_f \delta_f = k_j \delta_j$, where all k's and δ's have a positive sign. If a service load F is applied in a direction to open the joint, as from gas pressure on the head, then the fastener is extended a further amount δ, developing a force $F_f = F_i + k_f \delta$ (Fig. 7.20(c)). The joint force is decreased to $F_j = F_i - k_j \delta$. These forces are shown at E in Fig. 7.20(c). In Fig. 7.20(d) the three forces are shown acting on the head, and they are in equilibrium. This is written and solved for deflection, thus

$$F + F_j - F_f = 0$$

By substitution,

$$F + (F_i - k_j \delta) - (F_i + k_f \delta) = 0$$

and

$$\delta = \frac{F}{k_f + k_j} \tag{7.11}$$

Thus

$$F_f = F_i + \frac{k_f}{k_f + k_j} F \tag{7.12a}$$

$$F_j = F_i - \frac{k_j}{k_f + k_j} F \tag{7.12b}$$

The division of the external load is shown at E in Fig. 7.20(c). Here $k_f < k_j$ so the increase in fastener load is less than the decrease in gasket load. If the service load is pulsating between zero and F, the fastener and joint loads vary as shown in Fig. 7.20(e), which is in line with Fig. 7.20(c). Without a gasket it is possible to have $k_f \ll k_j$. By Eqs. (7.12), F_f becomes nearly equal to F_i. Then, the service load has negligible effect on the fastener load, and the latter is essentially under static loading. This is particularly beneficial, since it prevents fatigue failure of the bolt, to which it is vulnerable because of the stress concentration at its threads and under its head.

The stiffness k_b of one bolt is approximately $k_b = A_b E/l$, where A_b is the shank area and l is the length between the head and the nut. There may be N bolts per joint, and in series with each bolt there may be a spring, sleeve, or flexible washer of stiffness k_s. The total fastener deflection is $\delta_f = \delta_b + \delta_s$, or

$$\frac{F_f}{k_f} = \frac{F_f}{N k_b} + \frac{F_f}{N k_s}$$

whence

$$k_f = \frac{Nk_b\,k_s}{k_b + k_s} \tag{7.13}$$

If there is no spring or sleeve, then $k_s = \infty$, and $k_f = Nk_b$. Similarly, deflection of the parts joined may consist of a gasket deflection plus that of N small tubular, perhaps cylindrical, volumes of metal surrounding the bolts. With subscripts g and e designating the gasket and each bolt enclosure, respectively, $\delta_j = \delta_g + \delta_e$, or

$$\frac{F_j}{k_j} = \frac{F_j}{k_g} + \frac{F_j}{Nk_e}$$

whence

$$k_j = \frac{Nk_g\,k_e}{k_g + Nk_e} \tag{7.14}$$

If the stiffness Nk_e of the enclosures is much greater than k_g, then $k_j \to k_g$.

Fastener stiffness may be decreased by increasing bolt length, by reducing its shank diameter to or below that at the root of the threads, and by using a coil spring or Belleville-type spring washer in series with the bolt. A thin tubular sleeve placed between the bolt head and the first clamped surface not only adds its own flexibility but adds the flexibility of an increased length of bolt (Prob. 7.40). Joint stiffness k_j is greatly increased by avoiding soft gaskets and grinding flat the surface to be joined. If fluids are present, leakage may be prevented by an oil-proof shellac to fill the grooves in the finished surface or by a thin, soft copper sheet, which is forced into the grooves. Under higher pressures, an O-ring or Quad-ring, placed in a groove cut into a circumference or into one of the two flat surfaces joined will allow the high stiffness of metal-to-metal contact while preventing leakage by its self-sealing ability (Fig. 7.21).

Initial tension is commonly given by a torque wrench. However, it measures not only the torque to stretch the bolt but also the friction torque along the threads and under the nut, and errors of ±20% may occur unless all conditions such as surface finish, lubrication, and cleanliness are similar. More accurate tensioning has been obtained by tightening once with a wrench to seat the parts and compress the burrs, then loosening completely and retightening "finger tight," then turning the nut by a wrench through a

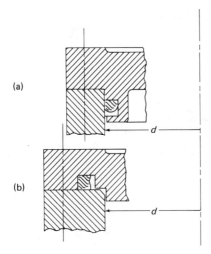

(a)

(b)

FIGURE 7.21. Metal-to-metal joint, sealed by an O-ring or Quad-ring. (a) On a cylindrical surface. (b) On the flat surface of the joint.

precalculated angle to give the tension desired. Another method is to tighten finger tight with the bolts heated to a temperature from which contraction will give the desired tension when cold. Bolts and washers are available with built-in sensors that indicate a degree of tightness. More accurate methods include hydraulic stretching under controlled pressure, electric-resistance strain gages, and micrometer measurement of the change in bolt length when the head end and the thread end are accessible and properly prepared for accurate measurement.

7.12 MECHANICALLY INDUCED YIELDING

The preceding assembly methods are done elastically. Deformation during the manufacturing process, sometimes an inherent part of a cold-forming operation, may give favorable residual stresses if the part is loaded in service in the same direction as when forming. For example, when a spring is being coiled from straight wire, the bending stresses are like those of Fig. 7.15, leaving a residual stress C-C. The latter is transferred to Fig. 7.22. If the spring is to be used as a torsion spring, i.e., one which is twisted about the axis of the coil, the service stress is a bending stress. If the moment is applied in the direction that winds up the spring, i.e., in the same direction as when coiling, it gives calculated stresses D-D and net stresses in service, E-E. A similar favorable effect occurs for crane hooks that are *proof tested*, i.e., overloaded beyond yielding before being put into service, and for helical compression springs that are *preset* by yielding in torsion to a desired length.

Favorable residual stresses are obtained in thick cylinders and guns by autofrettage (self-hooping) and in disks by overspeeding. After the final machining the ends of the gun are plugged and hydraulic pressure is applied in sufficient amount to cause permanent enlargement. The enlargement may be limited by encasing the gun in a heavy cylinder. Turbine disks may be rotated in the factory at higher speeds than the speeds that will be used in service and of sufficient amount to cause yielding in the inner portion of the disk. Some equations are developed in Section 8.12 for the cases of autofrettage and overspeeding. The stresses are different (Fig. 8.20), but both cases may be illustrated by the disk of Fig. 7.23. For the factory operations, line A-A represents the stress if the action were completely elastic and line B-B represents the actual stress as limited by S_y. Upon removal of the pressure or speed the remaining stress is line C-C, compressive in the

FIGURE 7.22. Bending stresses in a cold-wound torsion spring: *C-C* residual stress after coiling, *D-D* stress calculated from service torque applied in direction of coiling. *E-E* net stress in service, with a major reduction at the surfaces.

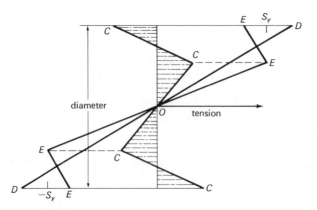

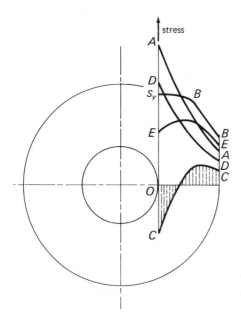

FIGURE 7.23. Distributions of the tangentially directed stresses in a uniform thin disk. During prestressing by overspeeding, calculated elastic stress *A-A* and actual elastic-plastic stress *B-B*. Resultant distribution of residual stress *C-C*. During service, elastic stress *D-D*, calculated from the speed of rotation and the net stress *E-E*.

inner region. In service the stress, line *D-D*, due to lower pressures or speeds is superimposed to give a net, more favorable stress, line *E-E*.

A thin but highly effective surface layer of compressive stress may be induced by cold-rolling, coining, and peening processes. It is seen that these processes work-harden an outer layer and squeeze or force it into smaller volume, thus causing compressive stresses to remain, together with minor tensile stresses in adjacent interior layers. Since the compressive layer is readily obtained all around, these processes are suitable for reversing loads and rotating components where the stress varies between tension and compression. The processes must be carefully controlled in respect to roller pressures and feeds, shot size and speed, depth of stressing, etc., for which extensive information is available in engineering books and periodicals.[15] Experimental development may be desirable for optimum results on new shapes, particularly when high production is anticipated.[16]

Cold-rolling is applied primarily to cylindrical and other shapes that can be rotated, such as threads and shaft fillets.[17] One of the earliest developments was in the strengthening of wheel-seating surfaces of railway axles, sometimes doubling their strength against fatigue failures initiated by fretting[18] (Section 5.12). A fixture containing spring-

[15] J. O. Almen and P. H. Black, *Residual Stresses and Fatigue in Metals*, New York: McGraw-Hill, 1963; S. S. Manson, "Metal Damage—Mechanism, Detection, Avoidance, and Repair," ASTM Special Publication 495, American Society for Testing and Materials, pp. 307–326, 1971; J. A. Scott, "The Influence of Processing and Metallurgical Factors on Fatigue," *Metal Fatigue—Theory and Design*, A. F. Madayag (Ed.), New York: John Wiley & Sons, Inc., pp. 66–106, 1969; H. O. Fuchs, "Techniques of Surface Stressing to Avoid Fatigue," *Metal Fatigue*, G. Sines and J. L. Waisman (Eds.), New York: McGraw-Hill, 1959; O. J. Horger, *ASME Handbook: Metals Engineering Design*, 2nd ed., New York: McGraw-Hill, 1965. More specific data may be obtained through the bibliographies in these books. Current information may be found in the journals of the American Society for Metals, American Society for Testing and Materials, Society of Automotive Engineers, and the Society for Experimental Stress Analysis.

[16] For a story of development, see e.g., R. O. Marklewitz, "The Tangential Rolling of Crankshaft Fillets," *General Motors Engineering J.*, Vol. II, 3rd Quarter. pp. 42–46, 1964.

[17] The usual cold-*drawing* process carried out in mills to produce cold-finished bars may produce residual *tensile* stresses unless the reduction per pass is small. See O. J. Horger in *Handbook of Experimental Stress Analysis*, M. Hetenyi (Ed.), New York: John Wiley & Sons, Inc., pp. 506–508, 1950.

[18] O. J. Horger, *Handbook of Experimental Stress Analysis*, M. Hetenyi (Ed.), New York: John Wiley & Sons, Inc., pp. 544–547, 1950; also O. J. Horger, "Stressing Axles by Cold Rolling," in *Surface Stressing of Metals*, American Society for Metals, 1947.

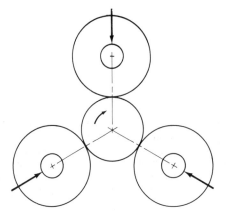

FIGURE 7.24. A method of cold-rolling cylindrical surfaces.

loaded rollers may be arranged to apply pressures to the shaft or axle (Fig. 7.24). The shape, size, and pressure of the roller and the yield strength of the shaft determine the depth of penetration, which can be calculated. The fixture shown in Fig. 7.24 may be attached to the carriage of a lathe and made to slowly traverse the desired lengths of the shaft while the latter is rotated by the lathe head. Special machines may be used for high production, and a number of different "tools" or roller assemblies have been used, particularly to roll fillets in crankshafts. (See Problem 11.19.) Thread rolling of bolts and screws has long been part of a forming process that not only forms but strengthens the threads by deformation and grain flow around the roots and by inducing compressive residual stresses.

Coining or *ballizing* of holes, also called *ball drifting*, is a manufacturing process of forcing a hard, tungsten carbide or AISI 52100 steel, slightly oversize ball through a hole in a plate, bushing, or tubing to give the holes final size and a fine finish. The length of the hole may be from 1/20 to 10 times its diameter. The machine is often set up for a high production of small parts with unskilled labor. An incidental result is that the process increases hardness, hence wear resistance, and induces around the hole a compressive residual stress that is usually advantageous, as in roller-chain links. The links are highly stressed in pulsating tension with a concentration of the stress at and near the hole surfaces. With the compressive stress from ball drifting, the net tensile stress in service is decreased, and failure is minimized.

7.13 PEENING

Peening is the most widely used method for prestressing by mechanically induced yielding. By the impact of rounded striking objects, the surface is deformed in a multitude of shallow dimples, which in trying to expand put the surface under compression. *Hammer peening*, usually by an air-driven tool with a rounded end, is useful on limited areas, such as a weld in a shaft or on areas found weakened by corrosion, plating, decarburization, or minor fatigue damage. With a hard spherical end to the tool, the depth of the compressed layer, which may be of the order of $5/16$ in, is about the same as the diameter of the surface dimples. The maximum residual compressive stress, which occurs below the surface, is about half the yield strength of the strain-hardened region.

Shot peening is done on steels by the high-velocity impingement of small, round, steel or chilled cast-iron shot with diameters from 0.007 to 0.175 in. The compressed layer has a depth from a few thousandths to a few hundredths of an inch, less than with hammer peening, but roughly proportioned to the shot size used and its velocity. Again

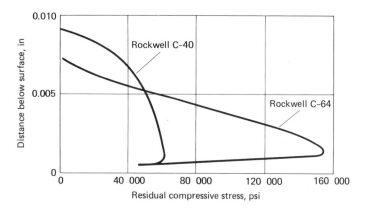

FIGURE 7.25. Magnitude and depth of residual stresses induced in steels of two different hardnesses by shot peening. (J. O. Almen, "Fatigue Weakness of Surfaces," *Product Engineering*, Vol. 21, No. 11, p. 117, November 1950. Copyright © Morgan-Grampian, Inc. Used with its permission.)

the residual stress produced is about half of the strain-hardened yield strength. A typical stress distribution is shown in Fig. 7.25.

Shot peening is extensively used because it may be applied with minimum cost to most metals and shapes, except some interior ones. On soft metals, glass beads may be used. Helical springs are commonly shot peened, with up to a 60% increase in allowable stress under pulsating loads. Part of the improvement may be due to the removal of the weakening longitudinal scratches left from the wire-drawing operation. Similarly, coarse-machined and coarse-ground surfaces are smoothed and improved by shot peening, which may be a more economical method than producing a final finish by machining or grinding. Peening is not used on bearing and other closely fitting surfaces where high precision is required. A final grinding for accuracy after peening would remove part or all of the residual stress. Machines are available for the automatic and continuous peening of small- and medium-size parts moving on a conveyor or turntable through the blast. Any fine parts such as oil holes, threads, and bearing surfaces must be masked by a resilient material.[19] The shot is either blown and directed by nozzles or it is thrown by impeller wheels.

The process may be controlled by test strips held flat and passed through the shot stream at the same rate and peened on one side. After removal, the strip has a curvature, the arc height of which can be measured with a dial indicator. Most commonly used is a standard strip of hardness Rockwell C44 to C50. This is placed in an Almen specimen gage for measurement and the arc height is the Almen intensity.

The benefits of residual stresses are somewhat limited for reversed loading and for high temperatures. In Fig. 7.26, a part is shown with initial compressive residual stresses OC_i and $O'C_i'$ on top and bottom, respectively, and with the small internal tensile stresses omitted. If the moment M which is first applied in service gives a calculated stress DD', the net tensile stress on the bottom is $O'E'$, the algebraic sum of $O'D'$ and $O'C_i'$. On the top, however, the net stress is not the sum of OC_i and OD, but instead is limited to the compressive yield strength, i.e., $OB = -S_y$. With removal of the moment, stress OD is subtracted, leaving a reduced compressive residual stress OC on top. If the moment next applied is a completely reversed $-M$, giving the calculated stress $D''D'''$, the net

---

[19] H. O. Fuchs, *Metal Fatigue*, G. Sines and J. L. Waisman, (Eds.), New York: McGraw-Hill, pp. 204–211, 1959.

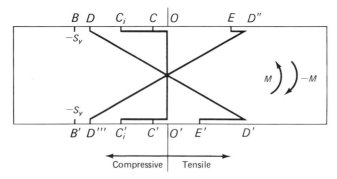

FIGURE 7.26. Reduction of residual stresses by high reversed loading. Beam with initial compressive residual stresses OC_i and $O'C_i'$, subjected to moments M, then $-M$, with corresponding calculated stresses DD' and $D''D'''$ such that $|OD + OC_i| > |-S_y|$ and $|O'D''' + O'C_i'| > |-S_y|$. The resulting, reduced, residual stresses are OC and $O'C'$.

tensile stress on top is OE, i.e., OD'' reduced only by the smaller residual OC. On the bottom the net compressive stress is limited to $O'B'$, the yield-strength value. Upon removal of moment $-M$, the beam is left with reduced residual stresses $O'C'$ and OC. Hence, theoretically the beneficial effect of prestressing, as by peening, has been reduced in one cycle. Thus the sum of the residual stress and the calculated compressive service stress should be limited to the compressive yield strength. Although half the strain-hardened yield strength is attainable by peening, it does not always need to be this to give a net stress in tension which is less than the endurance strength, a value less than the yield strength. Residual stresses from peening and other methods are also relaxed or annealed out if the temperatures are sufficiently high. However, depending on the material and the temperature, peening and other prestressing can be beneficial at elevated temperatures.

It is worth noting that the detrimental effect of *tensile* residual stresses can be reduced by an action similar to that described by Fig. 7.26, but involving the tensile yield strength. In the straightened-shaft investigation of Section 7.8 and Fig. 7.15, residual tensile stresses as high as 100 000 psi were measured, but these were relieved to 48 000 psi after 500 000 cycles of reversed bending. Also, the stress condition due only to the moment M in Fig. 7.26 is satisfactory if there is no reversal, i.e., if moment M is unidirectional and either static or pulsating. Tensile stress is never greater than $O'E'$. In fact, a compressive residual stress $O'C_i'$ close to the yield-strength value can be obtained by *strain peening*, which is peening done while the surface is held in tensile strain. Of course, this process should be used only for primarily unidirectional loadings in service. Thus it has been found suitable for prestressing automotive spring leaves.

7.14 THERMALLY INDUCED YIELDING

The rule given in the discussion of detrimental residual stresses (Section 7.8), namely, "the last region to cool is in tension" may be put to advantage by making the surface the *first* place to cool. This is done by heating a solid piece throughout to a *uniform* temperature, then cooling rapidly, as in water. The surface cools first, developing tensile stresses. Yielding must occur between it and the hotter and weaker interior, which is compressed. As the surface cools, it develops a higher yield strength, and it then resists the contraction of the slower cooling interior and is elastically compressed by it. The stresses of surface and interior change sign, leaving favorable compressive stresses on the surface and tensile stresses in the interior. Yielding must occur at the higher temperatures, or both surface and interior will return to the same original length, free of stress.

If the piece is a hollow tube or gun cylinder, subject to high, interior tensile stress in service, it may be beneficially prestressed by cooling from the interior outward, as by a spray. Heating at two or more spots in a line parallel to and adjacent to a weld causes yielding within the spots during heating and tension after cooling. If properly done, this may put the weld into favorable compression in its lengthwise directions.[20] Aluminum alloys and also ferrous metals which are not hardenable by quenching are among those that may be thermally strained to advantage. Glass is similarly strengthened, using air jets for cooling. Since the surface of glass is weak in tension but strong in compression, tempered glass with its surface of residual compressive stresses is three or four times stronger than the glass in an annealed condition. Chemical treatments of the glass surface can also develop residual compressive stresses.

7.15 TRANSFORMATION OF INTERNAL STRUCTURE

When a through-hardening steel is quenched in a liquid, the austenitic solution transforms wholly or partially into the hard and strong constituent martensite, which has lower density and higher volume. The average volume expansion is said to be about 1.5%, corresponding to a linear expansion of 0.5%, a strain of 0.005, and a biaxial stress of 200 000 psi on the surface. Although the interior is the last to cool, it is also the last to transform, which tends to expand the already strengthened exterior, leaving it with high-tensile residual stresses and possibly quenching cracks. It is common practice to temper to $1000° - 1200°F$ following quenching. This removes most of the residual stresses.

Since the rule of Section 7.14 does not hold for quench-hardened ferrous alloys, it might be replaced with, "the last region to transform is in compression." This is put to good use by several methods of control. In a process called Marstressing,[21] parts are heated in an atmosphere of ammonia, causing the diffusion of nitrogen into the surface. This lowers the temperature of transformation of the surface, and during quenching and cooling, the interior transforms first and the exterior transforms last. Methods more commonly used harden only a region at and near the surface. Some methods are based on an exterior heating that is rapid enough to limit the temperature necessary for the formation of austenite to a region of limited depth. This is followed immediately by quenching, and the only martensite to form is at and near the surface. Its attempt at expansion against the cooler interior may leave the surface region with peak compressive stresses up to 100 000 psi. Between it and the nonheated core there may be a layer which was heated high enough to yield readily, and this layer may be left with substantial tensile stresses.

In *flame hardening* the heating is usually done with a fixture, often with multiple jets and a neutral acetylene flame, to give an even heating of the entire area to be hardened. The part may rotate if its configuration allows, or the flame may be guided to follow a contour such as those of cams and gear teeth. In *induction hardening* closely placed magnetic coils induce large currents and heating in the surface region. The fixtures may have multiple holes for symmetrical, liquid spraying immediately after the heating stops. The two methods are popular for selective hardening, and warping of the larger unheated regions is avoided. The hardening of journal surfaces and of gear teeth should include the adjacent fillets for maximum benefit from residual stresses.

Other transformation methods do not use through-hardening steels. At one time there

---

[20] T. R. Gurney, *Fatigue of Welded Structures*, London: Cambridge University Press, 1968.
[21] Patented by General Motors Corporation.

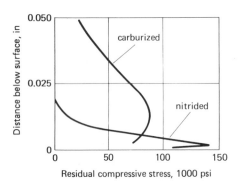

FIGURE 7.27. Residual compressive stresses developed in nitrided and carburized steel specimens by the expansion of the case during hardening transformations. (J. O. Almen, "Fatigue
Weakness of Surfaces," *Product Engineering,* Vol. 21, No. 11, p. 117, November 1950. Copyright © Morgan-Grampian, Inc. Used with its permission.)

was surprise that some *shallow-hardening steels*, those with medium-carbon content in
plain-carbon and low-alloy-content steels, when quenched all over gave higher fatigue
strength than the richer, through-hardening low-alloy steels in sections of medium size
or larger. It was discovered that transformation occurred only at and in the region near
the surfaces of the less expensive, supposedly weaker steels, leaving compressive residual
stresses at the surface. This is sometimes called *shell hardening*.

In *carburizing*, the matrix is a low-carbon-content steel (0.15−0.20%C) which does
not appreciably harden with quenching. The case has a high-carbon content (0.70−
0.90%C), in depths of 0.03−0.07 in, obtained by soaking in carbon-rich gas or solids
for 4 to 8 hours. Only the case becomes martensitic, and high-compressive residual
stresses and a high-inherent strength combine to give high-fatigue resistance. Low-tem-
pering temperatures of 300°−400°F are used, and they do not greatly reduce the residual
stresses. Carburizing is usually done for high wear and pitting resistance on the rolling
and rubbing surfaces of gear teeth and cams. Its extension to the stress-concentration
tensile-stressed regions at the tooth fillets should not appreciably increase the manufacturing expense of this high-cost process.

Nitriding gives an even harder but thinner case, typically 0.02 in deep. Nitrogen,
disassociated from ammonia, combines at about 1000°F (540°C) with some constituents
of steels to form hard nitrides with a theoretical, high-linear expansion of about 2%.
Aluminum is the unique constituent added to several steels known as nitralloys, but
regular steels may be used, usually with 0.20−0.40%C. Hardness and compressive
surface stresses are obtained by cooling to room temperature without quenching. The
long (typically 50−90 hours) and carefully controlled process is expensive, but it gives
larger residual stresses and hardness than carburizing. Residual stresses are compared in
Fig. 7.27.

7.16 SUMMARY OF PRESTRESSING METHODS: CHOICES AND CONDITIONS

Many methods have been discussed for avoiding tensile surface stresses and of obtaining
the generally favorable compressive stresses. A choice between methods is not difficult
when all the limitations of loading, shape, cost, reliability, and manufacturing facilities
are known. In the group dependent on *assembly*, the cost of obtaining favorable prestress
may not be any higher, and obtaining it is a matter of foresight and ingenuity, together
with proper specifications and control in manufacturing and assembly. In general, the

forces to preload must be in a direction *opposite* to the forces in service, as illustrated by the leaf spring, flange, and compound cylinder. An exception is the loading of fasteners for joints. *Some* of the prestressing by *mechanically induced yielding* is inherent in the forming process or in proof testing. Then it is only important that the loading in service be applied in the *same* direction as when forming or testing, as illustrated by torsion springs, autofrettage, and crane hooks. The foregoing are useful only when the load is unidirectional, either static or pulsating.

Where service stresses *alternate* between equal tensile and compressive values, a uniform compressive stress is required on opposite faces for a beam or all around for a shaft that rotates. Of course, parts so stressed may also be used for unidirectional loadings. All-around compressive stressing calls for *cold-working* of the surfaces, *thermal quenching*, or *transformation* of internal structure. Cold-rolling is principally for shafts, and it is an extra process sometimes requiring special fixtures. Ball drifting is restricted to strengthening around holes, but it is simply done or automated. It may replace a final sizing operation. Hammer peening is principally for corrective action on individual pieces. Shot peening requires some special equipment, but it is suitable for quantity processing of many shapes. It may replace other final finishing operations. Some experimentation may be required to obtain good results from these several surface-deforming operations.

In a thermal treatment applicable to nontransformable materials, rapid cooling from a uniform elevated temperature gives compressive residual stresses to the first regions to cool, provided yield strengths are exceeded at the higher temperatures during the cooling. Transformable materials must be treated by methods such that the desired region of compressive stress is the last to transform or the only region to transform. In a through-hardening steel, the surface transformation to martensite can be delayed by a nitrogen treatment called Marstressing. Otherwise, a through-hardening steel is tempered at a high temperature to relieve surface-quenching stresses that are tensile. In general, to obtain compressive surface stresses, a shallow heating and/or a shallow transformation are carried out. The former may be done on medium-carbon-content steels by quenching following locally applied, rapid flame or induction heating. Through-heating followed by quenching gives a shallow transformation in certain medium-carbon steels in sizes that are not deep hardening. Carburizing adds a high-carbon case to a low-carbon steel and, upon quenching, a martensite transformation in the surface region only. In the nitriding of steels, growth and compressive stress develop in the thin, nitrogen-penetrated case, and cooling is done without a quench from a temperature below the critical. In many cases the choice between the several thermal treatments may be based on size, material, available equipment, and on hardness requirements rather than on the degree of compressive stressing. The latter is then a bonus that is not always recognized.

PROBLEMS

Sections 7.3 *and* 7.4

7.1   The following values are reported for a high-temperature alloy steel, all at 1020°F.

Time (hours)	Rupture Strength (psi)	Creep Strength for 1% Elongation (psi)
1 000	38 000	
10 000	34 000	21 300
100 000	23 000	14 200

On a chart of strain vs log time, sketch freehand in relative position and shape five curves representing these stresses. For the lowest two stresses and neglecting the initial phase, what is the creep rate in percent per 1000 hours? In strain per hour?

7.2   The stress in order to rupture a certain 304H stainless steel at 1200°F is 21 500 psi for 1000 hours life and 16 500 psi for 10 000 hours life. Predict the rupture stress for 100 000 hours life.

7.3   In Fig. 7.3 let the temperature be 600°C and stress curves B, C, and D be 40, 60, and 75 MPa, respectively. Determine equations for each curve and by interpolation obtain the stress that will give 2.5% creep in one year. *Ans.* 64 MPa.

7.4   Solve Prob. 7.3 by obtaining the constants for an equation relating strain rate to stress, then estimating ϵ_a and calculating σ for 2.5% creep in one year.

7.5   (a) Obtain an equation for strain rate in terms of stress for the material of Fig. 7.8 at 800°F. *Ans.* $\dot{\epsilon}_0 = 1.72 \times 10^{-33} \, \sigma^{6.05} \, \text{h}^{-1}$. (b) Determine the increase in length of a 200 in long, 1-in-diameter pipe-hanger rod when first heated from 70°F and after 100 000 hours, if its load is steady at 20 000 lb, $\alpha = 6.3 \times 10^{-6} \, °\text{F}^{-1}$ and $E_{800} = 25 \times 10^6$ psi.

7.6   For the material of Fig. 7.8 at 1200°F, obtain an equation for strain rate in terms of stress and determine the stress to limit the creep to 1% in 100 000 hours.

7.7   For a stainless steel with columbium added, designated 316 + Cb, the creep rate at 730°C is given as 0.009% per 1000 hours at a stress of 18.5 MPa and 0.026% per 1000 hours at a stress of 32.5 MPa. Moduli E are 150 GPa at 730°C and 207 GPa at 20°C. Determine the creep constants n and B when the creep is measured in (mm/mm)/hour and estimate the stress to limit creep to 0.15% in 10 000 hours. *Ans.* $n = 1.88$, $B = 3.73 \times 10^{-10}$ $\text{h}^{-1}(\text{MPa})^{-1.88}$, $\sigma = 24.3$ MPa.

7.8   In a steel-mill reactor for converting methane to hydrogen, some 30-ft long vertical tubes of high Cr-Ni steel are to be hung from their tops and guided at their bottoms. They must be installed with a clearance between the latter and the reactor floor to allow for initial deflection, thermal expansion, and subsequent creep. Otherwise, if they reach the floor they will be partially supported by it and may buckle under their own weight at the high temperature of operation, 1600°F. The tube coefficient of expansion is $11.1 \times 10^{-6} \, °\text{F}^{-1}$, $E = 20.0 \times 10^6$ psi, and the unit weight is 0.285 lb/in³. The creep constants are $n = 1.3$ and $B = 8.0 \times 10^{-10} \, \text{h}^{-1} \, (\text{lb/in}^2)^{-1.3}$. Take the average temperature of the reactor wall to be 350°F and its coefficient of expansion to be $6.3 \times 10^{-6} \, °\text{F}^{-1}$. Determine the minimum clearance required below the tube when it is installed at 70°F for 5-year life. *Ans.* 8.04 in.

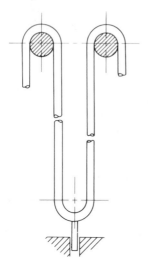

7.9   Concrete and plastics are viscoelastic materials, and to express the effect of creep it is common to use an *effective modulus* E_{eff} which decreases with the increase of time and temperature and sometimes with environment. Then E_{eff} is used in elastic equations such as $\epsilon = \sigma/E_{eff}$, instead of the modulus E of an initial, short-time test. To determine E_{eff} for plastics, an equation which is essentially $E_{eff} = E(b/t)^m$ has been proposed by MacLeod.[22] For polyethelene, the constants are $m = 0.040$ and $b = 1.0 \times 10^{-5}$ h at a temperature of 23°C, a relative humidity of 50%, with time t in hours. (a) Determine the effective modulus after one year and after five years. (b) Determine the increase in radius of a 100-mm-diameter pipe, 6 mm thick, with $E = 131$ MPa and an internal pressure of 0.70 MPa when put into service and after one year. (From Prob. 8.18 of Chapter 8, $\Delta d = 2pR^2/Et$, where R is the radius and t is the thickness.)

Section 7.5

7.10   Derive Eqs. (7.6) and (7.7), starting with Eq. (7.5).

7.11   A bolt with the material and temperature of Example 7.1 is to retain 60% of its initial tightening stress after four years. $E = 25 \times 10^6$ psi. What maximum tightening stress may be used?

7.12   A bolt, 125 mm long between head and nut, is made of a 12% Cr steel, with creep constants $n = 4.4$, $B = 1.35 \times 10^{-18}$ h^{-1}(MPa)$^{-4.4}$, and $E = 158.6$ GPa at 450°C and 207 GPa at 20°C. The tightening stress is not to drop below 75% of its initial value after 100 000 hours. Determine the allowable initial stress. What should be the deflection and stress when tightening cold? If the bolt is removed after 100 000 hours, how much will its length have increased? *Ans.* 145.6 MPa; 0.115 mm, 190 MPa; 0.029 mm.

7.13   For 9-in-long bolts at 800°F, made from the carbon steel of Fig. 7.8, determine a reasonable initial stress for this temperature if the time between tightenings is to be five years. Use data from Prob. 7.5, including the constants in the equation of the given answer. What should be the deflection and stress when tightening cold? If the bolt is removed after five years, by how much will its length have increased? *Ans.* 10 870 psi; 0.0039 in, 13 000 psi; 0.0008 in.

7.14   Use of 316-Cb stainless steel is proposed for bolts operating at 730°C. Use the data and answers given in Prob. 7.7 and an initial stress of 6.0 MPa. How often must the bolts be retightened if the stress is not allowed to drop below 4.0 MPa? Does the material seem suitable for this application?

Section 7.6

7.15   Derive an equation for the thermal stress in a block of dimensions l_x, l_y, and l_z when expansion is prevented in all directions. *Ans.* $-E\alpha(T_2 - T_1)/(1 - 2\nu)$.

7.16   Calculate the stress increases per 100°F temperature rise for the restricted rod and plate of Figs. 7.11 and 7.12 and the block of Prob. 7.15 when the material is steel, $E = 30 \times 10^6$ psi and $\alpha = 6.3 \times 10^{-6}$ °F^{-1}.

7.17   Same requirement as Prob. 7.16 except for each 100°C rise, with $E = 207$ GPa and $\alpha = 11.35 \times 10^{-6}$ °C^{-1}. *Ans.* 235, 336, and 588 MPa.

---

[22] A. A. MacLeod, "Design of Plastic Structures for Complex Static Stress Systems," *Ind. Eng. Chem.*, Vol. 47, p. 1319, 1955. Also, J. H. Faupel, "Creep and Stress-Rupture Behavior of Rigid PVC Pipe," *Modern Plastics*, Vol. 35, No. 11, p. 120 and No. 12, p. 132, 1958.

7.18 Approximately what difference in temperature between the walls of a cylinder or clamped plate will result in the yield strength being exceeded when the material is low-carbon (soft) steel with $S_y = 36\ 000$ psi and with the other constants given in Prob. 7.16?

7.19 A two-ply axially symmetrical tube of length l is made at temperature T_0 by bonding together two tubes of different materials and cross-sectional areas with an insulating adhesive cement. In service the inner ply 1 is heated to temperature T_1 and the outer ply 2 to temperature T_2. Derive equations for the axial thermal stress in each ply. Under what circumstances will the stresses be zero? *Ans.* $\sigma_1 = E_1[\alpha_2(T_2 - T_0) - \alpha_1(T_1 - T_0)]/(1 + A_1E_1/A_2E_2)$, $\sigma_2 = -(A_1/A_2)\sigma_1$. (b) With very thin walls may the same equations be used for tangential stresses?

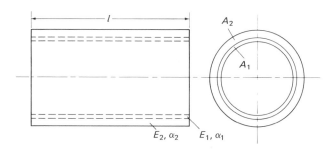

7.20 This problem is a continuation of Prob. 7.19. A composite cylinder consists of two tubes 400 mm long, one of bronze, 37.5 mm inside diameter and 44 mm outside diameter, fitted inside, bonded to, and insulated from a steel tube of 50 mm outside diameter. They are joined, stress free, at 25°C. In service, the inner tube is heated to 145°C and the outer tube heated to 45°C. For the bronze, $E = 103$ GPa and $\alpha = 17.7 \times 10^{-6}$ °C^{-1}. For the steel, $E = 207$ GPa and $\alpha = 11.35 \times 10^{-6}$ °C^{-1}. Calculate the axial stresses in each tube and the increase in length of the composite cylinder.

7.21 It is estimated that a cool liquid poured into the heated, cast iron pot, shown in section, will at first cool the bottom more or less uniformly to a temperature 100°F lower than the relatively massive side wall. $E = 15.0 \times 10^6$ psi, $\nu = 0.22$, and $\alpha = 5.9 \times 10^{-6}$ °F^{-1}. Estimate the thermal stress.

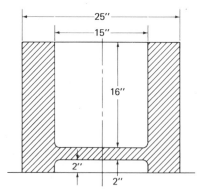

7.22 One method of controlling the clamping force in a bolt at room temperature is to heat the bolt to a calculated temperature, insert it in the parts to be joined, which may be assumed to be relatively larger, barely tighten the nut (as by a gloved hand), and allow it to cool. A steel bolt 10 mm in diameter and 100 mm long between the head and the nut is to have an initial tightening force of 30 000 N. Shop temperature is 25°C. At what temperature should the bolt be inserted? $E = 207$ GPa and $\alpha = 11.35 \times 10^{-6}$ °C^{-1}.

7.23 A common method for assembling a metal ring or tire on a wheel or cylinder is to machine its inner diameter to interfere with the cylinder diameter. It is then heated until it can be slipped over the cylinder, and a "shrink fit" occurs on cooling. If the "interference" in diameter is to be 0.003 mm per millimeter of diameter, to what minimum temperature should an aluminum ring 8 mm in thickness and 12 mm wide be heated to slip it over a steel disk 130 mm in diameter, which is at a temperature of 25°C? (a) What is the subsequent stress? *Ans.* 213 MPa. (b) What is the stress if the shaft is then operated at −50°C? at 150°C? For aluminum, $E = 71.0$ GPa and $\alpha = 24.7 \times 10^{-6}$ °C^{-1}, for steel, $E = 207$ GPa and $\alpha = 11.35 \times 10^{-6}$ °C^{-1}. *Ans.* 284, 94.5 MPa.

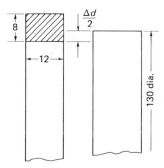

Section 7.7

7.24 Discuss the following for good thermal design, including minimization of stresses. (a) A brake disk (Prob. 3.41) which is used on an airplane wheel becomes excessively hot during the landing runs. Provision for additional, integral material for a "heat reservoir" has been proposed, adding it as a thin, wide rim just outside the pads. Comment. Suggest alternatives. (b) A neutron detector consists of an 18 ft long, small diameter, evacuated tube in which there is a fine central wire running along most of its length. The tension in the wire must be limited as well as maintained at all times (no slack). The detector is lowered into an atomic reactor, where the tube heats rapidly and the wire lags because of the vacuum; then, sometime after all parts reach a temperature of 600°F, the detector is removed to the room, with a temperature of 80°F, where the tube cools rapidly and the wire lags. What is the maximum difference in tube and wire expansions at any one time, if both are of metal with α averaging 7.0×10^{-6} °F^{-1}? Sketch a method for avoiding large changes in tension in the wire. (c) An auxiliary, such as a pump or small generator, runs somewhat cooler than the engine or turbine to which it is connected. Suggest how to avoid shaft stress and extra bearing loads.

7.25 Discuss the sketches for good thermal design, including minimization of stresses. (a) A circular steel diaphragm is clamped between two, much thicker, aluminum rings. It is used in a shock tube, where it is intended that the diaphragm burst only when the pressure behind it builds up to a desired value. It bursts prematurely, caused, it is believed, by thermal stresses from the heat in the gas. Explain and suggest a simple change. (b) There is a steep temperature gradient where a pipe or nozzle which carries a cold fluid is welded directly to the thicker wall of a heated pressure vessel, through which it enters. Sketch a "thermal sleeve," or extra elongated tubular member, which will reduce the gradient as well as add flexibility. (c) During the cooling period after casting, the thin, radial spokes of a pulley of brittle material tend to crack away from the rim or hub, shown as cast. Explain and suggest a different spoke configuration and improvements in cooling practice.

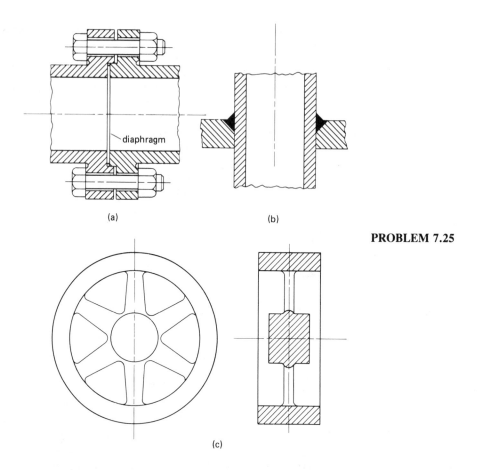

(a)

(b)

PROBLEM 7.25

(c)

Section 7.8

7.26 A long, 9-in-diameter lineshaft for a wire-rope drive was installed in a textile mill in the year 1900. It was driven by wire-ropes from a water turbine with water from a pond and stream beside the mill building. The lineshaft ran in bearings on several cast-iron pedestals supported by the mill floor, with columns below it and a brick wall near it. It ran smoothly and almost continuously without any problems for 54 years, then fractured at a fillet adjacent to a 60-in-diameter rope sheave. The sheave then broke loose and damage to the mill was extensive. Examination showed that the fracture was of the fatigue type, and that there was no preceding bearing failure nor cracks in the pedestals. After so many years what could have happened to cause the failure? How many cycles does it usually take for a fatigue fracture? Describe the probable stress situation.

7.27 Two flat bars 12 in long, 1½ in wide, and ½ in thick are laid on a welding table and a butt weld of a V cross section is made between them. Sketch a change of shape that may occur. Sketch the stress distribution if a correction is made by bending. What shape of weld and steps in welding might have avoided the need for a correction?

7.28 Tire rims for trucks and buses have been made from flanged, straight strips of steel by rolling them into a circle and resistance welding the two ends together as in Fig. 7.16. After mounting a tire and the other flange, the tire is inflated and in service the tire is bulged a bit once per revolution. Fatigue cracks have been discovered, starting in the weld at the fillet (Fig. 7.16), and progressing around the rim along the fillet of the flange. Describe the several stresses that contribute to failure and suggest corrective measures.

7.29 (a) Commonly, failures at the junction of a bolt head and its shank show that the fatigue crack spread from one point on the periphery. What does this indicate about the assembly

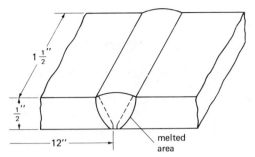

PROBLEM 7.27

or the accuracy of machining? (b) It has sometimes been found that a failure in an electro-plated part originated at a "burn," the result of arcing between the piece and an electrode. Explain the probable stress situation.

Section 7.9

7.30 A simple but semidestructive and approximate method of determining the nature and the magnitude of residual surface stresses is to spray the area with Stresscoat or other brittle lacquer which cracks or crazes when extended by the displacement of the metal to which it adheres. A hole about ⅛ in diameter by ⅛ in deep is made with a sharp drill causing a relief of stress in the metal part. Sketch the general nature of the crack pattern which would indicate that the residual stress present before drilling had been (a) compressive and (b) tensile.

7.31 (a) If metal is planed off the top surface of a flat plate and it bows upward in the middle, what kind of residual stress was present there? Explain. (b) If a hollow cylinder is machined externally on a lathe and its inside diameter increases, what kind of tangential, residual stress was present on the external surface?

7.32 (a) As a cold-drawn rectangular bar is cut along its central, axial plane by a thin saw, the separated halves bend away from each other. Sketch the distribution of stress before cutting. (b) If a hollow cylinder is bored internally and its outside diameter decreases, what stress was present on the inside surface?

7.33 Most residual stress considerations are made by assuming "perfect plasticity," i.e., a horizontal line at height S_y on the stress-strain diagram of the material, and no work hardening. Suppose the stress-strain diagram is better represented by one of slope E, followed by one of reduced slope E', somewhat characteristic of high-strength steels. Redraw the diagram of Fig. 7.15 for this case. Are residual stresses larger or smaller?

Section 7.10

7.34 For the top leaf of the spring of Fig. 7.18 sketch the relative values and distribution of the prestress, service stress, and net stress, labeling them according to the key given in Section 7.9.

7.35 A commercial airliner had an engine fire, the flames burning off a wing before a landing could be made. When the engine was found, examination showed that fatigue cracks had originated at several places in the fillet around a cylinder at its flange (Fig. 7.17). These cracks had joined, allowing an ignited gasoline and air mixture to escape. Overhaul records showed that the cylinder had been in service only six hours on this engine. It had been on another engine where the discovery was made during a routine check that eight of its sixteen bolts had failed. The airline's rules required that any cylinder with more than two failed bolts must be returned to the manufacturer since the flat, machined face of the flange might have become warped and wavy. This rule had been ignored by an inspector. (a) Explain carefully why the cylinder should fail when installed on the second engine. (b) Sketch the probable appearance that enabled identification of the originating points of fatigue failure.

7.36 The airplane bracket shown failed at point *A* under a service pull *P*. It was found that insufficient chamfer on the bracket at *C* had caused interference and there was bending when the bolts at *B* and *B'* were tightened. (a) Explain the failure. (b) If the chamfer is increased to eliminate the interference, what slight change in the 90° angle between the sides of the bracket would give a beneficial stressing at installation. Explain.

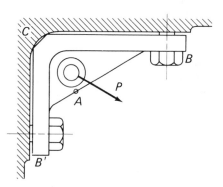

7.37 Obtain information on prestressed concrete beams and with sketches discuss purpose, rod location, how and where done, stress distribution, poststressing vs prestressing, etc. What advantage is given poststressing over prestressing by the fact that concrete creeps at normal temperatures?

Section 7.11

7.38 The steel cylinder head and cylinder of a compressor are joined by six steel bolts with a 1-in-shank diameter. They pass through rigid flanges and a gasket with a combined thickness of 4 in. The bolts are initially tightened to a force of 20 000 lb each. The stress area at a bolt thread is 0.663 in². The gasket has a 9-in outside diameter and a 6-in inside diameter. It deflects 0.012 in when all the bolts are tightened. Neglect deflection of the bolt enclosures. In service a pressure of 1000 psi is repeatedly applied and removed, acting on a 6-in-diameter area of the head. Determine the maximum and minimum nominal bolt stresses (not considering stress concentration) and gasket pressures. Compare the amplitudes of the variation of bolt load and external load, basing each on its average value. *Ans.* Bolt 35 700, 30 200 psi; gasket 3400, 3220 psi: 8.4%, 100%.

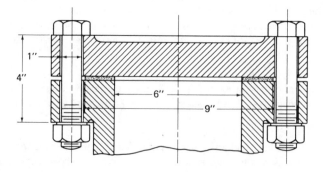

7.39 The maximum load on the cap of a small-steel-connecting rod is 3500 lb, from inertia forces at maximum speed with the engine throttle closed. The two steel bolts are initially tightened to 1.5 times the service load on each bolt. They are turned down to 0.300 in diameter over an effective length of 1.50 in between the head and the nut. Each boss around a bolt is considered to have an effective area three times that of the bolt over the same length. (a)

Determine the initial and maximum forces on the bolts and the minimum force on the joint. Determine the deflections and construct a force vs deflection diagram. (b) If the stress area at the root of the threads is 0.0878 in², the fatigue-stress-concentration factor is 3.5 (including notch sensitivity), the yield strength is 70 000 psi, and the corrected endurance limit is 45 000 psi, what is the factor of safety?

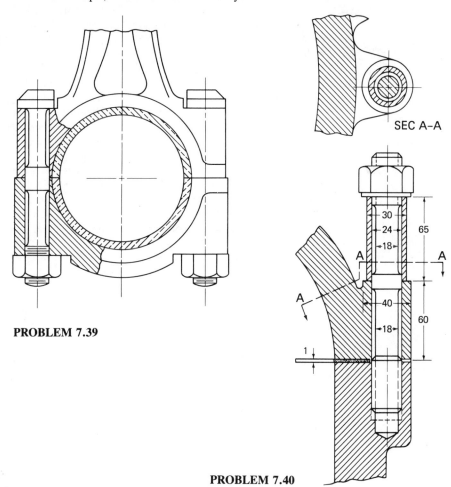

PROBLEM 7.39

PROBLEM 7.40

7.40   The cast-iron head of a cylinder is fastened by 24 stud bolts, the dimensions of which are shown. The gasket has a thickness of 1.0 mm, and its composition is chosen so that it is reduced to 0.75 mm when the bolts are tightened. To add fastener flexibility the steel bolt is lengthened, and a thin steel sleeve is placed between the nut and the surface of the boss. An estimate of the stiffness of the head beneath the boss should be made and included in the joint stiffness. A maximum service load on the head of 1.303 MN is expected, and each bolt is to be initially tightened to 1.5 times this load when divided between all the bolts. Base the bolt calculations on the reduced shank area and neglect the enlargement where the bolt is fitted between the boss and the sleeve. E_{steel} = 207 GPa, E_{CI} = 100 GPa. What should be the yield strength of the bolt and sleeve material for a factor of safety of 1.4? What is the percentage loss in gasket pressure when the service load is applied? *Ans.* 598 MPa, 33%.

Section 7.12

7.41   In a crane hook (Fig. 9.6(a)), the bending moment is largest at Section *B-B*. For simplicity assume a hook with a rectangular cross section and a linear distribution of elastic bending stress. By sketches determine the general distribution of residual stress at the section after the hook has been proof tested, i.e., yielded by an overload. Then, show the stress distribution in service under a rated load. Label with symbols corresponding to those of the text.

7.42 Compression-type helical springs with squared ends, e.g., Fig. 12.3(a), are frequently "set," i.e., brought to a desired, shorter length, by applying a load sufficient to cause yielding of the wire in torsion. For simplicity assume a linear distribution of shear stress. On a diameter of a circle representing the cross section of the wire, show by sketches of shear-stress distribution the development of residual stress and its effect under service loads. Label with symbols corresponding to those of the text.

7.43 A straight rod is uniform in diameter except midway of its length, where there is a circular groove for a retaining ring. In service the rod will be loaded by a pulsating (zero to maximum) tensile load. Overloading the rod statically before putting it into service has been found to improve its fatigue strength at the groove. Explain this phenomenon with sketches. Is the strength improved at sections away from the groove? Explain.

7.44 Some retaining rings or "snap rings" are formed cold by winding from straight and pre-tempered wire. Their principal stresses in service occur when they are opened up to slide over a shaft or closed up to slide inside a hole, both to reach a groove where they must remain. It is important that there be no permanent change in dimension. For the same ring diameters and net stress, which application can have the deeper groove? Explain with sketches.

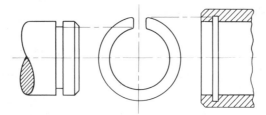

7.45 A number of fatigue failures of some rather expensive high-strength heat-treated bolts are being reported. The failures occur at the fillet where the shank joins the head. Quite a few of the bolts are still in stock. Suggest and explain a possible "cure" that might be made.

7.46 Sketch the distribution and label maximum values for the residual stresses in a beam that is bent sufficiently to place all parts of its rectangular cross section in a plastic condition, then is released. Note: The moment of the imaginary all-elastic-stress distribution must equal that of the all-plastic distribution.

7.47 Determine the distribution of residual stresses in a circular shaft that has been given a torque sufficient to yield it completely throughout. The equations of Example 5.5(b) and Section 12.2 may be helpful. *Ans.* $\pm S_y/3$ max.

7.48 If a solid shaft, elastically stressed, is to be replaced by a hollow shaft that has been prestressed as in Prob. 7.47, what should be the ratio of its inside to outside diameters if its outside diameter is the same as that of the shaft it replaces? What is the saving in weight? *Ans.* 0.63, 40%.

7.49 It has been suggested that shafts be built up of a large number of concentric tubes, first welded together at one end, then welded together at the other end after each tube is separately pretwisted an amount sufficient to bring it to its yield strength.[23] For an infinite number of

[23] W. E. Cawley, "The Prestressed Shaft," *Machine Design*, Vol. 33, pp. 99–101, Feb. 2, 1961.

tubes this gives the same result as Prob. 7.47. For a 50-in-long shaft with a 5 in outside diameter made of 9 steel tubes each 0.25 in thick and a central bar of 0.25 in radius, all with $S_{sy} = 50\,000$ psi, through what angle must the outer tube be twisted? the inner bar? *Ans.* 5°, 50°.

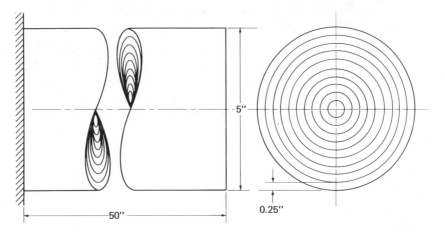

Section 7.13

7.50 Fine cracks are seen to be developing for a short length along the root on the tension side of several gear teeth of two large, herringbone gears used in a mine hoist. The teeth are approximately 1 in high and 30 in long, with a pitch-circle thickness (t_p of Fig. 11.16) of 0.785 in. There are more than 300 teeth in each gear. If the cracks progress on these teeth and form on others, the mine must be shut down for three months before replacements can be manufactured, shipped, and installed. Describe how to save the gears, at least until new ones can be obtained. What would you do about the cracks already formed?

7.51 It is proposed to make a leaf spring from a single leaf. After heat treatment it will be preset to nearly final shape by yielding with an overload, then shot-peened on the residual, compressive-stress side. Sketch the spring as formed in the typical curved shape of a leaf spring (Fig. 7.18), then describe with sketches the stress changes during the first and second prestressing operations and when finally installed and supporting a vehicle.

7.52 In a strain-peening operation a spring leaf is held under a bending moment such that the tensile and compressive stresses on the surfaces are $0.5S_y$ and $-0.5S_y$, respectively. The tensile side is shot peened until the stress at and below the surface for a short distance becomes $-0.5S_y$. Sketch this condition and the stress distribution after release of the bending moment.

7.53 Why, in general, must shot peening be the final finishing operation? Why is it sometimes done *before* electroplating with hard coatings?

7.54 What general variation in velocity, size, and material of the shot would you suggest for peening small coil springs, a large badly decarburized part, sheet steel, aluminum pistons, and flat brass springs?

7.55 Your company has sandblasting equipment for cleaning forgings and castings but no shot-peening equipment. Repeated failure of a part after machining makes some peening necessary. Would you suggest sandblasting? If so, what restrictions would you apply? Would you expect the same effectiveness?

Section 7.14

7.56 To explain beneficial prestressing by the thermally induced yielding process, sketch separately on three diameters of a cylinder in relative positions the temperature and stress distributions (a) just before rapid cooling, (b) during cooling, and (c) after cooling, labeling the stress lines with the previously used designations *A-A*, *B-B*, *C-C*, and S_y (at elevated temperature) as appropriate.

7.57 An I-beam or girder of ordinary structural steel is strengthened by welding a preheated cover plate of high-strength low-alloy steel to the bottom flange. The cover plate is welded at one end of the beam, insulated from the beam and preheated along its midlength until the other end has lengthened a predetermined amount, then is welded at that end. Explain the increase of load-carrying ability. Why is it more economical than simply welding a high-strength cover plate without preheating it?

7.58 A plate is attached to a rectangular bar, which crosses it, by using fillet welds as shown. The bar is loaded in tension, and the stress concentration at the ends of the welds may cause failure there. Sketch the locations for spot heating that may prevent this. Show and explain the distribution of residual stresses in the bar.

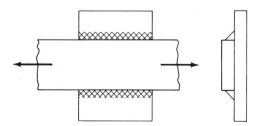

7.59 For the welds of Fig. 7.16 and Prob. 7.28, sketch the locations for spot heating that may be beneficial in preventing fatigue failure starting in the weld at the fillet. Sketch and explain the stress distribution which you would hope to attain across the weld and spots.

Section 7.15

7.60 Confirm the relationships between expansion, strain, and stress stated for steels in the first paragraph of Section 7.15.

7.61 A shaft is forged and machined with a flange at one end for the attachment of a gear. Adjacent to it is the journal for a plain bearing which it is proposed to surface harden because of uncertain lubrication by grease. Recommend and describe a method (a) if only one shaft is to be done and (b) if a hundred shafts are to be done and uniformity between them is important. Would you confine the hardening to the bearing length or would you extend it beyond? What must be the general composition of the material of the shaft?

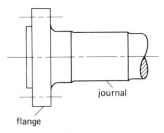

journal

flange

7.62 In an automotive, differential gear train between the rear axles, the pinion on the end of the propeller shaft and the ring gear get continuous service and wear. Maximum-tooth-endurance strength and hardness to a depth of $^1/_{32}$ in are wanted. What material and process or processes do you recommend and on what surfaces?

7.63 A shaft of 50-mm diameter has been made from AISI 4340 steel. This is a machinery steel with a relatively high-alloy content. An engineer proposes to substitute another oil-quenching alloy steel, AISI 8640, with the same carbon content, 0.40%, but less alloy content and less expensive. He claims it may have as high or higher fatigue strength in this size. Justify his claim.

Section 7.16

7.64 When a thick, hollow cylinder is subjected in service to *external* pressure, the tangential stress distribution is as indicated by line *D-D* in Fig. 7.19 but is inverted, as is σ_t in Fig. 8.12(b). The stresses are compressive and maximum at the interior wall. By sketches on several circle diagrams determine whether or not this maximum compressive stress can be reduced in magnitude if a prestressing operation is first completed by (a) hydraulic pressure applied internally, (b) hydraulic pressure applied externally, and (c) substituting two cylinders for one and assembling them with an initial interference fit, but no plastic deformation.

7.65 Roller-chain links sometimes develop fatigue cracks starting at the holes and breaking through to the outside, as shown. A pin or bushing is press fitted in the hole, and there is a pull P. Describe the nature of the stress and suggest two methods of prestressing. The methods of prestressing must be inexpensive and adapted to manufacture by the millions, using relatively unskilled labor.

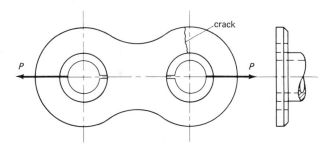

crack

P   P

7.66 Suggest and describe a method for strengthening plexiglass, which is a transparent plastic with low yield strength.

7.67 Fatigue failures in service are reported to be occurring at the roots of ground threads of some rather expensive, accurate bolts. Explain the weakness and suggest a correction that could be made to the bolts still in stock.

7.68 A long, low-stressed piston rod of a reciprocating pump needs a harder surface to minimize wear where it rubs in passing through the packing of the stuffing box. What material and process do you recommend?

7.69 Suggest methods and devices for heat relieving residual stresses in the circumferential welds used to join sections of large pipelines in the field. If less reliability is needed, what faster methods might be beneficial? Explain.

7.70 In an investigation of the breaking of several aircraft landing-wheel axles, evidence of severe grinding was found. Grinding for accuracy was necessary as a final operation following a quench and temper. Because the axle was simple, the grinding was done rapidly on a centerless grinding machine. Describe what happens if the surface temperature during grinding is (a) below the temperature at which the tempering was done, (b) between the tempering temperature and the critical range, and (c) above the critical in the austenitic range. Suggest corrections to the grinding operation and also suggest means of correcting the damage done.

7.71 It is sometimes necessary for time saving or economy to build up the damaged surface of a shaft by turning it down or cutting away, then building up with weld metal oversize, then turning down to size. Comment on this practice and its possible consequences. What additional steps would you recommend? Give alternatives.

7.72 Fine machining generally leaves a surface with a very thin layer of compressive stresses. There are numerous methods for further finishing; some involve *rubbing* like polishing, burnishing, and honing; some involve the *removal* of metal such as electropolishing, chemical milling, and electrochemical machining; and some involve the *addition* of a coating, such as by electroplating. Discuss which of these methods are likely to reinforce and which to destroy any favorable residual stresses left by fine machining. What precautions may be necessary with some? What is likely to be the difference in fatigue properties of a ductile steel coated with a hard material such as nickel and chromium and coated with a soft material such as copper?

III

FORCE, STRESS, AND DEFLECTION ANALYSES FOR LOAD-CARRYING COMPONENTS

8

Axially Symmetrical Loading
Vessels, Rotors, and Fits

8.1 INTRODUCTION

This chapter is concerned with mechanical structures and machine components that have an axis of symmetry in geometry and loading. Included are containers, pressure vessels, tubes, disks, rotors, and shaft assemblies. Special topics include filament-wound vessels and hoses, ASME Code pressure vessels, storage tanks, optimum multiple-layer cylinders and dies, loosening of fits under rotation, limit-analysis design and beneficial prestressing by plastic yielding (autofrettage), a uniform-strength rotating disk, thermal stresses, and plane strain in long cylinders. With the advent of nuclear plants, propulsion to outer space, and undersea exploration, much attention is being given to the analysis, design, and construction of pressure vessels and rotating propulsion devices.

The subject matter may be divided into two principal parts by the requirements of equation derivation. If the radially measured wall thickness is small compared with the inside diameter, say less than 5%, the significant stress is often assumed uniform across the thickness, with a simplification in the derivations. The structure is called a *shell*, or if basically of cylindrical shape, a *thin-walled cylinder*. If it is short, it is a *ring*. These structures will be treated first.

Across a wall of thicker proportion, the stresses may vary considerably. The derivation must start with the equilibrium of forces on an element within the material, an equation of compatibility must be satisfied, and boundary conditions must be used. The structure is called a *thick-walled vessel, cylinder,* or *rotor*, and one that is short is called a *disk*. For their treatment here, a single equation of equilibrium and a general solution to the differential equation will be derived, one which will provide for loading by boundary pressures, distributed centrifugal forces, and thermal differences.

8.2 THE MEMBRANE EQUATION FOR SHELLS

Balloons, parachutes, and tire inner tubes are membranes. Tangential or *membrane* stresses, uniform across the thickness, may be high, but the walls are so thin that no appreciable bending or shear stresses may be carried. When the walls are thicker and there can be appreciable resistance to forces that cause bending and shear stresses, the structure becomes a shell. The thickness is still small enough for the stresses caused by

FIGURE 8.1. Shell or vessel with an axis of symmetry and meridians from pole to pole.

fluid pressure to be considered uniform across the wall thickness, and they are still called *membrane stresses*. However, it may be necessary to consider local bending and shear stresses such as those that occur at supporting rings or where a change of shape occurs in a vessel between a cylindrical center and spherical heads. In most cases, these stresses, while significant, are smaller than the membrane stresses, and they are called *secondary stresses*. Secondary stresses at rings and heads are treated in Section 10.7, using the equations of continuous elastic support. When the shell is used to support major transverse and concentrated loads, it becomes a curved plate and a subject for more advanced treatment.[1]

A shell of general shape with an axis of symmetry Z-Z is shown in Fig. 8.1.[2] A line formed by the intersection of any axial plane with the surface of the shell is called a meridian, a number of which are shown. The intersection of the surface and any plane normal to the axis might be called a latitude circle. A portion of the shell is shown in a cross section of meridian lines in Fig. 8.2(a). The position of an element dS of the shell is defined by coordinates z and r. The normal to the surface makes an angle θ with the axis of symmetry. The radii of curvature, shown on the cross-sectional view in Fig. 8.2(a) and more clearly in Fig. 8.2(b) of the element, are R_m defining the radius of curvature of the meridian at θ and, to a tangent at S, the radius R_t lying in the plane which is normal at S to the meridian and to its plane. The two radii appear coincident in Fig. 8.2(a), but have different centers O_m and O_t, the latter lying on the axis of symmetry. If R_t is rotated so that it passes through all the surface points at axial position z or angle θ, it generates a conical surface.

The lengths of the sides of the element dS in Fig. 8.2(b) are shown in terms of increments $d\theta$ and $d\phi$ of position angles θ and ϕ, respectively. The wall thickness is t and the pressure is p. The forces on the sides are given in terms of their areas and the stresses acting upon them, namely, a stress σ_m directed along a meridian and a stress σ_t normal to σ_m and directed along a tangent to a latitude circle. An equation of equilibrium

[1] S. Timoshenko and S. Woinowsky-Krieger, *Theory of Plates and Shells*, 2nd ed., New York: McGraw-Hill, 1959; W. Flügge, *Stresses in Shells*, 2nd ed., Berlin: Springer-Verlag, 1973.

[2] Although devised to illustrate several basic shapes, this shell is not greatly different than the six, 60-ft diameter, 61-ft-high cylindrical tanks with ellipsoidal tops and conical bottoms, installed in 1965 to fit deep in the hull and over the deck of the ship, *Jules Verne*, to transport liquefied natural gas (LNG) at cryogenic temperatures.

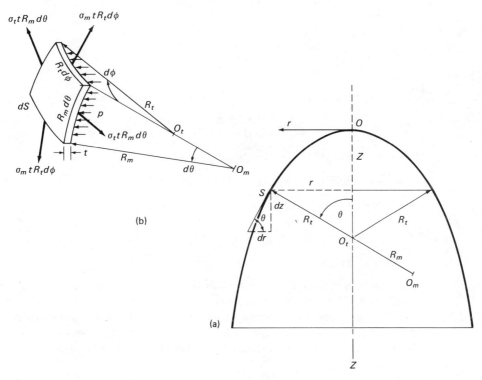

FIGURE 8.2. Top of shell of Fig. 8.1. (a) Section view with locating dimensions z, r, and θ, radii of curvature R_m and R_t, and corresponding centers O_m and O_t. (b) Shell element dS showing pressure p and the elastic forces $\sigma_m tR_t\,d\phi$ and $\sigma_t tR_m\,d\theta$.

is obtained by summing components of forces normal to the surface, thus

$$\Sigma F_n = 0; \quad -2(\sigma_t tR_m d\theta)\frac{d\phi}{2} - 2(\sigma_m tR_t d\phi)\frac{d\theta}{2} + p(R_m d\theta)(R_t d\phi) = 0 \qquad \text{(a)}$$

whence

$$-\sigma_t tR_m - \sigma_m tR_t + pR_mR_t = 0 \qquad \text{(b)}$$

and

$$\frac{\sigma_t}{R_t} + \frac{\sigma_m}{R_m} = \frac{p}{t} \qquad (8.1)$$

Equation (8.1) is called the *membrane equation* of the shell.

The values of the two radii are readily found if the surface of revolution may be expressed in the form $z = f(r)$. Then $\tan \theta = dz/dr$ (Fig. 8.2(a)), and

$$R_t = \frac{r}{\sin \theta} = \frac{r}{\sin\left[\tan^{-1}(dz/dr)\right]} \qquad (8.2)$$

The other radius is found from the equation for the radius of curvatures in any plane of r and z, namely

$$R_m = \frac{\left[1 + (dz/dr)^2\right]^{3/2}}{d^2z/dr^2} \qquad (8.3)$$

Where the distance to a center of curvature O_m is measured away from the axis of

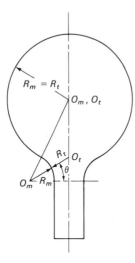

FIGURE 8.3. Vessel with an external radius R_m at fillet joining a cylinder to a sphere.

symmetry, as at the fillet of the tank of Fig. 8.3, the direction of the force components $\sigma_m t R_t d\phi(d\theta/2)$ in Fig. 8.2(b) is reversed. This changes the sign of the second term in Eq. (a). The change is taken into account in Eq. (8.1), and in Eq. (8.5) derived from it, by substituting a negative value for R_m. Likewise, if the pressure is external to the vessel, the pressure force term is reversed and p should be given a negative value.

8.3 THIN PRESSURE VESSELS

The treatment in this section is applicable to thin-walled containers, pipes, and power cylinders when the fluid pressure on all walls may be considered uniform. The case of variable pressure due to the weight of liquid or static head is treated under tanks in Section 8.4.

In addition to the membrane equation a second, necessary equation is that obtained by a summation of the forces acting upon a free body bounded by a radial plane (Fig. 8.4). The total loading in the axial direction is fluid pressure p times the area projected normal to the axis, πr^2, where r is the radius at the plane. The total resisting membrane force in the direction of the axis is the meridional stress σ_m multiplied by the area on which it acts, $2\pi rt$, then by $\sin \theta$ for projection along the axis. Thus

$$p\pi r^2 - 2\pi rt\sigma_m \sin \theta = 0 \tag{c}$$

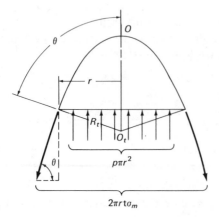

FIGURE 8.4. Free-body diagram of a symmetrical portion of a pressure vessel.

Substitution of $r = R_t \sin \theta$ for meridional stress gives

$$\sigma_m = \frac{pR_t}{2t} \tag{8.4}$$

Substitution from Eq. (8.4) into Eq. (8.1) gives for the tangential stress

$$\sigma_t = \frac{pR_t}{2t}\left(2 - \frac{R_t}{R_m}\right) = \sigma_m\left(2 - \frac{R_t}{R_m}\right) \tag{8.5}$$

There is a radially directed stress at the pressure-side surface, and it is equal in value to the pressure. It decreases to zero on the unpressurized surface. However, the pressure that can be carried in a thin-walled vessel is quite small in value compared with that of the tangential stress, so the radial stress may be considered negligible and written

$$\sigma_r = 0 \tag{8.6}$$

This stress σ_r must not be forgotten when the maximum shear stress is being determined.

The radii for several common shapes may be determined by inspection. Thus for a cylindrical portion of radius R, the central part of the vessels of Figs. 8.1 and 8.5, the radii are $R_m = \infty$ and $R_t = \text{constant} = R$. Then from Eqs. (8.4) and (8.5) the meridional or longitudinal stress σ_m and the tangential or circumferential stress σ_t for closed cylinders are, respectively,

$$\sigma_m = \frac{pR}{2t} \quad \text{and} \quad \sigma_t = \frac{pR}{t} \tag{8.7}$$

For a sphere, $R_m = R_t = R$, and

$$\sigma_m = \sigma_t = \frac{pR}{2t} \tag{8.8}$$

In both cylinders and spheres these membrane stresses are independent of location, except at attachments, heads, and other restraints.

Although the sphere is more favorably stressed, the cylinder is readily fabricated by rolling plates to a single radius of curvature and joining the plates by welds, as along the lines of Fig. 8.5. Thus the cylinder is by far the most common basic shape for pressure

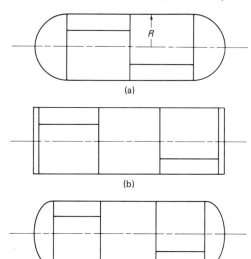

(a)

(b)

(c)

FIGURE 8.5. Common cylindrical vessels showing location of longitudinal and circumferential joints. (a) Spherical heads. (b) Flat heads. (c) Dished heads. (Note relative volumetric capacity for the same length.)

vessels. The second of Eqs. (8.7) gives the higher stress and determines the thickness. When the equation is solved for thickness t, a factor η, called joint efficiency, is commonly introduced for possible decrease of strength at the joint. The allowable stress is a material strength S divided by a factor of safety n and multiplied by η_t, where the subscript indicates the tangential direction of stress. Substitution into the second of Eqs. (8.7) of $\eta_t S/n$ for σ and of inside half-diameter $D/2$ for R gives the equation for thickness

$$t = \frac{pDn}{2\eta_t S} \tag{8.9}$$

Codes[3] specify the value of η_t as a function of the type of joint, the control of its fabrication, and the nature of the application and operation. A *longitudinal joint*, one which runs parallel to the cylinder's axis (Fig. 8.5), is usually butt welded, and a typical allowed efficiency is 0.90.

The *circumferential* or *girth joints* need not be as strong. If in the first of Eqs. (8.7) the meridional stress σ_m is set equal to $\eta_m S/n$, and the resulting equation for t is equated to that of Eq. (8.9), then $t = pDn/4\eta_m S = pDn/2\eta_t S$, whence $\eta_m = 0.5\eta_t$. Thus, lap joints with efficiencies less than 0.90 are used for circumferential joints of certain thin-walled vessels, and the time and cost for preparation and welding of the joints is less than for butt joints. However, such use is restricted to the less critical applications.

For end closures or heads of cylindrical vessels (Fig. 8.5), hemispherical shapes are the least stressed (Eq. (8.8)), and they may have the smallest thickness. However, they may be more costly to forge, and, if space is restricted, they may require a reduction in the cylindrical length and total volume. Flat heads act as plates in bending, thus are unfavorably stressed, and are suitable for low-pressure applications only. Compromise *dished heads* are ellipsoidal or similar in shape, and their thickness must be approximately that of the cylindrical portion.[4]

Example 8.1

The design of a vessel is commonly based on the maximum shear-stress theory of failure. It would be ideal if the head could be shaped so that the shear stress there would be everywhere the same and equal to that in the cylindrical portion. Derive an equation for such a shape,[5] keeping in mind that the meridional radius in the head near its junction with the cylinder should be somewhat less than R of the cylinder if the overall axial length is to be reasonable (Fig. 8.6).

Solution

In the *cylinder*, $\sigma_t > \sigma_m > \sigma_r$ with σ_t and σ_m tensile and $\sigma_r = 0$. Hence by Eqs. (8.6) and (8.7), $\tau_{max} = (\sigma_t - \sigma_r)/2 = pR/2t$. In the *head*, if R_m is made less than half of R_t, Eq. (8.5) indicates that σ_t is negative, and the maximum shear stress, from Eqs. (8.4) and (8.5), is

$$\tau_{max} = \frac{\sigma_m - \sigma_t}{2} = \frac{pR_t}{4t}\left(\frac{R_t}{R_m} - 1\right) \tag{8.10}$$

This shear stress in the head is equated to the constant shear stress in the cylinder to give

---

[3] *ASME Boiler and Pressure Vessel Code—An American National Standard*, revised and reissued every three years, as in 1977, and published by the American Society of Mechanical Engineers, New York.

[4] *ASME Boiler and Pressure Vessel Code—An American National Standard*. (See footnote 3.)

[5] First solved by Biezeno in 1922.

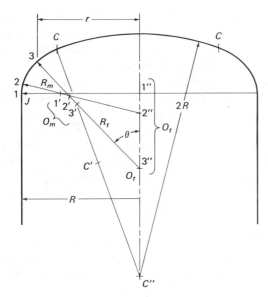

FIGURE 8.6. Graphical determination of the shape for the head of a pressure vessel of constant shear stress throughout (Example 8.1).

an equation relating R_m and R_t in the head, with thickness t the same in each, thus

$$\frac{pR_t}{4t}\left(\frac{R_t}{R_m} - 1\right) = \frac{pR}{2t}$$

and

$$R_m = \frac{R_t^2}{R_t + 2R} \tag{8.11}$$

At junction J of Fig. 8.6, $R_t = R$, whence by Eq. (8.11), $R_m = R_t/3 = R/3$. This satisfies the condition that $R_m < R_t/2$. Proceeding, a short, tangent arc 1-2 of radius $R_m = R/3$ is drawn with center $1'$. Then a line is drawn from 2 through $1'$ to $2''$ on the axis of symmetry, and the length 2-$2''$ is measured to give R_{t2}. The value R_{m2} is calculated from R_{t2} and Eq. (8.11). The value R_{m2} is laid off as 2-$2'$, with arc center $2'$ on line 2-$2''$ so that the arc 2-3 will be tangent to arc 1-2. The arc is drawn and the process is continued at 3 and beyond until a point C is reached where $R_t = C''C = 2R$, whence by Eq. (8.11), $(R_m)_c = R = C'C$. The process may not be continued beyond C because R_m would become larger than $R_t/2$, and the maximum shear stress would no longer be given by Eq. (8.10). A spherical cap with radius $C''C = R_m = R_t = 2R$ is blended in at C. Since $\sigma_m = \sigma_t = pR_m/2t = pR/t$ for the spherical portion, by Eq. (8.8), and σ_r is negligible, $\tau_{\max} = (\sigma_m - \sigma_r)/2 = pR/2t$ over the cap, the same shear stress as everywhere else.[6]   ////

Because of the geometrical discontinuities at J and C, there will be additional, usually minor, secondary stresses. If the head and cylinder were separately pressurized, they would not expand equally at J. In the vessel they must expand equally. Some resulting stresses in the region of the joint between a hemispherical head and cylinder are determined in Section 10.7 and Example 10.9.

[6] For further study see R. A. Struble, "Biezeno Pressure Vessel Heads," *J. Appl. Mech.*, Vol. 23 (*Trans. ASME*, Vol. 78) pp. 642–645, 1956; also, G. A. Hoffman, "Minimum-Weight Proportions of Pressure-Vessel Heads," *Trans. ASME*, Ser. E, Vol. 84, pp. 662–669, 1962. The latter paper includes Biezeno, torispherical, and ellipsoidal heads, some with variable thickness, and lists further references.

Other discontinuities exist in a vessel. Openings are required for the entrance and exit of the fluid at nozzles, and sometimes are required by manholes for inspection and cleaning. Openings are reinforced with extra thicknesses of plate and preferably are placed in the lower-stressed heads. A cylindrical vessel subject to *external* pressure is subject to buckling or collapse, since a small flat area once formed is stressed in bending, which increases the area and hence the "beam length" and bending moment. This is an unstable condition, and extra thicknesses or stiffening rings may be needed. Details and calculations for reinforcements and for supports are given in codes such as that of the ASME.[7] The applicable code, according to law, should always be consulted for design details and material specifications. More carefully determined analyses and design criteria are given for nuclear vessels.[8] Codes frequently make minor empirical modifications in equations to give added thickness for handling stresses, corrosion, threading, out of roundness; and in cast pipes and cylinders, for eccentricity of cores and hence nonuniform thickness.

A sphere is the most favorably stressed shape for a vessel (Eq. (8.8)). It is used for extremely high-pressure operations and is also used in space vehicles and missiles for the storage of liquefied gases at lower pressures but with lightweight thin walls. Spheres also have the greatest buckling resistance, and singly or in tandem they form the pressure-carrying structure and living space in most deep-submergence vehicles for oceanography.[9] The space between the spheres and the fairing is at local seawater pressure and is used for batteries, pumps, ballast, etc.

8.4 TANKS AND GRAVITY LOADS

Not only is the pressure variable, but the weight of the liquid contributes appreciably to the total stress. The support constitutes a major discontinuity, and equations must be written for the sections above and below the support (Fig. 8.7(a)). The pressure is $p = \gamma h$, where γ is the unit weight and h is the head or height of the liquid surface above the section in question. Note that in writing an equation for the lower surfaces of a vessel, it may be simpler to take the Z axis as positive upward [cf. the axis for the top surface of the vessel of Fig. 8.2(a)].

Above the support. There are three vertical forces acting on the shell above plane *A-A* (Fig. 8.7(b)). These are the weight W of the liquid and the shell above the plane, the upward pressure force $p\pi r^2$ from the liquid below the plane, and the axial component F_z of the elastic force in the shell. The latter is the product of the meridional stress σ_m, the area $2\pi rt$ on which it acts, where t is wall thickness, and $\sin\theta$. Thus $F_z = 2\pi rt\sigma_m \sin\theta$.

For equilibrium

$$-2\pi rt\sigma_m \sin\theta - W + p\pi r^2 = 0$$

whence

$$\sigma_m = \frac{p\pi r^2 - W}{2\pi rt \sin\theta} \tag{8.12}$$

The tangential stress is then found from the membrane equation, Eq. (8.1).

---

[7] *ASME Boiler and Pressure Vessel Code—An American National Standard.* (See footnote 3.)

[8] *ASME Boiler and Pressure Vessel Code*, Section III, on nuclear power-plant components, American Society of Mechanical Engineers, New York. For some stress analyses, see S.S. Gill (Ed.), *The Stress Analysis of Pressure Vessels and Pressure Vessel Components*, New York: Pergamon Press, 1970.

[9] For an analysis of intersecting spheres and many other details of vessel design see, e.g., J. F. Harvey, *Theory and Design of Modern Pressure Vessels*, New York: Van Nostrand Reinhold Company, 1974.

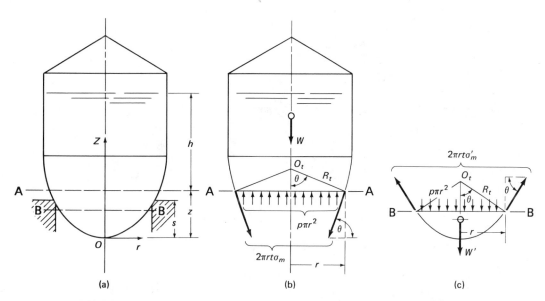

FIGURE 8.7. A tank with support. (a) Complete tank. (b) Free-body diagram of tank above plane A-A. (c) Free-body diagram of tank below plane B-B.

Below the support. Let the weight of the liquid and shell below plane B-B (Fig. 8.7(c)), be W' and the stresses be σ_m' and σ_t'. Then for equilibrium

$$+ \ 2\pi r t \sigma_m' \sin \theta - W' - p\pi r^2 = 0$$

and

$$\sigma_m' = \frac{p\pi r^2 + W'}{2\pi rt \sin \theta} \tag{8.13}$$

At the support, the difference in the axial components of the elastic forces, is

$$(\sigma_m' - \sigma_m) \ 2\pi rt \sin \theta = p\pi r^2 + W' - p\pi r^2 + W = W' + W = W_{\text{total}}$$

which was to be expected since the supporting ring must push upward with a force equal to the total weight. Hence the numerical value of the meridional stress immediately below the support is larger than that just above it. However, the meridional stress σ_m just above the support is often compressive, while it is always tensile below it. The radial stress is negligible, and if σ_t is tensile above the support, then $\tau = (\sigma_t - \sigma_m)/2$, and the maximum shear stress may occur just above the supporting ring.

Note carefully that W in Eq. (8.12) is the weight *above* a section for which the stress is calculated, and that W' in Eq. (8.13) is the weight *below* a section for which the stress is calculated. Also, where there is a change in geometric shape, or a restriction as at the supports, secondary stresses may be expected. Such stresses are determined in Example 10.8 of Section 10.7 for a cylindrical tank with a flat bottom.

Example 8.2

The conical tank shown in Fig. 8.8 is supported at a height s and filled with liquid of unit weight γ to a level l. (a) Derive equations for the stresses below the liquid level in terms of the half-angle α at the apex of the cone. (b) Determine the location and magnitude of the maximum values of the stresses.

Solution

(a) It will be advantageous to measure z from the apex of the cone as shown. Then

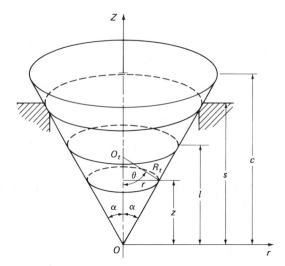

FIGURE 8.8. Conical tank of Example 8.2.

pressure $p = \gamma(l - z)$, $r = z \tan \alpha$, and $\sin \theta = \cos \alpha$. The volume of a cone is one-third its base area times its altitude, so the weight below any section of altitude z is $W' = (\frac{1}{3})\gamma\pi z^3 \tan^2 \alpha$. From Eq. (8.13)

$$\sigma_m' = \frac{p\pi r^2 + W'}{2\pi rt \sin \theta} = \frac{\gamma(l - z) \pi (z \tan \alpha)^2 + (\frac{1}{3})\gamma\pi z^3 \tan^2 \alpha}{2\pi(z \tan \alpha)t \cos \alpha}$$

$$= \frac{\gamma \tan \alpha}{2t \cos \alpha} \left[(l - z)z + \frac{z^2}{3} \right] = \frac{\gamma z(3l - 2z) \tan \alpha}{6t \cos \alpha} \tag{8.14a}$$

From the membrane equation, Eq. (8.1), and since $R_m = \infty$ and $R_t = r/\sin \theta = z \tan \alpha/\cos \alpha$

$$\sigma_t' = R_t \left(\frac{p}{t} - \frac{\sigma_m'}{R_m} \right) = \frac{z \tan \alpha}{\cos \alpha} \left[\frac{\gamma(l - z)}{t} - \frac{\sigma_m'}{\infty} \right] = \frac{\gamma z(l - z) \tan \alpha}{t \cos \alpha} \tag{8.14b}$$

Since $z \leq l$, σ_m' and σ_t' are always positive or zero. Since $\sigma_r' = 0$, the maximum shear stress may be either $\sigma_m'/2$ or $\sigma_t'/2$.

(b) The location of a maximum value of stress is found by setting its first derivative equal to zero. Thus

$$\frac{d\sigma_m'}{dz} = 0 = \frac{d}{dz} (3lz - 2z^2) = 3l - 4z, \quad \text{and} \quad z = \frac{3}{4} l$$

whence

$$(\sigma_m')_{\max} = \gamma(\frac{3}{4} l)(3l - \frac{3}{2} l) \tan \alpha/6t \cos \alpha = \frac{3\gamma l^2 \tan \alpha}{16t \cos \alpha} \tag{8.15a}$$

Also

$$\frac{d\sigma_t'}{dz} = 0 = \frac{d}{dz} (lz - z^2) = l - 2z, \quad \text{and} \quad z = \frac{l}{2}$$

whence

$$(\sigma_t')_{\max} = \gamma(\frac{l}{2})(l - \frac{l}{2}) \tan \alpha/t \cos \alpha = \frac{\gamma l^2 \tan \alpha}{4t \cos \alpha} \tag{8.15b}$$

The largest tensile stress, $(\sigma_t')_{max}$, and the largest shear stress, $\tau'_{max} = (\sigma_t')_{max}/2$, occur at half the liquid height l. At the liquid level, $z = l$ and $\sigma_m' = \gamma l^2 \tan \alpha/6t \cos \alpha$ and $\sigma_t' = 0$. At the bottom $z = 0$ and $\sigma_m' = \sigma_t' = 0$.   ////

8.5 FILAMENT-WOUND CYLINDERS

For years, strong, flexible, pressure hose has been made by winding wire over a plastomer core. The demand for lightweight vessels and thrust chambers in spacecraft, rockets, and airborne vehicles has led to the wrapping of wires about metallic-walled containers; it has also led to the forming of vessels from fiberglass and other high-strength filaments by winding them over a mandrel, followed by impregnation of the windings with a plastic and removal of the mandrel in pieces.

The material of the usual vessel fabricated from steel plates does not have its strength increased by quench heat treatments. This strengthening cannot be done after forming because of the vessel's bulk and hollowness, and the strengthening cannot be done before forming because of the increased resistance to cold shaping and the possible loss of strength in any subsequent welding. However, smaller cases can be forged and treated, and as much as 315 000 psi (2170 MPa) tensile strength can be obtained in 18% nickel steel. Steel wire can be strengthened by heat treatments and cold drawing, and a tensile strength of 250 000 psi (1720 MPa) in stranded form is typical for carbon steel. A vessel may be wrapped with this wire. Groups of glass fibers called *roving* may have a strength of 150 000 psi (1035 MPa). However, their density is lower than that of steel, and a composite glass-fiber vessel may weigh one-half to two-thirds as much as the best monolithic steel or titanium vessel. Modulus of elasticity is considerably lower, and the buckling strength may be less.

In an open-ended vessel, such as a rocket-engine thrust chamber, the stresses are mainly tangential, and the filaments are wound circumferentially. On closed-end composite vessels, various loop patterns are used, depending upon the shape of the heads, location and size of the nozzles, minimization of crossover points, etc. A helical winding is normally used to wrap the cylindrical portion, continue over the dished head, and return. Additional circumferential layers may be added to the cylindrical portion if necessary.

A cylinder wrapped by a filament at a helix angle ψ is shown in Fig. 8.9(a). A part of its upper half is taken as a free body and is shown in Fig. 8.9(b). Its length is the *lead* $l = \pi D/\tan \psi$, the axial advance made by one filament in 360° of wrap. If F is the tensile force in the filament, its tangential component acting across an axial plane is $F \sin \psi$. Its axial component acting across a radial plane is $F \cos \psi$. Let the number of filament starts in each hand, i.e., to the right and to the left, be $N/2$ per unit of length. Then there are Nl filaments per lead length to resist separation across an axial section. There is the same number across a radial section of circumferential length πD. The forces on the free body in Fig. 8.9(b) are shown in its end view. The pressure is $\int_0^\pi pl(rd\theta) \sin \theta = pDl$ or the pressure times the projected area. The forces on the portion of the cylinder cut out by a radial plane are shown in Fig. 8.9(c). The equations of equilibrium are thus

$$\Sigma F_y = 0; \quad -2(Nl)(F \sin \psi) + pDl = 0$$

$$\Sigma F_z = 0; \quad (Nl)(F \cos \psi) - p\pi D^2/4 = 0$$

whence $F \sin \psi = pD/2N$ and $F \cos \psi = p\pi D^2/4Nl$.

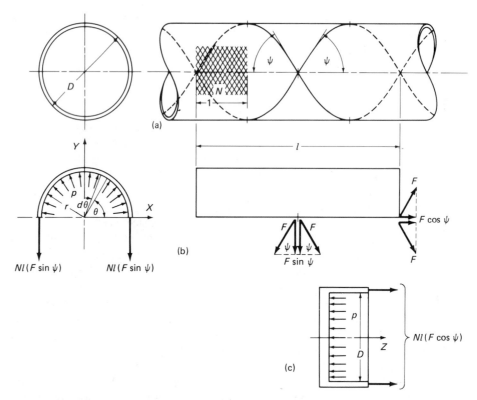

FIGURE 8.9. Filament-wrapped cylinder. (a) Helix angle ψ and lead l. (b) Free-body diagram of upper half. (c) Free-body diagram of a section cut out by a radial plane.

By division

$$\tan \psi = \frac{F \sin \psi}{F \cos \psi} = \frac{2l}{\pi D} = \frac{2(\pi D/\tan \psi)}{\pi D} = \frac{2}{\tan \psi} \tag{8.16}$$

or $\tan^2 \psi = 2$, $\tan \psi = \sqrt{2}$, and $\psi = 54.7°$. This is the condition for helical wrappings only to support an internal pressure without distortion of the wrapping. It is the angle at which the filament is wrapped on high-pressure hose.[10] By the additional use of circumferential wrapping, the angle of the helical wrapping may be decreased for convenience in wrapping, to avoid vessel irregularities in its path, etc. The ratio of the thickness t_c of the circumferential layer to that of the helical layer t_h when ψ is less than $54.7°$ should be

$$\frac{t_c}{t_h} = \frac{S_h}{S_c} (3 \cos^2 \psi - 1) \tag{8.17}$$

where S_c and S_h are respective tensile strengths (stresses) of the circumferential and the helical filament materials.[11]

[10] E. N. Ipiotos, "Calculate the Burst Strength of Braided Rubber Hose from Design Characteristics," *General Motors Engineering J.*, Vol. 6, No. 3, pp. 52–54, No. 4, pp. 44–48, 1959.

[11] R. L. Stedfeld and C. T. Hoover, "Design and Fabrication Techniques for Filament-Wound Pressure Vessels," *General Motors Engineering J.*, Vol. 13, No. 4, pp. 24–31, 1966; also, "Fiberglass-Reinforced Plastic Pressure Vessels," Section X of the *ASME Boiler and Pressure Vessel Code—An American National Standard*, American Society of Mechanical Engineers, New York.

8.6 THE GENERAL EQUATION OF EQUILIBRIUM FOR THICK CYLINDERS: THE SPECIAL CASE OF A DISK OF UNIFORM STRENGTH

The following derivations are for use with axially symmetric configurations and loadings, including those of pressure and centrifugal forces. In Fig. 8.10, the element has a variable width l, a mass $\rho l r\, d\phi\, dr$, radial faces of area $l\, dr$, and arc faces of lengths $r\, d\phi$ and $(r + dr)d\phi$. There are tangentially directed stresses σ_t and radially directed stresses σ_r. Rotational speed is ω rad/s, and centrifugal force $dm\, r\omega^2 = (\rho l r\, d\phi\, dr)r\omega^2$. The forces are shown acting on the element, and for equilibrium of their radial components

$$- 2(\sigma_t l\, dr)\frac{d\phi}{2} - \sigma_r l r\, d\phi + \left[\sigma_r l r\, d\phi + \sigma_r l\, dr\, d\phi \right.$$

$$+ r\frac{d}{dr}(\sigma_r l)\, dr\, d\phi + \frac{d}{dr}(\sigma_r l)\, (dr)^2 d\phi \Bigg]$$

$$+ (\rho l r\, d\phi\, dr)r\omega^2 = 0$$

With cancellation of the second term by the third, the disappearance of the third-order differential in the limit, and division through by $dr\, d\phi$, the following *equation of equilibrium* is obtained

$$- \sigma_t l + \sigma_r l + r\frac{d}{dr}(\sigma_r l) + \rho\omega^2 r^2 l = 0 \tag{8.18}$$

Equation (8.18) has several unknowns, and in general some elastic and geometrical relationships must be used for its solution. One might say that it is statically indeterminate.

FIGURE 8.10. Axially symmetric body of variable width l. Dimensions and forces on an element at radius r.

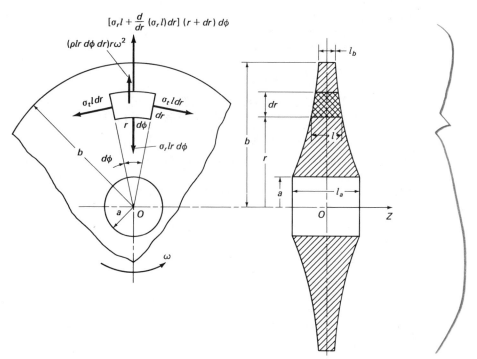

In the special case of $\sigma_t = \sigma_r =$ constant, the first two terms of Eq. (8.18) cancel out, and it reduces to an equation in one variable, l. This situation is physically possible if a uniform stress σ_b or pressure p_o is applied to the periphery of the disk, a condition approximated by the centrifugal forces from closely spaced blades. In addition the disk must be without a central hole so that the same radial stress may be transmitted throughout. The width l_b at the rim (Fig. 8.10), can be determined on the basis of the peripheral forces and a suitable allowable stress σ_b. Then Eq. (8.18) becomes

$$\sigma_b r \frac{dl}{dr} + \rho\omega^2 r^2 l = 0$$

whence

$$\frac{dl}{l} = -\frac{\rho\omega^2 r \, dr}{\sigma_b}$$

By integration

$$\ln l = -\frac{\rho\omega^2 r^2}{2\sigma_b} + \ln C$$

whence

$$l = Ce^{-\rho\omega^2 r^2/2\sigma_b}$$

The constant of integration is determined to make $l = l_b$ at $r = b$. There results

$$l = l_b e^{\rho\omega^2(b^2 - r^2)/2\sigma_b} \tag{8.19}$$

Radius r_a is zero (Fig. 8.10), and the ratio of center to rim widths is $l_a/l_b = e^{\rho\omega^2 b^2/2\sigma_b}$.

The foregoing solid-disk configuration is approximated by forging the disk integral with its shaft or by providing central, raised, flat surfaces to which flanged shafts are bolted. If the center-to-rim width ratio is larger than about 3.0, corresponding to a value larger than 1.1 for the exponent $\rho(\omega b)^2/2\sigma_b$, the shape may be impractical or uneconomical in axial space and in manufacture. Hence the constant-strength disk is restricted to the smaller values of peripheral velocity ωb, or to materials suitable for a high stress σ_b.

8.7 STRAIN IN CYLINDRICAL COORDINATES: COMPATIBILITY: PLANE STRESS VS PLANE STRAIN

The relationships between strain and stress are similar to those in Cartesian coordinates, Eqs. (5.1). Temperature strains will be included so that thermal stresses may be treated in Section 8.14. The unrestricted, unit-thermal expansion at a point is the same in all directions, αT, where α is the coefficient of thermal expansion and $T = T(r)$ is a function of radius only. If the axial direction is Z,

$$\epsilon_r = \alpha T + \frac{1}{E}\left[\sigma_r - \nu(\sigma_t + \sigma_z)\right]$$

$$\epsilon_t = \alpha T + \frac{1}{E}\left[\sigma_t - \nu(\sigma_r + \sigma_z)\right) \tag{8.20}$$

$$\epsilon_z = \alpha T + \frac{1}{E}\left[\sigma_z - \nu(\sigma_r + \sigma_t)\right]$$

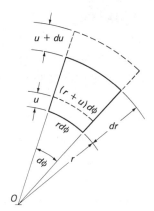

FIGURE 8.11. Radial displacement of an element in an enlarged view.

Radial strain ϵ_r and tangential strain ϵ_t are related to radial displacement u for compatibility under conditions of axial symmetry. In Fig. 8.11 an element in a radial plane is shown with a positive or outward displacement u at its inner boundary and $u + du$ at its outer boundary, corresponding to a positive or tensile strain and stress. The unit changes of radial length and of tangential length are seen to be, respectively, the strains

$$\epsilon_r = \frac{(u + du) - u}{dr} = \frac{du}{dr}$$

$$\epsilon_t = \frac{(r + u)d\phi - r\,d\phi}{r\,d\phi} = \frac{u}{r}$$

(8.21)

An assumption must be made to obtain a relation between axial strain ϵ_z and radial displacement u, also between σ_r or σ_t and u. One assumption is that, under radial loading only, the axial stress σ_z is zero. Then the only principal stresses are σ_r and σ_t, and they lie in radial planes. Hence the condition is designated *plane stress*. The stresses are obtained in terms of ϵ_r and ϵ_t by the simultaneous solution of the first two of Eqs. (8.20) and in terms of u by substitution from Eqs. (8.21). This is done in the derivation of Eq. (8.22). By the third of Eqs. (8.20), axial strain ϵ_z may then be expressed in terms of u and included in the compatibility group, Eqs. (8.21).

Except for thermal expansion, the axial strain is caused by the Poisson's effect and the radial displacement. Since the latter varies with radius, in general the axial strains and displacements also vary with radius. If the cylinder is short and there are no external forces on its end surfaces, then the axial stresses that are developed are negligible, i.e., $\sigma_z = 0$. Thus the equations of plane stress are applicable both to the short cylinders and to the disks. These equations are developed in the sections that immediately follow.

In long cylinders, except near the ends, the internal restraint to large differences in axial displacements results in an axial stress σ_z, and in a condition which approaches that of radial sections remaining plane. The radial displacements and radial and tangential stresses can be determined by letting $\epsilon_z = 0$ in Eqs. (8.20). This condition is known as *plane strain*. Equations for long cylinders under rotation and under radially symmetrical temperature distributions are developed in Sections 8.13 and 8.14. It is shown in Section 8.9 that the equations for a short cylinder under pressure loadings also apply to a long cylinder because the axial strain ϵ_z is constant, i.e., independent of radius.

8.8 PLANE STRESS EQUATIONS FOR GENERAL AND FOR CONSTANT WIDTHS

Substitution into the first two of Eqs. (8.20) of the condition of plane stress, $\sigma_z = 0$, followed by their simultaneous solution for σ_r and σ_t and the substitution of the conditions of compatibility, Eqs. (8.21), give

$$\sigma_r = \frac{E}{1-\nu^2}\left[\frac{du}{dr} + \nu\,\frac{u}{r} - (1+\nu)\alpha T\right]$$

$$\sigma_t = \frac{E}{1-\nu^2}\left[\frac{u}{r} + \nu\,\frac{du}{dr} - (1+\nu)\alpha T\right] \tag{8.22}$$

These stresses are substituted into the equation of equilibrium, Eq. (8.18), to obtain a differential equation in two dependent variables u and l, and one independent variable r, or

$$\frac{d^2u}{dr^2} + \left(\frac{1}{r} + \frac{1}{l}\frac{dl}{dr}\right)\frac{du}{dr} - \left(\frac{1}{r^2} - \frac{\nu}{rl}\frac{dl}{dr}\right)u$$

$$= (1+\nu)\alpha\,\frac{dT}{dr} + (1+\nu)\frac{\alpha T}{l}\frac{dl}{dr} - \frac{(1-\nu^2)\rho\omega^2}{E}\,r \tag{8.23}$$

If the width l can be expressed as a function of radius r, Eq. (8.23) is reduced to one dependent variable u. A profile for which a solution can be readily obtained is hyperbolic, or $l = (a/r)^k l_a$, i.e., $lr^k = a^k l_a = $ constant, where l_a is the width at the inside radius a (Fig. 8.10). Since the width becomes infinite at $r = 0$, a central hole is required. This profile is useful for rotating disks.[12] Width l_a at the hole and width l_b at the rim may be chosen from space and manufacturing considerations and the value of k determined to give a smooth profile between the rim and the hole. Up to a limit, the larger the ratio l_a/l_b, the larger is the value of k and the more uniform is the stress. When $k = 0$, then $l = l_a$, and the disk has constant width.

Constant-length pressure cylinders and sleeves are in wide usage. Derivations for them and for a constant-width rotating disk are readily made by letting l be constant in the differential equation, Eq. (8.23). Then $dl/dr = 0$, and the equation reduces to

$$\frac{d^2u}{dr^2} + \frac{1}{r}\frac{du}{dr} - \frac{u}{r^2} = (1+\nu)\alpha\,\frac{dT}{dr} - Nr \tag{8.24}$$

where $N = (1-\nu^2)\rho\omega^2/E$. This equation may be solved explicity for u. The sum of the second and third terms is recognized as the derivative of u/r with respect to r. Thus the left-hand side of the equation may be rewritten and factored as

$$\frac{d^2u}{dr^2} + \frac{d}{dr}\left(\frac{u}{r}\right) = \frac{d}{dr}\left(\frac{du}{dr} + \frac{u}{r}\right) = \frac{d}{dr}\left[\frac{1}{r}\left(r\,\frac{du}{dr} + u\right)\right] \tag{a}$$

The sum in parentheses is the derivative of (ru) with respect to r. Hence Eq. (8.24) becomes

$$\frac{d}{dr}\left[\frac{1}{r}\frac{d}{dr}(ru)\right] = (1+\nu)\alpha\,\frac{dT}{dr} - Nr \tag{b}$$

---

[12] For derivations and a discussion of the hyperbolic disk, see, e.g., J. P. Den Hartog, *Advanced Strength of Materials*, New York: McGraw-Hill, pp. 61–65, 1952; A. Stodola, *Steam and Gas Turbines*, Transl. 6th ed., New York: McGraw-Hill, 1927.

By integration with respect to r

$$\frac{1}{r}\frac{d}{dr}(ru) = (1 + \nu)\alpha T - \frac{Nr^2}{2} + C_1{}'$$

or

$$\frac{d}{dr}(ru) = (1 + \nu)\alpha Tr - \frac{Nr^3}{2} + C_1{}'r \tag{c}$$

By a second integration, where a is the inner radius

$$ru = (1 + \nu)\alpha \int_a^r Tr \, dr - \frac{Nr^4}{8} + C_1{}'\frac{r^2}{2} + C_2 \tag{d}$$

Division by r, substitution for N, and replacement of $C_1{}'/2$ by C_1 gives

$$u = \frac{(1 + \nu)\alpha}{r} \int_a^r Tr \, dr - \frac{(1 - \nu^2)\rho\omega^2}{8E} r^3 + C_1 r + \frac{C_2}{r} \tag{8.25}$$

The stresses σ_r and σ_t may now be written in terms of r and the constants of integration. Substitution from Eq. (8.25) into Eqs. (8.22) gives

$$\sigma_r = -\frac{\alpha E}{r^2} \int_a^r Tr \, dr - \frac{3 + \nu}{8}\rho\omega^2 r^2 + \frac{EC_1}{1 - \nu} - \frac{EC_2}{(1 + \nu)r^2}$$

$$\sigma_t = +\frac{\alpha E}{r^2} \int_a^r Tr \, dr - \alpha ET - \frac{1 + 3\nu}{8}\rho\omega^2 r^2 + \frac{EC_1}{1 - \nu} + \frac{EC_2}{(1 + \nu)r^2} \tag{8.26}$$

Two values of σ_r or one each of σ_r and u are generally known at the boundaries. For example, in the most common pressure vessel with internal pressure p_i, the radial stress at inner radius a is $\sigma_r = -p_i$, since a compressive force or stress must oppose the pressure on each unit of interior surface. At outer radius b, $\sigma_r = 0$. Hence the values of C_1 and C_2 may be determined by the two boundary conditions and the solution completed.

There are advantages in having separate equations for each of the loading conditions of pressure, rotation, and temperature variation. For conditions of pressure and rotation Table 8.1 lists for each the boundary conditions and the stress and displacement equations. These equations or their numerical values may be superimposed when the corresponding loading conditions are superimposed. Significant properties indicated by the equations are discussed in the next section, together with axial and shear stresses in closed-end cylinders. Discussion of the temperature terms of Eqs. (8.25) and (8.26) is postponed to Section 8.14, where the subject of thermal stresses is treated.

8.9 PRESSURE CYLINDERS, ROTATING THIN DISKS, AND SPHERES

With both internal and external pressures on cylinders, the tangential stress is largest at the interior surface and smallest at the external surface, as may be seen by inspection of the equations of Table 8.1 or the graphs of Fig. 8.12. The maximum radial stress occurs at that boundary where the pressure is applied, is equal to it and compressive, and is always less than the maximum tangential stress. The stresses, plotted as a ratio to applied pressure, are shown in Fig. 8.12 for a rather thick cylinder in order to give a large variation from interior to exterior.

It should be noted from the table that the boundary conditions for a solid disk are of a different type than those for a hollow disk. In addition to the external boundary con-

TABLE 8.1. Cylinders of Uniform Length—Loading, Boundary Conditions, Stresses and Displacements. ρ = density, ν = Poisson's ratio, E = modulus of elasticity

Loading	Boundary Conditions	Stresses and Displacement
(1) Internal pressure p_i	at $r = a$, $\sigma_r = -p_i$ at $r = b$, $\sigma_r = 0$	$\sigma_r = -p_i \dfrac{a^2}{b^2 - a^2}\left(\dfrac{b^2}{r^2} - 1\right)$, max $\sigma_r = -p_i$ at $r = a$ $\sigma_t = p_i \dfrac{a^2}{b^2 - a^2}\left(\dfrac{b^2}{r^2} + 1\right)$, max $\sigma_t = p_i \dfrac{b^2 + a^2}{b^2 - a^2}$ at $r = a$ $u = p_i \dfrac{r}{E}\dfrac{a^2}{b^2 - a^2}\left[(1 - v) + (1 + v)\dfrac{b^2}{r^2}\right]$
(2) External pressure p_o	at $r = 0$, $\sigma_r = 0$ at $r = b$, $\sigma_r = -p_o$	$\sigma_r = -p_o \dfrac{b^2}{b^2 - a^2}\left(1 - \dfrac{a^2}{r^2}\right)$, max $\sigma_r = -p_o$ at $r = b$ $\sigma_t = -p_o \dfrac{b^2}{b^2 - a^2}\left(1 + \dfrac{a^2}{r^2}\right)$, max $\sigma_t = -p_o \dfrac{2b^2}{b^2 - a^2}$ at $r = a$ $u = -p_o \dfrac{r}{E}\dfrac{b^2}{b^2 - a^2}\left[(1 - v) + (1 + v)\dfrac{a^2}{r^2}\right]$
(3) Thin uniform disk. Rotation ω.	at $r = a$, $\sigma_r = 0$ at $r = b$, $\sigma_r = 0$	$\sigma_r = \rho\omega^2 \dfrac{3 + v}{8}\left(b^2 + a^2 - \dfrac{a^2 b^2}{r^2} - r^2\right)$ max $\sigma_r = \rho\omega^2 \dfrac{3 + v}{8}(b - a)^2$ at $r = \sqrt{ab}$ $\sigma_t = \rho\omega^2 \dfrac{3 + v}{8}\left(b^2 + a^2 + \dfrac{a^2 b^2}{r^2} - \dfrac{1 + 3v}{3 + v}r^2\right)$ max $\sigma_t = \dfrac{\rho\omega^2}{4}\left[(3 + v)b^2 + (1 - v)a^2\right]$ at $r = a$ $u = \rho\omega^2 \dfrac{r}{E}\dfrac{(3 + v)(1 - v)}{8}\left(b^2 + a^2 + \dfrac{1 + v}{1 - v}\dfrac{a^2 b^2}{r^2} - \dfrac{1 + v}{3 + v}r^2\right)$
(4) Solid, thin uniform disk. Rotation ω and external pressure p_o.	at $r = 0$, $u = 0$ at $r = b$, $\sigma_r = -p_o$	$\sigma_r = -p_o + \rho\omega^2 \dfrac{3 + v}{8}(b^2 - r^2)$ max $\sigma_r = -p_o + \rho\omega^2 \dfrac{3 + v}{8}b^2$ at $r = 0$ $\sigma_t = -p_o + \rho\omega^2 \dfrac{3 + v}{8}\left(b^2 - \dfrac{1 + 3v}{3 + v}r^2\right)$ max σ_t = max σ_r at $r = 0$ $u = \dfrac{r}{E}(1 - v)\left\{-p_o + \dfrac{\rho\omega^2}{8}\left[(3 + v)b^2 - (1 + v)r^2\right]\right\}$

dition, the equations are derived for the solid disk from either the condition that displacement u is zero at $r = 0$, or that C_2 in Eqs. (8.25) and (8.26) must be zero if the results are to be finite at $r = 0$.

For the derivations the assumption of *plane stress* was made, based upon short cylinders. However, an inspection of Eqs. (8.26) indicates that when there is pressure only

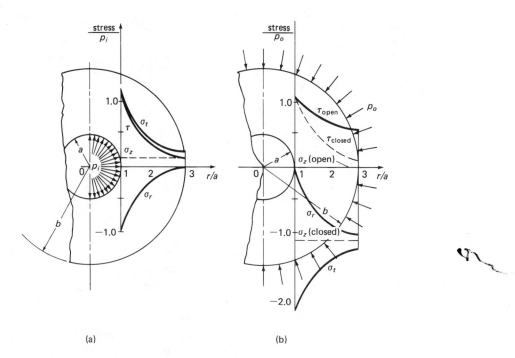

FIGURE 8.12. Tangential stresses σ_t, radial stresses σ_r, axial stresses σ_z, and major shear stresses τ in cylinders. (a) With internal pressure p_i. (b) With external pressure p_o. Radius ratio $b/a = 3.0$. Ordinates are the nondimensional ratios of stress to applied pressure. Stresses σ_z and τ are indicated by dotted lines for closed ends and by full lines for open ends. Pressures p_o and p_i have the same value.

and no temperature differential T and rotation ω, the sum $(\sigma_r + \sigma_t)$ equals $2EC_1/(1 - \nu)$, which is a constant. Since $\sigma_z = 0$, the axial strain, by Eq. (8.20), is

$$\epsilon_z = \frac{1}{E}\left[-\nu(\sigma_r + \sigma_t)\right] = -\frac{2C_1\nu}{1 - \nu} \tag{8.27}$$

which is constant and independent of radius. This means that any two radial sections remain plane while moving toward or away from each other, regardless of length, a case of plane strain as well as plane stress. Thus for the pressure loadings, the equations of Table 8.1 are applicable to all cylinders, whether short or long.

When at least one end of the cylinder is open, as in a gun barrel, there are no axial stresses. When both ends are closed, an axially directed stress σ_z is superimposed. Away from the ends the stress is uniform and equal to the product of pressure and the projected head area on which it acts, divided by the area of the resisting cross section, $\pi(b^2 - a^2)$.

For internal and external pressures, respectively, the stresses are

$$\sigma_z = p_i \frac{a^2}{b^2 - a^2} \quad \text{and} \quad \sigma_z = -p_o \frac{b^2}{b^2 - a^2} \tag{8.28}$$

It is seen by inspection that both these axial stresses are numerically smaller than the corresponding tangential stresses. They are plotted in Fig. 8.12(a) and (b).

Thus with *internal pressure*, σ_t is tensile, σ_r is compressive, and σ_z is either zero or tensile, i.e., $\sigma_t > \sigma_z \geq 0 > \sigma_r$. The major shear stress is

$$\tau = \frac{\sigma_t - \sigma_r}{2} = \frac{p_i a^2 b^2}{(b^2 - a^2)r^2} \tag{8.29}$$

The major shear stress has a maximum value of $p_i b^2/(b^2 - a^2)$ at $r = a$. Twice τ is sometimes used as the *stress intensity* S, i.e., $S = 2\tau = \sigma_t - \sigma_r$. With *external pressure*, σ_t and σ_r are compressive, and σ_z is either zero or compressive. With open ends, $|\sigma_t| > |\sigma_r| > |\sigma_z| = 0$. With closed ends, $|\sigma_t| > |\sigma_z| > |\sigma_r|$. The major shear stresses are, respectively,

$$\tau_{\text{open}} = \left| \frac{\sigma_t - 0}{2} \right| = \frac{p_o b^2}{2(b^2 - a^2)} \left(1 + \frac{a^2}{r^2} \right) \tag{8.30a}$$

$$\tau_{\text{closed}} = \left| \frac{\sigma_t - \sigma_r}{2} \right| = \frac{p_o a^2 b^2}{(b^2 - a^2) r^2} \tag{8.30b}$$

Both stresses have the same maximum value of $p_o b^2/(b^2 - a^2)$, at $r = a$. The several shear stresses are added to Fig. 8.12(a) and (b).

For internal pressure, the ratio of the maximum shear stress, Eq. (8.29), to the corresponding maximum tangential stress in Table 8.1 is $R = b^2/(b^2 + a^2)$, a value greater than one-half. Since the shear yield *strength* may be only one-half of the normal-stress yield strength in ductile materials, it is expected that yielding will be initiated in shear. The shear stress is critical, and it is only logical to use it and an allowable shear stress as a basis for the design of vessels with internal pressure. However, the equations in some codes for the design of pressure vessels use the maximum tangential stress together with the *tensile strength* of the material, since the latter is readily available from tests and is better understood. Since the ratios R of the maximum shear to tangential stresses vary but little within the usual range of values of b/a for pressure vessels, the designed dimensions may come out about the same if the factor of safety is properly adjusted.

In the rotation of uniform disks with a hole, tangential stress σ_t is tensile and maximum at the hole, as can be seen by inspection of the equation in Table 8.1. Radial stress is zero at the boundaries and tensile elsewhere, with a maximum value of $\rho\omega^2(3 + \nu)$ $(b - a)^2/8$ at $r = \sqrt{ab}$ (Fig. 8.13). The significant shear stress is $\sigma_t/2$ which is maximum at the hole. When the hole radius $r = a$ approaches zero, the tangential stress approaches $\rho\omega^2(3 + \nu)b^2/4$ which is double the value at $r = 0$ in a solid rotating disk. This indicates a stress-concentration factor of 2.0 for a tiny, central hole.

Thick, hollow spheres are used as vessels in high-pressure applications, and either singly or in multiple, as the cores of deep-sea vehicles. They give lower stresses than

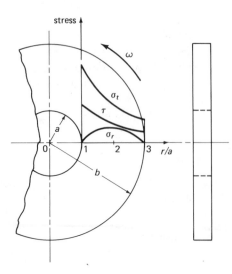

FIGURE 8.13. Stresses in a thin rotating disk of uniform thickness with a radius ratio $b/a = 3.0$.

other shapes, and, under external pressures, the largest resistance to buckling. Partial spheres are used in the heads of cylindrical vessels. The equations for radial and tangential stresses for an internal pressure p_i are .

$$\sigma_r = -\frac{p_i a^3(b^3 - r^3)}{r^3(b^3 - a^3)} \quad \text{and} \quad \sigma_t = \frac{p_i a^3(b^3 + 2r^3)}{2r^3(b^3 - a^3)} \tag{8.31}$$

and for an external pressure p_o,

$$\sigma_r = -\frac{p_o b^3(r^3 - a^3)}{r^3(b^3 - a^3)} \quad \text{and} \quad \sigma_t = -\frac{p_o b^3(a^3 + 2r^3)}{2r^3(b^3 - a^3)} \tag{8.32}$$

Example 8.3

A cylindrical vessel with hemispherical ends, Fig. 8.14, is to have a volume capacity of approximately 1.5 m³, a pressure of 20 MPa, and an overall length of not more than 2.5 m. The allowable shear stress is 70 MPa. (a) Determine the principal dimensions. (b) Calculate the maximum values of the several principal stresses and assess their significance relative to the shear stress.

Solution

(a) For the calculations, surfaces are assumed flush on the inside where the stresses are largest, although further consideration should be given to the abutment and tapers at the joint between unequal-thickness cylinders and heads.[13] With the allowance of a possible 100 mm for the thicknesses of two heads and of another 100 mm as the tolerance in dimensions, the inside length of the cylindrical portion is estimated to be $l = 2.50 - 0.20 - D = 2.30 - D$ meters (Fig. 8.14). The inside volume of each hemisphere is $(\frac{1}{2})(\pi D^3/6)$ and the volume of the vessel is $\pi D^3/6 + (\pi D^2/4)(2.30 - D) = 1.5$ m³, whence $D^3 - 6.9D^2 + 5.73 = 0$. The equation's one real root is 0.984 m, hence $D = 0.984$ and $a = D/2 = 0.492$ m. Also $\tau = 70$ MPa and $p_i = 20$ MPa.

For the *cylinder* at $r = a$, by Eq. (8.29),

$$70 = \frac{20b^2}{b^2 - 0.492^2},$$

whence $b = 0.582$ m. This indicates an outside diameter of $2b = 1.164$ m and a wall thickness of 90 mm.

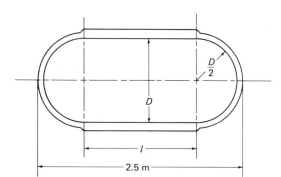

FIGURE 8.14. Pressure vessel of Example 8.3.

---

[13] The applicable pressure-vessel code should be consulted before the final dimensions and construction procedure are specified.

For the *sphere* at $r = a$, from Eqs. (8.31),

$$\tau_{max} = \frac{\sigma_t - \sigma_r}{2} = \frac{3p_i b^3}{4(b^3 - a^3)}$$

$$70 = \frac{3(20)b^3}{4(b^3 - 0.492^3)}$$

whence $b = 0.533$ m indicating an outside diameter $2b = 1.066$ m and a thickness of 41 mm. This is within the allowance made for it in length. To use a standard plate thickness, it may be necessary to modify the thickness and the outside diameter.[13]

(b) In the *cylinder* at $r = a$, from Table 8.1 and Eqs. (8.28),

$$(\sigma_r)_{max} = -p_i = -20 \text{ MPa}$$

$$(\sigma_t)_{max} = \frac{p_i(b^2 + a^2)}{b^2 - a^2} = \frac{20(0.582^2 + 0.492^2)}{0.582^2 - 0.492^2} = 120 \text{ MPa}$$

$$\sigma_z = \frac{p_i a^2}{b^2 - a^2} = \frac{20(0.492)^2}{0.582^2 - 0.492^2} = 50 \text{ MPa}$$

In the *sphere* at $r = a$

$$(\sigma_r)_{max} = -p_i = -20 \text{ MPa}$$

$$(\sigma_t)_{max} = \frac{p_i(b^3 + 2a^3)}{2(b^3 - a^3)} = \frac{20[0.533^3 + 2(0.492)^3]}{2(0.533^3 - 0.492^3)} = 120 \text{ MPa}$$

It is seen that the tangential stresses are equal in cylinder and sphere and less than twice the shear stress of 70 MPa, indicating that failure would initiate by shear yielding.   ////

Example 8.4

A thin steel disk is to be used at 14 000 rpm as a rotating blade for cutting blocks of paper. It will be of uniform thickness except where sharpened at the periphery. The outside diameter may be taken as 250 mm. The disk is mounted on a 56-mm-diameter part of a shaft (Fig. 8.15), and clamped as shown. Material is AISI 1060, unquenched, with $S_y = 480$ MPa. Ignore the friction of clamping and determine significant stresses and displacements.

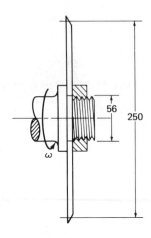

FIGURE 8.15. Rotating disk and mounting (Examples 8.4, 8.6, and 8.7).

Solution

Because the inner cylinder is solid there will be no confusion if we use the symbols of Table 8.1, a for the inner and b for the outer radius of the disk. From row 3, the tangential stress, with $\nu = 0.30$ as in steel, is

$$\sigma_t = 0.4125\, \rho\omega^2 (b^2 + a^2 + \frac{a^2 b^2}{r^2} - 0.5758 r^2) \tag{a}$$

If into this equation dimensions are substituted in mm and ω is substituted in rad^2/s^2 or s^{-2}, then stresses are obtained in MPa if density ρ is substituted in units of $\text{kg}\cdot\text{m/mm}^4$. Thus in Eq. (a) we have $(\text{kg}\cdot\text{m}\cdot\text{mm}^{-4})(\text{s}^{-2})(\text{mm}^2) = \text{kg}\cdot\text{m}\cdot\text{s}^{-2}\text{mm}^{-2} = \text{N}\cdot\text{mm}^{-2} = $ MPa. For steels, $\rho = 7.778 \times 10^3\ \text{kg/m}^3 = 7.778 \times 10^{-9}\ \text{kg}\cdot\text{m/mm}^4$. This system of units will be used for convenience in the stress equations of this and the following examples, justifying the otherwise awkward arrangement of units.

In this example $\omega = (14\,000)(2\pi/60) = 1466\ \text{rad/s}$, $\omega^2 = 2.149 \times 10^6\ \text{rad}^2/\text{s}^2$, and $\rho\omega^2 = 16.71 \times 10^{-3}\ \text{kg}\cdot\text{m}/(\text{s}^2\cdot\text{mm}^4)$. Also, $a = 56/2 = 28$ mm and $b = 250/2 = 125$ mm. Hence at $r = a$ and from Eq. (a)

$$(\sigma_t)_a = 0.4125(16.71 \times 10^{-3})(2b^2 + 0.4242 a^2)$$

$$= (6.893 \times 10^{-3})[2(125)^2 + 0.4242(28)^2] = 217.7\ \text{MPa} \tag{b}$$

and at $r = b$ and from Eq. (a)

$$(\sigma_t)_b = 0.4125(16.71 \times 10^{-3})(2a^2 + 0.4242 b^2)$$

$$= (6.893 \times 10^{-3})[2(28)^2 + 0.4242(125)^2] = 56.5\ \text{MPa} \tag{c}$$

The radial stresses are zero at a and b, and are maximum at $r = \sqrt{ab} = 59.2$ mm. From Table 8.1, the maximum value is

$$(\sigma_r)_{\text{max}} = \rho\omega^2 (3 + \nu)(b - a)^2/8$$

$$= 0.4125(16.71 \times 10^{-3})(125 - 28)^2 = 64.9\ \text{MPa (tensile)} \tag{d}$$

The maximum shear stress occurs at $r = a$ and is

$$\tau_{\text{max}} = \frac{\sigma_t - \sigma_r}{2} = \frac{\sigma_t}{2} = 108.9\ \text{MPa}$$

If the cutting forces are relatively small and the speed is steady, the loading may be considered static. The factor of safety by the maximum shear-stress theory of failure (Section 5.5) (and also by the maximum normal-stress theory in this case), is

$$n = (S_y/2)/\tau_{\text{max}} = (480/2)/108.9 = 2.20$$

If there are many starts and stops, the condition is one of pulsating stress and fatigue failure, and a lower value of n will be obtained by the methods of Section 6.3.

The displacement equation is obtained from Table 8.1. With $\nu = 0.30$, $E = 207$ GPa, and $\rho\omega^2 = 16.71 \times 10^{-3}$ for steel, the equation becomes

$$u = \frac{(16.71 \times 10^{-3})r}{207\,000} \frac{(3.3)(0.70)}{8} (b^2 + a^2 + \frac{1.3}{0.7}\frac{a^2 b^2}{r^2} - \frac{1.3}{3.3} r^2)$$

$$= (0.0233 \times 10^{-6})r(b^2 + a^2 + 1.857\frac{a^2 b^2}{r^2} - 0.394 r^2) \tag{e}$$

When $r = a$,

$$u = (0.0233 \times 10^{-6})a(2.857b^2 + 0.606a^2)$$

$$= (0.0233 \times 10^{-6})(28)\left[2.857(125)^2 + 0.606(28)^2\right] = 0.0294 \text{ mm}$$

When $r = b$,

$$u = (0.0233 \times 10^{-6})b(2.857a^2 + 0.606b^2)$$

$$= (0.0233 \times 10^{-6})(125)\left[2.857(28)^2 + 0.606(125)^2\right] = 0.0341 \text{ mm} \qquad \textit{////}$$

8.10 INTERFERENCE FITS

A cylinder or hub with a hole slightly smaller than the outside diameter of a second cylinder or shaft (Fig. 8.16(a)) may be assembled over it, as shown in Fig. 8.16(b), by axial force in a *force* or *press fit*, by preheating the outer cylinder in a *shrink fit*, by precooling the inner cylinder in an *expansion fit*, or by a combination of the fitting methods. A pressure p_f is developed at the fitted surfaces at radius n in Fig. 8.16(c). The sum of the absolute values of hole expansion and of shaft contraction equals the radial interference δ, or with the symbols of Fig. 8.16(a) and (b),

$$|u_2| + |u_1| = u_2 - u_1 = r_1 - r_2 = \delta = \Delta/2 \qquad (8.33)$$

where the subscripts 1 and 2 indicate the inner and outer cylinders, respectively, and $\Delta = 2\delta$ is the diametral interference. The negative sign is needed before u_1 to give it an additive value, since with external pressure the equation for u_1 gives a negative value to indicate an inward displacement.

Displacement equations from Table 8.1 are substituted into Eq. (8.33), using for the outer member $p_i = p_f$, $a = n$, $b = q$, $E = E_2$, and $\nu = \nu_2$; and for the shaft or inner member, $p_o = p_f$, $a = m$, $b = n$, $E = E_1$, and $\nu = \nu_1$. In these equations, the small initial difference between r_1 and r_2 is not significant, and n may taken as equal to either

FIGURE 8.16. Assembly of an interference fit. (a) Before assembly with initial radial interference δ between hole and shaft. (b) Assembled with new common radius n established. (c) Resulting pressure p_f at radius n, and stresses plotted nondimensionally.

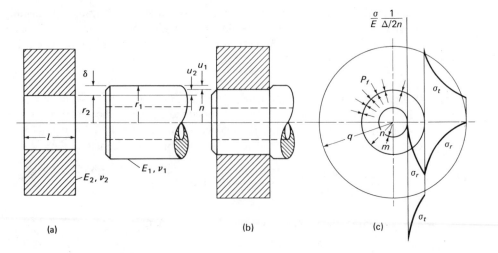

one or to their nominal values. Equation (8.33) becomes

$$\frac{2p_f n^3}{E_2(q^2 - n^2)}\left[(1 - \nu_2) + (1 + \nu_2)\frac{q^2}{n^2}\right]$$

$$+ \frac{2p_f n^3}{E_1(n^2 - m^2)}\left[(1 - \nu_1) + (1 + \nu_1)\frac{m^2}{n^2}\right] = 2\delta = \Delta \tag{8.34}$$

When the materials have the *same properties E and ν*, Eq. (8.34) reduces to

$$\frac{2p_f n}{E}\left[\frac{q^2 + n^2}{q^2 - n^2} + \frac{n^2 + m^2}{n^2 - m^2}\right] = 2\delta = \Delta \tag{8.35}$$

and when the *shaft is solid, m = 0*, giving the further simplification

$$\frac{4p_f n q^2}{E(q^2 - n^2)} = 2\delta = \Delta \tag{8.36}$$

If the length of the fitted surfaces is l and the coefficient of friction is μ, the torque that is developed just before slipping is the product of surface area, unit tangential force μp_f, and radius n, thus

$$T = (2\pi nl)(\mu p_f)n = 2\pi\mu p_f n^2 l \tag{8.37}$$

In a typical problem where the torque requirements and size are predetermined, one might first solve for the required fit pressure p_f, using, in Eq. (8.37) a value of μ reduced by a factor of safety. Then the required interference is determined from one of Eqs. (8.34) to (8.36). The diametral interference Δ should be specified since only hole and shaft *diameters* can be readily measured. The stresses may be determined by the equations of Table 8.1. A typical distribution is shown in Fig. 8.16(c), plotted to a nondimensional scale $(\sigma/E)/(\Delta/2n)$.

Example 8.5

Each driving axle of an electric freight locomotive supports a load of 44 200 lb on two wheels with a 42-in effective diameter (Fig. 8.17). A gear of 63 teeth and 31½-in pitch diameter on the axle is driven by a steel pinion of 16 teeth and 8-in pitch diameter on the shaft of an electric motor hung between the wheels. The pinion teeth are 1.12 in high on a root diameter of 7 in, with a short dedendum to avoid undercutting (Section 11.8). The pinion is bored with a hole of 5-in diameter and 6 in long. It is proposed to press this onto the motor shaft without keying. The minimum coefficient of friction at

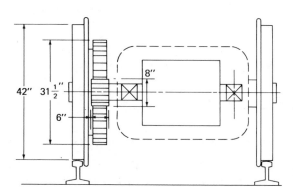

FIGURE 8.17. Press-fitted pinion on motor shaft and driven gear on axle of electric locomotive (Example 8.5).

this fit is expected to be 0.10, which is low because of the lubricant used when pressing. The maximum tractive force that can be developed at the wheel and rail corresponds to 35% adhesion, i.e., a friction coefficient of 0.35. (a) Determine the pressure, interference, and the mounting force required between the pinion and the shaft if there is to be a factor of safety of 3.33 against slipping in service. (b) What are the stresses in the pinion? (c) If a shrink fit is to be used, what change of temperature is required if the pinion is to be slipped over its shaft with a diametral clearance of 0.003 in? Expansion coefficient $\alpha = 6.3 \times 10^{-6} \ {}^\circ F^{-1}$.

Solution

At maximum adhesion the tangential force at the two wheels totals $0.35(44\ 200) = 15\ 470$ lb, and the force at the gear teeth is $(42/31.5)(15\ 470) = 20\ 630$ lb. Thus the torque on the pinion is $(8/2)(20\ 630) = 82\ 500$ lb·in.

(a) *Fit pressure and interference.* The fit radius is $n = {}^5/_2 = 2.5$ in and $l = 6$ in. In Eq. (8.37) the factor of safety is here applied to the product of the coefficient and the pressure, the most uncertain data. Thus

$$T = 2\pi\mu p_f\, n^2 l$$

$$82\ 500 = 2\pi(0.10p_f/3.33)(2.5)^2(6)$$

whence the fit pressure required is $p_f = 11\ 660$ psi.

The teeth have some stiffening effect, and to use Eq. (8.36) we must estimate the outside radius q of an equivalent cylinder. The teeth are widest at their base, and a radius to 20 or 25% of the tooth height may be reasonable. Accordingly, $q = (7/2) + 0.25 = 3.75$ in and from Eq. (8.36)

$$\Delta = \frac{4p_f\, nq^2}{E(q^2 - n^2)} = \frac{4(11\ 660)(2.50)(3.75)^2}{(30 \times 10^6)(3.75^2 - 2.50^2)} = 0.0070 \text{ in}$$

Usually, the hole is bored to nominal size plus tolerance, say 5.0000/5.0008 in, and the shaft will be turned to 5.0070/5.0078 in, for an average interference of 0.0070 in. By selective assembly, i.e., the larger holes over the larger shafts, etc., the variation in fit need not be as large as might appear.

The axial force to assemble the pinion is estimated from the product of unit-axial resistance μp_f and contact area $2\pi nl$ at the end of the pressing operation, namely,

$$F = 2\pi\mu p_f nl = 2\pi(0.10)(11\ 660)(2.50)(6) = 110\ 000 \text{ lb}$$

(b) *Stress at hole.* The maximum stresses are given in the first row of Table 8.1. There, $a = n = 2.50$, $b = q = 3.75$ in, and $p_i = p_f = 11\ 660$ psi. Hence

$$\sigma_r = -p_i = -11\ 660 \text{ psi}$$

$$\sigma_t = p_i \frac{b^2 + a^2}{b^2 - a^2} = (11\ 660) \frac{3.75^2 + 2.50^2}{3.75^2 - 2.50^2} = 30\ 320 \text{ psi}$$

$$\tau_{max} = \frac{\sigma_t - \sigma_r}{2} = \frac{30\ 320 - (-11\ 660)}{2} = 21\ 000 \text{ psi}$$

Shear is the most significant stress, since it is larger than half of the larger normal stress.

(c) *Shrink fit.* The required dimensional change at diameter $2n$ is $\Delta + 0.003 = 0.007 + 0.003 = 0.010$ in. This is equated to $\alpha(2n)\ \Delta T = (6.3 \times 10^{-6})(5.0)\ \Delta T$, and solved to give $\Delta T = 317°F$. The temperature change may be obtained by heating the pinion or by a combination of pinion heating and shaft cooling.   ////

In many applications where normal torque is expected to be carried by friction at the fit, it is not uncommon to add a key to care for unexpected overloads. A small allowance for the reduction of the clamping action by the keyway slot might be made in the interference Δ or pressure p_f. Also, the nominal stress is increased, although the stress concentration may not be significant in a steel hub because of the static nature of the stress.

There is a sharp increase in the pressure and stress near the ends of a hub or sleeve that is fitted over a longer shaft, (Fig. 8.18 (a)). This is because a short length of the shaft on either side adds resistance to deformation. This stress concentration decreases the fatigue resistance of the shaft, but advantage is taken of the increased pressure in a method that makes assembly easy on shafts of large diameter. Oil is introduced between the surfaces at a pressure larger than the uniform-fit pressure but less than that of the end peaks. This trapped oil slightly expands the outer member and separates the metal surfaces except near the ends, greatly reducing the resistance to axial motion. In Fig. 8.18 the method is shown applied to a thick sleeve used as a coupling between two shafts.[14] The thin, inner sleeve acts as a seal, and its taper minimizes the axial motion required of the heavy sleeve to make or free the fit. The changes in pressure during *removal* of the coupling are illustrated. During removal a temporary stop may be necessary to limit motion of the sleeve to the left.

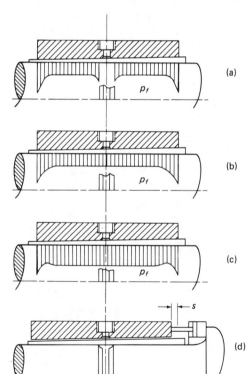

(a)

(b)

(c)

(d)

FIGURE 8.18. Stages in the removal of the ring of an SKF "OK" coupling. (a) Before removal, with fit pressure shown acting on sleeve. (b) Removal begins with pressure of oil from pump brought up to fit pressure. (c) Oil pressure greater than fit pressure, separating surfaces except at ends. (d) Ring freed by moving a short distance s, with axial force provided by an annular hydraulic cylinder if needed, as on large shafts. (Courtesy, of SKF Industries, Inc.)

---

[14] The SKF "OK" coupling, SKF Industries, Inc., Philadelphia, Pennsylvania.

Effect of speed. For shafts rotating at high speeds allowance may be necessary for the decrease in pressure at the fit due to radial expansion of the assembled members, the outer member more than the inner. The effective interference is decreased. Hence from δ there must be subtracted a net displacement $[(u_2)_\omega - (u_1)_\omega]$ which represents the gap that would develop between the surfaces of the two cylinders rotating separately at speed ω, if at standstill they had the same radius. Accordingly, in Eq. (8.33),

$$u_2 - u_1 = \delta - [(u_2)_\omega - (u_1)_\omega] \tag{8.38}$$

For the case of a solid shaft and the same materials, and with displacements from Table 8.1, Eq. (8.38) becomes

$$\frac{4p_f\, nq^2}{E(q^2 - n^2)} + \frac{3 + \nu}{2E}\, \rho\omega^2 nq^2 = 2\delta = \Delta \tag{8.39}$$

The pressure at the fit is a function of ω. Without rotation and with $\omega = 0$, Eq. (8.39) reduces to Eq. (8.36). The speed ω_0 that just completely loosens the fit is found by substituting $p_f = 0$, whence

$$\omega_0 = \sqrt{\frac{2E\Delta}{(3 + \nu)\rho nq^2}} \tag{8.40}$$

Example 8.6

(a) What is the minimum diametral interference with which the disk of Example 8.4 and Fig. 8.15 should be assembled if it is not to loosen and, hence, possibly become eccentric at its operating speed? (b) If the interference is made 25% larger than this, what are the stresses at the inner surface of the disk at assembly?

Solution

(a) Equation (8.40) governs, i.e., $p_f = 0$. With transformation to the symbols a and b

$$\Delta = \frac{3 + \nu}{2E}\, \rho\omega_0^2 nq^2 = \frac{3 + \nu}{2E}\, \rho\omega_0^2 ab^2$$

From Example 8.4, $\rho\omega_0^2 = 16.71 \times 10^{-3}$ kg·m/(s²·mm⁴), and

$$\Delta = 2\delta = \frac{3.3(16.71 \times 10^{-3})(28)(125)^2}{2(207\,000)} = 0.0582 \text{ mm}$$

The radial interference, $\delta = \Delta/2 = 0.0291$ mm, is slightly less than the radial deflection 0.0294 mm, calculated in Example 8.4. The difference is the shaft expansion under rotation, which is only 0.0003 mm.

(b) The new interference at assembly is $\Delta = 1.25(0.0582) = 0.0728$ mm. From Eq. (8.36) with $n = a$ and $q = b$, the fit pressure is

$$p_f = \frac{E\Delta(b^2 - a^2)}{4ab^2} = \frac{(207\,000)(0.0728)(125^2 - 28^2)}{4(28)(125)^2} = 127.8 \text{ MPa}$$

Radial stress $(\sigma_r)_n = -127.8$ MPa, and the tangential stress, from row 1 of Table 8.1, is

$$(\sigma_t)_n = p_f\frac{b^2 + a^2}{b^2 - a^2} = (127.8)\frac{125^2 + 28^2}{125^2 - 28^2} = 141.3 \text{ MPa}$$

The maximum shear stress is

$$\tau_{max} = \frac{(\sigma_t)_n - (\sigma_r)_n}{2} = \frac{141.3 + 127.8}{2} = 134.6 \text{ MPa} \qquad \qquad ////$$

8.11 COMPOUND CYLINDERS

Section 7.10 illustrates the effect of shrinking one cylinder over another to induce compressive residual stresses in the inner cylinder and to distribute the tangential and shear stresses more evenly when the assembly is in service under internal pressure. The tangential stresses are shown in Fig. 7.19—the stresses from the fit by lines C-C, from the pressure by lines D-D, and the net stresses by lines E-E. In the following, equations are derived for the shear stresses in each cylinder, then equations for the optimum thicknesses, i.e., those giving the least weight of material, and, finally, equations for the corresponding radial interference and fit pressure.

When fitted together under pressure p_f, the inner cylinder is externally loaded and the outer cylinder is internally loaded at their common radius n (Fig. 8.16). The maximum tangential and shear stresses for each cylinder occur at its inner surface. From Table 8.1 for the externally loaded inner cylinder at $r = m$ and from Eq. (8.29) for the internally loaded outer cylinder at $r = n$, the shear stresses are $-p_f n^2/(n^2 - m^2)$ and $p_f q^2/(q^2 - n^2)$, respectively. When a fluid pressure p is applied in service at radius m, the resulting stresses are distributed as though through a single cylinder of inner radius m and outer radius q. Accordingly, from Eq. (8.29), the shear stresses at radii m and n are $pq^2/(q^2 - m^2)$ and $pm^2q^2/(q^2 - m^2)n^2$, respectively. The principal stresses from internal pressure and from the fit act on the same planes, and their maximum shear stresses act on planes at 45° to the principal planes. The sense of each shear stress is given by its sign. Therefore, the shear stresses may be superimposed for the inner cylinder to give

$$\tau_{max} = -\frac{p_f n^2}{n^2 - m^2} + \frac{pq^2}{q^2 - m^2} \qquad \qquad (8.41a)$$

and for the outer cylinder

$$\tau_{max} = \frac{p_f q^2}{q^2 - n^2} + \frac{pm^2q^2}{(q^2 - m^2)n^2}. \qquad \qquad (8.41b)$$

These equations give the maximum shear stresses for a closed-end cylinder only if the axial stress, Eqs. (8.28), is less than the net tangential stresses. For the optimum dimensions that follow, it may be proved that this is always true. (See Prob. 8.93.)

The required volume or cylinder bore determines the inner radius m. Optimum values of n and q are those values that give the smallest q, hence the least material for the cylinder. This occurs when the net maximum shear stresses, those which occur at the inner surface of each section, are equal. This is obtained by setting the two shear equations, Eqs. (8.41), equal to the same allowable shear stress τ_0, then eliminating p_f between them, and solving for q. Finally q is minimized as a function of n by setting $dq/dn = 0$, solving for n and then a. It is found that optimum proportions are obtained when $mq = n^2$, and the equations for n and q in terms of the internal pressure p and the allowable shear stress τ_0 of the particular material are

$$n = m \sqrt{\frac{1}{1 - p/2\tau_0}} \quad \text{and} \quad q = \frac{m}{1 - p/2\tau_0} \qquad \qquad (8.42)$$

The required pressure at the fit and the initial radial interference δ are[15]

$$p_f = \frac{p^2}{2(4\tau_0 - p)} \quad \text{and} \quad \delta = \frac{mp}{E\sqrt{1 - p/2\tau_0}} \tag{8.43}$$

The higher cost of assembling compound cylinders is justified in some pressure-vessel applications, either for extreme pressures or for minimum weight.[16] Charts are available for optimizing two layers[17] and for selecting numbers and optimizing diameters of up to five layers.[18] The latter set of charts is based upon the Mises criterion for yielding failure (Section 5.6). Tests show that it more nearly represents the condition of yielding than does the maximum shear theory of failure.[19] Accurate fitting is avoided by filling the spaces between the cylinders with a fluid under controlled pressure, decreasing toward the exterior.[20] Cylinders are also built up by continuous wrapping in several layers with steel sheets under tension, thus compressing the inner layers.[21] In a "total-layered vessel" an inner, pressure-tight shell with any needed, special properties such as resistance to corrosion and hydrogen embrittlement, is covered at both the cylinder and head by several layers of high-strength preformed plate, each not more than ½ in (13 mm) thick. Each layer can be vented to the atmosphere for hydrogen escape through staggered holes. As a plate is held against the previous layer, it is tack welded, and subsequently, butt welded to join all plates in the course and to the preceding courses. The resultant shrinkage as the welds cool is claimed to have a beneficial prestressing effect on the inner layers.[22]

8.12 ELASTIC–PLASTIC STRAIN

The tangential stress and the shear stress are maximum on the inside surface of a pressure vessel or rotating disk (Figs. 8.12 and 8.13). It is at this location that yielding will first occur as the pressure or speed is increased. Further increase will create an inner plastic region and an outer elastic region with a radius e at their common boundary (Fig. 8.19). Since this condition may not constitute failure in some components, it may be used as a basis or limit, one to which a factor of safety may be applied to determine allowable design stresses. The principles of limit design were introduced in Section 5.8.

It is usually assumed that yielding proceeds either according to the maximum shear theory or according to the Mises criterion.[23] Perfect plasticity, as represented by Fig. 5.4, is also assumed. For a rotating disk, since $\sigma_t > \sigma_r > 0$ (Fig. 8.13), $\tau_{\max} = \sigma_t/2$. This is set equal to $S_{sy} = S_y/2$ in the maximum shear theory of failure. Hence $\sigma_t = S_y$, and it is constant in the plastic region (Fig. 8.20(a)). For a vessel with internal pressure

---

[15] G. J. Cloutier, "Equations and Charts for Stresses, Deflections, and Optimum Design of Thick-Walled Cylinders, Interference-Fit Assemblies, and Rotating Disks," M.S. Thesis, Cornell University, Ithaca, New York, 1959.

[16] W. R. D. Manning, "The Design of Compound Cylinders for High-Pressure Service," *Engineering*, Vol. 163, pp. 349–352, 1947.

[17] See footnote 15.

[18] R. L. Huddleston, "Design of Minimum Wall-Thickness Multiregion High Pressure Cylinders," ASME Paper 64-WA/PT8, New York, 1964.

[19] J. H. Faupel, "Yield and Bursting Characteristics of Heavy-Wall Cylinders," *Trans. ASME*, Vol. 78, pp. 1036–1064, 1956.

[20] I. Berman, "Design and Analysis of Commercial Pressure Vessels to 500 000 psi," *Trans. ASME, J. Basic Eng.*, Vol. 88, pp. 500–508, 1966.

[21] For theory of wire and ribbon-wound thick-wall vessels, see, e.g., J. H. Faupel, *Engineering Design*, New York: John Wiley & Sons, Inc., pp. 689–693, 1964.

[22] R. Pechacek, "Advanced Technology for Large Thick-Wall High-Pressure Vessels," *Mech. Eng.*, Vol. 99, No. 5, pp. 40–43, May 1977; also, J. S. McCabe and E. W. Rothrock, "Multilayer Vessels for High Pressures," *Mech. Eng.*, Vol. 93, No. 3, pp. 34–39, March 1971.

[23] For the latter, see, e.g., A. Nadai, *Theory of Flow and Fracture of Solids*, New York: McGraw-Hill, pp. 405–489, 1950.

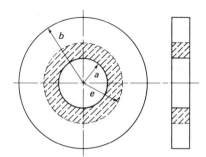

FIGURE 8.19. Disk with inner plastic region (shaded) extending from $r = a$ to $r = e$ and elastic region extending from $r = e$ to $r = b$.

and either closed or open ends, $\sigma_t > \sigma_z > \sigma_r$, as in Fig. 8.12(a), so that the maximum shear stress is $(\sigma_t - \sigma_r)/2$. Set equal to $S_y/2$, this gives $\sigma_t - \sigma_r = S_y$ or $\sigma_t = S_y + \sigma_r$ for the plastic region, as in Fig. 8.20(b).

Rotating disks. An equation for the *radial* stress of the plastic region may be obtained directly from the equation of equilibrium. Thus for a *rotating disk of constant width l*, and with the value S_y substituted for σ_t, Eq. (8.18) becomes,

$$r \frac{d\sigma_r}{dr} + \sigma_r = S_y - \rho\omega^2 r^2 \tag{8.44}$$

or

$$\frac{d}{dr}(r\sigma_r) = S_y - \rho\omega^2 r^2$$

Integrating,

$$r\sigma_r = S_y r - \frac{\rho\omega^2}{3} r^3 + C \qquad \text{or} \qquad \sigma_r = S_y - \frac{\rho\omega^2}{3} r^2 + \frac{C}{r}$$

FIGURE 8.20. Distribution of stresses in the plastic and elastic regions, plotted as ratios to S_y. (a) In a thin rotating disk with $b/a = 6$, $e/a = 4$, and $\omega^2 = 0.064 S_y/\rho a^2$. (b) In an internally pressurized cylinder with $b/a = 2$, $e/a = 1.67$, and $p_i/S_y = 0.664$.

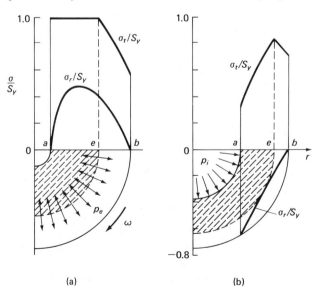

(a)   (b)

For a rotating disk with a central hole, $\sigma_r = 0$ at $r = a$; hence C is determined, and the two stresses of the plastic region become

$$(\sigma_r)_{\text{pl}} = S_y\left(1 - \frac{a}{r}\right) - \frac{\rho\omega^2 r^2}{3}\left(1 - \frac{a^3}{r^3}\right)$$

$$(\sigma_t)_{\text{pl}} = S_y$$

$$(8.45)$$

The radial stress σ_r is seen to be less than S_y and hence it is everywhere elastic in this region where σ_t is plastic.

For the elastic region the equations of Table 8.1 hold, except that the inner radius of the elastic region is e instead of a, and there is superimposed upon the stresses due to rotation ω the tensile radial stress of the plastic region acting at radius e. This is equivalent to a pull or negative pressure p_e at the inner boundary of the elastic region. Thus the equations of rows 1 and 3 in Table 8.1 are added to obtain σ_t and σ_r in terms of e and p_e. Expressions for e and p_e are found from the conditions of continuity at the plastic–elastic boundary where $r = e$, namely, $p_e = -(\sigma_r)_{\text{pl}} = -(\sigma_r)_{\text{el}}$ and $(\sigma_t)_{\text{el}} = (\sigma_t)_{\text{pl}} = S_y$. The values of e and p_e thus found are substituted into the elastic equations for σ_t and σ_r in order to complete them. The stress distribution for one case is shown in Fig. 8.20(a).

However, of particular interest and more easily found is the value of rotational speed ω_1 when yielding just begins at the inner boundary and also the value of ω_2 when the entire disk becomes plastic. The first speed ω_1 is found by setting the elastic equation for $(\sigma_t)_{\text{max}}$ equal to S_y. From row 3 of Table 8.1,

$$(\sigma_t)_{\text{max}} = \frac{\rho\omega_1^2}{4}\left[(3 + \nu)b^2 + (1 - \nu)a^2\right] = S_y$$

whence

$$\omega_1^2 = \frac{4S_y}{\rho b^2}\frac{1}{(3 + \nu) + (1 - \nu)(a/b)^2}$$

$$(8.46)$$

At the second speed ω_2, Eqs. (8.45) of the plastic region hold for the entire disk. The radial stress must be zero at the outer boundary. Substitution of $r = b$ gives

$$(\sigma_r)_{\text{pl}} = S_y\left(1 - \frac{a}{b}\right) - \frac{\rho\omega_2^2 b^2}{3}\left(1 - \frac{a^3}{b^3}\right) = 0$$

whence

$$\omega_2^2 = \frac{3S_y}{\rho b^2}\frac{1}{1 + a/b + (a/b)^2}$$

$$(8.47)$$

The ratio of the second speed to the first speed is the limit-design factor defined by Eq. (5.31). Thus

$$L_\omega = \frac{\omega_2}{\omega_1} = \sqrt{\frac{3}{4}\frac{(3 + \nu) + (1 - \nu)(a/b)^2}{1 + (a/b) + (a/b)^2}}$$

$$(8.48)$$

This factor is independent of the yield point of the material but not of Poisson's ratio. The values are not high; e.g., when the ratio of the hole to outside diameter of the disk is $a/b = \frac{1}{6}$, $L_\omega = 1.44$. Hence the plasticity of the material does not give much margin of safety for overspeeding. However, a satisfactory prestressing of the disk may be accomplished with very little overspeeding (Section 7.12).

Example 8.7

(a) At what speed will the disk of Example 8.4 become plastic to a radius of 52 mm, about one-fourth of the distance from the hole to the periphery? (b) What will be the residual stresses at the hole and the periphery following the operation of (a)? Are they favorable?

Solution

(a) A method and conditions at the elastic–plastic boundaries are given after Eqs. (8.45). Values $S_y = 480$ MPa and $\rho = 7.778 \times 10^{-9}$ kg·m/mm^4 are obtained from Example 8.4. From Eqs. (8.45), with $r = e$

$$(\sigma_r)_{\text{pl}} = 480\left(1 - \frac{28}{52}\right) - \frac{(7.778 \times 10^{-9})\omega^2(52)^2}{3}\left[1 - \left(\frac{28}{52}\right)^3\right] = -p_e$$

whence

$$p_e = -221.5 + (5.916 \times 10^{-6})\omega^2 \tag{a}$$

Also, at the inner boundary of the elastic region, by adding the maximum tangential stresses of rows 1 and 3 of Table 8.1, and by setting $a = e$ and $p_i = p_e$, we obtain

$$(\sigma_t)_{\text{el}} = p_e\frac{b^2 + e^2}{b^2 - e^2} + \frac{\rho\omega^2}{4}[(3 + \nu)b^2 + (1 - \nu)e^2] = (\sigma_t)_{\text{pl}} = S_y$$

$$p_e\frac{125^2 + 52^2}{125^2 - 52^2} + \frac{(7.778 \times 10^{-9})\omega^2}{4}[(3.3)(125)^2 + (0.70)(52)^2] = 480$$

whence

$$1.419p_e + (103.9 \times 10^{-6})\omega^2 = 480 \tag{b}$$

Substitution into Eq. (b) from Eq. (a) yields the speed ω, thus

$$-314.3 + (8.395 \times 10^{-6})\omega^2 + (103.9 \times 10^{-6})\omega^2 = 480$$

where $(112.3 \times 10^{-6})\omega^2 = 794.3$, $\omega^2 = 7.073 \times 10^6$, and $\omega = 2660$ rad/s (25 400 rpm) The tensile, radial stress or negative pressure p_e is, from Eq. (a),

$$p_e = -221.5 + (5.916 \times 10^{-6})(7.073 \times 10^6) = -179.7 \text{ MPa}$$

and

$$(\sigma_r)_e = -p_e = 179.7 \text{ MPa}$$

By adding the tangential stresses of rows 1 and 3 of Table 8.1, and setting $a = e$ and $r = b$, we obtain the tangential stress at the periphery of the elastic region

$$\begin{aligned}
(\sigma_t)_b &= p_e\frac{2e^2}{b^2 - e^2} + \frac{\rho\omega^2(3 + \nu)}{8}\left[b^2 + 2e^2 - \frac{1 + 3\nu}{3 + \nu}b^2\right] \\
&= (-179.7)\frac{2(52)^2}{125^2 - 52^2} + \\
&\quad \frac{(7.778 \times 10^{-9})(7.073 \times 10^6)(3.3)}{8}\left[1 + 2(52)^2 - \frac{1.9}{3.3}(125)^2\right] \\
&= -75.2 + 273.2 = 198 \text{ MPa}
\end{aligned}$$

(b) The residual stresses are obtained by subtracting a completely elastic condition from a plastic or plastic–elastic condition, both for the same speed. Equations (b) and (c) in the solution to Example 8.4 may be used to determine the former, but with a higher value of $\rho\omega^2$ coresponding to the value of ω^2 found under part (a). This gives $\rho\omega^2 = (7.778 \times 10^{-9})(7.073 \times 10^6) = 55.01 \times 10^{-3}$. Thus at $r = a = 28$ mm, and from Eq. (b) of Example 8.4

$$(\sigma_t)_a' = 0.4125(55.01 \times 10^{-3})(2b^2 + 0.4242a^2)$$

$$= (22.69 \times 10^{-3})\left[2 \times 125^2 + 0.4242 \times 28^2\right] = 716.6 \text{ MPa}$$

and from Eq. (c) of Example 8.4

$$(\sigma_t)_b' = 0.4125(55.01 \times 10^{-3})(2a^2 + 0.4242b^2)$$

$$= (22.69 \times 10^{-3})\left[2(28)^2 + 0.4242\,(125)^2\right] = 185.9 \text{ MPa}$$

Since $(\sigma_t)_a = S_y = 480$ MPa, the residual stress at the inside surface is

$$(\sigma_t)_a'' = (\sigma_t)_a - (\sigma_t)_a' = 480 - 716.6 = -236.6 \text{ MPa (compressive)}$$

and the residual stress at the periphery is

$$(\sigma_t)_b'' = (\sigma_t)_b - (\sigma_t)_b' = 198 - 185.9 = 12.1 \text{ MPa (tensile)}$$

These residual stresses are beneficial since the larger compressive stress subtracts from the highest tensile stress at $r = a$. The residual stress is not needed for the disk of Example 8.4, operating at 14 000 rpm, but if the operating speed were such that the stress due to rotation just reached the yield strength of 480 MPa, the net stress in a prestressed disk would be $480 - 236.6 = 243.4$ MPa.   ////

Internal pressure. For a cylinder of uniform width *l under internal pressure*, the equilibrium equation for an interior, plastic region is obtained from Eq. (8.18) by setting $\omega = 0$ and $\sigma_t = S_y + \sigma_r$, as previously found for open or closed ends. This gives

$$r\frac{d\sigma_r}{dr} = S_y \tag{8.49}$$

the solution of which is

$$\sigma_r = S_y \ln r + C$$

When $r = a$, $\sigma_r = -p_i$, whence $C = -p_i - S_y \ln a$ and

$$(\sigma_r)_{\text{pl}} = -p_i + S_y \ln \frac{r}{a}$$
$$(\sigma_t)_{\text{pl}} = -p_i + S_y(1 + \ln \frac{r}{a}) \tag{8.50}$$

The stress distribution for one case is shown nondimensionally in Fig. 8.20(b).

The pressure at which yielding begins is found by setting the elastic shear stress at $r = a$, from Eq. (8.29), equal to $S_y/2$, whence

$$(p_i)_1 = \frac{S_y(b^2 - a^2)}{2b^2} \tag{8.51}$$

The pressure at which the plastic region extends to the outer surface is found from the condition that $(\sigma_t)_{\text{pl}} = S_y$ at $r = b$. From Eqs. (8.50)

$$(p_i)_2 = S_y \ln \frac{b}{a} \tag{8.52}$$

The ratio of pressure at the completion of yielding to that at the beginning of yielding is

$$L_p = \frac{(p_i)_2}{(p_i)_1} = \frac{2(b/a)^2 \ln (b/a)}{(b/a)^2 - 1} \tag{8.53}$$

This ratio has the values 3.69, 1.85, and 1.10 corresponding to $b/a = 6$, 2, and 1.10, respectively. The first value is for comparison with the 1.44 of the rotating disk, but its b/a ratio of 6 is not typical of pressure vessels. The second value corresponds to a rather thick-walled vessel, and the third value is near the upper limit of thickness for a shell, where $t/d = 0.05$. Thus plasticity offers a fairly good margin of safety for a thick cylinder but not for a shell. However, inability to yield at nozzles and other irregularities may decrease the margin. Since the ASME Code for the design of *unfired pressure vessels* is based upon elastic stress only, a comparison with collapse pressure, that pressure which makes the vessel continue to yield without further increase in pressure, makes the Code look overconservative for thick-walled vessels and less conservative for thin-walled vessels.

The ASME Code for Class 1 *nuclear vessels* uses limit analysis as a basis for some allowable stress values. The general primary membrane stresses, the meridional and tangential stresses assumed to be uniformly distributed across a section, are set at a value below the yield strength in order to avoid gross yielding. However, combined *local* primary membrane and primary bending stresses are set at a limit below the collapse strength of the beam loaded in bending, considering the limit-design factor of the beam. (See Section 5.8.) Secondary stresses are defined as those developed by self-constraint, such as at a discontinuity between the head and cylinder or from nonuniform temperatures. They are self-limiting because minor yielding decreases them. After several applications and removal of a pressure sufficiently high to cause yielding, the material "shakes down" to an elastic condition, accompanied by the setting up of favorable residual stresses. The Code assures that this will happen by limiting the sum of the primary and secondary stresses, calculated as though elastic, to values no more than twice the yield strength. A third category of stress, which is the local stress from stress concentrations, is a possible source of failure from fatigue, brittle fracture, or stress corrosion. In this category, strains calculated on the basis of elastic action are compared with constant-strain low-cycle fatigue tests. The Code allows some stress limits to be exceeded if more exact limit analyses are made, for example, by including the strain-hardening effect of the material. Theoretical and engineering considerations in pressure-vessel design are found in the many publications of the American Society of Mechanical Engineers.[24]

8.13 THE PLANE-STRAIN SOLUTION: ROTATING CYLINDERS

In a very long or in an infinitely long cylinder with free ends, it is inconceivable that the axial displacements and strains would vary with the radius except near its ends. The material resists the large radial-plane distortion that would be required, and it does so by developing stresses in the axial direction. Plane sections away from the ends remain plane. Thus in contrast to the plane-stress condition of radial-plane only stresses and

---

[24] In addition to the ASME Code (see footnote 3), these include a Criteria Document for Section III, Division I—1980; *Pressure Vessel Technology*, a collection in several volumes of papers and discussions presented at international conferences—1969 in The Netherlands, 1973 in the United States, and 1977 in Japan; *Pressure Vessels and Piping: Design and Analysis—A Decade of Progress*, four volumes issued between 1972 and 1976; books on finite-element analyses, computer programs, etc.; and the quarterly transactions, *Journal of Pressure Vessel Technology*, ASME, 345 East 47th St., New York.

variable axial strain, as in a short cylinder or disk, there is a condition of radial-plane stresses and a variable axial stress together with uniform axial strain, known as *plane strain*. The open-end pressure cylinder is a special case of uniform axial strain *without* axial stress (Section 8.9).

At first it will be assumed that the ends of the cylinder are fixed between rigid blocks. Then $\epsilon_z = 0$ and from the third of Eqs. (8.20)

$$\sigma_z = -\alpha ET + \nu(\sigma_r + \sigma_t) \tag{8.54}$$

Substitution of this equation into the first two of Eqs. (8.20) gives

$$\epsilon_r = (1 + \nu)\alpha T + \frac{1}{E}[(1 - \nu^2)\sigma_r - \nu(1 + \nu)\sigma_t]$$

and

$$\epsilon_t = (1 + \nu)\alpha T + \frac{1}{E}[(1 - \nu^2)\sigma_t - \nu(1 + \nu)\sigma_r]$$

These equations may be rewritten in the form

$$\epsilon_r = (1 + \nu)\alpha T + \frac{1 - \nu^2}{E}\left(\sigma_r - \frac{\nu}{1 - \nu}\sigma_t\right)$$

$$\tag{8.55}$$

$$\epsilon_t = (1 + \nu)\alpha T + \frac{1 - \nu^2}{E}\left(\sigma_t - \frac{\nu}{1 - \nu}\sigma_r\right)$$

In the plane-*stress* case of $\sigma_z = 0$, the corresponding equations, Eqs. (8.20), become

$$\epsilon_r = \alpha T + \frac{1}{E}(\sigma_r - \nu\sigma_t)$$

and

$$\epsilon_t = \alpha T + \frac{1}{E}(\sigma_t - \nu\sigma_r)$$

A comparison of the ϵ equations of the plane-*strain* case with those of the plane-*stress* case shows that wherever α occurs it is accompanied by $(1 + \nu)$, i.e., α in the plane-stress case is replaced by $(1 + \nu)\alpha$. Similarly, E is replaced by $E/(1 - \nu^2)$ and ν, by $\nu/(1 - \nu)$. The other basic equations, Eqs. (8.18) and (8.21), which were used with Eqs. (8.20) to obtain the plane-stress results, express the action in any radial plane. They contain none of the preceding physical constants. These same basic equations are to be used with Eqs. (8.55) to determine the stress and displacement equations for plane strain.

As a consequence, we may transform the σ_r and σ_t equations of this plane-stress case into equations for the plane-strain case by making a similar replacement of the physical constants, thus avoiding another long derivation. By the transformation of Eqs. (8.26), the general plane-strain equations become

$$\sigma_r = -\frac{\alpha E}{(1 - \nu)r^2}\int_a^r Tr\, dr - \frac{3 - 2\nu}{8(1 - \nu)}\rho\omega^2 r^2$$

$$+ \frac{EC_1}{(1 + \nu)(1 - 2\nu)} - \frac{EC_2}{(1 + \nu)r^2} \tag{8.56}$$

and

$$\sigma_t = \frac{\alpha E}{(1 - \nu)r^2} \int_a^r Tr\, dr - \frac{\alpha ET}{1 - \nu} - \frac{1 + 2\nu}{8(1 - \nu)} \rho\omega^2 r^2$$

$$+ \frac{EC_1}{(1 + \nu)(1 - 2\nu)} + \frac{EC_2}{(1 + \nu)r^2} \tag{8.57}$$

The stress equations in rows 3 and 4 of Table 8.1 may be similarly changed. Thus for the rotation of a long cylinder of internal radius a and external radius b

$$\sigma_r = \rho\omega^2 \frac{3 - 2\nu}{8(1 - \nu)} \left(b^2 + a^2 - \frac{a^2 b^2}{r^2} - r^2 \right)$$

$$\sigma_t = \rho\omega^2 \frac{3 - 2\nu}{8(1 - \nu)} \left(b^2 + a^2 + \frac{a^2 b^2}{r^2} - \frac{1 + 2\nu}{3 - 2\nu} r^2 \right) \tag{8.58}$$

For the *pressure* cases of Table 8.1, the σ_r and σ_t equations do not contain the constants E, ν, and α, and, as was proven in Section 8.9, these two equations are valid for both plane stress and plane strain.

Axial stress will now be considered. Equation (8.54) expresses the condition of zero axial strain, and from it the foregoing equations were derived. By substitution into Eq. (8.54) from Eqs. (8.56) and (8.57), it is determined that for a cylinder with fixed ends

$$\sigma_z = -\frac{\alpha ET}{1 - \nu} - \frac{\nu}{2(1 - \nu)} \rho\omega^2 r^2 + \frac{2E\nu C_1}{(1 + \nu)(1 - 2\nu)} \tag{8.59}$$

Figure 8.21(b) shows a stress pattern caused by rotation only. This stress pattern and the equilibrium at the ends is maintained by holding the cylinder between fixed and rigid ends. This is seldom the case, particularly when there is rotation.

Now consider the ends to be freed. No force is available to maintain equilibrium, so the stress pattern must be altered to that of a different stress σ_z' such that, across any radial section *away from the ends*, the integral of stress with area, i.e., the net force across the section, is zero. From the elemental area of Fig. 8.21(a)

$$F_z' = \int_a^b \sigma_z'\, dA = 2\pi \int_a^b \sigma_z'\, r\, dr = 0 \tag{8.60}$$

Consider that release of the ends superimposes a stress system consisting of $\sigma_r = \sigma_t = 0$ and $\sigma_z = \sigma_0$ where σ_0 is constant across a radial section (Fig. 8.21(c)). The radial and

FIGURE 8.21. Axial stresses in a long rotating cylinder. (a) Elemental area of section. (b) Stress with ends held, σ_z. (c) Uniform stress σ_0. (d) Stress with free ends, $\sigma_z' = \sigma_z + \sigma_0$.

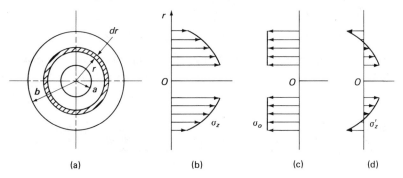

tangential stresses, Eqs. (8.56) and (8.57), are unaffected, but $\sigma_z' = \sigma_z + \sigma_0$, and because σ_0 is uniform, plane sections continue plane. The value of σ_0 is determined from Eq. (8.60) by the substitution of $(\sigma_z + \sigma_0)$ for σ_z' and then from Eq. (8.59) for σ_z, thus

$$\int_a^b (\sigma_z + \sigma_0) r \, dr$$

$$= -\frac{\alpha E}{1 - \nu} \int_a^b Tr \, dr - \frac{\nu \rho \omega^2}{2(1 - \nu)} \int_a^b r^3 \, dr + \left(\frac{2E\nu C_1}{(1 + \nu)(1 - 2\nu)} + \sigma_0 \right) \int_a^b r \, dr = 0 \quad \text{(a)}$$

from which a constant sum which includes σ_0 is found to be

$$\frac{2E\nu C_1}{(1 + \nu)(1 - 2\nu)} + \sigma_0 = \frac{2\alpha E}{(1 - \nu)(b^2 - a^2)} \int_a^b Tr \, dr + \frac{\nu \rho \omega^2 (b^2 + a^2)}{4(1 - \nu)} \quad \text{(b)}$$

When σ_0 is added to σ_z of Eq. (8.59) to give σ_z', this same constant sum appears. Its equivalent from Eq. (b) may be substituted to give

$$\sigma_z' = \sigma_z + \sigma_0$$

$$= \frac{\alpha E}{1 - \nu} \left(-T + \frac{2}{b^2 - a^2} \int_a^b Tr \, dr \right) + \frac{\nu \rho \omega^2}{4(1 - \nu)} (b^2 + a^2 - 2r^2) \quad (8.61)$$

Figure 8.21(d) shows a σ_z' distribution as the sum of distributions σ_z and σ_0 from Fig. 8.21(b) and (c), respectively. Stress σ_z' is accompanied by a uniform axial strain ϵ_z' and displacement w. This stress does not hold close to the ends, since at these free boundaries there can be no reaction, and σ_z' is zero at all radii. However, beginning a short distance from the ends, Eq. (8.61) does represent the axial stress, according to the principle of Saint-Venant.[25]

Example 8.8

If the disk of Example 8.4 were a cylinder 3 m long, what would be the stresses at the hole at the operating speed of 14 000 rpm?

Solution
From Eqs. (8.58), for a long cylinder, at $r = a$

$$(\sigma_t)_a = (16.71 \times 10^{-3}) \frac{3 - 0.60}{8(0.7)} \left[2b^2 + \left(1 - \frac{1.6}{2.4} \right) a^2 \right]$$

$$= (7.161 \times 10^{-3}) [2(125)^2 + 0.333(28)^2] = 225.7 \text{ MPa}$$

This result is slightly higher than the 217.7 MPa in the thin disk. At $r = b$

$$(\sigma_t)_b = (7.161 \times 10^{-3}) (2a^2 + 0.333b^2)$$

$$= (7.161 \times 10^{-3}) [2(28)^2 + 0.333(125)^2] = 48.5 \text{ MPa}$$

This result is lower than the 56.5 MPa at the periphery of the thin disk. The radial stresses are zero at radii a and b. The axial stresses, not present in the disk, from Eq.

---

[25] If the forces acting upon a small part of the surface are replaced by another statically equivalent system of forces acting on the same part of the surface, this redistribution of loading has a negligible effect on the stresses at distances that are large in comparison with the linear dimensions of the part of the surface on which the forces are changed.

(8.61) are

$$(\sigma_z{}')_a = \frac{0.3(16.71 \times 10^{-3})}{4(0.70)}(b^2 - a^2)$$

$$= (1.790 \times 10^{-3})(125^2 - 28^2) = 26.57 \text{ MPa}$$

$$(\sigma_z{}')_b = (1.790 \times 10^{-3})(a^2 - b^2)$$

$$= (1.790 \times 10^{-3})(28^2 - 125^2) = -26.57 \text{ MPa}$$

With the proportions given in this problem these stresses are not significant unless axial stresses occur from other sources.   ////

8.14 THERMAL STRESSES IN DISKS AND IN LONG CYLINDERS

In this section the general stress and deflection equations of Sections 8.8 and 8.13 are adapted for direct use with any function T that expresses a radial variation in temperature. This is done for both the plane-stress short-cylinder case and the plane-strain long-cylinder case. The equations are then applied to the common case of an internally heated long cylinder, and the results are charted.

At the boundaries $r = a$ and $r = b$, respectively, and without pressure and rotation, the first of Eqs. (8.26) becomes

$$(\sigma_r)_a = 0 = 0 + \frac{EC_1}{1 - \nu} - \frac{EC_2}{(1 + \nu)a^2}$$

and

$$(\sigma_r)_b = 0 = -\frac{\alpha E}{b^2}\int_a^b Tr\,dr + \frac{EC_1}{1 - \nu} - \frac{EC_2}{(1 + \nu)b^2}$$

By simultaneous solution

$$C_1 = \frac{\alpha(1 - \nu)}{b^2 - a^2}\int_a^b Tr\,dr$$

and

$$C_2 = \frac{\alpha(1 + \nu)a^2}{b^2 - a^2}\int_a^b Tr\,dr$$

The stresses σ_r and σ_t become

$$\sigma_r = \frac{\alpha E}{r^2}\left(\frac{r^2 - a^2}{b^2 - a^2}\int_a^b Tr\,dr - \int_a^r Tr\,dr\right) \tag{8.62}$$

$$\sigma_t = \frac{\alpha E}{r^2}\left(\frac{r^2 + a^2}{b^2 - a^2}\int_a^b Tr\,dr + \int_a^r Tr\,dr - Tr^2\right)$$

From Eq. (8.25)

$$u = \frac{\alpha}{r}\left(\frac{(1 - \nu)r^2 + (1 + \nu)a^2}{b^2 - a^2}\int_a^b Tr\,dr + (1 + \nu)\int_a^r Tr\,dr\right) \tag{8.63}$$

These equations also apply to a solid disk of uniform thickness with the substitution of $a = 0$.

Most turbine disks are tapered for more uniform strength, and this together with variations of α and E with temperature make it necessary to resort to finite difference and similar methods of computation. More detailed descriptions of this and other problems of a difficult nature are available elsewhere.[26]

For long cylinders, equations free of the constants of integration may be obtained from the general equations for plane strain, Eqs. (8.56) and (8.57), and the boundary conditions $\sigma_r = 0$ at $r = a$ and at $r = b$. They may be obtained more directly from the plane-stress solutions, Eqs. (8.62), by the transformation of physical constants described in Section 8.13. Thus transformed, the equations become

$$
\begin{aligned}
\sigma_r &= \frac{\alpha E}{(1-\nu)r^2}\left(\frac{r^2 - a^2}{b^2 - a^2}\int_a^b Tr\,dr - \int_a^r Tr\,dr\right) \\
\sigma_t &= \frac{\alpha E}{(1-\nu)r^2}\left(\frac{r^2 + a^2}{b^2 - a^2}\int_a^b Tr\,dr + \int_a^r Tr\,dr - Tr^2\right)
\end{aligned}
\tag{8.64}
$$

It may be noted that these two equations for thermal stress in a long cylinder differ from those for a short cylinder, Eqs. (8.62), by the multiplying factor $1/(1 - \nu)$ only, a value of 1.43 when $\nu = 0.30$.

Through long cylinders, such as heat-exchanger tubes, chimneys, and long rotors, the heat flow is essentially radial except near the ends. After a steady flow is established, the temperature at radius r is given by

$$
T = T_b + \frac{(T_a - T_b)\ln(b/r)}{\ln(b/a)}
\tag{8.65}
$$

Substitution from Eq. (8.65) into Eqs. (8.64), followed by integration, gives

$$
\sigma_r = \frac{\alpha E(T_a - T_b)}{2(1-\nu)\ln b/a}\left[\frac{a^2}{b^2 - a^2}\left(\frac{b^2}{r^2} - 1\right)\ln\frac{b}{a} - \ln\frac{b}{r}\right]
$$

$$
\sigma_t = \frac{\alpha E(T_a - T_b)}{2(1-\nu)\ln b/a}\left[1 - \frac{a^2}{b^2 - a^2}\left(\frac{b^2}{r^2} + 1\right)\ln\frac{b}{a} - \ln\frac{b}{r}\right]
\tag{8.66}
$$

By Eq. (8.61), the net axial stress for a free-end condition becomes

$$
\sigma_z{}' = \frac{\alpha E(T_a - T_b)}{2(1-\nu)\ln b/a}\left(1 - \frac{2a^2}{b^2 - a^2}\ln\frac{b}{a} - 2\ln\frac{b}{r}\right)
\tag{8.67}
$$

The distributions of the temperature and stresses are plotted nondimensionally in Fig. 8.22 for any cylinder with a radius ratio $b/a = 2.0$. The axial and tangential stresses at the outer surface are equal and *tensile*. This explains why internal heating may cause external cracks in chimneys and conduits of concrete and masonry since this material is weak in tension. The denominator of the stress ratio is $\alpha E(T_a - T_b)/2(1 - \nu)$, which for

[26] S. S. Manson, *Thermal Stress and Low Cycle Fatigue*, New York: McGraw-Hill, 1966; B. E. Gatewood, *Thermal Stresses, with Applications to Airplanes, Missiles, Turbines, and Nuclear Reactors*, New York: McGraw-Hill, 1957; B. A. Boley and J. H. Weiner, *Theory of Thermal Stresses*, New York: John Wiley & Sons, Inc., 1960.

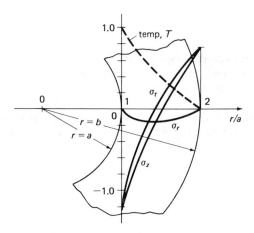

FIGURE 8.22. Distribution of temperature T and stresses σ in a long thick cylinder with radius ratio $b/a = 2.0$, steady radial heat flow, and $T_a > T_b$. Nondimensional ordinates are the temperature ratio $(T - T_b)/(T_a - T_b)$ and the stress ratio $\sigma/[\alpha E(T_a - T_b)/2(1 - \nu)]$. Stresses are σ_r radially, σ_t tangentially, and σ_z axially.

steels has a value of 13 500 psi for each 100°F (167.6 MPa for each 100°C) difference between inside and outside wall temperatures. For the radius ratio of the figure, the corresponding maximum tensile and compressive stresses are then 10 400 and 16 600 psi, respectively, for each 100°F.

When the wall is thin, as in the cylinder liner of an engine or compressor, the stress distribution is nearly linear, and the maximum stresses have a value nearly equal to αE $(T_a - T_b)/2(1 - \nu)$, as in Fig. 8.23 where $b/a = 1.1$. This is the stress value that is found in a flat plate[27] that is heated on one side while its edges are clamped to prevent bending of the plate. The bending of the wall of the cylinder is also prevented, but by its curvature, symmetry, and continuity. Near the ends of the cylinder, however, additional stresses occur if there is no inflexible end to oppose the moment of the stresses in the walls. For these free ends, there is locally a 25% increase in stress when the material is steel.

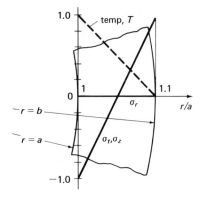

FIGURE 8.23. Distribution of temperature and stresses in a long thin cylinder with radius ratio $b/a = 1.1$, steady radial heat flow, and $T_a > T_b$. Ordinates and symbols are the same as in Fig. 8.22.

---

[27] See Eq. (10.53b) of Section 10.8 (also see Section 7.6). By definition the thickness of a plate is small compared to its other dimensions.

PROBLEMS

Section 8.3

8.1 Confirm the statement in the last paragraph of Section 8.2, regarding substitution of a negative value for R_m, by deriving the equivalent of Eqs. (8.1), (8.4), and (8.5) for the fillet section of Fig. 8.3, with θ taken as shown thereon.

8.2 The nose section for a space vehicle is essentially a cone, with dimensions as shown and internal pressure p. Determine equations for the stresses at any position z away from the apex and base. *Ans.* $\sigma_m = (pzD/4ht)\sqrt{1 + D^2/4h^2}$, $\sigma_t = 2\sigma_m$.

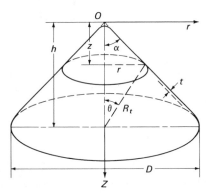

8.3 A paraboloid of revolution is to be used as the shape for a transparent observation capsule under a pressure differential p. (a) Taking the equation of a parabola to be $z = ar^2$, derive equations for the meridional and tangential stresses. *Ans.* $\sigma_t = (p/4at)(1 + 8a^2r^2)(1 + 4a^2r^2)^{-1/2}$. (b) Determine the numerical value of the constant a for a capsule which will fit around your head with a few inches to spare and will end about 3 in below your eye level. (c) Calculate the required thickness if the capsule is to be used on an underwater vehicle at a maximum depth of 650 ft and the allowable compressive stress in the glass is 5000 psi. The weight of seawater is approximately 64 lb/ft^3.

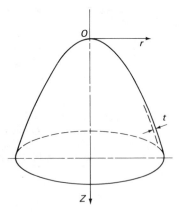

8.4 Derive the cylinder and sphere equations, Eqs. (8.7) and (8.8), directly from free-body sections through a cylinder and sphere, respectively.

8.5 What is the maximum tensile stress in the football of the sketch if the air pressure is 12 psi and the thickness is $^3/_{32}$ in?

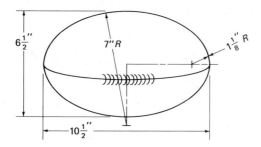

8.6 A pipeline has an average diameter of 36 in and carries gas at a pressure of 750 psi. Joint efficiency is 85%, yield strength is 40 000 psi, and the factor of safety is 2.0. What is the minimum required thickness? *Ans.* 0.794 in.

8.7 The high-pressure cylindrical tank in an SF_6 gas-enclosed electrical circuit breaker and switching substation may have a pressure of 2.0 MPa and a diameter of 600 mm. If the yield strength is 240 MPa, joint efficiency is 90%, and the factor of safety 1.75, what is the minimum thickness for the steel?

8.8 The forged, steel cylinder for an oil-hydraulic actuator giving a thrust of 27 kN has an inside diameter of 40 mm and a wall thickness of 4 mm. What should be the yield strength of the material for a factor of safety of 2.0?

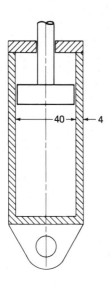

8.9 Derive an equation for the stress in a thin rotating ring. What is its limiting tangential velocity for a factor of safety of 5.0 if it is made of cast iron with a tensile strength of 25 000 psi and a unit weight of 0.26 lb/in³? *Ans.* 13 620 ft/min.

8.10 (a) Write equations for failure of a cylindrical pressure vessel according to the Mises criterion—in the cylinder at radius R and in the head at any radii R_m and R_t. (b) What is the failure value of σ_t in terms of S_y—in the cylinder and in a hemispherical head? Compare with the value given by the maximum shear theory of failure.

8.11 A condition of pure shear is not easy to obtain for testing purposes. One method is to subject a small, thin hollow cylinder simultaneously to internal fluid pressure and external axial loading. Explain and give some details of actuation (a) for a static-yield test and (b) for a fatigue test if there is available a hydraulic-power unit with controllable or programmed, rapidly changing pressures.

8.12 A partial torus of circular cross section is commonly used as a tube bend in piping systems, heat exchangers, and penstocks, and a complete torus may be used for gas storage where it fits available space. (a) Derive equations for the meridional, tangential, and maximum shear stresses in terms of position radius r. The horizontal ring defined by the shaded sections of the figure is suggested as a suitable free body for obtaining stress σ_m, and the membrane equation, Eq. (8.1), is valid. *Ans.* $\sigma_m = (pR/2t)(r + \bar{r})/r$, $\sigma_t = pR/2t$. (b) Study the stress variation with position, compare with a straight tube with which a partial torus might be joined, and plot the ratio of maximum normal stresses in torus and straight tube against radius ratio $\bar{r}/R$ values from 2 to 15 to show the curvature effect.

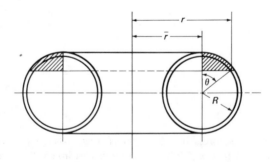

8.13 Water flows from an elevation of 1500 ft through a penstock that ends with a 90° turn into the casing of a water turbine. It has a cross section of 48-in diameter, and the inside radius at the turn is 150 in. What should be the thickness of a steel plate of yield strength 40 000 psi if a welded joint comes at the inside of the curve, its efficiency is 90%, and the factor of safety is 2.0. See Prob. 8.12. *Ans.* 0.936 in.

8.14 To form the living space of an underwater vehicle to operate at depths to 1000 m, two interconnected spheres of 2.5 m outside diameter are joined, the joints forming an included angle of 60°. A tank with a volume of 1 m³ is to surround the vehicle as a torus nested between the spheres. It is filled with air at 4-MPa pressure before submergence. All the air may be used before resurfacing. The material of the spheres and torus is steel with a yield strength of 650 MPa. The factor of safety is to be 2.0. Seawater weighs about 10 kN/m³. (a) What is the required thickness of the spheres? (b) Draw the spheres to scale and determine a size for the torus with a bit of clearance between them. The volume of a torus (Prob. 8.12) is $(2\pi \bar{r})(\pi R^2)$. What wall thickness is required?

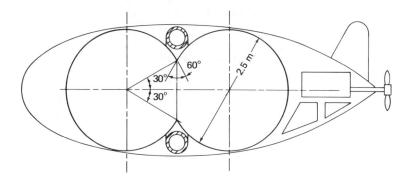

8.15 Ellipsoidal ends of thickness t are proposed for a cylindrical pressure vessel. The equation of the ellipse is $(z/l)^2 + (r/R)^2 = 1$, where the origin is at the center of the ellipse and the semiaxes are dome height $z = l$ and cylinder radius $r = R$, with $l < R$. Derive equations for the meridional and tangential stresses in the head. *Ans.* $\sigma_m = (p/2tl) [R^4 - r^2(R^2 - l^2)]^{1/2}$, $\sigma_t = (p/2tl) [R^4 - 2r^2(R^2 - l^2)] [R^4 - r^2(R^2 - l^2)]^{-1/2}$.

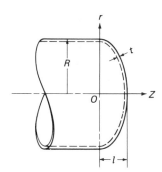

8.16 With the results of Prob. 8.15 determine the maximum and minimum values of σ_m and σ_t and also of the shear stress τ. At what l/R ratio does σ_t change sign at R? Is this reasonable? Compare the values for the head with those of the cylinder, particularly where they join. What must happen there? The ASME Code for unfired vessels discusses an ellipsoidal head with $l = D/4$ and gives for the thickness, $t = pD/(2\eta\sigma - 0.2p)$, where η is a joint efficiency and $0.2p$ accounts for the decrease of stress uniformity with the increase of p and t. Compare with your results.

8.17 Design a constant shear-stress head for a cylindrical vessel of 24-in diameter. If constructed graphically on a letter-size sheet, use one-quarter scale and draw one-half of the head only. List the radii at the junction and at the top and also the height of the head.

8.18 Derive equations for the radial expansion under uniform pressure p of (a) a cylinder with open ends, (b) a cylinder with closed ends (except for expansion near the ends), and (c) a sphere. Refer to Eqs. (8.6) through (8.8) and (5.1) as necessary. *Ans. pR^2/Et, $pR^2(2 - \nu)/2Et$, $pR^2(1 - \nu)/2Et$.*

8.19 (a) If the constant shear-stress vessel of Example 8.1 has a diameter of 24 in, carries a pressure of 600 psi, and allows a shear stress of 10 000 psi, what should be the thickness of the steel plate? (b) If the head and cylinder were pressurized separately, what would be their separate expansions in the plane where they are joined? In order to minimize secondary stresses at their junction it is proposed to enlarge one by force or by differential heating while they are being welded together. Which one and by how much? When do the resulting residual stresses disappear?

8.20 What is the increase in diameter of the rotating ring of Prob. 8.9 at its limiting tangential velocity if its diameter is 20 in? For cast iron of the given strength, $E = 9.7 \times 10^6$ psi.

8.21 Forged-steel tires may be shrink fitted onto cast-steel locomotive-driving wheels and replaced when worn. An initial interference of about $^1/_{64}$ in per foot of fit diameter has been used. The interference is removed by heating the tire, followed by assembly. The coefficient of expansion for steel is 6.3×10^{-6} °F^{-1}. A tire for an electric locomotive has a 42 in outside diameter, 1½ in thick, 4 in wide, and unflanged. Determine the temperature to which it must be heated for assembly and the stress and pressure on the tire after it cools. *Ans. $\sigma = 37\,600$ psi.*

8.22 (a) Derive an equation for the maximum stress that occurs in a thin 360°, retaining ring when it is snapped in place on a solid cylinder by pushing it over and beyond a raised annular bead on the cylinder, which gives a radial interference δ. (b) Derive an equation for the axial force required if the ramp angle is α and friction is neglected.[28] (c) A ring is made of soft plastic with $E = 350$ MPa, $\alpha = 15°$, $\beta = 40°$, b = 5.5 mm, $t_{avg} = 2$ mm, $d = 25$ mm, and $\delta = 1$ mm. What are the maximum stress and the axial snap-in and snap-out forces?

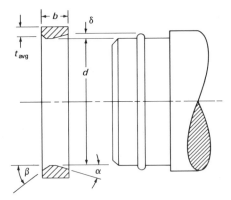

Section 8.4

8.23 For the conical tank of Fig. 8.8 determine (a) equations for the stresses *above* the liquid level but below the support. *Ans. $\sigma_m' = \gamma l^3 \tan \alpha/(6tz \cos \alpha)$, $\sigma_t' = 0$.* (b) Determine the location and magnitude of their maximum values.

---

[28] For a discussion of snap devices, see W. W. Chow, "Snap-Fit Design," *Mech. Eng.*, Vol. 99, No. 7, pp. 35–41, July 1977.

8.24 Stresses were determined in Example 8.2 and in Prob. 8.23 for a conical tank supported above its liquid level. (a) From these stresses derive equations for the tangential (hoop) *strains* at any location z, both above and below the liquid level. (b) Determine the location and magnitude of their maximum values if $\nu = 0.30$. Is there a place where $\epsilon_t = 0$, suitable for the attachment of a fitting? *Ans.* Above the liquid level: $-0.30\gamma l^2 \tan \alpha/(6Et \cos \alpha)$. Below the liquid level; $0.20\ \gamma l^2 \tan \alpha/(Et \cos \alpha)$, $0.945l$.

8.25 The conical tank of Fig. 8.8 is filled with liquid to its top, at height c. (a) Derive equations for the stresses above and for the stresses below the support. (b) Determine the location and magnitude of their maximum values.

8.26 A cylindrical tank with a vertical axis of symmetry is filled with a liquid. The tank is supported by a ring at one-third of the way from the bottom. Derive equations for the stresses above and below the supporting ring, expressing them in terms of unit weight γ and depth h below the water level. Discuss the variation of the stresses and compare them with a closed cylinder under constant pressure $p = \gamma h$. A process has been developed for rolling plate with a thickness varying from one side to the other. How could such plate be used advantageously?

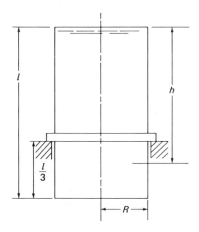

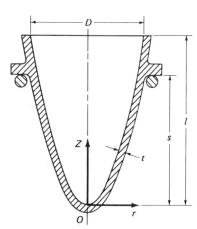

PROBLEM 8.26   **PROBLEM 8.28**

8.27 For increased capacity, a cylindrical shell of height c' is placed above the cone of Prob. 8.25. In what ways does this affect the stress equations for the cone? Compare the tangential strains of the cone and cylinder where they are joined. What develops there?

8.28 An open pot to hold heavy liquid metal of unit weight γ is a paraboloid, with a surface given by $z = ar^2$. It is supported at a level s, and it may be filled to the top at a level l. (a) Derive equations for the principal stresses below the support. *Ans.*

$$\sigma_m' = (\gamma/8at)(2l - z)(1 + 4az)^{1/2},$$
$$\sigma_t' = (\gamma/8at)\left[4(l - z)(1 + 4az)^{1/2} - (2l - z)(1 + 4az)^{-1/2}\right]$$

(b) Based on these stresses and an allowable of 1.0 MPa, what should be the minimum thickness of a uniform wall between the support and bottom if $\gamma = 103$ kN/m³, $l = 330$ mm, $D = 230$ mm, and $s = 255$ mm? What other considerations determine the thickness?

8.29 (a) A thin, hemispherical dish of thickness t is supported at its top by a horizontal ring and filled with liquid of unit weight γ. Derive the equations for the principal stresses. (b) Determine the regions in which they are tensile and in which they are compressive. Write an equation for the shear stress in each region. Determine the shear stresses at the top, bottom, and transition locations.

8.30 A spherical tank filled with liquid of unit weight γ is supported by a horizontal ring somewhere above its base. Stresses are to be found in terms of angle θ measured from the top center. Derive equations for the principal stresses in the region *above* the support. Note: the volume of a spherical cap is $(\pi z^2/3)(3R - z)$, where R is the radius of the sphere and z is the height of the cap. *Ans.* $\sigma_m = B(1 + 2 \cos \theta)$, $\sigma_t = B(5 + 4 \cos \theta)$, $\sigma_r = 0$, where $B = (\gamma R^2/6t)\,[(1 - \cos \theta)/(1 + \cos \theta)]$.

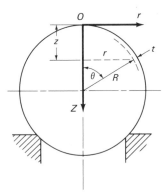

8.31 Same information as Prob. 8.30 but for the region *below* the support. Check the equations at the bottom central location against those for a sphere under uniform pressure. *Ans.*

$$\sigma_m' = (\gamma R^2/6t)(5 - 5 \cos \theta + 2 \cos^2 \theta)/(1 - \cos \theta),$$
$$\sigma_t' = (\gamma R^2/6t)(1 - 7 \cos \theta + 4 \cos^2 \theta)/(1 - \cos \theta), \quad \sigma_r' = 0.$$

8.32 (a) Examine the results of Prob. 8.30 and 8.31, and indicate the angular regions in which the stresses are tensile and those in which they are compressive. For each region write an equation for the maximum shear stress at angle θ. (b) The foregoing equations have been obtained without reference to the angle at which the support is located. Examine the shear equations for the regions above and for the regions below the support for the trends in value. Determine if there is an angle at which the support may be located such that the shear stress just above the support equals the shear stress just below it. Is this a best location? Compare the values there with those at a more easily constructed support, say at $\theta = 150°$. Is the cost saving for the other location worthwhile?

8.33 Investigate the relationships between R_m, R_t, σ_m, σ_t, and p at the bottom center of a tank at which the surface is curving, axis symmetric, and continuous.

8.34 A constant pressure may be superimposed on the liquid in a tank. Write equations for the net normal stress (a) if a neutral gas under pressure p' fills the space above the liquid level in the conical tank of Example 8.2 and (b) if above the sphere of Prob. 8.30 there is a standpipe with liquid at height h above the top of the sphere.

8.35 A tank of constant and equal meridional and tangential stress is built in the shape of a drop of liquid on a horizontal, frictionless surface, the drop having constant surface tension and a flat bottom. (a) If a radius $R_m = R_t$ is assumed for the top center of the tank, estimated from the desired volume capacity, then the constant thickness for the shell is calculated from the membrane equation and the pressure at the top, and a graphical construction is started by swinging an arc of this radius. If a radius of 12 m is assumed, the pressure is 0.15 MPa, and the constant stress is to be 90 MPa, what is the required thickness? *Ans.* 10 mm. (b) A graphical construction proceeds very much like that in Example 8.1 and Fig. 8.6 except that R_{t2} at the end of the top arc equals R_{t1} and centers 2", 1", and 1' coincide. The membrane equation is solved for R_m in general, then successively used to calculate an

R_m from measured R_t and calculated increased p, then constructing a new R_t. The first arc may be wide, followed by successively shorter ones in order to keep the pressure increments about equal. With the data of part (a) and water with $\gamma = 9800$ N/m³, carry out the construction for one side of the tank. If you plot on letter-size paper, use a scale of 1 cm = 1 m. Report the width and height of the tank.

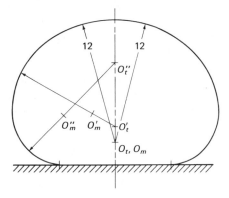

Section 8.5

8.36 (a) Show that each equation, ΣF_y and ΣF_z, preceding Eq. (8.16) gives the same equation for N in terms of ultimate breaking force F_u per filament and a factor of safety n, provided that the correct helix angle is used. (b) For a hose, if the steel-wire diameter is 0.010 in, $S_u = 350\,000$ psi, $n = 2.0$, $p = 3000$ psi, and $D = \tfrac{3}{4}$ in, how many wires per inch per hand should there be?

8.37 To the cylinder of Fig. 8.9 add N_c *circumferential* filaments per inch of axial length, with each filament requiring a force F_c to rupture. Designate the total number of helical windings (both hands) as N_h and the force to rupture each filament, F_h. Derive an equation for the ratio N_c/N_h for any angle ψ of the helical windings. *Ans.* $(F_h/F_c)(3 \cos^2 \psi - 1)/\sin \psi$.

8.38 When there are many filaments of small diameter, as with glass fibers, it is necessary to specify the thicknesses t_c and t_h of the circumferential and helical layers in terms of the tensile strengths of the material, S_c and S_h, respectively. The cross section of each filament may be assumed circular or rectangular. Convert the answer of Prob. 8.37 to an equation for t_c/t_h and check it against Eq. (8.17).

8.39 (a) Derive Eq. (8.17) directly from the tensile strengths S_c and S_h of the materials. *Suggestion:* first obtain for the helical layer the equivalent meridional and tangential strengths in terms of S_h and then use the free bodies of Fig. 8.9 with the force of the circumferential windings added. With $S_c = S_h$ check the equation for $t_c = 0$ and for $\psi = 0°$. What is the helix angle when $t_c = t_h$?

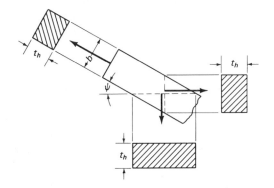

Section 8.6

8.40 Sketch three dimensionally a differential element of a thick-spherical-pressure vessel and label the forces. Derive the equation of equilibrium. *Ans.* $-2\sigma_t + 2\sigma_r + r(d\sigma_r/dr) = 0$.

8.41 A steel-turbine disk of uniform strength has a periphery of 12-in diameter, to which are attached 43 uniformly spaced blades each weighing 0.23 lb and having a center of mass at a radius of 7.3 in. The speed is 7200 rpm and the stress is to be 14 000 psi. Determine the width of the disk at its periphery, at its theoretical center, and at a radius of one-third the peripheral radius, where it will be flanged.

8.42 Same information as Prob. 8.41 except a disk periphery of 600-mm diameter, 83 blades each weighing 5.4 N with a mass center at 385-mm radius, a speed of 5000 rpm, and a stress of 115 MPa. *Ans.* $l_b = 22.3$ mm, $l_o = 51.3$ mm, $l_{100} = 46.8$ mm.

8.43 A cylindrical bar of length L and diameter D, with $L >> D$, is rotated at speed ω about an axis that is normal to and halfway along its length. (a) Write the equation of equilibrium of an element and derive an equation for the stress. (b) What is the maximum stress in the bar when its tip speed is 600 ft/s, material is steel, and $\gamma = 0.281$ lb/in³?

8.44 The casing of a rotating, hydraulic-torque converter, such as that of Fig. 4.4 is a vessel where the liquid pressure increases with radius. Derive an equation for the total liquid force on the cover or platelike end of the reducer for a rotational speed of ω rad/s. This is the force, plus any which may be due to a uniform-charging pressure, to be used in the calculation of the number and size of bolts to hold the cover.

Section 8.7

8.45 For a case of plane *stress*, derive from Eqs. (8.20) and (8.21) equations for the radial and tangential stresses in terms of u and du/dr.

8.46 For a case of plane *strain*, derive from Eqs. (8.20) equations for the radial and tangential strains in terms of σ_r and σ_t. *Ans.* Eqs. (8.55).

Section 8.8

8.47 The differential equations in u, Eqs. (8.23) and (8.24), were obtained by eliminating the variables σ_r and σ_t. A corresponding equation may be obtained in σ_r or σ_t. (a) Take $l = 1$ and $T = 0$ for simplification, adjust the equilibrium equation, Eq. (8.18), to include a term with $(r\sigma_r)$, and derive the differential equation in terms of $(r\sigma_r)$. *Ans.* $r^2(r\sigma_r)'' + r(r\sigma_r)' - (r\sigma_r) + (3 + \nu)\rho\omega^2 r^3 = 0$, where $(r\sigma_r)''$ and $(r\sigma_r)'$ are, respectively, the second and first derivatives of $(r\sigma_r)$ with respect to r. (b) Solve this equation without the last term by substituting $(r\sigma_r) = r^n$, solving for n, then finding a particular solution to account for the last term. Compare with σ_r in Eqs. (8.26).

8.48 For a rotating disk of hyperbolic profile, as defined after Eq. (8.23), (a) derive a general solution for u, corresponding to Eq. (8.25) for a uniform disk. *Ans.* $u = C_1 r^{n_1} + C_2 r^{n_2} - Nr^3/[8 - (3 + \nu)k]$ where $n_1 = (k/2) + \sqrt{(k/2)^2 + \nu k + 1}$, $n_2 = (k/2) - \sqrt{(k/2)^2 + \nu k + 1}$. (b) Write equations for σ_r and σ_t in terms of the constants of integration.

8.49 Refer to Prob. 8.40. (a) Derive a differential equation in one variable u for the sphere. Compare with Eq. (8.24). (b) Solve for u by the substitution $u = r^n$ and compare the result with Eq. (8.25). *Ans.* $u = C_1 r + C_2/r^2$.

8.50 Continue Prob. 8.49, obtaining σ_r and σ_t for a sphere with internal pressure p_i. Compare with Eqs. (8.31).

8.51 For a uniform flat disk with a hole, rotating at speed ω, evaluate the constants of integration, and check the equations for σ_r, σ_t, and u of Table 8.1.

8.52 Same requirements as Prob. 8.51 but for a flat disk with no hole.

8.53 A plastic ring of modulus E, inner radius a, and outer radius b is pressed onto a steel shaft of much greater rigidity. Before pressing, the inside diameter of the ring was smaller than the shaft diameter by an amount Δ. Write the boundary conditions and determine the constants of integration and the maximum tangential stress.

8.54 A plastic ring of modulus E just fits over a steel shaft of radius a and much greater rigidity. Subsequently, a uniform pressure is applied to the periphery of the ring at radius b. Write the boundary conditions and determine the constants of integration and an equation for the displacement of the periphery.

8.55 For a thin flat disk without boundary or body forces, evaluate the constants of integration and write equations for σ_r, σ_t, and u when the disk is given a temperature distribution $T = T(r)$.

Section 8.9

8.56 In Section 8.1 it was indicated that if the wall thickness is less than 5% of the diameter, a vessel is often treated as a shell. What is the percentage error in the calculation of the thickness of a cylinder at this thickness ratio?

8.57 (a) The ASME Code, Section VIII, in its rules for unfired vessels for internal pressure, has specified that when $t > R_i/2$, the wall thicknesses be calculated by the equation $t = (Z^{1/2} - 1)R_i$, where $Z = (\sigma e + p)/(\sigma e - p)$, R_i is the inside radius, e is the longitudinal-joint efficiency, and σ is the allowable normal stress. Show that, except for joint efficiency, this equation is the same as the equation for maximum tangential stress in Table 8.1. (b) It has been further specified that when $t \le R_i/2$, the thickness be calculated by $t = pR_i/(\sigma e - 0.6p)$. Compare this equation with the thin-cylinder equation. What is the effect of the modification ($0.6p$) as pressure increases? Used at its upper limit, what value does it give for stress? Compare this result with the result from the equation of part (a). Does the modified thin-cylinder equation appear safe?

8.58 From Table 8.1 write equations for the tangential, radial, and shear stresses in a thick cylinder subjected simultaneously to internal pressure p_i and external pressure p_o. Rearrange to the extent possible in terms of the pressure difference ($p_i - p_o$). What stress is indicated when $p_i = p_o$? Is there much error in the usual practice of using gage pressure when only one pressure is specified?

8.59 (a) Prove the inequalities preceding Eqs. (8.30). Derive these external-pressure shear-stress equations from Table 8.1. Put them into nondimensional form and plot them for a cylinder with $b/a = 1.5$. Compare Fig. 8.12(b).

8.60 Put Eq. (8.29) into nondimensional form. Derive and plot an equation for the ratio p_i/τ_0 vs $K = b/a$, where p_i is the allowable internal pressure and τ_0 is the allowable shear stress. What is the maximum possible value of p_i/τ_0? If the efficiency of a cylinder is considered to be this ratio divided by its weight, what are the proportions of the most efficient cylinder, the least efficient cylinder? Suggest possible better ways of increasing the pressure capacity of a vessel.

8.61 Derive an equation for the allowable internal pressure p_i based on the Mises criterion (Section 5.6) and the cylinder radii (a) for open ends and (b) for closed ends. (c) Compare with a similar equation based on the maximum shear-stress theory of failure. *Ans.* (a) 0.577 $A/\sqrt{b^4 + a^4/3}$, (b) $0.577A/b^2$, vs (c) $0.500A/b^2$, where $A = (S_y/n)(b^2 - a^2)$.

8.62 Write an equation for shear stress in an internally pressurized spherical vessel, then put it and the radial and tangential stress equations into nondimensional form and plot the ratio of each stress to pressure for a sphere with $b/a = 1.5$.

8.63 The inside radius a of an internally pressurized vessel is usually determined by the required volume. Derive for a cylinder and for a sphere equations for the required external radius b in terms of a, the pressure p_i, and the allowable shear stress τ. *Ans.* Cylinder $b = a\,[\tau/(\tau - p_i)]^{1/2}$, sphere $b = a[4\tau/(4\tau - 3p_i)]^{1/3}$. (b) Show that if these values are used, the cylinder and spherical head of a vessel have the same maximum tangential stress. (c) Confirm the foregoing by applying them to the calculation of the dimensions of Example 8.3.

8.64 A cylindrical vessel with spherical ends has an inside diameter of 20 in and an overall length of 15 ft. It operates in a chemical-plant process with an internal pressure of 5000 psi. At the operating temperature the shear stress is limited to 8000 psi to prevent excessive creep. Refer to Prob. 8.63 and determine the outer diameters and the volume capacity. *Ans.* Cylinder 32.7 in, sphere 24.7 in, 30.65 ft³.

8.65 The cylinder of a slow-speed reciprocating-piston boiler-feedwater pump has an inside diameter of 250 mm and operates against a pressure of 8.25 MPa. Cast iron with a tensile strength of 175 MPa is used. For a factor of safety of 4.0 what should be the minimum wall thickness?

8.66 A self-propelled exploration vessel with the supporting structure consisting of a sphere with an inside diameter of 75 in is to operate at a maximum depth of 15 000 ft. It is proposed to use an aluminium alloy with a compressive yield strength of 25 000 psi. Seawater has a specific weight of about 64.0 lb/ft³. For a factor of safety of 1.25, what should be the wall thickness based on shear stress? Calculate the weight and state if the resulting sphere is feasible.

8.67 A naval gun with a bore of 100 mm must sustain a maximum pressure of 200 MPa. What must be the yield strength of the material if weight limits the outside diameter to 300 mm at the breech end, the maximum shear theory of failure is used, and the factor of safety is 1.3. *Ans.* τ_{max} = 225 MPa, S_y = 585 MPa.

8.68 Confirm the several statements in Section 8.9 relative to the stresses in a rotating, uniform (flat) disk.

8.69 The limiting speed of a *solid*-rotating disk may be given as a simple peripheral velocity corresponding to an allowable stress. (a) What are the limiting velocities in m/s and in ft/min for a steel forging ($\rho = 7.778$ Mg/m³, $\nu = 0.30$) with an allowable tensile stress of 120 MPa (17 400 psi)? *Ans.* 193.4 m/s. (b) What are the limiting velocities for an aluminum forging ($\rho = 2.80$ Mg/m³, $\nu = 0.33$) with an allowable tensile stress of 80 MPa (11 600 psi)?

8.70 (a) Make the equations of Table 8.1 for the stresses in both hollow and solid rotating disks nondimensional by writing them as functions of peripheral velocity $V = b\omega$ and of radius ratios, a/b, a/r, etc. (b) Write an equation and construct a chart to show the variation of the ratio $[\rho(3 + \nu)/8]\,(V^2/\sigma_0)$ from $a/b = 0$ (solid disk) to $a/b = 1.0$, where σ_0 is the allowable tensile stress and $\nu = 0.30$. Comment. Limiting speeds for some disks, such as grinding wheels, are specified without regard to hole diameter provided the hole diameter is less than one-third of the peripheral diameter. Is this reasonable?

8.71 A resin-bonded grinding wheel of 24-in diameter is used for heavy-stock removal at a surface speed of 12 500 ft/min. It is mounted on a 3-in-diameter spindle. Before use, it is spin tested at 1.5 times the operating speed, according to a safety code. Unit weight is 210 lb/ft³ and $\nu = 0.20$. Determine the maximum tensile stress.

8.72 A circular saw blade consists of a thin disk 600 mm in diameter on which are mounted 96 equally spaced tool-steel inserts, each weighing 0.75 N. It is placed over an 80-mm-diameter spindle that rotates at 3000 rpm. The center of mass of each insert is at the periphery of the disk. What disk thickness is required if the yield strength of the material is 250 MPa and the factor of safety is 2.5?

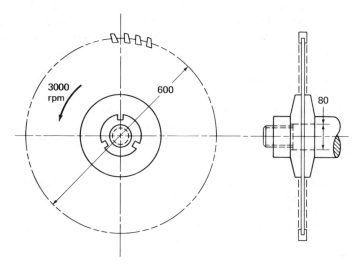

8.73 The rotor of a short electrical machine has an outer diameter of 32 in, a hole of 6-in diameter, but is slotted radially to a depth of 5 inches in order to hold windings. Consider the windings to be of the same density as the steel between the slots, none of which is able to carry tangential stresses, and assume the centrifugal forces from both to be uniformly distributed. If the speed is 3600 rpm, what are the unit external pull p_o and the maximum stresses at the hole of the rotor? *Ans.* $p_o = 8670$ psi, $\sigma_t = 29\ 200$ psi.

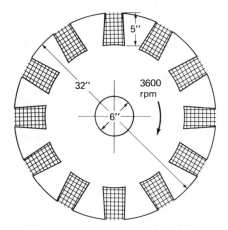

Section 8.10

8.74 Justify a handbook statement, based upon shear-stress considerations, that a cylindrical steel hub assembled on a solid-steel shaft with an interference of 0.001 in per inch of shaft diameter will produce a maximum equivalent stress of 30 000 psi. *Equivalent stress* is a tensile stress twice the shear stress, as in the tensile test.

8.75   A rule of thumb for the outside diameter of a cast-iron hub is to make it twice the shaft diameter. If the shaft is steel and solid, and the stiffening effect of spokes can be neglected, determine the maximum tensile stress in the hub corresponding to an interference of 0.001 in per inch of shaft diameter. $E_{C.I.} = 15 \times 10^6$, $E_{st} = 30 \times 10^6$, $\nu_{C.I.} = 0.25$, $\nu_{st} = 0.30$.

8.76   In the ANSI listings, the tightest fit suggested for high-grade cast-iron external members (on a shaft) is an FN2 medium-drive fit. For a nominal diameter of 4½ in, the tolerances give a hole range of 4.5000 to 4.5014 in and a shaft range of 4.5039 to 4.5030 in. If the hub outside diameter is 8 in, what is the tensile stress in the cast iron if the smallest hole is assembled over the largest shaft? If selective assembly is used and the smallest hole is matched with the smallest shaft, etc.? See Prob. 8.75 for constants.

8.77   A shrink fit is made of a steel sleeve with hole dimensions 3.0000 to 3.0012 in on a steel shaft with dimensions 3.0072 to 3.0060 in (ANSI—FN5). (a) What are the equivalent stresses if the sleeve with the smallest hole is assembled on the largest shaft? If selective assembly is used and the smallest sleeve is matched with the smallest shaft? See Prob. 8.74 for a quick formula. (b) What torque is developed with the selective assembly and a friction coefficient of 0.10?

8.78   A hardened-steel pinion with a yield strength of 690 MPa, a bore of 75 mm, an equivalent outside diameter of 150 mm, and a length of 100 mm must transmit a torque of 2100 N·m after being shrink fitted to a steel shaft. The minimum expected coefficient of friction is 0.10, but use a factor of safety of 3.0. (a) What interference is needed? To what temperature should the pinion be heated? Expansion coefficient α is 11.5×10^{-6} °C^{-1}. (b) Is the resulting stress reasonable?

8.79   SKF Industries has recommended for its hydraulically placed couplings (Fig. 8.18) an interference of 0.003 in per inch of shaft diameter. Assume that the thin tapered sleeve behaves as part of the shaft and neglect its thickness. A steel ring 16¾ in long having an outside diameter of 13 in with a gap of ¾ in between shafts has been used to couple 8-in marine-propeller steel shafts delivering 1200 hp at 300 rpm. (a) Determine the fit pressure corresponding to this interference. (b) What are the *design* coefficient of friction and the maximum shear and equivalent tensile stresses? *Ans.* $\mu = 0.0112$, $\tau = 45\ 200$ psi. (c) If the taper is 1:100, how far must the ring be pushed along during assembly?

8.80   To hold a sliding rod in any desired position, a tightly fitted sleeve surrounding the rod is proposed. The hold is released by introducing hydraulic fluid under pressure between the rod and sleeve, thus expanding the sleeve. Assume a given rod diameter of 1¾ in, an axial holding force of 5000 lb, a sleeve thickness of ⅛ in, and a hydraulic pressure of 3000 psi. The sleeve when expanded must clear the rod by 0.0005 in all around. Appropriate seals will prevent leakage. Assume a coefficient of friction of 0.10. Determine the interference and the length of the sleeve required.

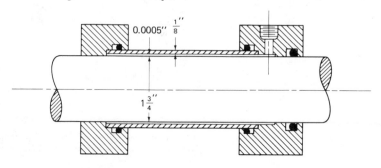

8.81   Binding between rings and rollers in a roller bearing may result from too heavy a fit or too wide a tolerance. Take the plain inner ring of a size 314 cylindrical-roller bearing, of width 35 mm and race diameter 90 mm, assembled on a shaft of 70 mm with an ABEC no. 1 fit (similar to Fig. 11.22(a), except the outer ring is the grooved one). Number 1 is the fit for general applications, but with the widest tolerances. It is possible to have a diametral interference of 0.03 mm. What is the enlargement of the raceway?

8.82   A steel bushing thinly lined with soft-bearing material is pressed into a cast-iron hub. Nominal diameters of the bushing are 3 in and $3^5/_{16}$ in. Nominal diameters of the hub are $3^5/_{16}$ in. and $5\frac{1}{2}$ in. The average diametral interference is 0.0065 in. See Prob. 8.75 for constants. (a) What is the decrease in inner diameter of the bushing? If this decreases the clearance too much for proper lubrication, how might this be anticipated and provided for? *Ans.* 0.00224 in. (b) What is the maximum tensile stress in the C.I.? Why is this reasonable although the unit interference is double that recommended for C.I. in Prob. 8.76?

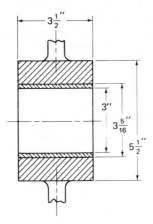

8.83   It has been found that locating shoulders near the ends of long-turbine shafts change relative position sufficiently during the shrinkage of turbine disks to require remachining after disk assembly. It is proposed to make allowance for this in the initially machined distance l between the shoulders at A and A'. The disks are located continuously along the lengths c only, which have outside and inside diameters D and d, respectively. Let $c = 2750$ mm, $D = 450$ mm, and $d = 250$ mm. Fit pressure is 69.0 MPa and $E = 207\ 000$ MPa. What should be the machined distance l if its final dimension must be 8230 mm? *Ans.* 8228.4 mm.

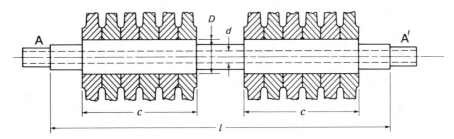

8.84   Same information as Prob. 8.83 except that $c = 88$ in, $D = 16$ in, $d = 8$ in, $p_f = 20\ 000$ psi, and the desired l is 260 in.

8.85   Because of necessary tolerances on hole and shaft dimensions, the tightness of a cylinder pressed onto a shaft is often in doubt. What measurement with a micrometer, made before and after assembly, can be used to indirectly determine the conditions at the fit? (a) Derive an equation for the pressure and torque capacity in terms of this measurement and the cylinder dimensions, for the general case of a hollow shaft. *Ans.* $p = EU_o(q^2 - n^2)/4n^2q$, where U_o is the increase in outside diameter. (b) Derive a general equation for this measurement in terms of the desired interference Δ and then simplify for a solid shaft. *Ans.* $q(n^2 - m^2)\Delta/n(q^2 - m^2)$.

8.86 From Table 8.1 derive Eq. (8.39) in detail.

8.87 A flat steel disk of 400 mm outside diameter, 75 mm hole diameter, and 60 mm width is used as a flywheel at 5000 rpm. A minimum interference of 0.001 mm per millimeter of hole diameter is desired at operating speed. What should be the interference at assembly? $E = 207$ GPa.

8.88 Tooth pitch-line velocities in steel gears seldom exceed 7500 ft/min. (a) Determine the percent loss at this speed in a fit pressure of 15 000 psi if the effective outside diameter can be taken as the pitch-line diameter, the gear is of uniform thickness, and the ratio of effective outside diameter to the diameter of the solid steel shaft is 3.0? (b) What is the maximum stress due to rotation only? (c) What do you conclude about these calculations for most small machine parts?

8.89 Interference fits may loosen with changes of temperature if different metals are used. Determine the interference required for assembly at 85°F if a die-cast aluminum-alloy pinion on a steel shaft in an airborne instrument is to have a torque capacity of 12 lb·in at −55°F. Equivalent pinion diameter is 0.50 in, its length 0.188 in, and shaft diameter 0.157 in, and $\mu = 0.15$. Expansion coefficients are 13.3×10^{-6} °F^{-1} for aluminum and 6.3×10^{-6} °F^{-1} for steel. For aluminum $E = 10.2 \times 10^6$ psi.

Section 8.11

8.90 (a) From Table 8.1 derive equations for the maximum net tangential and radial stresses for both the inner and outer cylinders of a compound cylinder with any proportions, where $q > n > m$. Dimensioned sketches for the separate conditions to be superimposed will be useful. *Ans.* (partial): $(\sigma_t)_m = -2p_f n^2/(n^2 - m^2) + p(q^2 + m^2)/(q^2 - m^2)$, $(\sigma_t)_n = p_f(q^2 + n^2)/(q^2 - n^2) + pm^2(q^2 + n^2)/n^2(q^2 - m^2)$. (b) From the results of (a), together with consideration of the axial stresses in both open and closed cylinders, derive the equations for the maximum shear stresses and compare them with Eqs. (8.41).

8.91 (a) From Eqs. (8.41) find that, for equal maximum shear stresses in both cylinders, $q/m = (n/m)(2 - p/\tau_0 - m^2/n^2)^{-1/2}$, then derive Eqs. (8.42) for optimum radii. *Suggestion:* Before optimizing, let $y = q/m$, $x = n/m$, and $k = p/\tau_0$. (b) Prove that on a chart of log q/m vs log n/m, lines for different ratios p/τ_0 will all have a minimum value along a straight line that has a slope of 2.0.

8.92 Derive Eqs. (8.43) in detail, starting with Eqs. (8.42).

8.93 Using the optimum, compound-cylinder dimensions and fit pressure of Section 8.11 and the answers given to Prob. 8.90(a), obtain $(\sigma_t)_m$, $(\sigma_t)_n$, and σ_z in terms of τ_0 and p and show that $(\sigma_t)_n > (\sigma_t)_m > \sigma_z$. Justify the application of the equations of Section 8.11 to closed-end cylinders as well as to open-end cylinders.

8.94 The gun of Prob. 8.67 is to be made as a composite cylinder. Use the same value of τ_{max} and determine the dimensions for least weight, plus the fit pressure and interference. What is the percentage relative weight, assuming that dimension changes along the entire length are in the same proportion.?

8.95 The vessel of Prob. 8.64 is to be redesigned as an optimum, two-layer-composite cylinder. Determine the plate thicknesses and the fit pressure and interference. From the answer given for the single cylinder, determine the percentage relative weight. *Ans.* Thicknesses 2.06, 2.49 in.

8.96 Rather than purchase plates of two thicknesses to give optimum proportions when only a single vessel is to be made (Prob. 8.95), if may be more economical to use and purchase plates of the same thickness, if the weight increase is small. Determine the required dimensions and compare. The equation in Prob. 8.91 may be used.

8.97 Dies used in the metal-extrusion process are subject to very high-internal pressures, and a compound cylinder giving a compressive prestressing to the inner surface is desirable. Also, a harder, better wearing, and replaceable inner cylinder may be used. Proposed dimensions for a die are shown, with the prestress established by forcing the inner cylinder a calculable axial distance along the slightly tapered fit. An initial tangential compressive stress of 100 000 psi is desired at the 1.000-in diameter. (a) Calculate the fit pressure and equivalent interference and the pressing distance h required to bring it about. *Ans.* $h = 0.123$ in. (b) The pressure on the inside surface of the die during the extrusion process will be 150 000 psi. What tensile strength and hardness are required of the outer cylinder for a factor of safety of 2.0 based on the ultimate tensile strength? (This problem and sketch were adapted from C.R. Bradlee, "Determine the Interference Fit and Resulting Stresses in the Design of a Cold Extrusion Die," *General Motors Eng. J.*, Vol. 2, No. 4, pp. 38–39 and No. 5, pp. 51–52, 1955.)

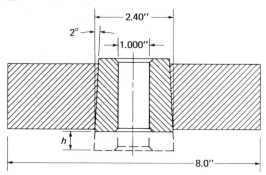

8.98 It may be shown that for a pressure vessel with inside radius r_i and outside radius r_o, divided into any number of layers N, the ratio of outer radius r_{n+1} to inner radius r_n of any one layer should be $r_{n+1}/r_n = K^{1/N}$, where $K = r_o/r_i$, if each layer is to have the same maximum shear stress. (a) Derive an equation for the maximum shear stress when there is a constant difference, $\Delta p = p_n - p_{n+1}$, in the fit pressures acting on all layers. *Ans.* $\tau = p_i K^{2/N}/N(K^{2/N} - 1)$. (b) Check the radius ratio and the shear-stress equations against those in the text for two layers.

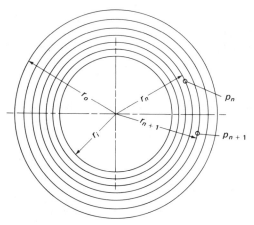

Section 8.12

8.99 (a) For the disk of Examples 8.4 and 8.7 determine the residual stresses at its hole and periphery after rotating it at the speed that causes it to become completely plastic. Assume no buckling when slowing down. (b) If the disk is subsequently operated at 21 000 rpm, what is the factor of safety based on $S_y = 480$ MPa? What is unsatisfactory? (c) What is the factor of safety if the residual stresses of Example 8.7(b) are used?

8.100 A flat disk has an outside diameter of 16 in and a hole of 6 in. A residual compressive stress of 15 000 psi is desired at the hole. Yield strength is 40 000 psi. At what speed should it be rotated? What is the net stress at the hole if the operational speed is 8400 rpm?

8.101 Derive an equation for the internal pressure p_i that will give a plastic region extending to any radius e in a thick cylinder. Check it at the extreme positions. *Ans.* $p_i = (S_y/2b^2)$ $[b^2(1 + 2 \ln e/a) - e^2]$.

8.102 A cylinder has internal and external diameters of 12 in and 16 in, respectively. After manufacture it will be overpressurized to give yielding to its midthickness. Yield strength is 40 000 psi. The cylinder will be used in service at an internal pressure of 7500 psi. The equation of Prob. 8.101 may be used. Determine the residual tangential stress and the net tangential stress in service at (a) the hole. *Ans.* -9600, 17 180 psi. (b) at the periphery and at the midthickness. What change would you suggest?

8.103 A hollow cylinder with open ends is subjected to an *external* pressure p_o. Assume perfect plasticity and that yielding proceeds according to the maximum shear theory, with the yield strength in compression equal to $-S_y$. (a) Prove the inequality that expresses the relative magnitude of the principal stresses and then determine an equation for σ_t in the plastic region. (b) Derive an equation for the corresponding radial stress σ_r. *Ans.* $\sigma_r = -S_y(1 - a/r)$.

8.104 (a) For a sphere with internal pressure and elastic conditions, Eqs. (8.31), determine the radii between which $\sigma_t > \sigma_r$ and $\sigma_r > \sigma_t$. (b) For the more representative, thinner-wall vessels, derive equations for σ_r and σ_t in the plastic region. The equation of Prob. 8.40 may be used. *Ans.* $\sigma_r = -p + 2S_y \ln r/a$, $\sigma_t = S_y + \sigma_r$. (c) Determine an equation for the limit-design factor and its value for the thickest sphere for which it holds. Compare its value for $b/a = 1.10$ with that of a cylinder, given after Eq. (8.53).

8.105 Derive equations for σ_t and σ_r in the plastic region of a rotating hyperbolic disk with a profile $l = al_a/r$, where the symbols are defined after Eq. (8.23).

Section 8.13

8.106 Transform the displacement equation, Eq. (8.25), for plane stress into a general form for the case of plane strain, to correspond with Eqs. (8.56) and (8.57).

8.107 From the equations derived for the stresses in a long cylinder with free ends, derive an equation for axial strain under rotation only. Does it satisfy the conditions for plane strain?

8.108 Extend Example 8.8 to determine (a) the location and magnitude of the maximum radial stress and to compare it with that of the disk of Example 8.4; (b) the change in cylinder lengths at the outer and inner surfaces, neglecting the end effects. Compare the two and explain.

8.109 Determine the unit external pull p_o and the safety in the long rotor of a turbogenerator at 1800 rpm that has a continuous cylindrical region of hole diameter 12 in and outer diameter 32 in, outside of which there is a slotted region of outer diameter 52 in, similar to Prob.

8.73. There are windings in the slots with the same density as the steel of the retaining fingers between the slots. Assume that fingers plus windings stress the inner region uniformly. Yield strength is 50 000 psi. *Ans.* $p_0 = 7265$ psi, factor of safety $= 2.22$.

8.110 (a) Obtain by transformation the stress and displacement equations for a *solid* rotating long cylinder. (b) Consider all principal stresses and write an equation for the maximum shear and for the limiting peripheral velocity based on an allowable normal stress $\sigma_0 = 2\tau_0$. Determine its value for steel and compare with that for a thin disk (Prob. 8.69). *Ans.* 219 m/s if $\sigma_0 = 120$ MPa.

Section 8.14

8.111 (a) A thin solid disk of uniform thickness is heated such that the variation in temperature is linear, given by $T = T_0 + kr$, where T_0 and k are constants. Determine equations for σ_r, σ_t, and u. What are the locations and magnitudes of their maximum values? (b) From the equations of (a) obtain the corresponding equations for a long cylinder. Compare their maximum values with those for the disk. *Ans.* $E\alpha kb/3(1 - \nu)$.

8.112 Determine a bounding-value formula for the maximum radial and tangential stresses developed in a nonrotating, solid, circular flat disk that has been heated rapidly to give a central "hot spot" of uniform temperature T_1 and radius n. The remainder of the disk stays "cold" with uniform temperature T_0. *Suggestion:* Consider use of an interference-fit equation. Also, simplify the equations for a very tiny spot.

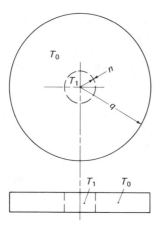

8.113 For the temperature distribution of Eq. (8.65) derive Eqs. (8.66) and (8.67) in detail.

8.114 (a) Derive equations for the location and magnitude of the maximum radial stress in a long cylinder under the temperature variations of Eq. (8.65). (b) Put them into nondimensional form and compare the results for $b/a = 2.0$ with $(\sigma_r)_{max}$ in Fig. 8.22.

8.115 Determine the thermal stress in a steel cylinder with internal and external diameters of 300 and 400 mm, respectively, and a temperature difference $T_a - T_b = 100°C$. E $= 207\ 000$ MPa and $\alpha = 11.5 \times 10^{-6}\ °C^{-1}$.

8.116 (a) Derive an equation for the axial strain from the equations derived for stresses with the temperature variation of Eq. (8.65). Does the resulting equation satisfy the condition for plane strain? (b) What is the increase in the length of a 40-ft steel cylinder for each 100°F of temperature difference $(T_a - T_b)$ if $b/a = 2.0$? Compare this increase with the increase if the heating were uniform for the average amount of 50°F, $\alpha = 6.3 \times 10^{-6}\ °F^{-1}$. *Ans.* 0.117 in vs 0.151 in.

9

Components in Flexure I

Beam and Shaft Shapes, Forces and Moments,
Bolted and Welded Connections,
Deflections by Moment Area and Graphical Methods,
Superposition, and
Statically Indeterminate Beams and Shafts

9.1 INTRODUCTION

Many, probably the majority, of stressed components in machines and structures are loaded in a manner that results in some flexing or bending. This usually gives a non-uniform distribution of stress, both along and across the member. The deflections and changes in initial shape may be small and go unnoticed, as in a machine-tool shaft and base, or these deflections and changes may be readily seen or felt, as in the cushioning action of a leaf spring, the vibration of a shaft, or the normal flexing of an airplane wing. The deflection may be a useful function of the component, as in a valve spring or a vibration isolator, or it may be highly undesirable as in a machine tool or turbine.

Principal dimensions must be based upon safe stresses suitable for the material and the condition of operation. Forces and moments can usually be determined first from statics, followed by calculations of stress and deflection. However, forces, moments, and deflections are interdependent when the condition of support is redundant; it is a statically indeterminate condition.

Table 9.1 lists some groups of mechanical components whose design is aided by the application of one or more of the flexure equations or methods of analysis. Their names generally come from their function or shape, but their loading or one of their loadings is that of a "beam" or "plate." This is not always apparent because of the many different methods of applying the loads and providing support through rigid, pivoting, or rotating connections. Also, the components have shapes suitable for multiple functions and shapes suitable for their processing and assembly. Some flexure loads are dynamic, the result of the component's motion. In pursuit of the theory and its application, the student who is unfamiliar with these components may have the opportunity to learn about their construction and functioning, for example, the camshaft of Fig. 9.1.

Some elementary theory and equations from first courses in statics and strength of materials are repeated here in condensed form as a review and as a basis for problems of application and selection. Some pointers and guidance are given for these applications to typical machine-component sections and tapers, to bolted and welded parts, and to the noncoplanar loading of shafts. For the flexure of straight but stepped beams the moment-of-area method is presented as well as the equally useful graphical method. The super-

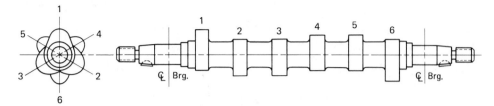

FIGURE 9.1. A six-lobed camshaft used for a fuel injection pump. A typical "beam" with moving loads.

TABLE 9.1. Some Components in Flexure

Rail, channel, bracket, rocker panel, pipeline, bolster.
Cylinder head, turbine casing, bulkhead, flange, planer table.
Lever, crank, connecting rod, torque arm, walking beam, sheave pin, hook, eyebolt.
Shaft, camshaft, axle, spindle, gear tooth, flywheel rim, turbine blade.
Flat spring, spring clip, retaining ring, piston ring, Belleville spring washer, wave washer, wire rope.

position of simple beams is applied to overhung shafts, brackets, and the double canti-lever, then to statically indeterminate cases such as simple frames, beams with reinforcement, beams in parallel (gear teeth), and a three-bearing shaft. Application of the moment area and graphical methods to statically indeterminate cases is also indicated.

The review of beam theory will also be useful as an introduction to Chapter 10. In Chapter 10 strain-energy methods are applied to determine the deflection of rings, tubes, and loops. Equations for thick curved beams are developed and applied to hooks and offset links. The theory for beams with distributed elastic support is developed and applied to rails and to secondary stresses in pressure vessels and pipes. Plate theory is developed for general application, then principally applied to circular plates as in pistons, housings, tank bottoms, and Belleville springs. A table is provided for the superposition of simple plates to more complicated loadings and supports, including statically indeterminate cases.

9.2 STRESSES IN STRAIGHT BEAMS

In the beam of Fig. 9.2 each element of area $dA = b\,dy$ contributes a force $\sigma\,dA$ to the total normal force on the section and a moment $(\sigma\,dA)y$ to the total internal resisting moment. In the absence of an external axial force on the beam, the net normal force over the section must be zero, i.e.,

$$\int_A \sigma\,dA = 0 \tag{a}$$

The applied external moment M and the internal moment from the stresses must be in equilibrium, i.e.,

$$M + \int_A (\sigma\,dA)y = 0 \tag{b}$$

It is observed that plane sections YZ remain plane during bending and rotate about a fixed axis OZ, the neutral axis. Therefore, displacements and strains are proportional to the

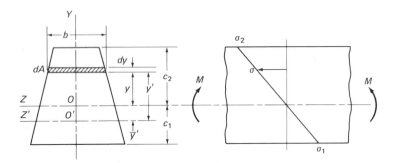

FIGURE 9.2. Distribution of strain and elastic stress in a beam.

distance y from this axis, and when the action is elastic, so are the stresses. Thus in terms of the stress σ_2 in an extreme fiber,

$$\sigma = \frac{y}{c_2} \sigma_2 \tag{c}$$

and its substitution into the equation of equilibrium of forces, Eq. (a), gives the equation

$$\int_A \sigma \, dA = \frac{\sigma_2}{c_2} \int_A y \, dA = 0 \quad \text{or} \quad \int_A y \, dA = 0 \tag{d}$$

The neutral axis may be located by the distance $\bar{y}'$ from any conveniently located baseline or second axis $O'Z'$. Then $y' = \bar{y}' + y$. This equation integrated on both sides relative to dA, with substitution from Eq. (d), gives

$$\int_A y' \, dA = \int_A (\bar{y}' + y) \, dA = \bar{y}' \int_A dA + \int_A y \, dA = \bar{y}'A$$

whence

$$\bar{y}' = \frac{1}{A} \int_A y' \, dA = \frac{1}{A} \int_{-c_1}^{c_2} y'b \, dy = \frac{1}{A} \sum_{i=1}^{i=n} \bar{y}_i' A_i \tag{9.1}$$

where $y = -c_1$ and $y = c_2$ are the locations of the extreme fibers. This is also the equation for location of the centroidal axis of the area. Thus the neutral axis and the centroidal axis are identical in a straight beam. The integral term of Eq. (9.1) may be used when b is constant or can be expressed as a function of y. The summation term may be used with composite sections, where $\bar{y}_i'$ is the distance from $O'Z'$ to the known centroidal axis of each area A_i.

From Eqs. (b) and (c),

$$M = -\int \sigma y \, dA = -\frac{\sigma_2}{c_2} \int y^2 \, dA \tag{e}$$

The integral is the moment of inertia of area A about its neutral and centroidal axis

$$I = \int_A y^2 \, dA = \int_{-c_1}^{c_2} y^2 b \, dy \tag{9.2}$$

For composite sections, a known moment of inertia of area I_i'' about the centroidal axis of area A_i may be transferred to a single convenient axis $O'Z'$ by the parallel-axis equation. Thus for n areas,

$$I' = \sum_{i=1}^{i=n} I'_i = \sum_{i=1}^{i=n} [I''_i + A_i(\bar{y}'_i)^2]$$ (9.3)

where $\bar{y}'_i$ is the distance from $O'Z'$ to the centroidal axis of A_i.

Then the moment of inertia I of the composite area A about its centroidal axis is

$$I = I' - A(\bar{y}')^2$$ (9.4)

where $\bar{y}'$ is the distance from $O'Z'$ to OZ, given by Eq. (9.1). (See Example 9.2 for an application of these equations.)

By use of Eq. (9.2), Eq. (e) may now be rewritten and solved for the stress σ_2. A similar equation may be written for the stress σ_1; thus

$$\sigma_2 = -\frac{c_2}{I}M \quad \text{and} \quad \sigma_1 = +\frac{c_1}{I}M$$ (f)

The positive sign indicates a positive stress and the negative sign indicates a compressive stress. One may be more significant than the other as when $|c_1| > |c_2|$, or when the material is weaker in tension than in compression, or when there is a higher stress concentration factor K at fiber 1. Subscripts and signs are customarily dropped, tension or compression being found by inspection, so that

$$\sigma = K\frac{c}{I}M = K\frac{M}{I/c}$$ (9.5)

The factor K is added here so that it will not be forgotten. ($K = 1.0$ for no stress concentration at the section.) It is understood that c is the distance to the extreme fiber in question. When the neutral axis is a central axis, $c_1 = c_2$, and I/c is named the section modulus Z.

Most beams are loaded by transverse forces, and these are resisted by internal shear forces V and shear stresses on transverse planes (Fig. 9.3(a) and (c)). Bending stresses act normal to these planes. Fortunately, there is little need to combine the two stresses, since shear stresses are zero at the extreme fibers and maximum at or near the neutral axis (Fig. 9.3(b)). Moreover, the shear stresses can usually be ignored in metal beams of uniform section that are wide at the neutral axis, such as those with solid circular, elliptical, or rectangular sections. Then, the numerical value of the maximum shear stress is much smaller than that of the maximum bending stress unless the affected length is very short, of the same order as the beam's depth. The shearing stress is more significant at a step in diameter at the end of a short supporting length, as in Fig. 9.3(d), which is discussed in Example 9.1. In long, rectangular beams of wood, a material that is weak in shear between longitudinal fibers, the longitudinal shear stress which always accompanies and is equal to the transverse shear stress, even though small, may cause cracks to form along or near the neutral plane. Shear stresses should also be considered in any beams made by bonding together two or more strips.

Also, in long metal beams with thin webs and wide flanges, such as H-beams and built-up girders, and perhaps in some hollow box sections, the shear stress may be high enough for consideration because of the relatively small area at the neutral axis.

Shear stresses accompanying bending may be expressed as a product of a shape factor α and the average stress, thus

$$\tau = \alpha\frac{V}{A}$$ (9.6)

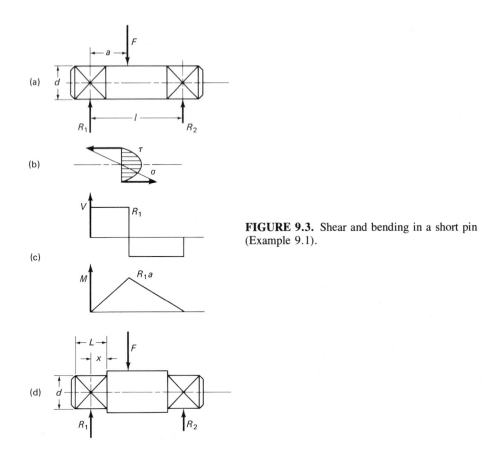

FIGURE 9.3. Shear and bending in a short pin (Example 9.1).

For example $\alpha = 3/2$ for a solid rectangular section, and $\alpha = 4/3$ for a circular one.

The allowable shear stress may be taken as one-half the allowable normal stress according to the maximum shear-stress theory of failure (Section 5.5).

Example 9.1

A pin of uniform diameter is loaded as in Fig. 9.3(a). Determine the distance a beyond which it is not necessary to check the shear stress.

Solution

Load, stresses, and shear and moment diagrams are shown in Figs. 9.3(a) through (c). It is unnecessary to check maximum shear stress when it is less than one-half the bending stress. The maximum shear stress is $(4/3)R_1/A = 16R_1/3\pi d^2$, Eq. (9.6), and since $I/c = \pi d^3/32$, the maximum bending stress is $M/(I/c) = 32R_1a/\pi d^3$, Eq. (9.5). Then

$$\tau \le \frac{\sigma}{2}, \quad \frac{16R_1}{3\pi d^2} \le \frac{1}{2}\frac{32R_1a}{\pi d^3}, \quad \text{whence} \quad a \ge \frac{d}{3} \tag{g}$$

Thus for a pin with central loading or $a = l/2$, the shear stress may be neglected when the span l between concentrated reactions is $\ge 2d/3$. This is a short beam, indeed!   ////

For the shouldered pin of Fig. 9.3(d), the moment at the shoulder is R_1x. The distance x replaces a in the criterion Eq. (g) in Example 9.1, so that if the reaction is considered centered along the supporting length L, then $x = L/2$, and $L \ge 2d/3$. If the reaction is considered concentrated at $x = L/3$, as in Fig. 9.11, then $L \geq d$. Thus a check on shear at the shoulder may be needed when L/d is less than 1.0.

Pins which are partially confined, such as the one in Fig. 9.11, are usually checked for bending stresses, as though bending were unconfined. They are then checked for shear stresses as though the pin were rigidly gripped by the adjoining parts and the only deformation is by shear through the pin. Since there is no bending under such conditions, the average stress is calculated. Thus

$$\tau_{avg} = \frac{V}{A} \tag{9.7}$$

where V is the shear force at each section of area A. In Fig. 9.11, it is "double shear," occurring simultaneously at the two sections of the pin where it enters the central rod, and then $V = F/2$ and $\tau_{avg} = V/2A$. The "confinement" effect is increased by shoulders, as in Figs. 9.6(a) and 9.10(a), thus making this calculation more significant.

Whereas transverse shear stress and bending stress are not additive in their effects, the deflections from shear and bending are overall effects that are additive and significant in short beams. For example, the flexibility of gear teeth, which determines the sharing of the load between two pairs in contact, is significantly affected by shear deflections, as well as by bending deflections and local deformation at the contacting surfaces (see Section 9.11 preceding Example 9.12).

The foregoing equations and comments may be applied with sufficient accuracy to stresses in curved beams provided the beams are "thin," so that the lengths of the inner and outer fibers differ only in the order of 10%. If the cause of the bending is an offset load P parallel with the bending stresses, it may be necessary to add an uniform stress P/A to the bending stress Mc/I. For "thick" curved beams, such as the hook of Fig. 9.6(a), see Section 10.4.

9.3 TYPICAL BEAM SHAPES: SECTION MODULI

The circular cross section is found most frequently in critically stressed parts such as shafts because it is the most uniformly stressed if torsion is also present, its continuous surface is readily and accurately formed by various processes, and holes may be readily and accurately bored in the parts to be fitted over shafts. In shafts formed hollow by extrusion, piercing and rolling, or by boring out the low-stressed interior, the same stress and rigidity are obtained by small increases of outside diameter.

Solid and hollow circular and rectangular sections of equal strength are shown in Fig. 9.4, together with their area (weight) and diameter or height ratios. There is a lower limit to the thinness of the wall, set by local buckling in bending or torsion.

Many special sections may be used in mechanical equipment, depending upon the application and the method of manufacture. The lever or rocker arm shown in Fig. 9.5(a) will have a hollow rectangular section at its pivot pin (Section A-A), and if cast, probably an elliptical section along its length. A connecting rod (Fig. 9.5(b)) will have an I-section with well-rounded corners, because it gives the required strength in two directions with minimum weight, and is readily forged. Structured shapes[1] such as I-beams, H-beams, and channels (Fig. 9.5(c)), the products of a rolling mill, are readily available, weldable, and relatively inexpensive, so are used for machine bases and frames. Steel tubing is common in aircraft frames and engine supports. In Fig. 9.5(d) two channels or a rolled

---

[1] Section properties for numerous standard sizes may be found in handbooks.

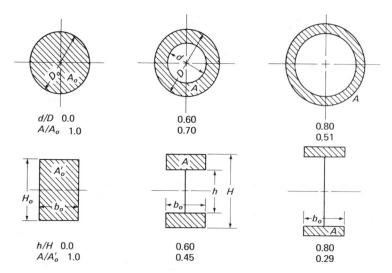

d/D	0.0	0.60	0.80
A/A_o	1.0	0.70	0.51

h/H	0.0	0.60	0.80
A/A_o'	1.0	0.45	0.29

FIGURE 9.4. Some areas of equal section modulus and bending strength. For the solid areas, rectangle height H_0 = circle diameter D_0, and the area ratio $A_0'/A_0 = 0.75$. For the hollow areas, an area ratio relates the area (weight) to the corresponding solid area. The hollow rectangles approximate I-beams.

sheet, welded along the neutral axis of bending, form "box-sections" of minimum weight.

The shape may be governed by the functional and retention requirements. The hook for a crane hoist may be supported by a trunnion or crosspiece between side straps of a traveling sheave block (Fig. 9.6(a)), with pin ends for vertical-plane rocking of the hook and with a central recess for retention of a thrust-type ball bearing to allow rotation of the hook-suspended load. Thus the beam cross section at the location of maximum bending moment consists of two L-shaped areas (Fig. 9.6(b)), with the narrow portion coming from either a recess down into or a ring raised above the top surface of the trunnion. This is similar to the rib or T-section at Fig. 9.6(c) with the L's back to back. The hook itself, a thick curved beam, will have a modified trapezoidal section (Fig. 9.6(d)) to make more uniform the two outer fiber stresses (see Section 10.4), as well as to minimize the bend in the ring or rope slung over it. The section of a rocker panel of a passenger car, giving stiffness to the body as well as acting as a doorsill, may be formed by bending flat sheet steel to the shape of Fig. 9.6(e).

Many of the foregoing are "composite sections." Their moment of inertia of area may be determined by adding or subtracting the values for several well-known geometric shapes (see Eqs. (9.1), (9.3), (9.4), and Example 9.2). One is warned not to add or subtract *section moduli*. Each section modulus has been obtained from its own extreme fiber distance c, whereas there is only one c for the composite section. The composite I/c is the composite I divided by that one extreme fiber distance.[2]

A retention ring raised above the surface of the trunnion (Fig. 9.6(b)) or a rib on a flat wall adds to the moment of inertia I and hence to the stiffness of the trunnion. However, there is a height e beyond which strength, proportional to I/c, begins to decrease, i.e., extreme fiber distance c increases more rapidly than I. The thickness t

---

[2] Thus formulas for the sections of Fig. 9.4 are

$$Z = \frac{\pi}{64} \frac{D^4 - d^4}{D/2} = \frac{\pi D^3}{32}\left[1 - \left(\frac{d}{D}\right)^4\right] \quad \text{and} \quad Z = \frac{b}{12} \frac{H^3 - h^3}{H/2} = \frac{bH^2}{6}\left[1 - \left(\frac{h}{H}\right)^3\right].$$

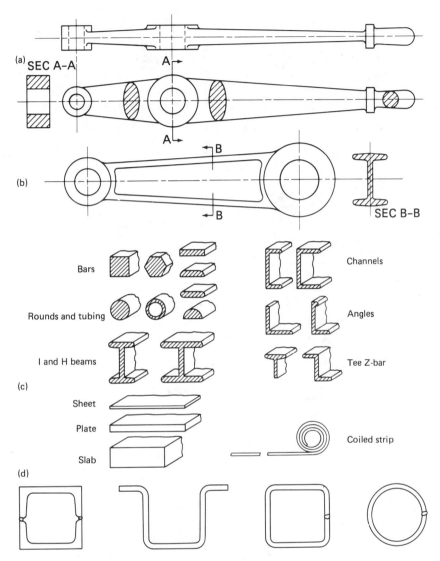

FIGURE 9.5. Some beam shapes determined by the method of processing. (a) Cast lever of elliptical section. (b) Forged connecting rod of H section. (c) Standard rolled shapes from steel mills. (d) Welded and cold formed shapes.

of the ring is important. If it covers the entire width b of the section, strength is increased indefinitely as e increases, but as it approaches zero, any increase of e weakens the section.

Example 9.2

Design the trunnion of Fig. 9.6(a) and (b). The hook is rated for a load of 45 kN, and the trunnion material is forged from AISI 1040, then normalized, resulting in a yield stress of 350 MPa. Span $s = 125$ mm, pin lengths $L = 19$ mm. A ball-thrust bearing allows rotation of the hook for positioning the load. Its lower race and a third or bottom ring have matching spherical surfaces to allow some degree of self-alignment of the hook with the bearing load. A bearing size selected for this load and service has internationally standardized dimensions, with 69 mm for the outside diameter of the bottom ring and 40 mm for the inside diameter of the upper ring.

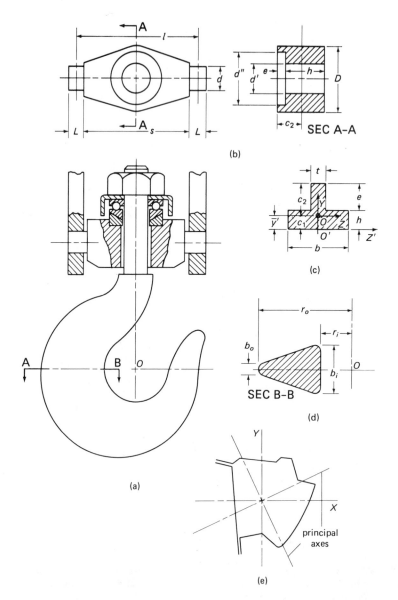

FIGURE 9.6. Some functional beam shapes. (a) Hook and swivel in lower portion of a traveling block. (b) Top and section views of the trunnion of (a), showing recess for ball bearing (Example 9.2). (c) A ribbed or T section, equivalent to the recessed section for the purpose of analysis. (d) Modified trapezoidal section of the hook of (a). (e) Section of a rocker panel. (From G. W. Ropers, "Analyze the Rocker Panel Section of an Automobile Body to Determine if Deflection is Within Allowable Limits," *General Motors Engineering J.*, Vol. 11, No. 2, pp. 50–53, No. 3, pp. 62–65, 1964. Reproduced with permission of General Motors Corporation.)

Solution

(a) *Choice of design stress:* In small-hoist service we can expect the rated or maximum load to be applied infrequently and without reversal. Values of stress concentration are not known for this shape, and they may not be significant. Hence, we shall design for rated load, with yield stress as the base, as though service were static, and with the reasonably higher factor of safety $n = 3.0$, because of the danger in case of a failure and because of our ignorance of the normal load and overload conditions. In all probability, any failure would be a yielding only, perhaps noticed by a binding at the bearing pins.

The design or allowable stresses are then, in bending,

$S_y/n = 350/3 = 116.7$ MPa

and for the average shear stress,

$\tau = (1/2)(116.7) = 58.3$ MPa

(b) *Pin diameters:* Equation (9.7)

$\tau = R/A = R/(\pi d^2/4)$

where reaction $R = F/2$, and $F = 45\ 000$ N, whence

$$d = \sqrt{\frac{2F}{\pi \tau}} = \sqrt{\frac{2(45\ 000)}{\pi(58.3)}} = 22.2 \text{ mm}$$

(c) *Properties of section:* The solution will be more general if we equate Section A-A of Fig. 9.6(b) to the T-section of Fig. 9.6(c), where $b = D - d'$ and $t = D - d''$. Its area is $A = bh + te$. The distance $\bar{y}'$ of the centroidal axis from the baseline $O'Z'$ is found from Eq. (9.1),

$$A\bar{y}' = \frac{h}{2}(bh) + \left(h + \frac{e}{2}\right)(te)$$

$$\bar{y} = \frac{bh^2 + 2teh + te^2}{2(bh + te)} \tag{a}$$

The distances from the neutral axis to the extreme fibers are $c_1 = \bar{y}$ and

$$c_2 = h + e - \bar{y} = \frac{bh^2 + 2beh + te^2}{2(bh + te)}$$

Since in the numerators of c_2 and c_1, $2\ beh > 2teh$, $c_2 > c_1$, and since $\sigma = Mc/I$, therefore $\sigma_2 > \sigma_1$. About the baseline $O'Z'$, Eq. (9.3),

$$I' = \frac{1}{3}bh^3 + \frac{1}{12}te^3 + (te)\left(h + \frac{e}{2}\right)^2 = \frac{1}{3}[bh^3 + te^3 + 3teh(h + e)]$$

About the centroidal axis OZ the moment of inertia of area is found, Eq. (9.4), as

$$I = I' - A\bar{y}'^2 = \frac{b^2h^4 + 4bteh^3 + 6bte^2h^2 + 4bte^3h + t^2e^4}{12(bh + te)} \tag{9.8}$$

The ratio I/c_2 becomes

$$\frac{I}{c_2} = \frac{b^2h^4 + 4bteh^3 + 6bte^2h^2 + 4bte^3h + t^2e^4}{6(bh^2 + 2beh + te^2)} \tag{9.9}$$

(d) *Dimensions of section:* A ring on the trunnion of thickness 4 mm and height $e = 5$ mm should suffice for retention of the bottom bearing ring and not interfere with operation of the self-aligning feature. With $d'' = 69$ mm in Fig. 9.6(b), $D = 69 + 2(4) = 77$ mm. Also for the self-alignment, a clearance of 5 mm is needed around the 40-mm shank of the hook, so that the hole diameter is $d' = 45$ mm. Then for Fig. 9.6(c), $b = D - d' = 77 - 45 = 32$ mm. The effective beam length l, taken to the center of the bearings, is $s + 2(L/2) = 125 + 19 = 144$ mm. With perfectly rigid bearing rings and trunnion, the load would be applied uniformly along a circular area, giving a concave, nonuniform pressure distribution along the central length of the beam. Deflection adds to this nonuniformity. Hence a customary assumption of uniform distribution, in this case 45 000 N along the 69-mm diameter of the ring, will be conservative or "on the safe side." The loading and bending moment diagrams are like those of Fig. 9.10(g), using the dotted line. The peak moment is safely approximated by considering half the distributed load F concentrated at a distance $d''/4$ to the left of the center of the beam. Then, together with the bearing reactions R,

$$M_{max} = R\left(\frac{l}{2}\right) - \frac{F}{2}\left(\frac{d''}{4}\right) = \frac{45\ 000}{2}\left(\frac{144}{2}\right) - \frac{45\ 000}{2}\left(\frac{69}{4}\right) = 1232\ \text{N}\cdot\text{m}.$$

Hence from $\sigma = Mc/I$, the I/c required of the section, is

$$\frac{I}{c} = \frac{M}{\sigma} = \frac{1232\ \text{N}\cdot\text{m}}{116.7 \times 10^6\ \text{N/m}^2} = 10.56 \times 10^{-6}\ \text{m}^3 = 10\ 560\ \text{mm}^3$$

This can be set equal to I/c_2 of Eq. (9.9), giving an equation in h^4. As a first approximation we may neglect the effect of the small raised ring, setting $t = e = 0$. This gives, as expected, $I/c_2 = bh^2/6$, and with $b = 32$ mm and $I/c_2 = 10\ 560$ mm³, $h = 44.5$ mm, a value lower than required. We notice that with t and e having values about one-tenth that of h, the last two terms of the numerator and the last term of the denominator of Eq. (9.9) will be numerically quite small relative to the others and may be neglected. This allows factoring to simplify the equation to

$$\frac{I}{c_2} = \frac{h(bh^2 + 4teh + 6te^2)}{6(h + 2e)} \tag{b}$$

With $b = 32$, $t = 4$, and $e = 5$ mm,

$$\frac{I}{c_2} = \frac{h(32h^2 + 80h + 600)}{6(h + 10)}$$

With I/c_2 set equal to the required value of 10 560 mm³ this is a cubic equation in h. Its solution, by trial or otherwise, indicates a value $h = 48$ mm for the height of the main part of the cross section.   ////

A circular shaft with a single noncircular portion, such as a milled flat, may have vibrations induced at speeds equal to half the normal critical speed. The variable flexibility that causes this can be made uniform in a keyed shaft (Fig. 9.7(a)) by using three keyways or a spline (Figs. 9.7(b) and (c)). A shaft that requires two sets of irregularities,

FIGURE 9.7. Some shaft sections. (a) Single keyway. (b) Three keyways. (c) Splines. (d) Slotted for coils of wire.

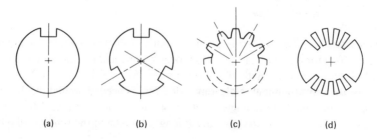

(a)   (b)   (c)   (d)

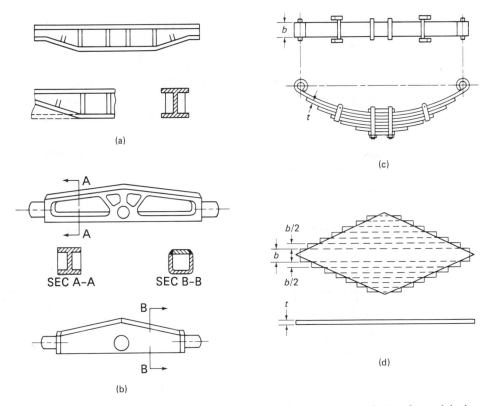

FIGURE 9.8. Some tapered beams of approximately uniform strength. (a) Bolster for truck body. (b) Equalizers, as a casting with section *A-A* or as a weldment with section *B-B*. (c) Leaf spring. (d) Dotted lines at half-widths *b*/2 of the spring of (c), showing how it approximates a beam of triangular shape.

such as the slots for windings in a two-pole generator (Fig. 9.7(d)), will run rough when going through half its critical speed.[3]

The section modulus is seldom made constant along the entire length of a beam-loaded member of a machine. In the lever and connecting rod of Figs. 9.5(a) and (b), the elliptical and the I-sections taper with the bending moment and then blend into the hollow rectangular sections. A heavy H-beam, such as the bolster beam for a truck (Fig. 9.8(a)), may have part of its web cut out with a torch and then the flange may be bent over to meet the web and welded to it. This saves weight where the moment and stress are lower and increases the payload. The centrally pivoted, load-equalizing lever of Fig. 9.8(b) similarly tapers off toward the locations of low moment, whether built as a casting with an I-section *A-A* or as a weldment with a box section *B-B*. Leaf springs are tapered in width of depth or are clustered in leaves of different lengths, as for vehicle body suspension (Fig. 9.8(c)), the leaves approximating a beam of constant stress (Fig. 9.8(d)) when laid side by side. This then gives maximum flexibility for a given weight. Shafts are generally stepped in a series of cylindrical portions ending in shoulders, against which added parts such as gears and ball bearings are accurately located and clamped (Figs. 9.10 and 9.12). The diameters usually increase toward the location of maximum moment.

---

[3] J. P. Den Hartog, *Mechanical Vibrations*, 4th ed., New York: McGraw-Hill, pp. 338–339, 1956.

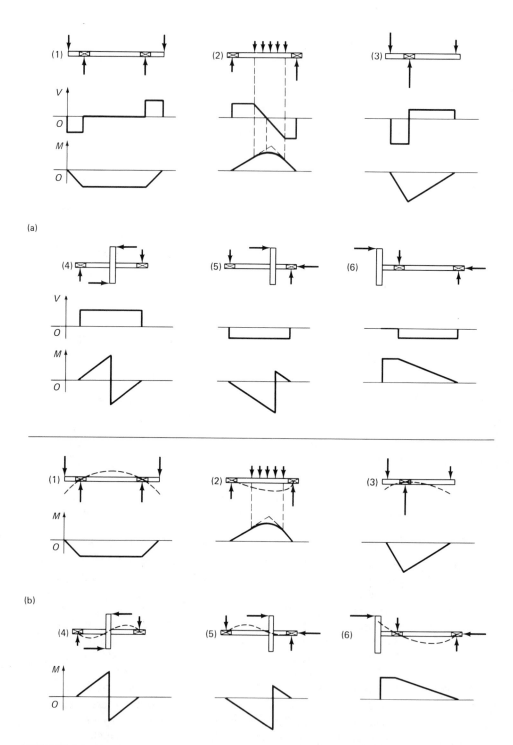

FIGURE 9.9. Bending moment diagrams for common shaft loadings. Their construction: (a) Top half of page, with the aid of shear diagrams. (b) Bottom half of page, with the aid of force and deflection diagrams. Bearing supports are indicated by an ×-mark.

9.4 DETERMINATE FORCES AND MOMENTS

Forces and moments are said to be statically determinate when the reaction forces may be found by the equilibrium relationships $\Sigma F = 0$ and $\Sigma M = 0$ only, appropriately applied as many times as there are unknowns. Any inertia forces may be treated as reversed effective forces and included in the equations above. When the elastic conditions of slope and deflection must also be used, the condition is said to be statically indeterminate (see Section 9.11). This latter condition occurs, for example, in a shaft that has more than two bearings, and in a beam when there is a built-in condition plus an outboard support.

Moment diagrams may be constructed with the aid of shear diagrams or with the aid of force and deflection relationships. If $q = q(x)$ represents the force per unit length where the load is distributed along the beam, then the shear at any section x is $V = \int_0^x q\, dx + C$. The constant must be evaluated from the reactions and other concentrated loads that usually occur along a beam. A second integration gives the moment distribution, $M = \int_0^x V\, dx$. Where shear V is constant, $M = Vx + C$ and the slope of the moment diagram is constant, as for portions of all the beams of Fig. 9.9(a). Where there is a discontinuity in the shear diagram, there is a discontinuity in the slope of the moment diagram, as in (1) and (3) of Fig. 9.9(a). Where a moment is applied, as in (4) and (5) of Fig. 9.9(a) there is no change in the shear and no change in the slope of the moment diagram, although the moment diagram changes in value by the amount of the moment.

Usually the moment diagram may be constructed directly from the free-body force diagram, as in Fig. 9.9(b), which repeats the loadings of Fig. 9.9(a). The rule is that the bending moment at any section equals the moments of the external forces about that section plus the external moments, all taken on one side or the other of the section. For this purpose, distributed forces may be concentrated at their centroid. It is conventional to consider an external moment to the left of a section to be positive when its sense is clockwise. The general shape of the deflection curve (dotted lines of Fig. 9.9(b)) can be sketched by inspection and used as an aid. Where the deflection curve is concave upward, the moments are positive[4] A position of inflection in the deflection curve is a position of change in sign in the moment diagram, as in (4) of Fig. 9.9(b).

Because section moduli and stress-concentration factors K generally change along a shaft, the critically stressed sections usually occur at locations other than that of maximum bending moment. The entire moment diagram needs to be constructed, as in Figs. 9.10, 9.16, and 9.23, and stresses at several fillets or other locations of stress concentration need to be calculated and compared. An increase in moment may be matched by an increase in shaft diameter, but the fillet radii are often determined by other considerations, such as available space adjacent to a mounted member.

The stress intensity is greatest at a location about one-third of the way along the arc of the fillet from the smaller diameter end, and the published values of stress-concentration factors are for this location. The moment at this location may be measured on a diagram that has been drawn to scale. Otherwise, the moment calculated at the face of the shoulder, where a dimension for location is given, will usually be accurate enough.

Example 9.3

Each sheave of a large traveling block, (Fig. 9.10(a)), carries a load of 20 000 lb. Construct the moment diagrams assuming the loads to be concentrated and then assuming

---

[4] It is sometimes said that the moment is positive where the deflection curve "will hold water."

374

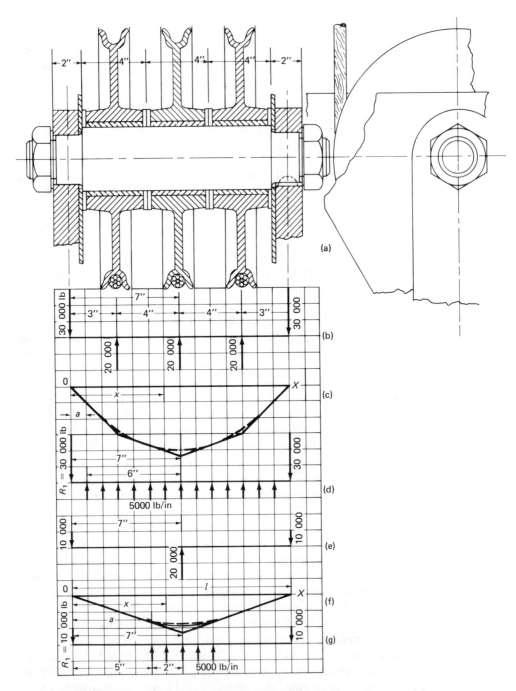

FIGURE 9.10. Comparison of bending moment diagrams from equivalent concentrated and distributed loadings (Example 9.3). (a) Sheaves, sheave pin, and rope in upper portion of a traveling block. *Three sheaves:* (b) Concentrated loading. (c) Moment diagram for concentrated loading in full lines, from distributed loading in dashed lines. (d) Distributed loading. *One centrally located sheave:* (e) Concentrated loading. (f) Moment diagram from concentrated loading in full lines, from distributed loading in dashed lines. (g) Distributed loading. Scales: space, 1 division = 1 in; force, 1 division = 10 000 lb; moment, 1 division = 30 000 lb·in.

the loads to be uniformly distributed. Compare the results. Do this for two cases (a) with three sheaves as shown and (b) with one sheave centrally located.

Solution

(a) The loading diagrams for the two assumptions are shown plotted to scale at Fig. 9.10(b) and (d). The peak moment for the first assumption is $M = 30\,000 \times 7 - 20\,000 \times 4 = 130\,000$ lb·in. In the second assumption, the uniform load covers the central 12 in of length, giving a distribution $q = 3(20\,000)/12 = 5000$ lb/in beginning at location $x = a$. The moment at any location x is $M = R_1 x - q(x - a)[(x - a)/2] = 30\,000x - 2500(x - 1)^2$ lb·in, giving a parabola with a peak value $M = 120\,000$ lb·in at $x = 7$ in. The two moment diagrams are plotted at Fig. 9.10(c). The assumption of uniform loading gives 7.7% less moment.

(b) With one central concentrated load (Fig. 9.10(e)), $M_{max} = (10\,000)(7) = 70\,000$ lb·in. With the load distributed, as in Fig. 9.10(g), $q = 5000$ lb/in, $a = 5$ in, and $M_{max} = 10\,000x - 2500(x - 5)^2 = 60\,000$ lb·in. The two moment diagrams are plotted at Fig. 9.10(f), and the moment is 14.3% lower for the assumption of uniform loading.　　　　////

In the preceding example, the moment at the shoulder was the same by both loading assumptions. For determining the diameter required for the center of the pin, the designer must decide which assumption to use. In new parts, any deflection of the pin will tend to concentrate forces at the ends of the central bushing, making the moment even less than that with uniform distribution. Wear of the bushing will redistribute the load toward a uniform one. From this, the assumption of uniform load seems safe enough. However, it is conceivable that under some conditions of operation, the sheaves will wobble and the bushings will wear to a bell shape which will give a distribution in between the uniformly distributed and the concentrated conditions.

When the span between supports is several times the length of the support, it is customary to assume the support reaction to be a concentrated force at the center of the support, and this was done in Example 9.3. It is "on the safe side," and any error is negligible. However, for very short spans, such as in pin-connected rods and links, the assumption of a nonuniformly distributed load may be advisable. When a shaft or pin is centrally loaded, its deflected shape has a slope at the supports that brings about line or point contact at the adjacent edges of the supports and rod, unless deformation or wear occurs. The greatly exaggerated pin clearance and deflection shown in Fig. 9.11(a) indicates why this occurs. The wear pattern (unloaded) in Fig. 9.11(b) is taken from an earth-moving machine's steering mechanism, which had been subjected to many load reversals. It shows how the concentrated forces of Fig. 9.11(a) may cause their own redistribution by deformation and wear.[5] Such a pattern suggests that a triangular load distribution (Fig. 9.11(c)) may be safely assumed at the supports. This is equivalent to a concentrated reaction one-third of the support length from its inner edge. Also, the wear pattern suggests that a minimum allowable deflection or slope of the pin should be the design criterion, resulting in a larger diameter, stiffer pin and a more uniform distribution of pressure.

Bent and curved beams are common in machines, e.g., a bell crank or rocker with arms at 90°, a C-frame, the automotive front-wheel spindle and yoke for steering (Fig.

---

[5] The pin is shown unloaded and straight in Fig. 9.11(b), but under one particular load it will deflect to fit the wear pattern of the bushing.

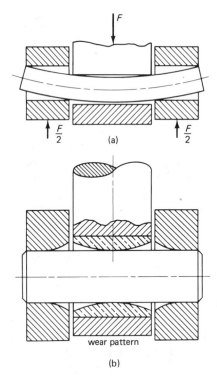

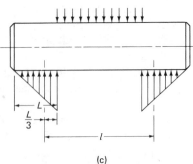

FIGURE 9.11. Short-pin connection between links. (a) Initial contact when new. (b) Wear pattern at the steering jack of an earth-moving machine. (Adapted from B. A. Slupek and F. L. Main, "Determine the Diameter and Length of a Pin for a Pin-Bushing Type Joint by Bending Stress Evaluation," *General Motors Engineering J.,* Vol. 3, No. 5, pp. 48–49, 1956, and Vol. 4, No. 1, pp. 59–61, 1957. Reproduced with permission of General Motors Corporation.). (c) Triangular loading assumption.

9.12), and the hook of Fig. 9.6(a). An otherwise straight link with a pull at each end may have a curved offset to avoid interference with a crossing shaft (Fig. 10.14). Bending stresses occur at the offset. A crankshaft may have one or more "throws" or cranks, not all in the same axial plane. In general, the moment distribution may be determined from free-body force diagrams in one or more coordinate planes. Figure 9.12(b) is a diagram for the vertical plane of the spindle and the effect of road-force components P and F. Not shown are a horizontal-plane diagram for the effects of component B and a combined diagram. Noncoplanar forces and moments are discussed in Section 9.6.

9.5 FORCES AT BOLTED AND WELDED CONNECTIONS

Brackets, yokes, and other cantilevered parts, as well as beams of all kinds, are commonly attached to similar or larger surfaces and walls by threaded fasteners, rivets, or welds (Figs. 9.13 and 9.14). The forces in these fasteners become determinate if it is assumed that the parts attached are very rigid relative to the fasteners. Then, moments cause the surfaces to move apart or be partially relieved of initial compression except at

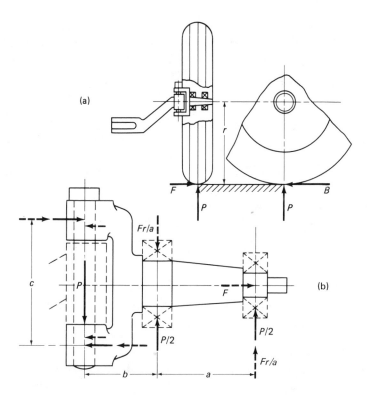

FIGURE 9.12. (a) Front wheel and spindle, showing road forces—normal force P, turning force F, and braking or driving force B. (b) Free body of spindle with forces from bearings, kingpin, and chassis.

the one edge about which they pivot (Fig. 9.13(b)). The axial deflections of the fasteners are proportional to their distances from this line of pivoting. Forces parallel to the wall cause equal sliding at each fastener, and if each fastener fills its hole, equal shearing action.

For brackets similar to those of Fig. 9.13, with the notations as shown in Fig. 9.13(b),

$$\frac{\delta_2}{\delta_1} = \frac{c_2}{c_1}, \quad \frac{\delta_3}{\delta_1} = \frac{c_3}{c_1}, \quad \text{etc.} \tag{a}$$

FIGURE 9.13. Bolted connection between relatively rigid parts. (a) Wall bracket as assembled. (b) Wall bracket as assumed for force analysis.

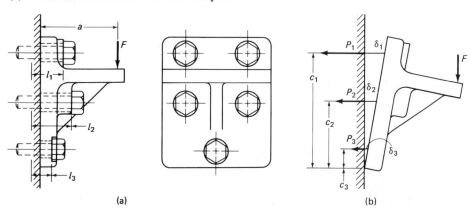

At each distance c, let there be N fasteners of one size, N' of another size, etc. If the effective axial length of each fastener is l or l', its modulus of elasticity E, and cross-sectional area A or A', the elastic forces developed, from $\delta = Pl/AE$, are

$$P_1 = \frac{A_1 E_1}{l_1}\delta_1, \quad P_1' = \frac{A_1' E_1'}{l_1'}\delta_1, \quad P_2 = \frac{A_2 E_2}{l_2}\delta_2 = \frac{A_2 E_2}{l_2}\frac{c_2}{c_1}\delta_1, \text{ etc.} \tag{b}$$

From the equation of moments, $\Sigma M_{\text{pivot}} = 0$,

$$(N_1 P_1 + N_1' P_1' + \cdots)c_1 + (N_2 P_2 + N_2' P_2' + \cdots)c_2 + \cdots - Fa = 0 \tag{c}$$

Substitution of Eqs. (b) into Eq. (c) will allow solution for the one unknown δ_1. Then substitution of δ_1 into Eqs. (b) will determine the forces. The stresses may then be found from the equation $\sigma = P/A$. When all fasteners are of the same material and size, and there are three rows, as in Fig. 9.13, Eqs. (c) and (b) yield

$$P_1 = \frac{Fa}{c_1} \frac{1}{N_1 + \left(\dfrac{c_2}{c_1}\right)^2 N_2 + \left(\dfrac{c_3}{c_1}\right)^2 N_3} \tag{9.10}$$

Equation 9.10 indicates that fasteners N_3 near the pivot line contribute very little toward reducing the load on the maximum-loaded ones. However, fasteners so located are very useful, since good design and assembly require a pretightening of threaded fasteners and, to the extent possible, of hot- and cold-headed rivets. With all bolts tightened to give normal forces several times the P_1 load, the friction force on the mating flat surfaces is more than sufficient to carry the parallel loads. Clearance holes are used so that neither hole location nor bolt shank diameter need have close tolerances, and the bolts do not carry and share the loads in shear. Some rivets are upset enough during heading to fill the holes, and they have a chance to share the load by shear stressing.

The stresses found from the preceding equations should be added to the estimated initial tightening stresses. The total stress should not exceed the yield strength at any time, and a suitable margin of safety should be used.

The assumptions and equations for bolt and rivet analysis may be applied to welded connections, provided the areas through which the welds might tear can be considered concentrated. Friction cannot be expected to resist shearing forces, so elastic shearing stresses should be combined with normal stresses. In most connections fillet welds are laid along the top and bottom of the beam, or along the sides, or both, either continuously or intermittently. The analysis will be indicated for each of these.

For the channel fitting of Fig. 9.14, moments will be taken about the lower weld, so that only the forces at the upper weld and the applied loads will appear in the equation. The free body is shown without contact with the wall except through the welds, and the weld forces are concentrated at their midheights $b/2$. All forces are shown resolved into components normal to and parallel to the wall (Fig. 9.14(a)).

The following equations may be solved for F_1' and F_2':

$$\Sigma M_2 = 0, \quad \left(c + \frac{b_1}{2} + \frac{b_2}{2}\right)F_1' - a(F\cos\theta) - \left(e + \frac{b_2}{2}\right)(F\sin\theta) = 0 \tag{9.11a}$$

$$\Sigma F_x = 0, \quad F_2' + F\sin\theta - F_1' = 0 \tag{9.11b}$$

If the "size" or leg dimensions b of the welds have not been chosen, a fair guess at their values will suffice for a solution of the first equation for F_1'. The displacement parallel to the wall of the relatively rigid fitting is the same at all welds, so the shear strain γ and corresponding stress τ will be the same for all. The shear *forces* developed

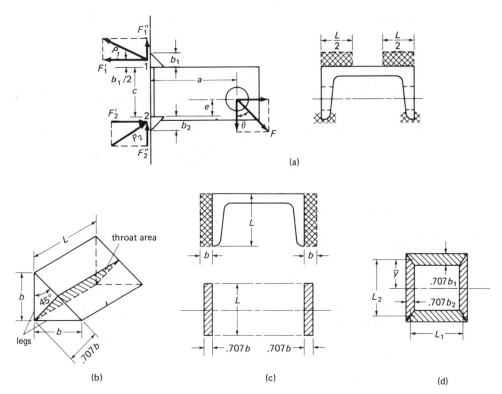

FIGURE 9.14. Fillet-weld connections between a beam and wall. (a) Welds normal to plane of bending. (b) Throat section. (c) Welds parallel to plane of bending. (d) Welds all around.

at the welds will be proportional to their areas A under shear stress τ so $F_1'' = A_1 \tau$ and $F_2'' = A_2 \tau$. From $\Sigma F_y = 0$,

$$F_1'' + F_2'' - F \cos \theta = A_1 \tau + A_2 \tau - F \cos \theta = 0.$$

This equation gives $\tau = F \cos \theta / (A_1 + A_2)$, which substituted into the equations for F_1'' and F_2'' yields

$$F_1'' = (F \cos \theta) \frac{A_1}{A_1 + A_2} \quad \text{and} \quad F_2'' = (F \cos \theta) \frac{A_2}{A_1 + A_2} \tag{9.12}$$

Common fillet weld dimensions are shown in Fig. 9.14(b). Either the leg areas bL or the throat area $bL \sin 45° = 0.707bL$ may be used in Eqs. (9.12). The resultant weld forces are

$$P_1 = \sqrt{(F_1')^2 + (F_1'')^2} \quad \text{and} \quad P_2 = \sqrt{(F_2')^2 + (F_2'')^2} \tag{9.13}$$

A fillet connection, because of its small size and discontinuous nature, cannot be given a very exact stress analysis. The welds most generally fail through their narrowest section or throat (Fig. 9.14(b)), regardless of the direction of the resultant force P. Hence it has been common practice, except when construction codes specify otherwise, to simply express the load-carrying capacity of the weld as the product of the throat area and an allowable safe stress σ_a, divided by a stress concentration factor K to be used when applied loads vary, namely,

$$P = \frac{(0.707bL)\sigma_a}{K} \tag{9.14}$$

Equation (9.14) may be equated to either of Eqs. (9.13) and solved for a dimension b or L. For static loading $K = 1.0$. For varying loads a value $K = 1.5$ has been suggested for transverse fillet welds, with the allowable stress for coated welding rod, 14 000 psi with static loading reduced to 5000 psi with dynamic loading.[6]

If equal fillet welds are laid along the side of the channel, (Fig. 9.14(c)), the two throat sections may be treated as though they were rectangular beam sections under bending. The moment of inertia of area of each is $I = (1/12) \times$ base $\times$ (altitude)$^3 = (1/12)(0.707b)L^3 = 0.059bL^3$. The distance c to the extreme fiber is taken to be $L/2$. The maximum bending stress is, by Eq. (9.5),

$$\sigma = K \frac{M}{2(I/c)} = K \frac{M(L/2)}{2(0.059bL^3)} = \frac{KM}{0.236bL^2} \tag{9.15}$$

The shear stress at the weld is $\tau = F_y/2A = F_y/2(0.707bL)$, which may be combined with σ if it is significantly large.

If welds are laid around the perimeter of an area (Fig. 9.14(d)), the moment of inertia of area may usually be found as for a composite section. The thickness of the throat is usually small compared with the distance from the centroid of area, and from Eq. (9.2), $I = \int y^2 dA = A\bar{y}^2$ each for the top and bottom welds. In Fig. 9.14(d), the welds are symmetrically placed, and $\bar{y} = L_2/2$. The composite section modulus from top and bottom welds and from two side welds is,

$$Z = \frac{I}{c} = \frac{2(0.707b_1L_1)(L_2/2)^2 + 2(1/12)(0.707b_2)L_2^3}{L_2/2}$$

$$= 0.236L_2(3b_1L_1 + b_2L_2) \tag{9.16}$$

In a long beam welded to two supports at its ends, the value of the moment on a connection is probably between that of a built-in and that of a freely supported end. An estimate may be made, based to some extent on the area of the welds.

9.6 NONCOPLANAR MOMENT ANALYSES

Some common power transmission elements—friction wheels, chain sprockets, belt pulleys, and gears—and the forces applied to them are shown in Fig. 9.15. Force on a gear tooth is considered concentrated at a pitch point. The force may be resolved into a tangential transmitted component F_t, a radial separating component F_s, so called because it tends to move the two gears apart, and an axial or thrust component F_a (Fig. 9.15(d) through (g)).[7] For their effect on shaft loads, friction forces are negligible in the plain spur, bevel, and helical drives shown, but not in the worm drive. In Fig. 9.15(g) the friction force μF_n acts in the direction of sliding of the worm thread. On the shaft the transmitted component causes a torque in a radial plane and a bending moment in an axial plane, the vertical plane of Fig. 9.15(d) and the horizontal planes of Fig. 9.15(e), (f), and (g). The separating and thrust components cause bending in the other coordinate axial plane. The thrust component also causes axially directed tensile or compressive stresses and a thrust reaction at one of the bearings. The external forces acting on several shaft-mounted components and the body forces are seldom parallel to each other, e.g.,

---

[6] C. H. Jennings, "Welding Design," *Trans. ASME*, Vol. 58, pp. 497–509, 1936.

[7] For the force resolution of spiral bevel, hypoid, and other gearing, see, e.g., *New Departure Handbook*, vol. II, New Departure, Bristol, Conn., or the *Timken Engineering Journal*, Section 1, The Timken Co., Canton, Ohio, also publications of the Gleason Works, Inc., Rochester, N.Y.

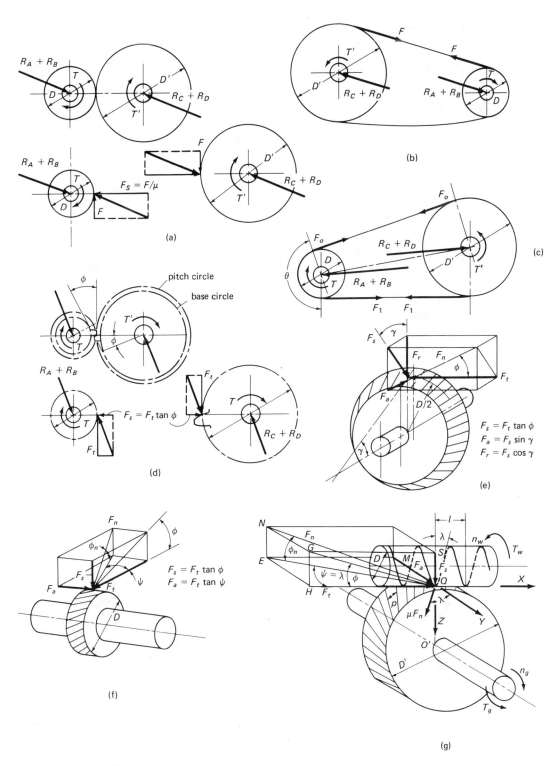

FIGURE 9.15. Some mechanical power-transmission components and applied forces. (a) Friction drive. (b) Chain drive. (c) Belt drive. (d) Spur gear drive. (e) Bevel gear. (f) Helical gear. (g) Worm gear. Total shaft bearing reactions are $R_A + R_B$ and $R_C + R_D$. Transmitted force F_t = torque/radius = $2T/D$, F_s = separating force, F_a = axial force, F_r = radial force, F_n = normal force, μF_n = friction force, ϕ = pressure angle, and ψ = helix angle.

the shafts of Figs. 9.16 and 9.17. It is obvious that most shafts in machines require a three-dimensional analysis.

For construction of the force diagrams of a shaft, the X-coordinate axis may be taken to coincide with the axis of rotation, and the Y and Z axes to be parallel to some one set of tangential and radial components, such as those of the spur gear of Fig. 9.16. All applied forces are resolved into x, y, and z components, and the components are projected onto the three coordinate planes. The unknown components of bearing reactions are added to complete the free-body diagrams for each plane. The bearing reactions are determined by equations of equilibrium, and the component moment diagrams are computed and plotted.

The first subscript of a force, reaction, or moment indicates its location along the shaft, and the second subscript indicates its direction, e.g., the reaction R_{3y} in Fig. 9.16(b). Moments acting in the XY plane, which are moments about a line parallel to the Z axis, are designated by the subscript z, the subscript also indicating the direction of a vector which could represent the moment, e.g., M_z and M_{1z} in Fig. 9.16(b). Similarly, the moments acting in the XZ plane of Fig. 9.16(c) are designated by the subscript y. For purpose of visualization, the XY planes of Figs. 9.16 and 9.17 may be considered as vertical planes, the XZ planes as horizontal, and the YZ planes as radial.

Each resultant radial bearing load and the resulting bending moment at one location x are, respectively,

$$R = \sqrt{R_y^2 + R_z^2} \quad \text{and} \quad M = \sqrt{M_z^2 + M_y^2} \tag{9.17}$$

The angular direction of the axial plane in which the moment acts is of little interest for a rotating shaft and for a circular spindle or pin which is not rotating. Therefore, although the resultant moment diagram is a warped surface in cylindrical coordinates M, ϕ, and x, it may be plotted in a plane with coordinates M and x and used to determine stress or dimensions.

Component moment diagrams are shown in Fig. 9.16(b) and (c) and a resultant moment diagram is shown in Fig. 9.16(d). Between zero end moments and the adjacent peaks, M has a linear relationship, plotted as a straight line, e.g.,

$$M = \sqrt{M_z^2 + M_y^2} = \sqrt{(M_{1z} x/x_1)^2 + (M_{1y} x/x_1)^2}$$

$$= (x/x_1) \sqrt{M_{1z}^2 + M_{1y}^2} = (x/x_1)M_1 .$$

It is usually sufficiently accurate to plot the resultant moment as a straight line between the other peaks, as between M_1 and M_2 or between M_2 and M_3. A torque diagram (Fig. 9.16(e)) may also be constructed.

Example 9.4

Between the energy-consuming punching strokes of a press, there are much longer, idle periods during which the work piece is being removed and replaced. Hence most of the energy may come from a flywheel by a rapid drop in its speed, and a small electric motor may be used to restore the speed before the next punching stroke. Figure 9.16 shows an arrangement of a flywheel, a pulley which is belt driven by a motor, and a pair of spur gears to reduce the crankshaft speed. A connecting rod, not shown, joins the crank to the punch ram. For the punching stroke, the torque diagram (Fig. 9.16(e)) shows 7/8 of the torque being supplied by the flywheel. At the pulley, $\mu = 0.35$, and the angle of wrap $\theta = 3.4$ rad. The pressure angle ϕ of the gear teeth is 20°. Determine the resultant bearing forces and moments.

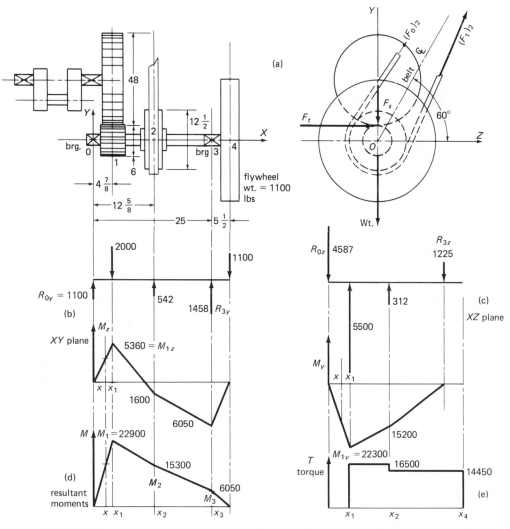

FIGURE 9.16. Force, moment, and torque diagrams for an intermediate shaft of a punch press (Example 9.4). The diagrams are not drawn to scale. Units are in inches and pounds.

Solution

At the pulley, from Fig. 9.15(c) and Eq. (3.42) and neglecting velocity effects, $F_1 = e^{\mu\phi} F_0 = e^{(0.35)(3.4)} F_0 = 3.29 F_0$, and the net belt pull which produces 1/8 of the torque is $F_1 - F_0 = 2.29 F_0$. Then

$$2.29 F_0 = \frac{(1/8)T}{D/2} = \frac{(1/8)(16\ 500)}{12.5/2}$$

whence $F_0 = 145$ lb and $F_1 = 478$ lb. The total belt pull $= F_0 + F_1 = 623$ lb, which has components 542 and 312 lb, shown on Fig. 9.16(b) and (c), the force diagrams for the XY and XZ planes.

At the 6-in pinion, transmitted force $F_z = 16\ 500/(6/2) = 5500$ lb; and the separating force F_s (Fig. 9.15(d)), is $F_s = F_t \tan \phi = 2000$ lb.

The moment diagrams are calculated from the force diagrams, and the peak values of

384

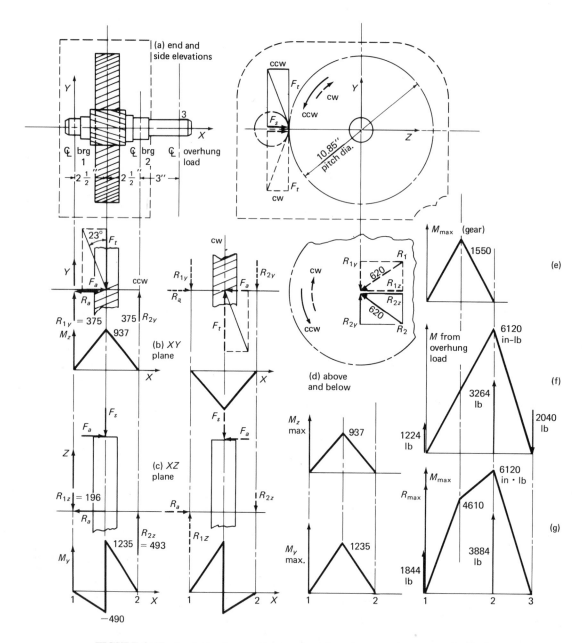

FIGURE 9.17. Analysis of component and resultant forces and moments for the low-speed (output) shaft of a helical gear unit, where the rotation may be in either direction, and the overhung load may have any radial direction (Example 9.5). (a) Elevation views of gear unit and forces for counterclockwise (ccw—full lines) and clockwise (cw--dotted lines) rotation of the larger gear. (b) From gear action only, XY-plane forces and moments for both counterclockwise and clockwise rotations of output shaft. (c) Same, but for XZ plane (view from bottom). (d) Maximum component and resultant bearing reactions and maximum component moment diagrams from gear action only. (e) Maximum resultant moment diagram from gear action only. (f) From overhung load only, forces and moments in one plane. (g) Direct summation of maximum possible forces and moments due separately to gear forces and overhung load. All forces and moments are to scale, except that in (f) and (g), the force scale is ¼ and the moment scale ½ of the preceding ones.

the resultant force and moment diagrams are calculated by Eqs. (9.17), e.g.,

$$R_{max} = R_0 = \sqrt{(1100)^2 + (-4587)^2} = 4710 \text{ lb}$$

$$M_{max} = M_1 = \sqrt{(5360)^2 + (-22\ 300)^2} = 22\ 900 \text{ lb·in}$$

The resultant moment diagram is plotted in Fig. 9.16(d), with straight lines between peak values. Together with the torque diagram (Fig. 9.16(e)), the problem is ready for bearing selection and for the calculation of stresses or diameters (Prob. 6.38).   ////

For theories of failure that combine the effects of bending moment and torque, or of normal and shear stresses, one may refer to Chapters 5 and 6, particularly Sections 5.5 through 5.7 for static loading, and Sections 6.4 through 6.8 for fluctuating loads and rotating shafts. Sections 5.9 through 5.12 provide information on stress concentration and fatigue strength.

When the place of load application changes in service, the maximum possible moment that can occur at a given fillet or critical section should be found. Often, the load location for this can be selected by inspection, as for the camshaft of Fig. 9.1 with its uniformly spaced lobes. For less symmetrical beams with traveling loads, the location may be found by writing an equation for the moment in terms of load position a, then taking $dM/da = 0$, solving it for a, and substituting this value into the moment equation.

A load may also change its radial direction, or this direction may not be known. A stock or commercial unit, such as a small or medium-size speed reducer, should be designed so that a customer may install it to rotate in either direction. Also, it should be designed to take a transverse load in any radial direction from a belt, chain, or spur gear mounted externally on the overhanging portion of the output shaft (Fig. 9.17). The analysis starts with only the loading on the internal gear considered, as if a pure torque were applied at the output end of the shaft. Moment diagrams are obtained for two component planes and for clockwise and counterclockwise rotation. Then, for each plane, the maximum values of moment, regardless of the direction of rotation, are plotted for every location along the shaft. The two diagrams are combined by Eqs. (9.17) to obtain a resultant moment diagram due to the gear only. The moment diagram due to the overhung load is then determined. The worst total moment at each location occurs when the moment of the overhung-load lines up with that due to gear forces only. Thus the final moment diagram is a direct addition of the diagram for the overhung load to the resultant moment diagram due to the gear only. Similarly, for the maximum possible load at a bearing, the load due to the overhung load is added directly to the maximum resultant load due to the gear.

Example 9.5
The gear unit of Fig. 9.17 is to be stocked in warehouses and sold "off the shelf" to any customer. Rated power is to be 15 hp, output speed is 232 rpm, and estimated maximum overhung load is 2040 lb. The helix and normal pressure angles are $\psi = 23°$ and $\phi_n = 20°$, respectively. Other dimensions are shown in the figure. Determine the maximum possible bearing loads and bending moments on the output shaft.

Solution
The torque is

$$T = \frac{63\ 000 \text{ hp}}{n} = \frac{(63\ 000)(15)}{232} = 4060 \text{ lb·in}$$

Forces on the helical gear are, by Fig. 9.15(f),

transmitted force $F_t = \dfrac{T}{D/2} = \dfrac{4060}{10.85/2} = 750$ lb

axial force $\qquad F_a = (750)\,\tan 23° = 318$ lb

separating force $\quad F_s = (750)\,\dfrac{\tan 20°}{\cos 23°} = 297$ lb

Forces F_a and F_t are seen in the XY plane (Fig. 9.17(b)), and F_a and F_s are seen in the XZ plane (Fig. 9.17(c)). One set of diagrams (force and moment) is required for counterclockwise rotation of the driven gear, a second set of diagrams in a column beside the first set is required for clockwise rotation. Forces F_t and F_a reverse but F_s does not. This causes a reversal or an exchange of the vectors representing bearing reactions, and a corresponding shift in the moment diagrams. Maximum component, bearing reactions from the gear action are added vectorially in the top diagram of Fig. 9.17(d). For the XZ plane, the right half of the moment diagram for counterclockwise rotation and the left half of the moment diagram for clockwise rotation are used to plot the maximum possible moment M_y in the bottom diagram of Fig. 9.17(d). This is combined with the similar diagram for M_z by Eqs. (9.17) to give the resultant diagram (Fig. 9.17(e)) for gear forces only. Figure 9.17(f) is readily constructed for the single, overhung load. The moment diagrams of Fig. 9.17(e) and (f) are added to give the maximum possible moments at each location (Fig. 9.17(g)), and the bearing reactions at locations 1 and 2 of Fig. 9.17(d) and (f) are added to give the maximum possible bearing loads. Figure 9.17(g) should be used in the design of the stock unit. By inspection, it is apparent that factors K, M, and I/c will combine to give the highest stresses at the fillets to the immediate right or left of bearing 2. For a final design and calculation, see Prob. 6.40. ////

9.7 DEFLECTION OF STRAIGHT BEAMS

A short section $\triangle s$ of a straight beam deforms to the shape shown by the dotted lines in Fig. 9.18. The fiber at position y is strained an amount $\triangle u = -y\,\triangle\theta$ so that strain ϵ_x is

$$\epsilon_x = \lim_{\triangle s \to 0} \frac{\triangle u}{\triangle s} = -y \lim_{\triangle s \to 0} \frac{\triangle\theta}{\triangle s} = -y\frac{d\theta}{ds} = -\frac{y}{\rho} \tag{9.18}$$

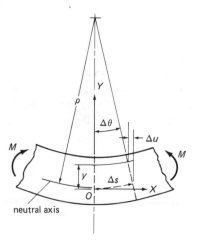

FIGURE 9.18. Curvature and strain in a beam under a moment M.

TABLE 9.2. Beam Relationships by Calculus

	By Integration	**By Differentiation (read upward)**
Distributed Loading	$q = q(x)$	$\dfrac{d^2}{dx^2}\left(EI\dfrac{d^2v}{dx^2}\right)$
Shear	$V = \displaystyle\int_0^x q\,dx + \cdot C_1$	$\dfrac{d}{dx}\left(EI\dfrac{d^2v}{dx^2}\right)$
Moment	$M = \displaystyle\int_0^x V\,dx + C_2$	$EI\dfrac{d^2v}{dx^2}$
Slope	$\theta = \displaystyle\int_0^x \dfrac{M}{EI}\,dx + C_3$	$\dfrac{dv}{dx}$
Deflection	$v = \displaystyle\int_0^x \theta\,dx + C_4$	$v = v(x)$

where ρ is the radius of curvature at the neutral axis and $ds = \rho\,d\theta$. Also, from Section 9.2, $\epsilon_x = \sigma_x/E = -(My/EI)$. Hence from Eq. (9.18).

$$\frac{M}{EI} = \frac{1}{\rho} \tag{9.19}$$

From Eq. (8.3), an approximation to the radius of curvature is $1/(d^2v/dx^2)$, where x and v are coordinates of a point on the deflection curve.[8] Hence

$$\frac{M}{EI} = \frac{1}{\rho} = \frac{d^2v}{dx^2} \tag{9.20}$$

The Eq. (9.20) relationship between moment and displacement allows the determination by integration of slope, $\theta = dv/dx$, and deflection v. All the relationships from distributed loading q to deflection v are listed in Table 9.2.

The integrations are readily made if the loading and section are uniform, or if they vary continuously so that q and I may be expressed analytically in x. With concentrated loads and/or stepped beams, a different expression for M/EI must be written as many times as there are discontinuities in loading or in cross section, and together with the necessary determination of the constants, the method of formal integration may be prohibitively long. This is particularly true with shafts, where shoulders are used to accurately locate and clamp mounted and loaded parts such as gears and bearings. Then, alternate methods of integration should be used, such as the moment-area method, graphical methods, or numerical integration by digital computer.

A convenient representation of the effort required to give a displacement at a particular location is the *spring constant* or *spring rate* k. For a beam it may take the form $k = P/v = \beta EI/l^3$ or $k_\theta = M/\theta = \gamma EI/l$, where β and γ are coefficients and θ is a slope. Constant k is also called the *stiffness influence coefficient* or simply *stiffness*. Its reciprocal is the flexibility influence coefficient α, used in Section 10.1. Stiffness k should not be confused with the *rigidity*, a product that allows a comparison of the effect of material and cross section for the same configuration or length. For beams, it is the *flexural rigidity EI*.

---

[8] Note that the term v for displacement or deflection in the direction of the y axis is consistent with that commonly used in the theory of elasticity (e.g., Section 5.2) and in some recent textbooks on mechanics of solids. Terms u, v, and w are the displacements in the X, Y, and Z directions, respectively, of a point at location x, y, z. If there were a moment M_y acting in the XZ plane about a line parallel to the Y axis, the equation corresponding to Eq. (9.20) would be written $M_y/EI = d^2w/dx^2$. However, in some books δ or y is used to indicate beam deflection.

9.8 MOMENT-OF-AREA METHOD FOR DEFLECTIONS

This method is particularly useful for obtaining the flexure of nonuniform beams and shafts, such as that of Fig. 9.19. Under its M/EI diagram (Fig. 9.19(d)), the area $dA = (M/EI)dx$ is, by Table 9.2, the change of slope $d\theta$ over the distance dx. The *integral* $\int_{x_1}^{x_2} (M/EI)dx$ is the area A_{12} between the M/EI line and its axis and hence, the change of slope between locations 1 and 2,

$$\theta_2 - \theta_1 = \theta_{21} = \int_{x_1}^{x_2} \frac{M}{EI} dx = A_{12} \tag{9.21}$$

If M/EI is negative, so is area A_{12}.

FIGURE 9.19. Moment-area method for slopes and deflections of a stepped shaft. (a) Shaft. (b) Force diagram. (c) Moment diagram. (d) M/EI diagram, moment M divided by rigidity EI. (e) Deviations t from a tangent at the left support, slopes θ, and deflections v. (f) Increment of deviation.

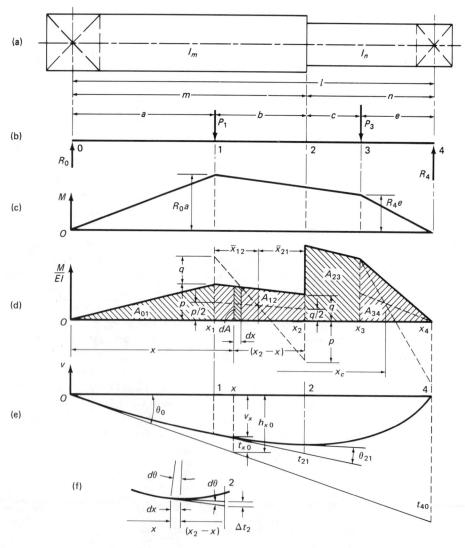

The distance to the deflected position at 2 from a line tangent at 1 will be called the tangential deviation t_{21}, read "the deviation at 2 from the tangent at 1." The contribution of the length dx to this deviation (Fig. 9.19(f)), is

$$\triangle t_2 = (x_2 - x)d\theta = (x_2 - x)dA = (x_2 - x)(M/EI)dx$$

The total deviation of the deflection curve at 2 from the tangent made at 1 (Fig. 9.19(e)), is

$$t_{21} = \int_{x_1}^{x_2} (x_2 - x)\left(\frac{M}{EI}\,dx\right) = \int_{x_1}^{x_2} (x_2 - x)dA = A_{12}\,\bar{x}_{21} \tag{9.22}$$

where $\bar{x}_{21}$ is the distance from 2 to the centroid of the area A_{12}, or the *centroidal distance*, which by definition, is

$$\bar{x}_{21} = \frac{\displaystyle\int_{x_1}^{x_2} (x_2 - x)dA}{A_{12}}$$

Thus the deviation t_{21} of the deflection curve at location 2 from the tangent at 1 is the moment about 2 of the area under the M/EI diagram between 1 and 2. Likewise, the deviation at 1 from the tangent at 2 is

$$t_{12} = A_{12}\,\bar{x}_{12} \tag{9.23}$$

where $\bar{x}_{12}$ is the distance to the centroid of area A_{12} measured from location 1. In general t_{12} differs from t_{21} because $\bar{x}_{12}$ differs from $\bar{x}_{21}$, except for rectangular areas.

If a beam is built in at its support, as in Fig. 9.20, the slope θ_0 of the tangent at location 0 is zero, so deviations such as t_{10} and t_{20} are also deflections v_1 and v_2, respectively. For other conditions of support, deflections must be found from the geometry of the beam. For a beam simply supported at two locations $x = 0$ and $x = x_4 = l$ (Fig. 9.19(e)), the deviation $t_{40} = A_{04}\,\bar{x}_{40}$ may be found by breaking A_{04} into several convenient areas and adding the moments about location 4 of each of the areas. By proportion the altitude of the triangle at x is $h_{x0} = (x/l)t_{40}$. The deviation t_{x0} is the moment of area A_{0x} about x, or $t_{x0} = A_{0x}\bar{x}_{x0}$. Finally, the beam deflection is $v_x = -h_{x0} + t_{x0}$. An alternative route (Fig. 9.21), is to use the tangent at the other support, location 4, finding first the deviation t_{04} and altitude $h_{x4} = [(l - x)/l]t_{04}$; then $t_{x4} = A_{x4}\bar{x}_{x4}$ and $v_x = -h_{x4} + t_{x4}$.

Example 9.6
Derive equations for the slopes at the bearings and the deflection at the shoulder of the shaft of Fig. 9.19.[9]

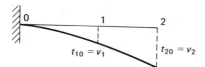

FIGURE 9.20. Deviations for a cantilever beam.

---

[9] Usually it is easier to work with numerical values and graphs throughout, but the derived equations will be useful in Section 9.10 and so they are derived here.

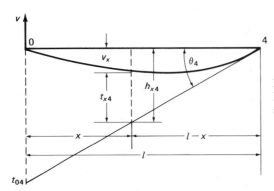

FIGURE 9.21. Alternate to Fig. 9.19(e). Deviations from a tangent at the right support.

Solution

The M/EI diagram of Fig. 9.19 is redrawn in Fig. 9.22(a). The centroid of a trapezoid is easily obtained graphically,[10] but not algebraically. Hence each trapezoid and adjacent triangle of Fig. 9.19 is replaced by two triangles by extending the top lines of the triangles to location 2. Thus the entire area to the left of location 2 consists of positive triangle A'_{02} and a negative triangle A'_{12}, with the effects of the latter to be subtracted as for any negative area of a moment diagram. Areas of the four triangles are listed in Table 9.3.

FIGURE 9.22. Diagrams and dimensions for the solution of a stepped shaft (Example 9.6). (a) The M/EI diagram of Fig. 9.19(d) converted into triangles. (b) Deflection curve and tangents.

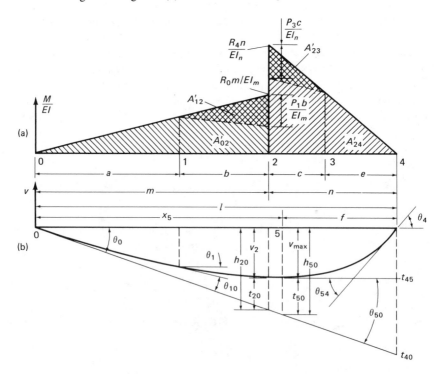

---

[10] The centroid of a trapezoid is readily located by the construction shown for A_{12} in Fig. 9.19. It is the point of intersection between a line of median altitude, and a line drawn between extensions q and p laid off in opposite senses from side altitudes p and q, respectively. The same construction for the triangle A_{34}, with $q = 0$, locates its centroid, which, of course, is at location $x_c = x_3 + (1/3)(x_4 - x_3)$.

TABLE 9.3. Tabulation of Areas and Moments of Fig. 9.22 for Example 9.6

Space	Area	Centroidal Distance from Location 4	Moment of Area about Location 4
A'_{02}	$\dfrac{R_0 m^2}{2EI_m}$	$n + \dfrac{m}{3}$	$\dfrac{R_0 m^2(m + 3n)}{6EI_m}$
A'_{12}	$-\dfrac{P_1 b^2}{2EI_m}$	$n + \dfrac{b}{3}$	$-\dfrac{P_1 b^2(b + 3n)}{6EI_m}$
A'_{24}	$\dfrac{R_4 n^2}{2EI_n}$	$\dfrac{2}{3} n$	$\dfrac{2R_4 n^3}{6EI_n}$
A'_{23}	$-\dfrac{P_3 c^2}{2EI_n}$	$e + \dfrac{2}{3} c$	$-\dfrac{P_3 c^2(2c + 3e)}{6EI_n}$

Also listed are the centroidal distances or moment arms measured from location 4 and their products with the areas, i.e., the moments of area about location 4. By Eq. 9.22, the sum of the moments gives the deviation at location 4 from the tangent at 0 in Fig. 9.22(b), or

$$t_{40} = \frac{1}{6E}\left[\frac{R_0 m^2(m + 3n) - P_1 b^2(b + 3n)}{I_m} + \frac{2R_4 n^3 - P_3 c^2(2c + 3e)}{I_n}\right] \tag{a}$$

In terms of t_{40}, the altitude h_{20} is

$$h_{20} = (m/l)t_{40} \tag{b}$$

The deviation t_{20} of Fig. 9.22(b) is obtained from areas A'_{02} and A'_{12}, each multiplied by its centroidal distance from x_2, or $m/3$ and $b/3$, respectively, whence

$$t_{20} = \frac{R_0 m^2}{2EI_m}\left(\frac{m}{3}\right) - \frac{P_1 b^2}{2EI_m}\left(\frac{b}{3}\right) = \frac{R_0 m^3 - P_1 b^3}{6EI_m} \tag{c}$$

The deflection at location 2 is the difference $v_2 = -h_{20} + t_{20}$. Substitution from Eqs. (a), (b), and (c), followed by factoring and the substitution of m for $(l - n)$, n for $(l - m)$, $(m - a)$ for b, and $(n - e)$ for c, yields the symbol-balanced equation

$$v_2 = -\frac{1}{6EI}\left\{\frac{n}{I_m}\left[2R_0 m^3 - P_1(2m^3 - 3am^2 + a^3)\right]\right.$$

$$\left. + \frac{m}{I_n}\left[2R_4 n^3 - P_3(2n^3 - 3en^2 + e^3)\right]\right\} \tag{d}$$

The equations of equilibrium and the force diagram (Fig. 9.19(b)) give

$$R_0 = \frac{P_1(l - a) + P_3 e}{l} \quad \text{and} \quad R_4 = \frac{P_1(a) + P_3(l - e)}{l} \tag{e}$$

Substitution for R_0 and R_4 in Eq. (d) and further rearrangement yields the following equation for deflection v_2

$$v_2 = -\frac{n\{P_1 a[m^2(l + 2n) - a^2 l] + 2P_3 em^3\}}{6EI_m l^2}$$

$$-\frac{m(I_m/I_n)\{2P_1 an^3 + P_3 e[n^2(l + 2m) - e^2 l]\}}{6EI_m l^2} \tag{9.24}$$

The slope at support location x_0 is, from Fig. 9.22(b),

$$\theta_0 = -\frac{t_{40}}{l} \tag{9.25}$$

The slope at any other location x is larger than θ_0 by the area of the M/EI diagram between x and x_0. At x_4, by the algebraic summation of all the areas in Table 9.3, area A_{40} is obtained, and

$$\theta_4 = \theta_0 + A_{40} = -\frac{t_{40}}{l} + \frac{1}{2E}\left[\frac{R_0 m^2 - P_1 b^2}{I_m} + \frac{R_4 n^2 - P_3 c^2}{I_n}\right]$$

and by substitution for R_0, R_4, b, and c

$$\theta_4 = -\frac{t_{40}}{l} + \frac{P_1 a[m(l+n) - al] + P_3 e m^2}{2EI_m l}$$

$$+ \frac{(I_m/I_n)\{P_1 a n^2 + P_3 e[n(l+m) - el]\}}{2EI_m l} \tag{9.26}$$

Designate the location of maximum deflection to be x_5 (Fig. 9.22). Since the slope here is zero, the changes of slope between here and the supports are

$$\theta_{50} = A_{50} = \theta_0 \quad \text{and} \quad \theta_{54} = A_{54} = \theta_4$$

Thus x_5 may be found by assuming successive locations for it and scaling the areas thus defined until one is found equal to the corresponding end slope, as previously determined. A second method is to write an equation $\theta_{50} = \theta_0$ or $\theta_{54} = \theta_4$ in terms of the unknown distance x_5, and then solve for it. Once the value of x_5 is obtained, a deviation t_{50} may be determined by moment of areas, and the maximum deflection by $v_{max} = v_5 = -h_{50} + t_{50}$ (Fig. 9.22). Alternately, since deflection is zero at location 4 and slope is zero at location 5, $v_{max} = v_5 = -t_{45}$. In a beam proportioned like that of Fig. 9.19, it is likely that location 5 will be very close to location 2, perhaps just to the right of location 2, and the calculation of v_{max} may be avoided, since the value of v_2 will be close enough to v_{max} for most practical uses.   ////

9.9 GRAPHICAL METHODS FOR SLOPE AND DEFLECTION

The shear, moment, slope, and deflection diagrams of beams may be obtained by successive numerical or graphical integrations, because of the relationships that are summarized in Table 9.2. One may start with a diagram of distributed loading q, but when moment diagrams are readily constructed from calculations, only the integrations for slope and deflection are made. It is with a stepped or tapered shaft, and with an M/EI diagram for a start, that the approximate methods are most valuable, as for Fig. 9.23. The M/EI diagram is constructed to a new scale below the moment diagram by dividing it by E and by the different moments of inertia of area I along each portion. This gives a jump in value at each step in the shaft.

In most graphical methods, the areas under the M/EI diagram are divided into a number of small sections that approximate parallel trapezoids, as by the dotted lines of Fig. 9.23(c). In one method, their areas are measured and their values tabulated and added to obtain successive summations. Each summation is plotted at the beam location corresponding to the end of the incremental area, and lines are drawn through all points

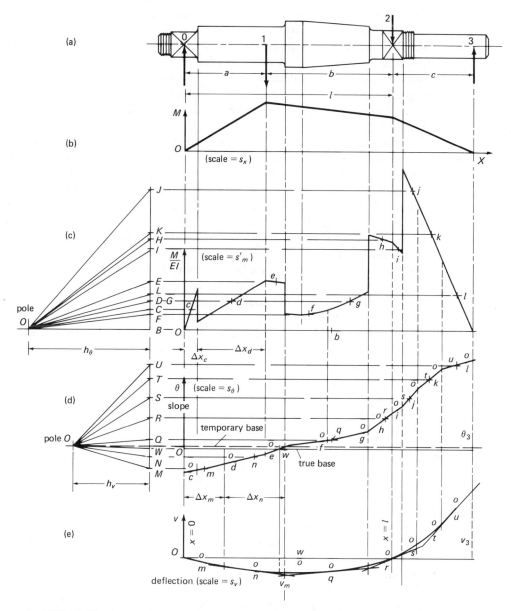

FIGURE 9.23. A graphical solution for the slope and deflection diagrams of a stepped and tapered shaft. (a) Shaft. (b) Moment diagram. (c) M/EI diagram and projections to its pole diagram OBJ. (d) Slope diagram and projections to its pole diagram OMU. (e) Deflection diagram.

to form the slope diagram. The process is repeated to obtain a deflection curve.[11] The method is essentially one of numerical integration, and it can be a basis for digital computer programming. A second method projects the mean heights of the several M/EI areas onto a normal line, to which rays are drawn from a pole O (Fig. 9.23(c)). The slope diagram (Fig. 9.23(d)) is constructed by drawing a series of lines parallel to

---

[11] R. M. Phelan, *Fundamentals of Mechanical Design*, 3rd ed., New York: McGraw-Hill, pp. 132–138, 1970; J. H. Faupel, *Engineering Design*, New York: John Wiley & Sons, Inc., pp. 127–130, 1964.

the successive rays. The construction is repeated to obtain the deflection curve (Fig. 9.23(e)). It is a completely graphical method except for the calculation of scales. It will be discussed in further detail on the following pages. A third method combines the graphical and numerical methods to perform a double integration directly from the M/EI diagram to the deflection curve. The *areas* rather than the heights of the M/EI trapezoids are laid off to scale on the pole diagram, and the lines are drawn parallel to its rays to form the deflection curve.[12] It is analogous to the moment-area method of Section 9.8.

Careful consideration should be given to the sectioning of the M/EI diagram. Along a portion of the shaft which is tapered or which carries a distributed load, the top boundary of the diagram is a curve. Under it, there must be enough divisions to make each section of the curve a reasonable approximation to a straight line, as in Fig. 9.23(c). Along portions where the diameters are small, the dividing of M by I increases the areas and hence their contributions relative to those of the portions where the diameters are larger. There is no need to make $\triangle x$ or the widths of the divisions the same. Accuracy is improved if the widths of the divisions are chosen by eye to make all the *areas* of about the same magnitude. A total of six to twelve areas is usually sufficient.

In the completely graphical method (Fig. 9.23), the mean height or mean value of M/EI for each division of the diagram is projected across to a normal line BJ. Thus point c projects to C, point d to D, etc., and finally l to L. The midpoint of the base of each area projects to B. A *pole O* should be placed beyond BJ at a distance such that rays drawn from it to the extreme points B and J will make with each other an angle of about 35 to 50°. All the rays are now drawn to complete the pole diagram. With pole O to the left and on the extension of the baseline of the M/EI diagram, the slope and deflection diagrams when constructed will have the correct signs or directions and the baseline of the slope diagram will be horizontal. An identification system such as the letters shown will help in avoiding errors.

The slope diagram is now constructed below the M/EI diagram. Across the first space division $\triangle x_c$, a line o/c is drawn parallel to ray OC. From its intersection with the extended right boundary of the division, a line o/d is drawn parallel to ray OD across space $\triangle x_d$, and so on for the length of the beam. The proof of this construction is given later. The zero axis of the slope diagram need not and cannot be located until the deflection curve is established, i.e., we need a "constant of integration" based upon a known location of zero deflection or slope. We do know that at the support locations $x = 0$ and $x = l$, the deflections are zero, and their difference is zero. Hence the definite integral $\triangle v = \int_0^l \theta \, dx$, where θ is slope, has a zero value, and the net area under the slope curve must be zero, i.e., the zero axis of slope is so located that the negative area and the positive area between supports have the same absolute value. If a temporary location for the horizontal line of zero slope is chosen by eye so that the two areas appear to be about equal,[13] then the axis of the deflection curve will turn out to be nearly horizontal, which is convenient but not necessary.

The slope diagram is now divided. With the desirability for somewhat equal areas considered, the widths $\triangle x$ may not be the same as in the M/EI diagram. A pole diagram is constructed and again the pole point O is placed to the left of the projections, on the extension of the baseline or temporary axis. Some rays have negative slopes and some

[12] This "force and funicular polygon" method is described in the third (1955) and earlier editions of V. M. Faires, *Design of Machine Elements*, New York: The Macmillan Company, 1955.

[13] Putting this zero line through the nearest point of discontinuity in the shear diagram will avoid an extra area and extra lines in the construction that follows.

rays have positive slopes, so a reasonable angle between the extreme rays may be larger than for the M/EI diagram, up to 90°. A succession of lines parallel to the rays are laid off below the slope diagram to form the deflection diagram.

The axis of the deflection diagram is the straight line drawn through its intersections with the two lines $x = 0$ and $x = l$, positions of known zero deflection. In general this axis will not be horizontal. Deflections should be measured off vertically from this inclined axis to the deflection curve. A smooth curve may be drawn *inside* the diagram, tangent to its several straight and discontinuous lines. This is done because if the integrand were divided into smaller and smaller areas, the lines forming the deflection diagram would bend inward at more and more frequent intervals, eliminating the out-standing corners, and approaching a smooth curve.

The true position of the baseline of the slope diagram is now found by drawing a ray from pole O, parallel to the axis or closing line o/w of the deflection diagram. Its intercept, point W on line MU, should be projected across to intercept the slope diagram at w. A line through w becomes the closing line of the slope polygon, and it is horizontal because it must be parallel to the ray OB of the upper pole diagram. Point w is at the location of zero slope to the deflection curve, and its projection downward to v_m locates the maximum deflection between supports. The largest slope and deflection for this particular shaft are θ_3 and v_3, respectively, occurring at the overhung position 3 (Fig. 9.23(d) and (e)).

A proof for the preceding graphical method will now be made, and formulas for the scales will be derived. In Fig. 9.24 the ordinate of the integrand diagram (Fig. 9.24(a)) is designated I_d. On the integral diagram (Fig. 9.24(c)), the ordinate is designated I_l. To make the proof more general, the pole O of diagram Fig. 9.24(b) is taken at a random position vertically relative to the projected points on line CG. Normal distance $OX = h$ is the *pole distance*, and it is the altitude of all the triangles in the pole diagram. On the integral diagram, mx_1 and px_2 are drawn parallel to OX, and pp' is drawn parallel to nq, which is part of baseline b/o and parallel to ray OB.

FIGURE 9.24. General proof of the graphical construction for integration. (a) Integrand I_d diagram. (b) Pole diagram. (c) Integral I_l diagram. Note scales s_x, s_d, and s_l.

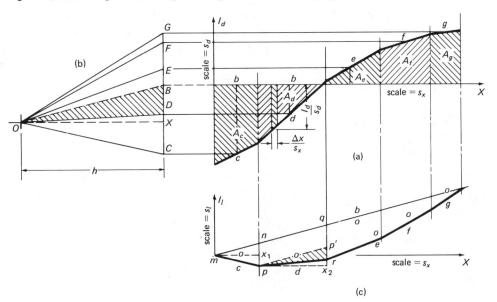

The vertical intercept np is found from similar triangles mnp and OBC, thus

$$\frac{np}{mx_1} = \frac{BC}{OX}, \quad \text{whence} \quad np = \frac{(mx_1)BC}{OX} = \frac{(mx_1)(bc)}{h} = \frac{A_c}{h}$$

since the area A_c of the integrand diagram is the product of its width mx_1 by its average height bc. The next intercept qr is found from similar triangles $pp'r$ and OBD, thus

$$\frac{p'r}{px_2} = \frac{BD}{OX}, \quad \text{whence} \quad p'r = \frac{(px_2)BD}{OX} = \frac{A_d}{h}$$

Now $qr = qp' + p'r = np + p'r = (1/h)(A_c + A_d)$. Similarly, any vertical intercept i has the value $(1/h)\sum_{n=c}^{n=i} A_n$, and the method of construction produces the integration desired. The proof requires that the integrand have a horizontal base, so the M/EI and slope diagrams must be plotted on a horizontal axis.

The scale of the integral diagram is a function of pole distance h, the scale s_d of the integrand, and the length scale s_x. If the real values of integrand and division width are I_d and $\triangle X$, respectively, the height and width on the diagram are I_d/s_d and $\triangle X/s_x$, and a division of the integrand diagram has the area $I_d \triangle X/s_d s_x$. If s_l is the scale of the integral and I_l its real value, then an intercept i as measured on the integral diagram has the value

$$\frac{I_l}{s_l} = \frac{1}{h}\sum_{n=c}^{n=i} A_n = \frac{1}{h}\sum_{n=c}^{n=i} \left(\frac{I_{dn}}{s_d}\right)\left(\frac{\triangle X_n}{s_x}\right) = \frac{1}{hs_d s_x}\sum_{n=c}^{n=i}(I_{dn}\triangle X_n)$$

Since the summation equals the numerator I_l of the first fraction, the denominators are also equal, and the integral scale is $s_l = hs_d s_x$. Let the scales of the M/EI, slope, and deflection diagrams be s_m', s_θ, and s_v, respectively, and the pole distances be h_θ and h_v, as shown in Fig. 9.23. Then, in summary

$$s_l = hs_d s_x, \quad s_\theta = h_\theta s_m' s_x, \quad s_v = h_v s_\theta s_x \tag{9.27}$$

The preceding graphical method may be applied to obtain the shear and moment diagrams when they are not readily obtained analytically, as when loading is irregularly spread along the beam. Any additional applied loads that are normally considered concentrated should be distributed, uniformly or otherwise, over a reasonable length of the beam. Then the loading diagram is analogous to the M/EI diagram. In the positioning of the axes of the diagrams, i.e., the "constants of integration," the construction for the shear diagram is somewhat analogous to that of the slope diagram, and that of the moment diagram to that of the deflection diagram.

9.10 SUPERPOSITION

By superposition, the net stress or strain at a point in an elastic body subjected to several forces and/or moments is determined by the algebraic addition of the several stresses or strains caused by the forces and moments acting separately. Stresses from centrifugal forces are superimposed on stresses from an interference fit in the rotors of Section 8.10, and thermal strains are superimposed on compressive strains in Section 7.6. Superposition is possible because the stresses and strains are linear functions of the disturbances. The slope and deflection equations for beams subject to small, elastic actions are linear functions of forces and their moments.

Linearity in beams is due to the linear differential equation $d^2v/dx^2 = M/EI$ (Eq. (9.20)). Its linearity follows from two linear relationships. One linear relationship is that between moment and curvature, $M/EI = 1/\rho$ (Eq. (9.19)). Modulus E is a constant only in the elastic range. Hence in the plastic range and for nonelastic materials, superposition is not possible. The other linear relationship is $1/\rho = d^2v/dx^2$, made linear by neglecting nonlinear terms in the expression for curvature (Eq. (8.3)), a reasonable assumption for small deflections. Hence superposition cannot be done if deflections are relatively large.

By superposition, solutions for beams with multiple loadings can be built up from solutions already made for simpler cases that have the same EI and support conditions. Thus any two or more loadings in Fig. 9.9, except for (3), and their corresponding deflection curves may be superimposed. These and the more extensive tables in handbooks are limited to uniform beams, usually of a single span, but loaded and supported in many different ways.[14] In this section, beams will be superimposed end to end to determine slopes and deflections for an overhung shaft, a double cantilever, and a bracket. In the next section, superposition will be used to determine statically indeterminate forces and moments, including those in a stepped, three-bearing shaft.

The beams and equations of Table 9.4 have been selected particularly for use in the method of superposition. Signs, if necessary, are to be assigned by inspection. The method itself will be developed by examples.

Example 9.7

As a first example of superposition we take the common case Fig. 9.25(a) of a shaft with one load P_1 somewhere between its bearings at locations 0 and 2 and one load P_3 *overhung*. It is an important case because unexpected wear of belts and gear teeth has been caused by excessive slope at outside mounted pulleys and gears, particularly where the two loads are oppositely directed, adding the slopes due to each load, as in Fig. 9.25. Determine equations for θ_3 and v_3.

Solution

Single load and corresponding moment diagrams are shown in Fig. 9.25(b) and (c). For the beam in Fig. 9.25(b), we find from Case 6 of Table 9.4 that slope $\theta_2' = P_1a(l^2 - a^2)/6EIl$, which is also the slope θ_3'. At the end of the straight portion of length c, deflection is $v_3' = c\theta_2' = P_1ac\,(l^2 - a^2)/6EIl$.

Finding nothing in Table 9.4 or in a handbook for the overhung loading of the beam in Fig. 9.25(c), we split the beam into two beams of single span, one a simply supported beam with moment $M_2'' = P_3c$ applied at its end, and the other a cantilever with the initial slope θ_2'' at its supported end. From Case 4 of Table 9.4, $\theta_2'' = M_2''l/3EI = P_3cl/3EI$, whence $\delta_3'' = P_3c^2l/3EI$. At the end of the cantilever, the deviations of slope and deflection from the tangent at location 2 are, respectively, $\theta_{32}'' = P_3c^2/2EI$ and $t_{32}'' = P_3c^3/3EI$, from Case 2 of Table 9.4. The slope and deflection at the end of the shaft due to the overhung load only are, respectively,

$$\theta_3'' = \theta_2'' + \theta_{32}'' = \frac{P_3c(2l + 3c)}{6EI} \quad \text{and} \quad v_3'' = \delta_3'' + t_{32}'' = \frac{P_3c^2(l + c)}{3EI} \tag{9.28}$$

The beams in Fig. 9.25(b) and (c) may now be superimposed to give for Fig. 9.25(a), the actual beam of this example,

$$\theta_3 = \theta_3' + \theta_3'' = \frac{P_1a(l^2 - a^2) + P_3cl(2l + 3c)}{6EIl} \tag{9.29a}$$

---

[14] Of particular reference value, R. J. Roark, *Formulas for Stress and Strain*, 4th ed., New York: McGraw-Hill, pp. 104–117, 1965, or R. J. Roark and W. C. Young, *Formulas for Stress and Strain*, 5th ed., New York: McGraw-Hill, pp. 96–108, 1975.

TABLE 9.4. Slopes and deflections of cantilever and simply supported beams (signs by inspection)

Loading and Diagram	Slope	Deflection
Cantilever Beams		
1. End moment M	$\theta = \dfrac{Mx}{EI}$ $\theta_B = \dfrac{ML}{EI}$	$v = \dfrac{Mx^2}{2EI}$ $v_B = \dfrac{ML^2}{2EI}$
2. End force F	$\theta = \dfrac{Fx(2L-x)}{2EI}$ $\theta_B = \dfrac{FL^2}{2EI}$	$v = \dfrac{Fx^2(3L-x)}{6EI}$ $v_B = \dfrac{FL^3}{3EI}$
3. Uniform load q	$\theta = \dfrac{Fx(3L^2 - 3Lx + x^2)}{6EIL}$ $\theta_B = \dfrac{FL^2}{6EI}$	$v = \dfrac{Fx^2(6L^2 - 4Lx + x^2)}{24EIL}$ $v_B = \dfrac{FL^3}{8EI}$
Simply Supported Beams		
4. End moment M_B	$\theta_A = \dfrac{M_B L}{6EI}$ $\theta_B = \dfrac{M_B L}{3EI}$	$v = \dfrac{M_B x(L^2 - x^2)}{6EIL}$ $v_{max} = 0.0642\,\dfrac{M_B L^2}{EI}$ at $x = 0.578L$
5. Central force F	$\theta = \dfrac{F(L^2 - 4x^2)}{16EI}$ $(x \leqslant L/2)$ $\theta_A = \theta_B = \dfrac{FL^2}{16EI}$	$v = \dfrac{F(3L^2 x - 4x^3)}{48EI}$ $(x \leqslant L/2)$ $v_{max} = \dfrac{FL^3}{48EI}$ at $x = L/2$

TABLE 9.4. *(continued)*

Loading and Diagram	Slope	Deflection
Simply Supported Beams		
6. Off-center force F	$\theta_A = \dfrac{Fb(L^2 - b^2)}{6EIL}$ $\theta_B = \dfrac{Fa(L^2 - a^2)}{6EIL}$	$v = \dfrac{Fbx(L^2 - b^2 - x^2)}{6EIL}$ $(x \leq a)$ $v_{max} = \dfrac{0.0642\,Fb(L^2 - b^2)^{3/2}}{EIL}$ at $x = 0.578\sqrt{L^2 - b^2}$
7. Uniform load q $F = qL$	$\theta = \dfrac{q(L^3 - 6Lx^2 + 4x^3)}{24EI}$ $\theta_A = \theta_B = \dfrac{FL^2}{24EI}$	$v = \dfrac{Fx(L^3 - 2Lx^2 + x^3)}{24EIL}$ $v_{max} = \dfrac{5FL^3}{384EI}$ at $x = L/2$

FIGURE 9.25. Superposition method applied to a shaft with an overhung load (Example 9.7). (a) Load and deflection diagram for the complete shaft. (b) Moment and deflection diagrams with only P_1 applied. (c) Moment and deflection diagrams with only P_3 applied.

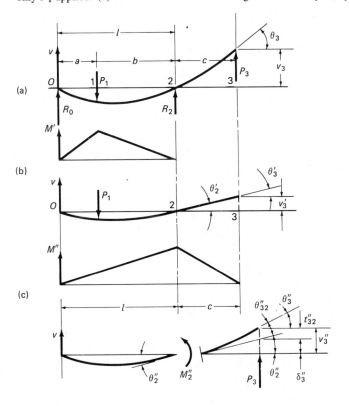

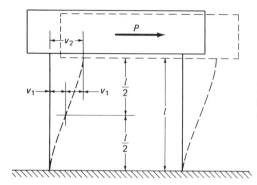

FIGURE 9.26. Flex glide. Double cantilever effect with parallel motion (Example 9.8).

and

$$v_3 = v_3' + v_3'' = \frac{P_1 ac(l^2 - a^2) + 2P_3 c^2 l(l + c)}{6EIl} \qquad (9.29b)$$

////

Example 9.8

In Fig. 9.26, a relatively rigid body is supported by two or more equal, flexible beams. If the beams are of flat spring stock, the device may be called a flex glide, a device for allowing small, essentially in-line motion without friction. The configuration is also that of the target and support leads of wire found in some electron tubes. There, sidewise motion of the target might be induced by vibration or impact. Derive equations for the motion and for the spring constant.

Solution

Each beam acts as a "double cantilever," or two cantilevers of length $l/2$, joined at their free ends. If sidewise force on the rigid body is P, and there are N beams, the deflection at their midpoints is, from Case 2 of Table 9.4, $v_1 = FL^3/3EI = (P/N) (l/2)^3/3EI = Pl^3/24NEI$. The motion v_2 of the rigid body and the stiffness or spring constant k are, respectively,

$$v_2 = 2v_1 = \frac{Pl^3}{12NEI} \quad \text{and} \quad k = \frac{P}{v_2} = \frac{12NEI}{l^3} \qquad (9.30)$$

////

Where beams are joined at angles, the joint is usually considered rigid within itself. Such beam joining by bolting or welding frequently occurs in structures, including box and inverted U structures or "bents."

Example 9.9

Derive equations for slope and deflection at the load P applied to the bracket or angle of Fig. 9.27.

Solution

The horizontal leg is a cantilever fixed to the free end of the vertical leg, which is also a cantilever. For a cantilever with an end couple, and for a cantilever with an end load, we obtain from Table 9.4, $\theta_1 = M_1 l/EI = Pcl/EI$, $\delta_2 = c\theta_1 = Pc^2 l/EI$, $t_{21} = Pc^3/3EI$, and $\theta_{21} = Pc^2/2EI$. Since at the end of the actual bracket, angle $\theta_2 = \theta_1 + \theta_{21}$ and deflection $v_2 = \delta_2 + t_{21}$, they are, respectively,

$$\theta_2 = Pc(2l + c)/2EI \quad \text{and} \quad v_2 = Pc^2(3l + c)/3EI \qquad (9.31)$$

////

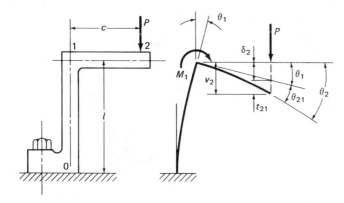

FIGURE 9.27. Angle beam, cantilevers end-to-end (Example 9.9).

9.11 STATICALLY INDETERMINATE FORCES AND MOMENTS

This is a condition of loading and support where the equations of equilibrium are insufficient by themselves for determining all the unknown reactions at the supports. At the supports there are one or more redundancies that can be removed without complete collapse of the body. For their determination it is necessary to use some condition or conditions of deformation. In beams, this means conditions of slope and/or deflection that may result in the rigidity EI appearing in the equations for the reaction forces and moments. The methods of solution will be developed by examples.

Example 9.10

The device of Fig. 9.28(a), if made of flat spring stock, is an elastic hinge or flex pivot, suitable for supporting a part subjected to small angular oscillations. It eliminates friction and lubrication, but it requires a torque T, which is proportional to the turning angle θ_1. In fact, the device may be used as an inexpensive torque meter. Determine the torsional spring constant k_t.

Solution

Pivot center 1 is essentially fixed in location because of the relatively high stiffness of the two legs in the directions of their axes, which are at 90°. Hence each leg is a beam fixed at one end and pivoted at the other. In the force diagram of one leg (Fig. 9.28(b)), the end moment M_1 is half of the applied torque, or $M_1 = T/2$. The normal force R_1 is the axial force on the other leg, and the axial force P_1 is the normal force on the other leg. Hence $P_1 = R_1$, since both legs are alike. From the equations of equilibrium, $P_0 = P_1 = R_1$, $R_0 = R_1$, and $M_1 + M_0 - R_1 l = 0$. There is one redundant reaction, i.e., either M_0 or R_1 could be removed without complete collapse. It is therefore necessary to consider the elastic conditions.

The flexure shown in the force diagram (Fig. 9.28(b)) may be considered the super-position of a simple beam (Fig. 9.28(c)) to which the known moment M_1 is applied, and a simple beam (Fig. 9.28(d)) to which the redundant moment M_0 is applied in direction and amount to give, in the superposition, zero slope at the built-in end of the beam. Then the condition for resolving the redundancy is

$$\theta_0 = \theta_0' + \theta_0'' = 0 \tag{9.32}$$

and this equation may be solved for the value of M_0. The axial forces P_0 and P_1 are omitted in Fig. 9.28(c) and (d) because they cause only small uniform compression, and their moment $P_1 \cdot v_{\max}$ is assumed to have a negligible effect on the flexure.

From Table 9.4 for a simple beam with a single end moment, substitution into Eq. (9.32) for the angle requirement gives $M_1 l/6EI - M_0 l/3EI = 0$, whence $M_0 = M_1/2$. At

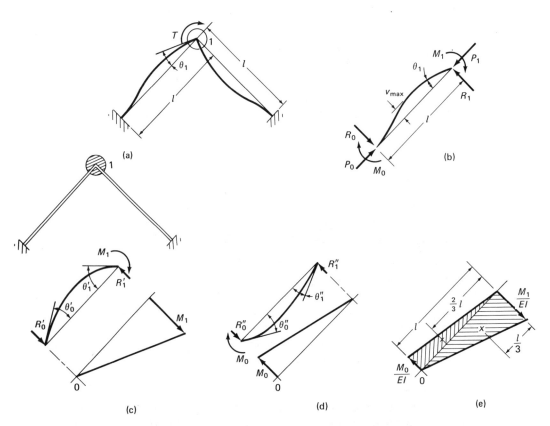

FIGURE 9.28. Flex pivot. (a) Configuration with torque T applied. (b) Free-body diagram of a single leg. Redundancy resolved (Example 9.10) by superposition of (c) with end moment M_1 applied and (d) with end moment M_0 applied. (e) Redundancy resolved by moment-area method.

the other end, slope θ_1 is $\theta_1 = \theta_1' + \theta_1'' = -M_1l/3EI + M_0l/6EI = -M_1l/4EI$, whence $M_1 = |4EI\theta_1/l|$. Reaction R_1 may be obtained by superposition or from the equilibrium of the actual beam, $M_1 + M_0 - R_1l = 0$, whence $M_1 + M_1/2 - R_1l = 0$, and $R_1 = 3M_1/2l$. For the two legs taken together, the torque is $T_1 = 2M_1$ and the "spring rate" or stiffness in twist is $k_t = T/\theta_1$, whence

$$T = \frac{8EI\theta_1}{l} \quad \text{and} \quad k_t = \frac{8EI}{l} \tag{9.33}$$

////

The solution of a statically indeterminate beam may often be obtained rather directly by the moment-of-area method (Section 9.8), without recourse to a table of simpler beams. For the beam of Fig. 9.28(b), the M/EI diagram consists of the two triangles (Fig. 9.28(e)), one related to the applied moment M_1 and one related to the redundant moment M_0, equivalent to the superposition of the M/EI diagrams for the beams of Fig. 9.28(c) and (d). The deviation of the deflection curve at location 1 from a tangent at location 0 is, from Eq. (9.22), $t_{10} = A_{01}\bar{x}_{10} = 0$, i.e., $(M_0l/2EI)(2l/3) - (M_1l/2EI)(l/3) = 0$, whence $M_0 = M_1/2$. From Eq. (9.21), $\theta_1 = \theta_0 + A_{01} = 0 + (M_0l/2EI - M_1l/2EI) = -M_1l/4EI$. For this particular indeterminate beam, the moment-area method is beautifully simple!

For the solution of many statically indeterminate beams, a condition of displacement is the basis for application of the equations of flexure, as in a three-bearing shaft (Fig.

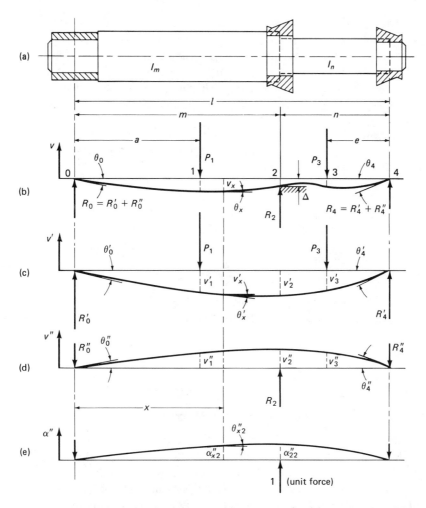

FIGURE 9.29. Three-bearing shaft with misalignment $\triangle$ of middle bearing. Redundant reaction R_2 determined by superposition (Example 9.11). (a) Approximation of the shaft by two diameters. (b) Actual loading. (c) Redundancy removed at location 2. (d) Loading by R_2 only. (e) Unit loading at location 2 for a graphical solution.

9.29).[15] To allow for possible misalignment of the bearings, a fixed displacement $\triangle$ is assumed at location 2 in the force-deflection sketch (Fig. 9.29(b)). The redundant reaction R_2 is removed for the beam of Fig. 9.29(c), and it is the only applied force in the beam of Fig. 9.29(d). The beams of Fig. 9.29(c) and (d) will be superimposed to give the deflection condition in the actual beam (Fig. 9.29(b)), and the equation for resolving the redundancy is

$$v_2' + v_2'' = \triangle \tag{9.34}$$

If no misalignment is expected, $\triangle = 0$.

---

[15] The shaft of Fig. 9.29 mounts gears centered at locations 1 and 3 in a large reduction unit. Three bearings are used to minimize deflection and slope at the gears and obtain uniform pressure and minimum wear at the mating teeth. The bearing seats can be through bored for alignment in a rigid housing. Three or more bearings may be required in long shafts, as for propellers in ships or for lineshafts in mills. In general, however, the use of more than two bearings is avoided because of alignment difficulties. Between two units with two bearings each, shafts are usually connected by flexible, misalignment couplings.

The analytical solution for the shaft of Fig. 9.29 is greatly simplified if the diameters of the short portions at the bearings are approximated by the adjacent diameters, as shown by the dotted lines in Fig. 9.29(a). The stiffening effect of fitted rings is uncertain, anyway, but it is sometimes approximated by adding to the shaft half or more of the ring thickness. With this dotted-line approximation, the shaft resembles that of Fig. 9.19(a) and Eqs. (9.24), (9.25), and (9.26) may be used as needed.

Example 9.11

Derive an equation for reaction R_2 of the simplified shaft of Fig. 9.29(a). Indicate the complete solution for the shaft.

Solution

Deflection v_2'' for the shaft of Fig. 9.29(d) can be obtained from Eq. (9.24) by substituting $P_1 = -R_2$, $P_3 = 0$, and $a = m$, whence

$$v_2'' = \frac{R_2 m^2 n^2 \left[m + n(I_m/I_n)\right]}{3EI_m l^2} \tag{9.35}$$

By substitution into Eq. (9.34) of v_2'' from Eq. (9.35) and v_2' from Eq. (9.24), a negative deflection, a solution for the redundant reaction R_2 is obtained, namely,

$$R_2 = \frac{P_1 a[m(l + 2n) - a^2(l/m)] + 2P_3 e m^2}{2mn\,[m + n(I_m/I_n)]}$$

$$+ \frac{(I_m/I_n)\,\{2P_1 a n^2 + P_3 e\,[n(l + 2m) - e^2\,(l/n)]\} - 6EI_m\,(l^2/mn)\,\triangle}{2mn\,[m + n(I_m/I_n)]} \tag{9.36}$$

With Eq. (9.36) solved for a numerical value, reactions R_0 and R_4 may be determined, followed by the calculations of moments and stresses at critical locations. End slopes are $\theta_0 = \theta_0' + \theta_0''$ and $\theta_4 = \theta_4' + \theta_4''$, obtained from Eq. (9.25), Eq. (a) preceding it, and Eq. (9.26). From the forces and reactions an M/EI diagram may be constructed, and other slopes may be obtained from it and Eq. (9.21). Deflections may be obtained by the moment-area methods (Eqs. (9.22) and (9.23), etc.).   ////

Redundant reactions and all other quantities may be obtained advantageously by the graphical method (Section 9.9), if the EI values of the beam and/or a distributed loading q vary along its length. Thus for the shaft of Fig. 9.29, without redundant reaction R_2, as in Fig. 9.29(c), deflection v_2' may be found graphically in the usual way from an M/EI diagram. The deflection v_2'' under force R_2 only (the beam of Fig. 9.29(d)) must equal $\triangle - v_2'$, from Eq. (9.34). A moment diagram and deflection curve can be constructed for any assumed force at location 2, say unity as in Fig. 9.29(e). Then the actual deflection is $R_2 \alpha_{22}''$, where α_{22}'' is scaled from the deflection curve and read, "the deflection at 2 due to unit load at 2." Reaction R_2 is found from the equality $R_2 \alpha_{22}'' = \triangle - v_2'$. Superimposed flexures at any location x (Fig. 9.29), with due regard for signs, become

$$v_x = v_x' + R_2 \alpha_{x2}'' \quad \text{and} \quad \theta_x = \theta_x' + R_2 \theta_{x2}'' \tag{9.37}$$

or the curve α'' may be redrawn to the same scale as that of v' and superimposed.

The final example involves local deformation, spacing errors, and the equality of deflections among several beams sharing the applied load. Gear teeth are short, tapered cantilever beams of involute profile, with varying locations of load application and variable bending and shear deflections v_b throughout their period of contact, from en-

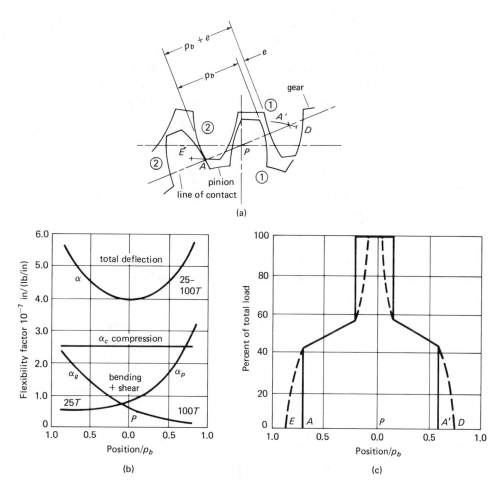

FIGURE 9.30. Division of load between gear teeth, a statically indeterminate case (Example 9.12). (a) Before loading, pair 2 in contact and pair 1 separated by error gap e. Line of contact of pair 2 is APA'. (b) Variation of tooth deflection with position along line of contact measured from pitch-point P. (After R. E. Peterson, "Load and Deflection Cycles in Gear Teeth", *Proc. 3rd Inter. Cong. for Applied Mech.* (1930) Vol. II. p. 388, published by P. A. Norstedt & Söner, Stockholm, Sweden, 1931.) (c) Percent of total load carried by any one pair of teeth *without* errors present. Full line for engagement at the kinematic point A (intersection of gear outside circle with line of contact) and disengagement at A'. Dotted line for premature engagement at E and delayed disengagement at D.

gagement E to disengagement D (Figs. 9.30(a), (b), and 11.14). Because of the convex surfaces, there is also compression v_c where they contact (Sections 11.4 and 11.9). This is added to the bending and shear deflections of the gear and pinion[16] teeth relative to their rims but measured along the line of contact. Their division by the transmitted force q per unit of tooth width gives a flexibility factor α for one pair of teeth, namely,

$$\alpha = \frac{v_c + (v_b)_g + (v_b)_p}{q} = \alpha_c + \alpha_g + \alpha_p \tag{9.38}$$

where the α values may be obtained from calculated charts like Fig. 9.30(b). Although

---

[16] The pinion is the smaller of any two gears in mesh.

v_c is not linear with load q, the sum is close to linear over the range of variation of q. Movement αq along the path of contact is equivalent in gear rotation to equal movement along the base circle, hence is a measure of relative rotation.

With the usual combinations of numbers of teeth and pressure angle, it is kinematically possible for two pairs of teeth to be in contact near the engagement and disengagement locations, and only one pair in the central or pitch-point region. The division of load between two pairs 1 and 2 depends upon their flexibility factors and the error e in base pitch p_b. This is the spacing error along the line of contact, shown under no load as an opening between the teeth of pair 1 (Fig. 9.30(a)).

Example 9.12

Derive equations for the load carried by each pair of teeth in terms of the flexibility factors.

Solution

If torque is applied to the pinion, say while the gear is locked, the relative motion of pinion to gear, measured along the line of contact, is

$$v = e + \alpha_1 q_1 = \alpha_2 q_2 \tag{9.39}$$

which is the elastic condition for resolving the redundancy. The total load transmitted per unit width is

$$q = q_1 + q_2 \tag{9.40}$$

From Eq. (9.39), $q_1 = (v - e)/\alpha_1$ and $q_2 = v/\alpha_2$. These values substituted into Eq. (9.40), followed by its solution for v, yield

$$v = \frac{\alpha_1 \alpha_2}{\alpha_1 + \alpha_2} q \left(1 + \frac{e}{\alpha_1 q} \right) \tag{9.41}$$

and the unit loads carried by teeth 1 and 2, respectively, are

$$q_1 = \frac{\alpha_2}{\alpha_1 + \alpha_2} q \left(1 - \frac{e}{\alpha_2 q} \right) \quad \text{and} \quad q_2 = \frac{\alpha_1}{\alpha_1 + \alpha_2} q \left(1 + \frac{e}{\alpha_1 q} \right) \tag{9.42}$$

From Eqs. (9.42) the load carried by one pair of teeth as it progresses along its path of contact AA' can be calculated and plotted, as shown by the solid lines of Fig. 9.30(c) for a tooth set without errors. ////

For common sizes of steel gears, the magnitude of error e for a zero value of q_1, i.e., an error beyond which no load sharing will occur, is of the order of half a thousandth of an inch. The need for accuracy in manufacturing is apparent. Another effect of deflection and errors is to cause premature engagement, in Fig. 9.30 a contacting of approaching teeth at E before they reach the kinematically correct engagement location A. This may cause gouging of the pinion flank, but the deflection may have a good effect in that it applies the load more gradually, with less shock. This is shown by the dashed lines of Fig. 9.30(c), a tooth set without errors and without consideration of dynamic effects.[17]

[17] The original work on load division was presented by the following: R. E. Peterson, "Load and Deflection Cycles in Gear Teeth," *Proc. 3rd Inter. Cong. for Applied Mech.*, *Vol. II*, P. A. Norstedt & Söner, Stockholm, Sweden, pp. 382–390, 1931; A. H. Burr, "Premature Engagement of Gear Teeth due to Deflection and Errors," M. S. Thesis, University of Pittsburgh, Pittsburgh, Pennsylvania, 1931; R. E. Peterson and A. H. Burr, "Theoretical Aspects of Tip Relief," Meeting of the American Gear Manufacturers Association, 1931; Further considerations were made by H. H. Richardson, "Dynamic Loading on High Speed Gearing," *Trans. 7th Conf. on Mech.*, Cleveland: Penton Publishing Company, pp. 146–160, 1962. With the increase of accuracies in gear-tooth cutting, load division can be considered an actuality.

PROBLEMS

Section 9.2

9.1 A portion of an aircraft fitting is shown. It consists of a clevis or yoke end that can rotate a small amount in bearing A to accommodate different alignments. There is a rod end or strut exerting a static force of 20 000 lb at an angle of 15°. A pin that fits tightly in the yoke joins it to the rod end. Some principal dimensions are shown. All parts are made from an alloy steel heat treated to a tensile strength of 145 000 psi, a yield strength of 110 000 psi, and a bearing (crushing) strength of 200 000 psi. Determine the factors of safety against failure by all possible methods at the apparently critical sections of the several parts.

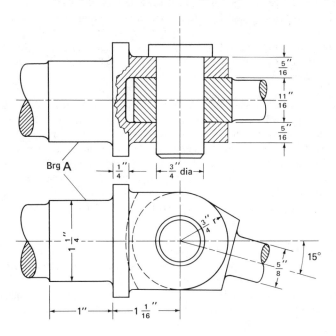

9.2 Design the cast-iron lever of Fig. 9.5(a), to carry an occasional maximum pull of 100 lb applied vertically near the end of the handle, a distance of 20 in from the center A-A of the pivot hole and 28 in from the center of the left hole. At the latter hole a vertical rod with a yoke is attached by a pin. Make the cross section for the handle circular, and for the two principal tapering lengths, an ellipse with a height twice its width (Area = $\pi bh/4$, $I/c = \pi bh^2/32$). The pivot-pin diameter at A-A is 1½ in. For ordinary gray cast iron, the minimum tensile strength is 20 000 psi and the shear strength is 22 000 psi. A reasonable factor of safety for this brittle material is 5.0. *Suggested procedure:* Choose a reasonable handle length and calculate the circle size at the circular shoulder. Calculate ellipse dimensions at the shoulder, at the left end, and at A-A as though extended. Make the width of the cylindrical hub at A-A slightly larger than the width of the adjacent ellipses so that it may be machined flat. Calculate a minimum outside diameter for the hub on the basis of bending stress. Calculate for the hub the bearing pressure per unit of projected area, and use it to determine the pin diameter at the left hole. The width between faces there may be made the same as at A-A, so that both cylinders may be faced by straddle milling in the same machining operation. Sketch the entire lever to approximate scale. Change any calculated dimensions as may seem desirable for appearance, casting, machining, etc.

9.3 Same instructions as Prob. 9.2, but the pull is 700 N, the distances from the pull are 400 and 750 mm, and the pivot-pin diameter is 40 mm. The tensile strength is 150 MPa and the shear strength is 165 MPa.

9.4 Design a cast-iron yoke for the left end of the lever of Prob. 9.2 or Prob. 9.3, as assigned.

Check the strength of the mild-steel connecting pin. A typical yoke is shown in the sketch of Prob. 9.1.

Section 9.3

9.5 The d/D and h/H ratios of Fig. 9.4 were calculated for equal section moduli. What is the effect on stiffness ratios I/I_0? *Suggestion:* Let $\beta = d/D$ and $\lambda = h/H$, determine the ratios D/D_0 and H/H_0 from equal moduli, then the ratios I/I_0 in terms of β or λ only. Calculate I/I_0 for each numerical value of β and λ in Fig. 9.4. *Ans.* 1.047 and 1.192 for $\beta = 0.60$ and 0.80, respectively.

9.6 (a) What is the percentage loss in nominal-load capacity based on an allowable stress, if through a shaft of diameter D there is bored a hole of diameter $d = D/2$? (b) What are the percentage losses in weight and in rigidity EI? (c) If the applied load is reduced by the percentage indicated in part (a), what is the percentage change in deflection that is proportional to P/EI? (d) If, instead, a steel of proportionally higher strength is chosen, what is the percentage change in deflection?

9.7 Because of troubles with lateral vibrations of a large, uniform circular shaft loaded only by its own weight, it is proposed to increase the natural frequency without a change of weight or length by replacing the solid shaft with a hollow one. (Refer to Fig. 9.4.) (a) Derive equations for the ratio ω/ω_0 of hollow shaft to solid shaft frequencies and for the ratio D_0/D, both in terms of diameter ratio $\beta = d/D$ and then solve for the required β in terms of frequency ratio. (Note: Frequencies are proportional to $\sqrt{k/m}$, where k is a stiffness constant proportional to EI/l^3 and m is the shaft mass W/g.) (b) What is the value of β and the percentage increase in outside diameter for a 50% increase in frequencies? *Ans.* 0.62, 27.5%.

9.8 (a) Determine an equation for the smallest section modulus of a circular shaft of diameter D that has had a radial hole of diameter d drilled through it for a pin, with $d \ll D$. If the shaft is rotating, what is likely to give a much larger decrease in the strength? (b) Same requirement as part (a), but a hole is drilled only halfway through to enable oil in a central axial channel to run outward to lubricate a bearing.

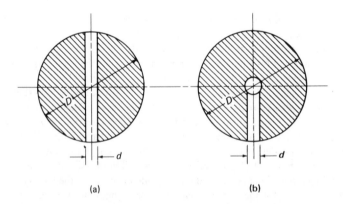

(a)   (b)

9.9 Two bars of square cross section are being considered as rails to support and guide a heavy sliding tool-head. One will be used flat and the other at (a) will be turned 45° to fit a 90° V-groove, as a guide. (a) Compare the two for strength and for rigidity, deriving any formulas needed. (b) To avoid interference with the apex of the groove and to provide an oil space, the top edge of the V-rail is removed by machining. If removed on both top and bottom as at (b), determine if there is an optimum height ratio h/H for maximum section modulus I/c. *Ans.* (a) $\sqrt{2}/2$, 1.0; (b) 8/9.

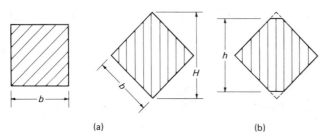

(a)   (b)

PROBLEM 9.9

9.10 An alternate shape for a rail of the preceding problem is shown. Determine equations for *I* and *I/c*.

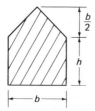

9.11 The connecting-rod bearing cap shown has a channel section for maximum rigidity with minimum weight. Its strength may be approximated by treating it as a straight beam of length equal to the distance between bolt centers and partially built in. Assume the worst case of a central concentrated load and a moment at the center halfway between a built-in condition ($M = Pl/8$) and a freely supported condition. For an allowable stress of 45 000 psi, what load may be taken by the cap? See Example 9.2 for equations that can be used to approximate the properties of the section.

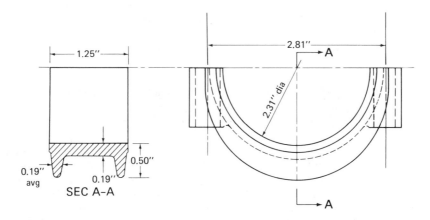

9.12 In electrical machines, a rotor may be built up out of a steel shaft surrounded by enameled (insulated) sheets of steel. The stack of sheets is squeezed together under pressure between heavy end plates. This has a stiffening effect on the assembly. Determine an equivalent diameter of steel shaft by equating bending rigidity, i.e., EI_{equiv} to the sum of the shaft rigidity and stack rigidity. Let the elastic modulus and outside diameter of the shaft be E

and d, respectively, and that of the stacks of punchings be E' and D. Ans. $d_{equiv} =$

$$\sqrt[4]{\frac{E'(D^4 - d^4)}{E} + d^4}.$$

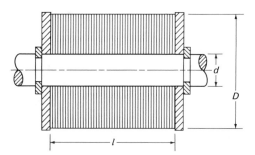

9.13 With reference to Prob. 9.12 it has been found by a compression test in a tensile-testing machine that the strain of the stack of sheets is related to the applied pressure and stress σ by the equation $\epsilon = a\sigma/(1 + b\sigma)$, where a and b depend upon the magnitude of pressure. In the stress range 100 psi $< \sigma <$ 340 psi, $a = 0.001\,75$ and $b = 0.030$. (a) Determine the slope of the stress-strain curve at an assembly pressure $\sigma = 300$ psi. Assume that subsequent changes in stress are small during deflection of the rotor, and take this slope as an average E'. (b) If $d = 3.0$ in, $d' = 12$ in, and $E = 30 \times 10^6$ psi, find the equivalent shaft diameter from the equation of Prob. 9.12. What is the percentage increase of stiffness?

9.14 Refer to Fig. 9.7(b) for three keyways spaced 120°, and prove that the centroid of the shaft section is at the center of the shaft circle and that the moment of inertia of area about any diameter making an angle θ with a diameter through the center of one of the keys is independent of this angle and hence is the same about all diameters. Assume that the area A of each keyway may be taken concentrated at its own centroid at a radius r from the shaft center. Calculate the value of I for the cross section in terms of A, r, and the shaft diameter D. Ans. $\pi D^4/64 - 3\,Ar^2/2$.

9.15 Same shaft and keyways as Prob. 9.14, but consider each keyway to be a rectangle of width b and height h, and use the inclined axis and transfer-of-axis theorems. By examination of the resulting expression for I, and the fact that a common size of keyway is given by $b = (1/4)D$ and $h = (1/8)D$, justify the assumption of Prob. 9.14.

Section 9.4

9.16 For each of the situations shown, sketch a loading diagram including reactions in relative magnitude to the applied loads. Superimpose a deflection curve upon it. Sketch a moment diagram below it. (As an example, see Fig. 9.9(b).)

9.17 The hub of a pulley or gear that is mounted on the end of an overhung shaft may be centered quite a distance from the nearest bearing. The sketch shows a locknut, cap, and seal between an inboard bearing and an outboard hub. This results in a large overhang and bending moment Fl if a typical stock pulley (dashed lines) is used. This may be reduced by the construction in full lines. It is not obvious to everyone that the moment on the shaft at the bearing then has the value Fc. Prepare a convincing proof, based not only on a free-body diagram of pulley and shaft taken together, but also on separate free-body diagrams of the pulley and shaft. Sketch the moment distribution along the shaft from hub center to bearing center and observe if the inward placement of F can be overdone. Discuss.

9.18 Flight maneuvers cause significant forces on engine-turbine shafts, from gyroscopic moments due to turning of the plane of rotation of the bladed wheels and from normal accel-

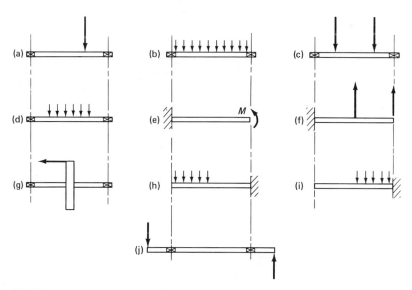

PROBLEM 9.16

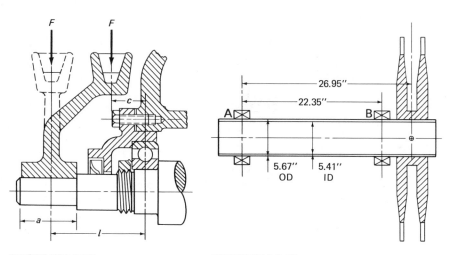

PROBLEM 9.17   **PROBLEM 9.18**

9.18   *(continued)*

eration forces during pullout from a dive. A yaw (turning about a vertical axis) gives a gyroscopic moment $\omega \Omega I_m$ about the pitch or transverse axis, where ω is the rotational velocity of the turbine shaft and Ω is the angular yaw velocity, both in rad/s, and I_m is the mass moment of inertia of the wheels. Dimensions of a shaft with overhung wheels are shown.[18] The weight of each wheel is $W = 85$ lb, its Wr^2 value is 29.66 lb·ft², the yaw velocity is 3.5 rad/s, and there is a possible vertical load of $10g$ due to a $10g$ acceleration during pullout from a dive. Rotational velocity is 11 000 rpm, and 16 560 hp is developed. Determine the bearing reactions, the maximum bending and torsional stresses, and the combined stresses. (It is assumed that these are significant because the maximum action is not frequently repeated.) *Ans.* $R_A = 4335$ lb, $R_B = 6035$ lb, $\tau_{\text{major}} = 22\,125$ psi.

[18] R. W. Lewis, "Determine the Maximum Tensile Stress in a Jet Aircraft Turbine Shaft with Overhung Turbine Wheels," *General Motors Engineering J.*, Vol. 2, No. 2, pp. 48–50, No. 3, pp. 40–43, 1955.

Section 9.5

9.19 A bracket similar to that of Fig. 9.13 has only three bolts, two in the top row and one below, at distances 6 and 2 in, respectively, from the bottom edge. The bottom bolt has half the length and twice the area of the other two. A vertical force of 1000 lb is applied 5 in from the wall. Find the forces in the three bolts that should be added to the tightening forces. *Ans.* 341 lb (top), 455 lb (bottom).

9.20 A dc motor delivers a torque of 2000 lb·in at its lowest speed. In its base there are holes for six equal bolts, two in each of three rows. The base is 14 in long with the rows spaced 1 in, 6 in, 6 in, and 1 in from one edge to the other. The motor weighs 450 lb, with its center of mass 8 in from the base. The base is attached to a vertical wall. What is the maximum bolt force, except for tightening, if the reaction torque and the gravity moment are in the same direction.

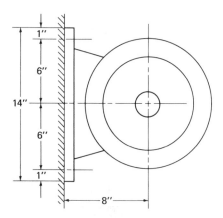

9.21 A cast-steel bracket, similar to that in Fig. 9.13 but without bolt holes, is loaded with an essentially static force of 10 000 lb at an angle $\theta = 30°$ with the vertical, as in Fig. 9.14(a). Dimensions *a*, *c*, and *e* (Fig. 9.14(a)) are, respectively, 4, 3¼, and 2 in. A ½-in fillet weld is proposed for the weld along the top of the plate and a ⅜ in fillet of the same length for the bottom, where the reaction is not so large. Sketch and determine a suitable length for the welds. *Ans.* 2¾ in (top), 2¼ in (bottom).

9.22 Same bracket and load as in Prob. 9.21, but the welds are of equal size and are placed along the entire length of each side of the plate, as in Fig. 9.14(c). Determine a suitable weld size.

9.23 Same bracket and load as in Prob. 9.21, but a weld of one leg size runs all around the plate, as in Fig. 9.14(d). Also the load continuously reverses and angle $\theta = 0°$. Suggest a suitable weld size. Width of the bracket is 2⅞ in.

9.24 A circular pin is joined to a plate by an all-around fillet weld, as at the ends of the equalizer bar of Fig. 9.8(b). Let the weld be small compared with the diameter of the pin, and assume that the force resisting bending is located along a circular *line* within the weld. Derive an equation for the maximum bending stress. *Ans.* $\sigma = 1.80M/bD_w{}^2$, where D_w is the diameter of the circle.

9.25 A large gear is fabricated by welding together three parts, a steel ring for a rim, a hollow cylindrical hub, and a plate or web between them. Teeth are cut in the rim at a pitch diameter of 40 in, at which a variable driving force of 2000 lb is applied. Determine the minimum total length of ⅜-in fillet welds that should be used along the 6-in-diameter surface of the hub to join it to the web.

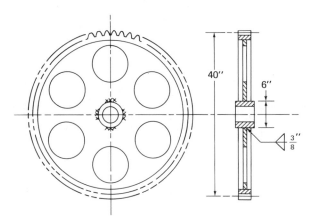

Section 9.6

9.26 Maximum road forces on a rear wheel of a motor vehicle are 1000 lb vertically, 600-lb-tractive force driving the wheel forward, and 300-lb sidewise on curves. Let overhang c be 3 in. Draw elevation and top views of the wheel and axle as a free body, with the 300-lb force directed to give the worst bending moment. Below them sketch the horizontal, vertical, and resultant moment diagrams, and a torque diagram. Label them with peak values. *Ans.* $(M_V)_{\max} = 7500$ lb·in, $(M_H)_{\max} = 1800$ lb·in.

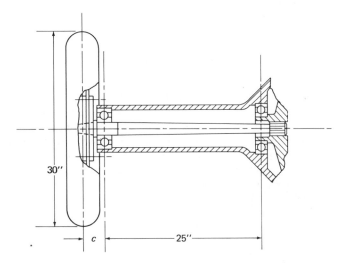

9.27 A front elevation and an end view of a shaft are shown. A pinion (shown by dashes) drives the spur gear, and power is taken off by a belt on a pulley that has twice the diameter of the gear. Tight-side belt pull F_1 is three times the loose-side pull F_0. (a) In terms of F_0 what are the values of total belt pull and of the force transmitted by the gear? (b) Draw end, elevation, and top views, and on all show the components of belt, gear, and bearing forces. Draw them carefully, approximately to scale, and correct in position and direction for free-body diagrams. (c) In line with the top and elevation views sketch the general shapes of the corresponding bending moment diagrams.

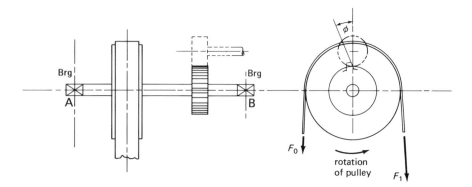

9.28 Draw three views of the idler gear and its shaft and on each show the locations and relative magnitudes of the tooth-force components and bearing reactions. In line with the top and elevation views sketch the shapes of the corresponding moment diagrams, including the resultant diagram.

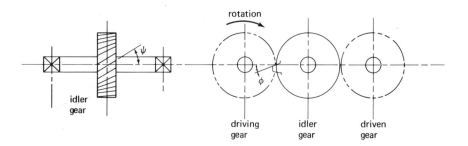

9.29 The high-speed driving shaft of the gear-reduction unit is retained radially and axially by bearings at *A* and *B*. (a) Sketch the directions of the helical teeth at *E*, *H*, and *F*, such that there can be zero net axial (thrust) force on the low-speed driven shaft. (b) Assuming symmetrically cut teeth, on a top view show the driven shaft displaced axially from a central position (in practice, by a few thousandths of an inch, as might occur during assembly with a thrust bearing). By force vectors show the unbalanced condition. (c) Show the best distribution of forces. What do you conclude should be the type or nature of the bearings at *C* and *D*?

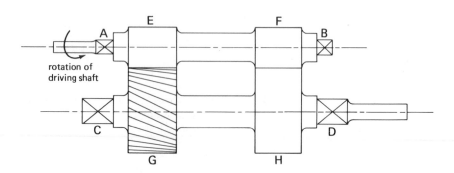

9.30 In a number of standard auto transmissions the second-speed drive is by helical gears from the engine shaft through a countershaft to the propeller shaft as shown in the sketch. Given are the values of the helix angle ψ_M on gear M and its hand (direction) on the sketch; also pressure angle ϕ of 20° (Fig. 9.15(f)) and pitch diameters D_M and D_Q. (a) Draw end, top, and front views of the countershaft. Make each a complete free-body diagram, starting with the end view. Sketch in a direction for the helix of gear Q to make the net end thrust (axial force) on the bearings of the countershaft equal to zero. (b) Derive an equation for the required helix angle ψ_Q.

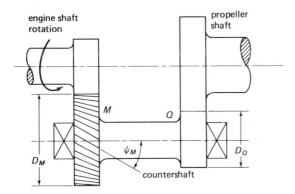

9.31 Gear A turns clockwise and drives gear B of the countershaft, and the other gear C of the countershaft drives gear E. The diameter of B is twice that of C. Helix angle $\psi_B = 45°$ and pressure angle $\phi = 23°$. The helix angle ψ_C is to have a value and hand such that there is no thrust on the bearings of the countershaft. (a) Sketch three views of it and show the force components on each approximately to scale. (b) Calculate the required value of ψ_C and show it with proper hand on the sketches. *Ans.* 26.6°, same hand as B.

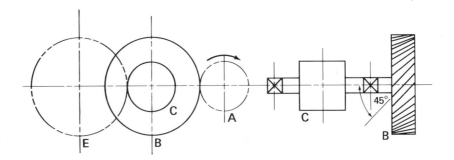

9.32 Bevel gear C rotates clockwise under a steady torque of 3750 lb·in. It and an equal mating gear E have straight teeth, are cut with a pressure angle of 20°, and have effective pitch diameters of 6.93 in. Helical gears G and H have pitch diameters of 5.00 in and 7.50 in, respectively, a helix angle of 30°, and a pressure angle of 20° in the plane of rotation (ϕ of Fig. 9.15(f)). (a) Determine the radial and thrust forces at the bearings A and B. (Ball bearings are selected according to the radial load, modified by the thrust. The latter can be taken at one bearing only.) *Ans.* At bearing A, 2250 lb radially; at bearing B, 825 lb radially and 585 lb axially. (b) Sketch component and resultant moment diagrams. Place the peak values on each one.

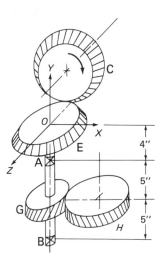

9.33 On the worm-driven gear of Fig. 9.15(g), two forces are shown acting at the pitch-point Q, a force F_n normal to the tooth's surface and a friction force μF_n in the direction of the sliding of the worm thread. The former makes a pressure angle ϕ_n and thread lead angle λ with coordinate planes. Force μF_n lies in a plane having a common tangency to the pitch cylinders of worm and gear. (a) Derive equations for the forces transmitted F_t, separating F_s, and axial F_a in terms of F_n, μ, and angles ϕ_n and λ, also for F_a in terms of F_t. (b) Since the torque turning the gear is $T_g = F_t D'/2$ and that overcome by the worm is $T_w = F_a D/2$ (Fig. 9.15(g)), obtain the T_w requirement in terms of T_g, diameters, angles, and the friction coefficient μ. (c) Write an equation for efficiency η of the drive. Note that efficiency is 100% when $\mu = 0$. *Ans.* $\eta = \{[\tan \lambda(\cos \phi_n - \mu \tan \lambda)]/(\cos \phi_n \tan \lambda + \mu)\} \times 100$ (percent). (d) For $\mu = 0.05$ and $\phi_n = 30°$, plot efficiency against lead angle from $\lambda = 0°$ to $\lambda = 90°$ and comment. (Note: worms may be cut with more than one thread.)

9.34 When the gear drives the worm of Fig. 9.15(g) and Prob. 9.33, the axial components of F_n and of μF_n from the gear act tangentially on the worm thread in opposite directions. (a) Show these on a radial plane through the worm. If the μF_n component is the larger one, there can be no motion of the worm unless torque is also applied to the worm shaft. This is known as a self-locking condition, an important one for holding a load when hoisting. State the condition for self-locking in terms of ϕ_n, λ, and μ. (b) Write an equation for the torque on the worm shaft required to lower the load. Compare it with T_w in Prob. 9.33. (c) From the condition in part (a), write an expression for the maximum efficiency of a self-locking drive. Plot it versus λ, and comment. *Ans.* $100[(1 - \tan^2 \lambda)/2]\%$.

9.35 Screws with special threads such as the Acme thread shown in axial section are used with a nut for the motion of translation. Consider the applicability of the analysis of the previous two problems, and write equations for the screw torques required for advancing and for retracting motion against a force F.

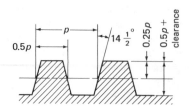

Section 9.7

9.36 Thin stainless-steel tubing of ornamental grade is being sold in competition with aluminum tubing for handrails, etc. For equal lengths and loads compare rigidity and strength of two tubes of 1.90-in outside diameter, one of stainless steel with a wall thickness of 0.065 in, and one of aluminum with a 0.142-in wall ($E_{s.s.} = 25 \times 10^6$, $E_{al} = 10 \times 10^6$ psi; $(S_y)_{s.s.} = 35\,000$, $(S_y)_{al} = 20\,000$ psi).

9.37 Steel pipes destined for an oil line on the ocean floor may be welded together, then wrapped around a large horizontal reel on the deck of a barge. When the pipe is laid, it unwraps slowly from this reel and goes off the rear end of the barge as it moves slowly forward. If high-strength pipe of 4.5 in outside diameter and of 4.026 in inside diameter is wrapped on a 150-ft-diameter reel, what is the bending stress? (Note: In pipe of other proportions the stress may be relieved by some flattening of the circular section. See the footnote to Example 10.5. But here, $K \approx 1.0$.)

9.38 It can be shown that the moment is constant along the entire length l of a spiral spring of many coils, clamped at its outer end, and wound loosely so that friction is not a factor.[19] Use Eq. (9.19) and $ds = \rho d\theta$ to show that the angular deflection measured by the number of turns N is related to the applied moment M by $N = 6Ml/\pi Ebh^3$, independent of radius R, where b is width and h is thickness. A restriction $h << R$ allows us to use the straight-beam sketch and theory of Fig. 9.18, although actually $\triangle\theta$ and M are measured relative to an initially curved and unstressed beam.

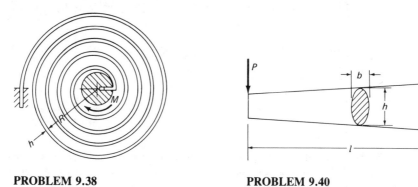

PROBLEM 9.38   **PROBLEM 9.40**

9.39 For a cantilever beam of constant cross section, acted upon by a uniform distributed load $q = -q_0$ along its entire length l, and with Table 9.2 as a basis, obtain by successive integrations the equations for slope and deflection at any position x measured from the free end of the beam.

9.40 Forged and cast brackets, rods, portions of shafts, and levers, such as in Fig. 9.5(a) and (b), are often made with a straight taper in one or both transverse dimensions. Take the case of a cantilever beam of elliptical cross section, tapering toward the left from its support, such that both height and width would decrease to zero at the same location. The beam is loaded at a distance l to the left of its support. Choose the zero location for x carefully, to simplify the expression for the variable moment of inertia of area and hence the integration. For an ellipse, $I = \pi bh^3/64$. (a) Derive an equation for slope as a function of location x. (b) Write the equation for slope under the load. *Ans.* $[32P/3\pi\kappa\lambda^3Ec^2\,(l + c)^3]\,[(l + c)^3 - (3l + c)c^2]$ where c is the distance between the load and the tapering-out point and κ and λ are the taper constants for breadth and height, respectively.

---

[19] For example, A. M. Wahl, *Mechanical Springs*, 2nd ed., New York: McGraw-Hill, pp. 300–302, 1963.

9.41 The steel camshaft of Fig. 9.1 for a fuel injection pump has a distance between bearing centers of 266 mm, divided into seven approximately equal spaces by the centerlines of the six cams. The lift is 9.2 mm, the plunger diameter is 13 mm, and the maximum injection pressure, occurring at highest speed, is 55 MPa. Assume that the resulting force acts on only one cam at a time, with no overlap. (a) Ignore the stiffening effect of the cams, assuming a uniform diameter throughout, and determine the diameter required to limit the deflection at any cam to 0.25 mm. $E = 207$ GPa. Deflection equations are in Table 9.4. (b) What will be the maximum nominal stress?

Section 9.8

9.42 Take a trapezoid with height h and base widths a and b and (a) derive an equation for the centroidal distance from the base of width a. Ans. $h(a + 2b)/3(a + b)$. (b) Prove the graphical construction for locating the centroid, as indicated in Footnote 10 and in Fig. 9.19(d).

9.43 For the shaft of Example 9.6 and Figs. 9.19 and 9.22, obtain in terms of θ_4 an equation for distance f that locates point 5, the place of maximum deflection in Fig. 9.22(b). Also obtain an equation for v_{max}.

9.44 By the moment of area method, determine equations for the slope and deflection at location 3, the end of the cantilever beam that has a step at 1 and a load P at 2. Ans. From triangles, $v_3 = -[P(a + b)^2/2EI_{01}][c + 2(a + b)/3] - (Pb^2/2E)(1/I_{13} - 1/I_{01})(c + 2b/3)$.

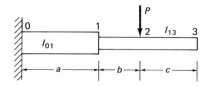

9.45 (a) By the moment of area method determine equations for the end slope and central deflection of a beam of uniform stress (Fig. 9.8(d)), of length l and maximum width B, simply supported at the ends and centrally loaded. (b) Adapt the foregoing to obtain an approximate equation for the deflection of a leaf spring with N leaves of width b (Fig. 9.8(c)).

9.46 The shaft of Fig. 9.9, Case (1), has a span between supports of length l and moment of section $2I$, and overhung lengths of $l/4$, with moment of section I. By the moment-area method, obtain equations for the slope and deflection at the overhung ends. Use the symmetry to make the solution easier. Ans. $\theta = \pm 3Pl^2/32EI$, $v = -Pl^3/48EI$.

9.47 Same requirement as Prob. 9.46, but the central span has a length $2a$ and moment of section $3I$, and the overhangs have a length a and moment of section I.

9.48 The shaft of Fig. 9.9, Case (5), is loaded by an axially directed thrust force, typical of the axial component of force on a helical gear. Number the bearing locations 0 and 2 and the gear location 1. Moment of inertia of area is I_0 from 0 to 1 over a length l_0, and is $I_2 = I_0/2$ from 1 to 2 over a length l_2. Let $l_0 + l_2 = l$ and gear diameter be D. Obtain formulas for the shaft deflection and slope at location 1 by the moment-area method. Ans. $v_1 = (PDl_0l_2/6EI_0l^2)(l_0^2 - 2l_2^2)$, $\theta_1 = -(PD/6EI_0l^2)(l_0^3 + 2l_2^3)$.

9.49 By the moment-area method obtain equations for the slope and deflection at the overhung gear, location 0.

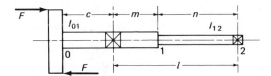

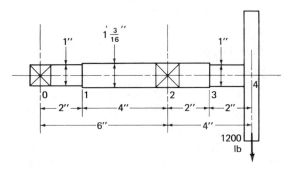

PROBLEM 9.49

9.50 For the steel shaft carefully sketch the moment and M/EI diagrams. Sketching to scale is suggested to minimize errors. By the moment-area method calculate the values of deflection and slope at the overhung load. *Ans.* -0.0234 in, $-0.429°$.

9.51 For a shaft loaded as in Fig. 9.9, Case (2), the length is 20 in, and the uniform loading of 500 lb/in begins at $x = 10$ in and continues to $x = 16$ in. Over the loaded portion, $I = 0.40$ in^4 but elsewhere, it is 0.25 in^4. Carefully sketch the M and M/EI diagrams, and by the moment-area method determine the slope at the right-hand bearing and the deflection at midpoint $x = 10$ in. Note: The centroid of a segment of a parabola is midway of its length c (measured along the x axis of the beam), and the area of the segment is $(2/3)hc$, where h is the height of the segment (measured in the M direction) at its midway point.

9.52 For the shaft of Prob. 9.51, determine the location and magnitude of the maximum deflection.

Section 9.9

9.53 By carefully drawn freehand sketches show the several steps for determining graphically the slope and deflection diagrams of the shaft of Prob. 9.49. Let I_{01}/I_{12} be approximately 3.0, and lay off $c = m = \frac{3}{4}$ in, $n = 1\frac{1}{4}$ in. Determine the zero axes for both diagrams, showing the construction clearly. Divide the largest area into two parts for better accuracy. Choose scales such that a standard-size letter sheet or sectioned paper is more or less filled.

9.54 Freehand, same instructions as Prob. 9.53, but for the shaft of Prob. 9.50. Let $I_{13} = 2I_{01} = 2I_{34}$.

9.55 To determine the lowest critical speed of the generator shaft, it is necessary to obtain the deflection curve when the shaft is loaded by its own weight. Start a solution, carefully sketching freehand a loading diagram, then graphically integrating to obtain the shear and moment diagrams. Sketch the appearance of the M/EI diagram but go no further. What must be done to the axis of this diagram before proceeding? Use scales such that a standard letter-size sheet is more or less filled.

9.56 Carefully sketch to show graphically how to solve for the deflection of a tapered cantilever beam of elliptical cross section, loaded at its free end. (An alternative for the analytical method of Prob. 9.40.)

9.57 By laying off force vectors on a pole diagram rather than projecting area midpoints, devise a means of obtaining the moment diagram and unknown reaction forces graphically. This is equivalent to projecting from a shear diagram, but with the zero axis and support reactions unknown. Try it out for some known simple cases, then apply it to the beam shown with four loads, including one overhung.

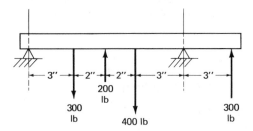

9.58 Using instruments and the graphical method, accurately determine the slope and deflection curves of the steel shaft. Calculate the scales and the maximum values of slope and deflection. Let $E = 207$ GPa. Suggested scales: For distance, 1 mm = 4 mm and for M/EI, 1 mm = 1×10^{-6} mm^{-1}.

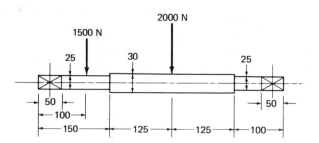

9.59 Same instructions as Prob. 9.58, but determine for the shaft shown the deflections at the three applied loads and the maximum slope at a bearing. Let $E = 30 \times 10^6$ psi. Suggested scales: For distance, 1 in = 8 in and for M/EI, 1 in = 50×10^{-6} in^{-1}. *Ans.* $v_{\max} = 0.0093$ in, $\phi_{\max} = 0.001\,15$ rad.

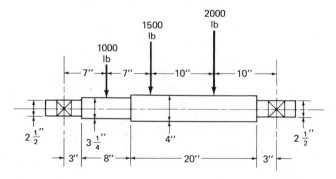

9.60 Same instructions as Prob. 9.58 but for the shaft of Prob. 9.50. Suggested scales: For distance, 1 in = 2.5 in and for M/EI, 1 in = 5×10^{-4} in^{-1}.

9.61 Reproduce the instrument-recorded acceleration vs time curve, plotting each space equal to ¼ in (quadrille paper suggested). Show how to obtain graphically the curve of displacement vs time from 0 to t_1. The body is at rest at $t = 0$. Obtain the zero axes and write the scales for the displacement and any other diagrams constructed. Careful freehand sketching may suffice.

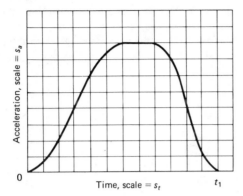

Section 9.10

9.62 By superposition determine equations for shaft slopes at bearings 0 and 2 in Fig. 9.25(a). Ans. $\theta_0 = [P_1 b(l^2 - b^2) + P_3 cl^2]/6EIl$.

9.63 The teeth of overhung pinions on the electric-motor shafts in rail car drives have been known to wear in a lengthwise taper. Excessive shaft slope causes nonuniform pressure between contacting teeth. See Fig. 8.17 and sketch to show that the loading is similar to that for Fig. 9.25(a). An alternative arrangement is to place the motor shaft behind the gear diametrically opposite. Apply the equations derived in Example 9.7 for Fig. 9.25 to compare the slopes of the two arrangements in ratio form. For this purpose, take the gear force equal to one-third of the armature weight W, and the dimensions as $a = b = l/2$ and $c = l/4$.

9.64 A uniform flat spring of length l is pivoted at its ends. A moment M is applied centrally, at $x = l/2$. By using Table 9.4 determine the angular stiffness constant M/θ for small oscillations. Ans. $\theta = -M_0 l/12EI$.

9.65 An example of the principle of elastic matching is a bearing sleeve overhung on its supporting wall in a direction such that the bearing reaction load R causes the wall to slope in the same sense as the shaft. This makes the pressure along the bearing more uniform. (a) Sketch to show the deflected shape of the wall near the bearing. (b) For a simply supported, centrally loaded shaft of length l and rigidity EI, determine an equation for the wall's angular stiffness k_t, where $k_t = R/\theta$, if the slope of the sleeve is to equal the slope of the shaft at the bearing.

9.66 By superposition derive equations for deflection and slope at the overhung end of the uniform shaft of Fig. 9.9, Case (6), where the bearing span is l, the overhang is c, and force F acts at radius $d/2$. Ans. $v = Fcd(2l + 3c)/12EI$.

9.67 A shaft is loaded as in Fig. 9.25(a), but it has steps in diameter. Assume that their effects can be approximated by two diameters, one of rigidity EI_t between locations 0 and 2, and one of rigidity EI_c between locations 2 and 3. (See Fig. 9.29 for a similar approximation.) By the method of superposition, derive equations for slope and deflection at location 3.

9.68 Same information as Prob. 9.67 but derive equations by the method of moment areas.

9.69 The pivot shown provides for small rotations by the flexing of three flat springs joining two hollow partial cylinders, indicated by the differently sloping section lines. Because the springs support the cylinders, there is a positive axis of rotation, and there is no friction, wear, and looseness. The closing of a gap g prevents overstressing of the springs. For small angles of rotation, it may be assumed that the bending moment M is constant along each spring, Section B-B. (a) Obtain general relationships between angle, externally applied torque T, stress, and deflection. Write an equation for the gap g to limit stress to a safe value σ_0. (b) If $h = 0.018$, $b = 0.16$ in, and $l = 0.33$ in, what should be the gap dimension to limit the stress to 150 000 psi? What is the corresponding torque at this limit of action? (Sketch adapted from Flexural Pivot manufactured by The Bendix Corporation with its permission.)

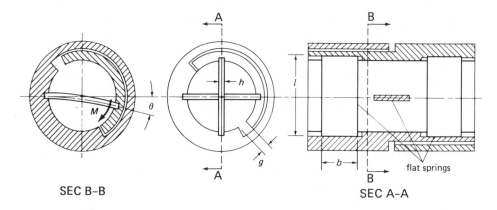

SEC B-B   SEC A-A

9.70 For a beam with a longitudinal axis in the X direction, derive an equation for the superposition at any location x of component slopes θ_{xy} and θ_{xz} that have been determined for the XY and XZ planes, respectively. Determine resultant slope θ and the angle ϕ made by its plane with the plane XY. Make the derivations first for large slopes, then by suitable approximations for the small slopes characteristic of beams and shafts, show that $\theta = \sqrt{\theta_{xy}^2 + \theta_{xz}^2}$ and $\phi = \tan^{-1}\theta_{xz}/\theta_{xy}$.

9.71 Refer to Prob. 9.26. Determine the maximum slope and deflection at the wheel when its overhang c is 5 in. What direction relative to the vertical is made by the plane of the resultant slope? Ignore the taper and use a uniform diameter of 1¼ in over the entire length of the steel axle. Use any available equations, such as Eqs. (9.29) and the answer to Prob. 9.66. *Ans.* $\theta = 1.89°$ at $\phi = 15.9°$, $v = 0.143$ in at $16.9°$.

9.72 (a) Assume that for a deflection analysis, the steel shaft of Fig. 9.17 can be approximated by a 1.5-in-diameter cylinder for the 5-in length between bearing centers 1 and 2, and by a 1.2-in-diameter cylinder for the 3-in extension to the center 3 of the overhung load, an approximation similar to that done on the shaft of Fig. 9.29(a). Determine the value of EI for each section of the shaft. (b) From the load and moment diagrams for planes XY and XZ and the overhung load, obtain separately for each plane the numerical values of shaft slope at position 3. (*Suggestion:* Use the shortest method available for each, such as superposition, the moment-area method, or equations from previous problems.) Of the two diagrams for different directions of rotation, choose the one to give the larger slope at position 3. (c) Add vectorially the slopes of planes XY and XZ and combine the resultant with slope due to overhung load to give the maximum possible slope.

9.73 Continue the previous problem to obtain the maximum possible deflection at location 3.

Section 9.11

9.74 A cantilevered support for a load P is reinforced by a hanger at location 1. By the method of superposition determine equations for the moment at location 0 and the deflection at location 2.

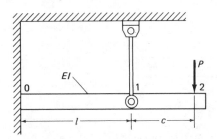

9.75 Determine the redundant force at B in terms of EI by the method of moment areas.

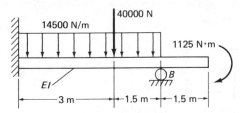

9.76 A cantilever beam of length l carries a concentrated load P at a distance a from its built-in end, and it is partially supported at its free end by a spring of stiffness k, installed without any initial compression. Determine the deflection of the beam at the spring by superposition and the bending moments at the load P and at the built-in end.

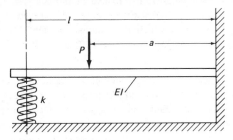

9.77 A flexible strip as it is bent over a cylinder gives a spring of increasing stiffness. Determine the instantaneous stiffness as a function of load P and cylinder diameter D, and by superposition determine an equation for the total deflection at the load. Assume small deflections. *Ans.* $v = [3l^2(PD)^2 - 4(EI)^2]/3D(PD)^2$.

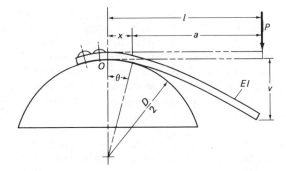

9.78   A leaf spring is reinforced by placing a second spring of the same section but of half-length below it. Assume point contact at location 1 and sketch the probable deflected shapes. Determine the redundant force at location 1 and equations for the deflections at locations 1 and 2. *Ans.* $R_1 = (5/4)P$, $v_2 = 0.203\ Pl^3/EI$.

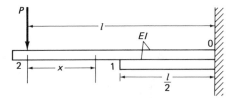

9.79   Small electronic components may be supported on bent wires built in at both ends, as shown. Of some interest are the stresses and deflections under inertia loads P. Derive equations for the redundant forces and moments and for the displacement of the component, which is to be considered rigid.

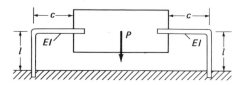

9.80   Given a shaft with the dimensions of Fig. 9.29 but with the supports in-line and the shaft loaded instead by a uniformly distributed force q_1 along length m and q_3 along length n (e.g., a shaft loaded by its own weight), also loaded by moments M_0 and M_4 at locations 0 and 4, respectively, in a direction to resist the changes of slope due to q_1 and q_3. Derive by superposition an equation from which the internal moment M_2 at location 2 may be determined. *Ans.* $(M_0 m/I_m) + 2M_2[(m/I_m) + (n/I_n)] + (M_4 n/I_n) = (q_1 m^3/4I_m) + (q_3 n^3/4I_n)$.

Note: This is known as a Theorem of Three Moments, because there are three moments in it. It may be solved directly for M_2 if M_0 and M_4 are known or are zero. If the two spans are part of a continuous beam with additional supports, M_0 and M_4 are the moments from adjacent spans. The equation can be written for each *pair* of spans in the entire beam, and if there are g spans, there will be $g - 1$ unknown moments, and $g - 1$ equations to be solved simultaneously.

9.81   For a uniform beam on three supports, without end moments and symmetrically placed and uniformly loaded ($m = n$, $q_1 = q_3 = q$), determine M_2 and the three-bearing reactions by using the three-moment equation in the answer to Prob. 9.80.

9.82   The shaft described in Prob. 9.80 has extensions of length c at each end. A downward force P is applied at the end of each extension. There are no other forces or distributed loads ($q_1 = q_3 = 0$). (a) From the three-moment equation given in the answer to Prob. 9.80, determine the value of M_2. (b) For the special case of $m = n = c$ determine the three bearing reactions, checking carefully for equilibrium of the entire shaft. Sketch the bending moment diagram for the shaft and compare maximum moment and force with those for a shaft with no central bearing (e.g., Case (1) of Fig. 9.9). *Ans.* $R_2 = 3P$.

9.83   Derive an equation for the relative deflection of the ends of the straight bimetallic strip when its temperature is increased by an amount T. Coefficients of expansion are α_1 and α_2 and moduli of elasticity, E_1 and E_2, respectively. *Suggestion:* Sketch the two materials separately (1) showing them straight and expanded different amounts at the higher temperature, (2) curved so that their adjacent surfaces will fit together, and acted upon at their ends by the necessary internal moments and axial forces to cause this configuration. Write equalities and solve for the redundancies, then deflection. *Ans.* $v = 6(\alpha_2 - \alpha_1)Tl^2E_1E_2/h(E_1^2 + E_2^2 + 14E_1E_2)$.

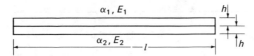

9.84 The use of three bearings to support a shaft is generally avoided because of the difficulty of placing them in alignment. Suppose that a uniform shaft of length l has bearings at $x = 0$ and $x = l$, and that a third bearing at midpoint $x = l/2$ is out of line by amount $\triangle$. There are parallel forces P applied at positions $x = l/4$ and $x = 3l/4$. By the method of superposition determine the bearing reactions and the maximum moment in terms of P, EI, $\triangle$, and l, (a) when both forces P are opposite in direction to $\triangle$, (b) in the same direction as $\triangle$. (c) Compare the reactions at the central bearing and the maximum moments for cases (a) and (b). *Ans.* (a) $R_0 = 5P/16 - 24\triangle EI/l^3$, $R_{l/2} = 11P/8 + 48\triangle EI/l^3$.

9.85 In order to give some central support to the shaft of Prob. 9.84, yet minimize the effect of misalignment, it is proposed to support the central bearing by a flexible structure of stiffness k. (a) For an installation made without misalignment and initial force at the central bearing determine equations for deflection and reaction there when the forces P are applied. (b) Determine the corresponding equations for deflection and reaction when there is also an initial misalignment $\triangle$ at the central bearing.

10

Components in Flexure II
Strain Energy, Thin and Thick Curved Beams,
Continuous Elastic Support, Plates

10.1 STRAIN ENERGY THEOREMS

Energy stored elastically is considered in some other sections of this book. In Section 5.2 the energy per unit volume at a point in a body is derived in terms of stress and strain components, then applied in Section 5.6 to the distortion-energy theory of failure. In Section 13.2 and Fig. 13.3, a study of the longitudinal impact on bars is based upon the energy stored from the application of a force F and corresponding deflection δ, namely,

$$U = \frac{F \cdot \delta}{2} = \frac{F^2 l}{2AE} = \frac{AE\delta^2}{2l} = \frac{\sigma^2 A^2}{2k} \tag{10.1}$$

where k is the stiffness constant AE/l. In this and the next section, the energy of flexure will be derived, together with theorems that will make it applicable to the determination of deflections and redundant reactions in statically loaded, straight and curved beams, including rings and pipe bends.

The energy stored elastically equals the work done during the deformation, and deformation increases linearly with the load. The work done on a beam by a moment that increases from zero to a value M_1 is the average moment multiplied by the angle of rotation, or $(M_1/2)\theta_1$, and it is represented by the shaded area on the M vs θ diagram (Fig. 10.1(a)). The moment is uniformly distributed and, hence, $\theta_1 = M_1 l/EI$, which is the area of the moment diagram below the beam times $1/EI$. Hence the work done and the stored energy are each

$$U = \frac{M_1\theta_1}{2} = \frac{M_1^2 l}{2EI} = \frac{EI\theta_1^2}{2l} \tag{10.2}$$

The work done in bending the beam of Fig. 10.1(b) is the average force times the deflection at the force, or $(P_1/2)v_1$. However, moment and stored energy vary along the beam, and integration is required to obtain the energy. The energy stored in an element of length dx is the average moment to rotate one side of the element relative to its other through the angle $d\theta$, or $(M/2)d\theta$. Now $d\theta = (1/\rho)dx$ (Fig. 10.1(b)). The curvature $1/\rho = M/EI = d^2v/dx^2$, from Eqs. (9.19) and (9.20). Hence the energy U stored in the beam

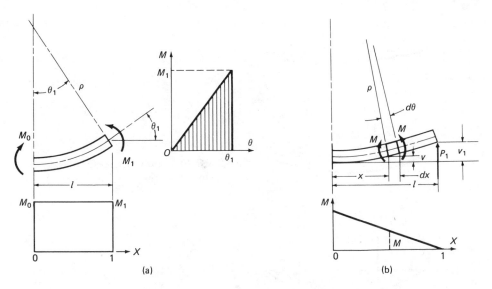

FIGURE 10.1. Energy in beams. (a) Uniform moment. (b) Variable moment.

may be expressed in terms of M or of d^2v/dx^2, as

$$U = \int \frac{M}{2}\, d\theta = \int \frac{M^2}{2EI}\, dx = \int \frac{EI}{2}\left(\frac{d^2v}{dx^2}\right)^2 dx \tag{10.3}$$

where the appropriate limits for the integration are to be applied. If EI and/or the equations for M differ for several parts of the beam or structure, a to b, b to c, etc., then

$$U = \int_a^b \frac{M_{ab}{}^2\, dx}{2E_{ab}I_{ab}} + \int_b^c \frac{M_{bc}{}^2 dx}{2E_{bc}I_{bc}} + \cdots \tag{10.4}$$

In structures and curved beams, bending may be accompanied by torsion which contributes to slope and deflection. The equation for energy stored by twisting with a torque T is developed in a similar manner, and is

$$U = \int \frac{T^2}{2GJ}\, dx = \int \frac{GJ}{2}\left(\frac{d\theta}{dx}\right)^2 dx \tag{10.5}$$

where θ is the angle of twist, G is the modulus of rigidity, and J is the polar moment of inertia of the cross-sectional area.

For the beam of Fig. 10.1(b), we may equate the work done, $P_1v_1/2$, to the energy stored, Eq. (10.3), and solve for deflection v_1. Since $M = P_1(l - x)$,

$$\frac{P_1v_1}{2} = \frac{P_1{}^2}{2EI}\int_0^l (l - x)^2\, dx = \frac{P_1{}^2 l^3}{6EI}$$

whence $v_1 = P_1 l^3/3EI$, which we recognize as the correct answer. A more useful and general method will be developed, however, and it will be done in three steps: influence coefficients, Maxwell's reciprocal theorem, and Castigliano's theorem.

An influence coefficient is the displacement of a body at one location due to a unit force applied at the same or some other location. For the loading of Fig. 10.2(a), α_{22} is the deflection at location 2 due to unit load at 2, and α_{12} is the deflection at location 1 due to unit load at 2. Loaded with unit force at location 1, as in Fig. 10.2(b), the deflection at 1 is α_{11} and the deflection at 2 is α_{21}. The value of a coefficient may be

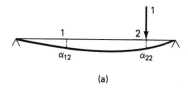

(a)

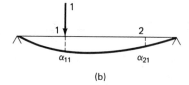

(b)

FIGURE 10.2. Influence coefficients for beam deflections.

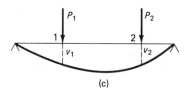

(c)

determined by substituting $P = 1$ in the equation for deflection, or it may be determined experimentally by applying a load, measuring the corresponding deflection at the desired position, and dividing deflection by load. In Fig. 10.2(c), the same beam is loaded by forces P_1 and P_2, and the "influence" of P_2 on the deflection at location 1 is the product $\alpha_{12}P_2$, etc. so that

$$v_1 = \alpha_{11}P_1 + \alpha_{12}P_2 \quad \text{and} \quad v_2 = \alpha_{21}P_1 + \alpha_{22}P_2 \tag{10.6}$$

Two work sequences are used to prove the theorem of reciprocity. In Fig. 10.3(a),

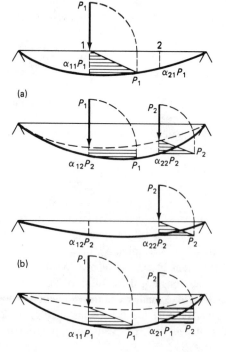

(a)

(b)

FIGURE 10.3. Work done on a beam (shaded areas) in sequence (a) and sequence (b) for proof of the theorem of reciprocity.

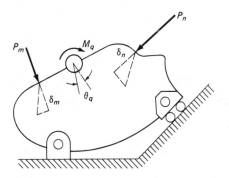

FIGURE 10.4. Generalized forces and corresponding displacements in an elastic body.

a force applied at location 1 is built up from zero to a value P_1, while doing the work $P_1 (\alpha_{11}P_1)/2$. This is followed by a buildup in P_2, which does the work $P_2 (\alpha_{22}P_2)/2$ at location 2, while the constant force P_1 moves through the distance $\alpha_{12}P_2$ and does the work $P_1(\alpha_{12}P_2)$. If there is no deflection at the supports, no work is done there. The total work done and the energy stored is

$$U = \frac{1}{2}\alpha_{11}P_1{}^2 + \frac{1}{2}\alpha_{22}P_2{}^2 + \alpha_{12}P_1P_2$$

In the sequence shown in Fig. 10.3(b), force P_2 is applied first, doing work $P_2(\alpha_{22}P_2)/2$, followed by P_1 with work $P_1(\alpha_{11}P_1)/2$, and additional work by P_2 of amount $P_2(\alpha_{21}P_1)$. The total work or energy is

$$U = \frac{1}{2}\alpha_{11}P_1{}^2 + \frac{1}{2}\alpha_{22}P_2{}^2 + \alpha_{21}P_1P_2$$

The final energy U must be the same by both sequences, which requires that

$$\alpha_{21} = \alpha_{12} \tag{10.7}$$

This is known as Maxwell's theorem of reciprocity. In words, Eq. (10.7) says that the deflection at location 2 due to unit load at location 1 is equal to the deflection at location 1 due to unit load at location 2. In pictures, α_{21} of Fig. 10.2(b) is shown equal to α_{12} of Fig. 10.2(a). The theorem applies to other loads and locations as well, and in general, $\alpha_{nm} = \alpha_{mn}$. It applies not only to beams, but to any elastic body, such as that in Fig. 10.4. Since the theorem was derived on the basis of work done, the displacement and all the coefficients α that apply at a particular location must be components in the direction of the force at that location. Also, "force" and "displacement" may be generalized to include "moment" and "angle of rotation," respectively.

With reference again to Figs. 10.3 and 10.4, let all the forces increase together, starting at zero and each at a uniform rate such that all forces reach their final value at the same time with the displacements increasing in proportion. The total work is

$$U = \frac{1}{2}P_1\delta_1 + \frac{1}{2}P_2\delta_2 + \cdots + \frac{1}{2}P_n\delta_n$$

which in terms of influence coefficients is

$$U = \frac{1}{2}P_1(\alpha_{11}P_1 + \alpha_{12}P_2 + \cdots + \alpha_{1n}P_n) + \frac{1}{2}P_2(\alpha_{21}P_1 + \alpha_{22}P_2$$

$$+ \cdots + \alpha_{2n}P_n) + \cdots + \frac{1}{2}P_n(\alpha_{n1}P_1 + \alpha_{n2}P_2 + \cdots + \alpha_{nn}P_n)$$

Differentiation with respect to one force, say P_2, yields

$$\frac{\partial U}{\partial P_2} = \frac{1}{2} P_1 \alpha_{12} + \frac{1}{2} (\alpha_{21} P_1 + 2\alpha_{22} P_2 + \cdots + \alpha_{2n} P_n) + \cdots + \frac{1}{2} P_n \alpha_{n2}$$

Note the one different term, $2\alpha_{22} P_2$, which comes from the differentiation of $\alpha_{22} P_2^2$. The reciprocity theorem is now applied such that all other terms occur in pairs, which may be added to give

$$\frac{\partial U}{\partial P_2} = \alpha_{21} P_1 + \alpha_{22} P_2 + \cdots + \alpha_{2n} P_n = \delta_2 \tag{10.8}$$

since the sum is by definition equal to the displacement at location 2. Generalized, this is Castigliano's theorem,[1] namely, that the partial derivative of the total strain energy in an elastic member or structure with respect to any of the external "forces" is the "displacement" of the point of application of that force in the direction of the force. In the form of equations,

$$\delta_n = \frac{\partial U}{\partial P_n} \quad \text{and} \quad \theta_n = \frac{\partial U}{\partial M_n} \tag{10.9}$$

These relationships have been widely used in the analysis of structures. They will be used here mainly in the analysis of curved beams and rings.

10.2 APPLICATIONS OF CASTIGLIANO'S THEOREM: THIN CURVED BEAMS

If there is more than one "force" in the moment equation, algebraic work is minimized if the differentiation of Eqs. (10.9) precedes the integration by Eq. (10.3). Thus Eqs. (10.9) with substitution from Eq. (10.3) become

$$\delta_n = \frac{\partial U}{\partial P_n} = \int \frac{1}{EI} \frac{\partial}{\partial P_n} \left(\frac{M^2}{2} \right) dx = \int \frac{M}{EI} \frac{\partial M}{\partial P_n} dx = \int \frac{M}{EI} \frac{\partial M}{\partial P_n} r \, d\phi \tag{10.10}$$

and

$$\theta_n = \frac{\partial U}{\partial M_n} = \int \frac{1}{EI} \frac{\partial}{\partial M_n} \left(\frac{M^2}{2} \right) dx = \int \frac{M}{EI} \frac{\partial M}{\partial M_n} dx = \int \frac{M}{EI} \frac{\partial M}{\partial M_n} r \, d\phi \tag{10.11}$$

The element of arc $r \, d\phi$ replaces dx for thin curved beams, where r is the radius to the centroidal axis. The rigidity EI is independent of P_n and M_n but in general is not independent of x. Only if EI is uniform may it be placed in front of the integral sign. In any case the integration may need to take place in several steps, as in Eq. (10.4), since the equation for M changes at every concentrated, external "force," whether applied or reactive. Then, from Eq. (10.10), with EI constant throughout,

$$\delta_n = \frac{\partial U}{\partial P_n} = \frac{1}{EI} \left[\int_a^b M_{ab} \frac{\partial M_{ab}}{\partial P_n} dx + \int_b^c M_{bc} \frac{\partial M_{bc}}{\partial P_n} dx + \cdots \right] \tag{10.12}$$

To obtain the slope at any location x_j where there is no external moment, a fictitious moment N_j is applied there and included in the moment equation. The derivative $\partial M / \partial N_j$ is taken, immediately after which, and only after which, substitution of zero for N_j is made in the equation for M. Then θ_j is found by Eq. (10.11). Deflections are similarly found. A fictitious force Q_j is introduced in the desired direction and location j of the

---

[1] For a careful delineation of Castigliano's theorem and a comparison of it with two other energy theorems, *virtual work*, and *least work*, see J. P. Den Hartog, *Advanced Strength of Materials*, New York: McGraw-Hill, Chapt. VII, p. 212, 1952. For an illustration of virtual work see the spring-clutch problem in Section 10.3.

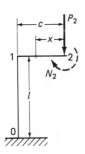

FIGURE 10.5. Angle beam with fictitious moment N_2 added to obtain slope θ_2 by the energy method (Example 10.1).

deflection, the moment equation is written, differentiation $\partial M/\partial Q_j$ is made, after which $Q_j = 0$ is substituted, and δ_j is found by Eq. (10.10).

Example 10.1
Determine the deflection and slope at the end of the angle or bracket of Example 9.9.

Solution
Figure 9.27 is redrawn in Figure 10.5, and in anticipation of solving for the slope at 2, we add a fictitious moment N_2. Then along the top leg, $M_{12} = P_2 x + N_2$, $\partial M_{12}/\partial P_2 = x$, and $\partial M_{12}/\partial N_2 = 1.0$; along the vertical leg, $M_{01} = P_2 c + N_2$, $\partial M_{01}/\partial P_2 = c$, and $\partial M_{01}/\partial N_2 = 1.0$. Now N_2 is set equal to zero, and by Eqs. (10.12) and (10.11)

$$\delta_2 = \frac{1}{EI}\left[\int_0^c (P_2 x)x\, dx + \int_0^l (P_2 c)c\, dx \right] = \frac{P_2}{EI}\left[\frac{c^3}{3} + c^2 l \right] = \frac{P_2 c^2}{3EI}(c + 3l)$$

and

$$\theta_2 = \frac{1}{EI}\left[\int_0^c (P_2 x)(1.0)\, dx + \int_0^l (P_2 c)(1.0)\, dx \right]$$

$$= \frac{P_2}{EI}\left[\frac{c^2}{2} + cl \right] = \frac{P_2 c}{2EI}(c + 2l)\ ,$$

which check with Eqs. (9.31).   ////

Example 10.2
The two electrically conductive springs of Fig. 10.6(a) clamp the pin of an exchangeable, printed-circuit board. The deflection of each spring at the contact point equals half the thickness t of the pin plus any initial compression δ_i at the assembly of the springs, or $\delta = t/2 + \delta_i$. Determine the contact force.

Solution
The undeflected position is represented in Fig. 10.6(b) by the dashed line with deflection δ exaggerated. For the straight leg of the spring, $M = Px$, and for the semicircle, $M = P(l + r \sin \phi)$. Hence from Eq. (10.12) with dx replaced by $r\, d\phi$ in the second term,

$$\delta = \frac{\partial U}{\partial P} = \frac{1}{EI}\left\{ \int_0^l (Px)\, x\, dx + \int_0^\pi P(l + r \sin \phi)(l + r \sin \phi)\, r\, d\phi \right\}$$

$$= \frac{P}{EI}\left\{ \left[\frac{x^3}{3} \right]_0^l + r\left[l^2\phi - 2lr \cos \phi + r^2\left(\frac{\phi}{2} - \frac{\sin 2\phi}{4} \right) \right]_0^\pi \right\}$$

$$= \frac{P}{EI}\left(\frac{l^3}{3} + \pi l^2 r + 4lr^2 + \frac{\pi r^3}{2} \right) = \frac{t}{2} + \delta_i$$

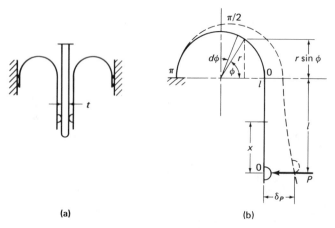

(a)   (b)

FIGURE 10.6. Electrical contact springs. (a) Pin clamped between two springs. (b) Dashed line position of free spring and solid line deflected position (Example 10.2).

Thus

$$P = \frac{EI\,(t/2 + \delta_i)}{l^3/3 + \pi l^2 r + 4lr^2 + \pi r^3/2} \tag{10.13}$$

Substitution of P and $\phi = \pi/2$ into the moment equation for the semicircular portion gives the maximum moment, from which the maximum stress may be determined.   ////

When forces are distributed, integration may be required to obtain the moment equations.

Example 10.3
Figure 10.7 shows a rotating split ring, such as a retaining ring on a shaft. Determine the moment equation and the radial expansion of the ring due to rotation.

Solution
The centrifugal force on an element of length $r\,d\psi$ and width b is $p_c br\,d\psi$, where p_c is a unit centrifugal force, given by Eq. (3.33a), and r is the mean radius. Its moment about point A at angle ϕ is $(p_c br\,d\psi)\,[r \sin(\phi - \psi)]$, whence

$$M = p_c br^2 \int_0^\phi \sin(\phi - \psi)\,d\psi = p_c br^2\,(1 - \cos\phi)$$

FIGURE 10.7. Rotating split retaining ring. (a) Fictitious force Q to determine displacement at $\phi = \pi/2$. (b) Free ring. (c) Ring in place in groove of shaft ($D > d$) (Example 10.3).

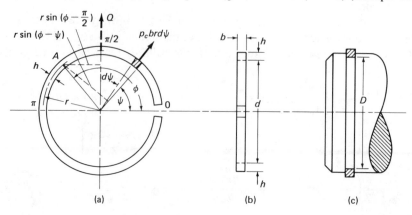

(a)   (b)   (c)

The expansion at $\phi = \pi/2$ is critical, so a fictitious radial force Q is added there. For $\pi/2 < \phi < \pi$, its moment about point A is $Qr \sin (\phi - \pi/2) = -Qr \cos \phi$. The total moment is $M' = p_c br^2 (1 - \cos \phi) - Qr \cos \phi$. By Eq. (10.12),

$$\delta_{\pi/2} = \frac{1}{EI} \left[\int_0^{\pi/2} M \frac{\partial M}{\partial Q} r \, d\phi + \int_{\pi/2}^{\pi} M' \frac{\partial M'}{\partial Q} r \, d\phi \right]$$

Now $\partial M/\partial Q = 0$, $\partial M'/\partial Q = -r \cos \phi$, and $Q = 0$, whence

$$\delta_{\pi/2} = \frac{p_c br^4}{EI} \int_{\pi/2}^{\pi} (\cos^2 \phi - \cos \phi) \, d\phi = \frac{p_c br^4}{EI} \left[\frac{\phi}{2} + \frac{\sin 2\phi}{4} - \sin \phi \right]_{\pi/2}^{\pi}$$

whence the radial expansion[2] at the 90° position is

$$\delta_{\pi/2} = \frac{p_c br^4}{EI} \left(\frac{\pi}{4} + 1 \right) \tag{10.14}$$

$////$

Energy methods may be applied for the solution of statically indeterminate problems. To write a moment equation use is made of some known or zero flexure, such as deduced by observation, and conditions of symmetry or temperature change, followed by a judicious separation of the member into similar parts. Thus for the circular ring of Fig. 10.8(a), the dotted-line sketch of the probable deflected shape indicates that there is no change of slope at $\phi = 0, 90, 180$, and 270 degrees because of symmetry of loading and resistance. At these locations the ring can be divided into two equal halves (Fig. 10.8(b)), each with energy U_π, or four equal quadrants (Fig. 10.8(c)), each with energy $U_{\pi/2}$. A moment equation may be written for either half or the quadrant in terms of only one unknown and statically indeterminate moment M_0 or M_0' and known force $P = F/2$. Use of P instead of $F/2$ will avoid confusion in taking derivatives relative to this force.

FIGURE 10.8. Circular ring with radial load F. (a) Deflected shape. (b) Analyzed by a semicircle. (c) Analyzed by a quadrant (Example 10.4).

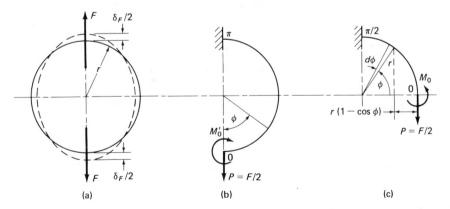

(a)   (b)   (c)

[2] A retaining ring, to locate and retain a ball bearing or other part, is usually expanded and slipped over the shaft and into a groove, on the smaller diameter of which it makes an interference fit, i.e., the ring's free (unstressed) inner diameter d was less than the groove diameter D (Fig. 10.7). High speeds will cause loss of this "fit," with resulting looseness and wear unless $(D-d)/2 \geq \delta_{\pi/2}$ of Eq. (10.14). Substitution for p_c from Eq. (3.33c), of $r = (D + h)/2$, $E = 30 \times 10^6$ psi and $\gamma = 0.281$ lb/in^3 (steel), and $I = bh^3/12$ yields as the loosening speed n (rpm),

$$n = (1.675 \times 10^6) \frac{h}{(D + h)^2} \sqrt{\frac{D - d}{D + h}} \tag{10.14a}$$

where dimensions are in inches. Derivation of the equivalent of Eqs. (10.14) and (10.14a) by integration of a differential equation with boundary conditions is given by R. L. Karr, *Product Eng.*, pp. 107–108, 125, October 29, 1962.

Example 10.4

Determine the moment M_0 and indicate how to find the extension of the ring of Fig. 10.8.

Solution

For the quadrant of Fig. 10.8(c), the moment at location ϕ is $M = M_0 - Pr(1 - \cos \phi)$. From Eq. (10.11), the equation for the slope θ_0 is written, then equated to zero, thus

$$
\begin{aligned}
\theta_0 &= \frac{\partial U_{\pi/2}}{\partial M_0} = \frac{1}{EI} \int M \frac{\partial M}{\partial M_0} r \, d\phi \\
&= \frac{1}{EI} \int_0^{\pi/2} [M_0 - Pr(1 - \cos \phi)] \, (1.0) \, r \, d\phi \\
&= \frac{r}{EI} \Big[(M_0 - Pr)\phi + Pr \sin \phi \Big]_0^{\pi/2} \\
&= \frac{r}{EI} \bigg[(M_0 - Pr) \frac{\pi}{2} + Pr] \bigg] = 0
\end{aligned}
$$

From this, moment M_0 is determined, then substituted into the moment equation, or

$$
M_0 = Pr\left(1 - \frac{2}{\pi}\right) \quad \text{and} \quad M = Pr\left(\cos \phi - \frac{2}{\pi}\right) = Fr\left(\frac{\cos \phi}{2} - \frac{1}{\pi}\right) \tag{10.15}
$$

At $\phi = \pi/2$, at the external load, M has its maximum absolute value, $M_{\max} = -Fr/\pi$. An inflection occurs when $M = 0$, or where $\cos \phi = 2/\pi$.

The amount that the loaded ends of the ring separate is twice the deflection at $\phi = 0$ in the one quadrant, or

$$
\delta_F = 2\delta_P = 2 \frac{\partial U_{\pi/2}}{\partial P} = \frac{2}{EI} \int_0^{\pi/2} M \frac{\partial M}{\partial P} r \, d\phi \tag{10.16}
$$

Also, since the energy in the entire ring is four times that in one quadrant, the separation may also be found as

$$
\delta_F = \frac{\partial U_{2\pi}}{\partial F} = 4 \frac{\partial U_{\pi/2}}{\partial F} = \frac{4}{EI} \int_0^{\pi/2} M \frac{\partial M}{\partial F} r \, d\phi \tag{10.17}
$$

////

Example 10.5

The flexible pipe loop of Fig. 10.9(a) might be placed in a steam line between a boiler and a turbine, to allow thermal expansion of the line without excessive binding and stressing. Determine the force and moment developed by a temperature increase T above that at installation.

Solution

Without restraint the ends would move apart by an amount $(2\sqrt{2}\, r)\alpha T$ where α is the coefficient of thermal expansion, and $(2\sqrt{2}\, r)$ is the distance between ends, as shown. The ends *are* constrained or fixed, so a force P and moment M_0 develop there. Because of symmetry, the top of the loop has no change of slope, and it is subjected to the same P and M_0. Hence for the half-loop free body of Fig. 10.9(b) and by $\Sigma M = 0$ taken about either one of the ends, $M_0 + M_0 - P(2 + \sqrt{2})r = 0$, whence $M_0 = (1 + \sqrt{2}/2)\, Pr$. The moment at any angular position ϕ between 0 and 135° is $M = M_0 - Pr(1 - \cos \phi) = Pr(\sqrt{2}/2 + \cos \phi)$. On the other hand, we might at the beginning have recognized

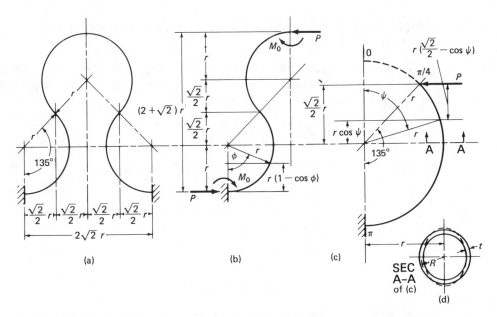

FIGURE 10.9. Pipe loop for thermal expansion. (a) Overall dimensions. (b) Half-loop as a free body. (c) Alternate 135° free body. (d) Deformation of the circular cross section (Example 10.5).

from the symmetry of bending that there is a point of inflection and zero moment at $\phi = 135°$, and then divided the loop into four 135° circular cantilevers (Fig. 10.9(c)). The moment equation for $\pi/4 < \psi < \pi$ is $M_\psi = Pr(\sqrt{2}/2 - \cos \psi)$.

The energy for the entire loop is four times that of one 135° cantilever, so $\delta_P = 4(\partial U/\partial P)$. Because the circular cross section of a curved pipe deforms to an elliptical shape (Fig. 10.9(d)), the pipe is more flexible than indicated by EI alone, and EI must be replaced by KEI,[3] where factor $K < 1.0$. If we choose to use M_ψ, then by Eq. (10.10),

$$\delta_P = \frac{4Pr^3}{KEI} \int_{\pi/4}^{\pi} \left(\frac{\sqrt{2}}{2} - \cos \psi \right)^2 d\psi = \frac{4Pr^3}{KEI} \left[\psi - \sqrt{2} \sin \psi + \frac{\sin 2\psi}{4} \right]_{\pi/4}^{\pi}$$

from which

$$\delta_P = \frac{3 (\pi + 1) Pr^3}{KEI} \tag{10.18}$$

The sum of thermal expansion and elastic deflection at the rigid supports must be zero, or

$$2 \sqrt{2} \, r\alpha T - \frac{3 (\pi + 1) Pr^3}{KEI} = 0$$

From this the force P is found to be

$$P = \frac{0.228 \, KE\alpha TI}{r^2} \tag{10.19}$$

and moment M_0 is found from $(1 + \sqrt{2}/2) \, Pr$. ////

---

[3] $K = (12\lambda^2 + 1)/(12\lambda^2 + 10)$, where $\lambda = tr/R^2$, t is the pipe wall thickness, R is the wall mean radius while circular, and r is the radius of curvature of centroidal plane, shown in Fig. 10.9(c) and (d). For a derivation, which is due to T. von Kármán, see e.g., J. P. Den Hartog, *Advanced Strength of Materials*, New York: McGraw-Hill, pp. 234–245, 1952. If $r/R = 10$ and $R/t = 10$, $\lambda = 1.0$ and $K = 0.59$. This considerably reduces the force P. However, high pressure and liquids may increase the stiffness.

10.3 ADDITIONAL RING ANALYSES: OUT-OF-PLANE LOADINGS: THE COIL CLUTCH

When loadings cause deflections normal to the plane of the ring, twist as well as bending may contribute to the deflection and both torque and bending moments will appear in the energy equation. Thus from Eqs. (10.3) and (10.4),

$$U = U_M + U_T = \int \frac{M^2}{2EI}\, dx + \int \frac{T^2}{2GJ}\, dx$$

Slope, twist, and deflection at location n are, respectively,

$$\theta_n = \frac{\partial U}{\partial M_n} = \int \frac{M}{EI} \frac{\partial M}{\partial M_n}\, dx + \int \frac{T}{GJ} \frac{\partial T}{\partial M_n}\, dx$$

$$\beta_n = \frac{\partial U}{\partial T_n} = \int \frac{M}{EI} \frac{\partial M}{\partial T_n}\, dx + \int \frac{T}{GJ} \frac{\partial T}{\partial T_n}\, dx \qquad (10.20)$$

$$\delta_n = \frac{\partial U}{\partial P_n} = \int \frac{M}{EI} \frac{\partial M}{\partial P_n}\, dx + \int \frac{T}{GJ} \frac{\partial T}{\partial P_n}\, dx$$

where M_n, T_n, and P_n are, respectively, the moment, torque, and force applied at location n.

Because of the radial symmetry of the engines they support, and because of a high strength/weight ratio, tubular circular rings are used in aircraft as supporting structures. The ring of Fig. 10.10 is a support and gimbal ring for pivoting the nozzle of a rocket motor to direct its thrust.[4] The net axial force applied to the ring both at the top and at

FIGURE 10.10. Gimbal ring for rocket motor nozzle. (a) Thrust forces F and deflection δ. (b) Forces P, bending moments M, and twisting moments T on a quadrant. (Adapted from R. B. Gausvik and W. E. Springer, "Analyze a Statically Indeterminate Ring by the Use of Castigliano's Theorem," *General Motors Engineering J.*, Vol. 10, No. 1, pp. 51–52, 1963, with permission of General Motors Corporation.)

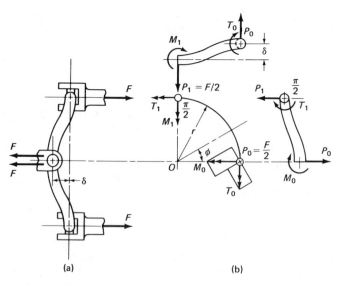

(a)   (b)

---

[4] This problem and two of the sketches of Fig. 10.10 are adapted from the article by R. B. Gausvik and W. E. Springer, "Analyze a Statically Indeterminate Ring by the Use of Castigliano's Theorem," *General Motors Engineering J.*, Vol. 10, No. 1, pp. 51–54 and No. 2, pp. 52–54, 1963.

the bottom is F in the direction of the gas flow. Equal and opposite reactive forces are exerted on the ring by the rocket body at two midway locations. Because of the rigidity of the clevises at the four pivots, neither slope measured circumferentially on the ring nor twist angle measured in a radial plane change significantly, i.e., they may be assumed to equal zero at the pivots. The ring is deflected out of its plane by the amount δ. This results in unknown bending and twisting moments at the pivots. These are pictured in the free body of a quadrant (Fig. 10.10(b)) together with forces $F/2$, since each F is shared by two quadrants. Two moments can be obtained in terms of the other two by the equations of equilibrium, but the latter, say M_0 and T_0, must be obtained from the known zero slopes and twist angles.

The bending moment and twisting moment equations are, respectively, for $0 \leqslant \phi \leqslant \pi/2$,

$$M = M_0 \cos \phi - P_0 r \sin \phi + T_0 \sin \phi \tag{10.21}$$

$$T = -M_0 \sin \phi + P_0 r (1 - \cos \phi) + T_0 \cos \phi$$

where P_0 rather than $F/2$ is used to avoid confusion with differentiation. The flexure conditions of zero slope θ_0 and zero twist β_0, plus deflection δ_0, are, respectively,

$$\theta_0 = \frac{\partial U}{\partial M_0} = 0, \quad \beta = \frac{\partial U}{\partial T_0} = 0, \quad \delta_0 = \frac{\partial U}{\partial P_0} \tag{10.22}$$

Substitution from Eqs. (10.20) and (10.21) is made in Eqs. (10.22). The first two are solved simultaneously for M_0 and T_0 in terms of $P_0 r$. The applied force F may be reintroduced by substituting $F/2$ for P_0. With M_0 and T_0 now known, deflection δ_0 may be found from Eqs. (10.22) and the maximum moment and torque from Eqs. (10.21), and then the stresses may be calculated.[5]

Section 3.9 describes a spring clutch that is assembled by forcing a helical coil over a cylinder that has a diameter ground slightly larger than the inside diameter of the free-standing coil. Its Eqs. (3.47) and (3.50) for pressure and stress, respectively, will be derived with the aid of an energy method which is essentially that of virtual work.[6] The dimensions of the free-standing coil are given with subscripts zero in Fig. 10.11(a). A length l along the centrodial arc subtends a central angle $\theta_0 = l/r_0$. When forced over the clutch drum, the spring fits on a radius R, larger than R_0 by the amount of the radial interference δ, such that $R = R_0 + \delta$ and $r = r_0 + \delta$, Fig. 10.11(b). The fiber of length

FIGURE 10.11. Coil-spring clutch. (a) Free coil. (b) Coil assembled over a cylinder. (c) Radial force vs elastic displacement.

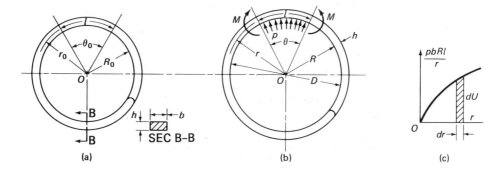

l subtends a smaller angle $\theta = l/r$, and the change of "slope" over the length l is $\theta_0 - \theta = (l/r_0) - (l/r)$.

An internal moment M and pressure p are developed. For a straight beam, which has zero initial curvature, moment M equals EI times the curvature developed (Eq. (9.19)). For a curved beam, the moment is EI times the *change* of curvature, or

$$M = EI\left(\frac{1}{r_0} - \frac{1}{r}\right) \tag{10.23}$$

From Eq. (10.2), which gives energy stored when the moment is uniform and θ_1 is the change of slope,

$$U = \frac{EI\theta_1^2}{2l} = \frac{EI\,(\theta_0 - \theta)^2}{2l} = \frac{EIl}{2}\left(\frac{1}{r_0} - \frac{1}{r}\right)^2 = \frac{EIl}{2r_0^2}\left(1 - \frac{r_0}{r}\right)^2 \tag{10.24}$$

The work done upon the coil must come from an external force. This is the pressure or unit force increasing from zero to value p and acting radially outward in the same direction as the displacement $R - R_0$ of the spring. The relationship between force and displacement, which we do not yet know, need not be linear (Fig. 10.11(c)), but it must be elastic, i.e., the same relationship must hold as force is removed so that all the energy is recovered. A small or "virtual" displacement dr caused by a *unit force p* does the work $p\,dr$. The force on the area $bR\theta$ of the coil does the work $dU = (p\,dr)\,(bR\theta) = (p\,dr)\,(bRl/r) = \dfrac{pbRl}{r}\,dr$ (Fig. 10.10(c)). From the foregoing equation,

$$p = \frac{r}{bRl}\frac{dU}{dr} \tag{10.25}$$

This determines the value of p, since the derivative dU/dr can be found from Eq. (10.24).

The derivation is completed as follows. From Eq. (10.24),

$$\frac{dU}{dr} = \frac{EIl}{2r_0^2}\left[2\left(1 - \frac{r_0}{r}\right)\left(\frac{r_0}{r_2}\right)\right] = \frac{EIl}{r_0 r^2}\left(1 - \frac{r_0}{r}\right) = \frac{EIl}{r_0 r^3}\,(r - r_0)$$

By substitution of dU/dr into Eq. (10.25),

$$p = \frac{EI\,(r - r_0)}{bRr_0 r^2} \tag{10.26}$$

For $(r - r_0)$ we may substitute $\delta = \triangle/2$, where $\triangle$ is the diametral interference, and for r we may substitute $(D + h)/2$ (Fig. 10.11(b)). If $\delta << r_0$ and $h << R$, we may make the approximations $r_0 \approx R \approx r = (D + h)/2$ in Eq. (10.26). With these substitutions,

$$p = \frac{8EI\triangle}{b\,(D + h)^4} \tag{10.27}$$

which with unit centrifugal force p_c subtracted, is Eq. (3.47) for an external coil clutch.[7] The bending moment may be similarly approximated. From Eq. (10.23),

$$M = \frac{EI}{r_0 r}\,(r - r_0) \approx \frac{EI(\triangle/2)}{[(D + h)/2]^2} \approx \frac{2EI\triangle}{(D + h)^2} \tag{10.28}$$

---

[7] For a reference to some original derivations, see footnotes to Section 3.9.

The bending stress is $\sigma = Mc/I$, where $c = h/2$, whence

$$\sigma = \frac{Eh\triangle}{(D + h)^2} \tag{10.29}$$

which is also Eq. (3.50).

10.4 STRESSES IN CURVED BEAMS

Plane sections through curved beams, as in straight beams, remain plane during bending deformations. However, were a section to rotate about its centroidal axis, the deformations would be proportional to their distance from this axis. Strain is deformation divided by fiber length, and the arc length between two radial sections decreases from the outside or convex surface to the inside or concave side (Fig. 10.12). Hence strains are greater toward the inside, and if the beam behaves elastically, the stresses are also greater. This unbalances the internal moment about the centroidal axis, so the neutral axis or axis of zero strain and stress must shift toward the center of curvature.

This curvature effect may often be negligible, as with curved beams having the proportions illustrated in Figs. 10.6 through 10.11. When the ratio of centroidal radius to radial thickness is 5:1, the error in maximum stress by straight beam formulas is 7 or 8% for most symmetrical sections. For smaller ratios, as in Figs. 10.12 through 10.14, curved beam formulas should certainly be applied. For ratios of the order of 3:1 or less, sections such as the trapezoidal and the T-section with the greater width on the inside, will be more economical and usually necessary. Also, tapering of the thickness is common such that the greatest thickness occurs at the section of maximum moment, as in the hook of Fig. 10.13.

The location of the neutral axis is found from the equilibrium of forces. In the beam portion subtended by angle ϕ (Fig. 10.12), the length l of a fiber ef at any radius r is $r\phi$. The section bfd rotates an amount θ about a neutral axis at radius r_n, and the fiber extension fg at f is $u = (r_n - r)\theta$. The stress is

$$\sigma = E\frac{u}{l} = E\frac{(r_n - r)\theta}{r\phi} = \frac{E\theta}{\phi}\left(\frac{r_n - r}{r}\right) \tag{10.30}$$

FIGURE 10.12. The shift $\overline{y}$ inward of the neutral axis from the centroidal axis, characteristic of a curved beam. Distribution of (a) strain and (b) stress in a curved beam of circular cross section.

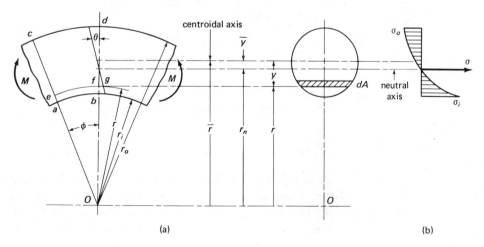

The requirement of force equilibrium is

$$\Sigma F = 0, \quad \int_A \sigma \, dA = \int_A \frac{E\theta}{\phi} \left(\frac{r_n - r}{r} \right) dA$$

$$= \frac{E\theta}{\phi} \left[r_n \int_A \frac{dA}{r} - \int_A dA \right] = 0$$

(a)

whence

$$r_n \int_A \frac{dA}{r} = \int_A dA = A$$

(b)

and

$$r_n = \frac{A}{\displaystyle\int_A \frac{dA}{r}}$$

(10.31)

The distance $\bar{y}$ from the centroidal to the neutral axis is $\bar{y} = \bar{r} - r_n$, and it is always measured inward from the centroidal axis.

The stress distribution may now be found from the equality of external and internal moments. The moment about the neutral axis of the force on element dA is $(r_n - r)$ $(\sigma \, dA)$. With integration and substitution for σ from Eq. (10.30)

$$M = \int_A (r_n - r) (\sigma \, dA)$$

$$= \frac{E\theta}{\phi} \int_A \frac{(r_n - r)^2}{r} dA$$

(c)

The integrand is expanded to

$$\int_A \frac{r_n{}^2 - 2r_n r + r^2}{r} dA = r_n \left[r_n \int_A \frac{dA}{r} - 2 \int_A dA \right] + \int_A r \, dA$$

The two terms within the bracket are equal to A and $-2A$, respectively, by Eq. (b), and $\int_A r \, dA = \bar{r} A$, by the definition of centroidal distance. Also, ratio $E\theta/\phi$ equals $\sigma r/(r_n - r)$, from Eq. (10.30). Hence the integrand becomes $r_n [A - 2A] + \bar{r} A = A(\bar{r} - r_n) = A\bar{y}$, and Eq. (c) becomes

$$M = \frac{\sigma r}{r_n - r} (A\bar{y}) \quad \text{or} \quad \sigma = \frac{M(r_n - r)}{A\bar{y}r}$$

(10.32)

A typical stress distribution is shown in Fig. 10.12(b).

The stresses at the inner and outer fibers are, respectively,

$$\sigma_i = \frac{M (r_n - r_i)}{A\bar{y}r_i} \quad \text{and} \quad \sigma_o = - \frac{M (r_o - r_n)}{A\bar{y}r_o}$$

(10.33)

where stress σ_i at the inner fibers is always positive and in tension if the moment tends to straighten out the beam, as in Fig. 10.12. Hence such a moment should be taken positive in Eqs. (10.33). Because $\bar{y}$ in these equations may be a small difference between the much larger radii, $\bar{r}$ and r_n, both must be accurately determined or there may be a

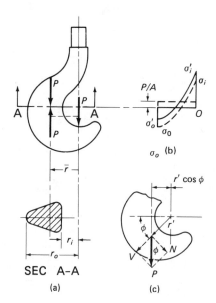

FIGURE 10.13. (a) Hook with typical trapezoidal section. (b) At Section A-A distribution of bending stress σ, direct stress P/A, and net stress σ'. (c) Section of hook at angle ϕ showing shear V and normal N components of applied force P.

large error in σ_i and σ_o.[8] Formulas for $\bar{y}$ or r_n for cross sections may be found in books, some of which will also list numerical values for the ratio of inner fiber stress to the corresponding stress in a straight beam.[9] For some shapes a graphical-numerical determination of properties is necessary.

Direct stresses and shear stresses may be significant when radii-to-thickness ratios are small. For the hook of Fig. 10.13 the applied force P is resolved into an equal, similarly directed force acting at the centroid of the section where maximum bending moment occurs, plus a moment $M = P\bar{r}$. Since the force acts at the centroid, the resulting stress P/A is uniformly distributed. It is because of this that the moment arm of M should always be taken to the centroid. The direct stresses are superimposed upon the bending stresses as illustrated in Fig. 10.13(b), and the net extreme fiber stresses are

$$\sigma_i' = \sigma_i + \frac{P}{A} \quad \text{and} \quad \sigma_o' = \sigma_o + \frac{P}{A} \tag{10.34}$$

where σ_i and σ_o are given by Eq. (10.33). For any other section, as in Fig. 10.13(c), the moment is $P(\bar{r}' \cos \phi)$, and the force is resolved into a normal component $N = P \cos \phi$ and shear component $V = P \sin \phi$.

Example 10.6

The dumbbell-shaped cross section of the forged aircraft link (Fig. 10.14) is offset to avoid interference with a crossing member. The outline of the section is well rounded to allow metal to flow readily to fill the die cavity. This makes the forging easier and

---

[8] Because of the chances for inaccuracy in small differences between large numbers, another method has been developed to obtain $\bar{y}$ in the curved beam equations. With reference to Fig. 10.12, let $r_n = \bar{r} - \bar{y}$ and $r = \bar{r} - y$ and substitute these values into the second integral of Eq. (a), which follows Eq. (10.30). The result is separated into two integrals, of which $\int y \, dA/(\bar{r} - y)$ is set equal to mA, where m is an area-modifying factor. The other integral $\int dA/(\bar{r} - y)$ is multiplied by $\bar{r}/\bar{r}$, after which $(-y + y)$ is added to the numerator. The result is separated into two integrals, the sum of which yields $(1 + m)$ ($A/\bar{r}$). Then, from the new form of Eq. (a), $\bar{y} = \bar{r}m/(1 + m)$, so that $\bar{y}$ is a *ratio* of $\bar{r}$ rather than a *difference* from it. Factor m may be found for a particular cross section from either of the two foregoing integrals. For details see, e.g., S. Timoshenko, *Strength of Materials*, 3rd ed., Princeton, New Jersey: D. Van Nostrand Company Company, Inc., p. 362, 1956.

[9] F. B. Seely and J. O. Smith, *Advanced Mechanics of Materials*, 2nd ed., New York: John Wiley & Sons, Inc., pp. 148−153, 1952; R. J. Roark and W. C. Young, *Formulas for Stress and Strain*, 5th ed., New York: McGraw-Hill, pp. 210−212, 1975.

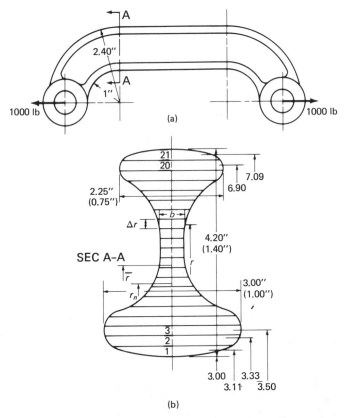

FIGURE 10.14. Graphical-numerical solution for a curved beam. (a) Offset link with 1000 lb load. (b) Section *A-A* enlarged to triple size and divided into elemental areas of $\triangle r = 0.20$ in width (shown here in reduced size). See Example 10.6 and Table 10.1.

more reliable but the engineer's analysis more difficult. The locations of maximum bending moment include the beginnings of the quadrants, so the analysis is that for a curved beam. Determine the properties of the cross section and the maximum inner and outer fiber stresses with a load of 1000 lb.

Solution

The section was drawn to triple size and divided into a convenient number of small strips or elements of equal thickness $\triangle r = 0.20$ in. This is shown to reduced size in Fig. 10.14(b). Samples of the measurements scaled from the drawing and their tabulation are given in Table 10.1. The following calculations are made from the summations of the table.

Area:

$$A = \int b\, dr = \triangle r \Sigma b = (0.20)(29.41) = 5.88 \text{ in}^2$$

Full-scale value, $A = 5.88/9 = 0.653$ in^2.

Radius to centroid:

$$\bar{r} = \frac{\int r\, dA}{A} = \frac{\int r\,(b\, dr)}{A} = \frac{\triangle r}{A}\Sigma(br) = \frac{\Sigma(br)}{\Sigma b} = \frac{142.58}{29.41} = 4.85 \text{ in}$$

Full-scale value, $\bar{r} = 4.85/3 = 1.62$ in.

TABLE 10.1. Determination of Properties for the Curved-Beam Section of the Link of Fig. 10.14

Element No. ($\triangle r = 0.20$ in)	Average Radius r (in) 3×	Average Width b (in) 3×	Product br (in²) 9×	Quotient b/r
1	3.11	2.00	6.22	0.643
2	3.33	2.83	9.42	0.851
3	3.50	2.98	10.43	0.852
.	.	.	.	.
.	.	.	.	.
20	6.90	2.20	15.20	0.319
21	7.09	1.72	12.20	0.243
Totals:		$\Sigma b = 29.41$	$\Sigma br = 142.58$	$\Sigma b/r = 6.597$

Radius to neutral axis:

$$r_n = \frac{A}{\int \frac{dA}{r}} = \frac{A}{\int \frac{b\,dr}{r}} = \frac{A}{\triangle r} \frac{1}{\Sigma(b/r)} = \frac{\Sigma b}{\Sigma(b/r)} = \frac{29.41}{6.597} = 4.46 \text{ in}$$

Full-scale value, $r_n = 4.46/3 = 1.49$ in.

Centroidal distance:

$$\bar{y} = \bar{r} - r_n = 1.62 - 1.49 = 0.13 \text{ in}$$

The need for accuracy in obtaining $\bar{r}$ and r_n is apparent, since their difference $\bar{y}$ is only 8% of their average value.

The applied load is 1000 lb and its moment about the centroid of the section is $P\bar{r} = 1620$ lb·in. The radii to the inner and outer fibers are $r_i = 1.00$ in and $r_o = 2.40$ in, as given. The net stresses are

$$\sigma_i' = \frac{(1620)(1.49-1.00)}{(0.653)(0.13)(1.00)} + \frac{1000}{0.653} = 9360 + 1530$$

$$= 10\ 890 \text{ psi (tensile)}$$

$$\sigma_o' = -\frac{(1620)(2.40-1.49)}{(0.653)(0.13)(2.40)} + \frac{1000}{0.653} = -7230 + 1530$$

$$= -5700 \text{ psi (compressive)} \hspace{3cm} ////$$

For an exact determination of the deflection of thick curved beams, deflection due to normal force N and shear force V should be included. These forces are illustrated in Fig. 10.13(c). The energy equation[10] is

$$U = \int \frac{N^2\,ds}{2AE} + \int \frac{\lambda V^2\,ds}{2AG} + \int \frac{M^2\,ds}{2AE\,\bar{y}\bar{r}} - \int \frac{MN\,ds}{AE\bar{r}} \hspace{2cm} (10.35)$$

where $ds = \bar{r}\,d\phi$ and $\lambda = \tau_{max}/\tau_{avg}$. Ratio λ is the ratio of the maximum to average transverse shear stresses in a beam, e.g., $\lambda = 1.5$ for a rectangular and $\lambda = 1.33$ for a

[10] S. Timoshenko, *Strength of Materials, Part I*, 3rd ed., Princeton, New Jersey: Van Nostrand, p. 382, 1955.

circular section. The deflection at load P is

$$\delta_P = \frac{\partial U}{\partial P} = \int \frac{N}{AE} \frac{\partial N}{\partial P} \, ds + \int \frac{\lambda V}{AG} \frac{\partial V}{\partial P} \, ds$$

$$+ \int \frac{M}{AE\bar{y}\bar{r}} \frac{\partial M}{\partial P} \, ds - \int \frac{M}{AE\bar{r}} = \frac{\partial N}{\partial P} \, ds - \int \frac{N}{AE\bar{r}} \frac{\partial M}{\partial P} \, ds \qquad (10.36)$$

In many cases the effect upon deflection of the normal and shear forces is negligible and the N and V terms may be neglected. Furthermore, the deflection due to moment is but little affected by a redistribution of stresses,[11] so that the deflection formulas of Section 10.2, based upon straight-beam stress distributions, usually may be used without much error. However, the maximum stress is considerably affected by the curvature, and the curved-beam stress equations, Eqs. (10.33) and (10.34), should always be used except under the large radii-to-thickness conditions discussed in the second paragraph of this section.

10.5 BEAMS WITH CONTINUOUS ELASTIC SUPPORT: THEORY

Instead of being concentrated at a few locations, the support forces or reactions on a beam may be distributed along its entire length. Examples include railroad rails, machines on continuous vibration-isolation mounting, and a beam supported by many crossing beams. The theory is adapted to determine secondary bending stresses in tubes and pressure vessels in the vicinity of reinforcing rings and end closures.

The differential equation that relates deflection v to distributed loading q and continuous elastic reaction q_R per unit length is, from Table 9.2 and Fig. 10.15,

$$EI \frac{d^4v}{dx^4} = q + q_R \qquad (a)$$

Unit force q_R is proportional and opposite in direction to deflection v. Hence $q_R = -kv$ where k, the proportionality or stiffness constant, is called the *foundation modulus*. With substitution and rearrangement, Eq. (a) becomes

$$\frac{d^4v}{dx^4} + \frac{k}{EI} v = \frac{d^4v}{dx^4} + 4\beta^4 v = \frac{q}{EI} \qquad (10.37)$$

where

$$\beta = \sqrt[4]{\frac{k}{4EI}} \qquad (10.38)$$

FIGURE 10.15. Deflection of a beam with distributed loading q and elastic support q_R per inch of length.

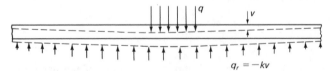

[11] F. B. Seely and J. O. Smith, *Advanced Mechanics of Materials*, 2nd ed., New York: John Wiley & Sons, Inc., p. 179, 1952; also A. P. Boresi, O. M. Sidebottom, F. B. Seely, and J. O. Smith, *Advanced Mechanics of Materials*, 3rd ed., New York: John Wiley & Sons, Inc., p. 357, 1978.

The product βx will appear frequently as a parameter in the solutions to Eq. (10.37). If the dimensions of force and length are F and L, respectively, then the dimension of q is F/L, the dimension of k is F/L^2, and the dimension of β is $1/L$. The *characteristic* β contains both the foundation modulus k and the beam rigidity EI. It will be seen to have an important influence on the shape of the deflection curve.

As with linear differential equations in vibration theory, the complementary function may be sought by the substitution of $q = 0$ and $v = e^{ax}$ into Eq. (10.37), thus

$$a^4 e^{ax} + 4\beta^4 e^{ax} = 0 \tag{b}$$

whence $a = \sqrt[4]{-4\beta^4} = \sqrt{2}\beta \sqrt[4]{-1}$. The four roots of $\sqrt[4]{-1}$ are $(1/\sqrt{2})(\pm 1 \pm i)$, where $i = \sqrt{-1}$, so $a = \beta(\pm 1 \pm i)$. The solution for the complementary function, containing the four values of parameter a and four constants of integration, is

$$v = e^{\beta x}(C_1 e^{i\beta x} + C_2 e^{-i\beta x}) + e^{-\beta x}(C_3 e^{i\beta x} + C_4 e^{-i\beta x}) \tag{c}$$

or alternately, in the equivalent form,

$$v = e^{\beta x}(A \cos \beta x + B \sin \beta x) + e^{-\beta x}(C \cos \beta x + D \sin \beta x) \tag{10.39}$$

Equation (10.39) and its derivatives hold for all spans of the beam between applied loads, whether concentrated or distributed. Most applications of interest have concentrated loads. If a single concentrated load is applied at location $x = 0$, and the beam begins there and extends a long distance to the right, it is known as a *semiinfinite* beam. Then the first half of Eq. (10.39) must be discarded, i.e., $A = B = 0$, for otherwise the amplitudes $Ae^{\beta x}$ and $Be^{\beta x}$ of successive sinusoidal waves increase indefinitely toward the right. The amplitudes $Ce^{-\beta x}$ and $De^{-\beta x}$ are exponentially "damped," in fact, they die out so rapidly that the theory may be applied with almost no error to rails and beams of normal proportions. For a beam extending only to the left of $x = 0$, only A and B would be retained. For a beam extending from the load in both directions, known as an *infinite* beam, as in Fig. 10.16, only the C and D terms are retained, and in the $e^{-\beta x}$ term, x is taken positive in both directions from $x = 0$.

The foregoing practice, which implies that $v = 0$ at the far end or ends, leaves only two integration constants to be determined. This is done from the known deflection, slope, moment, or shear conditions at $x = 0$. From Table 9.2, these are equal to or proportional to the following, respectively,

$$v = e^{-\beta x}(C \cos \beta x + D \sin \beta x) \tag{10.40a}$$

$$\theta = \frac{dv}{dx} = -\beta e^{-\beta x}(C \cos \beta x + D \sin \beta x) + e^{-\beta x}(-\beta C \sin \beta x + \beta D \cos \beta x)$$

$$= \beta e^{-\beta x}[(-C + D)\cos \beta x + (-C - D)\sin \beta x] \tag{10.40b}$$

$$\frac{M}{EI} = \frac{d^2 v}{dx^2} = \beta^2 e^{-\beta x}[-2D \cos \beta x + 2C \sin \beta x] \tag{10.40c}$$

$$\frac{V}{EI} = \frac{d^3 v}{dx^3} = \beta^3 e^{-\beta x}[2(C + D)\cos \beta x + 2(-C + D)\sin \beta x] \tag{10.40d}$$

For an infinite beam with a concentrated force P at $x = 0$ (Fig. 10.16), from symmetry the conditions at $x = 0$ are that the slope is $\theta = 0$ and the shear force is $V = -P/2$. Then, from Eq. (10.40b), $C = D$; and from Eq. (10.40d), $-P/2EI = \beta^3[2(C + D)]$, whence $C = D = -P/8EI(\beta^3) = -P\beta/8EI(\beta^4) = -P\beta/2k$ by Eq. (10.38). Equations (10.40), with substitution from the foregoing and from the definition of β, Eq. (10.38),

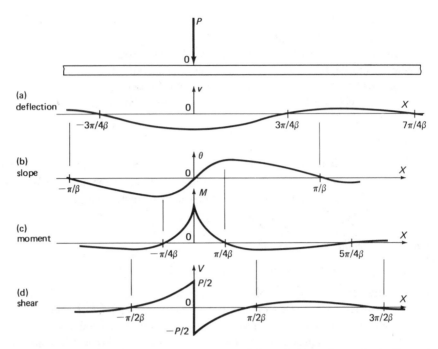

FIGURE 10.16. Deflection, slope, moment, and shear diagrams for an elastically supported beam of infinite length, with a concentrated load P applied at location $x = 0$.

become for an infinite beam with a concentrated force,

$$v = -\frac{P\beta}{2k} \left[e^{-\beta x} (\cos \beta x + \sin \beta x) \right] = -\frac{P\beta}{2k} \left[F_1 (\beta x) \right] \qquad (10.41a)$$

$$\theta = \frac{P\beta^2}{k} \left[e^{-\beta x} \sin \beta x \right] = \frac{P\beta^2}{k} \left[F_2 (\beta x) \right] \qquad (10.41b)$$

$$M = EI \frac{d^2 v}{dx^2} = \frac{EIP\beta^3}{k} \left[e^{-\beta x} (\cos \beta x - \sin \beta x) \right] = \frac{P}{4\beta} \left[F_3 (\beta x) \right] \qquad (10.41c)$$

$$V = EI \frac{d^3 v}{dx^3} = -\frac{EIP\beta^4}{k} \left[e^{-\beta x} \cos \beta x \right] = -\frac{P}{2} \left[F_4 (\beta x) \right] \qquad (10.41d)$$

The $F(\beta x)$ terms in Eqs. (10.41) replace the exponential and trigonometric products. These four functions of βx are defined in Table 10.2, and numerical values are calculated and listed for each, in βx steps of 0.10. The same functions appear in different ways in the solutions to other loadings, so that this and other tables[12] are quite useful. Each successive function is the derivative of the preceding one except for a multiplier, e.g., $F_4 = -(1/2\beta)F_3'$ and $F_1 = -(1/\beta)F_4'$. The curves of Fig. 10.16 to the right of the discontinuity at $x = 0$ are plots of these functions to different scales, and each successive curve shows the characteristics of a derivative, such as a zero value at a peak of the preceding one.

The several functions have a wavelength $\lambda = 2\pi/\beta$, i.e., for each increment of 2π in the parameter βx, the trigonometric terms repeat, although they are greatly reduced in amplitude by the $e^{-\beta x}$ multiplier. Of great interest, however, is the location of the

---

[12] More extensive tables appear in M. Hetenyi, *Beams on Elastic Foundation*, Ann Arbor, Michigan: University of Michigan Press, 1946.

first and sometimes the second crossing of the zero line. These are indicated for each curve of Fig. 10.16. The deflection, for example, is a function of F_1, which has a zero value when $\cos \beta x = -\sin \beta x$, or at $\beta x = 3\pi/4$ and $7\pi/4$. This occurs at $x = 3\pi/4\beta$ and $x = 7\pi/4\beta$, respectively. These distances, particularly the first one, should be somewhat less than the length of the actual beam if the theory is to be applied accurately. Also, when the elastic support does not sustain tensile force, such as the ground under a railroad rail, the beam tends to leave the support between $x = 3\pi/4\beta$ and $7\pi/4\beta$. Without force from the support, as required by the theory, there is a small error in the entire deflection surve. However, there may be adjacent applied loads from other wheels, or the weight of the beam itself, which when superimposed, maintain compression.

10.6 TABLES AND APPLICATIONS OF ELASTIC-SUPPORT EQUATIONS

The solutions by Eqs. (10.40) for concentrated loads and moments on infinite and semiinfinite beams are summarized in Table 10.3. The general shape of the deflection curve is shown for each case, and then the boundary conditions are listed. It is seen that all deflections, slopes, moments, and shear forces can be expressed as functions of the constants k and β, the loading, and the four $F(\beta x)$ functions. The regularity in the order of the functions going across the page, e.g., in Case 3, $F_3 \rightarrow F_4 \rightarrow F_1 \rightarrow F_2$, is a result of the relationship between their derivatives, stated in Section 10.5. The formulas hold for the positive or right side of $x = 0$. For infinite beams, the nature of the curves on the left side may be readily deduced from symmetry or continuity. Where there is a discontinuity, an otherwise continuous curve will be broken and displaced by the amount of the discontinuity, as in Fig. 10.16(d).

Example 10.7

An optical bench (Fig. 10.17), consists of a steel channel with a moment of area $I = 1.10$ in^4, supported by coil springs on 6-in centers to minimize its deflection and isolate it from certain vibrations. The springs are chosen to deflect ⅛ in under the weight of the beam, which is 13.0 lb/ft or 6.5 lb on each spring. In a particular test, light from a ruby laser is directed on a target. During the test, additional equipment weighing 35 lb is mounted at a position 30 in behind the laser. The resulting vertical deflection of the light beam at the target must be known so that data may be corrected accordingly.

Solution

The spring rate is $6.5/(⅛) = 52.0$ lb/in, and the foundation modulus is $k = 52.0/6 = 8.67$ lb/in per inch of length. From Eq. (10.38), the value of β is

FIGURE 10.17. Spring-supported optical bench (Example 10.7).

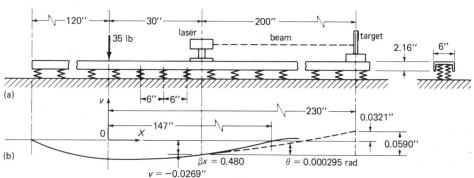

TABLE 10.2 Values of the $F(\beta x)$ Functions for Beams with Continuous Elastic Support

$F_1(\beta x) = e^{-\beta x}(\cos \beta x + \sin \beta x)$

$F_2(\beta x) = e^{-\beta x} \sin \beta x$

$F_3(\beta x) = e^{-\beta x}(\cos \beta x - \sin \beta x)$

$F_4(\beta x) = e^{-\beta x} \cos \beta x$

(βx)	$F_1(\beta x)$	$F_2(\beta x)$	$F_3(\beta x)$	$F_4(\beta x)$
0.00	1.0000	0.0000	1.0000	1.0000
0.10	0.9907	0.0903	0.8100	0.9003
0.20	0.9651	0.1627	0.6398	0.8024
0.30	0.9267	0.2189	0.4888	0.7077
0.40	0.8784	0.2610	0.3564	0.6174
0.50	0.8231	0.2908	0.2415	0.5323
0.60	0.7628	0.3099	0.1431	0.4530
0.70	0.6997	0.3199	0.0599	0.3798
0.80	0.6354	0.3223	−0.0093	0.3131
0.90	0.5712	0.3185	−0.0657	0.2527
1.00	0.5083	0.3096	−0.1108	0.1988
1.10	0.4476	0.2967	−0.1457	0.1510
1.20	0.3899	0.2807	−0.1716	0.1091
1.30	0.3355	0.2626	−0.1897	0.0729
1.40	0.2849	0.2430	−0.2011	0.0419
1.50	0.2384	0.2226	−0.2068	0.0158
1.60	0.1959	0.2018	−0.2077	−0.0059
1.70	0.1576	0.1812	−0.2047	−0.0235
1.80	0.1234	0.1610	−0.1985	−0.0376
1.90	0.0932	0.1415	−0.1899	−0.0484
2.00	0.0667	0.1231	−0.1794	−0.0563
2.10	0.0439	0.1057	−0.1675	−0.0618
2.20	0.0244	0.0896	−0.1548	−0.0652
2.30	0.0080	0.0748	−0.1416	−0.0668
2.40	−0.0056	0.0613	−0.1282	−0.0669
2.50	−0.0166	0.0491	−0.1149	−0.0658
2.60	−0.0254	0.0383	−0.1019	−0.0636
2.70	−0.0320	0.0287	−0.0895	−0.0608
2.80	−0.0369	0.0204	−0.0777	−0.0573
2.90	−0.0403	0.0132	−0.0666	−0.0534
3.00	−0.0423	0.0070	−0.0563	−0.0493
3.10	−0.0431	0.0019	−0.0469	−0.0450
3.20	−0.0431	−0.0024	−0.0383	−0.0407
3.30	−0.0422	−0.0058	−0.0306	−0.0364
3.40	−0.0408	−0.0085	−0.0237	−0.0323
3.50	−0.0389	−0.0106	−0.0177	−0.0283
3.60	−0.0366	−0.0121	−0.0124	−0.0245
3.70	−0.0341	−0.0131	−0.0079	−0.0210
3.80	−0.0314	−0.0137	−0.0040	−0.0177
3.90	−0.0286	−0.0139	−0.0008	−0.0147
4.00	−0.0258	−0.0139	+0.0019	−0.0120

TABLE 10.3. Beams with Continuous Elastic Support

Case Number	Loading and Deflection Diagram	Boundary Conditions	Deflection v	Slope θ dv/dx	Moment M $EI(d^2v/dx^2)$	Shear Force V $EI(d^3v/dx^3)$	Value of βx where $v = 0$
Infinite Beams 1. Force P		At $x = 0$, $dv/dx = 0$ $EI(d^3v/dx^3) = -\dfrac{P}{2}$ At $x \to \infty$, $v = 0$	$-\dfrac{P\beta}{2k} F_1$	$\dfrac{P\beta^2}{k} F_2$	$\dfrac{P}{4\beta} F_3$	$-\dfrac{P}{2} F_4$	$\dfrac{3\pi}{4},\ \dfrac{7\pi}{4}$
2. Moment M_0		At $x = 0$, $v = 0$ $EI\dfrac{d^2v}{dx^2} = \dfrac{M_0}{2}$ At $x \to \infty$, $v = 0$	$-\dfrac{M_0}{k} \beta^2 F_2$	$-\dfrac{M_0}{k} \beta^3 F_3$	$\dfrac{M_0}{2} F_4$	$-\dfrac{M_0\beta}{2} F_1$	$0,\ \pi,\ 2\pi$
Semiinfinite Beams 3. Moment M_0		At $x = 0$, $EI(d^2v/dx^2) = -M_0$ $EI(d^3v/dx^3) = 0$ At $x \to \infty$, $v = 0$	$\dfrac{2\beta^2 M_0}{k} F_3$	$\dfrac{M_0}{\beta EI} F_4$ $= -\dfrac{4\beta^3}{k} M_0 F_4$	$-M_0 F_1$	$2\beta M_0 F_2$	$\dfrac{\pi}{4},\ \dfrac{5\pi}{4}$
4. Force P		At $x = 0$, $EI(d^2v/dx^2) = 0$ $EI(d^3v/dx^3) = -P$ At $x \to \infty$, $v = 0$	$-\dfrac{2\beta P}{k} F_4$	$+\dfrac{2\beta^2 P}{k} F_1$	$-\dfrac{P}{\beta} F_2$	$-P F_3$	$\dfrac{\pi}{2},\ \dfrac{3\pi}{2}$
5. Slope $\theta_0 = 0$			Same as first row, except $EI(d^3v/dx^3) = -Q = -\dfrac{P}{2}$, so substitute $2Q$ for P.				

449

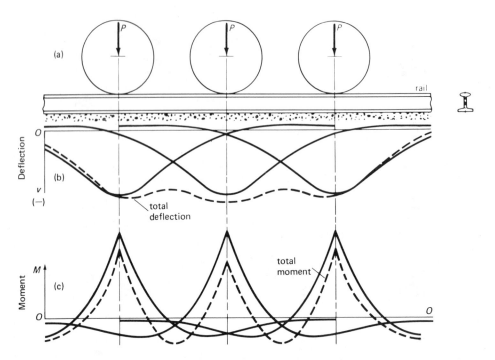

FIGURE 10.18. Superposition of rail deflections and moments from individual wheels (solid lines) to give total deflections and moments (dashed lines). (a) Loading. (b) Deflections. (c) Moments.

$\sqrt[4]{8.67/4(30 \times 10^6)(1.10)}$ = 0.0160. The deflection first becomes zero at $x = 3\pi/4\beta$ = 147 in from the load, so the results should be reasonably accurate. With $x = 0$ at the load, the parameter at the laser is $\beta x = (0.016)(30) = 0.480$. There, both the deflection and slope change the position of the light spot, the geometry of which is shown in Fig. 10.17(b). The following calculations are made, with F_1 and F_2 values interpolated from Table 10.2.

$$v = -\frac{P\beta}{2k}F_1 = -\frac{(35)(0.016)}{2(8.67)}(0.834) = -0.0269 \text{ in}$$

$$\theta = \frac{P\beta^2}{k}F_2 = \frac{(35)(0.016)^2}{8.67}(0.285) = 0.000\ 295 \text{ rad}$$

$$v_{\text{target}} = -0.0269 + (0.000\ 295)(200) = +0.0321 \text{ in} \qquad ////$$

The results of several loadings may be superimposed. If they occur at different locations, there must be an $x = 0$ axis for each location, i.e., the several deflection curves are displaced along the X axis. A sketch of the curves will aid in determining likely positions for maximum effects, as for the three-wheel rail loading of Fig. 10.18. The dashed lines are the superimposed curves. The maximum deflection occurs under the center wheel, and it is larger than the deflection from one wheel alone. The maximum bending moment occurs at the outer wheels, and it is less than the moment from one wheel alone.

The complete solution of the differential equation, Eq. (10.37), requires a particular integral. If a uniform loading $-q_0$ extends from $x = 0$ to the right in an infinite beam (Fig. 10.19), the particular value $v = -q_0/k$ satisfies the equation, and it will be added to the deflection equation Eq. (10.40a). The same equation without $-q_0/k$ and with

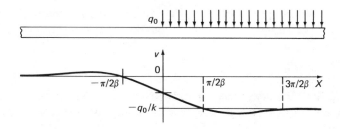

FIGURE 10.19. Uniformly distributed loading on positive half of an elastically supported beam.

different constants of integration holds for all values of x to the left of $x = 0$. At $x = 0$, both equations must give the same deflection, slope, moment, and shear. These four conditions determine the four constants of integration. The deflection equations become

$$v_{\text{right}} = -\frac{q_0}{k}\left[1 - \frac{1}{2}F_4(\beta x)\right] \quad \text{and} \quad v_{\text{left}} = -\frac{q_0}{k}\left[\frac{1}{2}F_4(\beta x)\right] \tag{10.42}$$

The deflection curve is shown below the beam in Fig. 10.19. Solutions have been made for other distributed loadings. Also, solutions are available for beams of finite length[13] and for beams with a variable foundation modulus k and with variable flexural rigidity EI.

10.7 CYLINDERS RESTRICTED IN THEIR DEFORMATION

An important application of the theory for beams on continuous elastic support is for the analysis of thin-walled tubes, cylindrical tanks and pressure vessels, and hollow rotors with loads that are symmetric about the axis. Moments generally arise from a restriction to radial deformation such as given by reinforcing rings and flanges, at the joints of tubes with more rigid containers, and end closures or supports for vessels. The source of loading may be internal pressure, centrifugal force, or thermal expansion.

When a thin, hollow cylinder is subjected to a pressure p, the tangential stress is $\sigma_t = pr/t$ given by Eq. (8.7), and the radial displacement is $u = r\epsilon_t$, given by Eq. (8.21). Since $\epsilon_t = \sigma_t/E$, displacement $u = pr^2/Et$. Next considered is a strip of unit width (Fig. 10.20(a)), the displacement of which is resisted by the remainder of the cylinder acting as an elastic support. Let the resistance per unit length be $q_R = -ku$, where k is the

FIGURE 10.20. Thin cylinder consisting of "beams" with elastic support. (a) Derivation of stiffness constant k. (b) "Beam" stiffening due to prevention of sidewise deformation. (c) Displacement of cylinder by a narrow band of force P per unit of circumference.

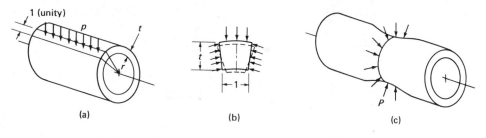

(a) (b) (c)

[13] M. Hetenyi, *Beams on Elastic Foundation*, Ann Arbor, Michigan: University of Michigan Press, 1946; J. P. Den Hartog, *Advanced Strength of Materials*, New York: McGraw-Hill, pp. 141–170, 1952; S. Timoshenko, *Strength of Materials, Part II*, 3rd ed., Princeton, New Jersey: Van Nostrand, pp. 1–25, 1956.

stiffness constant, as for the beam support of Fig. 10.15. For equilibrium of the strip, $q_R = -p$. Substitution of this and of $u = pr^2/Et$ into the preceding equation yields

$$k = \frac{Et}{r^2} \tag{10.43}$$

By the foregoing k has been determined as a characteristic of the supporting medium. As in Fig. 10.15 and Section 10.5 it may be used at any location along a cylinder, whether loaded or not.

For the unit width "beam" of Fig. 10.20(a) and (b), the moment of inertia of area is $I = t^3/12$, where t is the wall thickness of the cylinder. However, the rigidity EI is greater than that of the beams previously considered by the factor $1/(1 - \nu^2)$. This is because the rotation of the sides or Poisson's effect that would occur in a normal beam (dotted line of Fig. 10.20(b)), cannot occur here if the roundness of the tube is preserved. This is the same effect and factor used in the plate theory of Section 10.8. Substitution for k and EI in Eq. (10.38) gives

$$\beta = \sqrt[4]{\frac{Et}{r^2} \bigg/ \frac{4E(t^3/12)}{1 - \nu^2}} = \sqrt[4]{\frac{3(1 - \nu^2)}{r^2 t^2}} = \frac{1.285}{\sqrt{rt}} \tag{10.44}$$

the last form when $\nu = 0.30$. The formulas and data of the tables may now be used. When the cylinder of Fig. 10.20(c) is loaded by a narrow band of radial force P per unit length of circumference, the deflection curve is that of Case 1 of Table 10.3. The first location of zero deflection is at $\beta x = 3\pi/4$, or at $x = 3\pi \sqrt{rt}/(4 \times 1.285) = 1.834 \sqrt{rt}$ $= 1.834r\sqrt{t/r}$. If t/r is $1/16$, $x = 0.458r$. The "die-out" occurs even sooner for the other loadings of Table 10.3. Most practical cylinders will be longer than this, so the theory will hold well.[14]

Example 10.8

Figure 10.21 shows a cylindrical steel tank with a flat bottom, in common use for the storage of liquids. The bottom circular plate of the tank will not expand much radially, and may be thick enough and reinforced so that the cylindrical wall may be considered "fixed" or "built in" at its lower end. Determine the maximum bending stress in the wall of the tank.

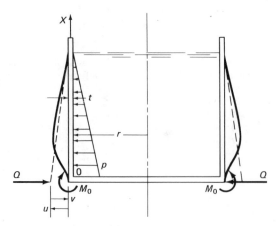

FIGURE 10.21. Displacements, forces, and moments on a cylindrical, flat-bottomed tank for liquid storage (Example 10.8).

---

[14] For many details and examples of cylindrical shells, see S. Timoshenko and S. Woinowsky-Krieger, *Theory of Plates and Shells*, New York: McGraw-Hill, pp. 446–532, 1959.

Solution

If unrestrained, the cylinder will expand in proportion to the depth of the fluid, as shown by the dotted line, an amount $u = pr^2/Et$, by Eqs. (8.7) and (8.21). To return and hold the cylinder in its restrained position, a moment M_0 and force Q per unit length of circumference are required. This occurs at location $x = 0$ of a semiinfinite beam (Case 5 of Table 10.3), and the deflection is negative. Since the net displacement $\triangle r$ is zero,

$$\triangle r = u + v = \frac{pr^2}{Et} - \frac{(2Q)\beta}{2k} F_1 = 0$$

Substitution of $F_1 = 1.0$ at $x = 0$ and of k from Eq. (10.43) gives $2Q = 2p/\beta$. The unit moment at $x = 0$, where $F_3 = 1.0$, is

$$M_0 = \frac{2Q}{4\beta} F_3 = \frac{2p}{\beta} \frac{1.0}{4\beta} = \frac{p}{2\beta^2} = \frac{p}{2(1.285)^2/rt} = 0.303prt \tag{10.45}$$

The stress due to this moment is $\sigma_0 = M_0 c/I = 6M_0/t^2 = 1.82pr/t$. This is the only normal stress at this location. Higher up away from the bottom plate, there will be a circumferential stress equal to pr/t, by Eq. (8.7). Hence the bending stress is relatively large here. Actually, it may be somewhat less because of flexure in the tank bottom near the corner.   ////

Example 10.9

A pressure vessel consists of a cylinder to which are welded hemispherical heads of the same wall thickness. Determine the maximum stresses under a pressure p.

Solution

In an unrestrained cylinder, the membrane stresses and expansions would be $\sigma_t = pr/t$, $\sigma_x = pr/2t$, and $u_1 = r\epsilon_t = r(\sigma_t - \nu\sigma_x)/E = (2 - \nu) pr^2/2Et$ (Fig. 10.22). In the heads they would be $\sigma_t = \sigma_x = pr/2t$ and $u_2 = (1 - \nu) pr^2/2Et$. The total radial displacement to bring an unrestrained head and cylinder together is the difference $u_1 - u_2 = pr^2/2Et$ (Fig. 10.22). For the cylinder this is Case 4 of Table 10.3. Function F_4 and the deflection are first equal to zero when $\beta x = \pi/2$ or, from Eq. (10.44), when $x = 1.22 \sqrt{rt}$. If $t/r = \frac{1}{25}$, $x = r/4$; if $t/r = \frac{1}{100}$, $x = r/8$. This is not very far around the curve of the sphere, so the stiffness of the sphere and cylinder is assumed to be the same and uniform in the area between $+x$ and $-x$. Then they will equally divide the total necessary displacements or $v_{x=0} = pr^2/4Et$. Table 10.3 gives the deflection for Case 4

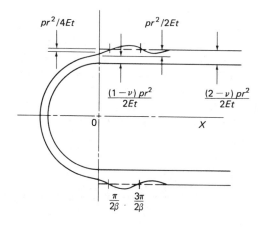

FIGURE 10.22. Displacements under internal pressure p of a cylindrical vessel with hemispherical ends (Example 10.9).

as $2\beta P F_4/k$. At $x = 0$, $F_4 = 1.0$, so $v_{x=0} = 2\beta P/k = 2\beta P r^2/Et$. Equated, the two expressions for $v_{x=0}$ yield $P = p/8\beta$.

The moment for Case 4 is $M = P F_2/\beta = p F_2/8\beta^2$. The bending stress is $\sigma_b = 6M/t^2 = 3p F_2/4t^2\beta^2 = 0.454 pr F_2/t$, where substitution for β is made from Eq. (10.44). This bending stress adds directly to the axial membrane stress $pr/2t$ to give a total axial stress

$$\sigma_x = \frac{pr}{2t} + 0.454\,\frac{pr F_2}{t} = (0.50 + 0.454 F_2)\frac{pr}{t} \tag{10.46a}$$

The maximum value of F_2 is 0.3224, at $\beta x = \pi/4$, so $(\sigma_x)_{max} = 0.646 pr/t$. The tangential stress consists of three parts, the membrane stress pr/t of the unrestrained cylinder, the increment of membrane stress from the radial deformation caused by bending, or $E\epsilon_t = Ev/r$ by Eq. (8.21), and a Poisson's ratio effect[15] $v\sigma_b$. For $0 \leqslant x \leqslant \pi/2$, deflection v is negative and equal to $-2\beta P F_4/k$. Thus

$$\sigma_t = \frac{pr}{t} + \frac{Ev}{r} + v\sigma_b = \frac{pr}{t} - \frac{2E\beta P F_4}{rk} + \frac{3v p F_2}{4t^2\beta^2}$$

With substitution of the preceding values of β, P, and k and of $v = 0.30$, the total tangential stress becomes

$$\sigma_t = (1 - 0.25 F_4 + 0.1363 F_2)\frac{pr}{t} \tag{10.46b}$$

This stress will be maximum between $x = \pi/2\beta$ and $x = 3\pi/2\beta$, the zero-deflection locations of Fig. 10.22. A cut and try solution using Table 10.2 indicates that the maximum value is $(\sigma_t)_{max} = 1.032 pr/t$, occurring at $x = 1.85/\beta$.   ////

10.8 FLAT PLATE THEORY IN RECTILINEAR COORDINATES

The plate problem is a three-dimensional one, in many respects analogous to the straight-beam problem, but complicated by a restriction to sidewise expansion and by twist from unsymmetrical loadings. The general differential equation will be derived in rectilinear coordinates, but its direct application will be limited here to a few simple cases. The differential equation will then be derived in cylindrical coordinates for symmetrically loaded circular plates, for which solutions are more readily made. In addition, tables are developed and solutions made by superposition.

The geometry of the deflected plate is expressed in terms of two coordinate curvatures and a "twist." The coordinate system commonly used for plates (Fig. 10.23), takes w as the deflection, positive in the upward direction and parallel to Z in the X, Y, Z system of locations. Analogous to straight beams of Section 9.7 and Eq. (9.20), the curvature is $1/\rho_x = \partial^2 w/\partial x^2$ in the XZ plane and $1/\rho_y = \partial^2 w/\partial y^2$ in the YZ plane. The twist is measured by the rate of increase in one direction of the slope in the other direction, or $1/\rho_t = (\partial/\partial y)(\partial w/\partial x)$, illustrated in Fig. 10.23. Corresponding to Eq. (9.18) and Fig. 10.24, the bending strains are $\epsilon_x = -z/\rho_x$ and $\epsilon_y = -z/\rho_y$, and the shear strain is $\gamma_{xy} = 2z(\partial^2 w/\partial x\,\partial y) = 2z/\rho_t$.[16] From Eqs. (5.1) and $\sigma_z = 0$, $\epsilon_x = (\sigma_x - v\sigma_y)/E$ and $\epsilon_y = (\sigma_y - v\sigma_x)/E$. These solved simultaneously, followed by substitution, yield σ_x and σ_y.

---

[15] Described for a *flat* rectangular plate under Example 10.10 of Section 10.8.

[16] For a derivation of γ_{xy}, e.g., see S. Timoshenko and S. Woinowsky-Krieger, *Theory of Plates and Shells*, 2nd ed., New York: McGraw-Hill, pp. 40–42, 1959. or J. P. Den Hartog, *Advanced Strength of Materials*, New York: McGraw-Hill, pp. 105–106, 1952.

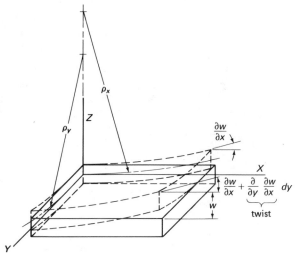

FIGURE 10.23. Plate coordinates, deflection, curvatures, and twist.

Together with stress τ_{xy} due to twist, they are

$$\sigma_x = \frac{E}{1 - \nu^2}(\epsilon_x + \nu\epsilon_y) = -\frac{Ez}{1 - \nu^2}\left(\frac{\partial^2 w}{\partial x^2} + \nu\frac{\partial^2 w}{\partial y^2}\right)$$

$$\sigma_y = \frac{E}{1 - \nu^2}(\epsilon_y + \nu\epsilon_x) = -\frac{Ez}{1 - \nu^2}\left(\frac{\partial^2 w}{\partial y^2} + \nu\frac{\partial^2 w}{\partial x^2}\right) \qquad (10.47)$$

$$\tau_{xy} = G\gamma_{xy} = 2Gz\frac{\partial^2 w}{\partial x\,\partial y}$$

Analgous to Eq. (b) of Section 9.2 for a beam, the internal resisting moment developed by a section per unit of length in the Y direction is, from Fig. 10.25, and by substitution from Eqs. (10.47),

$$M_x = -\int_{-t/2}^{+t/2}\sigma_x z\,dA = \frac{E}{1 - \nu^2}\left(\frac{\partial^2 w}{\partial x^2} + \nu\frac{\partial^2 w}{\partial y^2}\right)\int_{-t/2}^{+t/2}z^2 dz \qquad (a)$$

The integral equals I in a beam of unit width, and its product with $E/(1 - \nu^2)$ is called the plate modulus or flexural rigidity D. Like EI in a beam, it is a measure of the rigidity. Unit-length moment M_y and twisting moment T_{xy} are similarly obtained, and for T_{xy} the

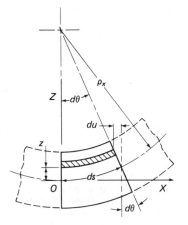

FIGURE 10.24. Strain ϵ_x in a plate. ($\epsilon_x = -du/ds = -z\,d\theta/\rho_x\,d\theta = -z/\rho_x$.)

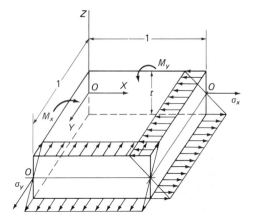

FIGURE 10.25. Components M_x and M_y of bending moment and σ_x and σ_y of stress in a plate.

substitution $G = E/2(1 + \nu)$ is made by Eq. (5.3). Summarized, the equations[17] are

$$D = \frac{E}{(1 - \nu^2)} \int_{-t/2}^{+t/2} z^2 \, dz = \frac{Et^3}{12(1 - \nu^2)} \tag{10.48}$$

and

$$M_x = D \left(\frac{\partial^2 w}{\partial x^2} + \nu \frac{\partial^2 w}{\partial y^2} \right)$$

$$M_y = D \left(\frac{\partial^2 w}{\partial y^2} + \nu \frac{\partial^2 w}{\partial x^2} \right) \tag{10.49}$$

$$T_{xy} = \frac{2Gt^3}{12} \frac{\partial^2 w}{\partial x \, \partial y} = D(1 - \nu) \frac{\partial^2 w}{\partial x \, \partial y}$$

Figure 10.26 shows the several forces and moments acting on a plate element of thickness t and breadths dx and dy. These include shear based on unit-length forces V_x and V_y.[17] Moments will be taken about the left side, Y-axis edge, and the moment arm of the shear force is dx. Then, the equation for moments in the XZ plane is

$$-M_x \, dy + \left(M_x + \frac{\partial M_x}{\partial x} \, dx \right) dy - T_{yx} \, dx + \left(T_{yx} + \frac{\partial T_{yx}}{\partial y} \, dy \right) dx$$

$$- \left(V_x + \frac{\partial V_x}{\partial x} \, dx \right) dy \, dx + (p \, dx \, dy) \frac{dx}{2} = 0 \tag{b}$$

This may be solved for V_x with terms of order higher than $dx \, dy$ neglected. A similar equation for moments in the YZ plane will yield V_y. Together, they are

$$V_x = \frac{\partial M_x}{\partial x} + \frac{\partial T_{yx}}{\partial y}$$

$$V_y = \frac{\partial M_y}{\partial y} + \frac{\partial T_{xy}}{\partial x} \tag{10.50}$$

---

[17] In plate theory it is customary to label a moment with a single subscript that sufficiently indicates the plane in which it acts, the same subscript as for the curvature and bending stress; thus M_x gives ρ_x and σ_x in plane XZ. However, the subscript of a shear force, which is always in the Z direction, and the first subscript of a twisting moment indicate the direction of the normal to the plane in which they act, thus V_x and T_{xy} on plane YZ (cf. τ_{xy} and τ_{yx} of Fig. 5.1).

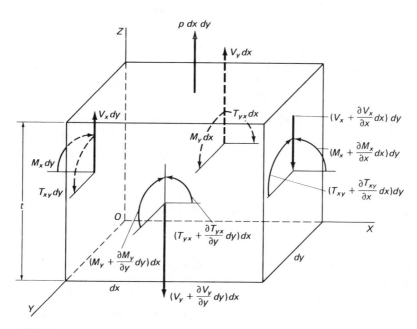

FIGURE 10.26. Forces and moments for equilibrium under bending of the element ($t\,dx\,dy$) of a plate of thickness t.

Equilibrium of forces in the Z direction requires that

$$V_x\,dy - \left(V_x + \frac{\partial V_x}{\partial x}\,dx\right)\,dy + V_y\,dx - \left(V_y + \frac{\partial V_y}{\partial y}\,dy\right)\,dx + p\,dx\,dy = 0 \qquad \text{(c)}$$

where p is a pressure in the positive direction. This equation yields[18]

$$\frac{\partial V_x}{\partial x} + \frac{\partial V_y}{\partial y} = p \qquad \text{(d)}$$

Substitution from Eqs. (10.50) and (10.49) gives the differential equation for the plate[18]

$$\frac{\partial^4 w}{\partial x^4} + 2\frac{\partial^4 w}{\partial x^2 \partial y^2} + \frac{\partial^4 w}{\partial y^4}$$

$$= \left(\frac{\partial^2}{\partial x^2} + \frac{\partial^2}{\partial y^2}\right)\left(\frac{\partial^2 w}{\partial x^2} + \frac{\partial^2 w}{\partial y^2}\right) = \nabla^2(\nabla^2 w) = \frac{p}{D} \qquad (10.51)$$

Functions for deflection w that satisfy Eq. (10.51) may be differentiated and used in Eqs. (10.49) to give equations for the moments. Then, since the section modulus for a rectangular section of unit width is $t^2/6$, the maximum bending stresses are

$$\sigma_x = \frac{M_x}{I/c} = \frac{6M_x}{t^2} \quad \text{and} \quad \sigma_y = \frac{6M_y}{t^2} \qquad (10.52)$$

In general, solutions of Eq. (10.51) to meet specific load and boundary conditions are difficult to obtain. Some have been obtained by the superposition of trial solutions $w = f(x,y)$. Two simple ones are tried in the examples that follow. The Fourier series has been very useful. There is a tendency for the corners of freely supported rectangular

---

[18] It is suggested that Eqs. (10.49), (10.50), (d), and (10.51) be compared with the corresponding derivative forms of Table 9.2.

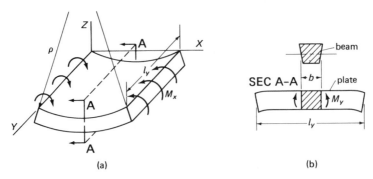

FIGURE 10.27. (a) Plate bent by moments M_x per unit length along opposite edges. (b) Section A-A comparing the restraint and moment M_y developed in a plate, except at its free edges, with the Poisson's ratio distortion in a beam (Example 10.10).

plates to rise off their supports, and the theory and equations do not hold unless this is prevented. In spite of the difficulties many solutions for rectangular and other straight-sided plates have been made and many collected in tables.[19] For derivations, discussions, and tests, specialized books and the journals of professional societies are available.[20]

Example 10.10

Study the function $w = Cx^2$ as a solution to the general plate equation, Eq. (10.51).

Solution

$\nabla^2 w = 2C$ and $\nabla^2 (\nabla^2 w) = 0$, so p must be zero, by Eq. (10.51). Curvatures are $1/\rho_x = \partial^2 w/\partial x^2 = 2C$ and $1/\rho_y = \partial^2 w/\partial y^2 = 0$. Furthermore, $M_x = D(\partial^2 w/\partial x^2) = 2CD$, $M_y = 2\nu CD$, and $T_{xy} = 0$. Thus the equations describe a plate loaded only by a uniform moment M_x along the y edges (Fig. 10.27), giving a deflected surface that is a cylindrical trough. Since $C = M_x/2D$, $w = M_x x^2/2D = 6(1 - \nu^2) M_x x^2/Et^3$. The case differs in two respects from a narrow beam loaded by end moments. Since w is a function of $1 - \nu^2$, which equals 0.91 with $\nu = 0.30$, the deflection is smaller or the beam is stiffer by about 10 percent. Also, acting in transverse planes there is an internal moment $M_y = \nu M_x$ that is 30 percent of the applied moment M_x. Only near the edges where there is no external moment does this internal moment disappear and the edges turn (Fig. 10.27(b)). Away from the edges, it is plane strain, as discussed in Sections 8.7 and 8.13. In fact, any trial solution of w as a function of x alone will give similar effects. For any wide *beam*, regardless how it is loaded, $EI/(1 - \nu^2)$ should be substituted for EI.   ////

Example 10.11

A flat circular plate which is fixed or clamped along its outer edge is heated on one side to a temperature T above ambient. Assume the temperature gradient to be linear, with a coefficient of expansion α. Determine the deflection of the plate and the resulting stresses.

Solution

If the plate were free along its edge and heated on its top surface (Fig. 10.28(a)), the top fiber would expand radially an amount αTr, and all fibers would expand in proportion

---

[19] W. Griffel, *Plate Formulas*, New York: Frederick Ungar Publishing Company, 1968; R. J. Roark and W. C. Young, *Formulas for Stress and Strain*, 5th ed., New York: McGraw-Hill, 1975.

[20] S. Timoshenko and S. Woinowsky-Krieger, *Theory of Plates and Shells*, 2nd ed., New York: McGraw-Hill, 1959; and *J. Appl. Mech., Trans. ASME*.

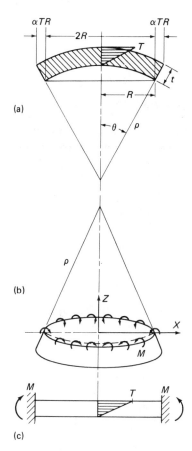

FIGURE 10.28. Displacements (exaggerated) of a flat circular plate. (a) With edges free and top surface heated to temperature T. (b) With edges free but subjected to uniform edge moments M. (c) Conditions (a) and (b) superimposed to represent the plate heated but clamped at its outer edge (Example 10.11).

to their increase of temperature. The plate becomes spherical, and by proportional triangles and for $\rho \gg t$, $R/\rho = \alpha TR/t$, whence the curvature $1/\rho = -\alpha T/t$. The slope at the edge is $\theta = -\alpha TR/t$. However, the edge of the plate is clamped and the rotation or slope at the edge is zero, with a moment required to hold it. Hence we must superimpose an elastic condition where edge moments give an edge slope equal but opposite to that of the free, heated plate. We recall that in the previous example a function $w = Cx^2$ gave a rectangular plate with moments along two sides. It appears that a function $w = Cr^2 = C(x^2 + y^2)$ might describe a circular plate with edge moments. Substitution of this function into Eq. (10.51), etc. and use of the ∇ notation of Eq. (10.51) gives $\nabla^2 w = 4C$, $\nabla^4 w = 0 = p$, $1/\rho_x = 1/\rho_y = 1/\rho = 2C$; also, by Eqs. (10.49), $M_x = M_y = M = 2DC(1 + v)$, $T_{xy} = 0$, and $V_x = V_y = 0$. Also, the deflection w and slope θ are

$$w = Cr^2 = Mr^2/2(1 + v)D \quad \text{and} \quad \theta = dw/dr = 2Cr = Mr/(1 + v)D \qquad (10.53a)$$

This is the plate we want, with edge moments only (Fig. 10.28(b)). The sum of the slopes in Fig. 10.28(a) and (b) at the edge $r = R$ must be zero to give the clamped plate of Fig. 10.28(c). Hence $-\alpha TR/t + 2CR = 0$ and $C = \alpha T/2t$. The curvature is $1/\rho = 2C = \alpha T/t$, whereas that of the heated free-edge plate was $-\alpha T/t$. Hence superposition shows the clamped plate to be without deflection anywhere (Fig. 10.28(c)). The bending moment, as found, is

$$M = 2DC(1 + v) = \frac{2Et^3}{12(1 - v^2)} \frac{\alpha T}{2t} (1 + v)$$

$$= \frac{\alpha ETt^2}{12(1 - v)}$$

and by Eqs. (10.52), the maximum stress is[21]

$$\sigma = \frac{6M}{t^2} = \frac{\alpha ET}{2(1 - \nu)} \tag{10.53b}$$

////

10.9 SYMMETRICALLY LOADED FLAT CIRCULAR PLATES

All displacements will be symmetrical about the Z axis, so one coordinate, the radius r, will serve to designate the location of all moments and deflections. Curvature of the deflected plate at any point S (Fig. 10.29), is defined by two radii similar to those used for shells in Section 8.2 but much larger. A meridian can be drawn from pole O' through S, and $\rho_m = O_m S$ is the radius of curvature of the meridian at S. Tangential radius $\rho_t = O_t S$ lies in a plane normal to the meridian at S (perpendicular to the paper). Now, arc $s \approx$ radius r and arc $ds \approx \triangle r$, and we may write $r = \rho_t \theta$ and $\triangle r = \rho_m \triangle \theta$. Furthermore, as $\triangle r \to 0$, $\theta = \lim (\triangle w / \triangle r) = dw/dr$. Then the curvatures are

$$\frac{1}{\rho_t} = \frac{\theta}{r} = \frac{1}{r}\frac{dw}{dr} \quad \text{and} \quad \frac{1}{\rho_m} = \lim_{\triangle r \to 0} \frac{\triangle \theta}{\triangle r} = \frac{d\theta}{dr} = \frac{d^2 w}{dr^2} \tag{10.54}$$

The moments in Eqs. (10.49) are each the rigidity D multiplied by a sum which is the curvature in the same plane plus ν times the curvature in the normal plane. Similarly here, from Eqs. (10.54)

$$M_t = D\left(\frac{1}{\rho_t} + \frac{\nu}{\rho_m}\right) = D\left(\frac{\theta}{r} + \nu \frac{d\theta}{dr}\right) = D\left(\frac{1}{r}\frac{dw}{dr} + \nu \frac{d^2 w}{dr^2}\right)$$

$$M_m = D\left(\frac{1}{\rho_m} + \frac{\nu}{\rho_t}\right) = D\left(\frac{d\theta}{dr} + \nu \frac{\theta}{r}\right) = D\left(\frac{d^2 w}{dr^2} + \frac{\nu}{r}\frac{dw}{dr}\right) \tag{10.55}$$

The twists are zero because of symmetry. Similarly, shear forces on the radial-plane sides of an element (Fig. 10.30), must also be zero. Equilibrium of the element requires that the sum of the moments acting in or parallel to its central radial plane be zero. This sum includes two components of the couples $M_t dr$ acting on the sides of the element.

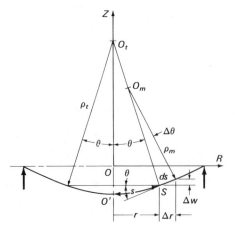

FIGURE 10.29. Radii of curvature for a symmetrically loaded circular plate.

---

[21] This is the stress approached in the wall of a thin cylinder heated on the interior (Fig. 8.23), and it is half the thermal stress in a uniformly heated plate constrained on all edges, Eq. (7.9).

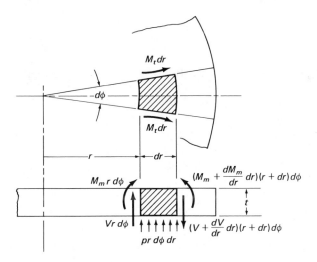

FIGURE 10.30. Forces and moments for equilibrium of an element of a circular plate.

Hence about the inner edge of the element, the moment equation is

$$-M_m r \, d\phi + \left(M_m + \frac{dM}{dr} \, dr \right)(r + dr) \, d\phi - 2(M_t \, dr) \frac{d\phi}{2}$$

$$- \left[\left(V + \frac{dV}{dr} \, dr \right)(r + dr) \, d\phi \right] dr + (pr \, d\phi \, dr) \frac{dr}{2} = 0, \qquad \text{(a)}$$

and with terms above the order $dr \, d\phi$ neglected, it reduces to

$$M_m + r \frac{dM_m}{dr} - M_t - Vr = 0 \qquad (10.56)$$

Equation (10.56), Eqs. (10.55), and an equation of force equilibrium for the element of Fig. 10.30 may be combined into a $\nabla^2(\nabla^2 w) = p/D$ form, corresponding to Eq. (10.51), but with $\nabla^2 = (1/r)(d/dr) + d^2/dr^2$. Its general solution[22] is

$$w = C_1 + C_2 \ln r + C_3 r^2 + C_4 r^2 \ln r + \frac{p_0 r^4}{64D} \qquad (10.57)$$

However, substitution into Eq. (10.56) of the moments of Eqs. (10.55) in terms of slope θ yields a simpler equation,

$$\frac{d^2\theta}{dr^2} + \frac{1}{r} \frac{d\theta}{dr} - \frac{\theta}{r^2} = \frac{V}{D} \qquad (10.58)$$

It is more readily solved, and it will serve the purposes of this section.

Figure 10.31 shows a central, cylindrical portion of radius r as a free body supporting a uniform pressure p_0 and a force P, either centrally concentrated or distributed uniformly along the edge of a central hole of radius a. Equilibrium of forces requires that $2\pi r V + \pi p_0 (r^2 - a^2) + P = 0$, where V is shear force per unit length. Let $P' = P - \pi p_0 a^2$.

---

[22] J. P. Den Hartog, *Advanced Strength of Materials*, New York: McGraw-Hill, p. 122, 1952.

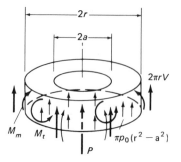

FIGURE 10.31. Forces and moments on a central, cylindrical portion of a symmetrically loaded circular plate.

Then

$$V = -\frac{1}{2}\left(p_0 r + \frac{P'}{\pi r}\right) \tag{b}$$

This may be substituted on the right-hand side of Eq. (10.58). The left-hand side is similar to Eq. (8.24), and it may be rewritten in the form of Eq. (b) which follows Eq. (8.24). Equation (10.58) becomes

$$\frac{d}{dr}\left[\frac{1}{r}\frac{d}{dr}(r\theta)\right] = -\frac{1}{2D}\left(p_0 r + \frac{P'}{\pi r}\right) \tag{c}$$

Slope θ is obtained by integrating twice, thus

$$\frac{1}{r}\frac{d}{dr}(r\theta) = -\frac{1}{2D}\left(\frac{p_0 r^2}{2} + \frac{P'}{\pi}\ln r\right) + C_1$$

$$\frac{d}{dr}(r\theta) = -\frac{p_0 r^3}{4D} - \frac{P'r\ln r}{2\pi D} + C_1 r \tag{d}$$

$$r\theta = -\frac{p_0 r^4}{16D} - \frac{P'r^2}{2\pi D}\left(\frac{\ln r}{2} - \frac{1}{4}\right) + \frac{C_1 r^2}{2} + C_2$$

$$\theta = -\frac{p_0 r^3}{16D} - \frac{P'r}{8\pi D}(2\ln r - 1) + \frac{C_1 r}{2} + \frac{C_2}{r} \tag{10.59}$$

where $P' = P - \pi p_0 a^2$. The deflection is

$$w = \int \theta \cdot dr = -\frac{p_0 r^4}{64D} - \frac{P'r^2}{8\pi D}(\ln r - 1) + \frac{C_1 r^2}{4} + C_2 \ln r + C_3 \tag{10.60}$$

All symmetrically loaded plates without holes will have zero slope at the middle and zero deflection at the supports. From Eq. (10.59) and $\theta_{r=0} = 0$, it is required that $C_2 = 0$. Since $a = 0$, $P' = P$ and Eq. (10.59) becomes

$$\theta = -\frac{p_0 r^3}{16D} - \frac{Pr}{8\pi D}(2\ln r - 1) + \frac{C_1 r}{2} \tag{10.61}$$

From Eq. (10.60), and $w_{r=b} = 0$,

$$0 = -\frac{p_0 b^4}{64D} - \frac{Pb^2}{8\pi D}(\ln b - 1) + \frac{C_1 b^2}{4} + C_3$$

From this, the value of C_3 is substituted into Eq. (10.60), whence

$$w = \frac{p_0}{64D}(b^4 - r^4) + \frac{P}{8\pi D}\left[b^2(\ln b - 1) - r^2(\ln r - 1) \right]$$
$$- \frac{C_1}{4}(b^2 - r^2) \tag{10.62}$$

The curvature in the meridional plane is

$$\frac{1}{\rho_m} = \frac{d\theta}{dr} = -\frac{3p_0 r^2}{16D} - \frac{P}{8\pi D}(2\ln r + 1) + \frac{C_1}{2}$$

and the moment from the second of Eqs. (10.55) is

$$M_m = -\frac{3+\nu}{16}p_0 r^2 - \frac{P}{8\pi}\left[2(1+\nu)\ln r + (1-\nu) \right] + \frac{(1+\nu)D}{2}C_1 \tag{10.63}$$

The constant C_1 is determined from one of two cases of support. If the edges are simply supported, then no moment M_m can exist there, so $(M_m)_{r=b} = 0$. If the edges are clamped, the slope is zero, so $\theta_{r=b} = 0$. Also, the preceding equations contain two cases of loading, p_0 and P. To keep the equation from becoming unwieldy, the cases will be separately treated. Thus four separate cases may be obtained from Eqs. (10.61) through (10.63).

Example 10.12
Determine moments, deflection, and slope for a circular plate simply supported at its edge and uniformly loaded.

Solution
Only the p_0 and C_1 terms are retained in the equations. From $(M_m)_{r=b} = 0$,

$$0 = -\frac{3+\nu}{16}p_0 b^2 + \frac{(1+\nu)D}{2}C_1 \quad \text{and} \quad C_1 = \frac{(3+\nu)p_0 b^2}{8(1+\nu)D}$$

whence moments M_m and M_t are found to be

$$M_m = \frac{3+\nu}{16}p_0(b^2 - r^2) \tag{10.64}$$

and

$$M_t = \frac{p_0}{16}\left[(3+\nu)b^2 - (1+3\nu)r^2 \right] = M_m + \frac{p_0(1-\nu)r^2}{8} \tag{10.65}$$

Deflection w is

$$w = -\frac{p_0(b^2 - r^2)}{64(1+\nu)D}\left[(5+\nu)b^2 - (1+\nu)r^2 \right] \tag{10.66}$$

At the center, deflection and moment are maximum, and $M_t = M_m = M_{\max}$,

$$M_{\max} = \frac{3+\nu}{16}p_0 b^2 \quad \text{and} \quad w_{\max} = -\frac{5+\nu}{64(1+\nu)D}p_0 b^4 \tag{10.67}$$

The slope is obtained from Eq. (10.61), thus

$$\theta = \frac{p_0 r}{16D}\left(\frac{3+\nu}{1+\nu}b^2 - r^2 \right) \tag{10.68}$$

Along the edge, slope is maximum, and although $M_m = 0$, M_t is about 43 percent of the maximum moment, respectively,

$$\theta_{\max} = \frac{p_0 b^3}{8(1 + \nu)D} \quad \text{and} \quad (M_t)_{\text{edge}} = \frac{(1 - \nu)p_0 b^2}{8} \tag{10.69}$$

////

10.10 TABLES FOR CIRCULAR PLATES: SUPERPOSITION AND STATICAL INDETERMINACY

Generally, we are interested only in the maximum values of stress and deflection. Much time can be saved with a table of the basic cases. From it other cases can be obtained by superposition, including statically indeterminate ones. We shall start Table 10.4 with the cases already solved, converting the moments into stresses by the equation $\sigma = 6M/t^2$, and substituting $Et^3/12(1 - \nu^2)$ for D. Case 1 is the plate of Fig. 10.28(b) and Eq. (10.53a) of Example 10.11. Case 5 is the uniformly loaded, edge-supported plate of Example 10.12 and Eqs. (10.67). Case 9 is similar to Case 5 except that the edges are fixed and $\theta_{r=b} = 0$. It may be solved in the same way, beginning with Eq. (10.61) and setting $\theta = 0$ at $r = b$ to obtain the value of C_1. It may also be solved by the superposition of Cases 5 and 1 (Fig. 10.32), setting the sum of their edge slopes to zero. Thus from Eqs. (10.69) and (10.53a)

$$\theta_{r=b} = \frac{p_0 b^3}{8(1 + \nu)D} + \frac{M_1 b}{(1 + \nu)D} = 0$$

whence $M_1 = -p_0 b^2/8$, which is uniform over the plate of Case 1. It is also the moment for the superimposed case, since the edge moment is zero for Case 5. Hence $(M_9)_{r=b} = -p_0 b^2/8$. For the center of the plate, we may either superimpose from Table 10.4 or use Eqs. (10.67) and (10.53a). Using the latter, together with the uniform moment M_1 and Fig. 10.32,

$$(M_9)_{r=0} = (M_5)_{r=0} + M_1 = \frac{3 + \nu}{16} p_0 b^2 - \frac{1}{8} p_0 b^2 = \frac{1 + \nu}{16} p_0 b^2$$

and

$$w_{r=0} = w_5 + w_1 = \frac{(5 + \nu)p_0 b^4}{64(1 + \nu)D} + \left(-\frac{p_0 b^2}{8}\right) \frac{b^2}{2(1 + \nu)D} = \frac{p_0 b^4}{64D} = \frac{3(1 - \nu)^2 p_0 b^4}{16Et^3}$$

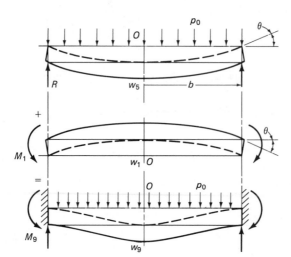

FIGURE 10.32. Superposition of Cases 5 and 1 to obtain Case 9 of Table 10.4.

The moment at the center is less than that at the edge, and at the edge $\sigma_{r=b} = 6M_{r=b}/t^2$ $= 3p_0b^2/4t^2$. Stress $\sigma_{r=b}$ and deflection $w_{r=0}$ are recorded as maximum values for Case 9 in Table 10.4.

Cases 2 and 7 for a central, concentrated load may be solved by using Eqs. (10.61) through (10.63). The maximum stress and deflection may also be obtained by letting the distributed load radius c of Cases 6 and 10 be very small, so that the c^2 terms are negligible. However, the logarithmic term approaches infinity as c approaches zero. An unrealistic, concentrated, point load gives an unrealistic stress. Therefore, the radius c of the load applicator must be used in the stress equations. However, local surface stresses will predominate on the applicator side of the plate, in many cases causing local yielding. A more extensive analysis of the far side has given a different approximate equation for the stress, which becomes practically independent of the ratio $2c/t$ as c becomes very small. It is listed as the second stress equation in rows 2 and 7. If the load is one of compression, these stresses will be tensile and hence the more significant ones.

The plates of Cases 3 and 8 are loaded by a force P uniformly distributed along a ring of radius c, as by loading with a tube. One method of solution depends on first finding equations for a plate with a central hole loaded by moments around the edge of the hole. The solid plate is then divided into two parts for analysis—an outer portion loaded around its inner edge at radius c by a moment and the applied forces, and an inner cap loaded only by moments (Case 1). Their slopes at $r = c$ must be equal, and this determines the moment there. Maximum deflection equals the deflection at $r = c$ plus the deflection of the cap. Case 4 follows. In Case 3 the unit load on the ring per unit length of arc is $q = P/2\pi c$. On an element of length $c\,d\phi$, the load is $dP = qc\,d\phi$. The contribution of this element dP to the deflection at the center is the same as that of another element located anywhere else on the ring, because of symmetry. Therefore, the elements may be redistributed around the ring in any way, including concentrating them on a small area, without affecting the total deflection at the center. The maximum deflection will occur closer to the center of loading, but it will not be much larger.

Cases 6 and 10 for a uniform central area of loading may be obtained from Cases 3 and 8, respectively. Let a pressure p_0 be distributed uniformly over an elemental ring of radius r, circumference $2\pi r$, and width dr. Then $2\pi p_0 r\,dr$ may be substituted for P, and r for c, in the equations for deflection and stress of Cases 3 and 8, followed by integration with respect to r from 0 to c. The results are given in the table.

Formulas for circular plates *with central holes* may be derived in various ways, from the general solutions for deflection, Eqs. (10.57) or (10.60) and their derivatives, and by superposition. The equations are sometimes rather lengthy, and they may be reduced by numerical substitution for Poisson's ratio ν and the ratio of outside to inside radius b/a. This has been done for maximum values, using $\nu = 0.30$ and combining all numerical values into coefficients.[23] Cases 11 through 18 with formulas and coefficients are listed in Table 10.5(A), and numerical values of the coefficients are found in Table 10.5(B) for several ratios of radii b/a. Rigid cores or rims are used to represent a fixed or built-in condition. This condition may be brought about, for example, by a hub or a heavy rim in a machine part. The several cases may be superimposed to solve additional cases,[24] including statically indeterminate ones, such as in the following example.

---

[23] Due to A. M. Wahl and G. Lobo, Jr., "Stresses and Deflections in Flat Circular Plates with Central Holes," *Trans. ASME*, Vol. 52, No. 1, pp. 29–43, 1930.

[24] For more complete tables of the same and of additional cases see W. Griffel, *Plate Formulas*, New York: Frederick Ungar Publishing Company, 1968; and R. J. Roark and W. C. Young, *Formulas for Stress and Strain*, 5th ed., New York: McGraw-Hill, 1975. Readers who consult these and other references should note that for plates it is common to use a and b as the outer and inner radii, respectively. On the contrary, for thick cylinders, it is common to use b and a, respectively. The author has chosen consistency for this book, using a for the hole and b for the periphery in both Chapters 8 and 10.

TABLE 10.4. Solid Circular Plates

Notation: a = hole radius, b = plate (outside) radius, c = load radius, t = plate thickness, p_0 = uniform pressure loading, P = total applied load, centrally concentrated, or uniformly distributed along or between concentric circles. M = moment per unit circular length, σ = max stress = $6M/t^2$, D = plate modulus = $Et^3/12(1 - \nu^2)$ = 0.091 Et^3 when $\nu = 0.3$.

Limitations: All decimal coefficients determined with $\nu = 0.3$. Thickness $t << 2b$. Accuracy may be unacceptable if the deflection is greater than $0.25t$; see "large deflection" or membrane theory.

Case. Loading and Deflection Diagrams	Maximum Bending Stress σ, Deflection w at Center, Slope θ at Edge
1. No support. Uniform moment along edge, M per unit length M ⤸ ⤹ M $2b$	$\sigma = 6\dfrac{M}{t^2}$ (uniform) $w = 6(1-\nu)\dfrac{Mb^2}{Et^3}$ $\theta = 12(1-\nu)\dfrac{Mb}{Et^3}$
2. Edge supported, central load P concentrated over a small circle of radius c $(c << b)$ $2c$ ⟶ P b	$\sigma \approx \dfrac{3}{2\pi}\dfrac{P}{t^2}\left[1 + (1+\nu)\ln\dfrac{b}{c}\right]$ (at center, on loaded side) $\sigma = \dfrac{P}{t^2}\left[\dfrac{3}{2\pi} + (1+\nu)(0.485\ln\dfrac{b}{t} + 0.52)\right]$ (at center, opposite loaded side) $w = \dfrac{3(1-\nu)(3+\nu)}{4\pi}\dfrac{Pb^2}{Et^3}$ $\theta = \dfrac{3(1-\nu)}{\pi}\dfrac{Pb}{Et^3}$
3. Edge supported, total load P uniformly distributed along a concentric circle of radius c c ⟶ c P b	$\sigma = \dfrac{3}{4\pi}\dfrac{P}{t^2}\left[(1-\nu)(1-\dfrac{c^2}{b^2}) + 2(1+\nu)\ln\dfrac{b}{c}\right]$ (at center) $w = \dfrac{3(1-\nu)}{4\pi}\dfrac{P}{Et^3}\left[(3+\nu)(b^2-c^2) - 2(1+\nu)c^2\ln\dfrac{b}{c}\right]$ $\theta = \dfrac{3(1-\nu)}{\pi}\dfrac{Pb}{Et^3}(1-\dfrac{c^2}{b^2})$
4. Total load P concentrated or distributed in any manner along a concentric circle of radius c c P b	Deflection w at center same as Case 3 for edge supported, same as Case 8 for edge fixed
5. Edge supported, uniform pressure P_0 on entire one side of plate p_0 ↓↓↓↓↓↓↓↓↓↓↓↓↓ b	$\sigma = \dfrac{3(3+\nu)}{8}\dfrac{p_0 b^2}{t^2}$ (at center) $w = \dfrac{3(1-\nu)(5+\nu)}{16}\dfrac{p_0 b^4}{Et^3}$ $\theta = \dfrac{3(1-\nu)}{2}\dfrac{p_0 b^3}{Et^3}$

TABLE 10.4. Solid Circular Plates *(continued)*

Case. Loading and Deflection Diagrams	Maximum Bending Stress σ, Deflection w at Center, Slope θ at Edge
6. Edge supported, total load P uniformly distributed within a concentric circle of radius c 	$\sigma = \dfrac{3}{8\pi} \dfrac{P}{t^2} \left[4 - (1 - \nu)\dfrac{c^2}{b^2} + 4(1 + \nu) \ln \dfrac{b}{c} \right]$ (at center) $w = \dfrac{3(1 - \nu)}{16\pi} \dfrac{P}{Et^3} \left[4(3 + \nu)b^2 - (7 + 3\nu)c^2 - 4(1 + \nu)c^2 \ln \dfrac{b}{c} \right]$ $\theta = \dfrac{3(1 - \nu)}{2\pi} \dfrac{Pb}{Et^3} \left(2 - \dfrac{c^2}{b^2} \right)$
7. Edge fixed, central load P concentrated over a small circle of radius c ($c \ll b$) 	$\sigma = \dfrac{3(1 + \nu)}{2\pi} \dfrac{P}{t^2} \ln \dfrac{b}{c}$ (at center, on loaded side) $\sigma = \dfrac{P}{t^2} \left[(1 + \nu)(0.485 \ln \dfrac{b}{t} + 0.52) \right]$ (at center, on opposite side) $w = \dfrac{3(1 - \nu^2)}{4\pi} \dfrac{Pb^2}{Et^3}$
8. Edge fixed, total load P uniformly distributed along a concentric circle of radius c 	$\sigma = \dfrac{3(1 + \nu)}{4\pi} \dfrac{P}{t^2} \left[\dfrac{c^2}{b^2} + 2 \ln \dfrac{b}{c} - 1 \right]$ (at center) $\sigma = \dfrac{3}{2\pi} \dfrac{P}{t^2} \left(1 - \dfrac{c^2}{b^2} \right)$ (at edge) $w = \dfrac{3(1 - \nu^2)}{4\pi} \dfrac{P}{Et^3} \left[b^2 - c^2 - 2c^2 \ln \dfrac{b}{c} \right]$
9. Edge fixed, uniform pressure p_0 on entire one side of plate 	$\sigma = \dfrac{3}{4} \dfrac{p_0 b^2}{t^2}$ (at edge) $w = \dfrac{3(1 - \nu^2)}{16} \dfrac{p_0 b^4}{Et^3}$
10. Edge fixed, total load P uniformly distributed within a concentric circle of radius c 	$\sigma = \dfrac{3(1 + \nu)}{8\pi} \dfrac{P}{t^2} \left[\dfrac{c^2}{b^2} + 4 \ln \dfrac{b}{c} \right]$ (at center) $\sigma = \dfrac{3}{4\pi} \dfrac{P}{t^2} \left(2 - \dfrac{c^2}{b^2} \right)$ (at edge) $w = \dfrac{3(1 - \nu^2)}{4\pi} \dfrac{P}{Et^3} \left[b^2 - \dfrac{3}{4}c^2 - c^2 \ln \dfrac{b}{c} \right]$

468

TABLE 10.5(A). Circular Plates with Concentric Holes

Notation: a = hole radius, b = plate (outside) radius, c = load radius, t = plate thickness, p_0 = uniform pressure loading, P = total applied load, centrally concentrated, or uniformly distributed along or between concentric circles. M = moment per unit circular length, σ = max stress = $6M/t^2$, D = plate modulus = $Et^3/12(1 - \nu^2)$ = 0.091 Et^3 when $\nu = 0.3$.

Limitations: All decimal coefficients determined with $\nu = 0.3$. Thickness $t \ll 2b$. Accuracy may be unacceptable if the deflection is greater than $0.25t$; see "large deflection" or membrane theory.

Case	Loading and Deflection Diagrams	Maximum Stress σ	Relative Deflection w of Edges
11. Inner edge fixed, total load P distributed uniformly along outer edge		$K_{11}\dfrac{P}{t^2}$	$H_{11}\dfrac{Pb^2}{Et^3}$
12. Inner edge fixed, uniform pressure p_0 on plate surface		$K_{12}\dfrac{p_0b^2}{t^2}$	$H_{12}\dfrac{p_0b^4}{Et^3}$
13. Inner edge supported, total load P distributed uniformly along outer edge		$K_{13}\dfrac{P}{t^2}$	$H_{13}\dfrac{Pb^2}{Et^3}$
14. Inner edge supported, uniform pressure p_0 on plate surface		$K_{14}\dfrac{p_0b^2}{t^2}$	$H_{14}\dfrac{p_0b^4}{Et^3}$
15. Outer edge supported, uniform pressure p_0 on plate surface		$K_{15}\dfrac{p_0b^2}{t^2}$	$H_{15}\dfrac{p_0b^4}{Et^3}$
16. Outer edge supported, inner-edge rotation prevented, uniform pressure p_0 on plate surface		$K_{16}\dfrac{p_0b^2}{t^2}$	$H_{16}\dfrac{p_0b^4}{Et^3}$
17. Inner edge fixed, outer-edge rotation prevented, total load P distributed uniformly along outer edge		$K_{17}\dfrac{P}{t^2}$	$H_{17}\dfrac{Pb^2}{Et^3}$
18. Inner edge fixed, rotation of outer edge prevented, uniform pressure p_0 on plate surface		$K_{18}\dfrac{p_0b^2}{t^2}$	$H_{18}\dfrac{p_0b^4}{Et^3}$

TABLE 10.5(B).[a] Values of Stress Coefficients K and Deflection Coefficients H for Cases 11 through 18 in Table 10.5(A)

Notation: a = hole radius, b = plate (outside) radius, c = load radius, t = plate thickness, p_0 = uniform pressure loading, P = total applied load, centrally concentrated, or uniformly distributed along or between concentric circles. M = moment per unit circular length, σ = max stress = $6M/t^2$, D = plate modulus = $Et^3/12(1 - \nu^2)$ = $0.091\,Et^3$ when ν = 0.3.

Limitations: All decimal coefficients determined with ν = 0.3. Thickness $t \ll 2b$. Accuracy may be unacceptable if the deflection is greater than $0.25t$; see "large deflection" or membrane theory.

b/a	1.25	1.50	2.00	3.00	4.00	5.00
K_{11}	0.227	0.428	0.753	1.205	1.514	1.745
H_{11}	0.0051	0.0249	0.0877	0.209	0.293	0.350
K_{12}	0.135	0.410	1.04	2.15	2.99	3.69
H_{12}	0.0023	0.0183	0.0938	0.2925	0.448	0.564
K_{13}	1.10	1.26	1.48	1.88	2.17	2.34
H_{13}	0.341	0.519	0.672	0.734	0.724	0.704
K_{14}	0.66	1.19	2.04	3.34	4.30	5.10
H_{14}	0.202	0.491	0.902	1.220	1.300	1.310
K_{15}	0.592	0.976	1.440	1.880	2.080	2.19
H_{15}	0.1841	0.4139	0.6640	0.8237	0.8296	0.813
K_{16}	0.122	0.336	0.74	1.21	1.45	1.59
H_{16}	0.0034	0.0313	0.1250	0.291	0.417	0.492
K_{17}	0.115	0.220	0.405	0.703	0.933	1.13
H_{17}	0.0013	0.0064	0.0237	0.0619	0.0923	0.114
K_{18}	0.090	0.273	0.71	1.54	2.23	2.80
H_{18}	0.0008	0.0062	0.0329	0.1096	0.1792	0.2338

[a] Data rearranged from paper by A. M. Wahl and G. Lobo, Jr., "Stresses and Deflections in Flat Circular Plates with Central Holes," *Trans. ASME*, Vol. 52, No. 1, pp. APM 29–43, 1930.

Example 10.13

A uniformly loaded plate with a central hole is simply supported at both its inner and outer edges (Fig. 10.33(c)). It is used in a high-pressure device, and the forces and stresses are to be minimized. Determine how much one support should be raised above the other in order to equalize the forces on the supports.

Solution

Case 14 of Table 10.5(A) shows inner-edge support and a pressure load. Case 13 shows inner-edge support and a load around the outer edge. If we superimpose Case 13 with a load $P_{13} = -P_{14}/2$, then the net loads P at inner and outer edges will be equal and of value $P_{14}/2$. Case 14 is reproduced in Fig. 10.33(a). The deflection of the outer edge from the support position is w_{14} downward. Addition of a force P to its outer edge (Fig. 10.33(b)), decreases the edge deflection of Fig. 10.33(a) by the amount w_{13}, and the distance between supports becomes $h = w_{14} - w_{13}$ (Fig. 10.33(c)). We start by putting the formula for w_{14} in terms of P_{14} by writing $p_0 = P_{14}/\pi(b^2 - a^2)$. Thus from Table 10.5(A), $w_{14} = H_{14}p_0b^4/Et^3 = H_{14}P_{14}b^4/\pi(b^2 - a^2)Et^3$, and $w_{13} = H_{13}Pb^2/Et^3$.

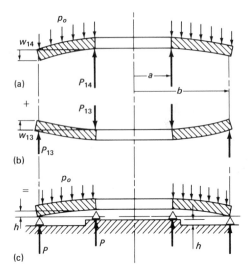

FIGURE 10.33. Uniformly loaded circular plate supported at inner and outer edges. Inner support higher than outer support to give equal reactions. Indeterminacy resolved by superposition of Cases 14 and 13 (Example 10.13).

Now $P = P_{14}/2$, so $w_{13} = H_{13}P_{14}b^2/2Et^3$. The required distance between supports is

$$h = w_{14} - w_{13} = \frac{P_{14}}{Et^3}\left[\frac{H_{14}b^4}{\pi(b^2 - a^2)} - \frac{H_{13}b^2}{2}\right]$$

$$= \frac{P_{14}b^4}{\pi(b^2 - a^2)Et^3}\left[H_{14} - \frac{\pi H_{13}(b^2 - a^2)}{2b^2}\right]$$

$$= H_{14}\frac{p_0 b^4}{Et^3}\left[1 - \frac{\pi H_{13}}{2H_{14}}\left(1 - \frac{a^2}{b^2}\right)\right]$$

$$= w_{14}\left[1 - \frac{\pi H_{13}}{2H_{14}}\left(1 - \frac{a^2}{b^2}\right)\right] \tag{10.70}$$

This distance is independent of the applied pressure p_0, so the equality of reactions is maintained during varying pressures. From Table 10.5(B), when $b/a = 2$, $h = 0.122w_{14}$ and when $b/a = 5$, $h = 0.190w_{14}$.   ////

If the deflection calculated by any of the equations of Table 10.4 exceeds about 25 percent of the plate thickness, then it should be recalculated using a "large deflection" formula, since there is uniform or membrane stressing of the plate in addition to stressing by bending. Deflections are nonlinear and increase more slowly with increase of load and deflection, i.e., the plate becomes stiffer. Special books on plates may be consulted for the theory and for exact and approximate equations and charts.[25] To obtain relatively large *linear* deflections, as needed in diaphragms for pressure-indicating instruments, circular corrugations may be used.

PROBLEMS

Section 10.1.

10.1   (a) Using Example 9.7, write the influence coefficients α_{31} and α_{33} for the beam of Fig. 9.25(a). (b) Determine α_{11} from Table 9.4, and using the theorem of reciprocity, write an

---

[25] S. Timoshenko and S. Woinowsky-Krieger, *Theory of Plates and Shells*, 2nd ed., New York: McGraw-Hill, 1959.

equation for the deflection of the beam at location 1. *Ans.* $v_1 = [2P_1a^2 (l - a)^2 + P_3ac$
$(l^2 - a^2)]/6EIl$.

10.2   In Table 8.1, obtain the displacement for Case (2) from the displacement given for Case
(1) by using the reciprocity theorem. *Note:* The *force* causing the displacement of a surface
is that acting over the entire surface.

Section 10.2

10.3   Derive the equations of Problem 9.48 by using the energy method.

10.4   Determine equations for the slope and deflection at the load on the stepped cantilever. Use
the energy method.

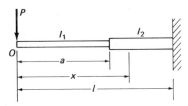

10.5   Determine an equation for the deflection at the midlength location of the uniformly loaded
cantilever beam shown. Use the energy method.

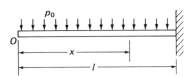

10.6   A helical coil of wire (Fig. 12.3(a) or 12.4(a), becomes a spring for torsion when a torque
or twisting moment M_t is applied to the coil about its axis. To the wire, M_t is a bending
moment, constant along its active length. Let the wire diameter be d, the mean diameter
of the coil D, and the number of active turns N. Write an equation for the energy stored
and differentiate to find the angular twist and the torsional spring constant k_t. *Ans.* $k_t =$
$Ed^4/64DN$ moment per radian.

10.7   A split uniform circular ring is used as a spring clamp. Assume a 360° extent and small
deflections, and derive an equation in terms of EI for the clamping force corresponding
to an opening $\triangle$.

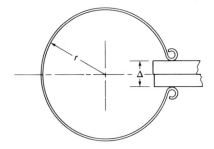

10.8   A 180° ring is used as a spring clamp. Assume small deflections and derive an equation in terms of EI for the clamping force developed by an extension $\triangle$.

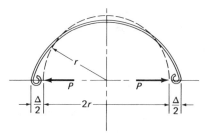

10.9   Partial loops of circular and rectangular cross section are often used as spring clips or clamps. (a) Assuming small deflections and a uniform radius r, derive an equation in terms of EI for the clamping force developed by an extension $\triangle$. *Suggestion:* Make use of what is known about the place of symmetry. *Ans.* $P = EI\triangle/\alpha r^3$, where $\alpha = (\theta/2)$ $[1 + 2 \cos^2 (\theta/2)] - (3/2) \sin \theta$. (b) Compare values of α for $\theta = 90°$, $\theta = 180°$, and $\theta = 360°$.

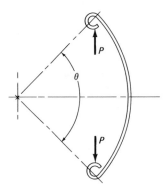

10.10   (a) The partial loop or thin circular beam is centrally loaded. One or both ends may slide on a frictionless surface. Determine an equation in terms of Fr^3/EI and θ for the deflection under the load. (b) The loop deflection should be of the same order of magnitude as for a straight beam of length l. Check your answer to part (a) by a comparison for the extreme case of $\theta = 180°$. Do you think that, in general, a somewhat more shallow loop may be treated as a straight beam?

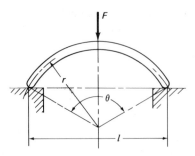

10.11   The flexible support proposed for an extraterrestrial vehicle consists of four wheels and semicircular springs, readily unfolded from the body of the vehicle after landing and then clamped in the position shown. Determine in terms of EI, vehicle load W, and spring radius r: (a) the downward deflection of the supported body and (b) the horizontal motion of a wheel due to vehicle load only.

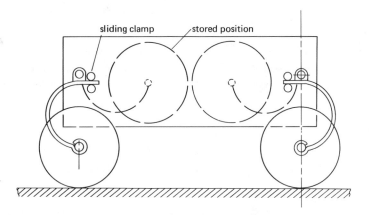

10.12 Same information as Prob. 10.11 except that the springs are longer, subtending an arc of 270° instead of 180°.

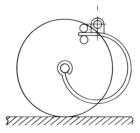

10.13 The coil spring has a relatively rigid rod brazed to it. A force P is applied to the rod at a point C coinciding with the center of the coil, and it may be at any angle α. Consider that the coil extends for 360°. Determine equations for the deflection of point C in the directions of the force and at 90° to the force. What interesting properties does the spring have?

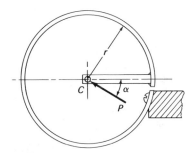

10.14 Determine an equation for the motion at the gap in a split ring when subjected to uniformly distributed pressure such as acts externally on a piston ring or to centrifugal force such as on the ring of Fig. 10.7. *Ans.* $3\pi pbr^4/EI$.

10.15 (a) Complete the solution of Eq. (10.16) or Eq. (10.17) of Example 10.4 to obtain in terms of Fr^3/EI the extension δ_F of the ring of Fig. 10.8(a). (b) If the ring is 75 mm in mean diameter, with a square cross section 3 mm × 3 mm what force may be applied if the allowable stress is 400 MPa? What is the corresponding extension of the ring?

10.16 (a) Derive an equation in terms of Fr^3/EI for the crosswise contraction of the ring of Prob. 10.15. *Ans.* $-0.137Fr^3/EI$. (b) Using the numerical data of Prob. 10.15(b), determine the values of the force and contraction.

10.17 (a) A thin circular ring is centrally loaded and hinged to immovable supports as shown. Determine the horizontal component of the reaction at a support in terms of load P and determine the maximum value of bending moment. (b) Determine the deflection at the load.

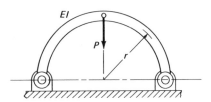

10.18 The statically indeterminate frame consists of one semicircular and two straight sections, the latter built-in at the base. Determine the moments and transverse forces at the base and write equations of bending moment for each section. *Ans.* $0.099Fr$, $0.104F$; straight, $F(0.099r - 0.104x)$; quadrant, $Fr(0.391 - 0.104 \sin \phi - 0.500 \cos \phi)$.

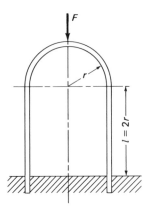

10.19 Determine the deflection at F for the frame of Prob. 10.18.

Section 10.3

10.20 Split rings are sometimes twisted and bent out of their plane by forces normal to their plane. (A lock washer (Prob. 12.56) when tightened is forced back into a plane.) Take a ring with a full circle, such as that of Fig. 10.7, subject it at the sides of the split to two oppositely directed forces, and derive an equation for the out-of-plane deflection there. Assume the forces are in line. *Ans.* $\pi Pr^3 [1/EI + 3/GJ]$, where I is about the neutral axis of bending.

10.21 A curved cantilever beam is semicircular, built-in at one end and loaded at the other by a force normal to its plane. (a) Derive equations for the deflection and slope at the load. (b) Compare deflection and slope for the curved cantilever with a straight one of length $2r$ when both have a cross section of diameter d.

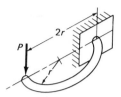

PROBLEM 10.21

10.22   Same requirement as Prob 10.21(a) but for a beam with a quarter turn.

10.23   Starting with Eqs. (10.21), determine the values of M_0 and T_0 for the gimbal ring of Fig. 10.10. Take $G = E/2(1 + \nu)$. Ans. $M_0 = 0.333\ Fr$.

10.24   The device which will hold the door of a passenger car in an open position includes an S-shaped, circular-wire spring. The spring is loosely supported at three locations as shown. As the door reaches its wide-open position, a small serrated wheel on the door rolls over and past the top end of the spring while deflecting it a small amount. It springs back, and the wheel acts as a detent. Determine the force developed and the maximum bending and twisting moments in the wire if it has the dimensions shown and a diameter of $^9/_{32}$ in, and the deflection at force P is $^3/_{16}$ in. An approximate solution which considers only the two principal lengths, taken as straight along their entirety, is suggested.

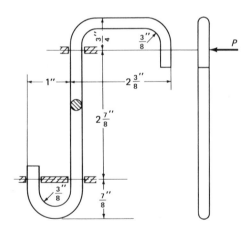

10.25   The coil of a coil spring clutch fits *inside* of a drum, as in Prob 3.69. Derive equations for the pressure and stress similar to Eqs. (10.27) and (10.29) for an external coil, using the method of Sec. 10.3. Ans. $p = 8EI\triangle/b(D - h)^4$.

Section 10.4

10.26   Show that the radius r_n of the neutral surface in a thick curved beam when the cross section is a rectangle is $r_n = (r_o - r_i)/\ln(r_o/r_i)$.

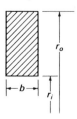

10.27 Show that for a curved beam of trapezoidal section, of widths b_i and b_o at the inner and outer radii, respectively,

$$r_n = \frac{(r_o - r_i)(b_i + b_o)/2}{\dfrac{b_i r_o - b_o r_i}{r_o - r_i} \ln \dfrac{r_o}{r_i} - (b_i - b_o)} .$$

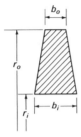

10.28 A T section for a curved beam has the dimensions shown. Show that the neutral surface has a radius

$$r_n = \frac{b_i (r_j - r_i) + b_o (r_o - r_j)}{b_i \ln r_j/r_i + b_o \ln r_o/r_j} .$$

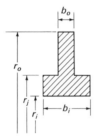

10.29 A hollow box section for the curved throat of a machine frame is made of formed and welded plate. Write an equation for the radius r_n of the neutral surface.

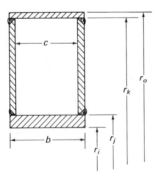

PROBLEM 10.29

PROBLEM 10.30

SEC A-A

10.30 An offset connecting link has the dimensions and loading shown. Determine the maximum stresses on the inside and outside fibers.

10.31 Determine the maximum stresses in the inside and outside fibers of the C clamp. *Ans.* 15 800, −6600 psi,

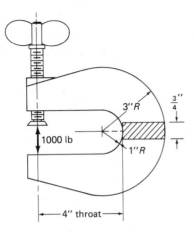

10.32 A photoelastic model for a curved beam is made of a special plastic plate with a thickness of 6 mm. What force P may be applied without exceeding an allowable stress of 7 MPa at any position?

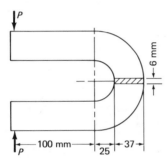

10.33 The C hook or "hairpin" hook shown is used in rolling mills for lifting and transporting large coils of wire and rods. The hook of a crane is put into one of the notches of the top piece or "eye," its location selected according to the number and position of the coils, such that the hook tilts slightly counterclockwise. With the dimensions shown and with a stress of 20 000 psi and a uniform distribution of the coils along the 3½ foot length, what should be the width of a rectangular cross section for a capacity of 2½ tons?

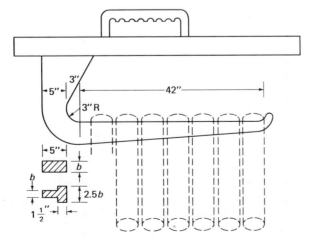

10.34 Same information as Prob. 10.33, but a T section as shown, with the width of the flange 2.5 times the width of the web. Use the equation for r_n in Prob. 10.28. *Ans.* $r_n = 4.5626$ in, $\bar{r} = 4.96$ in, $b = 1.22$ in.

10.35 Determine the maximum stress at the inside and outside surfaces of a hook, such as in Fig. 9.6 (a), but with the trapezoidal cross section shown here. The load is 45 000 N. Use the equation for r_n given in Prob. 10.27.

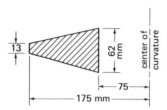

10.36 (a) Derive the expression for $\bar{y}$ in a thick curved beam by the method outlined in footnote 8 of Section 10.4. (b) Derive m and $\bar{y}$ for a rectangular cross section by either or both of the integrals in footnote 8. Let h be the radial height of the rectangle. *Ans.* $\bar{y} = \bar{r}[1 - (h/\bar{r})/\ln(r_o/r_i)]$. (c) Using the answer to part (b), calculate $\bar{y}$ to four decimal places for the section of Prob. 10.31.

10.37 With reference to Prob. 10.36, determine equations for m and $\bar{y}$ of a thick curved beam of circular cross section and diameter d. A transformation of the variable by letting $y = \bar{r} - r$ may be necessary to carry out the integration. *Ans.* $m = (1/4)(d^2/4\bar{r}^2) + (1/8)(d^2/4\bar{r}^2)^2 + (5/64)(d^2/4\bar{r}^2)^3$.

10.38 A circular section of 4-in diameter is proposed for the C hook of Prob. 10.33. Using the equations of Prob. 10.37, calculate the maximum stress.

10.39 A simple hook for a large coil of wire is formed as shown from 1-in-diameter steel rod. What is its capacity based on a stress of 15 000 psi? Use the equations of Prob. 10.37.

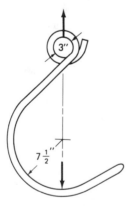

10.40 (a) A crane hook like that in Fig. 9.6(a) has a modified trapezoidal section with the dimensions shown. Draw the section to full scale and determine the radius r_n of the neutral surface by the graphical-numerical method of Example 10.6. Divisions of $\triangle r = \frac{1}{4}$ in should suffice. Careful work with sharp lines is necessary to make the small difference $\bar{y} = \bar{r} - r_n$ reasonably accurate. (b) Determine the capacity of the hook based on an allowable stress of 20 000 psi.

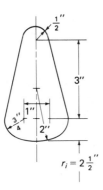

PROBLEM 10.40

10.41   The usual "torsion spring" is a helical coil of wire to the ends of which a moment M_t about the coil's axis is applied. If closely wound, the wire is a thick curved beam in bending. (a) For wire of rectangular cross section, width b and radial height h, write an equation for stress in terms of h, r_n, and the mean coil diameter D. (b) Expand $\ln(r_o/r_i)$ $= \ln x$ into the first two terms of a series. Put the result in terms of $D/h = C$, the *spring index* and substitute into the expression for $\bar{y}$ given in the answer to Prob. 10.36(b) to show that $\bar{y} \approx \bar{r}/(3C^2 + 1)$. Then determine that the answer to part (a) may be written $\sigma = K(6M/bh^2)$, where $K = (3C^2 - C + 1)/3C(C - 1)$.

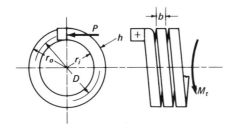

Sections 10.5 *and* 10.6

10.42   Verify that deflection v as given by Eq. (10.39) satisfies the differential equation, Eq. (10.37), when $q = 0$.

10.43   Determine the first derivatives of the functions at the top of Table 10.2. Rearrange to show that each successive function is the derivative of the preceding one except for a multiplier. (*Note*: the function "preceding" F_1 is F_4.) *Ans.* $F_1' = -2\beta F_2$, $F_2' = \beta F_3$, $F_3' = -2\beta F_4$, $F_4' = -\beta F_1$.

10.44   Check the values of βx for $v = 0$ for all the cases in the last column of Table 10.3.

10.45   Starting with Eqs. (10.40), derive the deflection, slope, moment, and shear force equations for an elastically supported beam of infinite extent when loaded at $x = 0$ by a moment M_0 (Case 2 of Table 10.3).

10.46   With reference to the loading of an elastically supported beam by three forces P (Fig. 10.18), determine the spacing between the loads that will result in the minimum bending moments under the loads, also the value of these moments. Express them in terms of P, k, and EI.

10.47   Calculate the maximum moment in the rail for a three-wheel loading as shown in Fig. 10.18 if each wheel carries 50 000 lb, the distance between wheels is 60 in, the foundation modulus is 1600 psi, and I for the steel rail is 86.0 in⁴.

10.48   In Prob. 10.47, the surfaces of the three wheels may not be horizontally in the same line. Determine the amount of misalignment of the middle wheel that would just relieve it of carrying any load when the end wheels are sharing the total load by carrying 75 000 lb each.

10.49   Derive Eqs. (10.42) for an infinite beam that is uniformly loaded over half its extent (Fig. 10.19). Sketch and describe the deflection curve, making use of answers to Prob. 10.43.

10.50   Obtain the bending moment equations for the beam of Fig. 10.19 by differentiating Eqs. (10.42). Make use of the relationships between the $F(\beta x)$ functions and their derivatives given in the answer to Prob. 10.43. Sketch the moment diagram and justify it from the shape of the deflection curve.

Section 10.7

10.51   Pipes are connected to pressure vessels through relatively thick and inflexible nozzles or flanges. For a standard steel pipe with an *outside* diameter of 10.75 in and a wall thickness of 0.593 in, carrying fluid at a pressure of 1150 psi, determine the radial displacement at some distance from the flange. At what two locations closest to the flange do these same displacements appear? What is the maximum value of stress in the pipe and where does it occur? *Ans.* 16 800 psi.

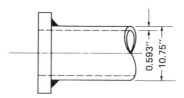

10.52   Rework Example 10.8 assuming that the cylinder is free to rotate but not to expand where it joins the tank bottom plate. Compare values with those from the assumption of no rotation, made in Example 10.8.

10.53   (a) Confirm Eq. (10.46b) by making the indicated substitution in the equation preceding it. (b) Confirm the statement following Eq. (10.46b) regarding the magnitude and location of the maximum tangential stress. What is the value of x in terms of r if $t/r = \frac{1}{25}$?

10.54   For the cylinder of Example 10.9 determine expressions in terms of pr/t for the factors of safety by the maximum energy of distortion theories of failure for steady loading, at two locations, $x = \pi/4\beta$ and $x = 1.85/\beta$, the respective locations of the maximum axial stress and the maximum tangential stress.

10.55   A coil retaining ring is shrunk onto a heavy generator rotor, as shown. The diametral interference at the shoulder is 2.5 mm before heating and assembly. Determine the tangential and the axial stresses in the ring at the edge of the fit. All parts are steel ($E = 207$ GPa). Neglect any contraction of the heavy rotor.

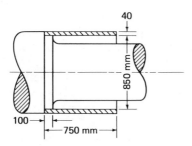

PROBLEM 10.55

10.56 A long pipe is firmly fitted at its ends by rollers, as shown. Show that raising the temperature by T degrees increases the maximum bending stress by the amount $0.586\ \alpha ET$, where α is the coefficient of thermal expansion. Is this form typical of other thermal stress results?

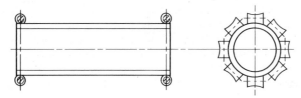

10.57 A free-standing thin cylinder is heated such that the temperature changes linearly in the radial direction by an amount T. Just as stresses in a thin curved beam may be considered linearly distributed as in a straight one, those in a thin cylinder may be taken as in a thin flat plate, from Section 7.6 or Eq. (10.53b). (a) Write equations for the maximum stress, moment, and radial displacement at places well away from the ends. (b) At an end, no moment can exist, and the moment in part (a) is in effect cancelled by the application of an equal and opposite one. Find this case in Table 10.3, and write an equation for the radial deviation. Sketch the shape of the deflected cylinder and place dimensions upon it for two cases: (1) higher temperature at the outside surface and (2) higher temperature at the inside surface. *Ans.* Deviation at end is $\pm 0.275\ r\alpha T/(1 - \nu)$.

10.58 A group of closely spaced simply supported cross beams of length l, rigidity EI_c, and center-to-center spacing c serves as an elastic foundation for a much longer beam of rigidity EI, centrally placed and loaded. (a) Calculate the foundation modulus k and characteristic β for the system of beams. *Ans.* $\beta = \sqrt[4]{12I_c/cl^3I}$. (b) Write an equation for the deflection under the load and an equation for the load carried by the central cross beam in terms of the applied load P. *Ans.* $P_0 = (Pc/2)\sqrt[4]{12I_c/cl^3I}$. (c) For the special case $I = 1.5\ I_c$ and $c = l/20$, list the ratios to the central deflection of the deflections of the third, sixth, and twelfth beam to the right and left.

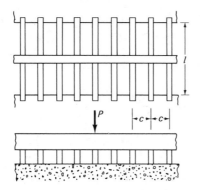

Section 10.8

10.59 Calculate the maximum thickness of a steel forming sheet that may be bent without yielding over an ellipsoid where its principal radii of curvature are 1000 and 2000 mm. The yield-point stress is 200 MPa and $E = 207$ GPa.

10.60 Show that the function $w = C(1 - x^2/a^2 - y^2/b^2)^2$ may satisfy the plate equation, Eq. (10.51), and evaluate the constant C. Describe the shape and the conditions at the boundary of this plate.

10.61 Investigate the function $w = Cxy$ as a description of plate deflection. Do not neglect to try all the equations (10.49).

10.62 With reference to Example 10.11 and the equations preceding Eq. (10.53a), put $1/\rho_y = -1/\rho_x$ and determine equations for moments, deflection, and slope. Sketch and describe the plate.

10.63 An approximate method that has been used to obtain a design equation for the *stress* in uniformly loaded square and rectangular plates is to assume the plate to be bent along a diagonal by a moment of the applied force acting at the centroid of pressure in the triangular half of the plate and a moment of the reactions centered along each side of the triangle. This gives an average stress along the diagonal. Sketch and dimension, and determine the stress equation for a square plate with sides of length b. *Ans.* $\sigma = pb^2/4t^2$.

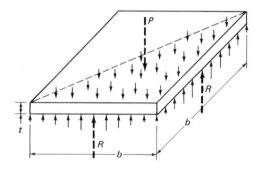

10.64 Same information Prob. 10.63 but for a rectangular plate with sides of length a and b. *Ans.* $\sigma = [a^2b^2/(a^2 + b^2)][p/2t^2]$.

Sections 10.9 *and* 10.10

10.65 For Case 2 of Table 10.4 the deflection at any radius r, with $c = 0$, is

$$w = (P/16\pi D)[(3 + \nu)(b^2 - r^2)/(1 + \nu) + 2r^2 \ln(r/b)].$$

Determine equations for the slope and the meridional-plane moment at any location r and compare θ_{max} and σ_{max} with values in the table.

10.66 Obtain an expression for moment M_m in terms of the constants of integration of Eq. (10.59).

$$Ans. \ M_m = -\frac{(3 + \nu) \, p_0 r^2}{16} - \frac{(1 + \nu)P'}{8\pi}(2 \ln r - 1)$$

$$-\frac{P'}{4\pi} + \frac{D(1 + \nu)C_1}{2} - \frac{D(1 - \nu)C_2}{r^2}$$

10.67 (a) To use the loaded-side stress equations of Cases 2 and 7 of Table 10.4 it is necessary to know or assume the radius of the loading circle. For one of these cases, compare the intensities of stress for c/b ratios of 0.10, 0.01, and 0.001. (b) For the case chosen and with a b/t ratio of 25, obtain the stress on the opposite side and compare with part (a).

10.68 Derive the first and third equations of Case 7 from Eqs. (10.61) through (10.63) and the boundary conditions.

10.69 Derive the maximum deflection equation of Case 6 from that of Case 3 by integration of an elemental ring of force, as briefly outlined in Section 10.10.

10.70 For the plates with holes of Cases 12, 14, 16, and 18 of Table 10.5(A) write the boundary conditions needed for determining the constants of integration, and indicate by number the equations to be used.

10.71 A circular plate supported at its outer edge is loaded by a moment M_a per unit length distributed uniformly around its concentric hole. Derive an equation for the moment M_m at any radius r. Use the moment equation given in the answer to Prob. 10.66. *Ans.* $M_m = M_a [(b^2/r^2 - 1)/(b^2/a^2 - 1)]$.

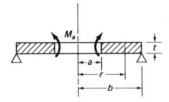

10.72 Derive the stress and deflection equations of Case 3 by the superposition of two other cases in the table.

10.73 For Case 7 of Table 10.4 derive all the equations for the stress and deflection at the center of the plate by superposition of two other cases in the table.

10.74 Same requirement as Prob. 10.73 but apply to Case 10.

10.75 Determine by superposition equations for maximum deflection and stress of a solid circular plate of radius b, freely supported at its edge and uniformly loaded over an area bounded by the edge and an inner circle of radius c.

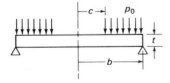

10.76 A solid circular plate is centrally supported over a very small area and loaded uniformly on the opposite side. Determine by superposition the equations for the deflection at the outer edge and stress at the center. *Ans.* $w = 3(1 - \nu)(7 + 3\nu)p_0 b^4/16Et^3$ and $\sigma = (3p_0 b^2/2t^2) [(1 + \nu) \ln (b/c) + (1 - \nu)/4]$.

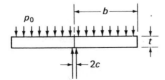

10.77 Welded construction is proposed for the piston of a double-acting pump, as shown. A pressure of 200 psi may act alternately on the two sides of the piston. If the allowable stress is 14 000 psi, what thickness is required of the disk if it is considered to be "built in" or fixed at the hub, but the stiffening effect of the outer ring is neglected? *Ans. t =* 0.963, say 1 in.

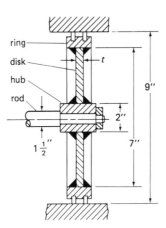

10.78 Same information as Prob. 10.77 except that the outer ring is considered very stiff.

10.79 A coned disk or *Belleville* spring consists of one or more dished plates, such as the two shown, alternately stacked in pairs so that all the disks are active. If the deflection-to-thickness ratio w/t and the "dish"-to-thickness ratio h/t are both less than 0.50, flat plate theory may be used with sufficient accuracy.[26] Design a safety-valve spring with a maximum compression corresponding to a load of 4000 lb and a stress of 100 000 psi. Determine: (a) thickness, and maximum dish and deflection if $b/a = 2.0$ and flat plate equations are to be suitable; (b) minimum desirable number of disks for a total deflection of 2 in; (c) required outside radius and diameter; (d) overall dimensions. What changes can be made to give less diameter?

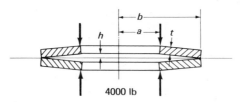

4000 lb

10.80 The bottom circular steel plate of an elevated water storage tank, similar to that of Fig. 10.21, is supported along its outer edge and in the center by a vertical pipe that comes through the plate and is welded to both sides of the plate. Assume both supports to be inflexible vertically. The outside diameters of the plate and pipe are 1.50 and 0.30 m, respectively, and water in the tank will have a maximum height of 2.50 m. (The unit weight of water is 9.807 kN/m³.) Recommend a thickness for the plate if the allowable

---

[26] If this restriction is removed, coned-disk springs of smaller size and nonlinear characteristics may be designed. Special equations and charts are available, e.g., A. M. Wahl, *Mechanical Springs*, 2nd ed., New York: McGraw-Hill, pp. 153–175, 1963.

stress is 100 MPa. (*Note*: A safe-side approximation is obtained by assuming that the outer walls give a freely supported condition.)

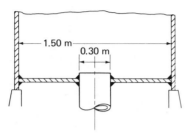

10.81 The turbine housing shown consists of one outer and two inner thick rings welded to circular flat plate ends. It contains steam at pressure p_0. Axial deflection at the inner rings must be limited to prevent rubbing at labyrinth seals (not shown). The plate deflection and stress cannot be solved by superposition from Table 10.5(A) because it has no equation for outer-edge fixed support, uniform loading, and inner-edge nonrotation. Hence start the solution using the general equations for circular plates. (a) Sketch the deflected shape of the plate, showing the applied loads. (b) Solve for P' and shear V or solve directly for V for equilibrium at radius r. (c) Write the boundary conditions and substitute them into the equations that would be used to solve for the constants of integration.

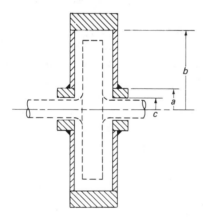

10.82 A perfectly flat, circular plate with a central hole is uniformly loaded over its top surface. The plate is supported on two narrow rings that are exactly at the same level, one around the outer edge and one around the hole. The ratio b/a is 4.0. Using Table 10.5 (A), make sketches and write an equation that resolves the indeterminacy. Find the reactions at both rings in terms of $p_0 b^2$.

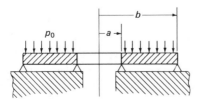

11

Surface Contacts and Wear

Concentrated and Distributed Forces; Spheres,
Cylinders, and General Shapes; Surface Failures and
Wear; Gear Teeth and Rolling-element Bearings

11.1 INTRODUCTION

There is a group of machine components whose functioning depends upon rolling and
sliding motion along surfaces while under load. Both surfaces are usually convex, so that
the area through which the load is transferred is very small, even after some surface
deformation, and the pressures and local stresses are very high. Unless logically designed
for the load and life expected of it, the component may fail by early general wear or by
local fatigue failure and the loss of portions of its surface by pitting, spalling, and scoring.
The magnitude of damage is a function of the materials and of the intensity of applied
load or pressure, as well as surface finish, lubrication, and relative sliding.

The intensity of load can be determined from equations which are functions of the
geometry of the surfaces, essentially the radii of curvature, and the elastic constants of
the materials. Larger radii and smaller moduli of elasticity give larger contact areas and
lower pressures. Careful alignment, smoother surfaces, and higher strength and oil viscosity minimize failures.

First in the chapter there are developed or presented the basic equations for contact
of points, knife-edge pivots, balls, and rollers, as well as for contact between more
general surfaces such as those of wheels on rails, three-dimensional cams and traction-
type transmissions. The nature of surface failures, the propagation of cracks, and the
effect of oil films are discussed, and design stresses and improvements are suggested.
Then particular attention is given to the design of gear teeth and rolling-element bearings.

11.2 CONCENTRATED AND DISTRIBUTED FORCES ON PLANE
SURFACES: BOUSSINESQ'S EQUATIONS

The theory of contact stresses and deformations is one of the more difficult topics in the
theory of elasticity. In this and the next section only the derivation for two spheres is
completed, but some other derivations are outlined. Stress and deflection formulas are
listed in Tables 11.1 and 11.2, and some discussion of their application is given.

Derivations start with forces applied to the plane boundaries of semiinfinite bodies,
i.e., bodies which extend indefinitely in all directions on one side of the plane. Theoretically this means that the stresses which radiate away from the applied forces and die

TABLE 11.1. Concentrated and Distributed Surface Forces

Notation: Z-axis normal to surface, z = depth, r = radial distance as sketched, P = force applied, l = length of contact line, P/l = force per unit length, σ = normal stress, τ = shear stress, s = resultant stress, w = surface deflection, E = modulus of elasticity, ν = Poisson's ratio, $\eta = (1 - \nu^2)/E$

Limitations: Dimensions of the body must be much larger than the locally affected portion.

Loading Case	Pictorial	Stresses and Deflections
1. Point		$q = \sqrt{\sigma_z^2 + \tau_{rz}^2} = \dfrac{3}{2\pi} \dfrac{P \cos^2\theta}{(r^2 + z^2)}$ $w = \dfrac{1-\nu^2}{\pi E} \dfrac{P}{r}$ at surface
2. Line		$\sigma = \dfrac{2}{\pi} \dfrac{(P/l)\cos\theta}{\sqrt{x^2 + z^2}}$
3. Knife edge or pivot		$\sigma_r = -\dfrac{(P/l)\cos\theta}{r\left(\alpha + \dfrac{1}{2}\sin 2\alpha\right)}$
4. Uniform distributed load p over circle of radius a.		With $\nu = 0.3$, at point O $\sigma_z = -p, \quad \sigma_r = \sigma_\theta = -0.8p$ and $w_{max} = \dfrac{2(1-\nu^2)pa}{E} = 2\eta pa$ $\tau_{max} = 0.33p$ at $z = 0.638\,a$
5. Rigid cylinder $(E_1 \gg E_2)$		$(\sigma_z)_{z=0} = -p = -\dfrac{P}{2\pi a\sqrt{a^2 - r^2}}$ $w = \dfrac{(1-\nu^2)P}{2E_2 a} = \dfrac{\eta_2 P}{2a}$

out rapidly are unaffected by any stresses from reaction forces or moments elsewhere on the body.

A concentrated force acts at point O in Case 1 of Table 11.1. At any point Q there is a resultant stress q on a plane perpendicular to OZ, directed through O and of magnitude inversely proportional to $(r^2 + z^2)$, or the square of the distance OQ from the point of load application. This is an indication of the rate at which stresses "die out." The deflection of the surface at a radial distance r is inversely proportional to r, hence, is a hyperbola asymptotic to axes OR and OZ. At the origin, the stresses and deflections theoretically become infinite, and one must imagine the material near O cut out, say, by a small hemispherical surface to which are applied distributed forces that are statically equivalent to the concentrated force P. Some such surface is obtained by yielding of the material.

An analogous case is concentrated loading along a line of length l (Case 2). Here, the force is P/l per unit length of the line. The result is a normal stress directed through the origin and inversely proportional to the *first* power of distance to the load, not fading out as rapidly. Again, the stress approaches infinite values near the load. Yielding followed by work hardening may limit the damage. Stresses in a knife or wedge, which might be used to apply the foregoing load, are given under Case 3. Note that the solution for Case 2 is obtained when $2\alpha = \pi$, or when the wedge becomes a plane.

In the deflection equation of Case 1 one may substitute for force P an expression that is the product of a pressure p and an elemental area, such as the shaded area of Fig. 11.1. This gives a deflection at any point M on the surface at a distance $r = s$ away from the element, namely,

$$dw = \frac{(1 - \nu^2)}{\pi E} \frac{p(s\, d\phi)\, ds}{s} = \frac{(1 - \nu^2)p\, ds\, d\phi}{\pi E}$$

where ν is Poisson's ratio. The total deflection at M is the superposition or integration over the loaded area of all the elemental deflections, namely,

$$w = \frac{(1 - \nu^2)}{\pi E} \int\int p\, ds\, d\phi = \frac{\eta}{\pi} \int\int p\, ds\, d\phi \tag{11.1}$$

where η is an elastic constant $(1 - \nu^2)/E$. If the pressure may be considered uniform, as from a fluid, and the loaded area is a circle, the resulting deflections, in terms of elliptic integrals, are given by two equations, one for M outside and one for M inside the circle.[1] The deflections at the center are given under Case 4 of Table 11.1. The stresses are also obtained by a superposition of elemental stresses for point loading. Shear stress is maximum below the surface.

If a rod in the form of a punch, die, or structural column is pressed against the surface of a relatively soft material, i.e., one with a modulus of elasticity much less than that

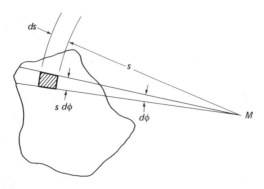

FIGURE 11.1. Sketch for obtaining deflection at any point M due to loading within an enclosed area (Eq. (11.1)).

---

[1] S. Timoshenko and J. N. Goodier, *Theory of Elasticity*, 3rd ed., New York: McGraw-Hill, pp. 403–409, 1970. The solution for this and for Case 1 are credited to J. Boussinesq, *Application des Potentiels à l'Etude de l'Equilibre et du Mouvement des Solides élastiques*, Paris: Gauthier-Villars, 1885.

of the rod, the rod may be considered rigid, and the distribution of deflection is initially known. For a circular section, with deflection w constant over the circle, the results, obtained by J. Boussinesq,[2] are listed in Case 5. The pressure p is least at the center, where it is 0.5 p_{avg}, and it is infinite at the edges. The resultant yielding at the edges is local and has little effect on the general distribution of pressure. For a given total load, the deflection is inversely proportional to the radius of the circle.

11.3 CONTACT BETWEEN TWO ELASTIC BODIES: HERTZ STRESSES FOR SPHERES

When two elastic bodies with convex surfaces, or one convex and one plane surface, or one convex and one concave surface are brought together in point or line contact and then loaded, local deformation will occur, and the point or line will enlarge into a surface of contact. In general, its area is bounded by an ellipse, which becomes a circle when the contacting bodies are spheres, and a narrow rectangle when they are cylinders with parallel axes. These cases differ from those of the preceding section in that there are *two* elastic members, and the pressure between them must be determined from their geometry and elastic properties.

The solutions for deformation, area of contact, pressure distribution, and stresses at the initial point of contact were made by Heinrich Hertz.[3] The maximum compressive stress, acting normal to the surface, is, of course, equal and opposite to the maximum pressure, and this is frequently called *the* Hertz stress. Equations and charts for the principal and shear stresses at any other point in the region were developed later by others.[4] Hertz and others have checked the theory experimentally using very smooth surfaces.

The assumption is made that the dimensions of the contact area are small relative to the radii of curvature and to the overall dimensions of the bodies. Thus the radii, though varying, may be taken as constant over the very small arcs subtending the contact area. Also, the deflection integral derived for a plane surface, Eq. (11.1), may be used with very minor error. This makes the stresses and their distribution the same in both contacting bodies.

The methods of solution will be illustrated by the case of two spheres of different material and radii R_1 and R_2. Figure 11.2 shows the spheres before and after loading, with the radius a of the contact area greatly exaggerated for clarity. Distance $z = R - R \cos \gamma \approx R - R(1 - \gamma^2/2 + \cdots) \approx R\gamma^2/2 \approx r^2/2R$ because $\cos \gamma$ may be expanded in series and the small angle $\gamma \approx r/R$. If points M_1 and M_2 in Fig. 11.2(a) fall within the contact area, their approach distance M_1M_2 is

$$z_1 + z_2 = \frac{r^2}{2}\left(\frac{1}{R_1} + \frac{1}{R_2}\right) = Br^2 \tag{11.2}$$

where B is a constant $(1/2)(1/R_1 + 1/R_2)$. If one surface is concave, as indicated by the

[2] J. Boussinesq, *Application des Potentiels à l'Etude de l'Equilibre et du Mouvement des Solides élastiques*, Paris: Gauthier-Villars, 1885.

[3] H. Hertz, "On the Contact of Elastic Solids," *J. Math.*, Vol. 92, pp. 156–171, 1881 (in German). For an English translation, see H. Hertz, *Miscellaneous Papers*, London: Macmillan & Company, Ltd., 1896. Of interest to students, the introduction to the book contains a brief history of the education and early professional life of this great physicist, much of it told by quotations from letters to his parents.

[4] Of the many papers, major contributions published in English are by H. R. Thomas and V. A. Hoersch, "Stresses Due to the Pressure of One Elastic Solid upon Another," *University of Illinois, Bulletin 212 of the Engineering Experimental Station*, July 15, 1930, and by H. L. Whittemore and S. N. Petrenko, "Friction and Carrying Capacity of Ball and Roller Bearings," Tech. Paper, Bureau of Standards 201, 1921.

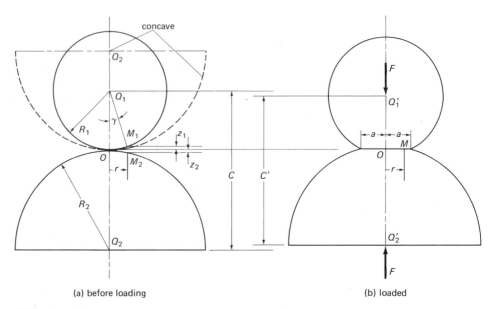

(a) before loading   (b) loaded

FIGURE 11.2. Compression between two spherical surfaces.

dotted line in Fig. 11.2(a), the distance is $z_1 - z_2 = (r^2/2)(1/R_1 - 1/R_2)$, which indicates that when contact is on the inside of a surface the numerical value of its radius is to be taken as negative in all equations derived from Eq. (11.2).

The approach between two relatively distant and strain-free points, such as Q_1 and Q_2, consists not only of the surface effect $z_1 + z_2$, but also of the approach of Q_1 and Q_2 relative to M_1 and M_2, respectively, which are the deformations w_1 and w_2 due to the as yet undetermined pressure over the contact area. The total approach or deflection δ, with substitution from Eqs. (11.1) and (11.2), is

$$\delta = (z_1 + z_2) + (w_1 + w_2) = Br^2 + (1/\pi)(\eta_1 + \eta_2) \iint p \, ds \, d\phi$$

where

$$\eta_1 = (1 - \nu_1^2)/E_1 \quad \text{and} \quad \eta_2 = (1 - \nu_2^2)/E_2$$

With rearrangement

$$\frac{\eta_1 + \eta_2}{\pi} \iint p \, ds \, d\phi = \delta - Br^2 \tag{11.3}$$

By symmetry, the area of contact must be bounded by a circle, say of radius a, and Fig. 11.3 is a special case of Fig. 11.1. A trial will show that Eq. (11.3) will be satisfied by a hemispherical pressure distribution over the circular area. Thus the peak pressure at center O is proportional to the radius a, or $p_0 = ca$. Then, the scale for plotting pressure is $c = p_0/a$. To find w_1 and w_2 at M in Eq. (11.3), an integration $\int p \, ds$ must first be made along a chord GH, which has the half-length $GN = \sqrt{a^2 - r^2 \sin^2 \phi}$. The pressure varies as a semicircle along this chord, and the integral equals the pressure scale c times the area A under the semicircle, or

$$\int p \, ds = cA = \frac{p_0}{a} \frac{\pi}{2} (a^2 - r^2 \sin^2 \phi)$$

By a rotation of line GH about M from $\phi = 0$ to $\phi = \pi/2$, half of the contact circle, the shaded area of Fig. 11.3, is covered. Doubling the integral completes the integration in

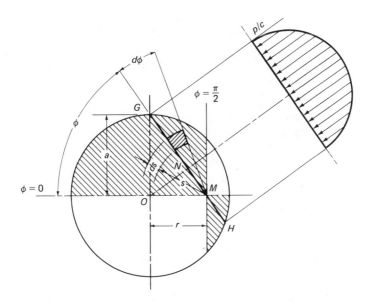

FIGURE 11.3. Distribution and integration of pressure over the circular contact area between spherical surfaces.

Eq. (11.3), namely,

$$\frac{p_0(\eta_1 + \eta_2)}{a} \int_0^{\pi/2} (a^2 - r^2 \sin^2 \phi)\, d\phi = \delta - Br^2$$

whence

$$\frac{p_0 \pi}{4a} (\eta_1 + \eta_2)(2a^2 - r^2) = \delta - Br^2 \tag{11.4}$$

Now the approach δ of centers Q_1 and Q_2 is independent of the particular points M and radius r chosen in the representation by which Eq. (11.4) was obtained. To make the equation independent of r, the two r^2 terms must be equal, whence it follows that the two constant terms are equal. The r^2 terms, equated and solved for a, yield the radius of the contact area

$$a = \frac{p_0 \pi (\eta_1 + \eta_2)}{4B} \tag{11.5}$$

The two constant terms when equated give

$$\delta = \frac{p_0 \pi (\eta_1 + \eta_2) a}{2} \tag{11.6}$$

The integral of the pressure over the contact area is equal to the force P by which the spheres are pressed together. This integral is the pressure scale times the volume under the hemispherical pressure plot, or

$$\frac{p_0}{a} \left(\frac{2}{3} \pi a^3 \right) = P$$

and the peak pressure has the value

$$p_0 = 1.5\, P/\pi a^2 = 1.5\, p_{\text{avg}} \tag{11.7}$$

Substitution of Eq. (11.7) and the value of B below Eq. (11.2) gives to Eqs. (11.5) and (11.6) the forms shown for Case 1 of Table 11.2. If both spheres have the same elastic modulus $E_1 = E_2 = E$, and Poisson's ratio is 0.30, a simplified set of equations is obtained. With a ball on a plane surface, $R_2 = \infty$, and with a ball in a concave spherical seat, R_2 is negative

It has taken all of this just to obtain the pressure distribution on the surfaces. All stresses can now be found by the superposition or integration of those obtained for a concentrated force acting on a semiinfinite body. Some results are given under Case 1 of Table 11.2. An unusual but not unexpected result is that pressure, stresses, and deflections are not linear functions of load P, but rather increase at a less rapid rate than P. This is because of the increase of the contact or supporting area as the load increases. Pressures, stresses, and deflections from several different loads cannot be superimposed because they are nonlinear with load.

11.4 CONTACT BETWEEN CYLINDERS AND BETWEEN BODIES OF GENERAL SHAPE

Equations for cylinders with parallel axes may be derived directly,[5] as done for spheres in Section 11.3. The contact area is a rectangle of width $2b$ and length l. The derivation starts with the stress for line contact (Case 2 of Table 11.1). Some results are shown under Case 2 of Table 11.2. Inspection of the equations for semiwidth b and peak pressure p_0 indicates that both increase as the *square root* of load P. The equations of the table except that for δ may be used for a cylinder on a plane by the substitution of infinity for R_2. The semiwidth b for a cylinder on a plane becomes $1.13\sqrt{(P/l)(\eta_1 + \eta_2)R_1}$.

Figure 11.4 shows for two parallel cylinders the variation of several stress ratios σ/p_0 and τ/p_0 as a function of depth ratio z/b for points along the Z axis, where the stresses have their largest values. All normal stresses are compressive, with σ_y and σ_z equal at the surface to the contact pressure p_0. Also significant is the maximum shear stress τ_{yz}, with a value of $0.304p_0$ at a depth $0.786b$. The variation of stresses along the Y axis at several depths z are also given by Radzimovsky,[6] based on equations of N. M. Belayev. Radzimovsky points out that the shear stress of greatest significance for rolling cycles, since it is completely reversed, is located at approximately $y = \pm 0.85b$ and $z = 0.5b$, with a value approximately $\pm 0.256p_0$.

Example 11.1
For the steel cams and flat-faced followers of Prob. 6.16 determine the minimum allowable tip radius if both cam and follower are made of AISI 8650 steel, hardened to give the yield-point stress of 150 000 psi required for shaft strength.

Solution
From steel charts this yield point is seen to correspond to an ultimate tensile stress of 170 000, with a hardness of about $170\,000/500 = 340$ Bhn. With Section 11.7 and Eq. 11.17 for guidance, we may choose $p_0 = -(\sigma)_{max} = 120\,000$ psi as a conservative value. From Prob. 6.16, $P = 120$ lb, $l = \frac{1}{2}$ in, $P/l = 240$ lb/in, R_1 (follower) $= \infty$, and

[5] M. F. Spotts, *Mechanical Design Analysis*, Englewood Cliffs, New Jersey: Prentice-Hall, pp. 166–171, 1964.

[6] E. I. Radzimovsky, "Stress Distribution and Strength Condition of Two Rolling Cylinders," *University of Illinois Engineering Experiment Station, Bulletin 408*, pp. 20–25, Feb. 1953.

TABLE 11.2. Contact between Two Elastic Bodies

Notation: a = major semiaxis of ellipse of contact (radius of circle of contact between spheres), b = minor semiaxis of ellipse, also half-width of rectangle of contact between parallel cylinders, where l = length of contact; P = load, P/l = load per unit length of parallel cylinders, p_0 = maximum pressure on surface, σ_c = compressive stress, σ_t = tensile stress, τ = shear stress, δ = approach of the bodies as a whole; E = modulus of elasticity, ν = Poisson's Ratio, and

$$\eta_1 = \frac{1 - \nu_1^2}{E_1} \quad \text{and} \quad \eta_2 = \frac{1 - \nu_2^2}{E_2}$$

(For steel, $\eta = 0.0303 \times 10^{-6}$ in²/lb or $\eta = 4.40 \times 10^{-6}$ mm²/N)

Loading case	Pictorial	Area, Pressure, Approach
1. Spheres or Sphere and Plane If the surface of the body 2 is concave (dash lines), take D_2 negative, if plane, take $D_2 = \infty$.		$a = 0.721\,[P\,(\eta_1 + \eta_2)\,D_1 D_2/(D_1 + D_2)]^{1/3} = c/2$ $p_0 = 1.5\,P/\pi a^2 = 1.5\,p_{avg} = -(\sigma_c)_{max}$ $\max \tau = \frac{1}{3}\,p_0$, at depth $0.638\,a$ $\max \sigma_t = (1 - 2\nu)\,p_0/3$, at radius a $\delta = 1.04\,[(\eta_1 + \eta_2)^2 P^2\,(D_1 + D_2)/D_1 D_2]^{1/3}$
2. Cylindrical Surfaces with Parallel Axes If the surface of body 2 is concave (dash lines) take R_2 negative. If plane, take $R_2 = \infty$ except for δ.		$b = 1.13\sqrt{(P/l)\,(\eta_1 + \eta_2)\,R_1 R_2/(R_1 + R_2)} = c/2$ $p_0 = 2P/\pi b l = 1.273\,p_{avg} = -(\sigma_c)_{max}$ If $\nu = 0.30$ $\max \tau = 0.304\,p_0$, at depth $0.786\,b$ If $\eta_1 = \eta_2 = \eta$ $\delta = 0.638\,(P/l)\,\eta\left[\dfrac{2}{3} + \ln\dfrac{2R_1}{b} + \ln\dfrac{2R_2}{b}\right]$
3. General Case At point of contact, Z-axis is the common normal. Minimum and maximum numerical values of the radii of curvature are respectively R_1 and R_1' for body 1, R_2 and R_2' for body 2. Angle between planes containing curvatures $1/R_1$ and $1/R_2$ is ψ. A radius is negative if the surface is concave in its plane.		$b = \beta\left[\dfrac{3P(\eta_1 + \eta_2)}{4\,(B + A)}\right]^{1/3}$ and $a = b/\kappa$ where β, κ, and λ are obtained from Fig. 11.6 and $B + A = \dfrac{1}{2}\left[\dfrac{1}{R_1} + \dfrac{1}{R_1'} + \dfrac{1}{R_2} + \dfrac{1}{R_2'}\right]$ $B - A = \dfrac{1}{2}\left\{\left[\dfrac{1}{R_1} - \dfrac{1}{R_1'}\right]^2 + \left[\dfrac{1}{R_2} - \dfrac{1}{R_2'}\right]^2 \right.$ $\left. + 2\left[\dfrac{1}{R_1} - \dfrac{1}{R_1'}\right]\left[\dfrac{1}{R_2} - \dfrac{1}{R_2'}\right]\cos 2\psi\right\}^{1/2}$ at $x = y = z = 0$ $p_0 = 1.5\,P/\pi ab = 1.5\,p_{avg} = -(\sigma_c)_{max}$ $\sigma_x = -2\nu p_0 - (1 - 2\nu)\,p_0\,\dfrac{b}{a + b}$ $\sigma_y = -2\nu p_0 - (1 - 2\nu)\,p_0\,\dfrac{a}{a + b}$ $\delta = \lambda\,[P^2\,(\eta_1 + \eta_2)^2\,(B + A)]^{1/3}$

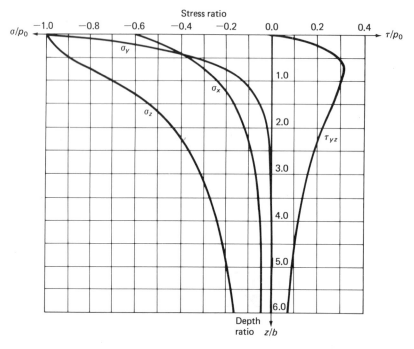

FIGURE 11.4. For two parallel cylinders in contact, variation of normal-stress and shear-stress ratios along the Z axis as a function of the ratio of depth to contact half-width b when $\nu = 0.30$. (Adapted from E. I. Radzimovsky, University of Illinois, *Engineering Experiment Station Bulletin 408*, p. 13, 1953, with permission of Engineering Publications, University of Illinois.)

from Case 2 of Table 11.2, $p_0 = 2P/\pi bl$. Then

$$b = \frac{2P}{\pi p_0 l} = \frac{2(120)}{\pi(120\ 000)(1/2)} = 0.001\ 273 \text{ in}$$

Also,

$$\eta_1 + \eta_2 = 2\eta = 0.0606 \times 10^{-6} \text{ in/lb}, \quad \frac{R_1 R_2}{R_1 + R_2} \rightarrow R_2,$$

and

$$b = 1.13\sqrt{(P/l)(\eta_1 + \eta_2)R_1 R_2/(R_1 + R_2)} = 1.13\sqrt{(240)(0.0606 \times 10^{-6})R_2}$$

Equating the two values of b and solving for R_2, we obtain

$$1.621 \times 10^{-6} = 1.277(14.54 \times 10^{-6})R_2, \quad R_2 = 0.0873 \text{ in}$$

A graphical layout shows that this radius is possible and reasonable. A larger value will reduce wear.　　　　////

Case 3 of Table 11.2 pictures a more general case of two bodies, each with one major and one minor plane of curvature at the initial point of contact. Axis Z is normal to the tangent plane XY, and thus the Z axis contains the centers of the radii of curvature. The minimum and maximum radii for body 1 are R_1 and R_1', respectively, lying in planes $Y_1 Z$ and $X_1 Z$. For body 2, they are R_2 and R_2', lying in planes $Y_2 Z$ and $X_2 Z$, respectively. The angle between the planes with the minimum radii or between those with the maximum radii is ψ. In the case of two crossed cylinders with axes at 90 deg, such as a car wheel on a rail (Fig. 11.5), $\psi = 90$ deg and $R_1' = R_2' = \infty$.

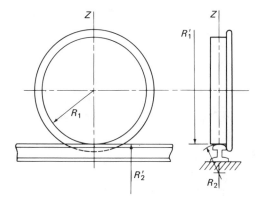

FIGURE 11.5. Crossed cylinders: wheel on railroad rail with a head radius R_2, $R_1' = R_2' = \infty$, and $\psi = 90°$ in Case 3 of Table 11.2.

Hertz solved this general case in his original paper.[7] Results may be presented in various ways. Here, two sums $(B + A)$ and $(B - A)$, obtained from the geometry and defined under Case 3 of Table 11.2, are taken as the basic parameters. The area of contact is an ellipse with a minor axis $2b$ and a major axis $2a$. The distribution of pressure is that of an ellipsoid built upon these axes, and the peak pressure is 1.5 times the average value $P/\pi ab$. The coefficients β and κ for the calculation of b and a, and λ for the calculation of the deflection δ, may be obtained from Fig. 11.6.[8] The abscissa

FIGURE 11.6. Coefficients in equations for contact between general elastic bodies (Case 3 of Table 11.2). (Based upon values calculated in the paper by H. L. Wittemore and S. N. Petrenko, "Friction and Carrying Capacity of Ball and Roller Bearings," Tech. Paper 201, Bureau of Standards, 1921.)

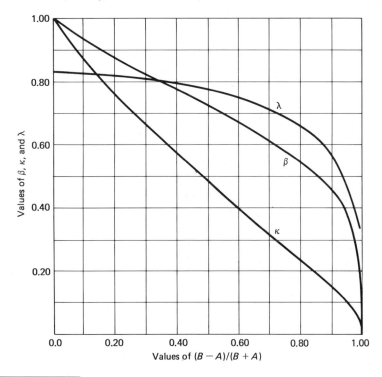

[7] See footnote 3.
[8] Based upon values calculated in the paper by H. L. Whittemore and S. N. Petrenko, "Friction and Carrying Capacity of Ball and Roller Bearings," Tech. Paper, Bureau of Standards 201, 1921.

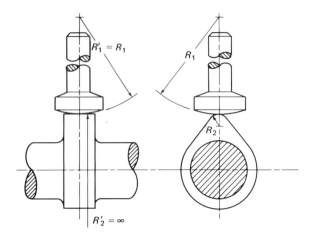

FIGURE 11.7. Cam and spherical-faced (mushroom) follower or tappet (Example 11.2).

is $(B - A)/(B + A)$, which has the limiting values of zero for two spheres and 1.0 when $\psi = 0$ and the radii R_1' and R_2' approach infinity together. However, for cylinders with parallel axes, the results are not usable in this form, and the contact area is a rectangle of known length, not an ellipse.

Example 11.2

A spherical-faced tappet of chilled cast iron is used as the follower in contact with a steel cam (Fig. 11.7). The face radius of the tappet is 10 in, and for it, $E = 20 \times 10^6$, $\nu = 0.25$. The minimum radius of the cam is 0.375 in, with a width 0.40 in. The load is 600 lb. Calculate stresses and deflection. Should the cam be carburized?

Solution

With the notation of Case 3 of Table 11.2 and the spherical surface designated by subscript 1, $R_1 = R_1' = 10$ in, $R_2 = 0.375$ in, $R_2' = \infty$. Hence

$$B + A = \frac{1}{2}\left(\frac{1}{10} + \frac{1}{10} + \frac{1}{0.375} + 0\right) = 1.433$$

and

$$B - A = \frac{1}{2}\left[\left(\frac{1}{10} - \frac{1}{10}\right)^2 + \left(\frac{1}{0.375} - 0\right)^2 + 2(0)\right]^{1/2} = 1.333$$

whence $(B - A)/(B + A) = 0.930$ and from Fig. 11.6, $\beta = 0.42$, $\kappa = 0.12$, and $\lambda = 0.51$. From the notation of Table 11.2,

$$\eta_1 = \frac{1 - \nu_1^2}{E_1} = \frac{1 - (0.25)^2}{20 \times 10^6} = 0.0469 \times 10^{-6},$$

$$\eta_2 = \frac{1 - \nu_2^2}{E_2} = \frac{1 - (0.30)^2}{30 \times 10^6} = 0.0303 \times 10^{-6}$$

The semiwidth of the elliptical area of contact is

$$b = \beta\left[\frac{3P(\eta_1 + \eta_2)}{4(B + A)}\right]^{1/3} = 0.42\left[\frac{3(600)(0.0469 + 0.0303)10^{-6}}{4(1.433)}\right]^{1/3} = 0.0122 \text{ in}$$

The semilength is

$$a = \frac{b}{\kappa} = \frac{0.0122}{0.12} = 0.101 \text{ in}$$

The maximum pressure is

$$\frac{1.5}{\pi ab} = \frac{1.5(600)}{\pi(0.101)(0.0122)} = 232\ 000 \text{ psi}$$

The deflection is

$$\delta = \lambda[P^2(\eta_1 + \eta_2)^2\ (B + A)]^{1/3}$$

$$= 0.51[(600)^2(0.0772 \times 10^{-6})^2(1.433)]^{1/3} = 0.000\ 74 \text{ in}$$

The stress of 232 000 psi may be satisfactory if a carburized cam is used (Section 11.7). With a flat-faced follower and a cam of length $l = 2a = 0.202$ in (the length of the ellipse), the *calculated* stress is less, but inaccuracies in machining, misalignment, and shaft deflection, particularly if the cam has a typical length of ⅜ or ½ in, may give edge contact and higher actual stress. This unscheduled increase cannot occur with a spherical face, which is frequently used with radii of 10 to 300 in.   ////

The principal stresses of the table occur at the center of the contact area, where they are maximum and compressive.[9] At the edge of the contact ellipse, the surface stresses in a radial direction (along lines through the center of contact) become tensile. Their magnitude is considerably less than that of the maximum compressive stresses, e.g., only $0.133p_0$ with two spheres and $\nu = 0.30$ by an equation of Case 1, Table 11.2, but the tensile stresses may have more significance in the initiation and propagation of fatigue cracks (Section 11.5). The circumferential stress is everywhere equal to the radial stress but of opposite sign, so there is a condition of pure shear. With the two spheres $\tau = 0.133p_0$.[10]

Forces applied tangentially to the surface, such as by friction, have a significant effect upon the nature and location of the stresses. For example, two of the three compressive, principal stresses immediately behind the tangential force are changed into tensile stresses. Also, the location of the maximum shear stress moves toward the surface and may be on it when the coefficient of friction exceeds 0.10. Results of a study made by J. O. Smith and C. K. Liu for a cylinder on a plane are available.[11]

11.5 SURFACE FAILURES

There are several kinds of surface failures and they differ in action and appearance. *Indentation*, a yielding from excessive pressure, may constitute failure in some machine components. Nonrotating but loaded ball bearings can be damaged in this way, particularly if vibration and therefore inertia force is added to dead weight and static loads.

[9] Equations and charts for stresses at points *along* the Z axis, plotting against *B/A*, were prepared by H. R. Thomas and V. A. Hoersch, ''Stresses Due to the Pressure of One Elastic Solid Upon Another,'' *University of Illinois, Bulletin 212 of the Engineering Experimental Station*, July 15, 1930. A more accessible source may be A. P. Boresi, O. M. Sidebottom, F. B. Seely, and J. O. Smith, *Advanced Mechanics of Materials*, 3rd ed., New York: John Wiley & Sons, Inc., Figure 14–6.1 (p. 595)—which is also plotted against *B/A*—1978.

[10] S. Timoshenko and J. N. Goodier, *Theory of Elasticity*, 3rd ed., New York: McGraw-Hill, pp. 414 and 418, 1970.

[11] J. O. Smith and C. K. Liu, ''Stresses Due to Tangential and Normal Loads on an Elastic Solid with Application to Some Contact Stress Problems,'' *J. Appl. Mech. Trans. ASME*, Vol. 75, pp. 157–166, 1953. Smith and Liu's work is summarized in R. C. Juvinall, *Stress, Strain, and Strength*, New York: McGraw-Hill, pp. 379–385, 1967.

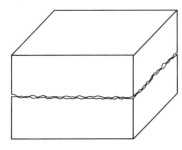

FIGURE 11.8. Surfaces in contact at asperities.

This may occur during the shipment of machinery and vehicles on freight cars or in devices that must stand in a ready status for infrequent and short-life operations. The phenomena is called *false brinelling*, named after the indentations made in the standard Brinell hardness test.

The term *wear* will be used here to describe a progressive loss of surface by shearing and tearing away of particles. This may be a flat spot, as when a locked wheel slides on a rail. More generally the wear is distributed over an entire active surface because of a combination of sliding and rolling action, as on gear teeth. It may occur in the presence of oil or grease where a hydrodynamic film for complete separation of metallic surfaces is not developed. On dry surfaces, it may consist of a flaking of oxides. If pressures are moderate, the wear may not be noticeable until looseness develops. The surfaces may even become polished, with machining and grinding marks disappearing. Removal of just the high spots may be a "wearing in" rather than the start of "wearing out." Large amounts of wear may result from misalignments and unanticipated deflections on only a portion of the surface provided to take the entire load. This has been observed on the teeth of gears mounted on insufficiently rigid shafts, particularly when the gear is overhung, outside the bearings (Example 9.7). Rapid wear may occur from insufficient lubrication, as on cam shafts, or from negligence in lubricating and protection from dirt.

One mechanism of wear begins with the shearing, compression, or flattening of asperities spaced randomly (Fig. 11.8) or regularly. Since their supporting area is very small compared with that on which the pressure calculation is based, they will be work hardened with possible interlocking of surfaces at planes of shear. Temperatures at these asperities may be high enough to cause flow or pressure welding where the surfaces are clean. Since the work-hardened material is stronger than the matrix, small particles may be torn out. If these particles continue to adhere to one of the surfaces, they will further damage adjacent areas of the other surface as sliding continues. This has been called *adhesive wear*.[12] If it is severe, it may be called *galling* or *scoring*. Overloaded and underlubricated gear teeth may score, with parallel gouges in the direction of sliding of the teeth into and out of mesh. If only the metal flows, having the appearance of having been melted, it is called *scuffing*.[12]

Abrasive wear occurs when foreign particles such as sand or grinding grit come between sliding surfaces. These may be soon rejected, but when one material is considerably softer then the other, the particles may become embedded in the softer material, gouging the harder one. On the contrary, asperities on a hard surface may gouge the softer surface. One reason for using white metal in the lining of plain bearings is to allow hard loose particles to become completely embedded in the soft metal, thus preventing the scoring of the shaft, a hard-to-replace component.

In another type of surface failure, *fatigue cracks* progress into and under the surface, and particles fall out of the surface. The holes remaining are called pits or spalls. This

---

[12] F. T. Barwell, "Surface Contact in Theory and Practice," *Proc. Inst. Mech. Eng.*, Vol. 175, pp. 853–878, 1961.

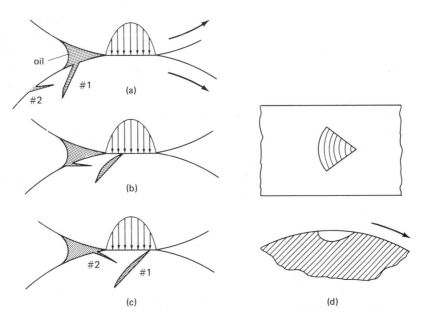

FIGURE 11.9. Progress of pitting originating at a surface crack.

pitting occurs on convex surfaces such as gear teeth, rolling-element bearings, and cams. In Sections 11.3 and 11.4, it was noted that the maximum shearing stress occurs below the surface of bodies in contact. Hence, at one time, it was strongly held that the crack forming a pit started at this point of maximum shear stress, then progressed outward.

Certain pure rolling tests, first made by A. H. Burr, then continued and reported by S. Way,[13] disclosed that the cracks commonly started at the surface and progressed only in the presence of oil, a particularly good penetrant that would fill any fine cracks present and act as a hydraulic wedge. S. Way subsequently showed by experimentation and sectioning that only cracks with their lips facing the approaching load would progress to failure.[13] In Fig. 11.9(a) and (b) surface crack #1, filled with oil, approaches the loading zone and has its lip sealed off. As the full length of the crack comes under the load in Fig. 11.9(c), oil in the crack cannot escape, and a high hydraulic pressure results. After repeated occurrence of this pressure, high stress from stress concentration along the root of the crack will cause the crack to spread by fatigue. Eventually, the crack will progress toward the surface, favoring the most highly stressed regions. In Fig. 11.9(d), the particle has fallen out, exposing a pit with typical lines of progressive cracking, radiating from the pointed lip. The pit may look much as though it were molded from a tiny sea shell, with an arrowhead point of origin. Pit depths may vary from a few thousandths of an inch to 1/32 in or more, with lengths from two to four times their depths.

Cracks facing away from the approaching zone of loading, such as #2, will not develop into pits. The root of the crack first reaches the loaded area (Fig. 11.9(c)), and oil in the crack is squeezed out by the time its lip is sealed off. A more viscous oil reduces or eliminates pitting, either by not penetrating into fine cracks, or by forming an oil film thick enough to prevent contact between asperities.

There are several possible causes for the initial surface crack, which only needs to be microscopic or submacroscopic. Machining and grinding are known to leave fine surface

---

[13] S. Way, "Pitting Due to Rolling Contact," *J. Appl. Mech., Trans. ASME*, Vol. 57, pp. A49–58, 1935; also, S. Way, "How to Reduce Surface Fatigue," *Machine Design*, Vol. 11, pp. 42–45, March 1939.

cracks, either from a tearing action or from thermal stresses. Polishing inhibits pitting, presumably by the removal of these cracks. Along the edge of spherical and elliptical contact areas a small tensile stress is present under static and pure rolling conditions. Tangential forces from sliding combined with rolling, as on gear teeth, add tensile stresses to the above and to the rectangular contact area of cylinders (Section 11.4). Surface inclusions at the tensile areas create stress concentrations and add to the chance that the repeated tensile stresses will initiate cracks. Sometimes a piece that has dropped out of a pit passes through the contact zone, making a shallow indentation probably with edge cracks. Handling nicks have also initiated pitting.

Sometimes the breaking out of material continues rapidly in a direction away from the arrowhead point of origin, increasing in width and length. It is then called *spalling*. Spalling occurs more often in rolling-element bearings then in gears, sometimes covering more than half the width of a bearing race.

Propagation of the crack from the surface is called a *point-surface origin* mode of failure by Littmann and Widner.[14] They also describe *inclusion-origin* failures. Inclusions are nonmetallic particles that are formed in and not eliminated from the melt in the refining process. They may be formed during the deoxidization of steel or by a reaction with the refractory of the container. The inclusion does not bond with the metal, so that essentially a cavity is present with a concentration of stress. Littmann and Widner detected inclusions before rolling tests by a magnetic particle method. Rolling was then interrupted at various stages of crack propagation, and radial cross sections were cut. A crack, starting at the inclusion, may propagate through the subsurface for some time, or the crack may head for the surface. If cracks to the surface form, further propagation may be by hydraulic action, with a final appearance like that from a point-surface origin. The damaged area is often large, a spall. It is well known that bearings made from vacuum-melted steel, and therefore a "cleaner," more oxide-free steel, are less likely to fail and may be given higher load ratings.

Three other types of failure are identified by Littmann and Widner[14] and illustrated by failure on heavily loaded roller bearings in test rigs. *Geometric stress concentration* occurs at the ends of a rectangular contact area, where the material is weaker without side support. A bit of misalignment, shaft slope, or taper error will shift much of the load to one of the ends. In *peeling*, fatigue cracks propagate over large areas but at depths of 0.0002 to 0.0005 in. It has been attributed to loss of hydrodynamic oil film, particularly when the surface finish has many asperities greater than the film thickness under the condition of service. *Subcase fatigue* occurs on carburized elements where the loads are heavy, the core is weak, and the case is thin relative to the radii of curvature in contact. Cracks initiate and propagate below the effective case depth, and cracks break through to the surface at several places, probably from a crushing of the case for lack of support.

11.6 OIL FILMS AND THEIR EFFECTS

There is theoretical and experimental proof from many recent investigations[15] that a hydrodynamic oil film separates rolling elements, or rolling and sliding elements, that have fine surface finishes. It is a problem in elastohydrodynamics, where the shape of the elastic surfaces of the disks and hence of the film depends upon the pressure distribution, and the pressure distribution depends upon the shape. The Hertzian shape is

---

[14] W. E. Littmann and R. L. Widner, "Propagation of Contact Fatigue from Surface and Subsurface Origins," *J. Basic Eng., Trans. ASME*, Vol. 88, pp. 624–636, 1966.

[15] For a summary and references see M. D. Hersey, *Theory and Research in Lubrication*, New York: John Wiley & Sons, Inc., pp. 322–341, 1966.

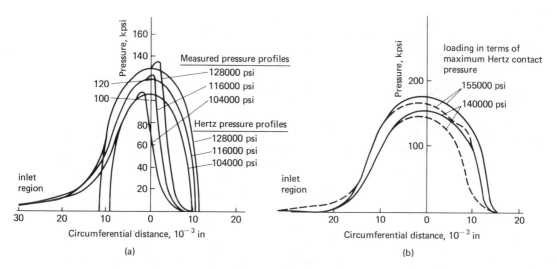

FIGURE 11.10. Measured film pressures between rolling disks. (a) Cylindrical disks, 1400 ft/min rolling speed, polyphenol ether as the lubricant. (b) Crowned disks, 2900 ft/min rolling speed, mineral oil (solid line) or polyphenol ether (dashed lines) as the lubricant. All disks 3-in diameter and polished to less than 1 μin rms. Temperature 115°F. Pressure transducer of manganin wire 1 μin thick by 0.002 in wide. (From J. W. Kannel, J. A. Walowit, J. C. Bell, and C. M. Allen, "The Determination of Stresses in Rolling-Contact Elements," *J. Lubrication Technol., Trans. ASME*, Vol. 89, pp. 453–465, 1967.)

therefore altered. Depending upon the velocity and viscosity, the pressure may begin to develop in the converging space in front of the Hertzian or static contact area at a distance of the order of the Hertzian width itself, perhaps 0.02 in, and may end early (Fig. 11.10(a)). Film thickness may be of the order of 30 μin.[16] It has sometimes been assumed that the spreading of the pressure area reduces the peak pressure, but the measurements of Fig. 11.10 show slight increases, a maximum of 12% with the mineral oil and crowned disks of Fig. 11.10(b). Calculations showed some changes, some smaller and some slightly larger, in the maximum shear stresses. However, the calculated Hertz pressure seems to remain as a good measure of the intensity of loading.

However, an oil film, unquestionably valuable when sliding is present, seems to be equally valuable for pure rolling conditions. Tallian et al.[17] suggested that fatigue life is sensitive to the metal-to-metal contact that occurs when asperities break through the oil film, and later proposed the ratio of film thickness to surface roughness as an important measure of the life of rolling-element bearings.[18] Given[19] used a rolling four-ball tester, AISI 52100 steel balls at Rockwell C64-65, and a rather high Hertz pressure of 680 000 psi to accelerate failure. He related his results to a film parameter Λ equal to the ratio of the average elastohydrodynamic film thickness to the composite roughness of the surfaces in contact. He concluded that premature surface-related fatigue failures can be expected when $\Lambda < 0.8$, marginal conditions with $0.9 < \Lambda < 1.5$, normal operation with $1.5 < \Lambda < 3.5$, and full film support and excess life expected with $\Lambda > 3.5$.

Syniuta and Corrow[20] have furnished evidence obtained on a four-ball tester that

[16] Data and figures from a paper by J. W. Kannel, J. A. Walowit, J. C. Bell, and C. M. Allen, "The Determination of Stresses in Rolling-Contact Elements," *J. Lubrication Tech., Trans. ASME*, Vol. 89, pp. 453–465, 1967.

[17] T. W. Tallian et al., "Lubricant Film in Rolling Contact of Rough Surfaces," *Trans. ASLE*, Vol. 7, pp. 109–126, 1964.

[18] R. R. Valori, L. B. Sibley, and T. W. Tallian, "Elastohydrodynamic Film Effects on the Load Life Behavior of Rolling Contacts," ASME Paper 65-LUBS-11, 1965.

[19] P. S. Given, "Rolling Contact Fatigue Rating for Lubrication Qualification," presented at a panel session on Bearings for Turbine Aircraft Engines, ASME Winter Annual Meeting, 1966.

[20] W. D. Syniuta and C. J. Corrow, "The Origin of Microcracks Leading to Rolling Contact Fatigue," ASME Paper 70-DE-46, Design Conf., May 1970.

surface cracks are *initiated* by corrosion or electrochemical effects, which are greater with a slightly corrosive lubricant and with moisture present than with a noncorrosive lubricant.

11.7 DESIGN VALUES AND PROCEDURE

The many investigations, some of which are reported in the preceding sections, have not produced a common basis on which materials, properties, component configuration, operating conditions, and theory may be combined to determine dimensions for a satisfactory life. The investigations indicate that much progress is being made, and they do furnish a guide to conditions and changes for improvement. Most surface-contact components are operating satisfactorily, and their selection is often based on a nominal Hertz pressure determined from experience with a particular component and material, or selection is made from manufacturers or manufacturers' associations tables based on tests and experience with their components.

The several types of stresses, failures, and their postulated causes, including those of subsurface origin, are all closely related to the maximum contact pressure calculated by Hertz equations. Each of the few values suggested in the following is therefore an allowable maximum Hertz pressure, which is also the maximum compressive stress, usually called the Hertz stress or the allowable contact stress. If a value seems large compared with other physical properties of the particular material, it is because it is a compressive stress and the other two principal stresses are compressive. The shear stresses and tensile stresses that may initiate failures are much smaller. Also, the materials used are often hardened for maximum strength. Radzimovsky[21] reports an endurance limit under pulsating (zero to maximum) compressive loading to be near or somewhat higher than the tensile test yield strength.

Rothbart[22] suggests allowable values for cams and followers, for several combinations of materials. For example, for a steel follower of hardness $51R_c$ and rolling contact, 79 000 psi with a nickel cast-iron cam of $35-40R_c$, and 250 000 psi with a carburized steel cam. Turkish,[23] in discussing automotive valve-gear design suggests that the stress not exceed 120 000 psi for flat-faced chilled cast-iron tappets (followers) on a steel cam, using no more than ½ in for cam width in the calculations because of possible camshaft deflection and misalignment. When the latter is avoided by using spherical-faced followers (Fig. 11.7), he suggests that the stress not exceed 190 000 psi using 30-in or 100-in radii for the spherical surfaces. Follower axes are often offset from the center of the cam, by which the follower is rotated to distribute the wear. If there is a possibility that rotation will not occur, Turkish suggests a reduction in the values cited. It is known that some manufacturers have used as high as 280 000 psi for carburized cams with chilled-iron tappets.

The American Gear Manufacturers Association lists for gear teeth the values given in Table 11.3. Its values are probably conservative. Considerably higher values are used for ball and roller bearings made of carburized and hardened AISI 4620 steel or fully hardened AISI 52100 steel. Combining the test results of several investigators, Lipson and Juvinall related Hertz contact stress to a 10% probability of failure after N cycles.[24] For case-hardened $(60R_c)$, high-quality spur gears the stress was 225 000 psi

[21] E. I. Radzimovsky, "Stress Distribution and Strength Condition of Two Rolling Cylinders," *University of Illinois Engineering Experiment Station Bulletin 408*, pp. 27–29, Feb. 1953.
[22] H. A. Rothbart, *Cams*, New York: John Wiley & Sons, Inc., pp. 227–282, 1956.
[23] M. C. Turkish, *Valve Gear Design*, Detroit, Michigan: Eaton Manufacturing Company, 1946.
[24] C. Lipson and R. C. Juvinall, *Handbook of Stress and Strength*, New York: The Macmillan Company, 1963.

TABLE 11.3. Allowable Contact Stress—AGMA 215.01[a]

Material	Minimum Surface Hardness	Stress, σ_c
Steel	Through hardened	
	180 Bhn	85–95 000
	240 Bhn	105–115 000
	300 Bhn	120–135 000
	360 Bhn	145–160 000
	440 Bhn	170–190 000
	Case carburized	
	$55R_c$	180–200 000
	$60R_c$	200–225 000
	Flame or induction hardened	
	$50R_c$	170–190 000
Cast iron		
AGMA Grade 20	—	50–60 000
AGMA Grade 30	175 Bhn	65–75 000
AGMA Grade 40	200 Bhn	75–85 000
Nodular iron	165–300 Bhn	10% less than the σ_c value of steel with the same hardness
Bronze	Tensile strength psi (minimum)	Stress, σ_c
Tin bronze		
AGMA 2C (10–12% Tin)	40 000	30 000
Aluminum bronze		
ASTM B 148-52 (Alloy 9C-H.T.)	90 000	65 000

[a] Extracted from AGMA Information Sheet: Surface Durability (Pitting) of Spur, Helical, Herringbone, and Bevel Gear Teeth, 1963 (Reaffirmed 1974), with the permission of the publisher, The American Gear Manufacturing Association, Arlington, Virginia.

for $N = 10^7$ and 120 000 for $N = 10^9$. For radial ball bearings the stress was 505 000 psi for $N = 10^7$ and 305 000 for $N = 10^9$. For overrunning clutches in automatic transmissions a stress of 650 000 psi in sprags and 810 000 in rollers has been reported.[25]

Caution must be exercised in applying and interpreting contact pressures or compressive stresses if a factor of safety is used. For most structural components the factor of safety may be taken for convenience as the ratio of the failure stress to the design or applied stress, which is the same in the equations as the ratio of failure load to design or applied load. But with surface contacts, there is a nonlinear relationship between load and pressure or stress. It may be determined from Table 11.2, since β and κ are functions of radii only, that $\sigma_{max} = p_0 = cP^{1/3}$ in the general case including spheres, and $\sigma_{max} = p_0 = c'P^{1/2}$ for two cylinders, where c and c' are constants. If it is the applied load that is in doubt, and n is the multiple of design load that will cause failure and S_e is the endurance limit above which failure occurs, then in the general case $S_e/\sigma = (P_{fail}/P)^{1/3} = n^{1/3}$, and $\sigma = S_e/n^{1/3}$. If n is 2.0 and $S_e = 100\ 000$ psi, the design stress can be as high as $100\ 000/1.26 = 79\ 300$ psi. On the other hand, if it is the strength or uniformity

[25] F. R. McFarland, "Practical Experience with Hertz Stresses," Appendix II of *Rolling Contact Phenomena*, J. B. Bidwell (Ed.), London: Elsevier Publishing Company, 1960.

of the material that is in doubt, the factor of safety should be applied to the failure stress S_e, and $\sigma = S_e/n$. In the example, $\sigma = 100\,000/2 = 50\,000$ psi.

The discussions of the foregoing sections may be summarized to indicate some directions for changes in contact-stress components by which their load or life may be increased. Of course, some of these may be impossible or impractical to implement in many situations. Directions are:

1. Larger radii or material of a lower modulus of elasticity to give larger contact area and lower stress.
2. Provision for careful alignment or minimum slope by deflection of parallel surfaces, or the provision of crowned surfaces as has been done for gear teeth, rollers of bearings, and cam followers.
3. "Cleaner" steels, with fewer entrapped oxides, as by vacuum melting.
4. Material and treatments to give higher hardness and strength at and near the surface, and if carburized, a sufficient case depth (at least somewhat greater than the depth to maximum shear stress) and a stronger core.
5. Smoother surfaces, more free of fine cracks, by polishing, by careful "wearing in," or by the avoidance of coarse machining and grinding and of nicks in handling.
6. Oil of higher viscosity and lower corrosiveness, free of moisture and in sufficient supply at the contacting surfaces. No lubricant on some surfaces with pure rolling and low velocity.
7. Provision by items 5 and 6 for increased film thickness to asperity–height ratio ($\Lambda > 1.5$).

11.8 GEAR KINEMATICS AND THE INVOLUTE TOOTH SHAPE

The design of gears includes the application in special forms of the contact stress and beam equations. This section is a brief survey of the kinematic requirements in preparation for the applications of Section 11.9. Only the basic spur gears, those with teeth parallel to the axis of rotation, will be discussed.

With a pair of toothed gears, rotational motion and power must be transmitted from the driving gear to the driven gear with a smooth and uniform positive motion and with only minor loss of power from friction. If the gears have different numbers of teeth, the magnitudes of the torque and of the rotational velocity are changed, the former in direct ratio and the latter in inverse ratio to the numbers of teeth. For kinematic analysis, a pair of gears may be considered to be a multilobed cam and a follower, with the lobes or teeth sufficiently close for the transfer of load from one lobe to the next, and with the lobes shaped to give a constant ratio of angular velocity in all positions of contact. The geometric requirement for constant ratio will first be found.

Two teeth of arbitrary shape are attached to the rotating disks of Fig. 11.11. Contact point Q on tooth 1 has a velocity V_{Q1} normal to O_1Q, and point Q on tooth 2 has a velocity V_{Q2} normal to O_2Q. These velocities must have the same component V_Q^n along the common normal to the surfaces at Q, and components V_{Q1}^t and V_{Q2}^t tangential to the surfaces. The relative or sliding velocity is $V_{Q1/Q2} = V_{Q1}^t - V_{Q2}^t$. Angular velocities are $\omega_1 = V_{Q1}/O_1Q = QM_1/O_1Q$ and $\omega_2 = V_{Q2}/O_2Q = QM_2/O_2Q$. If perpendiculars O_1T_1 and O_2T_2 are dropped from O_1 and O_2 onto the common normal, then from proportional triangles QM_1N and O_1QT_1, $QM_1/O_1Q = QN/O_1T_1$, and from triangles QM_2N and O_2QT_2, $QM_2/O_2Q = QN/O_2T_2$. The angular velocity ratio is then $\omega_2/\omega_1 = (QN/O_2T_2)/(QN/O_1T_1) = O_1T_1/O_2T_2$, which, from similar triangles O_1T_1P and O_2T_2P, becomes $\omega_2/\omega_1 = O_1P/O_2P$. Thus the velocity ratio equals the inverse ratio of the lengths into which the normal cuts the line joining the centers of rotation.

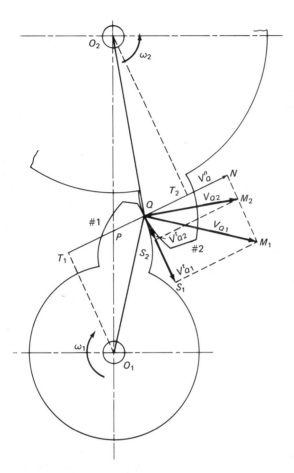

FIGURE 11.11. Velocity relations at contact point Q of two teeth on rotating disks.

As the disks rotate, the point of contact Q changes location. The point P, however, must not move if the ratio ω_2/ω_1 is to stay constant. Thus a requirement for the tooth profiles is that the common normals at all the points of contact intersect the line of centers at the same fixed point. It is called the *pitch point*, and its position is determined by the desired velocity ratio and the center distance. One may choose any reasonable shape for one profile and with the foregoing requirement, determine the other, conjugate profile. However, there are other requirements. It must be possible to economically and accurately form the cutting tools and cut the gears. Small errors in the center distance O_1O_2 should not affect the uniformity of contact. These requirements are only met well by an *involute* profile.

The involute is a curve generated by a tracing point on a cord as it is unwrapped from a cylinder. In Fig. 11.11 the cylinders will have the diameters of those circles to which the normal line PQ is tangent, namely, at points T_1 and T_2. These points are shown again in Fig. 11.12, together with generating cylinders of *base circle* diameter D_b attached to the cylindrical blanks on which the teeth are to be formed, one necessarily behind the other for our purposes here. The cord is wrapped around and fastened to each base cylinder, and as they are rotated, keeping the cord taut, the point Q on the cord traces out an involute surface on each blank. The progress of tracing is shown by the successive positions Q', Q, and Q''. Note that the instantaneous radii of curvature of the two profiles at contact point Q are T_1Q and T_2Q, that they change to values T_1Q' and T_2Q', T_1Q'' and T_2Q'', etc., but that the centers of curvature remain at T_1 and T_2.

The involute profile is limited inwardly by the base circle, where the radius of cur-

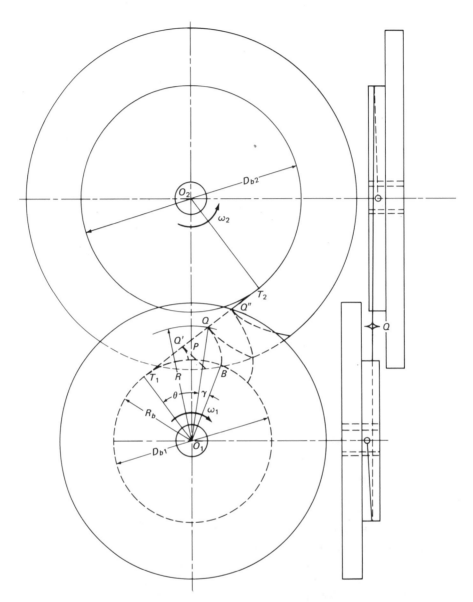

FIGURE 11.12. The generation of conjugate involutes on two gear blanks by a tracing point Q on a cord joining the base cylinders. Also, angles γ and θ, such that $\gamma = \tan \theta - \theta = \text{inv } \theta$.

vature of the involute becomes zero. The profiles are always perpendicular at Q to their instantaneous radii of curvature, which lie along line $T_1 T_2$. This line always intersects the line of centers $O_1 O_2$ at a fixed point P. Therefore, the requirement for a constant velocity ratio is satisfied.

In Fig. 11.12 let Q be any point on an involute profile, and let B be its origin on the base circle. Let the angle between radii $O_1 B$ and $O_1 Q$ be γ, and let the angle between radii $O_1 T_1$ and $O_1 Q$ be θ. Length $T_1 Q$ equals arc $T_1 B$. Hence $\angle T_1 O_1 B = (\text{arc } T_1 B)/O_1 T_1 = T_1 Q/O_1 T_1 = \tan \theta$, and

$$\gamma = \angle T_1 O_1 B - \theta = \tan \theta - \theta = \text{inv } \theta \tag{11.8}$$

where inv θ, read involute θ, is the difference between the tangent and the radian values of θ. The angle θ corresponding to point Q at radius R is found from $\theta = \cos^{-1}$

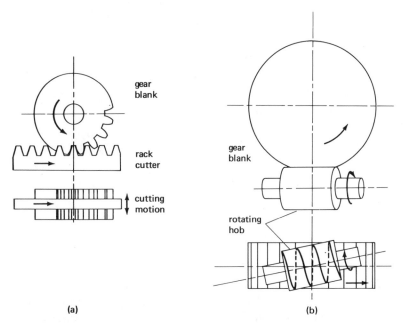

FIGURE 11.13. Generating action of rack-type cutting tools. (a) Cutter reciprocating in direction of axis of gear blank. (b) Schematic of rotating hob with cutting faces straight sided and normal to the thread helix.

$(O_1T_1/O_1Q) = \cos^{-1}(R_b/R)$, where R_b is the radius of the base circle. The involute function is extensively used in gear design and manufacture, particularly to determine tooth thicknesses at various radii. Tables are used to solve the inverse function, $\theta = \mathrm{inv}^{-1}\gamma$. In using inv θ, small differences of large numbers are involved. Since accuracy is important, tables of inv θ are available to six and more decimal places.[26]

If the teeth of one member are replaced by teeth with cutting edges we have a tool for cutting the other gear. In Fig. 11.13(a) the teeth have straight sides, normal to the path of generation, T_1PT_2 of Fig. 11.12. The cutter works by a rapid reciprocating motion parallel to the axis of the gear blank while slowly translating with the blank so that cutting occurs along path T_1PT_2. The straight-sided profile of the tool corresponds to that of a tooth on a gear with an infinite number of teeth, a profile formed by unwrapping a cord from a base cylinder of infinite radius. It is called a rack tooth, and this generation cutting of a gear with a straight-sided tool requires a special cutting machine that translates the tool and rotates the gear blank with the same velocities as though a rack were driving the gear. The straight sides of the cutter are themselves, of course, readily and accurately formed and ground to size. Instead of a rack, the tool may resemble a worm or screw, with its threads interrupted with cutting faces normal to the threads. This is called a hob (Fig. 11.13(b)), and to generate teeth, the cutting machine by means of other gears must force the gear blank and the hob to rotate together as in a worm-gear drive unit. Cutters resembling small circular gears are also used for the reciprocating generation cutting of gear teeth. Nongenerated teeth of somewhat less accuracy may be cut in an ordinary milling machine with an indexing fixture by form-milling cutters.

The circles drawn through pitch-point P are the *pitch circles*, with diameters D_1 and D_2 (Fig. 11.14). The circle into which the teeth are blended at their "root" is the *root*

[26] *Machinery's Handbook*, 20th ed., P. B. Schubert, Ed., New York: Industrial Press, Inc. (7 places); W. F. Vogel, *Involutometry and Trigonometry*, Michigan Tool Company, Detroit, Michigan (7 places), 1945; E. Buckingham, *Manual of Gear Design, Section One, Mathematical Tables*, New York: Industrial Press, Inc. (8 places), 1935.

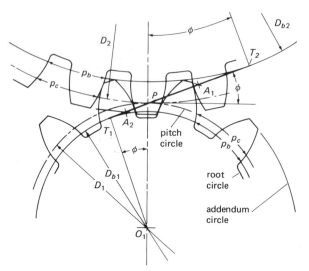

FIGURE 11.14. Geometry of meshing gears.

circle, and the circle at their tip is the *addendum circle*. Its diameter is closely toleranced since it is the outside diameter of the blank from which measurements for the depth of cutting are made. On the other hand, the pitch circles, which are never seen on the gears, are important in calculations. The spacing of the teeth along these circles is the circular pitch p_c, which is an arc length equal to the pitch-circle circumference divided by the number of teeth, or

$$p_c = \frac{\pi D_1}{N_1} = \frac{\pi D_2}{N_2} \tag{11.9}$$

The diameter of the base circle is seen to be $D_b = D \cos \phi$, from the construction in Fig. 11.14. The angle ϕ is called the *pressure angle* because the pressure force between the teeth is directed along $T_1 P T_2$ at an angle ϕ with the tooth velocity at P. The spacing between successive profiles on the base circle is the *base pitch* p_b. Since the same number of profiles intersect the base and pitch circles, and the diameters of the circles are in the ratio $\cos \phi$, it follows that

$$p_b = p_c \cos \phi \tag{11.10}$$

The tooth spacing along the normal line $T_1 T_2$ is also p_b, from the base circle from which it is unwrapped. Kinematic engagement of two teeth occurs where the addendum circle of the driven gear 2 intersects the normal line at point A_2, and they disengage at A_1 where the addendum circle of the driving gear 1 intersects. The *path of contact* or *line of action* is $A_2 A_1$. Kinematically, disengagement of one pair does not occur until some time after the following pair engages, and during this time there may be some sharing of the load, the amount depending upon spacing errors and tooth deflections.[27] A measure of this overlap and continuity is the contact ratio or average number of pairs of teeth in contact, namely,

$$m_c = \frac{A_2 A_1}{p_b} \tag{11.11}$$

Values of m_c between 1.25 and 1.40 are common in spur gears, those with teeth parallel to the axes of rotation. Values higher than 2.0 have been obtained by special design of

---

[27] See Example 9.12 and Fig. 9.30.

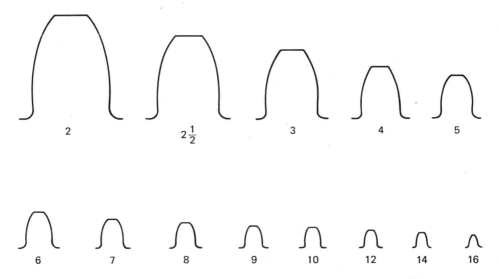

FIGURE 11.15. Some standard diametral pitches and comparative tooth sizes (full size).

the teeth, but the equivalent effect in division of load and quietness is more readily obtained by using helical gears with teeth skewed relative to the axes.

Tooth size depends upon the forces and the requirements for strength and durability. Tooth sizes are standardized for convenience and availability, and particularly for the economy of limiting the number of sizes of cutting tools. Standardization also provides for interchangeability. In the United States the standard system is established by the AGMA and the ANSI. It is based upon a *diametral pitch P*, which is defined as the number of teeth N per inch of pitch diameter D, i.e., $P = N/D$. From Eq. (11.9) it follows that $P = \pi/p_c$. It is an inverse relationship, the smaller the tooth the larger the diametral pitch (Fig. 11.15). Values used in power transmission include 1, 1¼, 1½, 1¾, 2, 2¼, 2½, 2¾, 3, 3½, 4, 5, 6, 7, 8, 9, 10, 12, 14, 16, and 18. Higher values of 20 to 200 are used for finer-pitch instrument and control system gears. Pressure angles ϕ are standardized at 20° and 25°, with 14½° available for replacement gears in an old standard.

In Europe and with metric units, the *module m* is used as a measure of tooth size. It is defined as the millimeters of diameter per tooth, or $m = D_m/N$, where D_m is the pitch diameter in millimeters. It is proportional to circular pitch and inversely proportional to diametral pitch. Thus $p_c = \pi D_m/N = \pi m$, and $m = D_m/N = 25.4D/N = 25.4/P$, where D is in inches. Values of m increase by steps of 0.1 from 0.3 to 1.0, by 0.25 to 4, by 0.5 to 7, by integers to 16, and by 2.0 to 24.

The nominal arc thicknesses of the teeth at the pitch circles must be the same for all gears of a given pitch if there is interchangeability, hence must be half the circular pitch p_c, or $t_p = p_c/2 = \pi/(2P)$. They may be a few thousandths of an inch less in thickness to provide a clearance, called backlash, between the teeth to avoid binding. Heights are also related to the pitch, that portion outside the pitch circle (Fig. 11.16), called the addendum, being $1/P$ in a full-depth system; and that portion inside, the dedendum, being $1.250/P$ for the 20° and 25° systems. The proportions are the same for the module system, with the addendum $= m$, and the dedendum, $1.250m$ in millimeters. The difference between addendum and dedendum provides a clearance at the root for a fillet. The smaller of two gears in contact is frequently called the *pinion*, particularly if it has fewer then 25 teeth. The base circle of a pinion with less than 18 teeth in the 20° system may overlap the addendum circle of the gear. The interference that would occur at the

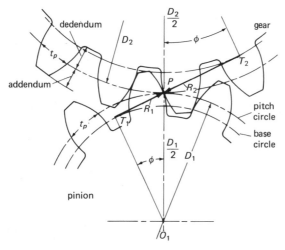

FIGURE 11.16. Nomenclature for gear teeth. Standard teeth shown, in contact at the pitch-point P.

noninvolute portion of the pinion's flank is eliminated when a straight-sided generating cutter is used, since it undercuts the pinion tooth near the root.

Two gears which have been cut by standard-system cutters may be assembled at a center distance slightly larger than $(N_1 + N_2)/2P$ in order to provide backlash, or because exact positioning is impractical. The base circles are inherent to the gears, once they are cut, so that spreading of the centers has the effect of increasing the inclination of their common tangent. In Fig. 11.17, with O_1 moved to O_1', the pressure angle ϕ increases to ϕ', and P moves to P'. Thus, the pitch diameters are increased, but the ratio of the segments cut from the line of centers is unchanged, and the velocity ratio is the same.

A further increase in center distance leaves a space which may be filled by a thicker and stronger pinion tooth. This is done by feeding the cutter into a larger diameter blank but no further than that position where undercutting will just begin to occur. This determines the assembled center distance. In another method of avoiding undercutting, the pinion is cut with a larger addendum and smaller dedendum, and the gear with a smaller addendum and larger dedendum. The center distance is unchanged. This is called a long-and-short-addendum system. The two foregoing types, plus others which may also give a better balance between strength and durability, are known as *nonstandard gearing*.[28]

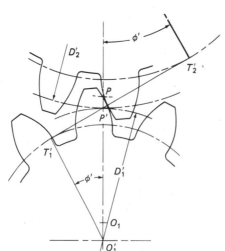

FIGURE 11.17. Center of gear 1 moved from O_1 to O_1', giving a new common tangent $T_1'T_2'$, pressure angle ϕ', and pitch-point P'.

---

[28] For more details, see for example, H. H. Mabie and F. W. Ocvirk, *Mechanisms and Dynamics of Machinery*, 3rd ed., New York: John Wiley & Sons, Inc., pp. 129–144, 1975 or SI Version, pp. 137–152, 1978; or J. E. Shigley, *Kinematic Analysis of Mechanisms*, 2nd ed., New York: McGraw-Hill, pp. 268–278, 1969.

Modifications of the tooth surface may be made. Because of unavoidable manufacturing and assembly errors and because of the slope inherent in the deflection of shafts, elements of tooth surfaces intended to be straight and parallel may be in misalignment, with mating teeth in contact over a portion at one end. A slight curvature or *crown* on the teeth will ensure a contact area near the center of their length, where the teeth are strongest. In gear-type couplings, a larger crown provides for larger misalignments. Another modification is related to the cantilever-beam deflection of the teeth themselves, a cause of premature engagement and noninvolute action (Point E, Fig. 9.30(c), Example 9.12 and the paragraph following it). The involute profile may be cut back very slightly at and near the tips of the tooth (tip relief) to avoid premature engagement and delayed disengagement. The profile modification may also be designed to give a more gradual rise to the load curve (Fig. 9.30(c)), reducing the impact and noise of engagement.

11.9 GEAR TOOTH LOADS AND SURFACE STRENGTH AND WEAR

The total force along the line of action is the torque T on the gear or pinion divided by its base circle radius (Fig. 11.18(a)), thus

$$F_n = \frac{T}{D_b/2} = \frac{2T}{D_b} \tag{11.12a}$$

For shaft and bearing analyses this force normal to the tooth surface is resolved into the transmitted force F_t and separating force F_s, respectively, normal and parallel to the line

FIGURE 11.18. (a) Normal force at tip of tooth translated to and resolved at pitch-point P. (b) Bending stress σ and uniform compressive stress F_s/A at root fillet.

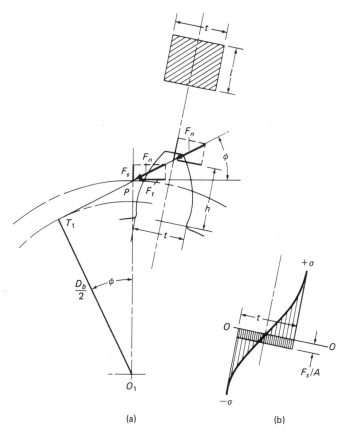

(a)   (b)

of centers as in Fig. 9.15 and 9.16. Since the pitch-circle diameter D equals $D_b/\cos \phi$,

$$F_t = F_n \cos \phi = \frac{2T}{D_b} \cos \phi = \frac{2T}{D} \qquad (11.12b)$$

$$F_s = F_n \sin \phi = F_t \tan \phi \qquad (11.12c)$$

Since gear teeth are lubricated, sliding forces are neglected.

Loads calculated by Eqs. (11.12) are static ones. Because of spacing errors and sudden deflections of the teeth, tooth loads are inherently dynamic. Their calculation should be based on the engagement velocity, tooth errors, and the flexibility and masses of the teeth, gears, and connected components. This problem has never been completely solved. Instead empirical factors are used based on pitch-line velocity V, manufacturing method, and materials. Two of the factors approved by the AGMA for steel gears are

$$K_V = \frac{50}{50 + \sqrt{V}} \quad \text{and} \quad K_V = \sqrt{\frac{78}{78 + \sqrt{V}}} \qquad (11.13)$$

where V is in feet per minute. The first factor is for use with carefully cut gears such as those generated by hobbing or gear shaping, the application of which is usually limited to $V < 4000$ ft/min. The second factor is to be used for highest accuracy gears such as those shaved, ground, or lapped. The dynamic load is then estimated by dividing the static load by a velocity factor, thus

$$F_{\text{dynamic}} = \frac{F_{\text{static}}}{K_V} \qquad (11.14)$$

The resistance to surface damage is determined from the Hertz equation for cylindrical surfaces (Case 2 of Table 11.2). The equations for maximum pressure p_0 and half-width b of the contact area combine to give for steels,

$$p_0 = \frac{1.273 \, F_{nd}}{2l \times 0.000\,277} \sqrt{\frac{l}{F_{nd}} \frac{R_1 + R_2}{R_1 R_2}} = 2300 \sqrt{\frac{F_{nd}}{l} \frac{R_1 + R_2}{R_1 R_2}}$$

Length l is the contact length of the teeth, and F_{nd} is the force normal to the tooth surface under dynamic conditions, obtained by substitution into Eq. (11.14) of F_n static from Eq. (11.12a) and a velocity factor such as given in Eqs. (11.13). Pitting usually occurs within the region on both sides of the pitch circles where only one pair of teeth is carrying the load. It is customary, then, to use the radii of curvature of the profiles at the pitch point, measured to P from their centers at T_1 and T_2. From Fig. 11.16, these are $R_1 = (D_1/2) \sin \phi$ and $R_2 = (D_2/2) \sin \phi$. Substitution gives

$$p_0 = 3250 \sqrt{\frac{F_{nd}}{l} \frac{D_1 + D_2}{D_1 D_2 \sin \phi}} \qquad (11.15)$$

Rearranged to solve for length, Eq. (11.15) becomes

$$l = \left(\frac{3250}{p_0} \right)^2 \frac{F_{nd}(D_1 + D_2)}{D_1 D_2 \sin \phi} \qquad (11.16)$$

The rpm and power or torque requirements are generally known. If the center distance has been approximated from other considerations, diameters D_1 and D_2 and dynamic load F_{nd} may be calculated. The allowable pressure p_0 may be selected from Table 11.3. A slightly more conservative value is

$$p_0 = 360 \, (\text{Bhn})_{\text{avg}} \qquad (11.17)$$

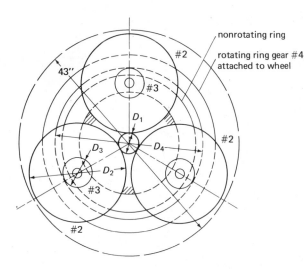

FIGURE 11.19. A double-reduction gear unit used for an individual-wheel drive (Examples 11.3 and 11.4).

where $(Bhn)_{avg}$ is the average of the Brinell hardness numbers of gear and pinion teeth. Less hardness is often specified for the gear because hardness is more difficult to obtain and cut in the larger member, and less is needed because each gear tooth makes fewer contacts then each pinion tooth. It is noteworthy that the surface-strength requirement represented by Eqs. (11.15) and (11.16) is independent of tooth size, i.e., diametral pitch P or module m. However, adjustments in dimensions or hardnesses may be necessary if the length l determined by Eq. (11.16) is excessive. Excessive length adds to the difficulty of obtaining parallel shafts and tooth surfaces for uniform, lengthwise distribution of pressure. An old rule suggested an upper limit of $12.5/P$ in inches for length l, but it is often exceeded.

Example 11.3

Each 6-ft-diameter wheel of a large off-the-road vehicle is to be individually driven by a 275-hp electric motor with a highest speed of 3500 rpm. All gears must lie within a cylindrical space 43 inches in diameter. The double-reduction arrangement of Fig. 11.19 is proposed. Pinion #1, attached to the motor, drives three gears #2 on fixed centers, with a speed reduction of approximately 4.5. The shafts of gears #2 carry pinions #3 which drive an internal-tooth ring gear #4 attached to the wheel. Pinion #1 will be allowed to "float," i.e., seek its own center, so that its torque will be transmitted equally to the three gears. To avoid quench-treatment difficulties and expense, the large gears #2 will be hardened to only 250 Bhn, after which they may be shaved for accuracy. The pinion will be quench treated for a hardness of 350 Bhn and finished by grinding. Use the 20° system. Determine dimensions for pinion #1 and gears #2.

Solution

The pinion torque is

$$T = \frac{63\ 000\ hp}{n} = \frac{(63\ 000)(275)}{3500} = 4950\ \text{lb} \cdot \text{in}$$

If we allow 1½ in for the addendums of two teeth, the maximum diameter that the several pitch circles may occupy is $43 - 1.5 = 41.5$ in. Then $41.5 = D_1 + 2D_2 = D_1 + 2(4.5D_1) = 10D_1$, where $D_1 = 4.15$ in and $D_2 = 4.15 \times 4.5 = 18.65$ in approximately. The base circle diameter for the pinion #1 is $D_{b1} = 4.15 \times \cos 20° = 3.90$ in. Then from Eqs. (11.11), (11.13), and (11.14), and with forces divided equally among the three gears #2,

$$F_n = \frac{T/3}{D_{b1}/2} = \frac{4950/3}{3.90/2} = 846 \text{ lb}$$

$$V = \frac{\pi D}{12} n = \frac{\pi(4.15)}{12}(3500) = 3800 \text{ ft/min}$$

$$K_V = \sqrt{\frac{78}{78 + \sqrt{V}}} = \sqrt{\frac{78}{78 + \sqrt{3800}}} = \sqrt{0.558} = 0.746$$

and

$$F_{nd} = \frac{F_n}{K_V} = \frac{846}{0.746} = 1134 \text{ lb}$$

From Eqs. (11.17) and (11.16),

$$p_0 = 360\left(\frac{250 + 350}{2}\right) = 108\,000 \text{ psi}$$

and

$$l = \left(\frac{3250}{108\,000}\right)^2 \frac{(1134)(4.15 + 18.65)}{(4.15)(18.65)(0.342)} = 0.885,$$

say ⅞ in. This problem is continued in Example 11.4.   ////

11.10 BENDING STRENGTH AND SIZE OF TEETH

The tooth is a cantilever beam, and even though it tapers, the nominal stress is maximum at the root. For repeated loading the effect is increased by the stress concentration at the fillet that joins the tooth to the main body of the gear. Because of uncertainty on sharing of loads between teeth,[29] it is conservative practice to assume that the full load F_n or F_{nd} acts at the tip of the tooth, as in Fig. 11.18(a). Its component causing bending is approximated by F_t, Eqs. (11.12), and the stress at the root of the rectangular section is

$$\sigma = K\frac{M}{I/c} = K\frac{F_t h}{lt^2/6} = \frac{6KF_t h}{lt^2} \tag{11.18}$$

The component F_s causes a uniform compressive stress (Fig. 11.18(b)), and since it only serves to reduce the net stress a small amount on the tensile side where fatigue cracks start, it has often been neglected.

For nonstandard teeth it may be necessary to use Eq. (11.18) as it stands, making a layout of the tooth to an enlarged scale to obtain the root thickness t. Charts of photoelastically determined stress-concentration factors are available as guides for determining K.[30] For standard tooth systems, Eq. (11.18) was rearranged by Wilfred Lewis to make it readily usable for determining tooth size and capacity, thus

$$F_t = \frac{\sigma lt^2}{6Kh} = \frac{\sigma lt^2 P}{6KhP} = \frac{\sigma l}{KP}\left(\frac{t^2 P}{6h}\right) = \frac{\sigma lY}{KP} \tag{11.19}$$

The quantity $Y = t^2 P/6h$ is nondimensional and independent of pitch P because t and h

---

[29] See Example 9.12.
[30] R. E. Peterson, *Stress Concentration Factors*, New York: John Wiley & Sons, Inc., pp. 249–251, 1974.

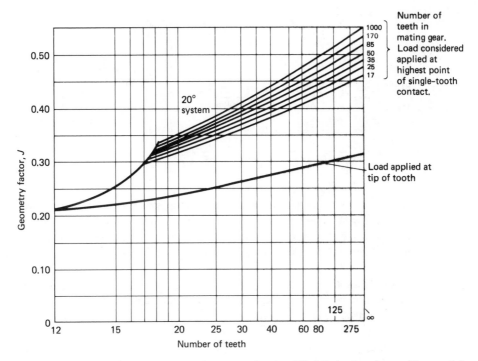

FIGURE 11.20. Spur-gear geometry factors J for the 20° full-depth system. (Extracted from AGMA Information Sheet—Strength of Spur, Helical, Herringbone, and Bevel Gear Teeth, AGMA 225.01, 1967, with permission of the publisher, The American Gear Manufacturers Association, Arlington, Virginia.)

are inversely proportional to P. It is a function only of the tooth system and the number of teeth. Lewis and others have drawn teeth to an enlarged scale, measured h and t, and calculated and tabulated Y. It is called the Lewis form factor.

The AGMA has improved on this by also determining stress-concentration factor K and combining it with Y into a single *geometry factor* J such that

$$F_t = \frac{\sigma l J}{P} \tag{11.20}$$

The J-factor also includes the beneficial effect of the F_s component, and by additional curves (Fig. 11.20), the effect of full load applied at a height corresponding to the beginning of theoretical single-pair contact. The discontinuity in this 20° chart is due to the undercutting of pinions with fewer than 18 teeth. Use of the lowest curve for tip loading is conservative practice, as previously discussed.

Equation (11.20) may be solved for pitch P, together with substitution of the dynamic load F_{td} and allowable stress σ_a, thus

$$P = \frac{\sigma_a l J}{F_{td}} \tag{11.21}$$

Since J is a function of tooth numbers, often as yet undetermined, a guess at its value based upon a reasonable number of teeth may be necessary for a first trial. Fortunately, it varies only a small amount. The allowable stress σ_a can be based upon a factor of safety and on yield strength and fatigue properties, measured or estimated, and the methods of Sections 5.11 and 5.14. The tooth loading is usually in a single direction and

FIGURE 11.21. Allowable bending stresses for steel-spur gears. (After AGMA 225.01, 1967, with permission of the publisher. See caption of Figure 11.20.)

pulsating, and only on an idler or intermediate gear is it completely reversed. An easier alternative for obtaining σ_a is to use the AGMA chart of Fig. 11.21, which appears to provide a factor of safety of 1.5 to 1.75. For idler gears, use 0.7 of chart values. For highest reliability, division of the chart values by 1.5 is suggested.

Example 11.4

Determine tooth size, numbers of teeth, and exact dimensions for the gears of Example 11.3 if full-depth teeth are used.

Solution

The dynamic transmitted load, Eqs. (11.12), is $F_{td} = F_{nd} \cos \phi = (1134) (0.940) = 1066$ lb. In order to obtain values of J to use in Eq. (11.21), we estimate 24 teeth for pinion #1; then $N_2 = 4.5N_1 = 108$ teeth, and from Fig. 11.20, $J_1 = 0.25$ and $J_2 = 0.30$ for tip loading. Based on Fig. 11.21, the allowable bending stresses are $\sigma_{a1} = 39\ 000$ psi for pinion #1 corresponding to 350 Bhn, and $\sigma_{a2} = 31\ 500$ psi corresponding to 250 Bhn. With reference to Eq. (11.21), l and F_{td} are the same for both gears, so pitch P is determined by the smaller of the products $\sigma_{a1}J_1$ and $\sigma_{a2}J_2$, indicating the weaker tooth. Now $\sigma_{a1}J_1 = 9750$ and $\sigma_{a2}J_2 = 9450$, so from Eq. (11.21),

$$P = \frac{(\sigma_a J)l}{F_{td}} = \frac{(9450)\ (7/8)}{1066} = 7.76$$

From the list of Section 11.8, we choose a standard pitch corresponding to the next stronger tooth, i.e., $P = 7$.

Integer numbers of teeth and the exact dimensions may now be calculated. Thus $N_1 = D_1 P = (4.15)\ (7) = 29.05$, use 29 teeth; $N_2 = 4.5N_1 = (4.5)\ (29) = 130.5$, use 130 teeth. A recalculation using J values corresponding to these new numbers of teeth will not change the result, $P = 7$. The nonfactorable "hunting ratio," $130/29 = 4.48$, is desirable since any one tooth on the pinion hunts out and contacts every tooth on the gear, supposedly giving uniform wear. Exact pitch diameters are $D_1 = 29/7 = 4.1429$ in and $D_2 = 130/7 = 18.5714$ in. The center distance is $(D_1 + D_2)/2 = 11.3572$ in., and the overall space swept has a diameter of $[D_1 + 2D_2 + 2$ (addendum)$] = 4.143 + 2(18.57) + 2(1/7) = 41.57$ in.   ////

Such well-balanced design as in Examples 11.3 and 11.4 is not always possible. In general, in continuous-duty drives and with heat-treated alloy steels, the surface-strength requirement predominates, and pitting and wear failures are not uncommon. Any root failures are nearly always at one end of a tooth, indicating misalignment, either at installation or by failure of a bearing. For occasional and intermittent operation at the maximum load with weaker materials, root fracture predominates, and wear and surface strength can sometimes be ignored.

The AGMA method for rating gear drives makes use of the basic Eqs. (11.15) and (11.20), but with further modification by additional factors for various conditions of usage.[31,32] The large field of other gear types—helical, bevel, worm, and others has not been discussed here, since the primary purpose is to illustrate the application of principles. The references and other sources may be consulted for these types of gear drives.[31,32,33]

11.11 ROLLING-ELEMENT BEARINGS: LOAD DIVISION, STRESS, AND DEFLECTION

The rolling elements are balls and rollers. They are assembled to roll between and in the raceway of an outer ring and the raceway of an inner ring (Fig. 11.22). In most cases, the balls or rollers are separated and equally spaced by a cage, also called a retaining ring or a separator. The rolling elements and raceways are shaped to take radial loads only (Fig. 11.22(a) and (b)); radial and axial (thrust) loads in various ratios and magnitudes (Fig. 11.22(c)–(g)); and axial load only (Fig. 11.22(h)).

In a radial-contact ball bearing the no-load contact points lie in the central, radial plane of symmetry (Fig. 11.22(c)). One type is filled with balls by offsetting one ring, the Conrad assembly, and the uniform groove provides support for axial as well as radial loads. It is the most popular type of rolling-element bearing. Another type has one side

FIGURE 11.22. Several types of rolling bearings in cross-sectional half-view. (a) Cylindrical-roller bearing. (b) Needle bearing. (c) Radial-contact deep-groove ball bearing. (d) Angular-contact ball bearings, a duplex pair, mounted face-to-face. (e) self-aligning ball bearing. (f) Tapered roller bearing (conical rollers). (g) Double-row spherical roller bearing, self-aligning with spherical outer raceway and barrel-shaped rollers. (h) Tapered roller thrust bearing.

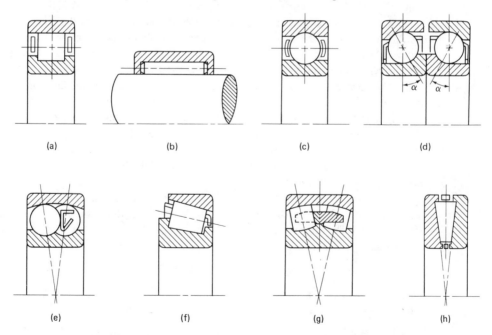

(a)   (b)   (c)   (d)

(e)   (f)   (g)   (h)

---

[31] See the various publications of the American Gear Manufacturers Association, Arlington, Virginia.

[32] For an excellent explanation of the AGMA relationships, see R. M. Phelan, *Fundamentals of Mechanical Design,* 3rd ed., New York: McGraw-Hill, pp. 365–379, 1970.

[33] E. Buckingham, *Analytical Mechanics of Gears,* New York: McGraw-Hill, 1949; D. W. Dudley (Ed.), *Gear Handbook,* New York: McGraw-Hill, 1962; J. E. Shigley, *Kinematic Analysis of Mechanisms,* 2nd ed., New York: McGraw-Hill, 1969; H. H. Mabie and F. W. Ocvirk, *Mechanisms and Dynamics of Machinery,* 3rd ed., New York: John Wiley & Sons, Inc., 1975.

of the groove interrupted by a filling slot so that more balls can be assembled for a higher radial capacity but with loss of almost all axial capacity. The bearings shown in Fig. 11.22(e) and (g) are self-aligning to accommodate misalignments of assembly and shaft slope due to deflection. In general, the roller bearings have more capacity in a given space than the ball bearings. Modifications of the bearings shown and other types are available, with or without integral shields and seals, and with extended rings, adapted for less-expensive assembly in numerous special applications. Also available are bearings of less-expensive construction, accuracy, and durability, suitable for price-competitive low-usage products.

Except below 20 mm, the nominal inside diameter (bore size) of the inner ring in millimeters equals the last two numbers of the bearing identification multiplied by 5. The preceding two numbers indicate a "series." In deep-groove ball bearings the numbers may be 18xx, 00xx, 02xx, 03xx, and 04xx in increasing order of ball and outer ring diameters and capacities. Thus designation 0207 indicates a series 02 bearing with a bore of 35 mm, 0307 a larger, series 03 bearing with the same bore. Each series is available in several grades of precison based on manufacturing tolerances, e.g., with designations ABEC-1, 3, 5, and 7 for ball bearings.

Usually the outer ring is supported by the housing of a machine and inserted with a snug fit for ease of assembly, and to allow its angular position to change very slowly or "creep." The inner ring is assembled on a shaft with a much tighter, press fit, necessary at this smaller diameter. The inner ring is conveniently and accurately located by a shoulder on the shaft, and it is further squared and locked by a threaded collar or nut that forces it against the shoulder. Shaft seat and shoulder are shown for the shafts of Figs. 9.1, 9.12(b), 9.17(a), and 9.23(a). Figure 9.6(a) shows a self-aligning, thrust-type, ball bearing so that a large crane hook and its load may be rotated and positioned by hand.[34]

Calculations for contact pressure and deflection are made with the equations of Table 11.2, Case 2 for cylindrical and tapered bearings, and Case 3 for the others. The outer raceways of ball bearings and the barrel-roller spherical bearing have a double concavity, the surface in one coordinate plane closely conforming to the rolling element. The inner ring of all bearings has a convex surface in one plane, and a smaller diameter, so that pressure and stress are higher than at the outer ring, except for the self-aligning ball bearing. With a purely thrust (axial) load all the rolling elements share the load equally. With a purely radial load about half the elements are loaded at any one time, with the largest share taken by an element as it rolls into line with the load vector.

To analyze the load sharing, it is assumed that each element is initially in contact with both rings (zero clearance) and that all deformation is local i.e., the circular shape of each ring is not deformed, a reasonable assumption since the rings are well supported. The inner ring is shown displaced an amount δ in the direction of the applied load F (Fig. 11.23). The normal deflections at positions 0, 1, 2, ... n of the elements are, respectively,

$$\delta, \quad \delta_1 = \delta \cos \phi_1, \quad \delta_2 = \delta \cos \phi_2, \ldots \delta_n = \delta \cos \phi_n \tag{a}$$

Only normal forces exist, and the corresponding force vectors are $P, P_1, P_2, \ldots P_n$. The equation of equilibrium in the direction of the applied force is

$$-F + P + 2P_1 \cos \phi_1 + 2P_2 \cos \phi_2 + \cdots + 2P_n \cos \phi_n = 0 \tag{b}$$

---

[34] For information on many practical aspects of selecting, installing, lubricating, maintaining, and trouble-shooting bearings, see D. F. Wilcock and E. R. Booser, *Bearing Design and Application*, New York: McGraw-Hill, 1957.

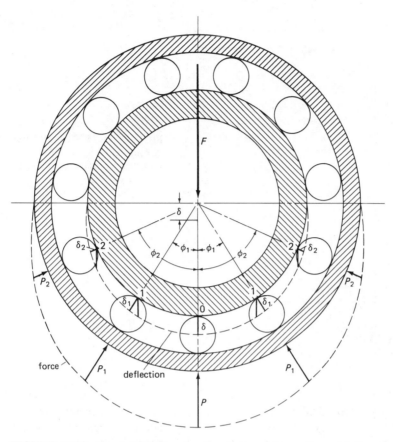

FIGURE 11.23. Local deflections (exaggerated) and forces (to scale) from the displacement of circular rings separated by equally spaced balls.

The relationships between deflections δ and forces P are found from Table 11.2, Case 3 for point-contact bearings. Terms η_1 and η_2 are elastic constants and terms λ and $(B+A)$ are geometric constants, different at each raceway, so at the inner raceway, $\delta_i = k_i P^{2/3}$ and at the outer raceway $\delta_o = k_o P^{2/3}$, where k_i and k_o combine all the constant terms. The total deflection locally is

$$\delta = \delta_i + \delta_o = (k_i + k_o)P^{2/3} = KP^{2/3} \qquad (c)$$

where $K = k_i + k_o$. Thus

$$P = \left(\frac{\delta}{K}\right)^{3/2}$$

$$P_1 = \left(\frac{\delta_1}{K}\right)^{3/2} = \left(\frac{\delta \cos \phi_1}{K}\right)^{3/2} = P(\cos \phi_1)^{3/2} \qquad (d)$$

$$P_2 = \left(\frac{\delta_2}{K}\right)^{3/2} = \left(\frac{\delta \cos \phi_2}{K}\right)^{3/2} = P(\cos \phi_2)^{3/2}$$

etc. Substitution into Eq. (b) yields

$$F = P[1 + 2(\cos \phi_1)^{5/2} + 2(\cos \phi_2)^{5/2} + \cdots] = PS = PN(S/N) \qquad (e)$$

where the sum S is constant and N is the number of rolling elements. This gives for the

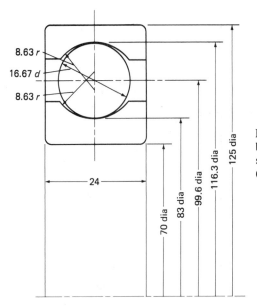

FIGURE 11.24. Size 214 deep-groove ball bearing. Dimensions in millimeters, with some interior radii and diameters estimated (Examples 11.5–11.7).

largest force P

$$P = (N/S)\frac{F}{N} \tag{11.22}$$

Stribeck[35] observed that for ball bearings (N/S) was nearly independent of the number of balls N and that it averaged 4.37. To account for diametral clearance the use of $(N/S) = 5$, or $P = 5F/N$ was proposed. This indicates that with 10 balls, half the load is taken by each ball as it passes through the applied-load vector.

Bearings may be rated for a *static capacity* C_s and for a basic *dynamic capacity* C. The static capacity is defined as that bearing load that will cause a permanent deformation at the maximum-loaded rolling element and at the weaker of the inner or outer raceway contacts of $0.0001d$, where d is the diameter of the rolling element. Larger deformations in the raceway may make the bearing noisy and cause vibration. Larger loads than the static capacity can be satisfactorily supported for a short life if the bearing is running continuously, because the deformations are uniform and smoothed out. The basic dynamic capacity, usually that load which will give a life of a million revolutions (only 9½ hours at 1750 rpm) is higher than the static capacity, but for longer life and other reasons, much lower loads are generally used. Life and dynamic capacity are discussed in Section 11.13.

Example 11.5
A size 214 deep-groove ball bearing is shown in Fig. 11.24 with its standardized surface dimensions and estimated interior dimensions. There are 11 balls with a 21/32 in (16.67 mm) diameter. Groove radii are typically 3 to 8% larger than ball radii to avoid sliding action under deformation. Determine maximum ball force and stress, and the relative displacement of the rings under the static rated load of 8300 lb (36 920 N).

Solution
Let the inner ring, ball, and outer ring dimensions be designated by the subcripts i, b, and o, respectively. Then for the inner ring and ball, $R_i = -8.63$ mm, $R_i' = 83/2 =$

---

[35] R. Stribeck, "Ball Bearings for Various Loads," *Trans. ASME*, Vol. 29, pp. 420–467, 1907.

41.5 mm, $R_b = R_b' = 16.67/2 = 8.34$ mm, and from Table 11.2,

$$B + A = \frac{1}{2}\left(\frac{1}{R_i} + \frac{1}{R_i'} + \frac{1}{R_b} + \frac{1}{R_b'}\right)$$

$$= \frac{1}{2}\left(-\frac{1}{8.63} + \frac{1}{41.5} + \frac{1}{8.34} + \frac{1}{8.34}\right) = 0.0741$$

$$B - A = \frac{1}{2}\left[\left(\frac{1}{R_i} - \frac{1}{R_i'}\right)^2 + 0\right]^{1/2} = \frac{1}{2}\left[\left(-\frac{1}{8.63} - \frac{1}{41.5}\right)^2\right]^{1/2} = 0.0700$$

$$\frac{B - A}{B + A} = \frac{0.0700}{0.0741} = 0.945$$

From Fig. 11.6, $\beta = 0.40$, $\kappa = 0.10$, $\lambda = 0.48$. From the Notation of Table 11.2, $\eta_i + \eta_b = 2(4.40 \times 10^{-6}) = 8.80 \times 10^{-6}$ mm²/N. By Eq. (11.22), with $N/S = 5$, the maximum force on a ball is

$$P = 5\frac{F}{N} = \frac{5(36\ 920)}{11} = 16\ 800\ \text{N}$$

The semiaxes of the contact ellipse at the inner ring are

$$b_i = \beta\left[\frac{3P(\eta_i + \eta_b)}{4(B + A)}\right]^{1/3} = 0.40\left[\frac{3(16\ 800)\ (8.80 \times 10^{-6})}{4(0.0741)}\right]^{1/3} = 0.458\ \text{mm}$$

$$a_i = b_i/\kappa = 0.458/0.10 = 4.58\ \text{mm}$$

and the ellipse is 0.92 mm by 9.16 mm. This width of contact area will cause some sliding in addition to rolling action. However, for higher speeds and long life, the design load, and hence width, will be smaller. The maximum pressure and compressive stress are

$$p_i = \frac{1.5P}{\pi a_i b_i} = \frac{1.5(16\ 800)}{\pi(4.58)(0.458)} = 3820\ \text{MPa (554 000 psi)}$$

The displacement at the inner ring is

$$\delta_i = \lambda\ [P^2(\eta_i + \eta_b)^2\ (B + A)]^{1/3}$$

$$= 0.48\ [(16\ 800)^2\ (8.80 \times 10^{-6})^2\ (0.0741)]^{1/3} = 0.056\ \text{mm (0.0022 in)}.$$

The pressure at the outer ring is 10% less because of greater conformity. The displacement, however, is only slightly less, 0.055 mm. The total displacement is $\delta = \delta_i + \delta_o = 0.111$ mm (0.0044 in). This value is much larger than the normal clearance and is not acceptable for higher speed operation nor where centering accuracy is needed, as with machine tools, where a much lower load is required.　　　　////

11.12 AXIAL LOADS AND ROTATIONAL EFFECTS IN BEARINGS

There are internally generated forces that may increase the normal forces on the elements and decrease the useful radial-load capacity of a bearing. A radial load applied to the bearings of Fig. 11.22(d)−(g) induces axial forces and, thereby, a larger normal force. Tapered bearings should be used in pairs with the rollers having opposite inclinations. In a deep-groove ball bearing (Fig. 11.22(c)), an applied axial load displaces the balls in the direction of the load a small amount, first by taking up clearance, then by deflection. A line through this new center of contact and the center of the ball makes a small *contact angle* α with the central radial plane. With sufficient axial load, the radial force on the balls as they pass through the load vector is reduced because clearance is eliminated and forces are better shared by the other balls. However, the axial load F_a, fortunately evenly divided between all balls as F_a/N, causes a much larger *normal* force $F_a/(N \sin \alpha)$, the wedging effect with a small contact angle. Deflection from any further increase of axial load increases the contact angle and decreases the force multiplier $1/\sin \alpha$.

If the axial load is of the order of magnitude of the radial load or greater, the use of angular-contact bearings should be considered. These are built with definite contact angles from 5 to 40°, and the angle is not effectively changed by deflection. Unless there is always an axial load in one direction they must be used in pairs, as in Fig. 11.22(d). More balls can be assembled, and by forcing the outer rings together an amount predetermined when grinding the ring widths, the bearings are effectively preloaded. This not only eliminates any "play" or looseness, but by placing the deflection due to external load further along on a flatter portion of the nonlinear deflection vs load curve, decreases the magnitude of this deflection and hence of the relative displacement of the rings. This together with high-axial-load capacity makes the arrangement popular for machine-tool spindles.

Analysis and experience has led to relatively simple equations and methods for combining the effects of radial and thrust loads. An equation standardized by the AFBMA and ANSI determines an equivalent radial load[36]

$$F_{\text{equiv}} = XF_r + YF_a \tag{11.23}$$

where F_r is the radial and F_a is the axial load. For application of the equation to radial-contact and angular-contact groove ball bearings the load ratio F_a/F_r must first be determined. When the ratio is small and does not exceed a certain fractional value e, the effect of axial force is ignored, and $X = 1.0$ and $Y = 0$. Factor e is a function of the intensity of axial loading of the particular bearing under consideration, the ratio F_a/C_s, where C_s is the static capacity of the bearing. When $F_a/F_r > e$, the beneficial effect of no clearance is reflected by the reducing factor X in the term XF_r ($X = 0.56$ for radial-contact groove ball bearings). The wedging effect of axial load is reflected in the multiplying factor Y ($Y > 1$) in the term YF_a.

Factor Y is also a function of load intensity F_a/C_s, and together with X and e, it may be determined from AFBMA tables. For radial-contact groove single- and double-row ball bearings they may be determined from Fig. 11.25. The equivalent radial load is then calculated by Eq. (11.23). In general reference must be made to publications of the AFBMA and the ANSI (see footnote 36) and to catalogs and manuals of individual bearing manufacturers to obtain values of the capacities and factors for radial and angular contact bearings, and for other types to which different equations and tables may apply.[37]

Example 11.6

Bearing 214 of Example 11.5 is rated for a static capacity $C_s = 8300$ lb. The inner ring rotates at constant speed. Determine the equivalent radial loads for two conditions: (a) $F_r = 2500$ lb and $F_a = 1500$ lb; (b) $F_r = 2000$ lb and $F_a = 300$ lb.

---

[36] "Load Ratings and Fatigue Life for Ball Bearings" (1972), Anti-Friction Bearing Manufacturers Association, AFBMA Standards, Section 9 or American National Standards Insttute, ANSI Standard B3.15. Also, "Load Ratings and Fatigue Life for Roller Bearings," AFBMA Standards, Section 11 or ANSI Standard B3.16. A ring rotation factor V was formerly included in the AFBMA equation. See Section 11.13 for a discussion.

[37] For more thorough and more extensive analyses of many phases of bearing design and investigation see T. A. Harris, *Rolling Bearing Analysis,* New York: John Wiley & Sons, Inc., 1966.

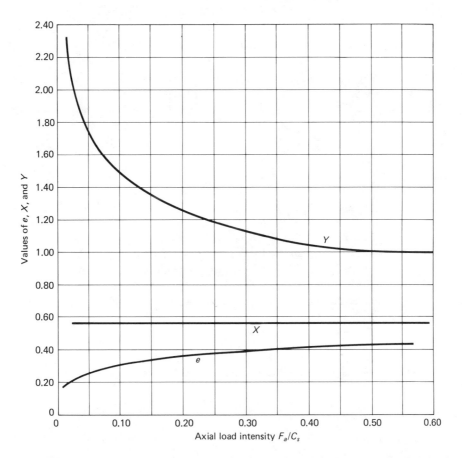

FIGURE 11.25. AFBMA factors e, X, and Y for single and double-row radial ball bearings for use with Eq. (11.23). *Note:* if $F_a/F_r \leq e$, use $X = 1.0$ and $Y = 0$. (Plotted from Table 2 of AFBMA Standard 9, June 1972 (ANSI B3.15-1972), "Load Ratings and Fatigue Life for Ball Bearings," with permission of the Anti-Friction Bearing Manufacturers Association, Arlington, Virginia.)

Solution

(a) $F_a/C_s = 1500/8300 = 0.18$. From Fig. 11.25, $e = 0.35$ and $Y = 1.29$. Then $F_a/F_r = 1500/2500 = 0.60$, which is greater than e, so from Eq. (11.23),

$$F_{\text{equiv}} = XF_r + YF_a = (0.56)(2500) + (1.29)(1500) = 3340 \text{ lb}$$

(b) $F_a/C_s = 300/8300 = 0.036$, $e = 0.23$, and $F_a/F_r = 0.15 < e$; hence $X = 1.0$ and $Y = 0$, so $F_{\text{equiv}} = F_r = 2000 \text{ lb}$ ////

When rotational speeds are above 5000 rpm, centrifugal and gyroscopic effects may be significant. Velocities and centrifugal force P_c for inner-ring rotation are shown in Fig. 11.26. They are higher for outer-ring rotation at the same rpm. Force P_c equals $mV_c^2/(D/2)$, where m is the mass of the ball. A calculation for a ball of Example 11.5 at an inner-ring rotation of 20 000 rpm gives $P_c = 717$ N from a 0.0189-kg ball.

The gyroscopic effect is greatest in thrust bearings, such as that of Fig. 11.27(a). The upper ring rotates at n rpm, the ball spins about axis X at ω rad/s, and the cage rotates about axis Z at Ω rad/s. Thus the spin axis of each ball precesses at a rate Ω, and a body-torque $T = I\omega\Omega$ will be developed. There will be sliding, heat, and wear unless this torque is exceeded by the friction torque available. A normal force P_n at angle α

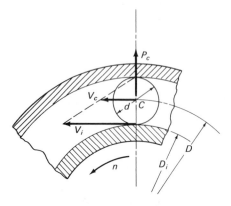

FIGURE 11.26. Velocities and centrifugal force in a radial bearing with inner-ring rotation.

(Fig. 11.27(b)), is developed by the applied load per ball F_a/N and a centrifugal force $P_c/2$. The maximum available friction torque is $(\mu P_n)(d/2) + (\mu P_n)(d/2) = \mu P_n d$, where μ is the coefficient of friction. According to Palmgren,[38] μ ranges from 0.02 at high speeds to 0.08 at low speeds. Thrust bearings and, in general, annular bearings, that have a smaller ball precession and gyroscopic effect, should be avoided for high-speed applications.

FIGURE 11.27. Precession of ball spin-axis in a ball thrust bearing. (a) Velocities. (b) Contact points and normal forces due to applied and centrifugal forces.

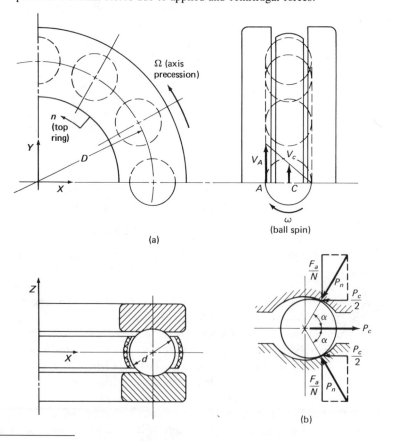

[38] A. Palmgren, *Ball and Roller Bearing Engineering*, 3rd ed., copyright 1959 by SKF Industries, Inc., Philadelphia, Pennsylvania.

A *Dn* factor is used as a measure of speed, where *D* is the bearing bore in millimeters and *n* is the shaft rpm. It is approximately proportional to centrifugal force at the rolling elements. Operation of radial ball bearings with *Dn* values above 0.2 or 0.3 $\times$ 10^6 requires special considerations, and values of 1.5 $\times$ 10^6 have been successfully used. Special, high-precision ball and cylindrical roller bearings are available. The cages are piloted on one of the rings for concentric, balanced operation, and bronze or composition cages are used to minimize scratching and friction of the highly polished rolling elements. The radial clearance may be increased to allow for thermal expansion of the inner ring. Spray and mist lubrication may be required.

11.13 BEARING LIFE, CAPACITY, AND VARIABLE LOAD

Despite careful control of materials and manufacturing accuracy, rolling bearings of the same size under controlled conditions, subjected to identical load, speed, lubrication, and environment, show a large variation in life from bearing to bearing. Some bearings in a large lot will have more than 20 times the life of others. This is shown in the plot of fatigue life vs probability of failure (Fig. 11.28). Most manufacturers publish combinations of load and life that will give a 10% probability of failure, i.e., 90% of the bearings thus loaded will survive beyond this life. It is known as L_{10} life. Published values have also been based on a median life L_{50}, that with a 50% probability of failure. In Fig. 11.28, $L_{50} \approx 5L_{10}$. However, only 50% of the bearings will survive this longer life, and the designer should make sure of the probability that he needs and the probability basis of the catalog he is using.

Rolling bearings fail with a removal by flaking, pitting, or spalling of material from the surfaces of the contacting elements, and they may become loose, rough, and noisy. Life is defined as the number of revolutions *L* until the first evidence of fatigue, such

FIGURE 11.28. Typical life distribution in rolling bearings. (Sketch courtesy of the SKF Industries, Inc.)

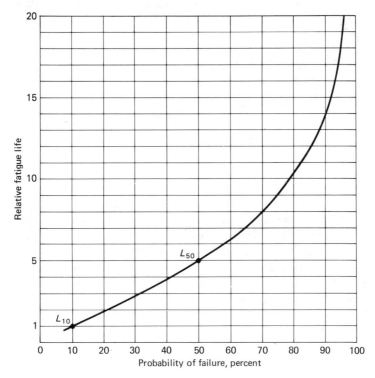

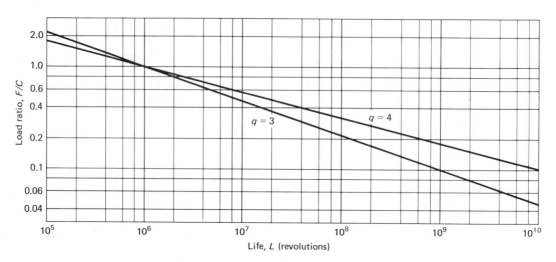

FIGURE 11.29. Logarithmic relationship between applied load and life in rolling bearings. Based on Eq. (11.26b), with $q = 3$, typical of point-contact elements and with $q = 4$, typical of line contact elements.

as a first crack, indicative of the beginning of pit formation. A plot of applied load F or, as in Fig. 11.29, of a load ratio F/C vs life L resembles the S–N charts of rotating-beam fatigue tests (Fig. 5.20 and 5.21). They both plot straight lines on a log-log chart, but the bearings have no endurance-limit load below which life will be infinite. The equation that fits the test results, relating life to load, is

$$L = \frac{c}{F^q} \tag{11.24}$$

where c and q are constants. This equation on a log-log chart becomes $\log F = \text{constant} - (1/q) \log L$, a straight line with a slope $(-1/q)$. The value generally assigned to ball and other point-contact bearings, such as spherical rollers, is $q = 3$; to line-contact bearings, $q = 4$; and sometimes $q = 10/3$ to modified line contact, as with crowned straight rollers. The nonlinearity between life and load is remarkable, e.g., for a ball bearing with $q = 3$, if the load is halved, the life is eight times as much. Conversely, overloading a bearing by even a small amount will reduce its life considerably. Equation (11.24), therefore, may be rewritten $LF^q = c = \text{constant}$ and as a proportionality,

$$\frac{L_2}{L_1} = \left(\frac{F_1}{F_2}\right)^q \tag{11.25}$$

For each type and size of bearing only a single value for load capacity need be established, an endurance value or *basic dynamic capacity* C corresponding to a chosen life and probability of failure. Then allowable loads for other lives may be calculated by Eq. (11.24). The basic life and probability approved by the AFBMA are 10^6 (one million) revolutions and 10% failure (90% survival), respectively. Substitution of these values into Eq. (11.24), gives $10^6 = c/C^q$, or $c = 10^6\, C^q$, which substituted back in Eq. (11.24), yields an equation for predicting the L_{10} life at a given load,

$$L_{10} = 10^6 \left(\frac{C}{F}\right)^q \tag{11.26a}$$

To determine the load that will give 90% chance of survival beyond a life L

$$F = C \left(\frac{10^6}{L_{10}} \right)^{1/q} \qquad (11.26b)$$

Equations (11.26) are charted in Fig. 11.29 by plotting F/C against L for $q = 3$ and $q = 4$. When $L = 10^6$, $F/C = 1.0$, and both lines pass through this point. Since the designer requires a value of C so that he may enter the table of a catalog and select a size of bearing with the next larger capacity, a most useful rearrangement is

$$C = F_{\text{equiv}} \left(\frac{L_{10}}{10^6} \right)^{1/q} \qquad (11.27)$$

where F_{equiv} is the equivalent radial load from Eq. (11.23). However, values of X and Y are needed for Eq. (11.23), and they must be estimated directly or the static capacity C_s must be estimated, for it is a property of the yet undetermined size of bearing. Figure 11.25 may be helpful in estimating.

Critical applications such as aircraft engines require higher reliability than 90%. Figure 11.30, due to Harris,[39] is an extension into the 0 to 10% region of Fig. 11.28. It gives the fraction which multiplied by the basic dynamic capacity will give a lower capacity C' for any probability of failure between 0.01 and 10%. Conversely, to enter capacity tables based on an L_{10} life in order to select a bearing for higher reliability, subtract the reliability from 100% to obtain the probability of failure. Enter Fig. 11.30, obtain the

FIGURE 11.30. Reduction in basic dynamic capacity required for increased reliability. (From T. A. Harris, "Predicting Bearing Reliability," *Machine Design*, Vol. 35, No. 1, pp. 129–132, 1963, with permission of the publisher, Penton Publishing Company, Cleveland, Ohio.)

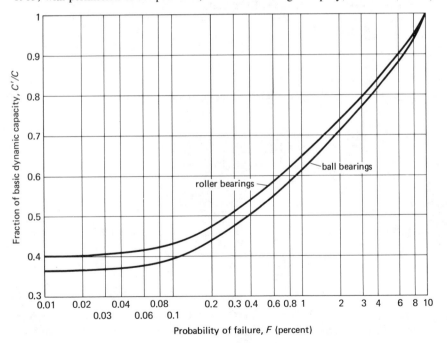

[39] T. A. Harris, "Predicting Bearing Reliability," *Machine Design*, Vol. 35, No. 1, pp. 129–132, January 3, 1963, or T. A. Harris, *Rolling Bearing Analysis*, New York: John Wiley & Sons, Inc., p. 414, 1966.

fraction C'/C, divide it into the value calculated by Eq. (11.27) and enter the capacity tables with this higher value.

Failure usually occurs first at the inner raceway, where the pressure and stresses are generally higher. If this raceway rotates through the load vector, then the entire circumference shares the maximum pressure. However, if the housing and outer ring rotate on a stationary shaft, as for the rope sheave of Fig. 9.10 with ball bearings used, there is no relative motion of the inner raceway and the load vector, and one small area takes the maximum load. This would seem to reduce life, and compensation for it in the calculations has been made by including a factor V in Eq. (11.23) so that it read $F_{equiv} = XVF_r + YF_a$, and by including V in the load ratio so that it read F_a/VF_r. Normally 1.0, V was changed to 1.2 for outer-ring rotation or for any other nonrotation of load vector to inner-ring, as by a centrifugal force from an unbalance on a rotating shaft. However, Harris determined equations for V based on loads and load capacities and states that a value of 1.2 for V for outer ring rotation exceeds theoretical requirements and is a conservative estimate.[40]

The concept of cumulative damage introduced for fatigue stress (Section 6.11) is applied to rolling bearings when the load varies or is cyclic. The total damage D, taken as 1.0, is the sum of several fractional damages $D_1, D_2, \ldots D_i$, as indicated in Eq. (6.35). Each fractional damage is the ratio of the number of revolutions N_i at load F_i to the number of revolutions to failure, life L_i, at that same load, namely,

$$\frac{N_1}{L_1} + \frac{N_2}{L_2} + \cdots + \frac{N_i}{L_i} = \Sigma \frac{N_i}{L_i} = 1.0 \tag{11.28}$$

Substitution of $L_i = c/F_i^q$ from Eq. (11.24) gives $\Sigma F_i^q N_i = c$, the constant for each particular bearing. Let $L = N_1 + N_2 + \cdots + N_i = \Sigma N_i$, the life of the bearing, and let F_m be a mean effective or equivalent single load. Also from Eq. (11.24), $c = F_m^q L$. Equating the two expressions for c yields

$$F_m = \left[\frac{\Sigma F_i^q N_i}{L} \right]^{1/q} \tag{11.29}$$

If each load repeats regularly for a time t_i and speed n_i, all within a period of time T and revolutions N, then it is convenient to substitute $n_i t_i$ for N_i and $N_T = \Sigma n_i t_i$ for L, thus

$$F_m = \left[\frac{\Sigma F_i^q N_i}{N_T} \right]^{1/q} = \left[\frac{\Sigma F_i^q n_i t_i}{\Sigma n_i t_i} \right]^{1/q} \tag{11.30}$$

If $n_i = n = $ constant,

$$F_m = \left[\frac{\Sigma F_i^q t_i}{\Sigma t_i} \right]^{1/q}$$

$$= \left[\frac{\Sigma F_i^q t_i}{T} \right]^{1/q} = \left[\Sigma F_i^q \frac{t_i}{T} \right]^{1/q} \tag{11.31}$$

The summations $\Sigma F_i^q N_i$ and $\Sigma F_i^q t_i$ are areas under a plot of F^q vs N and F^q vs t, respectively. If the plots are continuous curves and F may be written as a function of

---

[40] T. A. Harris, *Rolling Bearing Analysis,* New York: John Wiley & Sons, Inc., p. 363, 1966.

revolutions N in a period of N_T revolutions, or as a function of time t in a period T, then Eqs. (11.30) and (11.31) for cyclic action may be rewritten

$$F_m = \left[\frac{1}{N_T} \int_0^{N_T} F^q dN \right]^{1/q} = \left[\frac{1}{T} \int_0^T F^q dt \right]^{1/q} \tag{11.32}$$

A more general case of loading includes steady loads due to dead weight and belt and gear forces, rotating loads due to unbalance, and sinusoidal loads from inertia forces of reciprocation.[41] If there are both radial and axial loads in various amounts and ratios during the bearing life or cycle, then equivalent radial loads $(F_{equiv})_1$, $(F_{equiv})_2$, ... $(F_{equiv})_i$ should be calculated by Eq. (11.23) for each fractional part of the life or cycle and used in Eqs. (11.29) through (11.31) to obtain the mean effective load F_m. To select a bearing from tables based on an L_{10} life and the basic dynamic capacity C, F_m should be substituted for F_{equiv} in Eq. (11.27).

Example 11.7

Bearing 214 of Examples 11.5 and 11.6 has a basic dynamic capacity of 10 700 lb. It is loaded as in Example 11.6(a) during 0.3 of the time, as in Example 11.6(b) during 0.5 of the time, and it idles (no-load rotation) for the remaining time. Determine the mean effective load and the L_{10} life of the bearing.

Solution

From Example 11.6, $(F_{equiv})_1 = 3340$ lb and $(F_{equiv})_2 = 2000$ lb. Also, $t_1/T = 0.3$, $t_2/T = 0.5$, and $(F_{equiv})_3 = 0$ for 0.2 of the time. It is a radial bearing, so $q = 3$. From Eq. (11.31), the mean effective load is

$$F_m = \left[\Sigma F_i^3 \frac{t_i}{T} \right]^{1/3} = [(3340)^3 (0.3) + (2000)^3 (0.5) + 0]^{1/3} = 2475 \text{ lb}$$

From Eq. (11.26a), the L_{10} life is

$$L_{10} = 10^6 \left(\frac{C}{F} \right)^q = 10^6 \left(\frac{10\ 700}{2475} \right)^3 = 80.8 \times 10^6 \text{ revolutions} \qquad \text{////}$$

11.14 CLOSURE ON ROLLING-ELEMENT BEARINGS

So that one does not gain the impression that all rolling-element bearings are "standardized" and ready to be selected from catalogs of bearing specialists, it should be pointed out that there have been and will be more nonstandard developments by the specialists and by the customers and specialists cooperatively. Examples are rolling elements made from high-speed tool steels for gas-turbine bearings, hollow and drilled balls to decrease their inertia, rollers undercut at their ends for flexibility to reduce pressure and minimize geometric stress concentration (Section 11.5), outer raceways out of round a few thousandths of an inch to "pinch" or preload the rolling elements and prevent skidding, bearings with much lower friction torque for gyro-autopilots and other instruments, and very large and relatively narrow bearings for radar antenna systems. Sometimes the general manufacturer must design and develop his own rolling devices, as has been done

---

[41] For charts giving F_m as a fraction of total load with the individual loads in or out of phase, see T. A. Harris, *Rolling Bearing Analysis*, New York: John Wiley & Sons, Inc., pp. 405–409, 1966.

for cast-iron turntables, airborne turrets with plastic balls in magnesium rings, and rotary drills for oil-field drilling, with the rolling elements of the cutter disks coated by hard tungsten carbide. There are also a number of related devices using rolling elements, such as ball-guide bearings, ball screws, and speed-changing friction drives.

PROBLEMS

Section 11.2

11.1   Check the stress equation given for the knife edge of Table 11.1 by writing the equation of equilibrium of force across any cross section in terms of σ_r, substituting for it, and integrating.

11.2   Two methods of supporting a force P on the flat surface of a body of low modulus of elasticity are under consideration: (1) by a pad or hydraulic device that will give a constant pressure and (2) by a rod of much higher modulus, bearing directly on the surface. In both cases the supporting area has a radius $r = a$. (a) How much larger is one maximum deflection than the other? (b) Explain the result with sketches of deflection and stress. (c) If the loaded area and P are increased, keeping the average pressure the same, in what ratio to radius a does the deflection change, if at all? Is there any advantage of (1) over (2)?

Section 11.3

11.3   Starting with Eq. (11.1) and Fig. 11.3, derive an equation for the deflection when a constant pressure acts over a circular area of radius $r = a$ on the plane surface of a semi-infinite body (Case 4 of Table 11.1). Evaluate the deflection at $r/a = 0, 0.5, 1.0$. *Note:* A table of elliptic integral values will be needed.

Ans. $w = (4\eta p_0 a/\pi) \displaystyle\int_0^{\pi/2} \sqrt{1 - (r/a)^2 \sin^2 \phi}\, d\phi$.

11.4   Using the answer to Prob. 11.3 determine the average deflection over the circle loaded with uniform pressure and compare it with the deflection under the end of a rigid cylinder. *Note:* Such comparisons serve as checks on reasonableness of new equations. *Ans.* $w_{\text{avg}} = 0.54\eta P/a$.

11.5   Deflection *outside* the uniformly loaded circle in Prob. 11.3 may be found by directing all the chords through a point M located outside the loaded area in the figure (cf. Fig. 11.1), then proceeding as in Prob. 11.3 except indicating the limits of integration to be $-\phi_1$ and $+\phi_1$. Evaluation is then simplified by eliminating ϕ and substituting θ, where θ is the angle between chord GHM and radius OG. With corresponding limits of integration for θ, integration results in two elliptic integrals. Evaluate the deflection for point M at radius r for the ratios $a/r = 1.0, 0.5, 0.1$, and 0.0 and compare with the answer to Prob. 11.3.

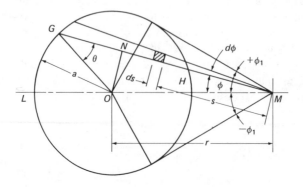

11.6 From Table 11.2 obtain a simplified equation for the contact-circle radius and maximum deflection at a steel ball on a flat surface. Compare values for soft wood ($E = 1.2 \times 10^6$) and steel ($E = 30 \times 10^6$) surfaces. Assume $\nu = 1/3$ for the wood. What is the definition of rolling friction? What do you conclude about the relative rolling friction of the two surfaces?

11.7 The ball end of a steel link is 10 mm in diameter. It is socketed in the hard-bronze bearing-alloy spherical seat, 10.1 mm in diameter, of a rocker arm in a valve-operating linkage. The maximum pressure is limited to 200 MPa (29 000 psi). For bronze, $E = 110$ GPa (16.0 × 10⁶ psi), and $\nu = 0.30$. Determine the maximum allowable load. *Ans. P =* 1690 *N.*

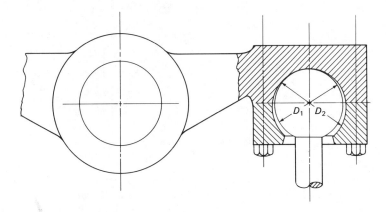

11.8 In the application of the equation for contact-circle radius a to a closely fitting ball and socket, as in Prob. 11.7, $D_2 \approx -D_1$ and $D_1 + D_2 = -c_d$, where c_d is the diametral clearance. Derive an approximate equation for load capacity in terms of clearance, allowable pressure, and ball diameter. Test this equation on Prob. 11.7, comparing answers. *Ans. $P = 1.29(\eta_1 + \eta_2)^2 p_0^3 D^4/c_d^2$.*

Section 11.4

11.9 Check by calculation the statement at the end of Example 11.2 that a flat-faced follower and cam of length 0.202 in will give a smaller *calculated* stress.

11.10 The steel cams of Prob. 6.16 rotate under flat-faced followers of chilled cast iron. What should be the minimum radius at the tips of the cams if the contact pressure is not to exceed 120 000 psi.

11.11 The roller-ramp clutch of Prob. 3.51 has an inside ring diameter of 2.90 in, a ramp angle $\alpha = 8.5°$, 10 rollers with diameters 0.375 in and length 0.625 in. The rated torque is 165 lb·ft. Determine the maximum pressure and compressive stress. Use the normal force given in the answer to Prob. 3.51, which force is the same on ring and ramp.

11.12 Carbon-steel rollers used under one end of a steel bridge to allow for thermal expansion may corrode and jam. This has been avoided without greatly increased initial cost by bonding a thin strip of stainless steel to the rolling surface. High-shearing stresses are avoided at the bond. For a flat plate on a 10-in-diameter roller with a ⅛ in thickness of stainless steel, loaded by 4000 lb/in of length, determine the shear stress at the bond and the location and magnitude of the maximum shear stress. Use Fig. 11.4 in the calculations.

11.13 A Geneva wheel has straight-sided radial slots entered by a rolling, cylindrical steel pin on a driving arm, with dimensions as shown. The wheel is nickel cast iron heat treated to R_c 35, and the steel pin is to have the same length as the diameter. The torque resistance of the wheel is 75 N·m. Use the allowable pressure suggested in Section 11.7 for a similar combination of materials in cams, and use $E = 138$ GPa and $\nu = 0.25$ for the hardened cast iron. Determine minimum pin dimensions.

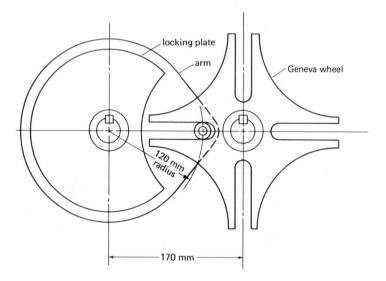

11.14 The pivot shown has a rounded point with a radius of 0.5 mm. It rests on a flat surface and supports a mass weighing 204 kg. Both surfaces are of hardened steel, and the allowable compressive stress at the point is 2000 MPa. (a) What length is required for the pivot? *Ans.* 36.1 mm. (b) The included angle is 18°. Determine the stress at a location 2 mm above the bottom. A large-scale sketch is suggested before using the equation of Table 11.1.

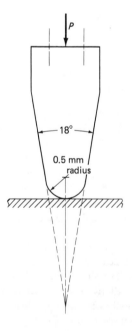

11.15 Calculate the maximum compressive stress and deflection for the steel wheel of Example 8.5 and Fig. 8.17 when on a rail with a head radius of 10 in. *Ans.* 180 000 psi, 0.004 08 in.

11.16 From Table 11.2 derive equations for the size and shape of the contact area and the maximum pressure and deflection for two cylinders of the same diameter and material (ν = 0.30), crossing at 90° and loaded by a force P.

11.17 Determine equations for shape of the contact area, pressure, and deflection for a sphere pressed against a cylinder of the same radius and material.

11.18 Ball bearings with thicker than normal outer rings and a crowned surface, case hardened, are available as cam roller followers. A follower with an outer-ring diameter of 1.50 in, a width of ⅝ in, and a circular crown with a height of 0.015 in operates on a steel-disk cam with a minimum radius of 0.50 in. Based on an allowable pressure of 250 000 psi, what is the surface load capacity?

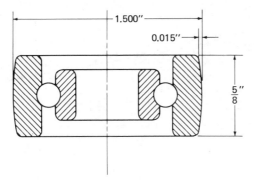

11.19 A lathe attachment is to be designed to induce beneficial residual stresses at shaft fillets by cold rolling, as shown in the sketch, a process discussed in Section 7.12. The hardened steel roller has a diameter of 2 in and a tip radius of ¼ in. For a steel shaft where $d = 3$ in, $D = 4$ in, and fillet radius $r = 5/16$ in, determine the minimum required roller force to produce yielding at the fillet surface where work hardening has raised the yield strength to 200 000 psi in compression: (a) when the roller is at the shaft end of the fillet and (b) when the roller's axis of rotation is turned 90° to work the shoulder end of the fillet.

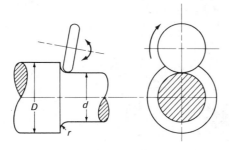

11.20 A proposed arrangement for a friction-type speed-changing mechanism is shown. It has tapered or conical steel elements in a carrier rotating about the axis of the mechanism at a speed of 1750 rpm and in rolling contact with the inside of a nonrotating but axially movable steel ring. (a) Determine the location and magnitude of the concentrated centrifugal force due to the conical part of the roller only. (b) Determine the maximum force on the ring assuming that the roller is free to pivot about the center of its bearing. (c) Determine the maximum compressive stress at the contact area. (A transmission using this principle of operation is made by the Graham Company, Menomonee Falls, Wisconsin.)

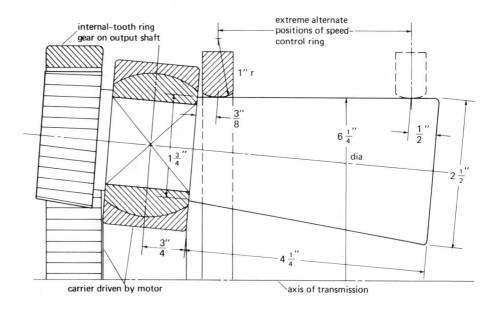

Section 11.8

11.21 We are given the arc thickness t of an involute gear tooth at one radius R. Obtain an equation for the arc thickness t' at any other radius R' in terms of involute functions. Also write an equation for the chordal thickness t_c' at radius R'. *Ans.* $t' = 2R'\left[(t/2R) + \text{inv}\,\theta - \text{inv}\,\theta'\right]$ and $t_c' = 2R' \sin(t'/2R')$ where $\theta = \cos^{-1}R_b/R$ and $\theta' = \cos^{-1}R_b/R'$.

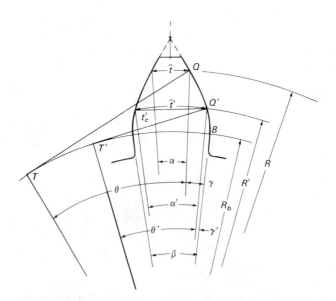

11.22 With the standard arc thickness at the pitch circle as a reference, determine equations for the arc thickness and for the chordal thickness of a tooth at its base circle. Use the answers to Prob. 11.21.

11.23 Same information as Prob. 11.22 except find thicknesses at the addendum circle.

11.24 Checks by the gear-tooth caliper shown are made during or after cutting. Member S is a vernier screw, which is preset to the desired depth for measuring chordal thickness t'_c. Using the answers to Prob. 11.21 write an equation for t'_c and h' in terms of any reference dimensions $\hat{t}$, R, and θ.

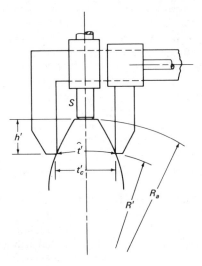

11.25 Use the answers and sketch of Prob. 11.21 to obtain an equation for the radius at which a standard tooth becomes pointed. Write the equation in terms of pressure angle ϕ, pitch P, and numbers of teeth N. Obtain a table of involute functions and calculate the radius for an 18 tooth, $5P$ tooth in the 20° system. Compare it with the standard tooth outside radius.

11.26 A gear in a machine has failed and you wish to replace it locally to save downtime. You count the number of teeth and measure the periphery with a tape and obtain 59 teeth and approximately 24 in, respectively. Assume a standard full-depth tooth and calculate the probable diametral pitch and the other dimensions to four decimal places. How could you determine which of several standard pressure angles was used?

11.27 A pair of rolling-mill drive gears built on the module system has $m = 8$, $N_1 = 38$, $N_2 = 103$, width $b = 150$ mm, and $\phi = 20°$, full-depth system. Some teeth have failed by pitting, and the gear must be replaced as soon as possible. Only diametral-pitch gears are available in the country. An increase of pitting resistance will be obtained by improved heat treatment and surface finish. The center distance is fixed by the existing bearing locations, but small changes may be made in tooth thickness, numbers of teeth, and width. Specify a suitable diametral pitch and numbers of teeth. What is the theoretical center distance for this new pair? (Operating on an extended center distance in the old bearings, any excessive gap between teeth can be filled by cutting the pinion teeth thicker than standard.)

Sections 11.9 *and* 11.10

11.28 If, as in Prob. 11.27, gears occupying about the same diametral space and width are substituted for the original ones, is there any change in surface-load capacity? In bending-load capacity? If the pitch P is somewhat larger, what dimension should be changed to maintain the same capacity?

11.29 The allowable surface capacity F of a pair of steel gears of a given hardness has been calculated. It is proposed to double the surface capacity of the gears by heat treating to a higher hardness. (a) By what percentage will the maximum contact pressure p_0 increase because of doubling the surface load? (b) By what percentage should the hardness be increased?

11.30 A spur pinion at 1000 rpm is to transmit 50 hp to a gear at 333 rpm. Other considerations require that the center distance between gears be approximately 10 in. Use steel at 350 Bhn for the pinion and at 250 Bhn for the gear, and Eq. (11.17) for the allowable pressure. Use 20° full-depth teeth, cut by hobbing, and assume that the load is sometimes applied at the tooth tips. Calculate tooth width and pitch, numbers of teeth, and pitch diameters. Are dimensions of reasonable proportions? *Ans. $F_t = 2170$ lb, $l = 1.63$ in, $P = 6$.*

11.31 The dimensions of the electric-locomotive gears of Example 8.5 and Fig. 8.17 were determined from space considerations. An analysis shows that the worst condition for the gears occurs when the speed of the axle is 128 rpm and the axle torque is 325 000 lb·in. The tooth form is 20° full depth, and the teeth will be ground. Assume that division of load will not occur with all tooth combinations. Determine minimum physical properties required by heat treatment of the material of the pinion and of the gear.

11.32 Most aircraft gears are case hardened, and they operate at high speeds to save size and weight in the power plant. One such gear for an accessory drive has the following properties: face width 0.80 in, 10 diametral pitch, 50 teeth, 20° full-depth system, alloy-steel carburized and oil quenched to give a hardness R_c 62 (680 Bhn) in the case, which extends over the fillet. The accuracy is high as by grinding and lapping, and load sharing may be expected. When this gear is meshing with an equal gear, what is its dynamic capacity based upon surface durability? On bending? What then is the allowable transmitted load and horsepower at 6000 rpm? *Ans. 3880 lb, 2065 lb, 335 hp.*

11.33 Determine the horsepower that may be transmitted by a 20° full-depth 18-tooth 12-pitch pinion in contact with a 35-tooth pinion at a pitch-line velocity of 4900 ft/min. The face width is ¾ in. The steel has been heat treated to a hardness over the tooth profile of 400 Bhn and at the fillets, to 350 Bhn, then ground and lapped, with division of load expected.

11.34 Sun, planet, and ring gears of a planetary transmission for aircraft use are shown, with numbers of teeth indicated. Teeth are alloy-steel carburized, including the root fillet, to a hardness R_c 61 (670 Bhn), and they have a diametral pitch of 10 in the 20° full-depth system, surface finished by grinding and lapping. The sun gear delivers 120 hp at 6300 rpm. (a) On a sketch draw in sun-gear pitch-line velocity vector V_1 and, in proportion, planet-carrier and planet-gear velocity V_{23} at center O_{23}. Determine an equation for tooth-engagement velocity V, which is the velocity of the two pitch circles rolling together relative to the "link" joining their centers, i.e., the carrier velocity at the pitch point of the sun and planet. What is its value here in feet per minute? *Ans. $V = V_1[1 - D_1/2(D_1 + D_2)] = 2640$ ft/min.* (b) Draw force vector diagrams of sun, planet, and ring gears and calculate the transmitted force at a tooth. Which one has teeth subject to the worst condition of fatigue? (c) Calculate the face width required for strength and for durability. Allow an extra 10% on the tooth loads for possible uneven distribution of force between the three planet gears. *Ans. 0.525 in.*

Section 11.11

11.35 Check Stribeck's observation given after Eq. (11.22) by calculating N/S for bearings with 7, 8, and 11 balls.

11.36 For the bearing of Example 11.5 determine the maximum pressure and displacement at the outer ring and compare them with those at the inner ring.

11.37 In a special series of bearings having a large ratio of diameter to radial thickness, one bearing has 44 balls with a diameter of ½ in and an inner raceway with a groove diameter

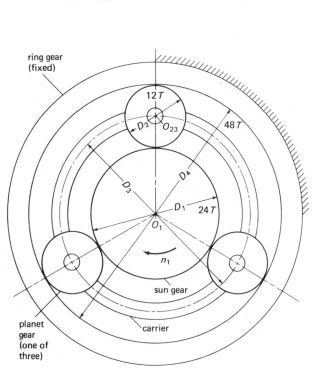

PROBLEM 11.34

of 10½ in (cf. Fig. 11.24). Assume a groove radius of 1.05 times the ball radius. The static and dynamic radial capacities are listed as 18 800 and 5100 lb, respectively, where dynamic life is based on 600 hours L_{10} life at 100 rpm. Determine the maximum compressive stress under the (a) static and (b) dynamic capacity loads.

11.38 Let a self-aligning ball bearing (Fig. 11.22(e)) have the same dimensions and number of balls as the bearing of Fig. 11.24, except that the outer raceway is spherical. What is the static capacity of the bearing if the maximum pressure is to be the same as found in Example 11.5? Compare the static capacities of the two bearings. *Ans.* 4615*N*.

11.39 Adapt the Hertz stress equations for ready use in finding the highest stress in the miniature-type, ball, pivot bearing with dimensions as shown. The load is F_a axially and there are N balls. (a) Obtain equations for $(B - A)/(B + A)$, etc. (Partial answer: $(B - A)/(B + A) = 1/[(2D/d \cos \alpha) + 1]$.) (b) Determine the maximum stress if there are 9 steel balls of diameter 0.10 in, a steel pivot with $D = 0.20$ in, $F_a = 6$ lb, and $2\alpha = 50°$.

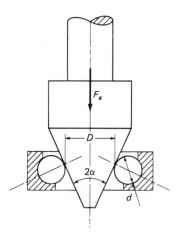

11.40 A platform for the accurate rotation of a heavy instrument about a vertical axis resembles a ball-thrust bearing (Fig. 11.27), but it has 38, 2-in-diameter steel balls and ASTM class 30 cast-iron supporting plates. The latter have ground raceways of 25-in average diameter and 1.0625-in radius. Adequate facilities for accurate grinding of the raceways were not available, and after installation the axis of rotation had a slight wobble. It was decided to grind the raceways "in place" with an abrasive slurry and balls, driving the top table slowly with a motor and friction wheel, and replacing the balls after the grinding, if necessary. How much weight can be placed on the table initially during the grinding process without exceeding a stress of 75 000 psi? Assume the equivalent of only half the balls being fully effective. Allow for the weight of the table itself. ($E_{CI} = 14 \times 10^6$ psi.)

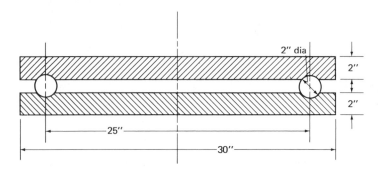

11.41 For self-protection against fighter planes, World War II bombing planes carried spherical or "ball" turrets, holding gun and gunner, in wing, nose, and tail positions. The turrets rotated within outside bearings, that were necessarily of large diameter. One type used hard plastic balls in magnesium raceways for low weight and inertia. Assume use of an acetal plastic or its equivalent with relatively high properties: $E = 410\ 000$ psi, $\nu = 0.25$ (estimated), endurance limit $S_e = 5000$ psi in bending, and a tensile strength to 10 000 psi. For magnesium $E = 6.5 \times 10^6$ and $\nu = 0.35$. If there are 330 balls of 3/8-in diameter in an inner raceway with a bottom diameter of 40 in and with a groove radius of 0.20 in, what is the stress for a force of 330 lb per bearing?

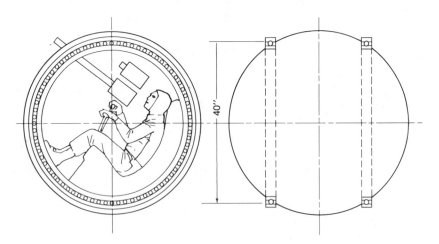

Section 11.12

11.42 For the inner ring of Example 11.5, plot to show the relationship $\delta = KP^{2/3}$ (Eq. (c) preceding Eq. (11.22)). Use several values of P between zero and the rated load. From the plot, explain why preloading minimizes the deflection changes during variable-load operation. Sketch to show several types of bearings and methods by which preloading may be done.

11.43 (a) If the inner ring of Fig. 11.24 and Example 11.5 rotates at 20 000 rpm and the outer ring is stationary, what centrifugal force acts against the outer ring at each ball? The density ρ of steel is 7820 kg/m³. (b) Same requirement as part (a) except it is the outer ring that rotates.

11.44 A thrust bearing (Fig. 11.27) of 70-mm bore has balls of 16.67-mm diameter, the same as in the radial bearing of Fig. 11.24 and Prob. 11.43, but the mean ball circle diameter is 97.5 mm, and there are 12 balls. One ring rotates at 10 000 rpm. (a) Determine the gyroscopic couple. *Ans.* $T = 0.845$ N·m. (b) The rated thrust load for 20 000 hours life at this speed is 2070 N. Include the effect of centrifugal force and determine the maximum available friction torque. Will there be skidding?

Section 11.13

11.45 For the bearing of Prob. 11.37 determine dynamic capacity for 10^6 cycles of L_{10} life.

11.46 What variation in L_{10} ball-bearing life can be expected in the machines of a certain model if customer usage varies from 30% underload to 30% overload?

11.47 A number of machines of the same size and type were designed for an L_{10} life with rather uncertain knowledge of the loads that might occur. Later reports indicate that about 20% of the ball bearings failed within the L_{10} design life, indicating that higher than design loads actually existed. In redesigning and selecting larger bearings, by how much should the basic dynamic capacity C be increased? *Ans.* 25%.

11.48 Suppose the catalog of your regular supplier of bearings rates them for an L_{10} life. In a certain application you have decided upon a suitable median or L_{50} life. What formula, similar to Eq. (11.27), but in terms of L_{50} should you use to obtain a value of C for entering the catalog? (a) For ball bearings? (b) For roller bearings? *Ans.* (a) $0.585 \, F_{equiv} \times (L_{50}/10^6)^{1/3}$.

11.49 A manufacturer rates his ball bearings on an L_{10} life and 500 hours at 1000 rpm. By what factor should you multiply his values to compare them with those of a manufacturer who uses the AFBMA rating? *Ans.* 3.11.

11.50 Fatigue life in bending is generally indicated by a plot of failure stress vs number of cycles, e.g., Fig. 5.20. However, fatigue life in rolling-element bearings is indicated by a plot of bearing load vs life in cycles (Fig. 11.29), and is represented by Eq. (11.24). It will be instructive to know how bearing life varies with maximum compressive stress σ and then to compare it with the variation in bending fatigue life. (a) For a ball bearing (Case 3 of Table 11.2), determine $p_0 = -\sigma_z$ in terms of load P and a combination elastic and geometric constant K that is independent of load. *Ans.* $\sigma_z = KP^{1/3}$. (b) Determine life L in terms of σ_z and an additional constant. *Ans.* $L = C'/\sigma_z^9$. (c) Write an equation for N in terms of S_f', from the lower line of Fig. 5.21, which is a log-log chart and typical of S–N diagrams. (d) Compare the exponents on the stress for the two cases. Is there reasonable correspondence?

11.51 Determine if a 321 deep-groove ball bearing with $C_s = 31\,500$ lb and $C_d = 32\,000$ lb is suitable for the following application: 500-lb radial load, 1000-lb thrust load, inner-ring rotation, a speed of 570 rpm, and 10 years of L_{10} life with continuous duty.

11.52 Use the answers to Prob. 9.32 and determine the basic dynamic capacity required for a ball bearing at B on the shaft. The speed is 275 rpm, and an L_{10} life corresponding to 10 years at 2000 hours per year is wanted. For the ball-bearing calculation of Eq. (11.23) and F_a/C_s, assume a static rating C_s of 7000 lb (based on a scan of the tables of bearings with an outside diameter limitation of 4 in). If a catalog is available, obtain the basic dynamic capacity for a roller bearing at A and select for locations A and B two or three bearings from different series, i.e., for different shaft dimensions.

11.53 The bearings of the speed-reducer output shaft of Fig. 9.17, Example 9.5, and Prob. 6.40 are to be selected for an L_{10} life of 5 years with 10 hours per day, full-load duty. Shaft speed is 232 rpm, and the possible maximum loads were determined to be 1844 lb radially and 318 lb axially on bearing #1, a ball bearing, and 3884 lb radially on bearing #2, a roller bearing. Estimate $C_s = 8000$ lb to obtain the equivalent radial load for the ball bearing. Determine the basic dynamic capacities required, using data from an AFMBA table or manufacturer's catalog for the roller bearing. *Ans.* #1, 11 680; #2, 15 500 lb.

11.54 Suppose that a reliability of 96% is wanted for the bearings of Prob. 11.53. Using the answers given, determine the values with which to enter the L_{10} basic capacity tables.

11.55 For the punch press of Fig. 9.16 and Example 9.4, the bearing at location 0 has a resultant radial load of 4710 lb during punching. Shaft speed is 480 rpm, and punching occupies 1/20 of the operating time. During the remaining time the loads on the shaft are essentially due to belt pull and flywheel, a resultant of 532 lb in another direction. This is easier on the outer raceway, but it still accumulates damage to the balls and inner raceway. The bearings are to have an L_{10} life for a 10-year operation, 300 days per year for two shifts or 16 hours per day. Is the 214 bearing of Fig. 11.24 and Example 11.5 suitable? Its basic ratings are 8300 lb static and 10 700 lb dynamic. If it is not suitable and a catalog is available, make several selections.

11.56 Same information as Prob. 11.55, but select a roller bearing from a catalog, probably a better selection than a ball bearing since some shock occurs at the beginning and end of the punching.

11.57 A load varies sinusoidally between zero and a maximum value F_{max} and back to zero every N_0 revolutions of a ball bearing. What fraction of F_{max} is the mean-effective load F_m? *Ans.* $F = (F_{max}/2)(1 - \cos 2\pi N/N_0)$, $F_m = 0.68F_{max}$.

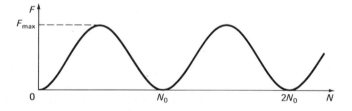

11.58 Same requirement as Prob. 11.57, except that a sinusoidal variation with amplitude F_a is superimposed upon a steady load of value $2F_a$.

11.59 The variable loading shown acts upon a bearing, repeating every N_0 cycles. (a) Determine a formula for the mean-effective load F_m in terms of F_{max} and exponent q. *Ans.* $F_m = F_{max}[(q + 3)/(4q + 4)]^{1/q}$. (b) Evaluate F_m for point-contact and for line-contact bearings. Compare the results with the mean load, $0.5F_{max}$.

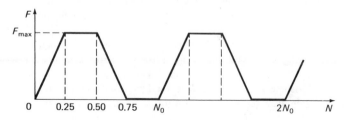

11.60   In the redesign of the sheave block of Fig. 9.10, needle bearings are being considered to replace the bronze bushings. They will require less attention to lubrication and wear, and speeds are slow. If the shaft is case hardened, the needles may run directly on it, and the space for the needles and outer ring will require only a minor modification of the core for casting the sheaves. Obtain a needle-bearing catalog, and select one or two bearings for each sheave according to dimension considerations and the manufacturer's recommendations. Figure 9.10 may be scaled. The shaft diameter, about 4 in, may be modified a small amount to fit the bearing, if necessary.

Section 11.14

11.61   A ball screw and nut assembly is needed as an actuator on a linkage for extending and retracting the wing flaps of an aircraft. The screw rotates at 960 rpm to drive a nut axially, without rotation, against a load of 9000 lb for a 62-in stroke. The required life is 12 000 cycles of two strokes per cycle. Tooling is available for a screw that has a 0.3636-in lead and a 1.625-in pitch diameter (to center of balls), using balls of 3/16-in diameter. Previous satisfactory designs have used a load of 116 lb per ball for a life of 10^6 impacts, which is the number of times a ball passes a given spot in the raceway of the nut. This has been determined from kinematics to be 11.75 per revolution. The nut may be designed with one, two, or three separate circuits each with 1½, 2½, or 3½ turns before returning the balls externally. The problem is to determine the number of balls required, and then the number of circuits and turns to contain this number of balls. (Adapted from G. A. Widmoyer, "Determine the Basic Design Specifications for a Ball Bearing Screw Assembly," *General Motors Engineering J.*, Vol. 2, No. 5, pp. 49–50 and No. 6, pp. 47–48, 1955 or Reprints, Prob. 11, by permission from the publisher, General Motors Corporation.

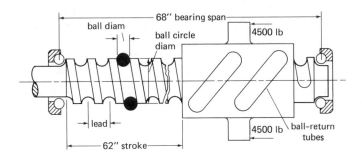

12

Torsion

Straight Circular Bars, Compression and Extension Springs, Noncircular Solid and Tubular Sections

12.1 INTRODUCTION

The elementary theory of the torsion of circular, solid and hollow bars, as used in the design of shafts and torsion bars is reviewed in this chapter. This theory, modified for the effect of curvature, is then applied to determine equations for the maximum stress and deflection in compression and extension springs wound from circular wire. Proportions and the design of springs for static and dynamic service is considered in some detail. Then, from the theory of elasticity more general equations for torsion are developed, equations applicable to bars of any cross section, and some analogous experimental methods are described. Approximate analytical solutions are made for some solid and tubular sections of interest in mechanical components, including air-foil sections and rectangular wire in springs.

Some methods and applications involving torsion occur in other parts of the book. These include limit design in Example 5.5 of Section 5.8, fluctuating stresses and theories of fatigue failure in Sections 6.2 through 6.6, shaft design in Sections 6.7 through 6.9, strain-energy methods in Sections 10.1 and 10.3, flat-plate theory in Section 10.8, torsional impact of shafts in Section 13.8, and longitudinal impact of helical springs in Section 13.9.

A spring is any element used primarily because of its elastic deflection. Thus there are springs to maintain force or torque, to support masses and minimize the transmission of vibration or shock, to store and deliver energy, to measure force or torque, to act as pivots or guides, and to act as retainers in assemblies. Many of these functions are obtained with convenience in attachment, in simple form, and with low expense by the elastic bending of thin beams and plates in numerous configurations. Thus there are cantilever springs, multiple-leaf springs, volute and spiral springs, loops, clips, spring washers, and retaining rings. Many are *flat springs*, of thin rectangular section flexible in only one direction, or are thin dished disks, but others are wound. The helical coil spring which is twisted about the axis of its coil is known as a *torsion spring*, but its action is one of bending. Theories and methods for bending analyses and specific applications to springs are found in the text and figures, problems, and answers of Chapters 9 and 10.[1] Other references are available for detailed treatments.[2]

12.2 TORSION OF STRAIGHT BARS OF CIRCULAR SECTION

When a twisting moment or torque M_t is applied to the ends of a solid or hollow circular bar, as in Fig. 12.1(a), the end rotates through an angle θ, plane cross sections remain plane, and the shearing strain $\gamma = \overline{aa}'/l = r\theta/l$, is proportional to radius r. On an element $dx\,dy\,dz$ (Fig. 12.1(b)), shear stress τ_{zx} on a radial plane face $dx\,dy$ will be directed tangentially, and on an axial plane face $dy\,dz$, τ_{xz} is parallel to the axis of the bar and equal to τ_{zx}, say τ. Because it is related to strain, the stress τ will be uniform over any elemental ring dA (Fig. 12.1(c)), and the external moment M_t equals the total internal moment at a section, namely,

$$M_t = \int_{r_i}^{r_o} r(\tau\,dA)$$

Within the elastic range, stress τ is proportional to strain γ and to radius r, and in terms of the maximum stress, $\tau = \tau_{\max} r/(d_o/2)$ (Fig. 12.1(c)). Hence

$$M_t = \frac{\tau_{\max}}{d_o/2} \int_{r_i}^{r_o} r^2\,dA \tag{12.1}$$

FIGURE 12.1. Cylindrical bar. (a) Twisted by moment M_t. (b) Shear distortion of an element. (c) Distribution of shear stress τ in a hollow bar.

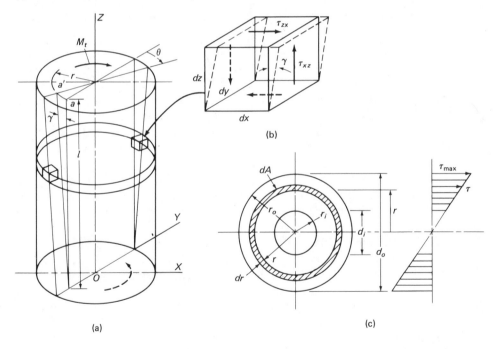

---

[1] Flat and leaf springs: Fig. 9.8(c) and (d), Probs. 9.64, 9.77, and 9.78; flex pivots and guides: Examples 9.8 and 9.10, Figs. 9.26 and 9.28, Probs. 9.69 and 9.79; bimetallic: Prob. 9.83; circular loops: Example 10.5, Fig. 10.9, Probs. 10.11 through 10.14; clips, including electrical: Example 10.2, Fig. 10.6, Probs. 10.7 through 10.9; retainers: Example 10.3, Fig. 10.7, Probs. 10.20 and 10.24; Belleville disks: Prob. 10.79; spiral: Prob. 9.83; torsion springs: Section 10.3, Fig. 10.11, Probs. 10.6, 10.25, and 10.41. The equation summaries in Tables 9.4, 10.4, and 10.5 may also be useful.

[2] A. M. Wahl, *Mechanical Springs*, 2nd ed., New York: McGraw-Hill, 1963.

The integral is the polar moment of inertia J of the cross-sectional area, and, since $dA = 2\pi r \, dr$

$$J = \int_{r_i}^{r_o} r^2 \, dA = 2\pi \int_{r_i}^{r_o} r^3 \, dr = \frac{\pi}{2}(r_o^4 - r_i^4) = \frac{\pi}{32}(d_o^4 - d_i^4) \tag{12.2}$$

Equation (12.1) solved with Eq. (12.2) for maximum shearing stress becomes

$$\tau_{\max} = \frac{(d_o/2)M_t}{J} = \frac{16M_t d_o}{\pi(d_o^4 - d_i^4)} \tag{12.3}$$

Shear strain γ is related to stress by $\gamma = \tau/G$, where G is the shear modulus of elasticity. The angle of twist for a small angle θ (Fig. 12.1(a)), is

$$\theta = \frac{aa'}{r} = \frac{\gamma l}{r} = \frac{\tau l}{Gr} = \frac{\tau_{\max} l}{G(d_o/2)} = \frac{M_t l}{JG} \tag{12.4}$$

where the last term is obtained by substitution for $\tau_{\max}$ from Eq. (12.3).

In the case of a solid section, $d_i = 0$ in Eqs. (12.2) and (12.3). With $d_o = d$, Eqs. (12.3) and (12.4) are frequently written

$$\tau_{\max} = \frac{M_t d}{2J} = \frac{16M_t}{\pi d^3} \quad \text{and} \quad \theta = \frac{32M_t l}{\pi d^4 G} \tag{12.5}$$

In the case of a bar with variable diameter and moment, $\tau_{\max}$ everywhere along its length is given by the first of Eqs. (12.5), where M_t and d are measured at the section in question. The increment of twist $d\theta$ contributed by an elemental length dz of round wire is, by analogy with Eqs. (12.4) and (12.5),

$$d\theta = \frac{M_t \, dz}{JG} \tag{12.6}$$

Substitution of the variables in terms of location z is followed by a summation or integration, as appropriate, to obtain θ between definite limits.

A *torsion bar* is a straight bar fixed at one end and twisted at the other end, where it is supported. Thus the stress is principally one of torsional shear. Figure 12.2 shows a typical bar for a vehicle wheel suspension. One of the upset-forged splined ends fits into a socket on the chassis, the other into the pivoted end of an arm. The arm is part of a linkage that allows the wheel to rise and fall in nearly parallel motion.[3] In a passenger automobile bar lengths may be of the order of 30 in, diameters 0.75 to 1.00 in, and twists 30° to 45°.

Attention is called to the analyses of twist using an energy method, in curved bars or in structures containing curved sections (Section 10.3 and Eq. (10.20), and in Probs. 10.20 through 10.24). Bending generally adds to the displacements that occur and is simultaneously treated by the energy method and equations. Attention is also called to the twist that may be induced by transverse loads on noncircular plates (Section 10.8).

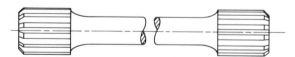

FIGURE 12.2. Torsion bar with splined ends (only partial length shown).

[3] See also Probs. 12.1 through 12.3.

12.3 STRESS IN COMPRESSION AND EXTENSION COILED SPRINGS

When loaded axially to give a decrease in length, the spring is a *compression spring*, and it has relatively simple ends, as in Fig. 12.3(a). When loaded to increase its length, it is an *extension spring*, and the ends are made with various loops or hooks, as in Fig. 12.3(b). These *cylindrical* springs with constant pitch and helix angle except at their ends are known as *helical springs*. Spring force plots against deflection in a straight line, and the springs are *linear*. An initial force is required to open the coils of the extension spring, where each coil is wound tightly against the previous one.

If the pitch or coil diameter varies, the spring may become nonlinear. In the conical spring of Fig. 12.3(c), the deflection is initially linear followed by a phase during which the coils successively close up, beginning with the most flexible, largest coil. This gradually increases the spring's stiffness or rate, which may be valuable for resisting overloads and large vibrations.

Coils are formed while cold by feeding a straight wire against a die or tool which curls and offsets it. Coils with wire larger than ⅜ to ½ in, depending on material, are wound hot on a grooved mandrel. The ratio of mean coil diameter to the wire diameter is the *spring index*. Springs of low index are difficult to coil, and those of high index have large variations in coil diameter. Indexes 4 to 15 are readily manufactured, but a range of 5½ to 9 is preferred for requirements of close tolerance and cyclic loading. The ends of compression springs may be left open (Fig. 12.3(c)), or the last turn at each end

FIGURE 12.3. Coiled springs. (a) Cylindrical compression spring with closed and ground ends, right-hand coiling. (b) Extension spring with full loops over center, bent up from end coils, left-hand coiling. (c) Conical compression spring with ends open and plain, right-hand coiling.

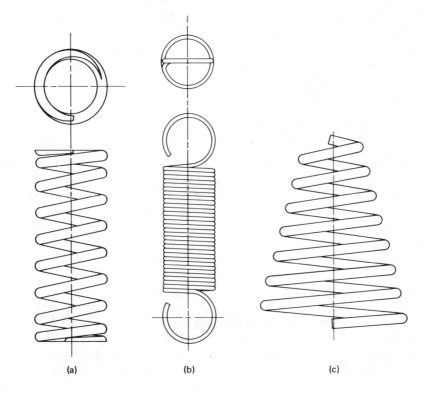

(a)   (b)   (c)

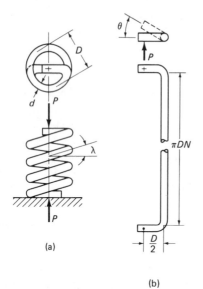

FIGURE 12.4. (a) Coiled cylindrical spring with crosswire end. (b) Wire unwound from the cylindrical spring. Note total twist θ in end view.

may be coiled closed against the adjacent one and ground flat or not. Springs with closed and ground ends (Fig. 12.3(a)), are most uniformly stressed and are generally used for cyclic and other critical applications.

The stress is primarily torsional shear. The forces distributed over the end of the coil are assumed to have a resultant P located at and directed along the coil's axis (Fig. 12.4(a)). This is resolved at the wire into a shearing force P applied at the centroid of the wire, at radius $D/2$, and a twisting moment around the axis of the wire of value $M_t = P(D/2) \cos \lambda$ where λ is the lead angle. With most coils, $\cos \lambda \approx 1.0$ and is omitted. With round wire, D is the mean diameter of the coil, and it is called the *coil diameter*. In terms of outside diameter (OD) it is $D = (\text{OD}) - d$, where d is the wire diameter. If the effect of direct shear and curvature on stress is neglected, the action is equivalent to the twisting of the straight bar of Fig. 12.4(b), which is unwound from the helix of Fig. 12.4(a). Equations (12.5) are applied to give the torsional shear stress at the surface of the wire

$$\tau_0 = \frac{16P(D/2)}{\pi d^3} = \frac{8PD}{\pi d^3} = \frac{8PC}{\pi d^2} \tag{12.7}$$

where, in the last term, $C = D/d$ is the *spring index*. Equation (12.7) applies to both compression and extension springs and whether D and C are constant or variable. For springs of rectangular and square wire see Section 12.10 and Example 12.4.

Equation (12.7) is sometimes used as written, but for most applications a correction factor should be applied. Most obviously, we have neglected the direct shear stress. Also, we have assumed the torsional shear stress to vary linearly across the section. However, in a curved bar (Fig. 12.5), and between any two radial sections bc and nm, the length of the inner fiber cm is less than that of the outer fiber bn. With the shear stress γ measured as in Fig. 12.1(a), then if the strain were linear, as shown on the end (circle) view of Fig. 12.5(a), the strain cc'/cm for the inner fiber is larger than the strain bb'/bn for the outer fiber. Since the shear stresses are proportional to these strains, the neutral axis of stress must shift inward from the centroid to maintain an equilibrium of internal forces. The stresses will be distributed somewhat as shown for τ_a in Fig. 12.5(b). This is similar to the case of *bending* in a curved beam (Fig. 10.12(b)).

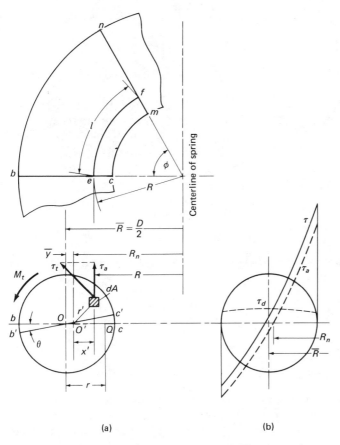

FIGURE 12.5. Effect of spring curvature. (a) Shift of neutral axis from O to O', with shear stress τ_t and axial component τ_a shown acting on an element dA. (b) Stresses along the diameter bc of (a) plotted as torsional stress τ_a (bold dashed line), direct shear stress τ_d (dashed line), and net stress τ (full line).

On the element dA at the end of fiber ef of arc length $l = R\phi$ (Fig. 12.5(a)), the torsional shear stress, by the third fraction of Eq. (12.4), is $\tau_t = Gr'\theta/l = Gr'\theta/R\phi$. This stress on dA is tangential in the radial plane, and its component parallel to the axis of the coil is $\tau_a = \tau_t(x'/r') = (Gr'\theta/R\phi)(R_n - R)/r'$ which reduces to $\tau_a = (G\theta/\phi)$ $(R_n - R)/R$, where R and R_n are radii of the coil measured to dA and to the neutral axis, respectively. In the absence of external force, the integral of $\tau_a\,dA$ over the section must be zero, i.e., $\int_A \tau_a\,dA = 0$.

These last two equations are similar to those derived for a thick curved bar in bending, Eq. (10.30) and Eq. (a) which follows it. Hence, the distance $\bar{y}$ of the neutral axis from the centroidal axis for torsion is identical to that for bending, the solution for which is outlined in Section 10.4 following Eqs. (10.31) and (10.33). Taking for $\bar{y}$ of a solid circular section a series approximation $d^2/8D$, A. M. Wahl obtained an equation for the torsional stress τ_a. To this he added the direct shear stress τ_d shown in Fig. 12.5(b), to obtain the net shear stress τ along the diameter normal to the axis of the spring.[4]

For the maximum net stress, occurring at the inner surface, Wahl presented his result

---

[4] See Prob. 12.14 for a suggestion for further derivation. For τ_d, Wahl used the slightly nonuniform stress given by the theory of elasticity, of value $1.23P/A$ at the surface.

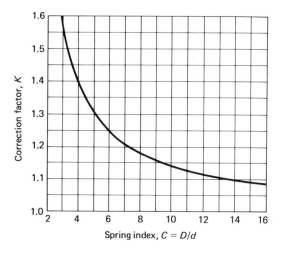

FIGURE 12.6. Wahl correction factor for curvature and direct shear, Eq. (12.9), for use in Eq. (12.8).

in the form of a stress *correction factor K* which is multiplied into Eq. (12.7).[5] Thus the *corrected stress* is

$$\tau = K\tau_0 = K\frac{8PD}{\pi d^3} = K\frac{8PC}{\pi d^2} \tag{12.8}$$

where

$$K = \frac{4C - 1}{4C - 4} + \frac{0.615}{C} \tag{12.9}$$

The first fraction in Eq. (12.9) accounts for the curvature and the second fraction accounts for the direct shear stress. Factor K is commonly called the *Wahl factor*. It is a function of spring index only, and its plot in Fig. 12.6 may be used for most calculations. A more exact theory shows it to be accurate within 2% for $C \geqslant 3$.

A correction factor that includes only the direct shear stress $P/A = 4P/\pi d^2$, taken as uniform over the section, is sometimes recommended for use in static applications for compression springs that have been well preset, giving surface residual stresses that are opposite in direction to the service stresses. Thus $\tau = 8PC/\pi d^2 + 4P/\pi d^2 = (1 + 0.5/C)(8PC/\pi d^2) = K_s(8PC/\pi d^2)$, where

$$K_s = 1 + \frac{0.5}{C} \tag{12.10}$$

It is stated in some references that the curvature-correction part of Eq. (12.9) is a localized stress, in the nature of a stress concentration. This is not apparent in Fig. 12.5(b), which is plotted for an index $C = 5.0$. There τ_a is greater than the nominal stress Eq. (12.7), for 12.5% of the wire radius.

12.4 DEFLECTION AND SPRING LENGTHS

DEFLECTION

The twisting of the wire causes the distance between coils and the length of the spring to change. This is indicated in the top view of the equivalent straight bar (Fig. 12.4(b)), as a rotation of the cross wire of length $D/2$ through an angle θ. With the same twist in a round coiled wire (Fig. 12.4(a)), the deflection or dip δ of the cross wire at the axis

---

[5] A. M. Wahl, "Stresses in Heavy Closely Coiled Helical Springs," *Trans. ASME*, Vol. 51, APM 51–17, 1929; also A. M. Wahl, *Mechanical Springs*, 2nd ed., New York: McGraw-Hill, pp. 231–235, 1963.

is $(D/2)\theta$. Let N be the number of active coils, those contributing to the deflection. Then, for a cylindrical spring, and by Eq. (12.4),

$$\delta = \frac{D}{2}\theta = \frac{D}{2}\left(\frac{M_t l}{JG}\right) = \frac{D}{2}\frac{(PD/2)(\pi DN)}{(\pi d^4/32)G} = \frac{8PD^3N}{Gd^4} = \frac{8PC^3N}{Gd} \tag{12.11a}$$

From Eq. (12.8), deflection may be written in terms of stress, thus

$$\delta = 8\left(\frac{\tau\pi d^3}{8KD}\right)\frac{D^3N}{Gd^4} = \frac{\pi\tau D^2N}{GKd} = \frac{\pi\tau C^2Nd}{GK} \tag{12.11b}$$

Also of significance is the force required for unit deflection, called the *spring rate* k, a measure of stiffness. By a rearrangement of Eqs. (12.11a), the rate for a cylindrical spring is

$$k = \frac{P}{\delta} = \frac{Gd^4}{8D^3N} = \frac{Gd}{8C^3N} \tag{12.12}$$

This last equation does not hold for extension springs, which require an initial force to open. Springs of square and rectangular wire are briefly treated in Section 12.10 and Example 12.4.

Deflection is not appreciably affected by the redistribution of stress, and Eqs. (12.11a) and (12.12) do not need a correction factor. The value of G used for spring steels is 11.5×10^6 psi (79.3×10^3 MPa). The value of N to use in the equations depends on the shape of the end coils and the total number N_t, as can be seen from Fig. 12.3(a). For a compression spring, the following relationships are used.

open, plain ends	$N = N_t$	(12.13a)
open, ground	$N = N_t - 1$	(12.13b)
closed ends	$N = N_t - N_j$	(12.13c)

where N_j is the number of inactive coils.

The value of N_j is not constant, since with increasing load and deflection each end coil becomes less active as it obtains increased support from the adjacent one. A. M. Wahl suggested an average value of 1.75. However, $N_j = 2.0$ is often used for design, sometimes a higher value, to give a larger N_t and free length L_f. The spring manufacturer requires the specification of a maximum permissible free length. For the problems of this book, unless otherwise stated, it is suggested that $N_j = 2.0$ be used to obtain N_t from N, followed by rounding N_t to an even ⅛ turn, which is a practical specification. This should be downward, unless the calculated value is only a bit less than an even ⅛.

SPRING LENGTHS

Figure 12.7 shows the spring of Fig. 12.3(a) in the solid condition. Its deflection curve is shown below it. There are four lengths of interest—the free length L_f when no force is acting and the active coils are spaced at a pitch distance p, the initial or installed length L_1, the maximum useful or operating length L_2, and the solid length or height L_s, when all coils are closed up. The corresponding deflections and forces $(0, 0)$, (δ_1, P_1), etc. are shown plotted below the spring. It is convenient to take the spring rate k, first defined by Eq. (12.12), as the ratio of any load increment to the corresponding deflection. Thus with reference to Fig. 12.7, one example is

$$k = \frac{\Delta P}{\Delta\delta} = \frac{P_2 - P_1}{\delta_2 - \delta_1} \tag{12.14}$$

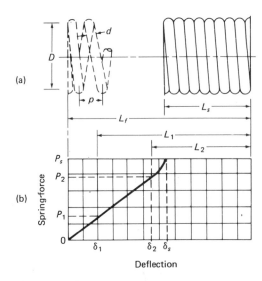

FIGURE 12.7. (a) Compression spring of Figure 12.3(a) shown closed solid. (b) Corresponding force vs deflection curve, with values of significant steps in reaching this condition.

There are several relationships between the various lengths. Since perfectly uniform spacing of the coils may not be attainable, pairs of coils begin to close up before the spring becomes completely solid. This increases spring rate or stiffness, as shown by the curved portion of the force-deflection plot of Fig. 12.7. Thus the pitch should be adjusted in calculations to give a solid deflection larger than the maximum operating deflection. Making $\delta_s = 1.15\delta_2$ is satisfactory, i.e., a *clash allowance* of 15%. This should prevent noisy and harmful action.

The solid length of a spring with *unground* ends is

$$L_s = (N_t + 1)d \tag{12.15}$$

With *closed* and *ground* ends, as may be seen in Fig. 12.7, the solid and free lengths are

$$L_s = N_t d = (N + N_j)d \tag{12.16}$$

$$L_f = L_s + \delta_s = L_s + 1.15\delta_2 \tag{12.17a}$$

also,

$$L_f = Np + N_j d \tag{12.17b}$$

where p is the pitch or center-to-center spacing of the free coils. From the three equations the pitch for a spring with closed and ground ends is found to be

$$p = d + \frac{\delta_s}{N} \tag{12.18}$$

BUCKLING

Relatively long compression springs should be checked for buckling. Buckling is a possibility for δ/L_f and L_f/D combinations to the right of the applicable line in Fig. 12.8. An end is *hinged* if it is pivoted or ball supported as in Fig. 12.9, although friction in the latter decreases the tendency to pivot. The top ends are *guided* in the several isolator springs supporting a machine base. An end is approximately *fixed* if it is firmly seated with closed and ground ends. A spring with factors in the unstable region should be

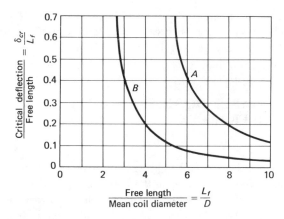

FIGURE 12.8. In a compression spring, ratio of critical deflection δ_{cr} for buckling to free length L_f. Critical load $P_{cr} = (\delta_{cr}/L_f)L_f k = k\delta_{cr}$. Buckling occurs to right and above curve A for both ends fixed, and to the right and above curve B for one end fixed and the other guided in parallel, sidewise motion (equivalent to both ends hinged). (Plot based on an equation due to J. A. Haringx; e.g., see Eq. 24−17 of A. M. Wahl, *Mechanical Springs,* 2nd ed., New York: McGraw-Hill, p. 282, 1963.)

redesigned, or supported internally or externally against sidewise buckling if friction along the coils is not a consideration.

SURGE

A longitudinal vibration that should be avoided is observed as a *surge* or pulse of compression passing through the coils to the ends, where it is reflected and returned. An initial surge is sustained if the natural frequency of the spring is close to the frequency of the repeated loading. The equation[6] for the lowest natural frequency in cycles per second is

$$f_n = \frac{14\ 150\ d}{ND^2}\ \text{Hz} \tag{12.19}$$

Vibration may also occur at whole multiples, such as 2, 3, and 4 times the lowest frequency. The spring should be redesigned to avoid these frequencies.

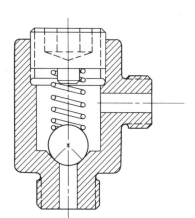

FIGURE 12.9. Pressure-relief valve with spring-loaded ball against a conical seat.

---

[6] See Prob. 13.33.

The stresses due to impact are treated in Section 13.9 and Example 13.4. Elevated temperatures may cause relaxation or loss of force, above 225°F (107°C) in hard-drawn wire, 325°F (163°C) in oil-tempered wire, and 450°F (232°C) in Cr–Va wire.[7] Stainless steels and inconel may be needed at higher temperatures.

12.5 EXTENSION SPRINGS, VARIABLE-DIAMETER SPRINGS, NESTED SPRINGS, AND TOLERANCES

EXTENSION SPRINGS

Extension springs are wound with each coil tight against its preceding coil (Fig. 12.3 (b)). An initial force P_1 is required to just open the coils, and the maximum force in operation is $P_2 = P_1 + k\delta$, where δ is the measurable deflection. The stress concentration at the bend where an end loop or hook joins the coil can be estimated to be r_c/r_i, where r_c is the radius of the wire's centerline along the bend and r_i is the radius of the inside fiber along the bend. It is recommended that this ratio not exceed 1.25. For fluctuating loading this stress-concentration factor should be substituted for the Wahl factor if it exceeds the latter. A bending-stress calculation is desirable for the loops, using the same stress-concentration factor. By bending, the loops add to the deflection, and two full loops are the equivalent of about one active coil. Thus if N is the desired number of active coils, the number between end loops should be $(N - 1)$.

The extension spring may be subjected to sidewise whip and wear of the loop at its place of attachment. A mechanical means of limiting the extension to a safe value is desirable. The coils cannot be strengthened by presetting, and shot peening is difficult. Maximum stress values may be 70% of those used for compression springs of the same material. For fluctuating loads the shorter and better supported compression spring is preferred. By use of a central pull rod or other mechanism, compression springs may be arranged to take tensile loads.

VARIABLE COIL DIAMETER

The centroidal radius $\overline{R}$ may be written as a function of location angle ϕ (Fig. 12.5(a)). Then, if the variation in diameter is small, the elemental length is $ds = \overline{R}\,d\phi$. The contribution of this element to axial deflection is $d\delta = \overline{R}\,d\theta$, and $d\theta$ is given by Eq. (12.6), with $dz = ds$. The twisting moment M_t is $P\overline{R}$, and the limits of integration are $\phi_0 = 0$ and $\phi_1 = 2\pi N$. Then

$$d\delta = \overline{R}\,d\theta = \overline{R}\,\frac{(P\overline{R})}{JG}(\overline{R}\,d\phi) = \frac{P}{JG}\overline{R}^3 d\phi$$

whence

$$\delta = \frac{P}{JG}\int_0^{2\pi N}\overline{R}^3\,d\phi \qquad (12.20a)$$

For a conical spring of constant pitch, small cone angle, and radii $\overline{R}_0$ and $\overline{R}_1$ at the two ends of the active part of the coil this yields

$$\delta = \frac{\pi PN}{2JG}(\overline{R}_0{}^2 + \overline{R}_1{}^2)(\overline{R}_0 + \overline{R}_1) \qquad (12.20b)$$

This equation is valid only is there is no decrease in the number of active coils N. In

---

[7] See Section 12.6 for spring materials.

a spring shaped like that of Fig. 12.3(c), the decrease may not be appreciable during a first phase of action. If the change in diameter per turn is sufficient to allow the spring to telescope, one coil within another without bottoming, the equation will be valid for larger deflections.

NESTED SPRINGS

Compression springs may be *nested*, i.e., two or more springs assembled one within another, to give maximum resistance in a given space. There must be a small clearance between the springs, and they should be alternately coiled right and left hand to prevent interlocking. If the springs are of different length, a stepped spring rate may be obtained, sometimes useful against overloads. For a constant rate, the several springs will have the same free lengths. The deflections are the same, and by inspection of Eq. (12.11b), it is readily seen that for any two springs in the nest, if C_2 is made equal to C_1, then $K_2 = K_1$, and if $N_2 d_2$ is made equal to $N_1 d_1$, then $\tau_2 = \tau_1$, which is a common design objective. With closed and ground ends, the solid lengths of the two springs, Eq. (12.16), will then differ by only the amount $N_j (d_1 - d_2)$.

In general, if #1 is the outer spring of the two and the diametral clearance is taken as a fraction η of the inside diameter of this spring, then $\eta (D_1 - d_1) = (D_1 - d_1) - (D_2 + d_2)$, whence

$$\frac{d_2}{d_1} = (1 - \eta) \frac{C_1 - 1}{C_2 + 1} \tag{12.21a}$$

From Eq. (12.11a), for the same deflection, the ratio between any two loads P_2/P_1 is found to be $(C_1/C_2)^3 (N_1/N_2) (d_2/d_1)$. If there are several springs, $P_1 + P_2 + P_3 + \cdots = P$. If there are only two springs, $P_1 + P_2 = P$, and

$$P_1 = \frac{P}{1 + (C_1/C_2)^3 (N_1/N_2) (d_2/d_1)} \tag{12.21b}$$

If τ_2 is made equal to τ_1 by making $C_2 = C_1$, $K_2 = K_1$, and $N_2 d_2 = N_1 d_1$ in Eq. (12.11b), then by Eqs. (12.21a) and (12.21b),

$$P_1 = \frac{(C + 1)^2 P}{(C + 1)^2 + (1 - \eta)^2 (C - 1)^2} \tag{12.21c}$$

If η and a single C and τ are chosen, then load P_1 is determined from Eq. (12.21c), the wire diameter d_1 is determined from Eq. (12.8), and the number of active coils N_1 is determined from Eqs. (12.11a) or (12.11b). Then, d_2 is determined by Eq. (12.21a), and $N_2 = N_1 (d_1/d_2)$. Some adjustment may be needed in order to use available wire sizes. If this results in a τ_2 slightly larger than τ_1, it is acceptable since wire *strength* increases with a decrease in its diameter. One may prefer initially to set $\tau_2 = \alpha \tau_1$ where α is an an approximate ratio, a small amount larger than 1.0, based on the charts of Sections 12.6 through 12.8. Then Eqs. (12.21a) and (12.21b) are still valid, but Eq. (12.21c) is no longer valid.

TOLERANCES

The tolerances that should be specified to control manufacturing variations in spring dimensions and properties should be broad if costs are to be reasonable. In explanation it may be noted that spring rate k varies with the fourth power of wire diameter, and inversely with the third power of coil diameter. This magnifies the individual variations, and a commercial tolerance for spring rate k is ± 8% with an index of 5 and ±6% with an index of 10.[8] Tolerances for dimensions are given in references.[9] Statistical studies of strength and dimensional variations and probabilities of failure have been made for large lots of identical springs.[10]

To accommodate variations in spring length and deflection, threaded adjustment may be provided in the housing, as in the relief valve of Fig. 12.9. The specification of desired spring characteristics and only those dimensions that are limited by available space will allow the manufacturer to make compensating changes in other dimensions or number of coils to best meet the characteristics desired.

12.6 SPRING MATERIALS AND TREATMENT

Anyone designing a spring for the first time may be surprised at the high values of torsional shear stress. There are several reasons why this is permissible. In most installations the deflection of the spring is limited by the motion of the mechanism or by stops, or, in compression springs, by closing of the coils. Therefore, an overload capability in the plastic range is not needed, and materials may be chosen that have nominal yield strengths of 85% or more of the ultimate strengths. Within its elastic range a spring is inherently shock resistant. Wire-to-wire strengths are highly uniform because the wire is symmetrical and small in section. Manufacture is a continuous operation that improves wire strength, increasingly so with a decrease in diameter (Fig. 12.10). In steels, high-carbon contents of 0.60% to 1.00% are chosen, giving maximum ultimate tensile strengths of 200 000 to 350 000 psi. Also, the stress can be accurately calculated, and surface treatments may be applied to increase fatigue resistance. Because of the foregoing favorable conditions, the allowable stresses can be high and the factor of safety, low.

The wire is strengthened and brought to size by cold drawing it through dies. Wires thus treated are *music wire* of 0.70–1.00% C and of highest quality and strength, available in the smaller sizes; also *hard-drawn* (HD) *wire* of 0.55–0.85% C, a cheaper, general-purpose spring steel. When wire of this carbon content is first cold drawn, then heated into austenitic range, quenched in oil, and tempered, often in molten lead, it is known as *oil-tempered* (OT) *wire*. In the alloy steels, hard-drawn or oil-tempered chrome-vanadium (AISI 6150) and chrome-silicon (AISI 9254) give higher strengths and resistance to relaxation at higher temperatures. These and the carbon-steel wires (except music) can be obtained in commercial quality, or in a better, valve-spring quality (VSQ) suitable for severe fatigue conditions. A chart of relative costs and a pricing analysis are available.[11]

Wires up to ⅜ in or more are wound cold, and the spring is greatly improved by a low-temperature heating to remove coiling strains. Some torsional elastic limits of springs are compared in Fig. 12.10. In larger sizes and particularly with severe coiling ($D/d <$ 5.0) the wire is drawn for size, annealed to soften, wound hot, and then quench treated. Because of the larger sections the alloy steels may be needed to make the heat treatment sufficiently responsive.

Other materials in a hardened condition are widely used for springs because of their noncorrosive and heat-resistant properties. All are more expensive than the foregoing steels and in order of increasing cost they are stainless steels, phosphor bronze, beryllium

[8] *Design and Application of Helical and Spiral Springs,* Handbook Supplement HS J795a, 4th ed., Society of Automotive Engineers, Warrendale, Pa., Fig. 23, 1973.

[9] *Design and Application of Helical and Spiral Springs,* Handbook Supplement HS J795a, 4th ed., Society of Automotive Engineers, Warrendale, Pa., pp. 37–45, 1973; or manufacturer's manuals such as *Design Handbook,* Associated Spring Corporation, Bristol, Conn.

[10] *Metals Handbook,* 9th ed., Vol. 1, American Society for Metals, Metals Park, Ohio, Fig. 7 (p. 296), Fig. 14 (p. 302), and Fig. 15 (p. 304), 1978.

[11] *Metals Handbook,* 9th ed., Vol. 1, American Society for Metals, Metals Park, Ohio, p. 305, 1978.

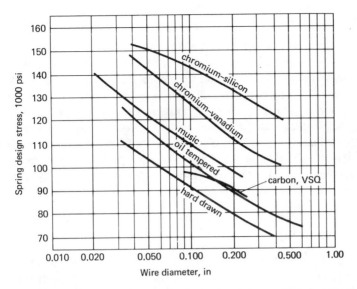

FIGURE 12.10. Torsional elastic limits (sometimes called design stresses—see footnote 13 of this chapter) vs wire diameter for nonset compression springs. (From H. J. Elmendorf, "Selection of Steels for Springs," *Metal Progress*, Vol. 73, Figure 6, p. 84, April 1958, with the addition of carbon VSQ, as in the *Metals Handbook*, 8th ed., Vol. 1, 1961, Figure 4, p. 163. Reproduced by permission, American Society for Metals, Metals Park, Ohio.)

copper, and inconel. Monel is used in saltwater applications, and the stainless steels in food and sanitary equipment. Copper alloys may be used for their electrical conductivity. For temperatures in the 500°F to 700°F range (260°C to 370°C), some of the stainless steels are suitable. For a temperature of 1100°F (595°C), inconel X750 is specified. For the plain carbon and alloy steels, corrosion protection by zinc or cadmium is suitable for the harder springs used statically, but electroplated coatings should be used only on springs with hardness less than R_c48. For the fatigue conditions of fluctuating loads, coatings are generally weakening (Sections 5.10 and 7.8).

There are two prestressing operations that add to the strength and durability of steel springs, and they are particularly valuable for fluctuating-load conditions. One prestressing operation is *shot peening* after coiling, which induces a layer of compressive residual stresses. Shot peening is described in Section 7.13. In a coiled spring the highest stressed portion of the wire, the inside of the coil, must be effectively reached. The residual stress reduces the net tensile stress, which, at an angle of 45° to the torsional stress, initiates a fatigue failure. In compression springs the second prestressing operation is *presetting*. The spring is forced to yield to the desired free length after it has been purposely wound with too large a pitch. This leaves the wire with torsional residual stresses of the opposite sign to those that will be imposed in service (Section 7.12 and Prob. 7.42).

12.7 STATIC STRENGTHS AND SPRING DESIGN

Periodicals, manuals of certain professional societies, and data sheets of spring manufacturers abound in tables, charts, nomographs for wire properties, and design shortcuts.[12] Exact treatments and terms may not be specified nor the test points shown, and to find a correlation between data is not easy. In the following a few methods for using data are discussed, and sufficient data are given for practice in obtaining spring dimensions.

---

[12] M. Massoud and L. Hubert, "Brief Survey of Spring Design Nomographs," Paper 76-DET-77, ASME, 1976. Significant nomographs are listed and evaluated. The authors conclude that many need updating.

TORSIONAL ELASTIC LIMITS

The torsional elastic limits are given in the chart of Fig. 12.10 as a function of material and wire size.[13] The chart is for nonset springs at room temperature and static conditions. For preset springs, values 25% higher are commonly used. When values from the chart are multiplied by 1.25, they compare well with those on a chart for preset springs given in the 1958 edition of an SAE manual on springs. These values are about 20% higher than the "allowable uncorrected torsional stress" for springs with moderate presetting, obtained by a method of the 1973 SAE manual. Therefore, it seems reasonable to use a factor of safety of 1.20 with the ASM values of Fig. 12.10 for nonset springs and with its values multiplied by 1.25 for preset springs. Higher factors of safety may be needed if relaxation must be completely avoided.

DETERMINATION OF WIRE SIZE

The design may be started by determining wire diameter with Eq. (12.8). For this calculation the load P is presumably known. If there is a space restriction on either the inside or outside coil diameter, the mean coil diameter D can be estimated. Without space restrictions, one may choose instead an average spring index C, say 7 or 8, one which avoids both a high value of K and excessive coil diameter. Then, a Wahl-factor K can be read from Fig. 12.6. The allowable stress τ depends on the wire diameter, as well as on the material and the factor of safety. If one is familiar with diameters used in similar designs, a good guess of diameter may be made in order to select a strength value from a chart. A poor guess may be sufficient. Allowable stress and Wahl-factor K do not change as rapidly with d as do the d^2 or d^3 terms of Eq. (12.8). Hence, with τ, K, and D or C substituted a calculation will give a first, approximate value for wire diameter d, followed by a second trial or check as necessary.

Example 12.1

A relief valve is attached to a hydraulic line. The valve proper closes against a conical seat and presents a circular area of $7/16$-in diameter to the fluid. Construction resembles that of the ball valve of Fig. 12.9, except that the valve is a conical disk with a guided stem. The valve is to open when the pressure reaches 225 psi, after which the fluid will act against a larger area of the valve, estimated to be equivalent to 0.675 inches in diameter. The valve shoulder for support of the spring can have a maximum diameter of 1 in. The desired lift is ¼ in. Determine the basic specifications for the spring.

Solution

Installed force $P_1 = pA = 225\pi(7/16)^2/4 = 33.8$ lb

Maximum useful force $P_2 = pA = 225\pi(0.675)^2/4 = 80.5$ lb

Solid force with $\delta_s = 1.15\delta_2$ $P_s = 1.15P_2 = 92.6$ lb

---

[13] The chart is due to H. J. Elmendorf, "Selection of Steels for Springs," *Metal Progress*, Vol. 73, pp. 80–84, April 1958. The curve for carbon valve-spring wire is added, as in the *Metals Handbook*, 8th ed., Vol. 1, American Society for Metals, Metals Park, Ohio, 1961, p. 168. From static tests on springs and spring wires, Elmendorf found definite ratios between spring torsional elastic limits and wire tensile strengths. Ratios were maximum values, based on optimum stress-relieving temperatures after coiling. The chart was constructed by multiplying the ratios by corresponding minimum tensile strengths specified in an AISI manual. The elastic limits are also designated "spring-design stresses." Since Elmendorf recommends their use with a factor of safety, it is assumed that a shear-stress *basis* for design is intended.

The shoulder limits the outside diameter of the coil to about 1 in. We must guess at a wire diameter in order to approximate the correction factor and the material strength. We recall seeing some small automobile-engine valves of about this outside diameter, with a wire diameter that was perhaps ⅛ in. We try this, making $D = OD - d = 1 - ⅛ = $ ⅞ in and $C = (⅞)/(⅛) = 7.0$. From Fig. 12.6, $K = 1.21$. Assuming the system to be well regulated and the relief valve seldom open, we have essentially static loading, and we choose readily available oil-quenched carbon-steel wire. We use the solid-length force of 92.6 lb, since the spring may be squeezed solid during installation, also, since no mechanical limitation to deflection is indicated. From Fig. 12.10 the elastic limit is about 95 000 psi, and with a factor of safety of 1.20, the allowable stress is 79 200 psi. In Eq. (12.8),

$$\tau = K\frac{8PD}{\pi d^3}, \qquad 79\ 200 = 1.21\frac{8(92.6)(7/8)}{\pi d^3}$$

whence

$$d = 0.1466 \text{ in}$$

The next standard wire size is W & M #9 (0.1483). If it is used, $D = 1.000 - 0.148 = 0.852$ in, $C = 0.852/0.148 = 5.75$, and $K = 1.265$. Solving Eq. (12.8) for τ, we obtain

$$\tau = 1.265\frac{8(92.6)(0.852)}{\pi(0.1483)^3} = 77\ 900 \text{ psi}$$

By Fig. 12.10 the elastic limit for this larger wire is about 93 000 psi so the factor of safety will be 93 000/77 900 = 1.19, close enough to the specified value, considering the several uncertainties. The next larger wire, 0.1563, gives an unnecessarily high factor, 1.62.

The spring rate, Eq. (12.15), must be

$$k = \frac{P_2 - P_1}{\delta_2 - \delta_1} = \frac{80.5 - 33.8}{1/4} = 187 \text{ lb/in}$$

and by rearranging Eq. (12.12), the active number of coils is

$$N = \frac{Gd}{8C^3k} = \frac{(11.5 \times 10^6)(0.1483)}{8(5.75)^3(187)} = 6.00$$

Hence, from Eqs. (12.13c), $N_t = N + N_j$, and $N_t = 8$ turns if two inactive coils are assumed.

The deflection corresponding to the maximum operating load is $\delta_2 = P_2/k = 80.5/187 = 0.43$ in. From Eqs. (12.16) through (12.18) the solid and free lengths and the pitch are, respectively,

$$L_s = N_t d = (8.0)(0.1483) = 1.19$$

$$L_f = L_s + 1.15\delta_2 = 1.19 + 1.15(0.43) = 1.68 \text{ in}$$

$$p = d + \frac{\delta_s}{N} = 0.1483 + \frac{1.15(0.43)}{6} = 0.231 \text{ in}$$

Since the ratio $L_f/D = 1.68/0.852 = 1.97$, there is no possibility of buckling (Fig. 12.8).   ////

12.8 FATIGUE STRENGTHS AND SPRING DESIGN

FATIGUE CHARTS

Spring failures under fluctuating loads are typical of failures in torsional shear. A crack starts at the surface on the inside of the coil and takes off at 45° to the radial plane of shear, in a direction perpendicular to the principal tensile stress. The loading during a fatigue test on a spring is unidirectional, usually from zero to a maximum value. The maximum torsional stress value for failure at a finite life or for endurance for infinite life is plotted along the ordinate, as in Fig. 12.11. From this point a straight line is drawn toward the ultimate torsional strength, stopping at its intersection with the horizontal line of stress at which permanent set begins. Thus it resembles the top, modified Goodman line of Fig. 6.3(b), obtained by complete reversal of stress. However, the 45° lines of Fig. 12.11 represent the minimum stress, not the mean stress of Fig. 6.3(b).

Figure 12.11(a) is plotted for four sizes of wire from S-N diagrams of fatigue tests run for 10^7 cycles, and its stress values presumably may be used with an appropriate factor of safety for infinite-life design. Figure 12.11(b), for two other wire sizes, compares failure stresses at finite lives of 10^4 and 10^5 cycles with values at 10^7 cycles. For other wire sizes at 10^7 cycles, interpolation between the given sizes of the two charts may be made. If the ratio ρ of minimum-to-maximum spring load is known, a straight line drawn from the origin at a slope of $1/\rho$ will intersect the maximum-stress lines at values that may be used, together with a factor of safety, for the different wire sizes. Figure 12.11 is for oil-tempered carbon steel only, but similar diagrams for other common wire materials are available from the same source.[14]

In another approach charts of maximum stress factor K_{s2} are plotted as a function of an initial (minimum) stress factor K_{s1}.[15] The factors are the ratios of the respective stresses to the minimum tensile strength of the wire. The resulting charts have the same general shape as those of Fig. 12.11. For the K_{s2} line of 10^7 cycles (infinite life) and for

FIGURE 12.11. Goodman-type diagrams of the range of torsional shear stress for helical springs made of oil-tempered wire (ASTM A229) of different diameters. (a) 10^7 cycles. (b) 10^4, 10^5, and 10^7 cycles. (Reproduced with permission, from *Metals Handbook*, 9th ed., Vol. 1, 1978, Figures 6 (c) and (h), p. 295, American Society for Metals, Metals Park, Ohio.)

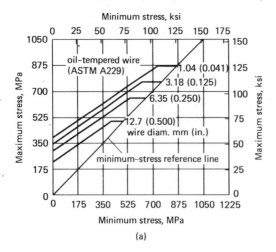

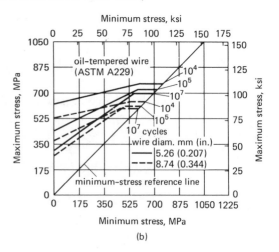

(a)   (b)

[14] *Metals Handbook*, 9th ed., Vol. 1, American Society for Metals, Metals Park, Ohio, 1978, Fig. 6 of the section "Steel Springs," pp. 283–308.

[15] *Design and Application of Helical and Spiral Springs*, Handbook Supplement HS J795a, 4th ed., 1973, Society of Automotive Engineers, Warrendale, Pa., pp. 28–29.

springs shot peened and preset, the following equation may be written,

$$K_{s2} = 0.36 + 0.49K_{s1} = 0.36/(1 - 0.49\rho) \qquad (12.22)$$

where ρ is the ratio $K_{s1}/K_{s2} = P_1/P_2$, and the upper limit for K_{s2} is 0.56 to avoid relaxation. Lines are available on the charts for 10^4, 10^5, and 10^7 cycles and for combinations of springs that are preset or not and shot peened or not.[15] All the lines are said to be applicable to music wire and valve-spring quality wire. A chart is also available for hot coiled springs that are shot peened, and preset.[16]

ULTIMATE TENSILE STRENGTHS

Tensile strength values S_u are needed for use with the foregoing factors. They are usually presented in tables or in a chart of strength vs the logarithm of wire diameter d. When plotted as log-strength vs log-diameter, fairly straight lines result. For each line there may be written an equation

$$S_u = \frac{B}{d^m} \qquad (12.23)$$

where m is the slope of the line and $B = S_{u0}d_0^m$ where (d_0, S_{u0}) is any convenient point on the line.[17] From a log-log replot of the minimum tensile strengths of the steels[18] for which Eq. (12.22) applies, the following equations are obtained for tensile strength. They are listed here together with the range of wire sizes, namely,

music wire $(0.004 \leqslant d \leqslant 0.250$ in$)$

$$S_u = (192 \times 10^3)/d^{0.151} \text{ psi} \qquad (12.24a)$$

HD, VSQ $(0.092 \leqslant d \leqslant 0.250$ in$)$ and
OT, VSQ $(0.062 \leqslant d \leqslant 0.250$ in$)$

$$S_u = (191 \times 10^3)/d^{0.0912} \text{ psi} \qquad (12.24b)$$

Cr-Va, OT and VSQ $(0.020 \leqslant d \leqslant 0.500$ in$)$

$$S_u = (169 \times 10^3)/d^{0.159} \text{ psi} \qquad (12.24c)$$

DETERMINATION OF WIRE SIZE

When maximum force P_2 and minimum force P_1 are known, the ratio $\rho = P_1/P_2$ is calculated, then K_{s2} is obtained by Eq. (12.22) for shot peened and preset springs. For other treatments K_{s2} may be read from the charts in the reference.[19] The appropriate equation for S_u is selected from Eqs. (12.24). Then the maximum stress for infinite life is

$$\tau_e = K_{s2}S_u \qquad (12.25)$$

and the allowable stress, for substitution into Eq. (12.8) is $\tau = \tau_e/n$, where n is the factor of safety. If spring index C is assumed or estimated to obtain factor K, then Eq. (12.8) may be solved directly for wire diameter d.

---

[16] See footnote 15, p. 55 of the reference.

[17] An equation of this type was presented by R. C. Johnson, *Optimum Design of Mechanical Elements*, New York: John Wiley & Sons, Inc., p. 481, 1961.

[18] *Design and Application of Helical and Spiral Springs*, Handbook Supplement HS J795a, 4th ed., Society of Automotive Engineers, Warrendale, Pa., Figs. 6A, 6B, 6C, and 6D, pp. 15–16, 1973.

[19] See footnote 18.

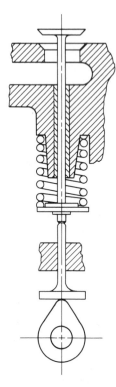

FIGURE 12.12. Spring-held poppet valve of the type used in internal combustion engines.

Example 12.2

A spring force of 90 lb holds a poppet valve closed in a large engine. In its open position (Fig. 12.12), the valve is lifted ⅝ in by a cam acting on the valve stem through an adjustable tappet or rocker arm (not shown). A spring force of 180 lb is permissible at maximum lift. No space restrictions have been specified. (a) Compare the sizes required using the oil-tempered carbon-steel wire of Fig. 12.11 with the Cr-Va, valve-spring-quality wire of Eqs. (12.24), preset and shot peened after coiling. (b) Use the smaller wire and determine the other spring dimensions.

Solution

Diameter calculations will be based upon the lift force $P_2 = 180$ lb since the spring can only be compressed to its solid length at installation, which is a static condition with a higher permissible stress. Choose an average, desirable spring index $C = 7.5$, making $K = 1.20$. The load ratio during operation is $\rho = P_1/P_2 = 90/180 = 0.50$.

(a) For the oil-tempered carbon-steel wire, the maximum endurance-stress value is obtained from Fig. 12.11(a) by drawing a line with a slope $1/\rho = 2.0$ from the origin. Based on our experience with Example 12.1 and in consideration of the fatigue conditions and larger loads of this example, we guess a wire diameter of 0.250 in and read a value of about 64 000 psi at the intersection of the two lines. Because of the cost of a failure in a large engine, a factor of safety of 1.33 is applied, giving $\tau = 48\,000$ psi. From Eq. (12.8),

$$\tau = K\frac{8PC}{\pi d^2}, \quad 48\,000 = (1.20)\frac{8(180)(7.5)}{\pi d^2}$$

whence $d = 0.293$ in. The next larger wire size must be used, W & M #0-1, with $d = 0.3065$ in. Using a reduced allowable stress of 45 000 psi, a second calculation shows this size to be sufficient. The outside diameter will be $d(C + 1) = 0.3065(7.5 + 1) = 2.60$ in.

For the Cr-Va wire the tensile strength is given by Eqs. (12.24c), namely, $S_u = (169 \times 10^3)/d^{0.159}$. From Eq. (12.22), for preset and shot-peened springs, for infinite life, and with $\rho = 0.50$,

$$K_{s2} = \frac{0.36}{1 - 0.49\rho} = \frac{0.36}{1 - 0.49(0.50)} = 0.477$$

By Eq. (12.25), the maximum stress for infinite life is

$$\tau_e = K_{s2} S_u = 0.477 \frac{169 \times 10^3}{d^{0.159}} = \frac{80.6 \times 10^3}{d^{0.159}}$$

Spring-quality wire is more uniform in strength, and a lower factor of safety is justified, say 1.20, than used for oil-tempered wire. For Eq. (12.8),

$$\tau = K \frac{8PC}{\pi d^2}, \quad \frac{80.6 \times 10^3}{(1.20)d^{0.159}} = (1.20) \frac{8(180)(7.5)}{\pi d^2}$$

whence

$$d^{1.841} = 0.0614 \quad \text{and} \quad d = 0.220 \text{ in.}$$

The next larger, standard wire size is W & M #4 (0.2253 in). The mean coil diameter is $D = Cd = (7.5)(0.2253) = 1.69$ in, and the outside diameter is $D + d = 1.69 + 0.23 = 1.92$ in.

Note that because of Eqs. (12.24) and no restrictions on coil diameter, the calculation for wire size was made without any guessing. The spring size and mass is considerably less than with oil-tempered carbon-steel wire, and it is resistant against relaxation up to 450°F vs 225°F for the carbon-steel wire. Hence the alloy-steel wire is far more suitable for this application.

(b) For a lift of ⅝ in, the spring ratio is

$$k = \frac{P_2 - P_1}{\delta_2 - \delta_1} = \frac{180 - 90}{5/8} = 144 \text{ lb/in}$$

The number of active coils is, by Eq. (12.12),

$$N = \frac{Gd^4}{8D^3 k} = \frac{(11.5 \times 10^6)(0.2253)^4}{8(1.69)^3(144)} = 5.33$$

From Eq. (12.13c) by taking the inactive coils to be a bit less than 2, we obtain a readily specified 7¼ total coils. The deflection at maximum lift, measured from the free length, is $\delta_2 = P_2/k = 180/144 = 1.25$ in. Then by Eqs. (12.16) through (12.18), with a 15% clash allowance,

$$L_s = N_t d = (7.25)(0.2253) = 1.63 \text{ in}$$

$$L_f = L_s + \delta_s = 1.63 + 1.15(1.25) = 3.07 \text{ in}$$

$$p = d \frac{\delta_s}{N} = 0.225 + \frac{1.15(1.25)}{5.33} = 0.495 \text{ in}$$

The ratio $L_f/D = 3.07/1.69 = 1.82$, and by Fig. 12.8, no buckling is expected. The fundamental frequency of vibration, Eq. (12.19), is

$$f_n = \frac{14\ 150\ d}{ND^2} = \frac{(14\ 150)(0.2253)}{(5.33)(1.69)^2} = 209 \text{ Hz}$$

This is high for large-engine disturbances, and surging need not be expected.   ////

GENERAL METHOD

Another basis for wire-size selection that has undoubtedly occurred to the reader is to use the Soderberg or modified Goodman lines of Fig. 6.3 and the equations of Section 6.3. Wire tensile strength, torsional ultimate or yield strengths, and reversed-torsion endurance limit must be known or estimated. However, the Soderberg or Goodman lines and equations are only empirical, based mainly on reversed-bending tests of ductile materials. It would seem preferable to use one of the foregoing methods, which are more closely related to the unidirectional tests and torsional data taken on actual springs. When such data are not available, as for new materials or for large wire diameters, then the more general methods of Section 6.3 may be necessary.[20]

12.9 THEORY OF TORSION FOR BARS OF ANY SECTION

Consider a short section of a bar of square cross section twisted by a torque M_t (Fig. 12.13).[21] For the elements $dx\ dy\ dz$ located centrally along the sides of the bar, the free or open face and the opposite face are seen distorted by the shear stresses and forces acting on the other four faces.[22] However, for elements lying along the corner edges of the bar, all six faces rotate without distortion because no shear forces can be developed along the two free faces. The trace of the plane joining these four elements becomes curved, as shown, and the plane itself warps.[23] In the theory that follows, it is assumed that the ends of the bar are free to warp and that the stresses and strain angles are uniform along the entire length of the bar.

FIGURE 12.13. Warping and shear stresses in a bar of square cross section when twisted.

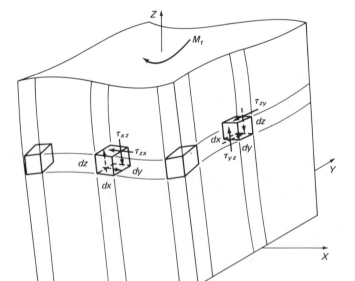

[20] See Probs. 12.40 and 12.41; also A. M. Wahl, *Mechanical Springs*, 2nd ed., New York: McGraw-Hill, 1963.

[21] For convenience Figs. 12.13 and 12.14 have been rotated from the *XYZ* position of Figs. 5.1 and 5.2.

[22] The first subscript on τ identifies the plane on which it acts by naming the direction of the normal to the plane. The second subscript is the direction of τ itself.

[23] The circle is the only fully symmetrical cross section, and all its surface elements distort and develop resisting shear forces in equal amounts. Therefore, there is not warping, and the assumption of Section 12.2, that plane sections remain plane, is justified.

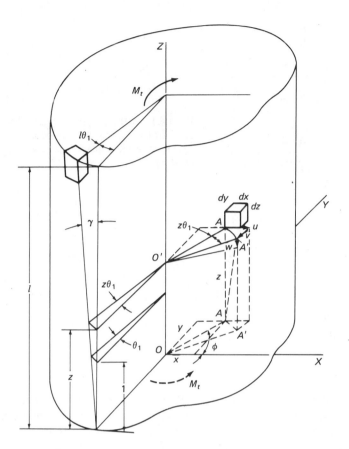

FIGURE 12.14. Displacements in a bar of any constant cross section when twisted.

Let the axis of the bar of any constant section (Fig. 12.14), be the center of twist. Upon application of a moment M_t, a point (x, y, z) of the element at A is displaced the amounts u, v, w in the X, Y, and Z directions, respectively. Seen projected for clarity onto the XY plane below, the line OA lies at an angle ϕ with the X axis, $\sin \phi = y/OA$, and $\cos \phi = x/OA$. Rotation in the XY plane at elevation z is through an angle $z\theta_1$ where θ_1 is the angle of twist per unit length of the bar. For small displacements, $AA' = (z\theta_1)OA$, and since $OA \sin \phi = y$ and $OA \cos \phi = x$, the X and Y components of AA' are, respectively,

$$u = AA' \sin \phi = z\theta_1 y \qquad\qquad (12.26)$$

$$v = - AA' \cos \phi = - z\theta_1 x$$

The displacement w is independent of position z and is some function $w = f(x, y)$ to be determined.[24]

For the element at A there is no change in angle between edges dx and dy of faces $dx\,dy$, so $\gamma_{xy} = \tau_{xy} = 0$. The changes for the $dx\,dz$ faces are shown on the XZ plane of Fig. 12.15. There are displacements u and w, and the shear strain γ_{xz} is the sum of angular changes $\partial w/\partial x$ in edge dx and $\partial u/\partial z$ in edge dz. This, together with a similar summation for the YZ plane gives the following relationships between strains and dis-

---

[24] These assumptions on displacements and the theory that follows are due to Saint-Venant.

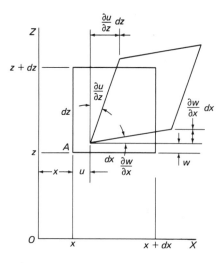

FIGURE 12.15. Projection onto the XZ plane of the element at A in Figure 12.14, showing displacements and distortion.

placements

$$\gamma_{xz} = \frac{\partial u}{\partial z} + \frac{\partial w}{\partial x}$$

$$\gamma_{yz} = \frac{\partial v}{\partial z} + \frac{\partial w}{\partial y}$$

(12.27)

No other strains exist, i.e., $\epsilon_x = \epsilon_y = \epsilon_z = 0$. Stresses are related to strains by Hooke's law, and from Eqs. (12.27), together with substitution for the $\partial/\partial z$ terms from Eqs. (12.26)

$$\tau_{xz} = G\gamma_{xz} = G\left(\theta_1 y + \frac{\partial w}{\partial x}\right)$$

$$\tau_{yz} = G\gamma_{yz} = G\left(-\theta_1 x + \frac{\partial w}{\partial y}\right)$$

(12.28)

The elastic forces and their increments are shown on the element of Fig. 12.16. From a summation in the Z direction the equation of equilibrium becomes

$$\frac{\partial \tau_{xz}}{\partial x} + \frac{\partial \tau_{yz}}{\partial y} = 0$$

(12.29)

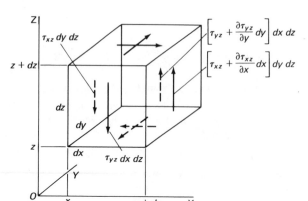

FIGURE 12.16. Shear forces on an element.

Saint-Venant replaced the *two* stresses by *one* stress function $\phi(x, y)$. To satisfy Eq. (12.29), such a function must be continuous and have the following properties

$$\tau_{xz} = -\frac{\partial \phi}{\partial y} \quad \text{and} \quad \tau_{yz} = +\frac{\partial \phi}{\partial x} \tag{12.30}$$

Function ϕ is now introduced into Eqs. (12.28) by substitution from Eqs. (12.30). Following this, an operation by $\partial/\partial y$ on the first and by $\partial/\partial x$ on the second of Eqs. (12.28) yields, respectively,

$$-\frac{\partial^2 \phi}{\partial y^2} = G\left(\theta_1 + \frac{\partial^2 w}{\partial y \partial x} \right)$$
$$\frac{\partial^2 \phi}{\partial x^2} = G\left(-\theta_1 + \frac{\partial^2 w}{\partial x \partial y} \right) \tag{12.31}$$

Since the warped surface described by w will be continuous over the cross section, $\partial^2 w/\partial y \partial x = \partial^2 w/\partial x \partial y$, and subtraction of the first equation from the second eliminates w and gives the differential equation

$$\frac{\partial^2 \phi}{\partial x^2} + \frac{\partial^2 \phi}{\partial y^2} = -2G\theta_1 \tag{12.32}$$

Since this equation is derived from the strain and displacement relationships that ensure continuity, Eqs. (12.26) through (12.28), it is the *equation of compatibility* in terms of the stress function.

CHARACTERISTICS OF THE ϕ-FUNCTION

Since $\tau_{zx} = \tau_{xz}$, Eqs. (12.30) indicate that the shear stress τ_{zx} in the X direction on the plane facing Z is the (negative) slope in the Y direction ($\partial \phi/\partial y$) of the function ϕ plotted above an XY base. Likewise, since $\tau_{zy} = \tau_{yz}$, the stress τ_{zy} in the Y direction on the same plane is the slope of the function in the X direction ($\partial \phi/\partial x$). At the Z axis or axis of twist there is no shear, and the ϕ-function must have zero slope there. We expect to find the maximum stress somewhere along the boundary of the section, so the maximum slope should be there. Hence we may think of the ϕ-function as the surface of a hill, with its apex at the Z axis, as in Fig. 12.17. Furthermore, it may be proven from Eqs. (12.30) that the shear-stress component in *any* direction on the plane facing Z equals the slope of the ϕ-surface in the *perpendicular* direction. Since a shear stress cannot exist on the surface of the bar, there can be no shear in a direction N normal to its surface.

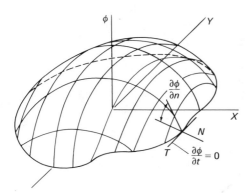

FIGURE 12.17. The ϕ-function "hill."

Along the boundary of the ϕ-hill in a tangent direction T, perpendicular to N, we have the slope $\partial\phi/\partial t$ equal to zero. Thus

$$\frac{\partial\phi}{\partial t} = \tau_{zn} = \tau_{nz} = 0 \tag{12.33}$$

where τ_{zn} would indicate a stress on the plane facing Z and in the direction normal to the boundary if the stress existed.

Equation (12.33) is a boundary condition that must be satisfied by the function ϕ, and it further defines the ϕ-hill. There can be no change in the value of ϕ as we move tangentially along the boundary, since the value of ϕ, the first integral of $(\partial\phi/\partial t)$, must be zero or a constant. Hence the base of the hill is flat. Also, with no tangential component, the slope at a point on the boundary must be maximum in a plane normal to the boundary, e.g., $\partial\phi/\partial n$ of Fig. 12.17. This means that the maximum stress is τ_{zt}, tangent to the boundary, equal to τ_{tz}, and of magnitude $\partial\phi/\partial n$.

To obtain a solution of the torsion problem for a bar of a given cross section, one must first devise an equation for ϕ that satisfies the boundary condition, Eq. (12.33), and the compatibility equation, Eq. (12.32). For the singly connected boundaries of solid bars, Eq. (12.33) is satisfied if the function has zero value along the boundary. For example, the stress function for an elliptical section with semiaxes a and b is

$$\phi = C\left(1 - \frac{x^2}{a^2} - \frac{y^2}{b^2}\right) \tag{12.34}$$

where C is a constant. The term in parentheses is the equation of an ellipse set equal to zero, so ϕ is 0 at all points (x, y) on the boundary.

One other relationship is needed to complete a solution by either the foregoing or by the analogy methods of the next section. This is for the torque that must be applied to produce the unit twist θ_1. Figure 12.18 shows the two shear-force components that act on an element $dx\,dy$ of the cross section. The torque is the moment of the forces integrated over the section

$$M_t = \int\int y[\tau_{zx}\,dx\,dy] - \int\int x[\tau_{zy}\,dx\,dy] \tag{a}$$

For the first term, by the first of Eqs. (12.30)

$$\int\int y\tau_{zx}\,dx\,dy = \int dx\int y\left(-\frac{\partial\phi}{\partial y}\right)dy \tag{b}$$

The second integral sign in Eq. (b) calls for integration along a strip of width dx with

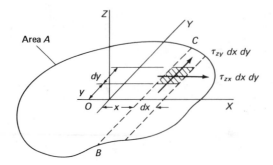

FIGURE 12.18. Determination of torque represented by the ϕ-hill.

x constant (Fig. 12.18). Hence

$$\int y(\partial\phi/\partial y)\,dy = \int_B^C y(d\phi/dy)\,dy = \int_B^C y\,d\phi$$

From this equation and by substitution from a differentiation by parts, $d(y\phi) = y\,d\phi + \phi\,dy$, the double integral of Eq. (b) becomes

$$-dx\int_B^C \left[d(y\phi) - \phi\,dy\right] = -\int dx\left[y\phi\right]_B^C + \int\int \phi\,dx\,dy$$

$$= -\int dx(y_C\,\phi_C - y_B\,\phi_B) + \int\int_A \phi\,dA \qquad (c)$$

If we take $\phi = 0$ everywhere along the border (see discussion following Eq. (12.33)), then $\phi_C = \phi_B = 0$, and the first term of Eq. (c) is zero. The second term, ϕ integrated over the area A, is the volume under the ϕ-hill. Thus the first term of Eq. (a) is determined. Integration of the second term of Eq. (a) by the same procedure will give the same result, with a sign that makes the torque of Eq. (a) equal to twice the volume under the ϕ-hill, or

$$M_t = 2\int_A\int \phi\,dA \qquad (12.35)$$

Example 12.3
Derive equations for stress and deflection in a bar of elliptical cross section.

Solution
The function of Eq. (12.34) satisfies Eq. (12.33), since there is no change in its value (zero) as we move tangentially along its boundary defined by $x^2/a^2 + y^2/b^2 = 1$ (Fig. 12.19). The function satisfies the second-order differentiations of the compatibility equation, Eq. (12.32), and a value for C is obtained by substituting Eq. (12.34) into it. Thus

$$-C\left(\frac{2}{a^2} + \frac{2}{b^2}\right) = -2G\theta_1 \quad \text{and} \quad C = \frac{a^2b^2}{a^2 + b^2}\,G\theta_1$$

whence

$$\phi = \frac{a^2b^2}{a^2 + b^2}\,G\theta_1\left(1 - \frac{x^2}{a^2} - \frac{y^2}{b^2}\right) = 0 \qquad (d)$$

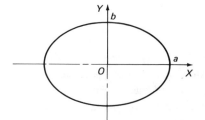

FIGURE 12.19. Bar of elliptical section in torsion (Example 12.3).

From Eq. (12.35) and Fig. 12.19

$$M_t = 2\int\int_A \phi \, dA = \frac{2a^2b^2G\theta_1}{a^2 + b^2}\left[\int\int_A dA - \frac{1}{a^2}\int\int x^2 \, dA - \frac{1}{b^2}\int\int y^2 \, dA\right] \tag{e}$$

Now from the properties of an ellipse

$$\int\int_A dA = A = \pi ab \qquad \int\int x^2 \, dA = I_y = \frac{\pi a^3 b}{4} \qquad \int\int y^2 \, dA = I_x = \frac{\pi ab^3}{4}$$

whence

$$M_t = \frac{\pi a^3 b^3 G\theta_1}{a^2 + b^2} \quad \text{and} \quad \theta_1 = \frac{M_t(a^2 + b^2)}{\pi a^3 b^3 G} \tag{12.36}$$

Function ϕ becomes

$$\phi = \frac{M_t}{\pi ab}\left(1 - \frac{x^2}{a^2} - \frac{y^2}{b^2}\right) \tag{f}$$

The stresses, Eqs. (12.30), are

$$\tau_{xz} = -\frac{\partial\phi}{\partial y} = \frac{2M_t y}{\pi ab^3}$$

$$\tau_{yz} = \frac{\partial\phi}{\partial x} = -\frac{2M_t x}{\pi a^3 b} \tag{12.37}$$

Their maximum values occur at $y = b$ and $x = a$, respectively,

$$(\tau_{xz})_{\max} = 2M_t/\pi ab^2 \quad \text{and} \quad (\tau_{yz})_{\max} = -2M_t/\pi a^2 b$$

of which $(\tau_{xz})_{\max}$ is the larger, occurring at the ends of the minor axis. If $a = b$, we have a circle with $(\tau_{xz})_{\max} = (\tau_{yz})_{\max} = 2M_t/\pi a^3 = 16M_t/\pi d^3$, in agreement with the first of Eqs. (12.5). Substitution into Eqs. (12.28) from Eqs. (12.36) and (12.37) and integration yield the displacement

$$w = \frac{M_t(a^2 - b^2)}{\pi a^3 b^3 G}xy \tag{g}$$

This describes the warping over the ends of the bar. If warping is prevented, as by rigid attachment of the bar to a much larger component at one end, the bar is stiffened, and Eqs. (12.36) and 12.37) do not apply. ////

APPROXIMATE METHOD FOR SOLID SECTIONS

The first of Eqs. (12.36) for the elliptical section may be rewritten as

$$M_t = \frac{A^4 G\theta_1}{4\pi^2 I_p} \approx \frac{A^4 G\theta_1}{40 I_p} \tag{12.38}$$

where A is the area of the section and $I_p = I_x + I_y$ is the polar moment of inertia. Saint-Venant discovered that this simple equation gave results accurate within 10% when applied to all the other sections for which he had made derivations, except some very elongated ones. Thus Eq. (12.38) may be useful for estimating the relationship between moment and twist in solid, squatty sections for which a strain function cannot be written but for which area and moments of inertia may be found analytically or graphically.

12.10 THE MEMBRANE ANALOGY AND THIN OPEN SECTIONS: RECTANGULAR SECTIONS

For many practical shapes, such as a channel beam, hollow airfoil sections, and a shaft with a rectangular keyway, ϕ-functions are not readily devised, and analogies, approximations, and experimental methods are useful. Equation (8.1), the membrane equation for a shell, may be rearranged to describe a thin elastic membrane of thickness t that is prestressed equally in all directions with a uniform tension T per unit width, then deflected a small amount by a fluid at pressure p. The stresses remain essentially unchanged and are $\sigma_t = \sigma_m = T/t$, and the curvatures for small deflections, by Eq. (8.3), are approximately $1/R_t = -(\partial^2 z/\partial x^2)$ and $1/R_m = -(\partial^2 z/\partial y^2)$ in rectilinear coordinates. Substitution into Eq. (8.1) gives

$$\frac{\partial^2 z}{\partial x^2} + \frac{\partial^2 z}{\partial y^2} = -\frac{p}{T} \tag{12.39}$$

THE MEMBRANE

Equation (12.39) is analogous to Eq. (12.32), and it suggests that a membrane be stretched over a flat plate, then clamped with a second plate that has an opening of the same size and shape as the cross section of the bar in torsion. If this assembly is the top side of a container, light air pressure may be applied to inflate the membrane. If the p/T of the membrane is adjusted to be numerically equal to the $2G\theta_1$ of the bar, the membrane will duplicate the ϕ-hill of Fig. 12.17. The slope of the membrane at a position (x, y) in a plane normal to the plate equals the shear stress in the bar at the same position but in a direction perpendicular to the plane of measurement, as discussed in the paragraphs following Eqs. (12.32) and (12.33). Twice the volume under the membrane equals the torque in the bar, by Eq. (12.35). The membrane tension before inflation must be sufficiently high so that its value will not change appreciably when deflected and so that T may be considered constant.[25] Two commonly used membranes are a thin rubber sheet and a soap film. Angles on the latter may be measured by an incident and reflected beam of light.[26]

Qualitative information may be deduced by visualizing the membrane. The steepest part occurs at the boundary nearest the top of the hill. Thus it occurs at the minor axis in an elliptical section and at the middle of each longer side in a rectangular section (Fig.

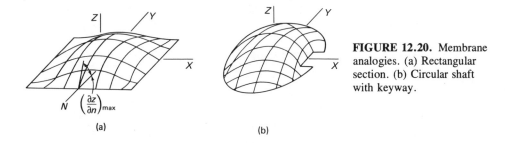

FIGURE 12.20. Membrane analogies. (a) Rectangular section. (b) Circular shaft with keyway.

(a)

(b)

---

[25] This membrane analogy was introduced by L. Prandtl in 1903.

[26] For details, see e.g., M. Hetenyi, (Ed.), *Handbook of Experimental Stress Analysis,* New York: John Wiley & Sons, Inc., pp. 725–736, 1950. In actual experiments, a circular hole is cut in the clamping plate, off to one side, so that the membrane there and at the section to be investigated are both subjected to the same convenient values of p and T. Then both will have the same $2G\theta_1$ values. For the circular section the stiffness $k_t = M_t/\theta$ and the M_t/τ ratios are known from Eqs. (12.5). The volumes and maximum slopes of each hill are measured, and the stiffness and the maximum stress for the section under investigation are obtained by proportions.

12.20(a)). The membrane is clamped in two perpendicular directions at an outside corner of this rectangle, making the slope zero over the 90-deg included angle. Shear stress is zero there. However, it is high at an inside corner, such as the one formed by a rectangular keyway cut into a circular shaft (Fig. 12.20 (b)). There the sloping membrane makes a 90-deg turn, with an infinite slope (and stress) when the corner is theoretically sharp.

THIN OPEN SECTIONS

From the membrane concept, approximate equations may be derived for long narrow sections and for thin hollow sections. For narrow sections of thickness c the membrane will be nearly uniform in height over most of its length, as indicated by the contour lines of the long rectangle of Fig. 12.21. If the drop at each end is neglected, the $\partial^2 z/\partial y^2$ term of Eq. (12.39) disappears and integration of the remainder, $d^2z/dx^2 = -p/T$ yields

$$\frac{dz}{dx} = -\frac{p}{T}x \tag{a}$$

The constant of integration is zero because the slope of the membrane is zero at mid-thickness $x = 0$. The maximum slope, at edges $x = \pm c/2$, is $(dz/dx)_{max} = \pm(p/T) \times (c/2)$. By the conversion of this maximum slope to maximum stress and of p/T to $2G\theta_1$

$$\tau_{max} = G\theta_1 c \tag{b}$$

Integration of Eq. (a) and evaluation of the constant of integration by setting $z = 0$ at $x = \pm c/2$ yields

$$z = \frac{p}{2T}\left(\frac{c^2}{4} - x^2\right) \tag{c}$$

a parabola. If the length of the section is L (not to be confused with bar length l) and the height z is given by Eq. (c), the volume V under the membrane becomes

$$V = L\int_{-c/2}^{+c/2} z\, dx = \frac{pL}{2T}\int_{-c/2}^{+c/2}\left(\frac{c^2}{4} - x^2\right)dx = \frac{p}{12T}(Lc^3) \tag{d}$$

The torque corresponds to twice this volume, Eq. (12.35). Thus substitution into Eq. (d) of $M_t/2$ for V and $2G\theta_1$ for p/T gives

$$M_t = \frac{G\theta_1 Lc^3}{3} \quad \text{or} \quad \theta_1 = \frac{3M_t}{GLc^3} \tag{12.40}$$

FIGURE 12.21. Membrane analogy for a narrow rectangular section (3 views).

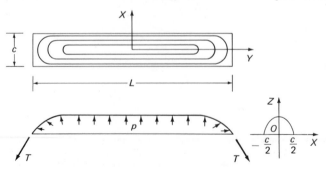

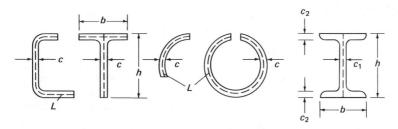

FIGURE 12.22. Thin, open sections to which Eqs. (12.40) through (12.42) apply. The term L is the total length of corresponding thickness c.

From Eqs. (b) and (12.40),

$$\tau_{max} = G\theta_1 c = \frac{3M_t}{Lc^2} \tag{12.41}$$

These equations are not only applicable to a narrow rectangle ($L \gg c$), but also to curved or bent narrow sections, as in Fig. 12.22, where L is the average length of thickness c, indicated by the dotted lines. Thus for the arcs, it is the length at midthickness, for the Tee, the total height h plus the length ($b - c$) of the flanges. However, if the section has several thicknesses c_1, c_2, etc., of length L_1, L_2, etc., the membrane volume of Eq. (d) is modified to $(p/12T)\,(L_1 c_1^3 + L_2 c_2^3 + \cdots)$. The torque and maximum stress are, respectively,

$$M_t = \frac{G\theta_1}{3}\,(L_1 c_1^3 + L_2 c_2^3 + \cdots) \tag{12.42a}$$

and

$$\tau_{max} = G\theta_1 c_{max} \tag{12.42b}$$

For the wide-flanged beam of Fig. 12.22, the summation in Eq. (12.42a) is $[hc_1^3 + 2(b - c_1)c_2^3]$. Again, outside corners are unstressed, and inside corners have stress concentrations that are functions of the radii.

RECTANGULAR SECTIONS

A more exact solution for a rectangle of any proportions was obtained in an infinite series. The results are usually presented as numerical values of coefficients η and λ (Table 12.1), to be multiplied into terms containing M_t and the dimensions h and b of

TABLE 12.1. Coefficients for Torsion of Bars of Rectangular Section.

Key: h/b is the ratio of side dimensions, η is the coefficient for shear stress, λ is the coefficient for unit twist for use in Eqs. (12.43)

h/b	η	λ	h/b	η	λ
1.0	4.81	7.11	4.0	3.55	3.56
1.2	4.57	6.02	5.0	3.44	3.44
1.5	4.33	5.10	6.0	3.34	3.34
2.0	4.07	4.37	8.0	3.26	3.26
2.5	3.87	4.02	10.0	3.20	3.20
3.0	3.74	3.80	∞	3.00	3.00

the rectangle. Thus

$$\tau = \frac{\eta M_t}{hb^2} \quad \text{and} \quad \theta_1 = \frac{\lambda M_t}{Ghb^3} \tag{12.43}$$

where $h \geq b$. It may be noted from the table that when h/b is ∞, $\eta = \lambda = 3.0$, and with L replacing h and c replacing b, Eqs. (12.43) become the "narrow rectangle" equations, Eqs. (12.41) and (12.40). The error in using the latter is indicated by Table 12.1, e.g., if $L/c = h/b = 10.0$, the error is $100(3.20 - 3.00)/3.20 = 6.25\%$ which is on the low side.

The second of Eqs. (12.43) may be rewritten $\theta_1 = M_t/G(hb^3/\lambda) = M_t/GJ_{\text{eff}}$ where J_{eff} is the effective polar moment of inertia. In equations where J has been used for sections in general, as in the energy-method derivations of Section 10.3, $J_{\text{eff}} = hb^3/\lambda$ should be substituted for J when the section is rectangular.

Example 12.4
Compare square with circular wire for coiled cylindrical springs.

Solution
For square wire $h/b = 1.0$. From Table 12.1, $\eta = 4.81$ and $\lambda = 7.11$. Let the mean coil diameter be D and let the number of active coils be N. Then, by Eqs. (12.43), Eq. (12.11a), and Fig. 12.4, and with the Wahl factor for square wire, K_{sq},

$$\tau_{\max} = K_{\text{sq}} \frac{\eta M_t}{hb^2} = K_{\text{sq}} \frac{(4.81)(PD/2)}{b^3} = 2.40 K_{\text{sq}} \frac{PD}{b^3}$$

$$\theta_1 = \frac{\lambda M_t}{Ghb^3} = \frac{7.11(PD/2)}{Gb^4} = 3.56 \frac{PD}{Gb^4}$$

$$\delta = \frac{D}{2}\theta = \frac{D}{2} l \theta_1 = \frac{D}{2}(\pi DN)\left(3.56 \frac{PD}{Gb^4}\right) = 5.58 \frac{PD^3N}{Gb^4}$$

$$k = \frac{P}{\delta} = 0.179 \frac{Gb^4}{D^3N}$$

The Wahl factor K_{sq} is only slightly less than K for circular wire and coils of the same index, and we shall take them to be the same in our comparison.[27]

With reference to Eqs. (12.8), (12.11a), and (12.12) for circular wire, for the same *load* and space, i.e., coil diameter, active coils, and $b = d$, the stress for square wire is 5.6% less, the deflection 30.2% less, the spring rate 43.2% more, and the weight 27% more. For the same *deflection* and space, the load, $P = k\delta$, is 43.2% more. Likewise, the energy stored, $P\delta/2$, is 43.2% more. The corresponding ratio of stresses is

$$\frac{\tau_{\text{sq}}}{\tau_{\text{circ}}} = \frac{2.40 K_{\text{sq}} P_{\text{sq}} D/b^3}{8KP_{\text{circ}} D/\pi d^3} = \frac{2.40\pi}{8}(1.432) = 1.35$$

i.e., the stress in the rectangular wire is 35% more. This may require a wire of higher-strength material and greater cost. It is seen that the purpose for using square wire is to give greater stiffness and energy storage in the same cylindrical space.   ////

---

[27] For values of K for square and rectangular wire, see A. M. Wahl, *Mechanical Springs*, 2nd ed., New York: McGraw-Hill, pp. 124–130, 1963; or *Design Handbook*, Associated Spring Corporation, Bristol, Conn.

12.11 HOLLOW SECTIONS:
ANALOGY FOR PLASTIC CONDITIONS: OTHER METHODS

THIN CLOSED SECTIONS

The equations for a thin strip or open section do not hold for a hollow, closed section or tube. The shear stress does not drop to zero halfway through the thickness, but rather remains nearly constant across a thin wall. In the analogy, the slope of the membrane is approximately constant across the wall, and the inner boundary lies in an elevated flat plane, since the inner wall has the same boundary condition of zero stress component in the normal direction as has the outer wall, Eq. (12.33). This condition is provided by a flat and stiff but weightless plate, bonded to the membrane (Fig. 12.23). The thickness c of the wall may vary around its periphery provided that it remains "thin." With variable thickness, however, the plate will tilt and it must be forced to remain horizontal to satisfy the condition that around its boundary the stress function ϕ is constant. It may be proved[28] that the membrane analogy gives correct results if the plate is made horizontal by applying a pure couple; thence, the equation of equilibrium is unaffected. With the membrane tension, pressure, and couple adjusted so that p/T equals $2G\theta_1$ and the membrane height is a constant w, then the slope w/c equals the shear stress τ, or $w = c\tau$. The volume under the membrane and plate is the height w multiplied by the area A within the dotted line of Fig. 12.23(a), or the mean of the inner and outer areas of the cross section. The torque M_t equals twice this volume, so $M_t = 2(wA) = 2c\tau A$, whence

$$\tau = \frac{M_t}{2cA} \tag{12.44}$$

The twist may be obtained from the equation of equilibrium of the membrane for the area A and midlength L, shown as a free body in Fig. 12.23(b). With the requirement of little change in T and hence small membrane height w, $\sin \lambda$ is nearly w/c and from $\Sigma F_z = 0$,

$$pA - \int_L (T\, dL)\frac{w}{c} = 0$$

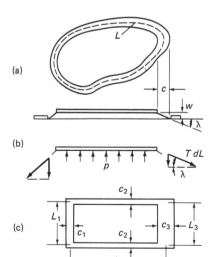

(a)

(b)

(c)

FIGURE 12.23. Membrane analogy for thin, hollow, closed sections. (a) General shape with varying wall thickness. (b) Equilibrium of membrane. (c) Wall with several lengths of different constant thicknesses.

---

[28] J. P. Den Hartog, *Advanced Strength of Materials*, New York: McGraw-Hill, pp. 25 and 33, 1952.

whence

$$\frac{p}{T} = \frac{1}{A} \int_L \frac{w}{c} \, dL \tag{a}$$

For the equivalent torsional case, with $2G\theta_1$ substituted for p/T and τ for slope w/c, Eq. (a) becomes

$$2G\theta_1 = \frac{1}{A} \int_L \tau \, dL = \frac{M_t}{2A^2} \int_L \frac{dL}{c} \tag{b}$$

where the substitution for τ is made from Eq. (12.44). From Eq. (b), the unit twist is

$$\theta_1 = \frac{M_t}{4GA^2} \int_L \frac{dL}{c} \tag{12.45}$$

For a constant thickness c,

$$\theta_1 = \frac{M_t L}{4GA^2 c} \tag{12.46}$$

For N parts each of length L_i and thickness c_i (Fig. 12.23(c)),

$$\theta_1 = \frac{M_t}{4GA^2} \sum_{i=1}^N \frac{L_i}{c_i} \tag{12.47}$$

The foregoing equations are applicable to box sections and to hollow airfoil and stream-line sections such as those commonly used in aircraft structures.

Methods and equations are also available for sections consisting of several cells formed by inner webs. In many cases the webs will increase the resistance to bending but will have very little effect on the *torsional* stiffness and strength. In the analogy a single plate is replaced by one for each cell, and the plates are joined by the same membrane. The plates may have the same heights if there is symmetry, and the membrane slope and corresponding stress are zero at the webs.[29]

Example 12.5

A flat steel strip 3 mm thick, 300 mm wide, and 1 m long is rolled up into a hollow rectangular tube, as shown in Fig. 12.24. What twisting moment may be applied if the shear stress is limited to 80 MPa and what is the corresponding deflection for (a) an open seam, (b) a welded seam, (c) a solid section? Compare these and the stiffnesses.

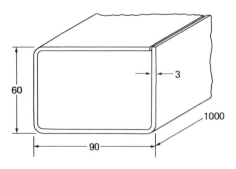

FIGURE 12.24. Rolled, rectangular section of tubing (Example 12.5).

---

[29] For a short treatment, see J. P. Den Hartog, *Advanced Strength of Materials*, New York: McGraw-Hill, 1952, pp. 28–30.

Solution

(a) *Open seam*

The length of the narrow strip is $L = 2[(90 - 3) + (60 - 3)] = 288$ mm. From Eqs. (12.41) and (12.40), respectively,

$$M_t = \frac{Lc^2\tau_{max}}{3} = \frac{(288)(3)^2(80)}{3} = 69\ 100\ \text{N}\cdot\text{mm}\ (69.1\ \text{N}\cdot\text{m})$$

$$\theta = \theta_1 l = \frac{3M_t l}{GLc^3} = \frac{3(69\ 100)(1000)}{(79.3 \times 10^3)(288)(3)^3} = 0.336\ \text{rad}$$

The stiffness is

$$k = \frac{M_t}{\theta} = \frac{69.1}{0.336} = 206\ \text{N}\cdot\text{m/rad}$$

(b) *Welded seam*

A welded seam is a closed section and Eqs. (12.44) and (12.46) apply. The mean area A in the membrane analogy, that enclosed by a dotted line drawn at the midthickness of the metal in Fig. 12.24, is $A = (60 - 3)(90 - 3) = 4960$ mm^2, and as before, $L = 288$ mm. From Eq. (12.44)

$$M_t = 2cA\tau = 2(3)(4960)(80)$$
$$= 2380 \times 10^3\ \text{N}\cdot\text{mm}\ (2380\ \text{N}\cdot\text{m})$$

This is 34 times that of the unwelded tube. From Eq. (12.46) the deflection is

$$\theta = \theta_1 l = \frac{M_t L l}{4GA^2 c} = \frac{(2380 \times 10^3)(288)(1000)}{4(79.3 \times 10^3)(4960)^2(3)} = 0.0293\ \text{rad}$$

This is 0.087 of that of the unwelded tube. The stiffness is

$$k = \frac{M_t}{\theta} = \frac{2380}{0.0293} = 81\ 200\ \text{N}\cdot\text{m/rad}$$

or 394 times that of the unwelded tube.

(c) *Solid section*

From Table 12.1 and Eqs. (12.43), for $h/b = 90/60 = 1.5$, $\eta = 4.33$ and $\lambda = 5.10$,

$$M_t = \frac{hb^2\tau}{\eta} = \frac{90(60)^2(80)}{4.33} = 5985 \times 10^3\ \text{N}\cdot\text{mm}\ (5985\ \text{N}\cdot\text{m})$$

This is only 2.5 times that of the hollow welded tube. From the second of Eqs. (12.43), the deflection is

$$\theta = \theta_1 l = \frac{\lambda M_t l}{Ghb^3} = \frac{(5.10)(5985 \times 10^3)(1000)}{(79.3 \times 10^3)(90)(60)^3} = 0.0198\ \text{rad}$$

This is 0.676 of that of the welded tube. The stiffness is

$$k = \frac{M_t}{\theta} = \frac{5985}{0.0198} = 302\ 000\ \text{N}\cdot\text{m/rad}$$

or 3.72 times that of the welded tube.   ////

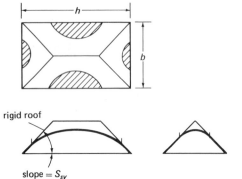

FIGURE 12.25. Constant-slope roof over the membrane to give the analogous plastic portion of an elastic-plastic torsional condition. The illustration is for a solid rectangular cross section.

PLASTIC ANALOGY

A shear yield strength is first reached and plastic flow begins at some point on the outside periphery of a section. As the torque is further increased, this yield stress spreads along the periphery and inward, and without further increase in value if ideal plasticity is assumed. In the analogy this is accomplished by a rigid sloping roof over the membrane which everywhere limits its slope to the equivalent of S_{sy}.[30] As the pressure builds up under the membrane of Fig. 12.25, representing a solid rectangular section, the membrane first contacts the roof at the midpoints of the long sides, then continues to spread as shown by the shaded areas. Thus for a completely plastic condition the membrane becomes a pyramid for a square cross section and a cone for a circular section. With a slope everywhere of S_{sy}, twice their volumes gives $M_t = S_{sy} b^3/3$ for the square section of side b and $2\pi S_{sy} R^3/3$ for the circular section of radius R (cf. Example 5.5).

OTHER METHODS

There are other analogies that are useful in the qualitative and in the experimental treatment of torsion. Kelvin's fluid-flow analogy, in an equation similar to Eq. (12.32), relates $G\theta_1$ to vorticity ω. The shear stresses are proportional to flow velocities. Thus one may easily visualize that fluid rotating within a rectangular section will have zero velocity at the center and corners and maximum velocities along the walls closest to the center or rotation. The velocities will be further increased by a notch or inclusion. Jacobsen's electrical analogy was applied to the torsion of a circular shaft with a filleted step or shoulder.[31]

Approximate equations have been obtained for the stress concentration at reentrant corners in both open and closed thin sections.[32] A method developed by W. Ritz for cross sections in general determines the stress function from the minimum condition of a certain integral that is obtained from the strain energy of the twisted bar. Trigonometric series as well as polynomials may be suitable for stress functions.[33]

[30] Due to A. Nadai, 1925. For a more complete description plus that of a sand-heap analogy, see A. Nadai, *Theory of Flow and Fracture of Solids*, New York: McGraw-Hill, pp. 494–505, 1950.

[31] J. P. Den Hartog, *Advanced Strength of Materials*, New York: McGraw-Hill, 1952. The chapter on torsion contains excellent descriptions of the flow and electrical analogies as well as torsion in general.

[32] S. Timoshenko and J. N. Goodier, *Theory of Elasticity*, 3rd ed., New York: McGraw-Hill, 1970.

[33] See footnote 32.

PROBLEMS

Section 12.2

12.1 In a particular passenger automobile an arm with an effective length of 12 in is attached to one end of a torsion bar, such as that of Fig. 12.2, the other end of which is anchored to the chassis. The end of the arm, which is connected to the wheel linkage, rises 6 in to a horizontal position when subjected to a vertical force of 750 lb. An additional rise of 4 in occurs under maximum impact loads. For through-hardened material heat treated to a hardness of $50R_c$ and shot peened and preset, an SAE manual suggests a maximum stress of 140 000 psi, which is high because of a limited number of cycles involving maximum deflection. Determine a suitable bar diameter and effective length. *Ans.* $d = 0.813$ in, *l* $= 28.6$ in.

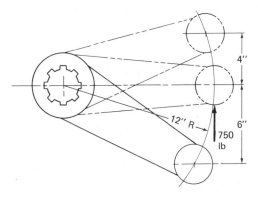

12.2 If the torsion bar of Prob. 12.1 is replaced by a hollow one with a diameter ratio of 0.70, what should be the diameters and length to give the same characteristics?

12.3 Torsion bars are frequently prestressed by a torque sufficient to cause yielding over an outer portion of the bar. Sketch for this condition an idealized diagram of shear stress vs radius. Determine the inner diameter of the plastic region when the limit-design factor (Section 5.8) is 1.20. *Ans.* 0.737*D*.

12.4 For the rotating system shown, an undamped vibration absorber consists of a disk on a steel shaft that is "tuned" to have the same natural frequency as the main shaft and mass to which it is attached. The absorber disk oscillates at the same frequency and opposite sense to create a torque to counteract the disturbance torque on the mass, thus preventing its vibration The torque amplitude is 850 N·m and the "spring constant" or stiffness required of the absorber shaft is 7400 N·m/rad. $G = 79.3 \times 10^3$ MPa. For compactness it is proposed to put it inside of the hollow main shaft, as shown. Available space requires that its outside diameter be 32 mm and the length be 585 mm. (a) Can the desired stiffness be obtained by making it hollow, and if so, with what inside diameter? *Ans.* 26.5 mm. (b) What will be the maximum angular deflection? (c) The enlarged end of the shaft is 41 mm in diameter and is joined to the 32-mm diameter by 5-mm fillets. For what endurance limit in shear should the material be chosen and treated if the factor of safety is 1.5?

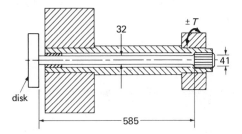

12.5   A stator subjected to a torque T is mounted on a shaft at a distance a from its left end and b from its right end. Both ends are clamped. (a) Derive equations for the torque on the two parts of the shaft if its diameter is uniform. (b) Derive equations for the diameters d_a and d_b to make the torques in the two parts of the shaft equal.

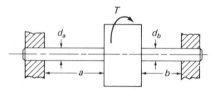

12.6   Derive an equation for the twist under torque M_t of that length l of a solid circular shaft which tapers linearly from a diameter d_1 to a diameter d_2 $(d_2 > d_1)$. Ans. $[32M_t l/3\pi G \times (d_2 - d_1)] [(1/d_1)^3 - (1/d_2)^3]$.

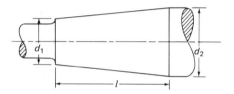

12.7   A long shaft of uniform diameter d is driven at one end. It has closely spaced gears along its length to drive the feed screws of a furnace grate. Consider this take-off torque to be uniformly distributed and derive equations for maximum stress and total twist.

12.8   Assume when deep drilling that there is a torque T_0 concentrated at the drilling end at $z = 0$ and additional torque from friction along the drill, its action decreasing uniformly from a value t_0 per unit of length at the drilling end to zero at surface level or location $z = a$. A drive unit is located above the surface at $z = L$. Derive equations for the maximum stress and total twist. Ans. $(16/\pi d^3)(T_0 + t_0 a/2)$, $(32/\pi d^4 G) [T_0 L + t_0 a(3L - a)/6]$.

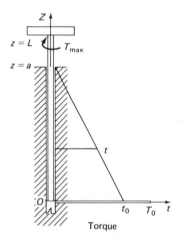

12.9   Determine the total twist between the gear and V-belt sheave for the shaft of Fig. 6.10 with the torque of Fig. 6.8. Neglect any effect of fillets and keyways. For the twist of the tapered length, see the answer to Prob. 12.6. Assume that the twist under the relatively rigid gear or sheave is determined by the average shaft torque between its center and the end of the keyway. Ans. 0.15°.

12.10   Determine the maximum twist between the sprockets of Prob. 6.21. Assume that attached hubs have no stiffening effect and that the twist under a sprocket is determined by the average shaft torque between the sprocket center and the adjacent shoulder.

12.11 For the intermediate shaft of the punch press of Example 9.4 and Fig. 9.16, calculate the twist and the maximum torsional stress between the pinion and the flywheel during the working stroke. The diameter of the shaft to the right of the pinion is $3^1/_{32}$ in, calculated in Prob. 6.38. Take the diameter under the pulley to be $2^7/_8$ in, between pulley and bearing #3 to be $2^5/_8$ in, and at bearing #3 to be $2^1/_4$ in. Any lengths needed may be scaled from Fig. 9.16(a).

12.12 Determine the maximum twist between coupling and crankpin in the crankshaft of Prob. 6.41. Include the effect of bending in the crankarm. To what displacement at the center of the crankpin would the twist and bending correspond if the coupling were held fixed? $G = 79.3 \, GPa$. *Ans.* 0.207 mm.

Section 12.3

12.13 Sketch schematically force vs deflection for a cylindrical compression spring with coils uniformly spaced (uniform pitch), for a cylindrical compression spring with nonuniform pitch, for a cylindrical extension spring, and for a conical spring. How can a conical spring be made to have a solid height (with all coils bottomed) equal to one wire diameter? How can a conical spring be made to give a constant spring rate?

12.14 From the equation for τ_a in the second paragraph following that for τ_0, Eq. (12.7), and from the approximation indicated for $\bar{y}$, show that at any radius R in Fig. 12.5(a), $\tau_a = \tau_0[(4D^2 - d^2 - 8RD)/8Rd]$. Assume that Eqs. (12.5) for the twist of straight bars hold for curved bars (confirmed experimentally), such that the substitution $l = l_{avg} = (D/2)\phi$ may be made in it to obtain the ratio θ/ϕ. Continue the derivation to confirm Eq. (12.8) and the Wahl correction factor, Eq. (12.9).

12.15 Let $r = OQ$ (Fig. 12.5(a)) be the location where $\tau_a = \tau_0$. From the τ_a equation of Prob. 12.14 derive an expression in terms of C for the ratio $r/(d/2)$. The result is the ratio of radius r to the wire radius $d/2$ above which the torsional shear stress τ_a is larger than the nominal maximum τ_0. Determine the value of the ratio for $C = 4, 6,$ and 8, with reference to the statement following Eq. (12.10).

Section 12.4

12.16 Derive Eqs. (12.11) for spring deflection by determining the strain energy from Eq. (10.5), then applying Castigliano's theorem.

12.17 Derive Eq. (12.18) in detail.

12.18 A company has in stock compression springs with squared and ground ends, an outside coil diameter of 56 mm, a wire diameter of 7 mm, 6.625 total turns, and a free length of 80 mm. Assume $1^3/_4$ inactive coils. The springs are being considered for a deflection of 25 mm. (a) Determine the maximum stress and force. *Ans.* 457 MPa, 1037 N. (b) What is the pitch? Will there be a chance that the coils will clash?

12.19 The steel compression spring of Fig. 12.3(a) is coiled from $1/_8$ in wire, has an outside coil diameter of $7/_8$ in, a free length of 2 in, and it has been preset. Obtain the number of coils by counting and assume $1^3/_4$ inactive coils. What is the maximum load and corresponding stress that may be carried if the coils are to be free from clashing? *Ans.* 87.3 lb, 106 700 psi. What are the spring rate, pitch, and solid length?

12.20 A small compression spring is coiled from 2-mm wire. It has an outside diameter of 10 mm, with 8 total turns and closed and ground ends. It is preset and used under a static load only, and it is proposed that a correction factor for only the direct shear stress be used. What should be the free length and pitch in order that the stress not exceed 900 MPa when the spring is compressed solid?

12.21 Derive an equation for the energy E_p that can be stored elastically by a helical spring of circular wire without exceeding a stress τ_0. (a) Show that the energy can be expressed in terms of stress and volume V of the active length of wire. What is the significance of the result? (b) Solve the result for a product of terms of N and d. This gives a useful form for design. What is its value for steels and $C = 8$? *Ans.* $Nd^3 = (3.27 \times 10^6)E_p/\tau^2$.

12.22 Coiled springs are used to gradually reduce the velocity of a striking body by temporarily and elastically absorbing its energy of motion. Sketch the shapes for the curves of the load vs displacement x and of the acceleration d^2x/dt^2 vs displacement. Derive equations for the velocity and displacement as functions of time t, spring rate k, body mass m, and striking velocity v_0. Sketch velocity and displacement vs time both before and after the body has been brought to zero velocity. Suggest mechanisms to minimize or even eliminate the rebound.

Section 12.5

12.23 The steel extension spring of Fig. 12.3(b) has an outside diameter of ⅝ in, and it is close coiled with an initial tension of 2.5 lb from #17 wire (0.054 in). The coiled length is 2.70 in, measured between wire centers at the start of the loops. What should be the maximum load and deflection if a torsional shear stress of 75 000 psi is not to be exceeded in the coils? *Ans.* $\delta = 3.62$ in. (b) What must be the minimum radius of the inside fiber at the bend where the loop joins the coils if the stress-concentration factor is not to exceed the K factor of the coils? *Ans.* 0.20 in.

12.24 The sketch shows an element SS' of the center line of the wire in a conical spring in a Y, ϕ, $\overline{R}$ cylindrical coordinate system, with the symbols corresponding to those of Fig. 12.5(a). The centroidal radius $\overline{R}$ is now a variable, changing in length by $d\overline{R} = -\tan \alpha \, dy$ while rising an amount $dy = (\overline{R} \, d\phi) \tan \lambda$, where α and λ are the half cone angle and helix lead angle, respectively. Since the pitch p is constant for easiest winding, the lead angle must increase as $\overline{R}$ decreases. Let the initial active radius at the base of the cone (the larger end) be $\overline{R}_0$ and at the other end let it be $\overline{R}_1$. (a) Derive an equation for $\overline{R}$ as a function of ϕ. *Ans.* $\overline{R}_0 - (p/2\pi)\phi \tan \alpha$. (b) Assume that the helix angle is large enough so that coils telescope, one through another, rather than gradually close. Derive Eq. (12.20b) for spring deflection δ under load P.

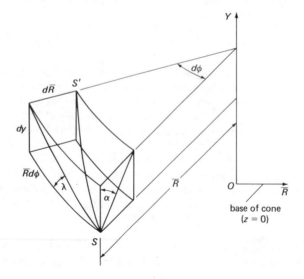

base of cone
$(z = 0)$

12.25 A spring of $^3/_{16}$ in wire has 4¼ active coils that are wound conically with a pitch of ¾ in and with a cone angle such that the coils may telescope with a radial clearance of $^1/_{32}$ in. At each end there is a closed and ground, inactive coil with a constant mean coil-radius, $\overline{R}_0$ at the bottom end and $\overline{R}_1$ at the top end. The bottom coil fits into a 3-in-diameter recess with a diametral clearance of $^1/_{16}$ in. Refer to Prob. 12.24 and Eq. (12.20b). (a) Determine the dimensions of the spring. (b) What load will cause the spring to deflect the amount of its active length? Is the stress reasonable? *Ans.* 176 lb.

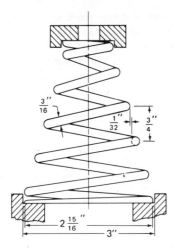

12.26 Confirm the several relationships stated for spring nests and derive Eqs. (12.21a), (12.21b), and (12.21c) in detail.

12.27 The load supported by the spring of Prob. 12.18 is found to be insufficient for the proposed application. Determine the dimensions of a second spring to be placed within the first with a clearance ratio $\eta = 0.10$. Assume that available wires come in whole and half millimeters only. Check the stress in the second spring and the solid heights. What total load may be carried? *Ans.* 1675 N or 1398 N, depending on the wire size chosen.

12.28 Coiled springs expand a small amount with compression, and sufficient clearance around them must be provided. Sketch one turn of wire with a pitch several times wire diameter, first open, then closed to its solid height. Assume that rotation of the coils about their axis is prevented by end restraints. (a) Derive an approximate equation for the percentage increase in coil diameter from the free to the solid configuration. (b) What is the expansion if $p = 4d$ and $D = 6d$?

Section 12.7

12.29 The stresses in the springs of Probs. 12.18, 12.19, and 12.20 are given in their statements or answers. Select materials appropriate for static loading and a factor of safety of 1.2.

12.30 The relief valve of Fig. 12.9 uses a spring-loaded ball against a conical seat of 90° included angle. The discharge capacity is to be 60 in³/s and the flow velocity is to be 100 ft/s. The discharge coefficient of the cylindrical hole is about 0.65. The ball should contact the cone a bit above its lower edge, and it should be an available, common fractional size. It should leave its seat when the pump pressure reaches 60 psi. With the pressure then acting against a larger projected area of the ball, the ball should rise to a position where the cylindrical area of the opening between the ball and the cone is equal to the cross-sectional area of the hole. (a) Determine the hole and ball diameters and ball lift. *Ans.* $^5/_{16}$, ½, and 0.087 in, respectively. (b) Determine the required spring rate. Assume a small spherical ring ground into the end of the cone so that when the valve is closed, the effective area has the diameter of the hole; also, assume that when the valve is open, the effective area has a diameter ¾ of the ball diameter. *Ans.* 23.3 lb/in. (c) Since adjustment is provided, make the solid deflection 30% larger than the normal useful deflection and select wire for the possibility of solid deflection. Use hard-drawn carbon wire, with spring preset, and a factor of safety of 1.4 to avoid any relaxation. Take the mean-coil diameter to be ⅜ in, and calculate the other spring dimensions, using a table of standard spring-wire sizes.

12.31 A study is being made of the feasibility of using helical springs for minimizing collision damage to an automobile. The automobile weighing 4000 lb loaded, and traveling at 5 miles per hour, is to be stopped in a distance of 5 in when striking a rigid wall askew, i.e., with one end of the bumper. Six springs will be placed between the chassis frame and a relatively stiff bumper, three at each end. Assume three springs operative for a hit askew and six for a square hit. The available space limits the outside spring diameter to 6 in. The springs will be coiled hot from Cr-Va wire, then heat treated, shot peened, and preset. They will have only a few maximum deflections (hopefully), so the loading may be considered static. Let the factor of safety be 1.2. (a) Determine the stopping force per spring for a hit askew. *Ans.* 5350 lb. (b) From this, determine the dimensions of the springs. *Ans.* $d = {}^{31}/_{32}$ in, $N_t = 11¼$, etc. (c) If the hit is square on, what is the deflection and force per spring? (d) What kind of mechanism might be used to prevent rebound? Is the use of helical springs a feasible method mechanically? Costwise?

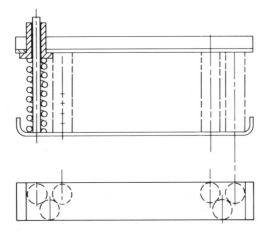

Section 12.8

12.32 Derive an equation for the 0.125-in line of Fig. 12.11(a) in terms of a range ratio $\rho = \tau_{min}/\tau_{max}$. *Ans.* $\tau_{max} = 50\,000/(1 - 0.67\rho)$ psi.

12.33 Same requirement as Prob. 12.32 but for the 5.26 mm, 10^7 cycles line of Fig. 12.11(b).

12.34 Derive an equation similar to Eq. (12.22) but for a spring preset and not shot peened. Tests at zero to maximum stress, repeated for 10^7 cycles, gave a K_{s2} factor of 0.24. An upper limit for K_{s2} to avoid relaxation is 0.56, at a K_{s1} value of 0.485.

12.35 Confirm Eq. (12.24c) by a calculation, without plotting, to obtain a straight line on a log-log chart. Two points are (0.033, 290 000) and (0.40, 195 000).

12.36 A single-disk clutch, such as that described and sketched in Prob. 3.48, is held in engagement by an axial force of 1000 lb produced by 10 symmetrically spaced and equally loaded, helical compression springs acting against the pressure plate. The motion of the pressure plate at disengagement is limited to 0.15 in with a corresponding force increase of 25%. Use the oil-temperd wire of Fig. 12.11, a spring index of 8.0, and a factor of safety of 1.2. (a) Determine the wire diameter and the total number of coils. (b) Determine the other spring dimensions.

12.37 A poppet valve similar to that of Fig. 12.12 is to be lifted 10 mm, with the spring force increasing from 155 to 380 N. The spring must clear the 21-mm valve-stem bushing and guide by at least 1 mm all around, and it must fit at one end into a recess of 35-mm diameter with a radial clearance of no more than 0.5 mm. Assume that spring wire is available only in whole and half millimeters. Use oil-tempered carbon-steel valve-spring-quality wire, with the spring shot peened and preset. Use a factor of safety of 1.20 and the method of direct solution for wire diameter, converting the appropriate equation, Eqs. (12.24), to megapascals. Determine the dimensions of a suitable spring.

12.38 Helical springs may be placed between two parts of a coupling or gear to give torsional flexibility and to change the natural vibration frequency of the shaft system. Shown is one of four springs centered on a circle of diameter D_s. Under variable torque the spring length may vary from the installed length L_1, as shown, to a length $(L_1 - \delta)$ at which the central pins or stops close to carry heavy overloads without flexibility. The spring should fit loosely over the pin diameters D_p, which act as guides. Let $D_s = 6$ in and $D_p = 1\frac{1}{8}$ in. Reliability is important, so use shot peened and preset springs, a factor of safety of 1.40, and oil-tempered carbon-steel wire of valve-spring quality. Design the coupling for a capacity of 1500 lb·in torque at the upper limit of spring action and 300 lb·in at the lower limit, and a torsional stiffness of 8400 lb·in/rad. (a) Determine the spring dimensions and the size of the gap between stop pins. (b) What length of 2-in-diameter shaft would be required to give the same torsional stiffness as the coupling?

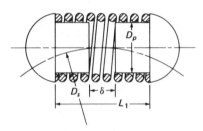

12.39 Adapt Soderberg's equation, Eq. (6.10), for use with springs, writing it in terms of τ_{max}, the range ratio $\rho = \tau_{min}/\tau_{max}$, and strengths S_y and S_u. Take $S_{sy}/S_y = 0.62$,[34] and use the customary relationship between S_e and S_u. Take $K_{fs} = 1.0$ and consider the Wahl correction factor to be part of τ_{max} and τ_{min}. Ans. $n = 1.24(S_y/\tau_{max})/[(1 + \rho) + 2(S_y/S_u)(1 - \rho)]$.

12.40 A large spring is needed for a fluctuation of 2.5 in between an initial load of 5000 lb and a maximum load of 10 000 lb. Ends are closed and ground. The factor of safety is to be 1.40. The shop's winding capabilities are limited to bars of 2-in diameter. The steel proposed is AISI 1095, stress relieved after winding hot, then oil quenched and tempered at 1000°F. A steel company's tables give tensile and yield strengths of 151 000 and 92 500 psi, respectively, for a 2-in round. Scale is removed after heat treatment and enough shot peening is done to compensate for all loss of surface fatigue strength. Use the Soderberg method adapted as in Prob. 12.39 to determine allowable stress, and calculate the maximum spring index if the 2-in bar is used. Determine the other spring dimensions and check for buckling and frequency.

---

[34] Based on the total-deformation-energy theory of failure, Prob. 5.32, and reported by C. Samónov by personal communication to have been found by German spring manufactures to give better agreement with test results on spring steels than the distortion energy theory.

12.41 In a small bowl mill for pulverizing materials, a base with motor, bearings, and shaft for a roller must be pivoted so that it can swing away to allow large or hard materials to pass. This is accomplished by spring loading the base. To provide sufficient force in the space available, two nested springs are advisable. The initial force will be adjustable, as low as zero when only the net weight at the roller is needed to pulverize. A spring rate of about 750 lb/in and a total deflection of about 1½ in from free length to solid length is desired. Available space restricts the outside diameter of the outer spring to 3 in, and it is anticipated that a small spring index, say $C = 5.0$, will be necessary. Use an approximate diametral clearance between springs corresponding to $\eta = 0.10$, as defined for Eq. (12.21a). Use Cr-Va wire of valve-spring quality, with the spring shot peened and preset, a factor of safety of 1.20, and the direct method of solving for wire diameter. Determine suitable dimensions for the two springs.

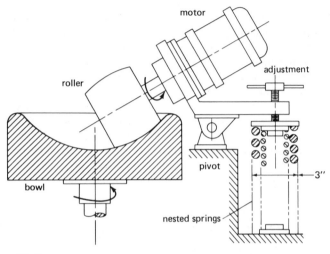

Section 12.9

12.42 Derive equations relating torsional moment, twist, and shearing stress for a shaft of circular cross section by writing a suitable stress function and using the methods of Section 12.9. What is the geometric shape of the ϕ-hill and analogous membrane?

12.43 Warping is measured by the displacement or warping function w. Show that it must satisfy the partial differential equation $\partial^2 w/\partial x^2 + \partial^2 w/\partial y^2 = 0$.

12.44 Derive an equation for the total twist in length of a bar of elliptical section that tapers only in height, from $2a_1$ at the small end to $2a_2$ at the large end. *Ans.* $[M_t l/\pi G b^3 (a_2 - a_1)]$ $\times [\ln a_2/a_1 - (b^2/2)(1/a_2{}^2 - 1/a_1{}^2)]$.

12.45 In the cast-iron lever of Fig. 9.5(a) the part to the left of A-A, which is redrawn here, may be twisted and bent by a force of 3750 N, acting on a pin inserted in the left-hand hole but offset 50 mm from the central plane of the lever. The distance from the center of the hole to the smallest elliptical section is 30 mm, and to the largest elliptical section the distance is 205 mm. The height of the ellipse varies from 50 to 76 mm, and it is everywhere twice the width. Determine the location and magnitude of the maximum shear and the maximum bending stresses. *Ans.* $\tau_{\max}$(torsional + direct shear) = 35.6 MPa, $\sigma_{\max}$ = 35.7 MPa.

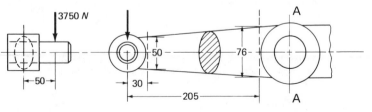

12.46 (a) Derive Eq. (12.38) from Eqs. (12.36). (b) Apply it to the 3:1 rectangle of Table 12.1 and Eqs. (12.43), and to the equilateral triangle of Prob. 12.49. Are the errors in the order that you might expect? (c) Apply Eq. (12.38) to determine the twist in a bar of semicircular section. Compare it with that of a circular section with the same radius and moment.

12.47 Determine by an approximate method the total twist in a steel hexagonal bar that is 800 mm long and measures 50 mm across the flats (twice the radius r of the inscribed circle), under a torsional moment of 1000 N·m. Compare the result to that for a circular bar with the diameter of the inscribed circle. The moment of inertia of area of any regular polygon about any line through its center is $I = (\text{Area})(12r^2 + a^2)/48$, where a is the length of a side and r is the radius of the inscribed circle.

12.48 (a) Determine a ϕ-function for the shaft with a semicircular keyway. Write separate equations for the large and the small circle in terms of x and y in the coordinate system shown. Combine them in such a way that ϕ has the value $-Cr^2$ for all values (x, y) on the large circle and $-2CRx$ for all values (x, y) on the small circle. Determine one additional term that has one or the other of these values, depending upon the circle substituted, thus cancelling out the remainder and satisfying the boundary condition for the ϕ-function. (b) Determine the value of C. (c) Write the equations for the shear stresses and for the maximum shear stress. *Ans.* $\tau_{xz} = G\theta_1[y - 2r^2Rxy/(x^2 + y^2)^2]$, $\tau_{yz} = -G\theta_1[x - R + r^2R/(x^2 + y^2) - 2r^2Rx^2/(x^2 + y^2)^2]$. (d) From the foregoing derive an equation for the torsional stress-concentration factor for a shallow, semicircular groove along a cylindrical shaft. What is its maximum value?

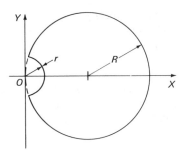

12.49 Analyze a shaft of equilateral triangular section under a torsional moment. (a) Write an equation for each of the three sides in the coordinate system shown and in terms of base width a and height h. (b) Multiply these three equations together, substitute for a in terms of h, and test the suitability of the product for a ϕ-function. *Ans.* $\phi = C[y^3 - hy^2 + (4/27)h^3 - x^2(3y + h)]$, suitable. (c) Derive equations for the shear stresses. (d) Derive a relationship between M_t and θ_1. *Ans.* $M_t = G\theta_1 ha^3/40$. (e) Obtain the maximum shear stress in terms of moment M_t. *Ans.* $20M_t/a^3$.

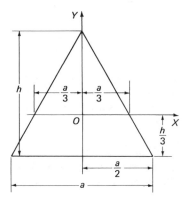

Section 12.10

12.50 Starting with the membrane equation, Eq. (12.39) derive in detail Eqs. (12.40) and (12.41) for narrow open sections.

12.51 A nominal 4-in 5.4 lb/ft steel channel (Fig. 9.5(c)), has a depth (height) of 4 in, a web thickness of 0.180 in, and flange widths of 1.58 in with an average thickness of 0.296 in. The area of the section is 1.56 in^2, and $G = 11.5 \times 10^6$ psi. Calculate the shear stress and twist in a 10-ft length under a twisting moment of 1000 lb·in. *Ans.* 9600 psi.

12.52 A nominal 8-in 20-lb/ft, wide-flange I-beam (Fig. 9.5(c)), has a depth (height) of 8.14 in, a web thickness of 0.248 in, a width of 5.268 in, and flange thicknesses of 0.378 in. The area of the section is 5.88 in^2. Calculate the shear stress and twist in a length of 10 ft under a twisting moment of 8000 lb·in.

12.53 Calculate the torsional stress and the unit twist in the leaves of the spring of Fig. 9.8(c), both adjacent to the end loop and at the shackle nearest the end loop. A twisting moment of 30 N·m is applied. The leaves are 50 mm wide, with the longest leaf 4 mm and the others 3 mm thick. $G = 79.3 \times 10^3$ MPa. Assume all leaves at the shackle are twisted through the same angle. *Ans.* At loop: 112.5 MPa, 0.355 rad/m; at shackle: 37.3 and 49.7 MPa, 0.157 rad/m.

12.54 A narrow flat bar of constant thickness c varies in width from a to b in a length l. Derive an equation for total twist under a torsional moment M_t. *Ans.* $[3M_t l/Gc^3(b - a)]\ln b/a$.

12.55 Write deflections in terms of stresses for coiled compression springs and compare energy-storage capacities of square wire with circular wire springs of the same dimensions and same stress. Assume Wahl factors to be the same.

12.56 The ends of a lock washer are deflected up to the amount of their thickness when closed in use. Consider a washer with an inside diameter of 1¼ in, ¼ in wide, and ³⁄₁₆ in high, of rectangular section. The deflection corresponding to loads P on the washer ends is given in the answer to Prob. 10.20. Calculate the maximum torsional stress. *Ans.* 95 200 psi.

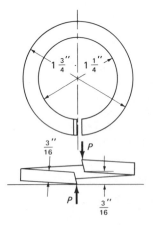

12.57 Compare the nominal value of polar moment of inertia J for a rectangular section with J_{eff} as determined in the paragraph following Eqs. (12.43). Calculate for h/b values of 1.0, 2.0, and 5.0. Explain physically why J_{eff} is less than J. What about a circular section?

Section 12.11

12.58 (a) For a thin circular tube of outside diameter D and thickness c, derive equations for the stress and stiffness by the approximate method of Section 12.11 and by the more exact method of Section 12.2. Compare for $D = 2$ in and $c = ⅛$ in. (b) Derive equations for the stress and stiffness of a thin, open, unwelded tube made by rolling sheet into a cylinder. For the same dimensions compare with the tube in part (a). (c) Find a small cardboard tube (that on which paper towels are wrapped is a good size) and feel the difference in resistance to twist before and after making a longitudinal cut. Note the relative motion along the cut and the warping at the ends.

12.59   Calculate the twist between the ends and the stress in the semicylindrical hollow tube of aluminum alloy ($G = 26.2$ GPa), subject to a torque of 900 N·m.

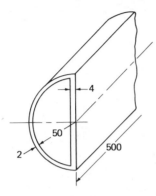

12.60   (a) Derive an equation for the stress and unit twist in a hollow, thin tube of uniform thickness c, formed in the shape of a hexagon. The outside dimension measured across the flats is d. *Ans.* $\tau = 0.577 M_t / c(d - c)^2$, $\theta_1 = 1.155\, M/Gc(d - c)^3$. (b) Compare the results with a circular ring of outside diameter d and thickness c.

12.61   A circular member is formed by seam welding two semicircular parts preformed from steel sheet, as shown. Calculate to determine if the projections appreciably change the torsional strength and stiffness in the manner that you anticipated.

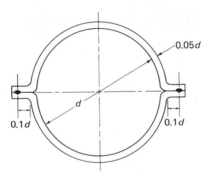

12.62   A hollow aluminum supporting member used in an airstream has a teardrop shape that can be approximated by the arcs and dimensions in millimeters as shown. If its length is one meter and a twisting moment of 1250 N·m is applied at one end, what are the stress and twist? ($G = 26.2$ GPa). *Ans.* 49.0 MPa, 3.30°.

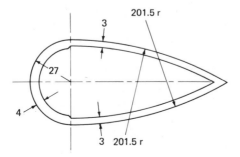

12.63 Determine the maximum stress and the twist in degrees for the 55-in-long steel rocker panel (door sill) of an automobile under a torsional moment of 10 000 lb·in. An outer panel of thickness 0.041 in is welded to an inner panel (vertical straight part on left) of thickness 0.047 in. Obtain the lengths by scaling, based on a full-size length of 3.25 in for the X-axis intercept. Approximate the area by dividing it into trapezoids and triangles. *Ans.* 12 600 psi, 2.7°. (Sketch reproduced with permission of General Motors Corporation. See Fig. 9.6(e).)

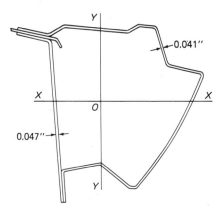

12.64 Derive the values for complete plasticity for square and circular sections, given in the paragraph following Example 12.5. To what limit-design factors do they correspond?

12.65 For a rod of thin, rectangular section ($c \ll L$) determine: (a) the twisting moment to just bring the points of maximum stress to the yield strength S_{sy}, (b) the twisting moment to complete yielding throughout, (c) the limit-design factor.

12.66 Derive an equation for the moment on the rectangular section of Fig. 12.25 to just produce complete plasticity. *Ans.* $S_{sy} b^2 (3h - b)/6$. If the width-to-thickness ratios h/b of the section are 2.0, 4.0, and ∞, what are the values of the limit-design factors?

12.67 Derive an equation for the twisting moment to give complete plasticity to the triangular section of Prob. 12.49, also the limit-design factor. The answers to Prob. 12.49 may be used. *Ans.* 1.67.

13

Impact

Energy and Wave Analyses, Longitudinal Impact and
Springs, Torsional Impact and Shafts

13.1 THE NATURE OF IMPACT

Impact occurs when two parts with different velocities meet or are suddenly connected; it also occurs when force is applied at a rapid rate. The resulting stresses depend upon the shape and mass of the stressed body, the masses of related bodies which may be considered rigid, the relative striking velocity or force, and the physical properties of the materials. As in static loading the impact may be directed to give uniform tensile or compressive stresses, shear stresses, or bending stresses. In machinery, longitudinal impact, shown schematically in Figs. 13.1(a) and 13.2(a), may occur in hoisting rope, linkages, hammer-type power tools, coupling-connected cars, and helical springs. Torsional impact, shown in Figs. 13.1(b) and 13.2(b), occurs in rotating shafts of punches and shears, in geared drives, at clutches and brakes, and in oilfield drill pipe. Bending impact occurs in shafts and in structural members such as beams, plates, and vessels.

One kind of impact consists of a striking or collision of bodies (Fig. 13.1), as in the engagement of jaw clutches, in hammering actions, and in the taking up of slack in rope and of clearance in bearings. This will be called *striking impact*. Another kind of impact consists of the rapid application of force or torque (Fig. 13.2), as in the sudden support of the weight of a body, and as friction torque in clutches and brakes. This will be called *force impact*. The two kinds of impact may occur simultaneously, as during the sudden connection of two members, one driven by a motor which has inertia and applied torque; or by dropping a mass onto a support, which is not only struck but also has suddenly applied to it the force of gravity or weight of the mass.

The stress must usually remain elastic in a machinery component if the component is to be used again or is to continue to run smoothly. In many structures much higher energy may be absorbed with permanent deformation and without fracture because of the large plastic yielding of which many ductile metals are capable. This is useful as protection against missiles, collisions, and accidental overloads. This chapter, however, will be concerned only with stresses in the elastic region.

When the striking body may be considered rigid and its mass is large compared with the mass of the elastic member, the kinetic energy of the mass before striking, modified for the mass of the elastic member, may be equated to the energy stored elastically at

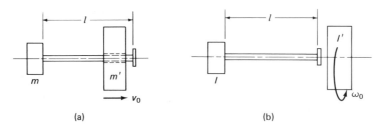

FIGURE 13.1. Striking impact. (a) Longitudinal. (b) Torsional.

maximum deflection (Fig. 13.3). This leads to an approximate equation for maximum stress. The energy method, treated in Section 13.2, is independent of time and position, and it assumes that stresses at all positions along the elastic member simultaneously reach peak values.

In a more exact method, the stress at any position is treated as a function of time, and waves of stress are found to sweep through the elastic material at a propagation rate that is often called its "speed of sound." This wave method, treated in Section 13.4 and onward, indicates higher stresses than does the energy method. This is because the energy is stored by strains and corresponding stresses that are nonuniform along the member, hence larger at some places than the uniform stress of the energy method. The wave method is correct for all ratios of the masses, but it is particularly useful at the smaller mass ratios of rigid body to elastic member, where the error from the energy method is large.

13.2 ENERGY METHOD: GENERAL

After an elastic member, such as a beam or the bars shown in Fig. 13.4, is struck by a rigid body of mass m', strains will travel through the elastic member, developing stress in it and a force on the rigid body, that will soon bring its velocity from its striking value of v_0 down to zero. The assumption is made in the energy method that all of the particles of the elastic member also have zero velocity at the same time. Then at this time, before rebound, it may be said that the initial kinetic energy, $m'v_0^2/2$, of the striking body has been converted into potential or elastic energy of the elastic member. The force, stress, and deflection have maximum values. If there is a linear relationship between the force F and deflection δ at the place of impact, then the elastic energy is $F_i\delta_i/2$, where the subscript i indicates a maximum value under impact. This is represented by the shaded area in Fig. 13.3. The linear relationship between force and deflection may be expressed by a stiffness constant $k = F/\delta$, or $\delta = F/k$, and the energy may be written as $F_i^2/2k$.

Since the particles of the elastic member must be put into motion in varying amounts, its mass must have some effect on the energy of motion. Allowance for this is made

FIGURE 13.2. Force impact. (a) Longitudinal. (b) Torsional.

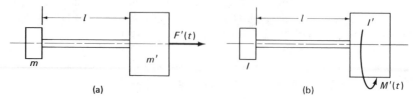

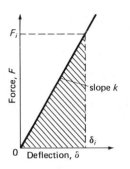

FIGURE 13.3. Force vs deflection in the elastic member, with the shaded area representing energy stored.

by multiplying the kinetic energy of the striking body by a correction factor η, that must be determined separately for each type of elastic member. If the effect is negligible, then $\eta = 1.0$. Thus for a body striking horizontally, as in Fig. 13.4(a), the energy is $\eta m' v_0^2 / 2$. The elastic energy may now be equated to the kinetic energy, or

$$\frac{F_i^2}{2k} = \frac{\eta m' v_0^2}{2} \tag{a}$$

whence

$$F_i = v_0 \sqrt{\eta m' k} \tag{13.1}$$

Corresponding to a load $W' = gm'$ applied gradually, we have a static stress σ_{st} and deflection δ_{st}. Under impact, let σ_i be the maximum stress corresponding to δ_i, the maximum deflection. With $m' = W'/g$ and $k = W'/\delta_{st}$, the term within the radical of Eq. (13.1) becomes $W'^2 \eta / g \delta_{st}$, and Eq. (13.1) may be rewritten as

$$\frac{F_i}{W'} = \frac{\sigma_i}{\sigma_{st}} = \frac{\delta_i}{\delta_{st}} = v_0 \sqrt{\frac{\eta}{g \delta_{st}}} \tag{13.2}$$

The last term is a factor by which the static load W', stress σ_{st}, or deflection δ_{st} may be multiplied to give the impact load, stress, or deflection, respectively, for any elastic member with linear characteristics, struck horizontally.

When a member is struck vertically, as is the bar in Fig. 13.4(b) or a beam on which a body is dropped from a height h, the body's energy includes an amount gained in

FIGURE 13.4. Striking of a supported elastic member by a body of mass m'. (a) Horizontally. (b) With vertical fall from a height h.

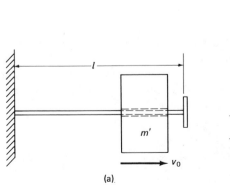

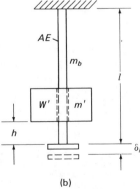

(a)   (b)

moving through the deflection distance δ_i. Thus Eq. (a) is replaced by

$$\frac{F_i^2}{2k} = \frac{\eta m' v_0^2}{2} + W'\delta_i = W'(\eta h + \delta_i) \tag{b}$$

since $v_0^2 = 2gh$. To obtain a ratio F_i/W', k is again replaced by W'/δ_{st}, and after rearrangement, there is a quadratic equation to be solved. Thus from Eq. (b),

$$F_i^2 = 2(W'/\delta_{st})W'(\eta h + \delta_i)$$

$$\left(\frac{F_i}{W'}\right)^2 = \frac{2\eta h}{\delta_{st}} + \frac{2\delta_i}{\delta_{st}} = \frac{2\eta h}{\delta_{st}} + 2\left(\frac{F_i}{W'}\right)$$

and

$$\left(\frac{F_i}{W'}\right)^2 - 2\left(\frac{F_i}{W'}\right) - \frac{2\eta h}{\delta_{st}} = 0$$

a quadratic equation in F_i/W'. Thus

$$\frac{F_i}{W'} = \frac{\sigma_i}{\sigma_{st}} = \frac{\delta_i}{\delta_{st}} = 1 + \sqrt{1 + \frac{2\eta h}{\delta_{st}}} \tag{13.3}$$

When $h = 0$, i.e., when the body is held just in contact with the elastic member but is not supported by it, and is then released, Eq. (13.3) gives $\sigma_i = 2\sigma_{st}$. From this comes the statement that a load suddenly applied doubles the deflection and stress. It is a case of force impact, under a constant gravity force. A wave analysis, such as that of Fig. 13.13, shows that the stress may be more than doubled.

13.3 ENERGY METHOD: PARTICULAR CASES

For the determination of correction factors the assumption is made that at any time following contact the velocities of the particles of the elastic member are proportional to their deflection under a static load applied at the place of impact. Thus for the bar of Fig. 13.5(a) with a cross-sectional area A, velocity u of an element of mass $\rho A\, dx$ at location x is $u = v_a(x/l)$, where v_a is the velocity of the mass m' immediately after striking. The total kinetic energy of the system is

$$E_k = \frac{m' v_a^2}{2} + \int_0^l \frac{\rho A\, dx}{2}\left(\frac{v_a x}{l}\right)^2 = \frac{v_a^2}{2}\left(m' + \frac{\rho A}{l^2}\int_0^l x^2\, dx\right)$$

$$= \frac{v_a^2}{2}\left(m' + \frac{\rho A l}{3}\right) = \frac{v_a^2}{2}\left(m' + \frac{m_b}{3}\right) \tag{a}$$

The energy is the same as if the bar were replaced by a massless spring and one-third of its mass were added to the striking mass m', as illustrated in Fig. 13.5(b).

The velocity v_a is obtained from the principle of conservation of momentum applied to the collision of two inelastic bodies of masses m' and $m_b/3$.[1] The momentum after impact (Fig. 13.5(b)) equals that before impact (Fig. 13.5(c)). Thus $(m' + m_b/3)v_a =$

---

[1] Some writers have equated $m'v_a + \int_0^l (\rho A\, dx)(v_a x/l)$ to $m'v_0$ and obtain $v_a = v_0 m'/(m' + m_b/2)$. This gives slightly lower values to σ_i. This was apparently done to obtain the correction factors listed in a book by R. J. Roark and W. C. Young, *Formulas for Stress and Strain*, 5th ed., New York: McGraw-Hill, pp. 580–581, 1975.

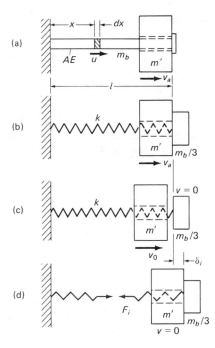

FIGURE 13.5. Consideration of mass of elastic member by the energy method. (a) Bar immediately after being struck, with struck end and mass moving at velocity v_a, element at x with velocity $u = v_a(x/l)$. (b) Equivalent concentrated bar mass $m_b/3$ moving with m' at velocity v_a. (c) Condition for momentum just before striking. (d) Maximum deflection, with maximum elastic energy stored and velocity zero throughout.

$m'v_0$, whence

$$v_a = v_0 \frac{m'}{m' + m_b/3} \tag{b}$$

The condition of zero velocity and maximum deflection and force is shown in Fig. 13.5(d).

Substitution from Eq. (b) into Eq. (a) gives for the kinetic energy

$$E_k = \frac{m'v_0^2}{2} \frac{m'}{m' + m_b/3} \tag{c}$$

A comparison with Eqs. (a) and (b) of Section 13.2 shows that the second fraction is the correction factor η, to be used in Eqs. (13.2) and (13.3). Thus for a bar

$$\eta = \frac{m'}{m' + m_b/3} = \frac{1}{1 + m_b/3m'} \tag{13.4}$$

For a beam it is assumed that its elements are set into transverse velocities that are proportional to the deflections of the elements under static loading at the place of impact (refer to Table 9.4). An equivalent mass is found by integration of the energy of an element over the length of the beam. This mass is added to the striking mass, and the equation of conservation of momentum is applied to find the velocity v_a. Then, from the equality of kinetic and elastic energies, a correction factor is found for use in Eqs. (13.2) and (13.3). For transverse impact at the end of a cantilever beam, it is

$$\eta = \frac{1}{1 + (33/140)(m_b/m')} \tag{13.5}$$

and for impact at the midlength of a simply supported beam, it is

$$\eta = \frac{1}{1 + (17/35)(m_b/m')} \tag{13.6}$$

where m_b is the total mass of the beam.

Equations for the horizontal impact of a bar will now be put into more fundamental terms for convenience and comparisons. Since $\sigma_{st} = W'/A$ and $\delta_{st} = W'l/AE$, Eq. (13.2) may be written

$$\sigma_i = \sigma_{st} v_0 \sqrt{\frac{\eta}{g\delta_{st}}} = \frac{W'v_0}{A} \sqrt{\frac{\eta AE}{gW'l}} = v_0 \sqrt{\frac{E(W'/g)\eta}{Al}} = v_0 \sqrt{E\rho} \sqrt{\frac{m'\eta}{\rho Al}} \tag{d}$$

It is shown in Section 13.5 and by Eq. (13.17) that $v_0\sqrt{E\rho}$ is the initial stress σ_0 at the struck end of the bar. Also, ρAl is the mass of the bar m_b. Let the ratio of masses $m_b/m' = \beta$. Then the correction factor, Eq. (13.4), is $\beta = 1/(1+\beta/3)$ and Eq. (d) becomes

$$\sigma_i = \sigma_0 \sqrt{\frac{1}{\beta}\left(\frac{1}{1+\beta/3}\right)} \tag{13.7}$$

Note that the maximum impact stress σ_i is a function only of striking velocity v_0, bar material properties E and ρ, and the mass ratio β.

Example 13.1
A steel bar 50 mm in diameter and 750 mm long is rigidly held at one end and struck at the other by a relatively rigid body weighing 5.0 kN. Predict the stress if the bar is horizontal and the striking body has a velocity of 0.5 m/s.

Solution
Area $A = (\pi/4)(50 \times 10^{-3})^2 = 1.964 \times 10^{-3}$ m². For steels, $E = 207$ GPa (207×10^9 kg/m·s²) and $\rho = 7.778 \times 10^3$ kg/m³ (Example 8.4). Also, $m' = 5000/9.8067 = 510$ kg. Hence

$$m_b = \rho Al = (7.778 \times 10^3)(1.964 \times 10^{-3})(0.750) = 11.46 \text{ kg}$$

$$\beta = m_b/m' = 11.46/510 = 0.0225 \quad \text{and} \quad 1/\beta = 44.5$$

$$\sigma_0 = v_0\sqrt{E\rho}$$

$$= 0.5 \text{ m/s} \sqrt{(207 \times 10^9 \text{ kg/m·s}^2)(7.778 \times 10^3 \text{ kg/m}^3)}$$

$$= 20.1 \times 10^6 \text{ kg/m·s}^2 \quad (20.1 \text{ MPa})$$

From Eq. (13.7)

$$\sigma_i = 20.1 \sqrt{(44.5)/(1 + 0.0225/3)} = 134 \text{ MPa}$$

Equation (13.37), obtained from a wave analysis, gives

$$\sigma = \sigma_0 R = \sigma_0(1 + \sqrt{(1/\beta) + (2/3)}) = 20.1(1 + \sqrt{44.5 + 0.67}) = 155 \text{ MPa}$$

It is seen that the energy method predicts a lower stress and that the error is much larger than that made by neglecting the correction factor for bar mass. Actually, the error may not be as large as apparent because local effects reduce the peak stresses predicted by the wave method. ////

It is shown in Section 13.8 that torsional impact is analogous to longitudinal impact. The initial shear stress, analogous to σ_0, is $\tau_0 = (\omega_0 d/2)\sqrt{G\rho}$, where ω_0 is the striking velocity in rad/s, d is the shaft diameter, and G is the shear modulus. The moment of inertia of the cylindrical shaft is $I_b = m_b d^2/8$. Moments of inertia I_b and I' correspond to m_b and m', respectively, and $\beta = I_b/I'$. Thus for the case of a shaft, fixed or attached

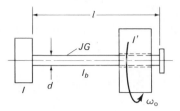

FIGURE 13.6. A body, with mass moment of inertia I' and rotational velocity ω_0 about to engage one end of a shaft, the other end of which carries an attached body with inertia I.

to a very large mass at one end and suddenly engaged by a rotating body through a jaw clutch at the other end, as in Fig. 13.1(b), Eq. (13.7) may be rewritten to give an approximation for maximum shear stress,

$$\tau_i = \tau_0 \sqrt{\frac{1}{\beta}\left(\frac{1}{1+\beta/3}\right)} \tag{13.8}$$

Again, attention is called to the more accurate Eq. (13.37).

A derivation in torsional terms will be made for the case shown in Fig. 13.6, where the nonstruck end carries an attached body of finite inertia I. The shaft has a rigidity JG, where J is the polar moment of inertia of the cross-sectional area, $\pi d^4/32$. An approximation is to assume that the energy before striking is converted into elastic energy of twist in the shaft plus the kinetic energy of motion of all bodies rotating together at some velocity ω_a, where $\omega_a < \omega_0$. From Eq. (12.3), $M_t = (2J/d)\tau_i$ and from Eqs. (12.4), $\theta = M_t l/JG = (2J\tau_i/d)l/JG = (2l/Gd)\tau_i$. The elastic energy is $M_t\theta/2$, and the energy balance becomes

$$\frac{2Jl}{Gd^2}\tau_i^2 + \frac{1}{2}(I + I_b + I')\omega_a^2 = \frac{1}{2}I'\omega_0^2 \tag{e}$$

From conservation of momentum, $(I + I_b + I')\omega_a = I'\omega_0$, whence

$$\omega_a = \frac{I'\omega_0}{I + I_b + I'} \tag{f}$$

By substitution for ω_a in Eq. (e) and its solution for stress

$$\tau_i^2 = I'\omega_0^2\left(1 - \frac{I'}{I + I_b + I'}\right)\frac{Gd^2}{4Jl}$$

and

$$\tau_i = \frac{\omega_0 d}{2}\sqrt{\frac{I'G}{Jl}}\sqrt{\frac{I + I_b}{I + I_b + I'}} \tag{g}$$

Two inertia ratios are introduced, $\alpha = I_b/I$ and $\lambda = I'/I$. The term under the second radical becomes $(1+\alpha)/(1+\alpha+\lambda)$. Now $I_b = \rho Jl$, and the term under the first radical becomes $I'G/(I_b/\rho) = G\rho(I'/I_b) = G\rho(\lambda/\alpha)$. Since $(\omega_0 d/2)\sqrt{G\rho}$ is the initial stress τ_0, the equation for maximum shear stress becomes

$$\tau_i = \tau_0 \sqrt{\frac{\lambda(1+\alpha)}{\alpha(1+\alpha+\lambda)}} \tag{13.9}$$

An equation for the corresponding case in longitudinal impact may be written by analogy, thus

$$\sigma_i = \sigma_0 \sqrt{\frac{\lambda(1 + \alpha)}{\alpha(1 + \alpha + \lambda)}}$$ (13.10)

where $\alpha = m_b/m$ and $\lambda = m'/m$. A correction for these two equations is proposed in Eq. (13.36) of Section 13.7.

13.4 LONGITUDINAL WAVES IN ELASTIC MEDIA

It is assumed that the action initiated in a long bar (Fig. 13.7) is such that the particles in any normal section move together through parallel planes. Their displacement u is a function of position x and time t. The change in length of the element is $(\partial u/\partial x)dx$, and the unit change or strain is $\partial u/\partial x$. Hence stress $\sigma = E(\partial u/\partial x)$ and the force on the section at x is $A\sigma = AE(\partial u/\partial x)$. Particle velocity is $\partial u/\partial t$, acceleration is $\partial^2 u/\partial t^2$, and element mass is $(A\ dx)\rho$, where ρ is mass density.

With stresses chosen to be positive or tensile, a summation of forces on the element (Fig. 13.7) is equated to mass times acceleration to give

$$AE\left[\frac{\partial u}{\partial x} + \frac{\partial}{\partial x}\left(\frac{\partial u}{\partial x}\right)dx\right] - AE\frac{\partial u}{\partial x} = A\ dx\ \rho\ \frac{\partial^2 u}{\partial t^2}$$ (a)

whence

$$\frac{\partial^2 u}{\partial t^2} = \frac{E}{\rho}\frac{\partial^2 u}{\partial x^2} = c^2\frac{\partial^2 u}{\partial x^2}$$ (13.11)

where $c = \sqrt{E/\rho}$.

The general solution of this partial differential equation, which can be verified by substitution into Eq. (13.11), is

$$u = f(x - ct) + f'(x + ct)$$ (13.12)

where the particular functions f and f' remain to be evaluated. The nature of the parameters $(x - ct)$ and $(x + ct)$ indicates that the displacement and its derivatives, strain and velocity, travel in waves. A value of the function at x and t moves unchanged in an increment of time Δt to a new position $(x + \Delta x)$, where Δx is an amount such that the value of the parameter is unchanged (Fig. 13.8). Thus for the parameter of function f, $(x + \Delta x) - c(t + \Delta t) = x - ct$, whence $\Delta x = c\Delta t$. The rate or velocity with which this function or disturbance moves is $\Delta x/\Delta t = c = \sqrt{E/\rho}$. It is commonly called the wave velocity or the speed of sound. In steels c has a value of 16 900 ft/s (5150 m/s), and it is much larger than the velocities $\partial u/\partial t$ of the particles, which never move very far relative to each other.

FIGURE 13.7. Displacements and forces on an element of an elastic bar.

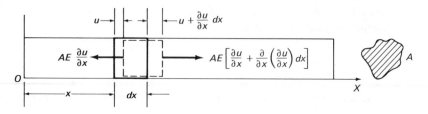

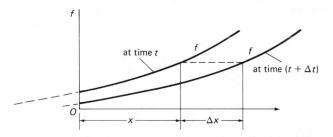

FIGURE 13.8. Travel $\triangle x$ of values of the function f in time $\triangle t$.

It is seen from the foregoing that the wave for function f has a positive velocity and that it continuously moves to the right. The parameter of function f' indicates that its wave velocity is $\Delta x/\Delta t = -c$, a negative velocity or continuous motion to the left. The displacement u is the sum of a function in a positive wave and a function in a negative wave, and so are its derivatives.

A relationship between stress and particle velocity in each wave will be obtained. For the positive wave designate the parameter of f as y, the stress as s, and the velocity as v. Then, $y = x - ct$ and

$$v = \frac{\partial f}{\partial t} = \frac{\partial f}{\partial y}\frac{\partial y}{\partial t} = \frac{\partial f}{\partial y}(-c) \tag{b}$$

and

$$s = E\frac{\partial f}{\partial x} = E\frac{\partial f}{\partial y}\frac{\partial y}{\partial x} = E\frac{\partial f}{\partial y}(1) \tag{c}$$

Their ratio is

$$\frac{v}{s} = \frac{-c}{E} = -\frac{\sqrt{E/\rho}}{E} = -\frac{1}{\sqrt{E\rho}}$$

Hence

$$v = -\frac{s}{\sqrt{E\rho}} \quad \text{and} \quad s = -v\sqrt{E\rho} \tag{13.13}$$

From this equation it is seen that in a positive wave, i.e., one propagated to the right, particles that move with a positive velocity v give a negative value of stress or a compression, as shown in Fig. 13.9(a). Likewise, particles moving in a negative direction give a positive value of stress or an extension.[2]

For a negative wave, designate the parameter of f' as z, the stress as s', and the velocity as v'. Then, $z = x + ct$ and

$$v' = \frac{\partial f'}{\partial t} = \frac{\partial f'}{\partial z}\frac{\partial z}{\partial t} = \frac{\partial f'}{\partial z}(+c)$$

$$s' = E\frac{\partial f'}{\partial x} = E\frac{\partial f'}{\partial z}\frac{\partial z}{\partial x} = E\frac{\partial f'}{\partial z}(1)$$

---

[2] A short treatment of wave propagation is given in a standard reference, S. Timoshenko and J. N. Goodier, *Theory of Elasticity*, 3rd ed., New York: McGraw-Hill, pp. 485–513, 1970. There, however, contrary to the practice in other areas of elasticity, compressive stresses are given a positive sign. Thus the equation corresponding to Eq. (13.13) carries a positive sign.

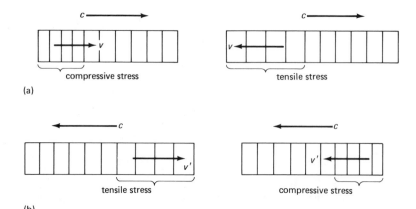

FIGURE 13.9. Strains and corresponding stresses. (a) In a positive wave, with propagation to the right. (b) In a negative wave, with propagation to the left.

Their ratio is

$$\frac{v'}{s'} = \frac{c}{E} = \frac{\sqrt{E/\rho}}{E} = \frac{1}{\sqrt{E\rho}}$$

whence

$$v' = \frac{s'}{\sqrt{E\rho}} \quad \text{and} \quad s' = v'\sqrt{E\rho} \tag{13.14}$$

Thus in a negative wave, positive-particle velocities give positive stresses, and negative-particle velocities give negative stresses, as illustrated in Fig. 13.9(b).

The total stress σ is the sum of the stresses in the two waves,

$$\sigma = E\frac{\partial u}{\partial x} = E\left(\frac{\partial f}{\partial x} + \frac{\partial f'}{\partial x}\right) = s + s' \tag{13.15}$$

Likewise, the net velocity V, in terms of stresses, is

$$V = \frac{\partial u}{\partial t} = \frac{\partial f}{\partial t} + \frac{\partial f'}{\partial t} = v + v' = \frac{1}{\sqrt{E\rho}}(-s + s') \tag{13.16}$$

Thus if we can sketch the wave-stress distributions s and s' at any time t, we may add them to obtain the total stress and subtract them to represent the net particle velocity. This is done in Figs. 13.10, 13.12, and 13.13.

13.5 INITIAL AND FIRST PERIOD STRESSES

When a rigid body strikes the end of an elastic bar, the particles of the bar in the plane of striking are instantly brought to the velocity of the body, and a single wave starts along the bar. If striking occurs by a body with mass m' at the positive end, at $x = l$ in Fig. 13.10(a), a negative wave is initiated. By Eq. (13.15), the initial stress is $\sigma_0 = s_0'$, and if the striking velocity is v_0, then by Eq. (13.14),

$$\sigma_0 = v_0\sqrt{E\rho} \tag{13.17}$$

This stress is shown in Fig. 13.10(b). It becomes the "wave front" stress, moving along

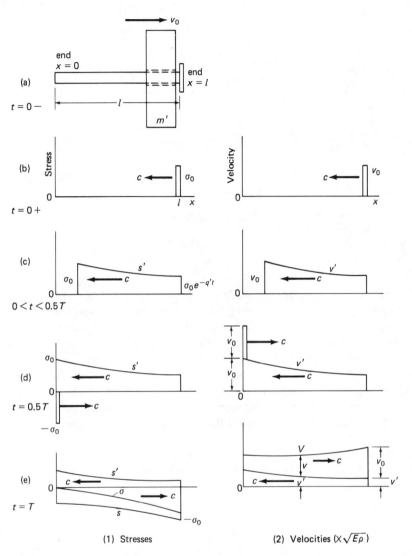

FIGURE 13.10. Striking impact by a mass m' on a bar with both ends free. (a) Configuration just before striking. From (b) to (e), stresses in column (1) and particle velocities (times $\sqrt{E\rho}$) in column (2). Time period $T = 2l/c$, where c is the propagation velocity. Striking at $t = 0$, reflection at $t = 0.5T$, and separation at $t = T$.

the bar (Fig. 13.10(c)). Behind it, an elastic force As' is developed, which acts against m' to reduce its velocity.[3]

The subsequent velocity, v' and stress s' at end $x = l$ are found from the equation that the force on m' equals its mass times acceleration. From Eqs. (13.14) and since acceleration is dv'/dt,

$$-As' = m'\frac{dv'}{dt} = \frac{m'}{\sqrt{E\rho}}\frac{ds'}{dt} \tag{a}$$

---

[3] The direction of striking has been taken in the positive x direction at the positive end of the bar in order to initially induce tensile stresses. These are indicated by a positive sign, the usual convention. In many references, striking is shown to occur in the positive direction at the *negative end*, probably because this is easily accomplished physically. However, the resulting compressive stresses are usually given a positive sign and plotted in an upward direction.

or

$$\frac{ds'}{dt} + \frac{A\sqrt{E\rho}}{m'} s' = \frac{ds'}{dt} + q's' = 0 \tag{b}$$

where $q' = A\sqrt{E\rho}/m'$. A multiplication through by $e^{q't}$, followed by integration gives

$$e^{q't}\frac{ds'}{dt} + q's'e^{q't} = 0 \tag{c}$$

and

$$s'e^{q't} = C \quad \text{or} \quad s' = Ce^{-q't} \tag{d}$$

where C is the constant of integration. At time $t = 0$, $s' = \sigma_0$, whence $C = \sigma_0$, and since no positive wave exists,

$$\sigma'_{x=l} = s' = \sigma_0 e^{-q't} \quad \text{and} \quad V'_{x=l} = v' = \frac{\sigma_0}{\sqrt{E\rho}} e^{-q't} \tag{13.18}$$

These decreasing exponential values initially occur at time t at end $x = l$, but they issue continuously to travel to the left at velocity c, trailing the wave front, as shown in Fig. 13.10(c) through (e) in the stress column (1) and the velocity column (2).

The wave-front stress σ_0 arrives at the nonstruck end, $x = 0$, at time $t = l/c = 0.5T$, where l is the length of the bar, c is the propagation velocity, and $0.5T$ is a half-period or interval of time for a wave to travel the length of the bar. The subsequent action depends upon the end conditions. If this end is "free" of attachment or mass, then no force or stress can be developed there. The total stress is $\sigma = 0$, and by Eq. (13.15) $s = -\sigma_0$, which is the front value of the wave of compressive stress which now starts to the right. The wave that continues to arrive, the *incident* wave, has the stress value developed at the other end at a time $0.5T$ earlier, or, by Eqs. (13.18), $\sigma_0 e^{-q'(t-0.5T)}$. By Eq. (13.15) with $\sigma = 0$, the stress in the *reflected* wave during the first full period at end $x = 0$ is

$$s = -\sigma_0 e^{-q'(t-0.5T)} \tag{e}$$

This is a positive wave s with a negative stress value, as plotted in Fig. 13.10(d) and (e). When the struck end $x = l$ is reached at the end of the period, time T, the total stress indicated, $\sigma = s + s'$, is negative or compressive. Unless m' has become attached, a force on the body to the right cannot be supported, and separation occurs.

The velocities are represented in a second column (Fig. 13.10(2)), by plotting $(-s + s')$ from Eq. (13.16). Note that the velocity is doubled at the free end, and at time $t = T$, the particle velocity of the struck end becomes $V = v + v'$, which is larger than the reduced velocity v' of the mass m'. Hence the end pulls away from the body, again indicating separation.

If the nonstruck end is fixed or attached to a rigid wall, i.e., $m = \infty$, no particle motion can occur there at any time, and Eq. (13.16) is applied to determine the stress in the positive wave, thus

$$V = 0 = \frac{1}{\sqrt{E\rho}} (-s + s')$$

and

$$s = s' = \sigma_0 e^{-q'(t-0.5T)} \tag{f}$$

giving

$$\sigma = s + s' = 2\sigma_0 e^{-q'(t-0.5T)} \tag{13.19}$$

at end $x = 0$ at times $0.5T \leq t < 1.5T$. This is a doubling or complete reflection of the impinging stress. Subsequent reflections at both ends may increase the stress considerably.

Another condition occurs when a wave front impinges upon a *finite*, attached mass m, as in Fig. 13.1(a). At the instant of impingement no instantaneous change of its velocity is possible, and the stress front is reflected in full. However, an elastic force now acts on the mass, and its velocity begins to change with an increment having the sign of the impinging stress s'. Thus from Eq. (13.16),

$$V = \frac{1}{\sqrt{E\rho}}(-s + s') \quad \text{and} \quad s = s' - V\sqrt{E\rho} \tag{13.20}$$

and the reflection of stress occurs in less than full magnitude. The determination of the subsequent action before separation is somewhat more complicated. It is discussed in the following section.

13.6 IMPACT ON A UNIFORM BAR WITH A RIGID BODY AT ONE END

In a case commonly treated, one end of the bar is rigidly attached to an *infinite* mass, and the other end is struck.[4] A more general case and one more representative of machinery components is that of Fig. 13.6, in which there is an attached *finite* mass at the nonstruck end. This case is treated in the following, and the results are applied to both longitudinal and torsional impact.[5]

It is apparent that several periods of wave travel may occur before the stress in the rod reaches a peak. Every T seconds the wave front arrives at one end of the bar, and because of this discontinuity a new equation is required for the stress in the wave leaving the end. Hence general equations are needed, good for any interval or period $n = 1, 2, 3, \ldots$. At end $x = l$, the times of the intervals are $0 \leq t < 1.0T$, $1.0T \leq t < 2.0T$, $\ldots$. At end $x = 0$ the times are $0.5T \leq t < 1.5T$, $1.5T \leq t < 2.5T$, $\ldots$. As before, stresses originating at $x = l$ and traveling in a negative direction are designated by a prime mark. A stress written as $s_n'(t - 0.5T)$ is read "the s_n' function of $(t - 0.5T)$", i.e., into the equation of stress issuing from end $x = l$ during one of its intervals n, there must be substituted for t the value $(t - 0.5T)$, where t is the time from the initiation of impact. Stress $s_n'(t - 0.5T)$ is that which arrives at end $x = 0$ at time t. It is added to the stress $s_n(t)$ that is leaving $x = 0$ to give the total stress there. Thus the time intervals, total stresses by Eq. (13.15), and net velocities by Eq. (13.16) are, respectively,

at end $x = 0$,

$$(n - 0.5)T \leq t < (n + 0.5)T$$

$$\sigma_n = s_n(t) + s_n'(t - 0.5T) \tag{13.21}$$

$$V_n = \frac{1}{\sqrt{E\rho}}[-s_n(t) + s_n'(t - 0.5T)]$$

[4] S. Timoshenko and J. N. Goodier, *Theory of Elasticity*, 3rd ed., New York: McGraw-Hill, pp. 485–513, 1970.
[5] A. H. Burr, "Longitudinal and Torsional Impact in a Uniform Bar with a Rigid Body at One End," *J. Appl. Mech.*, Vol. 17, No. 2, pp. 209–217, June 1950; or bound in rear of *Trans. ASME*, Vol. 72, 1950.

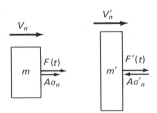

FIGURE 13.11. Forces acting on masses m and m', in contact with a bar at ends $x = 0$ and $x = l$, respectively, following striking and force impacts.

and at end $x = l$,

$$(n - 1)T \leqslant t < nT$$

$$\sigma_n' = s_{n-1}(t - 0.5T) + s_n'(t) \tag{13.22}$$

$$V_n' = \frac{1}{\sqrt{E\rho}}[-s_{n-1}(t - 0.5T) + s_n'(t)]$$

Figure 13.11 shows bodies with masses m and m' acted upon by the elastic forces from the bar, also by forces $F(t)$ and $F'(t)$, functions of time for the case of force impact. General equations of equilibrium for m and m' are, respectively,

$$A\sigma_n + F(t) = m\frac{dV_n}{dt}$$

$$-A\sigma_n' + F'(t) = m'\frac{dV_n'}{dt} \tag{13.23}$$

A solution for the first of Eqs. (13.23) is started with substitutions from Eqs. (13.21) for σ_n and V_n and the following representation,

$$q = \frac{A\sqrt{E\rho}}{m} \tag{a}$$

This gives

$$q[s_n(t) + s_n'(t - 0.5T)] + (q/A)F(t) = \frac{d}{dt}[-s_n(t) + s_n'(t - 0.5T)] \tag{b}$$

This is followed by a regrouping of terms, the addition of $-2qs_n'(t - 0.5T)$ to both sides, and the substitution of the following representations,

$$Q = -(q/A)F(t) - 2qs_n'(t - 0.5T) \tag{c}$$

$$y = s_n(t) - s_n'(t - 0.5T)$$

where Q and y are functions of time t. These actions yield the simple, first-order differential equation

$$\frac{dy}{dt} + qy = Q \tag{d}$$

Multiplication throughout by e^{qt}, followed by integration, gives

$$y = e^{-qt}\int e^{qt} Q \, dt + C_n e^{-qt} \tag{e}$$

The resubstitution of the terms for y and Q from Eq. (b) yields

$$s_n(t) = s_n'(t - 0.5T) - 2qe^{-qt} \int e^{qt} s_n'(t - 0.5T) \, dt$$

$$- (q/A)e^{-qt} \int e^{qt} F(t) \, dt + C_n e^{-qt} \tag{13.24}$$

This equation, the "recurrence formula" for end $x = 0$, enables one to find the stress s_n in the wave that is leaving end $x = 0$ during the new interval n. The formula includes the previously determined value of s_n' and the integration constant C_n. The constant C_n is determined from the condition that the initial value of s_n in the new interval n equals the final value of s_{n-1} in the preceding interval plus any wave-front stress σ_0. This, as previously stated, is because an instantaneous change of velocity of mass m is not possible, and by Eq. (13.16) the wave front must be reflected in full. The intervals begin at times $t = (n - 0.5)T$, so

$$s_n[(n - 0.5)T] = s_{n-1}[(n - 0.5)T] + \sigma_0 \tag{13.25}$$

At $x = l$ the intervals begin at times $t = (n - 1)T$. From the second of Eqs. (13.23) and by a method similar to the foregoing, there is obtained the recurrence equation and the condition for C_n', respectively,

$$s_n'(t) = s_{n-1}(t - 0.5T) - 2q'e^{-q't} \int e^{q't} s_{n-1}(t - 0.5T) \, dt$$

$$+ (q'/A)e^{-q't} \int e^{q't} F'(t) \, dt + C_n' e^{-q't} \tag{13.26}$$

and

$$s_n'[(n - 1)T] = s_{n-1}'[(n - 1)T] + \sigma_0 \tag{13.27}$$

where the value of q', as before, is $q' = A\sqrt{E\rho/m'}$.

13.7 APPLICATION TO STRIKING IMPACT: CHART SOLUTIONS: COMPARISONS

The stress for the first interval at end $x = l$ may be found by letting $n = 1$ and $s_0 = 0$ in Eqs. (13.26) and (13.27), or it is available from Eqs. (13.18), namely,

$$\sigma_1' = s_1' = \sigma_0 e^{-q't} \tag{13.28}$$

This is used for the stress value in the wave arriving at a time $0.5T$ later at end $x = 0$, and Eq. (13.24) for its first interval, $n = 1$, becomes

$$s_1 = \sigma_0 e^{-q'(t - 0.5T)} - 2qe^{-qt} \int e^{qt} \sigma_0 e^{-q'(t - 0.5T)} \, dt + C_1 e^{-qt}$$

Upon integration the middle term becomes $-\sigma_0[2q/(q - q')]e^{-q'(t - 0.5T)}$ and s_1 is

$$s_1 = -\sigma_0 K e^{-q'(t - 0.5T)} + C_1 e^{-qt} \tag{a}$$

where $K = (q + q')/(q - q')$. By Eq. (13.25), for the first interval with $n = 1$, $s_1(0.5T) = 0 + \sigma_0$, and with substitution from Eq. (a) with t set equal to $(n - 0.5)T = 0.5T$, we obtain $C_1 = \sigma_0(1 + K)e^{q(0.5T)}$. Equation (a) for the stress wave leaving end $x = 0$

becomes

$$s_1 = \sigma_0[(1 + K)e^{-q(t-0.5T)} - Ke^{-q'(t-0.5T)}] \tag{13.29}$$

The total stress σ_1 at $x = 0$ during the first interval there, is, by Eqs. (13.21),

$$\sigma_1 = s_1 + s_1'(t - 0.5T)$$
$$= \sigma_0[(1 + K)e^{-q(t-0.5T)} + (1 - K)e^{-q'(t-0.5T)}] \tag{13.30}$$

This is followed by the application of Eqs. (13.26) and (13.27) to obtain s_2' and σ_2', then by Eqs. (13.24) and (13.25) again to obtain s_2 and σ_2, and so on, alternately.

The quantities q and q' were chosen earlier as convenient abbreviations for several terms. However, their greater significance can be shown by the following manipulation,

$$q = \frac{A\sqrt{E\rho}}{m} = \frac{Al\rho}{m}\frac{\sqrt{E\rho}}{l\rho} = \frac{m_b}{m}\frac{\sqrt{E/\rho}}{l} = \alpha\frac{c}{l} = \frac{2\alpha}{T} \tag{13.31}$$

where α is the mass ratio m_b/m and $T = 2l/c$ is the period. Also,

$$q' = \frac{A\sqrt{E\rho}}{m'} = \frac{m}{m'}q = \frac{2\alpha}{\lambda T} \tag{13.32}$$

where λ is the mass ratio m'/m. Ratio K becomes $(\lambda + 1)/(\lambda - 1)$, and with the introduction of a time ratio $\zeta = t/T$, we may put Eqs. (13.28) through (13.30) into nondimensional forms. Let the symbols for total-stress ratio be $R = \sigma/\sigma_0$ and $R' = \sigma'/\sigma_0$ and for wave-stress ratio $S = s/\sigma_0$ and $S' = s'/\sigma_0$. Then

$$R_1' = \frac{\sigma_1'}{\sigma_0} = e^{-(2\alpha/\lambda)\zeta} \tag{13.33}$$

$$S_1 = \frac{s_1}{\sigma_0} = \frac{2\lambda}{\lambda - 1}e^{-2\alpha(\zeta-0.5)} - \frac{\lambda + 1}{\lambda - 1}e^{-(2\alpha/\lambda)(\zeta-0.5)} \tag{13.34}$$

$$R_1 = \frac{\sigma_1}{\sigma_0} = \frac{2\lambda}{\lambda - 1}e^{-2\alpha(\zeta-0.5)} - \frac{2}{\lambda - 1}e^{-(2\alpha/\lambda)(\zeta-0.5)} \tag{13.35}$$

It is seen that the first-interval stresses, and the stresses derived from them for subsequent intervals, are functions only of the initial stress σ_0 at impact, the two mass ratios α and λ, and the time ratio ζ. It is also true for the particle velocities, and nondimensional velocities $\bar{V} = V/v_0$ and $\bar{V}' = V'/v_0$ may be written from Eqs. (13.21) and (13.22).

To illustrate the wave action, Fig. 13.12 is plotted for the values $1/\alpha = m/m_b = 4$ and $\lambda = m'/m = 10$ at different time ratios $\zeta = t/T$. The stresses S and S' in the positive and negative waves, the total stress $R = S' + S$, and the velocity $\bar{V} = S' - S$ are shown. Note that the stresses are augmented at the beginnings of their several intervals. However, wave stress S at end $x = 0$ soon begins to drop rapidly, as should be expected considering that the mass m is one-tenth of mass m'. Its value becomes negative soon after $t/T = 2.0$, and that negative value which arrives at end $x = l$ at time $t/T = 2.85$ exactly cancels out the positive value in wave S', giving zero stress R in the bar. This is the moment of "rebound" or separation; henceforth the end of the bar moves more rapidly to the right than does the mass m'. It can be shown that subsequently there will be no stresses higher than those that have already occurred. The velocities corresponding to each stress picture are also plotted. Note that although there are discontinuities in the particle velocities for $0 < x < l$, there are no discontinuities at $x = 0$ and $x = l$ where the bar is in contact with the rigid masses m and m'.

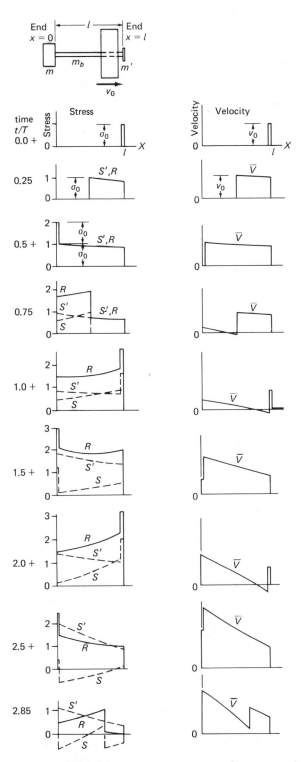

FIGURE 13.12. Stress and particle velocity profiles from impact to separation in a steel bar. Stress ratios are $R = \sigma/\sigma_0$, $S = s/\sigma_0$, and $S' = s'/\sigma_0$. Attached mass m, bar mass m_b, and striking mass m', as shown. Plotted for $m/m_b = 1/\alpha = 4.0$ and $m'/m = \lambda = 10.0$. (From A. H. Burr, "Longitudinal and Torsional Impact in a Uniform Bar with a Rigid Body at One End," *J. Appl. Mech.*, Vol. 17, No. 2, pp. 209–217, 1950; or *Trans. ASME*, Vol. 72, 1950, in rear.)

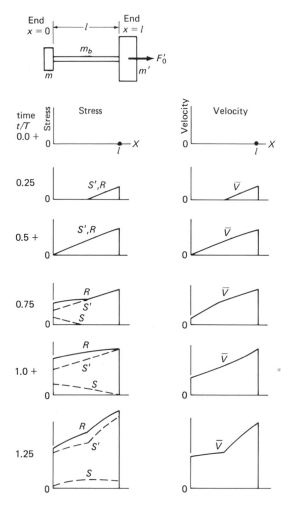

FIGURE 13.13. Stress and particle velocity profiles following the sudden application and maintenance of a constant force F_0' at end $x = l$ of a steel bar with masses at both ends. $R = \sigma/\sigma_{st}$. Same ratios $1/\alpha$ and λ and same source as Figure 13.12.

Provision was also made for *force impact* in the equilibrium equations, Eqs. (13.23), and the recurrence formulas, Eqs. (13.24) and (13.26). There is no wave front, and the formulas for determining the constants of integration, Eqs. (13.25) and (13.27) are used with σ_0 set equal to zero. This is illustrated in Fig. 13.13 for a constant force F_0' suddenly applied to the body of mass m' at end $x = l$. Note that the stress at this end and elsewhere builds up without discontinuity; there is no rigid striking mass that requires a sudden change in particle velocity. Stress ratios R, S, and S' are based on the static stress $\sigma_{st} = F_0'/A$.

For the case of striking velocity, the nondimensional stresses and velocities are now plotted as a function of time in Fig. 13.14(a) and (b) for ends $x = l$ and $x = 0$, respectively. Note that the total stress R at end $x = 0$ reaches a maximum value 3.14 times the initial stress at time $t/T = 1.5$; at end $x = l$, the maximum is 3.22 at time 2.0, both occurring upon arrival of a wave front. Again, note the continuity of end velocities $\bar{V}'$ and $\bar{V}$, with $\bar{V}'$ starting at 1.0 and dropping slowly, while $\bar{V}$ at contact with the smaller mass rises from zero to a maximum of 1.85 times the striking velocity.

The highest peak value from Fig. 13.14, namely $R' = 3.22$, is plotted in Fig. 13.15

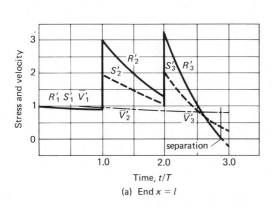
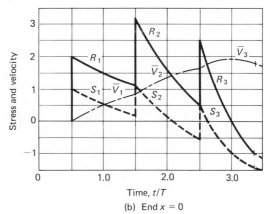

(a) End $x = l$

(b) End $x = 0$

FIGURE 13.14. For striking impact, variation with time of the total stress, wave stress, and velocity ratios. For a bar of mass m_b, with an attached mass m at end $x = 0$ and striking mass m' at end $x = l$. Plotted for $m/m_b = 1/\alpha = 4.0$, $m'/m = \lambda = 10.0$.

(a) At end $x = l$: $R' = \sigma'/\sigma_0$, $S' = s'/\sigma_0$, $\overline{V}' = V'/v_0$

(b) At end $x = 0$: $R = \sigma/\sigma_0$, $S = s/\sigma_0$, $\overline{V} = V/v_0$.

(From A. H. Burr, "Longitudinal and Torsional Impact in a Uniform Bar with a Rigid Body at One End," *J. Appl. Mech.*, Vol. 17, No. 2, pp. 209–217, 1950; or *Trans. ASME*, Vol. 72, 1950, in rear.)

together with the highest peaks from other combinations of $1/\alpha$ and λ to form a chart from which the maximum stress is readily found.[6] The chart is applicable not only to the longitudinal impact of bars, but, as will be shown in Sections 13.8 and 13.9, to the torsional impact of bars or shafts and to the longitudinal impact of helical springs. Formulas for initial stresses are given in the caption of Fig. 13.15, and an initial stress is multiplied by the stress ratio to give a maximum stress.

Example 13.2

In a soil-impacting device a tool consisting of a steel bar with a foot weighing 15 lb, a diameter of 1½ in, and a length of 16 in is repeatedly struck at its top end at a velocity of 50 in/s by a hammering body weighing 100 lb. Determine the maximum stress.

Solution

$$W_b = (\pi/4)(1.5)^2(16)(0.281) = 7.95 \text{ lb}$$

$$1/\alpha = m/m_b = W/W_b = 15/7.95 = 1.89$$

$$\lambda = m'/m = W'/W = 100/15 = 6.67$$

From the chart and its caption (Fig. 13.15),

$$R = 2.83 \quad \text{and} \quad \sigma_0 = 148v_0 = 148(50) = 7400 \text{ psi}$$

Hence

$$\sigma_{max} = R\sigma_0 = 2.83(7400) = 21\ 000 \text{ psi} \qquad\qquad \text{////}$$

---

[6] See footnote 5.

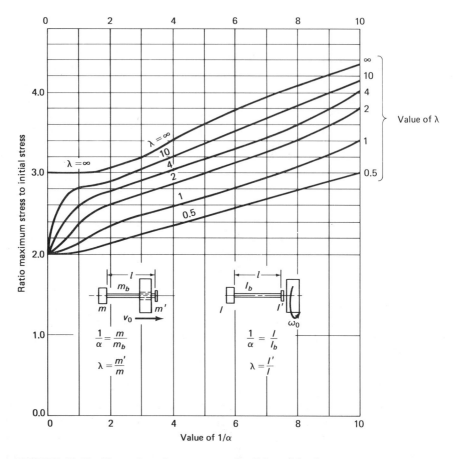

FIGURE 13.15. Chart of maximum stress ratios R in striking impact.

Longitudinal: $1/\alpha = m/m_b$, $\lambda = m'/m$, $\sigma_{max} = R\sigma_0$ where $\sigma_0 = v_0\sqrt{E\rho}$, v_0 is the striking velocity, E is the modulus of elasticity, and ρ is density. For steels, $\sigma_0 = 148v_0$ psi with v_0 in in/s, and $\sigma_0 = 40.2v_0$ MPa with v_0 in m/s.

Torsional: $1/\alpha = I/I_b$, $\lambda = I'/I$, $\tau_{max} = R\tau_0$, where, $\tau_0 = (\omega_0 d/2)\sqrt{G\rho}$, ω_0 is the striking velocity in rad/s, G is the modulus of rigidity, and d is the shaft diameter. For steels, $\tau_0 = 4.79n_0d$ psi with d in inches, and $\tau_0 = 1.30 \times 10^{-3}n_0d$ MPa with d in mm, both with the striking velocity n_0 in rpm.

(From A. H. Burr, "Longitudinal and Torsional Impact in a Uniform Bar with a Rigid Body at One End," *J. Appl. Mech.*, Vol. 17, No. 2, pp. 209–217, 1950; or *Trans. ASME*, Vol. 72, 1950, in rear.)

The equations and chart give conservative results for a uniform, symmetrical impact. This is because local effects, including deformation of the striking surfaces and any small flexibility in the bodies and at the attachment, will reduce the steepness of the wave front in the bar and the fullness of reflections and, therefore, the magnitude of the peak stresses. On the other hand, it may be difficult to prevent some eccentricity, which will superimpose bending stresses.

The $\lambda = 1$ and $\lambda = \infty$ lines of Fig. 13.15 are reproduced in Fig. 13.16 and compared with the dotted lines calculated from Eq. (13.10), derived for this case by the energy method. For the range of $1/\alpha$ in Fig. 13.16, the error is large by the energy method, a stress ratio as much as 1.2 in absolute value and on the unsafe side. Hence use of the chart of Fig. 13.15 is recommended.

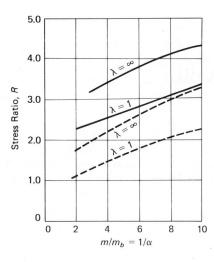

FIGURE 13.16. Stress ratios by wave method (full lines) vs energy method (dotted lines).

Values of $1/\alpha$ between 10 and 100 are common, particularly for the torsional impact of shafts. The chart has not been calculated for this region because many intervals occur before separation, and the equations become extremely long for the higher intervals. However, there is reason to believe that the difference between the stress ratios by the two methods approaches 1.0 as $1/\alpha$ increases. Hence for $1/\alpha > 10$ it is recommended that the energy equation, Eq. (13.10), be used with the addition of 1.1, thus

$$R \approx 1.1 + \sqrt{\frac{\lambda(1 + \alpha)}{\alpha(1 + \alpha + \lambda)}} \tag{13.36}$$

The case of a bar fixed at end $x = 0$, i.e., where $m \to \infty$, has been solved by the wave method for a wide range in the ratio $1/\beta = m'/m_b$. The maximum peak stresses give a discontinuous plot, that may be approximated very closely by the equation[7]

$$R = 1 + \sqrt{\frac{1}{\beta} + \frac{2}{3}} \tag{13.37}$$

A comparison with Eqs. (13.7) and (13.8) again shows a lowside error by the energy method. Hence use of Eq. (13.37) is suggested for this case, particularly for low $1/\beta$ ratios.

13.8 TORSIONAL IMPACT ON A SHAFT

From the equation relating twist to torque, Eq. (12.4), there may be written for variable twist, $M_t = JG(\partial\theta/\partial x)$, where J is the polar moment of inertia, Eq. (12.2), and G is the modulus of rigidity. The torques acting on an element of length dx are shown in Fig. 13.17(a). The moment of inertia of the element is $(J\,dx)\rho$. In the equation of equilibrium, net applied moment equals the moment of inertia multiplied by angular acceleration, or

$$JG\left(\frac{\partial\theta}{\partial x} + \frac{\partial^2\theta}{\partial x^2}\,dx\right) - JG\,\frac{\partial\theta}{\partial x} = J\,dx\,\rho\,\frac{\partial^2\theta}{\partial t^2} \tag{a}$$

---

[7] L. H. Donnell, "Longitudinal Wave Transmission and Impact," *Trans. ASME,* Vol. 52, Paper APM-52-14, pp. 153–167, 1930.

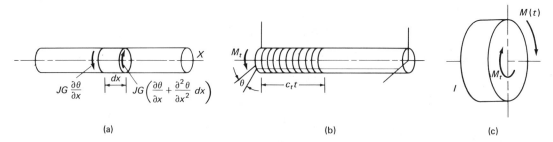

FIGURE 13.17. Torsional impact. (a) Torques on an element of a uniform shaft. (b) Rotation θ of end of shaft in time t. (c) Torques on attached body of moment of inertia I at left end of shaft, end $x = 0$.

whence

$$\frac{\partial^2 \theta}{\partial t^2} = \frac{G}{\rho} \frac{\partial^2 \theta}{\partial x^2} = c_t{}^2 \frac{\partial^2 \theta}{\partial x^2} \tag{13.38}$$

This equation is analogous to Eq. (13.11) for longitudinal impact. Hence the general solution is like Eqs. (13.12), or

$$\theta = f(x - c_t t) + f'(x + c_t t) \tag{13.39}$$

where $c_t = \sqrt{G/\rho}$ is the longitudinal velocity of torsional wave propagation. In steels its value is 10 500 ft/s (3200 m/s).

The relationship between particle velocity and stress may be obtained by the differentiation of Eq. (13.39), as was done to obtain Eqs. (13.13). A different method will be used here. Figure 13.17(b) represents a circular bar to which a constant torque M_t has been applied at one end for a time t. Hence the wave has traveled the distance $x = c_t t = \sqrt{G/\rho}\, t$ along the bar, and the end has rotated through the angle $\theta = M_t x/JG = M_t t/(J\sqrt{G\rho})$. From Eq. (12.4) $M_t = 2J\tau/d$, where d is the bar diameter. Thus the angular velocity ω of the end relative to the unstressed region becomes

$$\omega = \frac{\theta}{t} = \frac{(2J\tau/d)t}{J\sqrt{G\rho}\, t} = \frac{2\tau}{d\sqrt{G\rho}} \tag{13.40}$$

This is the equation that relates particle velocity to stress. If a rotating rigid body strikes with velocity ω_0, the end of the bar takes the velocity of the body, and by Eq. (13.40) the initial shear stress is

$$\tau_0 = (\omega_0 d/2)\sqrt{G\rho} \tag{13.41}$$

The torsional case of a bar with a mass attached at one end and struck at the other is shown in Fig. 13.1(b). The body that is attached to the bar is shown as a free body in Fig. 13.17(c), where I is its moment of inertia, $M(t)$ is an applied torque that is a function of time, and M_t is the elastic torque from the bar. Corresponding to Eqs. (13.21) and by the relationship of Eq. (13.40), the total shearing stress and net velocity are, respectively,

$$\tau_n = s_n(t) + s_n{}'(t - 0.5T)$$

and

$$\Omega_n = \frac{2}{d\sqrt{G\rho}} \left[-s_n(t) + s_n{}'(t - 0.5T) \right] \tag{13.42}$$

The equation of motion of I is

$$M_t + M(t) = \frac{2J\tau_n}{d} + M(t) = I\frac{d\Omega_n}{dt} \tag{13.43}$$

The substitution of Eqs. (13.42) into Eq. (13.43), and the representation of $J\sqrt{G\rho}/I$ by the symbol p, gives

$$p[s_n(t) + s_n'(t - 0.5T)] + (pd/2J)M(t) = \frac{d}{dt}[-s_n(t) + s_n'(t - 0.5T)] \tag{13.44}$$

This equation is identical in form with Eq. (b) of Section 13.6. Hence it has the same solution except for values of the coefficients. The symbol p may be obtained in terms of mass ratios, thus

$$p = \frac{J\sqrt{G\rho}}{I} = \frac{Jl\rho}{I}\frac{\sqrt{G\rho}}{l\rho} = \frac{I_b}{I}\frac{1}{l}\sqrt{\frac{G}{\rho}} = \alpha\frac{c_t}{l} = \frac{2\alpha}{T} \tag{13.45}$$

The last term is the same as that for q in Eq. (13.31) if α equals the ratio of moments of inertia I_b/I rather than masses. Likewise, from an equation of motion for body I' at end $x = l$, it may be shown that $q' = 2\alpha/\lambda T$, the same as for q' in Eq. (13.32), provided $\lambda = I'/I$ is the ratio of the moments of inertia of the end bodies. Hence the diagrams of Figs. 13.12, 13.13, and 13.14 and the chart of Fig. 13.15, also Eqs. (13.24) through (13.27), (13.36), and (13.37) apply to the torsional case as well. To use the chart of Fig. 13.15 it is only necessary to multiply R by the initial shear stress τ_0, Eq. (13.41).

Example 13.3

A steel flywheel consists of a disk 500 mm in diameter and 100 mm thick. While rotating at 250 rpm, it engages a stationary steel shaft 50 mm in diameter and 1 m long through a jaw clutch. At its far end there are gears with an equivalent moment of inertia of 0.040 kg·m². Determine the maximum stress.

Solution

The moments of inertia are $Jl\rho = (\pi/32)d^4l\rho$, and ρ for steel is 7.778×10^3 kg/m³ (Example 8.4). For the flywheel and shaft, respectively,

$$I' = (\pi/32)(0.5)^4(0.1)(7.778 \times 10^3) = 4.773 \text{ kg·m}^2$$

$$I_b = (\pi/32)(0.05)^4(1.0)(7.778 \times 10^3) = 0.004\ 773 \text{ kg·m}^2$$

$$\frac{1}{\alpha} = \frac{I}{I_b} = \frac{0.040}{0.004\ 773} = 8.38, \quad \lambda = \frac{I'}{I} = \frac{4.773}{0.040} = 119$$

From the chart and its caption (Fig. 13.15), $R = 4.15$ and $\tau_0 = 1.30 \times 10^{-3}n_0d = 1.30 \times 10^{-3}(250)(50) = 16.25$ MPa. Hence, $\tau_{max} = R\tau_0 = (4.15)(16.25) = 67.4$ MPa.

Although the modified, energy-method equation, Eq. (13.36), is intended for use when $1/\alpha > 10$, its correspondence here is interesting to note. With $\alpha = 1/8.38 = 0.119$ and $\lambda = 119$,

$$R = 1.1 + \sqrt{\frac{\lambda(1 + \alpha)}{\alpha(1 + \alpha + \lambda)}}$$

$$= 1.1 + \sqrt{\frac{119(1 + 0.119)}{0.119(1 + 0.119 + 119)}} = 1.1 + 3.05 = 4.15 \qquad ////$$

13.9 LONGITUDINAL IMPACT ON HELICAL SPRINGS

Springs together with means for damping are useful for stopping or limiting the motion of moving bodies, with a longer stroke and less shock than with some other devices. In mechanisms springs are often given very rapid displacements. Equations and charts obtained for the impact of bars may be applied but with different multiplication constants. Because the wave velocity is much smaller, in an average spring about one-hundredth of that for a straight bar in torsion, experimental results more closely duplicate the theory.

The spring-deflection equation, Eq. (12.11a), may be rearranged to give the axial force on a spring element of length dx (Fig. 13.18(a)). If the pitch of the coil is p, then the number of active coils in length l is $N = l/p$. With C the spring index D/d,

$$\delta = \frac{8PC^3N}{Gd} = \frac{8PC^3l}{Gpd} \quad \text{and} \quad P = \frac{Gpd}{8C^3}\frac{\delta}{l} \tag{a}$$

For an elemental length dx, $\delta/l = \partial u/\partial x$, and $P = (Gpd/8C^3)(\partial u/\partial x)$. If ϕ is the lead angle, then the length of wire per pitch distance p is $\pi D/\cos\phi$ (Fig. 13.18(b)), and per unit distance is $(\pi D/\cos\phi)/p$. The mass of the spring in axial distance dx is the product of wire density, cross-sectional area, and length, namely, $\rho(\pi d^2/4)(\pi D/p \cos\phi)\,dx = \rho\pi^2 Dd^2 dx/(4p \cos\phi)$. The equilibrium equation for the element (Fig. 13.18), is

$$\frac{Gpd}{8C^3}\left[\frac{\partial u}{\partial x} + \frac{\partial}{\partial x}\left(\frac{\partial u}{\partial x}\right)dx\right] - \frac{Gpd}{8C^3}\frac{\partial u}{\partial x} = \left(\frac{\rho\pi^2 Dd^2}{4p \cos\phi}dx\right)\frac{\partial^2 u}{\partial t^2}$$

whence, together with the substitution of D/d for one C,

$$\frac{\partial^2 u}{\partial t^2} = \frac{Gp^2 \cos\phi}{2\pi^2\rho C^3 Dd}\left(\frac{\partial^2 u}{\partial x^2}\right) = \frac{Gp^2 \cos\phi}{2\pi^2\rho C^2 D^2}\left(\frac{\partial^2 u}{\partial x^2}\right) = c^2\frac{\partial^2 u}{\partial x^2} \tag{13.46}$$

whence c, the speed of axial propagation of the wave, is

$$c = \frac{p}{\pi CD}\sqrt{\frac{G \cos\phi}{2\rho}} = \frac{\tan\phi}{C}\sqrt{\frac{G \cos\phi}{2\rho}} \tag{13.47}$$

since $p/\pi D = \tan\phi$ (Fig. 13.18(b)). For steels, $c = 2360\,p\sqrt{\cos\phi}/CD$ ft/s, or $c = 719\,p\sqrt{\cos\phi}/CD$ m/s.

When a wave travels an elemental distance $\Delta l = c\,\Delta t$ along the spring, there is a corresponding displacement Δu. The axial force, from Eq. (a), is $P = (Gpd/8C^3)$

FIGURE 13.18. Impact of a helical spring. (a) Dimensions and forces on an element of length dx. (b) Length of wire in pitch distance p.

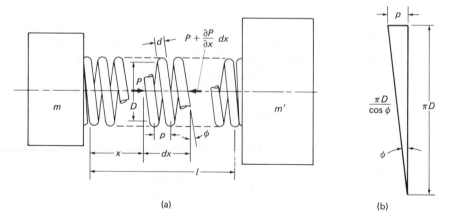

(a)   (b)

$(\Delta u/c\Delta t)$. In the limit, $\Delta u/\Delta t$ becomes particle velocity v, and the force, together with substitution from Eq. (13.47), becomes

$$P = \frac{Gpd}{8C^3}\left(\frac{v}{c}\right) = \frac{\pi vDd}{8C^2}\sqrt{\frac{2G\rho}{\cos\phi}} \tag{b}$$

The torsional shear stress in the spring wire is obtained with Eq. (12.8) by substituting from Eq. (b) and noting that $D = Cd$, thus

$$\tau = K\frac{8PC}{\pi d^2} = Kv\sqrt{\frac{2G\rho}{\cos\phi}} \tag{13.48}$$

With an initial striking velocity of v_0, the initial stress is

$$\tau_0 = Kv_0\sqrt{\frac{2G\rho}{\cos\phi}} \tag{13.49}$$

where K is the Wahl factor (Fig. 12.6), and $\phi = \tan^{-1}(p/\pi D)$. For steels, $\tau_0 = 129.4Kv_0/\sqrt{\cos\phi}$ psi when v_0 is in/s, $\tau_0 = 35.13Kv_0/\sqrt{\cos\phi}$ MPa, when v_0 is m/s. If the coils are initially closely wound, there is little error in taking $\cos\phi = 1.0$. For large displacements, ϕ should be considered a variable in Eq. (13.48).

The mass of the entire spring is equivalent to m_b in the longitudinal impact of a bar. From the dimensions following Eq. (a),

$$m_b = \rho\frac{\pi d^2}{4}\frac{\pi D}{p\cos\phi}Np = \frac{\pi^2\rho}{4}\frac{Dd^2N}{\cos\phi} \tag{13.50}$$

Wave action subsequent to striking proceeds in the same manner as analyzed for longitudinal impact of a bar. The chart of Fig. 13.15 and Eqs. (13.36) and (13.37) may be used. As before, $\lambda = m'/m$, $1/\alpha = m/m_b$, and $1/\beta = m'/m_b$, with m' the mass of the striking body and m, the mass of the attached body.

Example 13.4

(a) A shuttle weighing 1 lb to which a helical spring is attached is put into motion when the other end of the spring is hit at 100 in/s by a flywheel-driven striker. The allowable shear stress is 50 000 psi. Design a spring for the action. (b) If the shuttle is caught and stopped at the end of its travel by a spring attached to a heavy member, can an equal spring be used? Assume that the shuttle velocity is close to 100 in/s.

Solution

(a) We shall try a steel spring of average proportions, say $D/d = 8$ and $l/D = 4$, where l is the length of the active coils. It must be a compression spring with open coils. If we assume that the pitch p is $2d$, then $\phi = \tan^{-1}p/\pi D = \tan^{-1}2d/\pi(8d) = 4.55°$ and $\sqrt{\cos\phi} \approx 1.0$. Since $D = 8d$, $l = 4D = 32d$. Then the number of coils N is $l/p = 16$. From Eq. (13.50) the spring's weight, based on $\gamma = 0.281$ lb/in^3, is

$$W_b = gm_b = \frac{\pi^2\gamma}{4}\frac{Dd^2N}{\cos\phi} = \frac{\pi^2(0.281)}{4}\frac{(8d)(d^2)(16)}{1.0} = 88.75d^3$$

Now $v_0 = 100$ in/s, $W = 1$ lb, and $\lambda = \infty$ because of the large mass of the striking member. The Wahl factor K is 1.19, from Fig. 12.6. From the formula following Eq. (13.49),

$$\tau_0 = 129.4Kv_0 = (129.4)(1.19)(100) = 15\,400 \text{ psi}$$

Since the allowable is 50 000 psi, $R = \tau_{max}/\tau_0 = 50\,000/15\,400 = 3.25$, and together with $\lambda = \infty$, we read from the chart (Fig. 13.15), that $1/\alpha$ is 3.15. From its definition,

$$\frac{1}{\alpha} = \frac{m}{m_b} = \frac{W}{W_b}, \qquad 3.15 = \frac{1.0}{88.75 d^3}, \quad \text{and} \quad d = 0.153 \text{ in}$$

Then $D = 1.22$ in and $l = 4.90$ in. The weight of the active coils is 0.318 lb. The inactive coils at the ends move with the bodies there, and since their mass is relatively small, they are ignored.

(b) The second spring is attached to the stop, so $W = \infty$ and $W_b = 0.318$ lb. The energy method indicates that we might add one-third of the weight of the first spring to the 1-lb mass to which it is attached to give $W' = 1.0 + (1/3)(0.318) = 1.106$ lb. Then $1/\beta = W'/W_b = 1.106/0.318 = 3.48$. Since $\lambda = W'/W \approx 0 \ (m_b \to \infty)$, we must use Eq. (13.37), and

$$R = 1 + \sqrt{(1/\beta) + (2/3)} = 1 + \sqrt{3.48 + 0.667} = 3.04$$

$$\sigma_{\max} = R\sigma_0 = (3.04)(15\ 400) = 46\ 800 \text{ psi} < 50\ 000 \text{ psi}$$

An equal spring can be used.   ////

13.10 STRIKING OF TWO BARS: TRANSMISSION AND REFLECTION AT A STEP

When one bar with a velocity v_d strikes another that has a velocity v_c (Fig. 13.19(a)), their plane of contact will initially take a common, constant velocity v_e such that the forces started in the two bars are equal. Let the bars have unequal cross-sectional areas A_c, A_d and material constants ρ_c, ρ_d and E_c, E_d. The initial stresses depend on the relative velocities, and if $v_d > v_e > v_c$, then from Eq. (13.17),

$$(\sigma_c)_0 = (v_e - v_c)\sqrt{E_c\rho_c} \quad \text{and} \quad (\sigma_d)_0 = (v_d - v_e)\sqrt{E_d\rho_d} \tag{13.51}$$

FIGURE 13.19. Some special cases. (a) Striking of two different bars, free at both ends, with the connection at plane of impact arranged so that velocities and stresses are initially positive. (b) Striking of bars carrying attached masses at their far ends. (c) partial transmission τ_t' and reflection τ_r of an incident stress wave τ_i' at a step in a shaft under torsional impact.

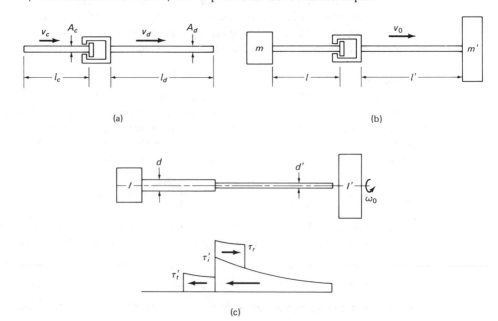

(a)   (b)

(c)

Let $\kappa = (A_c/A_d)\sqrt{E_c\rho_c/E_d\rho_d}$. With the forces on the two sides of the contact plane set equal, there is obtained the contact-plane velocity,

$$v_e = \frac{\kappa v_c + v_d}{\kappa + 1} \tag{13.52}$$

If the bars have the same section and material, then $\kappa = 1$ and v_e is the average velocity $(v_c + v_d)/2$.

From the plane of contact, waves of constant stress with the values of Eqs. (13.51) travel along the bars in opposite directions. If the far ends are free, the reflected waves are equal with opposite signs. If the bars are of equal length, both are free of stress at the end of the first period, with an exchange of velocities. If of unequal length, the shorter bar will be free of stress at the end of its first period, but separation will not occur until the end of the period for the longer bar. The latter will then carry a stress that is reflected back and forth within it until damped out.[8] For the case of attached, rigid masses m and m' at the far ends of the bars (Fig. 13.19(b)), a study has been made in a manner similar to that of Section 13.7,[9] also by a method of characteristics.[10]

A partial reflection and a partial transmission of stress occurs at a step or shoulder in a bar or shaft (Fig. 13.19(c)). The incident, reflected, and transmitted waves will be identified by subscripts i, r, and t, respectively, and, as before, a negative wave will have a prime mark. It is necessary that the forces or torques on both sides of the step be the same, also that the particle velocities be the same. Torque in terms of stress is $(\pi d^3/16)\tau$ (Eqs. (12.5)). With Eqs. (13.42) applied to a negative torsional wave, one which approaches the step from the right, we have

$$T_{\text{left}} = T_{\text{right}}, \quad \frac{\pi d^3}{16}\tau_t' = \frac{\pi d'^3}{16}(\tau_r + \tau_i') \tag{a}$$

also

$$\omega_{\text{left}} = \omega_{\text{right}}, \quad \frac{2\tau_t'}{d\sqrt{G\rho}} = \frac{2(-\tau_r + \tau_i')}{d'\sqrt{G\rho}} \tag{b}$$

Let $D = d/d'$ be the *step ratio*. Then Eqs. (a) and (b) become, respectively,

$$\tau_t' = (1/D^3)(\tau_r + \tau_i') \quad \text{and} \quad \tau_t' = D(-\tau_r + \tau_i') \tag{c}$$

Equated, these equations yield τ_r, then τ_t', as

$$\tau_r = \frac{D^4 - 1}{D^4 + 1}\tau_i' \quad \text{and} \quad \tau_t' = \frac{2D}{D^4 + 1}\tau_i' \tag{13.53}$$

By a similar analysis for a positive wave, one which approaches the step from the left, there is obtained

$$\tau_r' = -\frac{D^4 - 1}{D^4 + 1}\tau_i \quad \text{and} \quad \tau_t = \frac{2D^3}{D^4 + 1}\tau_i \tag{13.54}$$

These equations indicate that when an incident wave encounters an increase in section, the reflected stress has the same sign and when an incident wave encounters a decrease in section, the reflected stress has the opposite sign. The transmitted stress always has the same sign as the incident wave.

---

[8] For a more complete discussion of striking rods, see L. H. Donnell, "Longitudinal Wave Transmission and Impact," *Trans. ASME*, Vol. 52, Paper APM-52-14, pp. 153–167, 1930.

[9] See footnote 5.

[10] G. H. Garzón M., M.S. thesis, Universidad de los Andes, Bogotá, Colombia, 1972.

For this and other complicated cases and when there are many time intervals, it is necessary to use a digital computer to obtain results. Computation is facilitated by use of the method of characteristics,[11] which reduces the integration of partial differential equations to the integration of ordinary differential equations. Dalessandro[12] used this method to construct charts showing the effect of step location and step magnitude. For the same values of λ and $1/\alpha$, stress-ratio values R were somewhat larger than for the unstepped bar of Fig. 13.15. This shows that waves reflected from the step reinforce those reflected from the ends, particularly when the step is so located that the fronts of the waves arrive at the same time. The larger the step up and the closer its location to the struck end, the larger is the stress ratio.

In an experiment the stresses were smaller than those calculated, probably due to local action and delay at the clutch jaws and step. Observing that on the oscilloscope the wave front was not vertical, Dalessandro modified the computer program to feed in the same sloped front and thus obtained better agreement between computation and experimentation.

The stresses at a step are increased by stress concentration, and in the absence of actual test data, the appropriate theoretical factors, such as given by Figs. 5.26 and 5.28, may be applied. It is known that impact resistance is improved by decreasing stress concentration in the same way as is done in fatigue situations. A more gradual transition between diameters is likely to further decrease the peaks in the basic impact stresses. Also, the yield and ultimate strengths of metals are generally higher under rapidly applied loads. There are many uncertainties, and for an important application, a test program on the actual component may be advisable.

The literature on the subject of impact is fairly extensive. It includes the impact of beams, plastic waves, shock waves, dispersion of waves, internal friction, and fractures. Some publications are referenced below.[13]

PROBLEMS

Section 13.2

13.1   If the cylinder shown for Prob. 5.9 is used with a pneumatic hammer, the bolts may be subjected to an energy loading. (a) Two short bolts, one at each flange, with a shank length of 1½ in between the head and nut could replace each long bolt shown. Make calculations to compare the short bolts with a long bolt for ability to accept energy elastically. (b) For a more uniform maximum stress, the diameter of the shank may be reduced to that at the threads, where the root area is 0.551 in². For a long bolt compare the amount of energy it will accept elastically with that by a bolt with a full-diameter shank. *Ans.* 43% more.

13.2   Determine the elastic-energy capacity in terms of yield strength and volume: (a) of a bar stressed longitudinally and (b) of a simply supported rectangular beam loaded at its midlength. For the same energy loading what is the bulk and weight advantage in using higher strength steels, say with double the yield strength?

13.3   A horizontal bar of high-strength aluminum alloy consists of two cylindrical portions, as shown. The free end is struck centrally with a velocity of 0.6 m/s by a rigid body weighing

[11] S. H. Crandall, *Engineering Analysis,* New York: McGraw-Hill, pp. 352–358, 1956.

[12] J. A. Dalessandro, "Torsional Impact of a Stepped Shaft with a Rigid Body at One End," *J. Appl. Mech.*, Vol. 40, *Trans. ASME*, Vol. 95, Series E, Dec. 1973, pp. 1004–1008.

[13] W. Goldsmith, *Impact,* London: Edward Arnold, Ltd., 1960; H. Kolsky, *Stress Waves in Solids,* New York: Dover Publications, Inc., 1963. Varley, E. (Ed.), "The Propagation of Shock Waves in Solids," AMD—Vol. 17, Symp. *ASME,* New York, 1976.

98.1 N. For the aluminum, modulus $E = 71.0$ GPa. Predict the maximum stress, neglecting the mass of the bars ($\eta = 1$). *Ans.* 131 MPa.

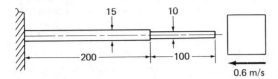

13.4 The bottom end of a vertical, massless bar is fixed. A rigid mass is placed symmetrically in contact with the top end but is supported by a cord such that the bar carries no force. The cord is suddenly cut. From an energy relationship determine equations for maximum force, stress, and deflection.

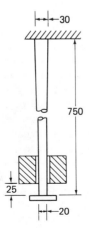

13.5 A 50-kg mass is dropped from a height of 25 mm onto the end of a steel rod that tapers from a 30-mm diameter to a 20-mm diameter in 750 mm. Neglect the mass of the rod and by the energy method predict the resulting maximum stress. The deflection of a tapered rod is given in Prob. 5.14.

Section 13.3

13.6 (a) Obtain v_a for a bar by integrating the momentum of its elements by the method indicated in footnote 1 of Section 13.3. (b) Derive the corresponding correction factor η and compare with that of Eq. (13.4).

13.7   Derive Eq. (13.5) for the correction factor with a cantilever beam, struck at its unsupported end.

13.8   Derive Eq. (13.6) for the correction factor with a simply supported beam, struck at mid-length.

13.9   Derive a correction factor for a bar when the end struck carries a rigid body of mass m. Compare with Eq. (13.4). Will the correction factor for the beams be similarly affected?

13.10   A part weighing 750 lb was accidentally dropped about ½ in while it was being lowered onto a machined way at its midlength. The way may be approximated by a beam 1 in wide by 3 in high by 60 in long with the ends freely supported. Its yield strength is 55 000 psi. Is it likely that the way was damaged by yielding, except locally at the surface?

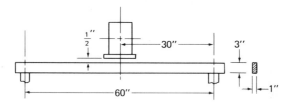

13.11   Same requirement as Prob. 13.10 but the beam is made of an aluminum alloy ($E = 71.0$ GPa and $S_y = 110$ MPa), has a section 50 mm by 20 mm, is 800 mm long, and was struck by a 50-kg mass, dropped about 20 mm.

13.12   Derive an equation for the correction factor η for the effect of the mass of a beam that is fixed at its ends and struck at its midlength by a rigid body. Deflection $u = (F/48EI)\, x^2 (4x - 3L)$ with $0 < x \leq L/2$.

13.13   Derive Eq. (13.8) directly in torsional terms rather than by analogy, first writing equations for elastic and kinetic energies and momentum.

13.14   Derive Eq. (g) preceding Eq. (13.10) directly in longitudinal-impact terms, first writing equations for elastic and kinetic energies and momentum.

Sections 13.4 and 13.5

13.15   In deep-well reciprocating pumps, e.g., Fig. 6.11, the stroke of the plunger at the bottom of the well and the stroke of the actuating crank at the top of the well have been observed to be different, sometimes with considerable loss of plunger stroke. To get an idea of the magnitude of the effects, take an oil-well pump with sucker rods of ¾ in diameter and 5000 ft long and a crank with a 30-in radius. With the plunger and crank both at bottom dead center at the start of upward motion, the one-way valve on the plunger closes to life a column of oil, and the plunger needs additional force to move. If the crank is moving at a rate of 20 rpm, through what angle does the crank move before it receives a signal to send back the additional force required to move the plunger? (It will be recognized that this is an oversimplification of the problem, particularly in its negligence of the weight of the sucker rods and their inertia.)

13.16   Prove by substitution that Eq. (13.12) is a solution of Eq. (13.11).

13.17   (a) Determine the velocity in ft/s and in m/s of longitudinal-wave propagation in steel bars.
(b) Determine the constant in the equation for initial stress for use with steel bars when v_0 is in in/s and v_0 is in m/s. *Ans.* Caption of Fig. 13.15.

13.18   Same requirement as Prob. 13.17 but for aluminum bars. $E = 10.3 \times 10^6$ psi (71.0 GPa), unit weight is 0.098 lb/in^3 (26.6 kN/m^3).

13.19   Describe with sketches the wave action, stresses, time of separation, and final velocity of a bar with free ends that is struck by a rigid body of infinite mass.

13.20   Describe with sketches similar to those of Fig. 13.10 the stress and velocity action in a bar with one end fixed and struck at the other by a rigid body of finite mass m'. Do this for one period.

Sections 13.6 and 13.7

13.21   Derive the recurrence formula, Eq. (13.24), from Eqs. (13.23).

13.22   Derive the recurrence formula, Eq. (13.26), from Eqs. (13.23).

13.23   What reflection relationship must take the place of the first of Eqs. (13.23) for the special case of a fixed end $x = 0$, where, in effect, the mass is infinite? Apply Eqs. (13.26) and (13.27) to the struck end, and with $F_t' = 0$, derive equations for wave stresses s_1' and s_2'. *Ans.* $s_1' = \sigma_0 e^{-q't}$, $s_2' = \sigma_0 \{e^{-q't} + [1 - 2q'(t - T)]e^{-q'(t-T)}\}$.

13.24   (a) Is the case of Prob. 13.4 one of striking impact or of force impact? What is the value of the initial stress? Of the force? Explain. (b) Write the equation that governs wave reflection at the bottom end of the bar. How does the equation differ at the top end? Sketch the stress waves s', s, and σ at $0.25T$, $0.75T$, and $1.25T$.

13.25   For the case of constant-force impact at end $x = l$, as shown at the top of Fig. 13.13, but for the special case of $m' = m$, i.e., $q' = q$, derive equations for the first time interval at each end. Recurrence equations, Eqs. (13.24) and (13.26), may be used. *Ans.* $s_1' = (F_0/A)(1 - e^{-qt})$, $s_1 = (F_0/A)\{-1 + [1 + 2q(t - 0.5T)e^{-q(t-0.5T)}]\}$.

13.26   A steel bar with a diameter of 2 in and a length of 30 in has a relatively rigid body weighing 150 lb attached at one end. It is struck axially at the other end by a relatively rigid body weighing 400 lb, traveling at 4 ft/s. Using the chart of Fig. 13.15, determine the maximum stress. *Ans.* 22 700 psi.

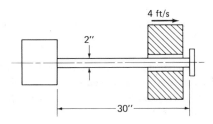

13.27   Same information as Prob. 13.26 except a diameter of 15 mm, a length of 250 mm, an attached body of 3.0 kg and a striking body of 30 kg at 1.5 m/s.

13.28   Same information as Prob. 13.26 except a diameter of 1 in, a length of 40 in, an attached mass of 0.45 lb·s^2/in, and a striking mass of 1.35 lb·s^2/in at 60 in/s.

Section 13.8

13.29   (a) Determine the longitudinal velocity in ft/s and in m/s for torsional-wave propagation in steel shafts. For steels, $G = 11.5 \times 10^6$ psi (79.3 GPa) and unit weight is 0.281 lb/in^3 (76.3 kN/m^3). (b) Determine the constant in the equation for initial shear stress for use with steel bars with angular velocity n_0 in rpm and d in inches, also in mm. *Ans.* Caption of Fig. 13.15.

13.30 Same requirements as Prob. 13.29 but for aluminum. $G = 3.80 \times 10^6$ psi (26.2 GPa) and unit weight is 0.098 lb/in³ (26.6 kN/m³).

13.31 A steel disk 10 in. diameter and 2 in wide while rotating at 200 rpm engages through a jaw clutch a stationary steel shaft of 1.5 in diameter and 30 in length, at the further end of which there are gears with an equivalent moment of inertia of 0.451 lb·in·s². Determine the maximum stress. *Ans.* 9585 psi.

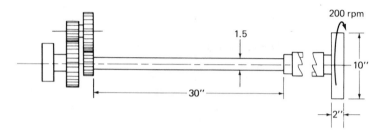

13.32 A rotating body with a moment of inertia of 6.0 kg·m² strikes a shaft with a diameter of 100 mm and length of 1600 mm at a relative velocity of 300 rpm. The shaft carries at its other end bodies with an equivalent single moment of inertia of 1.10 kg·m². Determine the maximum stress.

Section 13.9

13.33 The lowest natural frequency for the surging of spring coils between two fixed ends, Eq. (12.19), can be derived from the spring-wave velocity and period. Derive it for steel springs and compare the result.

13.34 (a) Confirm the constants for British Gravitational Units in the velocity and stress equations for steel springs, as given just after Eqs. (13.47) and (13.49). (b) Determine the velocity for a compression spring with $C = 8$ and $p = 3d$. Compare it with that for longitudinal and torsional waves in straight bars.

13.35 Obtain the constants in Eqs. (13.47) and (13.49) for use with steel springs in SI units, with v_0 in m/s. *Ans.* 719 and 35.13, respectively.

13.36 Derive by the energy method an equation for the shear stress τ_i from impact by a mass m' striking one end of a helical spring with the other end fixed. Ignore the mass of the coils and the lead angle ϕ. Show that the equation can be put in the form $\tau = \tau_0 \sqrt{1/\beta}$ and compare it with Eqs. (13.7) and (13.37).

13.37 A helical spring with ⁵/₁₆-in wire, a mean-coil diameter of 2.25 in, 10 total and 1¾ inactive coils, is struck at 100 in/s at one end by a body weighing 30 lb. Determine the impact stress.

13.38 Determine the number of active coils required for a helical spring with a wire diameter of 6 mm, a mean-coil diameter of 50 mm, with a body of 5.0-kg mass at one end, struck at the other by a body of 50-kg mass at 3.5 m/s. The allowable shear stress is 550 MPa. *Ans.* 19.

Section 13.10

13.39 Derive Eq. (13.52) for the contact-plane velocity of two striking bars.

13.40 A bar traveling at velocity v_0 strikes a second bar as in Fig. 13.19(a), but the bars are of the same material, section, and length and the left-hand bar is initially stationary. Describe the subsequent action with sketches of stress and velocity waves, as done in Fig. 13.10, but for $t = 0+$, $0 < t < 0.5T$, $0.5T < t < T$, and T. Relate to Eqs. (13.13) through (13.16). State the stress and velocity conditions after separation.

13.41 Same requirements as Prob. 13.40, but the striking bar is one-third longer than the other (Fig. 13.19(a)). If you have done Prob. 13.40, sketch only for $t = T$, $t = 1.1T$, $t = (4/3)T$ (at separation), and $t = (5/3)T+$, where l_c is the length of the shorter bar and $T = 2l_c/c$. Confirm the statement in the paragraph following Eq. (13.52).

13.42 Two bars have equal sections and material but different lengths l and l' (Fig. 13.19(b)). Bar l' strikes bar l with a positive velocity v_0. At the far ends, bar l has a rigid, attached body of mass m, and bar l' has one of mass m'. What is the value of the initial stress σ_0? For the total stresses σ_n and σ_n' in the bars at their attached bodies, write equations in terms of σ_0 and the incident and reflected stresses to correspond with the stress equations of Eqs. (13.21) and (13.22). Use periods, $T = 2l/c$ and $T' = 2l'/c$. Express the first two time intervals at each end in terms of T and T'.

13.43 Derive Eqs. (13.54) for the stresses reflected and transmitted at the step when a torsional wave approaches from the left.

13.44 Derive an equation for the stresses reflected and transmitted when a *longitudinal* wave approaches a step from the right. Let the area ratio be $a = A/A'$, where A' is the area on the right. *Ans.* $s_r = s_i'(a - 1)/(a + 1)$, $s_t' = 2s_i'/(a + 1)$.

13.45 Show that Eqs. (13.53) and (13.54) are the same expression if rewritten in terms of a step ratio δ which is defined as the ratio of the diameter beyond the step to that before the step, taken in the direction of the incident wave.

13.46 Determine numerically the fraction of an incident wave transmitted in a circular bar when the step is upward with a diameter ratio of 1.5: (a) in torsional impact and (b) in longitudinal impact. The answer to Prob. 13.44 may be used.

Name Index

An italicized page number indicates that the name appears on the page in a footnote only.

A

Aiyer, M. G. S. -R, *226*
Alison, N. L., *139, 145*
Allen, C. M., 501
Alley, R. P., *2*
Almen, J. O., *279, 281*, 284
Anno, J. N., 54
Argon, A. S., *259*
Auman, R. J., *82*

B

Bachman, R. W., *153*
Barwell, F. T., *498*
Belayev, N. M., 492
Bell, J. C., 501
Berman, I., *330*
Bidwell, J. B., *503*
Biezeno, C. B., *306, 307*
Black, P. H., *279*
Boley, B. A., *269, 340*
Booker, J. F., *23*, 52, 53
Booser, E. R., *18, 31*, 38, 39, 51, *518*
Boresi, A. P., *174, 444, 497*
Boussinesq, J., 486, *488, 489*
Boyd, J., *17*, 32, 46
Boyer, H. E., *193*
Bradlee, C. R., 357
Buckingham, E., *507, 517*
Bugbee, J. T., *153*
Burr, A. H., *2*, 40, *406*, 499, *601*, 605, 607, 608

C

Campbell, J., 52, 53
Carden, A. E., *193*
Cardullo, F. E., 34
Carroll, G. R., 58
Castigliano, A., 430
Cawley, W. E., *294*

D

Dalessandro, J. A., 616
DeHart, A. O., *47*, 53
Den Hartog, J. P., *316, 371, 430, 435, 451, 454, 461, 573, 574, 576*
Dodge, T. M., 229
Donnell, L. H., *609, 615*
Drucker, D. C., 108
DuBois, G. B., *34, 41*, 42, 43, 44, *49*
Dudley, D. W., *517*
Dundore, M. W., *153*
Durelli, A. J., *169*

E

Eichinger, A., 179
Eksergian, R., *142, 145, 149*
Elmendorf, H. J., 555, *556*
Elrod, H. G., Jr., 53
Ensign, C. R., *242*
Erdogan, F., *193*

F

Faires, V. M., *394*
Faucett, T. R., *198*
Faupel, J. H., *273, 287, 330, 393*
Fazekas, G. A. G., *82, 87*

Chase, T. P., *88*
Chen, F. Y., *174*
Chester, L. B., 229
Chow, W. W., *346*
Cloutier, G. J., *330*
Corrow, C. J., 501
Cox, H. L., 179
Crandall, S. H., *616*
Crooker, T. W., *193*
Cummings, H. N., 192

Subject Index

An italicized page number indicates that the subject appears on the page in a footnote only. For professional societies and organizations, only special publications such as codes, handbooks, and standards are indexed. Page numbers are included for those problems that contain equations of possible value, also for those figures among the "Problems" that show devices or concepts pertinent to the text.